Ultra-Wideband, Short-Pulse Electromagnetics 6

Ultra-Wideband, Short-Pulse Electromagnetics 6

Eric L. Mokole
Naval Research Laboratory
Washington, D.C.

Mark Kragalott
Naval Research Laboratory
Washington, D.C.

and

Karl R. Gerlach
Naval Research Laboratory
Washington, D.C.

Kluwer Academic / Plenum Publishers
New York, Boston, Dordrecht, London, Moscow

Library of Congress Cataloging-in-Publication Data

Ultra-wideband, short-pulse electromagnetics 6: UWB SP6/[edited by] Eric L. Mokole, Mark Kragalott, and Karl Gerlach.
p. cm.
"Proceedings of the Ultra-Wideband, Short-Pulse Electromagnetics 6, held at the American Electromagnetics 2002 conference, June 3–7, 2002 in Annapolis, Maryland—T.p. verso.
Includes bibliographical references and index.
ISBN 0-306-47481-6
1. Radar—Congresses. 2. Ultra-wideband devices—Congresses. 3. Signal processing—Congresses. I. Mokole, Eric L. II. Kragalott, Mark. III. Gerlach, Karl R. IV. International Conference on Ultra-Wideband, Short-Pulse Electromagnetics (6th: 2002: Annapolis, Md.)

TK6573 .U574 2003
621.3848—dc22

2003061086

Proceedings of the Ultra-Wideband, Short-Pulse Electromagnetics 6, held at the American Electromagnetics 2002 conference, June 3–7, 2002, in Annapolis, Maryland.

ISBN: 0-306-47481-6

233 Spring Street, New York, N.Y. 10013

http://www.wkap.com

10 9 8 7 6 5 4 3 2 1

A C.I.P. record for this book is available from the Library of Congress

Printed in the United States of America

PREFACE

The Sixth Conference on Ultra-Wideband, Short-Pulse Electromagnetics (UWB SP6), chaired by Eric Mokole of the United States Naval Research Laboratory (NRL) and hosted by the NRL and the United States Naval Academy (USNA), was held at the USNA in Annapolis Maryland (USA) from 3-7 June 2002. UWB SP6 was part of the AMEREM 2002 Symposium, chaired by Terence Wieting of the NRL. AMEREM 2002 continued the series of international conferences that were held in: Brooklyn New York at the Polytechnic University in 1992 and 1994; Albuquerque New Mexico in 1996 as part of AMEREM '96; Tel-Aviv Israel in 1998 as part of EUROEM '98; and Edinburgh Scotland in 2000 as part of EUROEM 2000. The next conference (UWB SP7) will be held from 12-16 July 2004 at Otto von Guericke University in Magdeburg Germany (EUROEM 2004) and will be chaired by Frank Sabath.

The purpose of these meetings is: to focus on advanced technologies for the generation, radiation, and detection of ultrawideband (UWB) short-pulse signals, taking into account their propagation about, scattering from, and coupling to targets and media of interest; to report on developments in supporting mathematical and numerical methods; and to describe current and potential future applications of the technology. The session topics of UWB-SP6 included electromagnetic theory, scattering, UWB antennas, UWB systems, ground penetrating radar (GPR), pulsed-power generation, time-domain computational electromagnetics, UWB compatibility, target detection and discrimination, propagation through dispersive media, and wavelet and multi-resolution techniques.

The Chairman and Editors wish to thank all of those involved in AMEREM 2002 for their assistance and participation, especially the contributors to this book and members of the National and International Committees and their supporting institutions. We gratefully acknowledge the cooperation and support of the following institutions: Office of Naval Research; US Naval Research Laboratory; Defense Threat Reduction Agency; Permanent HPEM Committee; SUMMA Foundation; Defense Research Establishment Ottawa, Canada; URSI Commissions B, E, and K; IEEE EMC Society; IEEE Antennas and Propagation Society; IEEE Electron Devices Society; US Air Force Office of Scientific Research; Air Force Research Laboratory–Directed Energy; US Army Research Laboratory; Los Alamos National Laboratory; DGA Centre d'Etudes de Gramat, France; MOD Wehrtechnische Dienstelle 81, Germany; Zurich Electromagnetic Compatibility Symposium; and EMA, Denver CO.

We also acknowledge with gratitude the rest of the UWB Committee: Robert Berkovits (Northrop Grumman), Larry Cohen (NRL), Frank Koye (Time Domain Corporation), Peter McCoy (USNA), Suren Samaddar (SFA, Inc.), Kim Scheff (NRL), and Kevin Wagner (NRL). We are especially indebted to Lynda Kelly (NRL), Daniel Nitsch (Wehrwissenschaftliches Institute für Schutztechnologien), Lisa Simpson (NRL), and Troy Stuart (NRL) for much hard work and many significant contributions towards organizing and running this conference and to Kevin Wagner and Brian Gold of NRL for logistical and editorial help in preparing this book. We also thank Peter McCoy (USNA) for providing excellent liaison with the USNA in organizing AMEREM 2002.

Eric L. Mokole
Mark Kragalott
Karl Gerlach

CONTENTS

ELECTROMAGNETIC THEORY

GROUP DELAY DESCRIPTION FOR BROADBAND PULSES
M. Ware, S.A. Glasgow, and J. Peatross 1

TWO-DIMENSIONAL NARROW-WAISTED GAUSSIAN BEAM ANALYSIS OF PULSED PROPAGATION FROM EXTENDED PLANAR ONE-DIMENSIONAL APERTURE FIELD DISTRIBUTIONS THROUGH PLANAR DIELECTRIC LAYERS
L.B. Felson and V. Galdi 11

AXIAL SCATTERING FROM THIN CONES
C.E. Baum 23

AN OBSERVATION CONCERNING THE ENTIRE FUNCTION IN SEM SCATTERING
C.E. Baum 41

TWO METHODS FOR PROFILE RECONSTRUCTION IN SCATTERING THEORY
F. Seydou and T. Seppanen 51

OBJECT IMPULSE RESPONSE EVALUATION FOR ULTRAWIDEBAND APPLICATION
V.I. Koshelev, V.T. Sarychev, and S.E. Shipilov 63

THE LIFE AND TIMES OF A SEA SPIKE: A STUDY OF WIDEBAND SEA CLUTTER AT HIGH RANGE RESOLUTIONS AND VERY LOW GRAZING ANGLES
F. Posner and K. Gerlach 75

A FULL WAVE APPROACH TO ELECTROMAGNETIC, SCATTERING BY NANOSTRUCTURES WITH SMALL AND LARGE SCALE PERIODIC AND RANDOM ROUGH SURFACES AND LATERAL FLUCTUATIONS IN THE PERMITTIVITY AND PERMEABILITY
E. Bahar 85

FRAME-BASED BEAM-SUMMATION ALGORITHMS FOR ULTRA WIDEBAND RADIATION FROM EXTENDED APERTURES PART I: FORMULATIONS IN THE MULTI-FREQUENCY DOMAIN
A. Shlivinski, E. Heyman, A. Boag, and C. Letrou 101

FRAME-BASED BEAM-SUMMATION ALGORITHMS FOR ULTRA WIDEBAND RADIATION FROM EXTENDED APERTURES PART II: TIME DOMAIN FORMULATION
A. Shlivinski, E. Heyman, A. Boag, and C. Letrou 113

DISPERSIVE MEDIA PROPAGATION

PULSE CENTROVELOCITY: ASYMPTOTIC AND FFT RESULTS
N.A. Cartwright and K.E. Oughstun 123

REVERSIBLE AND IRREVERSIBLE PROCESSES IN DISPERSIVE/DISSIPATIVE OPTICAL MEDIA: ELECTRO-MAGNETIC FREE ENERGY AND HEAT PRODUCTION
C. Broadbent, G. Hovhannisyan, M Clayton, J. Peatross, and S.A. Glasgow 131

FAST TIME-DOMAIN INTEGRAL EQUATION APPROACH FOR WIDE-BAND PULSE PROPAGATION IN DISPERSIVE MEDIA
E. Bleszynski, M. Bleszynski, and T. Jaroszewicz 143

EFFECT OF THE INVERSION SYMMETRY OF DISPERSION RELATIONS ON FOUR-WAVE INTERACTIONS IN PHOTONIC CRYSTALS
A. Babin and A. Figotin 159

A FINITE-DIFFERENCE-TIME-DOMAIN/PERFECTLY-MATCHED-LAYER ALGORITHM APPLIED FOR NONLINEAR DISPERSIVE MEDIA
S.J. Yakura 175

FOCUSING OF ULTRA-WIDEBAND COMPOSITE X-WAVES
A.M. Shaarawi and I.M. Besieris 193

MULTI-RESOLUTION HOMOGENIZATION (MRH) FOR SHORT PULSE FIELDS AND FOR FAULT INTERROGATION IN COMPLEX LAMINATES
V. Lomakin, B.Z. Steinberg, and E. Heyman 203

FOILAGE PENETRATION STRIPMAP SAR IMAGING USING UWB RANDOM NOISE WAVEFORM
R.M. Narayanan, X.Xu, and C. Kumru 215

TIME-DOMAIN COMPUTATIONAL TECHNIQUES

ITERATIVE SOLUTION OF FIELD PROBLEMS WITH A VARYING PHYSICAL PARAMETER
P.M. Zwamborn, A.G. Tijhuis, and M.C. van Beurden 223

A NEW FDTD SOLUTION METHOD WITHOUT A MARCHING-ON-IN-TIME SCHEME USING LAGUERRE POLYNOMIALS
Y. Chung, T.K. Sarkar, and B.H. Jung 231

TIME-DOMAIN INTEGRAL-EQUATION-BASED SOLVER FOR TRANSIENT AND BROADBAND PROBLEMS IN ELECTROMAGNETICS
A.O. Boryssenko and D.H. Schaubert 239

EFFICIENT MIXED FINITE ELEMENTS FOR MAXWELL'S EQUATIONS IN TIME DOMAIN
G. Cohen, X. Ferrieres, P. Monk, and S. Pernet 251

SOLUTION AND STABILITY ANALYSIS OF A NEW INTEGRAL EQUATION FOR THE TRANSIENT SCATTERING BY A FLAT, RECTANGULAR CONDUCTING PLATE
Jos G.A. van Riswick and A.G. Tijhuis 261

ANTENNAS

RESHAPING THE APERTURE FIELDS IN IMPULSE RADIATING ANTENNAS AND OSCILLATOR REFLECTOR ANTENNAS
J.S. Tyo 269

OPTIMIZATION OF IMPULSE RADIATING ANTENNAS
L.H. Bowen, E.G. Farr, C.E. Baum, and T.C. Tran 281

ELECTROMAGNETIC LENS DESIGN
A.P. Stone and C.E. Baum 291

LOCATION AND ORIENTATION OF ELECTRICALLY SMALL TRANSMITTING AND RECEIVING ANTENNA PAIRS WITH COMMON LINEAR POLARIZATION AND BEAM DIRECTION FOR MINIMAL MUTUAL COUPLING
C.E. Baum 299

THE ROLE OF MUTUAL COUPLING IN A 5:1 BANDWIDTH STRIPLINE NOTCH ARRAY
M. Kragalott, W.R. Pickles, and M.S. Kluskens 317

UWB ANTENNA CHARACTERIZATION AND OPTIMIZATION METHODOLOGIES
G.R. Salo and J.S. Gwynne 329

OPTIMIZED ULTRA-WIDEBAND RADIATION OF DIPOLE ANTENNAS WITH TRIANGLE DRIVING PULSES
A.O. Boryssenko and D.H. Schaubert 337

ANALYTICAL, NUMERICAL, AND EXPERIMENTAL RESULTS FOR WIDE-ANGLE BICONICAL ANTENNAS
E. Mokole, S. Samaddar, G. Sorenson, M. Kragalott, B. Gold, and M. Parent 345

ULTRAWIDEBAND TRANSMITTING ANTENNAS, ARRAYS, AND HIGH-POWER RADIATION SOURCES
V.I. Koshelev, Y.A. Andreev, Y.I. Buyanov, V.V. Plisko, and K.N. Sukhushin 357

PULSED POWER

RECENT DEVELOPMENTS IN ULTRA-WIDEBAND SOURCES AND ANTENNAS
W.D. Prather, C.E. Baum, J.M. Lehr, R.J. Torres, T.C. Tran, J.W. Burger, D.I. Lawry, and J.A. Gaudet 369

DESIGN AND DEVELOPMENT OF A HIGH VOLTAGE COAXIAL HYDROGEN SWITCH
J.W. Burger, C.E. Baum, W.D. Prather, R. J. Torres, T.C. Tran, M.D. Abdalla, M.C. Skipper, B.C. Cockreham, and D.R. Mclemore 381

PROGRESS IN DEVELOPMENT AND CHARACTERIZATION OF PHOTOCONDUCTIVELY-SWITCHED STACKED BLUMLEIN DEVICES PRODUCING HIGH POWER NANOSECOND PULSES
F. Davanloo, C. B. Collins, and F. J. Agee 391

THE EFFECTS OF PASSIVATION LAYERS ON THE RESPONSE OF PHTOCONDUCTIVE SWITCHES FOR ULTRA WIDEBAND HIGH POWER MICROWAVE APPLIACTIONS
N.E. Islam, G. Tzeremes, J. Gaudet, and E. Schamiloglu 401

NANOSECOND PULSED BREAKDOWN FOR POINT-PLANE GEOMETRIES AT MODERATE VOLTAGE
H. Krompholz, L. Hatfield, M. Haustein, J. Spears, and M. Kristiansen 409

DESIGN AND TESTING OF A COMPACT SUB-GW, SUBNANOSECOND PULSER
Y. Yankelevich, A. Pokryvailo, and M. Shapira 415

RESONANCE PULSE COMPRESSION AS AN EFFECTIVE METHOD FOR GENERATION OF UWB HIGH REPETITIVE MICROWAVE PULSES
P.Y. Chumerin, A.N. Didenko, S.A. Novikov, and Y.G. Yushkov 427

INFORMATION EXTRACTION

EXCITATION OF SUB-STRUCTURE RESONANCES USING A T-PULSE
Z. Ji, T.K. Sarkar, B.H. Jung, and C.E. Baum 435

EXPLOITING EARLY TIME RESPONSE USING THE HALF FOURIER TRANSFORM (HFT)
S. Jang, T.K. Sarkar, K. Kim, and C.E. Baum 443

DETECTION OF A BURIED OBJECT WITH PULSE-COMPRESSED WIRE ANTENNAS
S. Vossen, A.G. Tijhuis, E. Lepelaars, and P.M. Zwamborn 453

TARGET RECOGNITION VIA LATE TIME SCATTERING
K.J. Pascoe, W.D. Wood, P.S. Maybeck, and A.W. Wood 463

DETECTION OF WIRES USING UWB RADAR
G.N. Crisp, C.R. Thornhill, and G. Gooding-Williams 471

COMPARISON OF TIME, FREQUENCY AND TIME-FREQUENCY ANALYSIS OF BACK-SCATTERED ECHOES FROM RADAR TARGETS BURIED IN REALISTIC HALF-SPACES
G.C. Gaunaurd, H.C. Strifors, and A. Sullivan 481

DETECTION OF TARGETS BEHIND WALLS USING ULTRA WIDE BAND SHORT PULSE
W.A. Chamma and S. Kashyap 493

RADAR SYSTEMS

INTEGRATED DIGITAL UWB-RADAR
J. Sachs, P. Peyerl, P. Rauschenbach, F. Tkac, M. Kmec, and S. Crabbe 507

VECTOR RECEIVING ANTENNA FOR ULTRAWIDEBAND RADAR POLARIMETRY
V. Koshelev, E.V. Balzovsky, and Y.I. Buyanov 515

VERY HIGH VOLTAGE DIVIDER WITH LARGE BANDWIDTH
J. Wey and F. Bieth 523

AN UWB RADAR BASED STEALTHY 'LIE DETECTOR'
E.M. Staderini 537

ELECTROMAGNETIC COMPATIBILITY (EMC) OF ULTRAWIDEBAND (UWB) SHORT-PULSE SYSTEMS IN MILITARY ENVIRONMENTS
L.S. Cohen and A.H. Light 553

DESIGN AND TESTING OF A 1-3 GHZ SFCW GROUND PENETRATING RADAR
S. Peters, I. Jeffrey, C. Gilmore, D. Collins, and J. LoVetri 563

HIGH RESOLUTION IMAGING USING A SFCW GPR
G. Thomas, E. Corral, and J. LoVetri 573

RADAR DETECTION AND IMAGING OF BURIED OBJECTS
K. Scheff and P. Hansen 583

AUTHOR INDEX 591

SUBJECT INDEX 593

Ultra-Wideband, Short-Pulse Electromagnetics 6

GROUP DELAY DESCRIPTION FOR BROADBAND PULSES

M. Ware*, S. A. Glasgow†, and J. Peatross†

1. INTRODUCTION

The traditional concept of group delay usually arises in connection with an expansion of the phase delay for an electromagnetic pulse. In this context, the group delay function (evaluated at a single 'carrier' frequency) describes the time required for a pulse to traverse a given displacement. However, if the bandwidth of the pulse encompasses a substantial portion of a resonance structure the expansion fails to converge over the relevant range of frequencies. Because of this failure, traditional group delay suffers severe shortcomings when applied to broadband pulse propagation.[1-4] We recently introduced a method for describing pulse delay[5,6] in which the group delay function naturally arises. In contrast to the traditional formulation of group delay, this method retains validity for pulses of arbitrary bandwidth propagating in linear dielectrics (including cases where the spectrum overlaps resonances in the medium). In this work we give an overview of this method of description and demonstrate how it may be applied to gain insight into the behavior of broadband electromagnetic pulses. As an illustration we show how the method may be applied to the well-known precursor problem, where the traditional formulation of group delay fails.

2. PULSE DELAY TIME

To concretely discuss delay times, we first define the arrival time of a pulse at a point. Arrival times defined in terms of the advent of some pulse feature (such as the peak) at a point can be ambiguous, since the pulse shape can change dramatically during propagation (e.g., in a precursor field a pulse may evolve two peaks). In the

*National Institute of Standards and Technology, Gaithersburg MD.
†Brigham Young University, Provo UT.

present formulation we avoid this ambiguity by defining the arrival time as a time expectation[7] of the electromagnetic flux:

$$\langle t \rangle_{\mathbf{r}} \equiv \int_{-\infty}^{\infty} t\rho\left(\mathbf{r},t\right)\, \mathrm{d}t \tag{1}$$

where $\rho\left(\mathbf{r},t\right)$ is a normalized temporal distribution of the Poynting flux at $\mathbf{r}$:

$$\rho\left(\mathbf{r},t\right) \equiv \hat{\eta}\cdot\mathbf{S}\left(\mathbf{r},t\right) \Big/ \hat{\eta}\cdot \int_{-\infty}^{\infty} \mathbf{S}\left(\mathbf{r},t\right)\, \mathrm{d}t \tag{2}$$

The Poynting vector is defined as usual:

$$\mathbf{S}\left(\mathbf{r},t\right) \equiv \mathbf{E}\left(\mathbf{r},t\right) \times \frac{\mathbf{B}\left(\mathbf{r},t\right)}{\mu_0} \tag{3}$$

We assume that the propagation medium is neutral, nonconducting, and nonmagnetic. The unit vector $\hat{\eta}$ defines the direction in which the energy flow is detected. It plays a role in angularly dispersive systems,[8] but is unimportant in this work since we consider propagation in only one dimension.

Using this definition for arrival time, the delay Δt between the time when a pulse arrives at a point $\mathbf{r}_0$ and when it arrives at a point $\mathbf{r} \equiv \mathbf{r}_0 + \Delta\mathbf{r}$ is found naturally from

$$\Delta t \equiv \langle t \rangle_{\mathbf{r}} - \langle t \rangle_{\mathbf{r}_0} \tag{4}$$

After transforming Eq. (4) into the frequency domain and inserting the frequency solution to Maxwell's equations, we can write the delay time as the sum of two terms with intuitive interpretations:

$$\Delta t = \Delta t_G\left(\mathbf{r}\right) + \Delta t_R\left(\mathbf{r}_0\right) \tag{5}$$

For details on how this expression is obtained, see Refs. [5] and [6].

The first term in Eq. (5), the *net group delay*, is a spectral average of the group delay function weighted by the spectral intensity that is experienced at the final point $\mathbf{r}$:

$$\Delta t_G\left(\mathbf{r}\right) \equiv \int_{-\infty}^{\infty} \left(\frac{\partial\,\mathrm{Re}\,\mathbf{k}}{\partial\omega}\cdot\Delta\mathbf{r}\right) \rho\left(\mathbf{r},\omega\right)\, \mathrm{d}\omega \tag{6}$$

where $\rho\left(\mathbf{r},\omega\right)$ is given by

$$\rho\left(\mathbf{r},\omega\right) \equiv \hat{\eta}\cdot\mathbf{S}\left(\mathbf{r},\omega\right) \Big/ \hat{\eta}\cdot \int_{-\infty}^{\infty} \mathbf{S}\left(\mathbf{r},\omega\right)\, \mathrm{d}\omega \tag{7}$$

The frequency representation of the Poynting vector, $\mathbf{S}\left(\mathbf{r},\omega\right)$, is found from the Fourier transforms of the electric and magnetic fields in the usual manner. The distribution $\rho(\mathbf{r},\omega)$ is a spectral density function (i.e., it has units of $1/\omega$) which describes the fraction of the pulse's flux that may be associated with a given frequency range. Note that the distribution $\rho(\mathbf{r},\omega)$ is evaluated at the *final* point, so that only the spectral components that actually arrive at $\mathbf{r}$ contribute to the delay.

Equation (6) explicitly demonstrates the connection between the delay time for a pulse and the group delay function. Note the close resemblance between Eq. (1) and Eq. (6). Both are expectation integrals, the former being executed as a 'center-of-mass' integral in time, the later being executed in the frequency domain on the group delay function, $\partial\,\mathrm{Re}\,\mathbf{k}/\,\partial\omega\cdot\Delta\mathbf{r}$. In contrast to the traditional group delay, in Eq. (6) the group delay function evaluated at every frequency present in the final pulse influences the delay time as recorded by the time 'center-of-mass'.

The second term in Eq. (5) arises from a reshaping of the spectrum through absorption (or amplification). It is given by

$$\Delta t_R(\mathbf{r}_0) \equiv T\left[e^{-\,\mathrm{Im}\,\mathbf{k}\cdot\Delta\mathbf{r}}\mathbf{E}(\mathbf{r}_0,\omega)\right] - T\left[\mathbf{E}(\mathbf{r}_0,\omega)\right] \tag{8}$$

where T is the arrival time given in Eq. (1), but written as a function of the pulse spectrum rather than its temporal profile:

$$\langle t\rangle_{\mathbf{r}} = T\left[\mathbf{E}(\mathbf{r},\omega)\right] \tag{9}$$

where

$$T\left[\mathbf{E}(\mathbf{r},\omega)\right] \equiv -i\frac{\hat{\eta}\cdot\int\limits_{-\infty}^{\infty}\dfrac{\partial\mathbf{E}(\mathbf{r},\omega)}{\partial\omega}\times\dfrac{\mathbf{B}^{*}(\mathbf{r},\omega)}{\mu_0}\,d\omega}{\hat{\eta}\cdot\int\limits_{-\infty}^{\infty}\mathbf{S}(\mathbf{r},\omega)\,d\omega} \tag{10}$$

In Eq. (10), the function T depends only on the electric field since the magnetic field can be written in terms of the electric field using the complex index of refraction for the propagation medium.

The reshaping delay Δt_R is evaluated at $\mathbf{r}_0$, *before* propagation takes place. It is the difference between the pulse arrival time at $\mathbf{r}_0$ *without* and *with* the spectral amplitude that will be lost (in the absorbing case) during propagation. Since Δt_R uses the phase of the fields in the initial pulse, the reshaping delay is sensitive to how the pulse is organized. It measures the effect of absorption on a pulse and is sensitive to the pulse's state of chirp. If the pulse experiences roughly the same attenuation or amplification over its entire spectrum (as is the case for a pulse far from resonance or a sufficiently narrowband pulse near a resonance) the reshaping delay is minimal.

Equation (5) gives the delay between the 'center-of-mass' arrival times for any two points in a medium regardless of how the pulse evolves during transit. Since the form is obtained without approximation, it is valid for pulses with arbitrary bandwidths and for arbitrary propagation distances. This formalism does not attempt to define a specific 'group velocity' as being representative of the overall pulse propagation. Rather we have demonstrated that the delay between arrival times at two points is related to a linear superposition of group delays weighted by the pulse's spectral content, as in Eq. (6). By taking the limit of small separation a local velocity can be defined at each point along the propagation.[7] However, we prefer to consider the difference in arrival times at two discretely separated points since this is more directly applicable to experiment.

The formulation presented here can be thought of as a generalization of the traditional concept of group delay, and in the narrowband limit the delay time given by Eq. (5) reduces to the traditional expression derived using expansion techniques. In this limit the reshaping delay tends to zero, and the normalized spectral distribution approaches a delta function $\rho(\mathbf{r},\omega) \rightarrow \delta(\omega - \bar{\omega})$ where $\bar{\omega}$ is the central frequency of the pulse. The total delay is then calculated from Eq. (6) as $\Delta t = \partial \operatorname{Re} \mathbf{k} / \partial\omega|_{\bar{\omega}} \cdot \Delta\mathbf{r}$, in agreement with the well verified observation[9–13] that a pulse travels at the group velocity evaluated at $\bar{\omega}$ in the narrowband limit. It is important to note that group delay (in both broadband and narrowband cases) measures the delay between when pulse energy is *observed* at two points, and not necessarily the velocity at which pulse energy is *transported* from one point to the other. When a pulse is absorbed or amplified in the propagation medium, this delay does not necessarily indicate energy transport or information transfer since the propagation medium exchanges energy asymmetrically with the leading and trailing portions of a pulse.[14–16]

3. APPLICATION TO A BROADBAND PULSE

To illustrate the use of the formulation introduced above we consider a specific example of a broadband pulse whose spectrum is centered near a resonance in the medium. For this example we employ the Lorentz model with a single resonance frequency ω_0 and a damping frequency γ (please note, however, that the formulation presented above is independent of model). The complex index of refraction in this model is given by

$$[n(\omega)]^2 = 1 + \frac{f{\omega_p}^2}{{\omega_0}^2 - \omega^2 - i\gamma\omega}, \tag{11}$$

where ω_p is the plasma frequency and f is the oscillator strength. We choose the medium parameters as $\omega_0 = 50\gamma/7$, $\omega_p = 25\sqrt{5}\gamma/7$, and $f = 1$. (The parameters in this example closely correspond to those in Ref. [17].) The group delay function for these parameters is plotted in Fig. 1(a). If we ignore the frequencies near ω_0 (which will be absorbed after propagating a significant distance), Fig. 1(a) shows that low frequency spectral components have slightly longer delay times than high frequency components. This indicates that in the mature dispersion regime (i.e. when $\Delta\mathbf{r}$ is large) a pulse will be chirped such that high frequency components arrive at a point before low frequency components.

We consider a pulse with a Gaussian electric field at $\mathbf{r}_0$, given by

$$\mathbf{E}(\mathbf{r}_0, t) = \mathbf{E}_0 \exp\left(-t^2/\tau^2\right) \cos(\bar{\omega}t) \tag{12}$$

as it propagates through this medium. The pulse parameters are chosen as $\tau = 0.5/\gamma$ and $\bar{\omega} = 10\gamma$. This results in a pulse whose spectrum is broad compared to the resonance, and is centered slightly above ω_0. In this situation the traditional expansion method for group delay fails, since it predicts the delay time based on

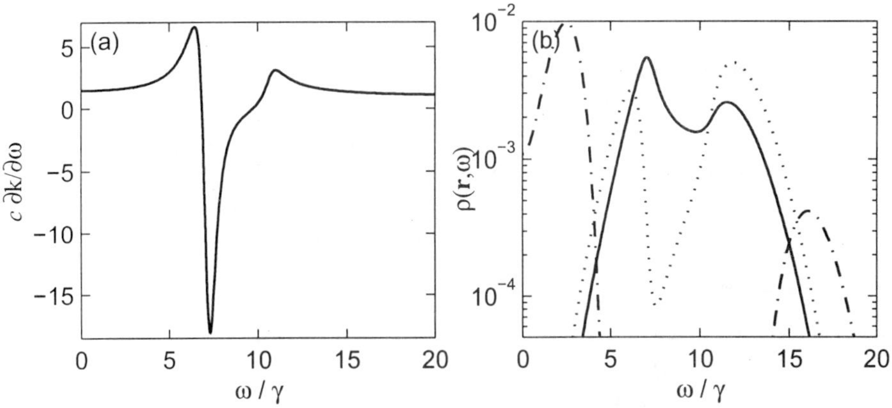

Figure 1. (a) Group delay function normalized by $\Delta\mathbf{r}/c$. (b) Normalized spectral distribution for propagation distances of 8.2×10^{-3} (solid), 0.82 (dotted), and 82 (dash-dot) absorption depths.

the group delay function evaluated at the central frequency in the *initial* pulse. However, in this example the energy associated with spectral components near $\bar{\omega}$ is quickly absorbed as the pulse propagates, so these spectral components do not influence the arrival at observation point located at points deeper in the medium. This shows why the prediction of propagation delay needs to be made using the spectrum that will survive propagation, as is done in Eq. (6).

Equation (5) can be used to predict the delay time for this pulse. Figure 1(b) shows the normalized spectral distribution $\rho(\mathbf{r}, \omega)$ for three values of $\Delta\mathbf{r}$. The relative net group delay $(c\Delta t_G/\Delta r)$ calculated using (6) is plotted in Fig. 2(a) as a function of the propagation depth. (The points at which the spectral distributions in Fig. 1(b) occur are marked with stars.) For small displacements, Δt_G is negative since the near-resonance components of the pulse spectrum still affect the expectation in Eq. (6). As the pulse propagates further into the medium and experiences absorption, the longer delay times of the off-resonance components dominate the expectation resulting in the eventual subluminal delay times generally observed for broadband pulses.

Even though the initial electric field in Eq. (12) is symmetric, the Poynting flux at $\mathbf{r}_0$ has a significant chirp. (This occurs because of the connection required by Maxwell's equations between **E** and **B** in a dielectric medium.) Thus, the reshaping delay has a noticeable effect on the total delay time at very shallow propagation depths (note the log scale on the propagation distance). At about one absorption depth the reshaping delay approaches a constant value, and the relative reshaping delay, $c\Delta t_R/\Delta r$, (plotted in Fig. 2(a)) goes quickly to zero as the propagation distance increases beyond this depth. At large propagation depths the total delay is dominated by the net group delay.

The total delay is given by the sum of the reshaping and net group delays as in Eq. (5). The relative total delay, $c\Delta t/\Delta r$, is plotted in Fig. 2(a). In the mature

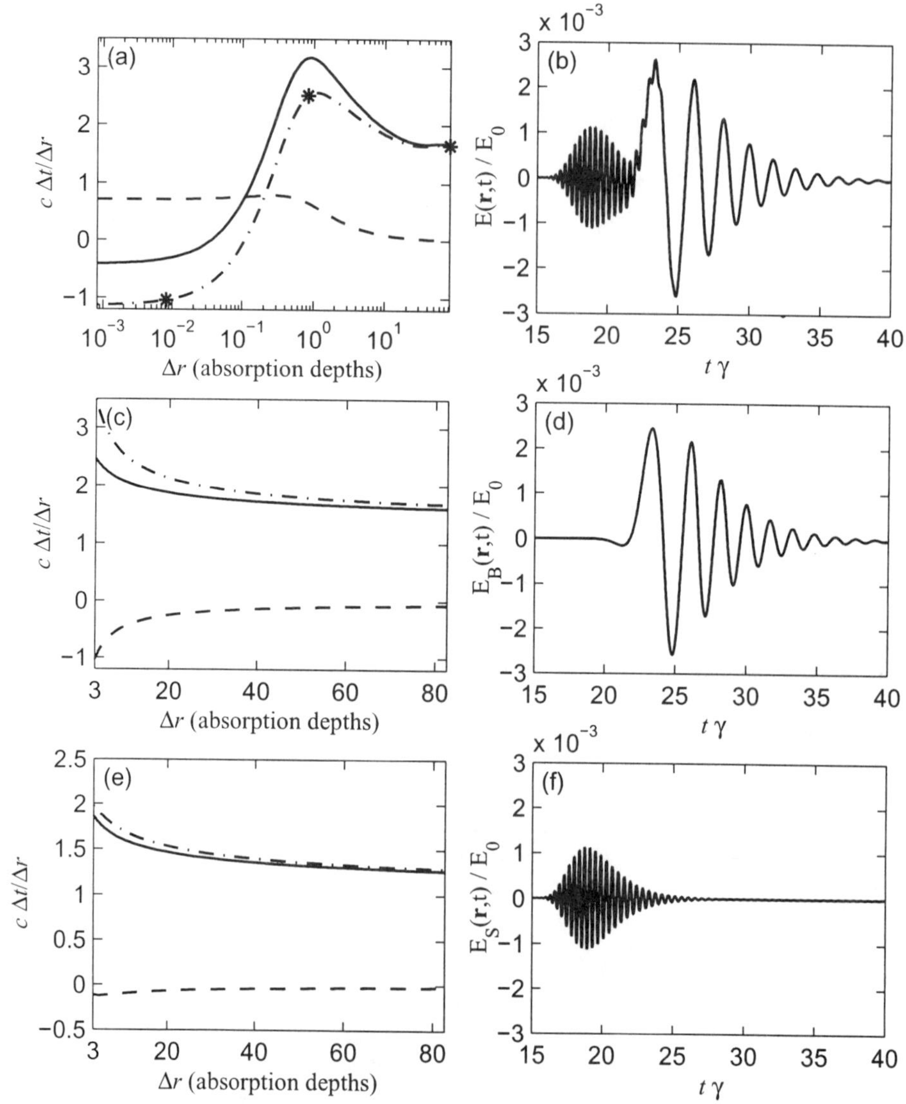

Figure 2. (a) Relative net group delay (dash-dot), reshaping delay (dash), and total delay (solid) as a function of propagation depth for the entire pulse. The stars indicate the absorption depths at which the distributions in Fig. 1(b) are evaluated. (b) Temporal profile of the pulse after propagating 82 absorption depths. (c) Relative net group delay (dash-dot), reshaping delay (dash), and total delay (solid) as a function of propagation depth for the low frequency component. (d) Temporal profile of the low frequency component after propagating 82 absorption depths. Figures (e)–(f) repeat (c)–(d) for the high frequency component.

dispersion regime the delay times settle into a stable situation where the delay of a pulse is given almost entirely by the net group delay. The reshaping delay plays a minimal role in this regime since absorption is small here (the spectral components that experience significant absorption having already been removed from the pulse). This behavior is easily observed in Fig. 2(a).

4. INDIVIDUAL PRECURSOR FIELDS

Figure 2(b) shows the temporal profile of the electric field for this pulse at 82 absorption depths into the medium. Notice that the high frequency components arrive first as would be predicted from the group delay function. For the pulse/medium parameters that we have used here, there are two distinct bumps in the final pulse. In the traditional asymptotic treatment of pulses in the mature dispersion limit, the high frequency component is referred to as the Sommerfeld precursor and the low frequency component is referred to as the Brillouin precursor. To this point we have considered only the delay of the pulse taken as a whole, so that the total delay is a weighted average of the delay times for the two individual precursor components. It is also interesting to consider the delay time of the two precursor fields separately.

Because the Fourier transform is a linear operation, we can separate the spectrum of a pulse propagating in a single resonance absorber (with resonance frequency ω_0) into two pieces as follows:

$$\mathbf{E}(\mathbf{r},\omega) = \mathbf{E}_B(\mathbf{r},\omega) + \mathbf{E}_S(\mathbf{r},\omega) \tag{13}$$

where

$$\mathbf{E}_B(\mathbf{r},\omega) = \begin{cases} \mathbf{E}(\mathbf{r},\omega) & , \quad |\omega| \le \omega_0 \\ 0 & , \quad |\omega| > \omega_0 \end{cases} \tag{14}$$

$$\mathbf{E}_S(\mathbf{r},\omega) = \begin{cases} 0 & , \quad |\omega| \le \omega_0 \\ \mathbf{E}(\mathbf{r},\omega) & , \quad |\omega| > \omega_0 \end{cases} \tag{15}$$

The temporal profile of the pulse is then given by

$$\mathbf{E}(\mathbf{r},t) = \mathbf{E}_S(\mathbf{r},t) + \mathbf{E}_B(\mathbf{r},t) \tag{16}$$

where the two terms in Eq. (16) are the inverse Fourier transforms of Eqs. (14) and (15). The term $\mathbf{E}_S(\mathbf{r},t)$ is then the 'above resonance' portion of the pulse, and $\mathbf{E}_B(\mathbf{r},t)$ is the 'below resonance' portion.

If a pulse's initial spectrum $\mathbf{E}(\mathbf{r}_0,\omega)$ has significant on-resonance amplitude, then $\mathbf{E}_S(\mathbf{r}_0,\omega)$ and $\mathbf{E}_B(\mathbf{r}_0,\omega)$ will have a significant discontinuity at ω_0. Because an abrupt discontinuity in frequency requires a broad temporal profile, $\mathbf{E}_S(\mathbf{r}_0,t)$ and $\mathbf{E}_B(\mathbf{r}_0,t)$ will be poorly localized. In this case the two terms have little significance individually except that when added together they produce the temporal profile of the total field. However, as the pulse propagates the on-resonance amplitude of $\mathbf{E}(\mathbf{r},\omega)$ decreases and the discontinuity becomes less important. Eventually the discontinuity becomes insignificant, and $\mathbf{E}_S(\mathbf{r},t)$ and $\mathbf{E}_B(\mathbf{r},t)$ represent two independent pulses. The delay times for these two pulses can be tracked separately using Eq. (5).

Figure 2(c) plots the relative delay times for the low frequency component $\mathbf{E}_B$ of the pulse considered in the preceding section. For this plot we place $\mathbf{r}_0$ (used to calculate Δt_R) three absorption depths beyond the point where the electric field is Gaussian so that the near-resonance portion of the pulse has been significantly attenuated. Figure 2(d) shows the temporal profile of the low frequency component at the same distance as Fig. 2(b). Figures 2(e)–(f) repeat these plots for the high frequency component $\mathbf{E}_S$. As with the total field, when propagation distance becomes large the reshaping delay becomes negligible and the total delay is given by the net group delay. At these large propagation distances the normalized spectral distribution for the low frequency component $\mathbf{E}_B$ becomes more tightly concentrated about $\omega = 0$ since frequency components near resonance experience the most absorption. Thus the delay time for $\mathbf{E}_B$ approaches the group delay at $\omega = 0$ in the limit of large propagation distance. The spectral distribution for the high frequency component, $\mathbf{E}_S$, shifts towards higher frequencies at large propagation distances as the near resonance components are absorbed. Thus, the high frequency component approaches an exactly luminal delay in the limit of large propagation distances.

5. CONCLUSION

In this article we have discussed a method for describing pulse delay which is valid for pulses of any bandwidth and showed how it may be applied in a broadband situation where precursor fields are observed. This method correctly describes the delay of a pulse taken in its entirety, and also the delay of the high and low frequency precursor components (i.e., the Sommerfeld and Brillouin precursors) individually. This method does not provide a full description of precursor fields, of course. However, when predicting delay times for a precursor fields, it has some advantages when compared to the asymptotic approach generally employed. The approach used here to study precursor behavior is independent of the model used to describe the medium. In addition, this method is easily generalizable to describe precursor fields in multiple resonance media. (For example, in a two resonance medium there are three precursor fields describing different spectral regions: one below the lower resonance, one above the higher resonance, and one between the two resonances.) Perhaps the most appealing advantage of the approach used here to analyze precursor behavior is its conceptual clarity; we are simply analyzing the behavior of different spectral portions of a pulse after it has propagated a large distance. This notion can sometimes be obscured by the details of the traditional approach.

REFERENCES

1. M. Born and E. Wolf, *Principles of Optics*, sixth ed. (Pergamon, Oxford, 1980).
2. J. D. Jackson, *Classical Electrodynamics*, 3rd ed. (Wiley, New York, 1998).
3. K. E. Oughstun and H. Xiao, Failure of the Quasimonochromatic Approximation for Ultrashort Pulse Propagation in a Dispersive, Attenuative Medium, Phys. Rev. Lett. **78,** 642–645 (1997).
4. H. Xiao and K. E. Oughstun, Failure of the group-velocity description for ultrawideband pulse propagation in a causally dispersive, absorptive dielectric, J. Opt. Soc. Am. B **16,** 1773–1785 (1999).
5. J. Peatross, S. A. Glasgow, and M. Ware, Average Energy Flow of Optical Pulses in Dispersive Media, Phys. Rev. Lett. **84,** 2370–2373 (2000).
6. M. Ware, S. A. Glasgow, and J. Peatross, The Role of Group Velocity in Tracking Field Energy in Linear Dielectrics, Opt. Express (2001).
7. R. L. Smith, The Velocities of Light, Am. J. Phys. **38,** 978–984 (1970).
8. M. Ware, W. E. Dibble, S. A. Glasgow, and J. Peatross, Energy Flow in Angularly Dispersive Optical Systems, J. Opt. Soc. Am. B **18,** 839–845 (2001).
9. R. Y. Chiao, Superluminal (but causal) propagation of wave packets in transparent media with inverted atomic populations, Phys. Rev. A **48,** R34–R37 (1993).
10. R. Y. Chiao and A. M. Steinberg, Tunneling Times and Superluminality, Progress in Optics **37,** 347–406 (Emil Wolf ed., Elsevier, Amsterdam, 1997).
11. S. Chu and S. Wong, Linear Pulse Propagation in an Absorbing Medium, Phys. Rev. Lett. **48,** 738–741 (1982).
12. E. L. Bolda, J. C. Garrison, and R. Y. Chiao, Optical pulse propagation at negative group velocities due to a nearby gain line, Phys. Rev. A **49,** 2938–2947 (1994).
13. C. G. B. Garrett and D. E. McCumber, Propagation of a Gaussian Light Pulse through an Anomalous Dispersion Medium, Phys. Rev. A **1,** 305–313 (1970).
14. S. A. Glasgow, M. Ware, and J. Peatross, Poynting's Theorem and Luminal Energy Transport Velocity in Causal Dielectrics, Phys. Rev. E **64,** 046610–1–046610–14 (2001).
15. J. Peatross, M. Ware, and S. A. Glasgow, The Role of the Instantaneous Spectrum in Pulse Propagation in Causal Linear Dielectrics, J. Opt. Soc. Am. A **18,** 1719–1725 (2001).
16. M. Ware, S. A. Glasgow, and J. Peatross, Energy Transport in Linear Dielectrics, Opt. Express, (2001).
17. K. E. Oughstun and C. M. Balictsis, Gaussian Pulse Propagation in a Dispersive, Absorbing Dielectric, Phys. Rev. Lett. **77,** 2210–2213 (1996).
18. K. E. Oughstun and G. C. Sherman, *Electromagnetic Pulse Propagation in Causal Dielectrics* (Springer-Verlag, New York, 1994).

TWO-DIMENSIONAL NARROW-WAISTED GAUSSIAN BEAM ANALYSIS OF PULSED PROPAGATION FROM EXTENDED PLANAR ONE-DIMENSIONAL APERTURE FIELD DISTRIBUTIONS THROUGH PLANAR DIELECTRIC LAYERS

Vincenzo Galdi and Leopold B. Felsen*

1. INTRODUCTION

Propagation of electromagnetic (EM) wavefields in the presence of layered dielectric media is a problem of long-standing interest, with applications to antenna radome design, guided propagation, etc. For this class of problems, most available analytic (rigorous and approximate) approaches are in the frequency domain (FD).[1,2] With a few exceptions,[3-7] time domain (TD) analysis is typically pursued via inversion from the FD. However, advances in ultra-wideband technology and the consequent more frequent use of space-time highly localized signals in communication and sensing systems motivate *direct* time domain (TD) analysis and processing, which is better matched to the short-pulse wave phenomenology and can therefore be expected to yield better numerical efficiency as well as deeper physical insight. These considerations have led to our investigation of a *direct* TD formulation of a previously developed FD Gabor-based narrow-waisted (NW) quasi-ray Gaussian beam (GB) algorithm for propagation of aperture-excited wave fields through planar dielectric layers. This approach is based on a NW-GB discretized decomposition of the aperture field distribution on the Gabor lattice and permits efficient *quasi-real complex ray tracing* (via the *complex source point* (CSP) method) of the thereby excited individual basis beams through the environment, with eventual recombination to synthesize the total field at the observer. In the FD, this approach has

* Vincenzo Galdi, Waves Group, Department of Engineering, University of Sannio, I-82100 Benevento, Italy. Leopold B. Felsen, Department of Aerospace and Mechanical Engineering and Department of Electrical and Computer Engineering, Boston University, Boston, Massachusetts 02215 (part-time). Also, University Professor Emeritus, Polytechnic University, Brooklyn, New York, 11201.

been applied successfully to propagation through arbitrarily shaped dielectric layers, for both two-dimensional (2-D) fields radiated by 1-D aperture distributions,[8,9] and 3-D fields radiated by 2-D aperture distributions.[10] Accurate predictions over calibrated parameter ranges have been obtained with modest computational effort. Basically, when dealing with electrically large domains, the NW-GB scheme preserves the favorable computational features of standard ray-optical techniques, *without* failing in typical ray-field transition regions.

In our TD analysis, to learn the rules, we have started with the simplest configuration involving 2-D fields excited by 1-D pulsed aperture distributions in the presence of a single planar dielectric layer. Using a procedure similar to that already developed in connection with TD discretization of pulsed aperture radiation in free space,[11] the FD narrow-waisted GB synthesis for the layer-transmitted field obtained by Maciel and Felsen[8] has been inverted analytically to the TD, yielding a *direct* closed form representation in terms of pulsed beam (PB) wavepackets. The detailed formulation, with several numerical examples, is presented in Galdi and Felsen.[12] In the present paper, we provide a compact review of the algorithm, and a number of *new numerical experiments* (not included in Galdi and Felsen[12]), involving abruptly-truncated and multilobe aperture field distributions.

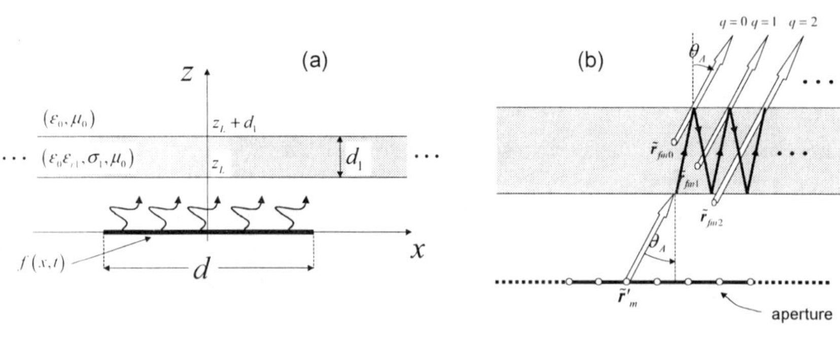

Figure 1. (a): Problem geometry and coordinate system. (b): Schematic interpretation of the *quasi-real* ray-tracing scheme pertaining to transmission of a single narrow-waisted basis beam through the layer. White arrows denote complex rays; black arrows denote real rays; "o" denotes the locations of actual and virtual (subscript *f*) complex source points.

2. STATEMENT OF THE PROBLEM

The problem geometry is illustrated in Fig. 1a. In a 2-D (x,z) reference system, an EM transient field with transverse magnetic (TM) polarization impinges on an infinite plane dielectric layer with lower interface at $z=z_L$, thickness d_1, relative permittivity

ε_{r1}, and electrical conductivity σ_1. Both ε_{r1} and σ_1 are assumed to be frequency independent, and attention is restricted to slightly lossy layers, i.e., $\sigma_1/(\omega\varepsilon_0\varepsilon_{r1}) \ll 1$, with ω being the angular frequency. The TD incident field e_y^i is generated by a 1-D truncated planar pulsed y-directed aperture field distribution which occupies the region $|x| \leq d/2$ at $z = 0$ where the space-time distribution of the incident tangential electric field is specified,

$$e_y^i(x, z = 0, t) = \begin{cases} f(x,t), & |x| \leq d/2, \\ 0, & |x| > d/2. \end{cases} \tag{1}$$

The assumed space-time aperture field distributions are separable with general nonlinear time delay,

$$f(x,t) = g(x)\, p\left[t - c_0^{-1}\phi(x)\right], \quad \phi(x) = x\sin\theta_A + \phi_{NL}(x), \tag{2}$$

where $g(x)$ is a spatial tapering function, $c_0 = (\varepsilon_0\mu_0)^{-1/2}$ is the free-space wavespeed, $p(t)$ is a time pulse with characteristic duration $T \ll d/c_0$, and $\phi(x)$ is a general nonlinear phase (delay) function, which is conveniently split into a linear part (parameterized by the real angle θ_A in Fig. 1b) plus a nonlinear remainder $\phi_{NL}(x)$. In computing the layer-transmitted EM field in the halfspace $z > z_L + d_1$, we restrict our attention to the y-directed electric field $e_y^t(x,z,t)$ from which all other field quantities of interest can be obtained via Maxwell's equations.

Throughout the paper, FD and TD quantities are identified by capital and lower-case letters, respectively. Moreover, the tilde ~ identifies dependence on analytically continued spatial source coordinates.

3. GABOR-BASED NARROW-WAISTED GAUSSIAN BEAM SYNTHESIS OF LAYER-TRANSMITTED FIELD

3.1 Frequency Domain Synthesis

As stated previously, the proposed PB synthesis for the layer-transmitted wave field is obtained via analytic Fourier inversion of the corresponding Gabor-based FD-GB synthesis. In the FD, the aperture field distribution is parameterized in terms of x-domain discretized m-indexed Gabor basis functions with narrow width $L_x \leq \lambda_0 \ll d$ (λ_0 being the free-space wavelength), centered on the Gabor lattice points $x_m = mL_x$; these initial conditions generate the NW, quasi-ray, CSP GBs. In Maciel and Felsen,[8] the *complex ray* machinery,[13,14] which allows *rigorous* tracking of the individual basis GBs through the layer, is shown to be simplified substantially for NW-GBs because this makes possible the use of efficient *quasi-real* ray tracing implemented through paraxial beam shooting. Basically, the single multi-hop GB is tracked along its real-ray axis via conventional real ray tracing; the slightly complex NW-GB spectrum of the emerging GB is accounted for approximately through augmentation of its on-axis (real ray) value by a complex phase correction in the (perpendicular) distance from the axis to the off-axis observer. In Galdi and Felsen,[12] a slightly different, though similar in spirit, approximation was developed, which facilitates analytic Fourier inversion to the TD. Here, we only give the final results, referring the reader to Galdi and Felsen[12] for details. The y-directed layer-transmitted field $E_y^t(x,z)$ is discretized approximately in terms of NW-GBs as

$$E_y^t(x,z) \approx \left(1-\mathrm{R}_{1A}^2\right) \sum_{|m|\le(d/2L_x)} A_m \sum_{q=0}^{N_q} \mathrm{R}_{1A}^{2q} \tilde{B}_m^{(q)}(x,z), \tag{3}$$

where $\tilde{B}_m^{(q)}$ are the GB propagators, and the (Gabor) expansion coefficients are obtained by sampling the FD counterpart of the aperture field distribution in Eq. (1),

$$A_m = \left(L_x/\sqrt{2}\right)^{1/2} g(x_m) P(\omega) \exp\left[ik_0\phi_{NL}(x_m)\right], \tag{4}$$

with $P(\omega)$ denoting the Fourier spectrum of the pulse $p(t)$, and $k_0 = \omega/c_0 = 2\pi/\lambda_0$ denoting the free-space wavenumber. Moreover, R_{1A} represents the plane-wave TM Fresnel reflection coefficient

$$\mathrm{R}_{1A} = \frac{\cos\theta_A - \sqrt{\varepsilon_{r1} - \sin^2\theta_A}}{\cos\theta_A + \sqrt{\varepsilon_{r1} - \sin^2\theta_A}}. \tag{5}$$

In Eq. (3), the q-index tags *roundtrip* reflections inside the dielectric layer (i.e., $2q$ *individual* reflections at each layer interface), with N_q denoting the total number of roundtrips retained. Apart from the explicit internal roundtrip reflection coefficient factor R_{1A}^{2q} in Eq. (3), the GB propagator $\tilde{B}_m^{(q)}$ represents the wavefield contribution at the observer, due to the m-th basis GB, after undergoing q internal roundtrip reflections, and is given by[12]

$$\tilde{B}_m^{(q)}(x,z) = -i2^{5/4}\left(z-\tilde{z}_{fq}\right)\sqrt{\frac{k_0 L_x}{8\pi \tilde{R}_{fmq}^3}} \exp\left[i\left(k_0\tilde{\Psi}_{mq} + \pi/4\right)\right], \tag{6}$$

where

$$\tilde{\psi}_{mq} = \tilde{R}_{fmq} + x_m \sin\theta_A + \frac{(1+2q)\left(\varepsilon_{r1}\cos^2\theta_t - \cos^2\theta_A\right)d_1}{\sqrt{\varepsilon_{r1}}\cos^3\theta_t} + ib, \tag{7}$$

and $\tilde{R}_{fmq}$ denotes the complex distance between the observation point (x,z) and the virtual complex source point $\tilde{\boldsymbol{r}}'_{fmq} \equiv \left(\tilde{x}_{fmq}, \tilde{z}_{fq}\right)$,

$$\tilde{R}_{fmq} = \sqrt{\left(x - \tilde{x}_{fmq}\right)^2 + \left(z - \tilde{z}_{fq}\right)^2}, \quad \operatorname{Re}\left(\tilde{R}_{fmq}\right) \geq 0, \tag{8}$$

$$\tilde{x}_{fmq} = x_m + \frac{(1+2q)\left(\sqrt{\varepsilon_{r1}}\sin\theta_t\cos^2\theta_t - \sin\theta_A\cos^2\theta_A\right)d_1}{\sqrt{\varepsilon_{r1}}\cos^3\theta_t} + ib\sin\theta_A, \tag{9}$$

$$\tilde{z}_{fq} = d_1 - \frac{(1+2q)d_1\cos^3\theta_A}{\sqrt{\varepsilon_{r1}}\cos^3\theta_t} + ib\cos\theta_A. \tag{10}$$

In Eqs. (7)-(10), the transmission angle θ_t is related to θ_A via Snell's law, $\sqrt{\varepsilon_{r1}}\sin\theta_t = \sin\theta_A$, whereas $b = \left(L_x\cos\theta_A\right)^2/\lambda_0$ is the GB-CSP displacement parameter (Fresnel length). Referring to Fig. 1b, the approximation in Eqs. (6)-(10) corresponds to tracing a ray along a *complex* trajectory (white arrow) from the CSP at $\boldsymbol{r}'_m \equiv \left(x_m + ib\sin\theta_A, d_1 + ib\cos\theta_A\right)$ to the intersection of the *real* beam axis with the *real* layer (lower) interface; from there, the path into the layer proceeds entirely in *real* configuration space, along the beam axis, undergoing q roundtrip reflections between the layer boundaries (black arrows). The last segment, which reaches the observer, proceeds in *complex* space, as a *complex ray* (white arrows) emanating from the complex virtual focus $\tilde{\boldsymbol{r}}'_{fmq}$. In real configuration space, these complex rays describe NW paraxial GBs.

3.2 Time Domain Synthesis

The FD GB synthesis for the layer-transmitted field can be transformed analytically to the TD using a procedure similar to that in Galdi and Felsen[11], yielding the approximate NW-PB synthesis[12]

$$e_y^t(x,z,t) \approx \left(1 - \mathrm{R}_{1A}^2\right) \sum_{|m| \leq (d/2L_x)} a_m \sum_{q=0}^{N_q} \mathrm{R}_{1A}^{2q} b_m^{(q)}(x,z,t-t_m), \tag{11}$$

where $b_m^{(q)}$ are the PB propagators, and the amplitude coefficients a_m and time delays t_m are obtained by sampling the aperture field distribution amplitude and nonlinear phasing, respectively, at lattice points x_m,

$$a_m = \left(L_x/\sqrt{2}\right)^{1/2} g(x_m), \quad t_m = \phi_{NL}(x_m)/c_0. \tag{12}$$

For Rayleigh pulses,

$$p(t) = P_0 \frac{d^j}{dt^j} \exp\left[-\left(\frac{t-T/2}{\varsigma}\right)^2\right], \tag{13}$$

and subject to certain constraints[12] which are validated subsequently by numerical experiments, the PB propagators $b_m^{(q)}$ can be evaluated analytically in closed form,

$$\begin{aligned} b_m^{(q)}(x,z,t) = \mathrm{Re}\Bigg\{(-i)^j \beta_{mq}\Bigg[&\bar{T}_{mq}\Gamma\left(\frac{3+2j}{4}\right)\mathrm{M}_1^{(j)}\left(\frac{\bar{t}_{mq}}{\bar{T}_{mq}}\right) \\ &-2i\bar{t}_{mq}\Gamma\left(\frac{5+2j}{4}\right)\mathrm{M}_2^{(j)}\left(\frac{\bar{t}_{mq}}{\bar{T}_{mq}}\right)\Bigg]\Bigg\}, \end{aligned} \tag{14}$$

where $\Gamma(\cdot)$ is the gamma function,[15] and

$$\mathrm{M}_1^{(j)}(t) = {}_1\mathrm{F}_1\left(\frac{3+2j}{4}, \frac{1}{2}, -t^2\right), \quad \mathrm{M}_2^{(j)}(t) = {}_1\mathrm{F}_1\left(\frac{5+2j}{4}, \frac{3}{2}, -t^2\right), \tag{15}$$

with ${}_1\mathrm{F}_1$ denoting the Kummer confluent hypergeometric function.[15] The functions $\mathrm{M}_{1,2}^{(j)}$ can be computed efficiently using the rapidly converging expansions in Galdi and Felsen.[12] Moreover,

$$\bar{t}_{mq} = t - \tau_{mq} - T/2, \quad \beta_{mq} = 2^{j+1/2}\pi^{-1/2}\left(\bar{T}_{mq}\right)^{-j-5/2} \Lambda_{mq} P_0 \varsigma T, \tag{16}$$

$$\bar{T}_{mq} = \sqrt{T_{mq}^2 + \varsigma^2 T^2}, \quad T_{mq} = \frac{L_x \cos\theta_A}{c_0}\left[\frac{2}{\pi}\left(1 - \frac{z_{bmq}}{R_{fmq}}\right)\right]^{1/2}, \tag{17}$$

$$z_{bmq} = \left(x - x_{fmq}\right)\sin\theta_A + \left(z - z_{fq}\right)\cos\theta_A, \tag{18}$$

$$\tau_{mq} = c_0^{-1}\left[R_{fmq} + x_m \sin\theta_A + \frac{(1+2q)\left(\varepsilon_{r1}\cos^2\theta_t - \cos^2\theta_A\right)d_1}{\sqrt{\varepsilon_{r1}}\cos^3\theta_t}\right], \tag{19}$$

$$\Lambda_{mq} = -i2^{5/4}\sqrt{\frac{L_x}{8\pi c_0}}\left(z - z_{fq}\right)R_{fmq}^{-3/2}\exp\left(-\alpha R_{Lq} + i\pi/4\right), \tag{20}$$

$$\alpha = \frac{\sigma_1}{2c_1\varepsilon_0\varepsilon_{r1}}, \quad R_{Lq} = (1+2q)d_1/\cos\theta_t, \quad R_{fmq} = \sqrt{\left(x - x_{fmq}\right)^2 + \left(z - z_{fq}\right)^2}, \tag{21}$$

$$x_{fmq} = x_m + \frac{(1+2q)\left(\sqrt{\varepsilon_{r1}}\sin\theta_t\cos^2\theta_t - \sin\theta_A\cos^2\theta_A\right)d_1}{\sqrt{\varepsilon_{r1}}\cos^3\theta_t}, \tag{22}$$

$$z_{fq} = d_1 - \frac{(1+2q)d_1\cos^3\theta_A}{\sqrt{\varepsilon_{r1}}\cos^3\theta_t}. \tag{23}$$

In Eq. (21), $c_1 = c_0/\sqrt{\varepsilon_{r1}}$ denotes the wavespeed in the dielectric layer.

In order to calibrate systematically the algorithm and estimate its range of applicability, it is useful to formalize the constraints arising from the sequence of approximations utilized in the analysis in terms of nondimensional estimators which embody the important (suitably scaled) problem parameters. We found that the most stringent overall constraint can be expressed as (see Galdi and Felsen[12] for details)

$$Q \equiv N_b^{-1}\sqrt{\frac{\kappa\cos^3\theta_A}{\chi}} \ll 1, \tag{24}$$

where the integer $N_b \equiv d/L_x$ represents the number of beams in the expansion (11), $\kappa = \Omega T/(2\pi)$ (with Ω denoting the pulse bandwidth), and $\chi = (z-d_1)/F_d$ is the distance to the observation plane, reduced by the layer thickness d_1 and scaled by the Fresnel distance of the aperture, $F_d = d^2/(c_0T)$. The nondimensional estimator Q contains a combination of useful physical problem parameters, through which the change in one can be compensated by a change in the others within the constraint. It is observed, for instance, that at *larger* observation distances z, *fewer* beams are required.

4. NUMERICAL RESULTS

An extensive series of numerical examples, including smoothly-tapered aperture field distributions with linear and nonlinear (e.g., focusing) phasing, have been presented in Galdi and Felsen[12]. Here, we focus on two interesting aperture distributions that have not been considered in Galdi and Felsen[12]. To validate and calibrate the PB synthesis, we use a reference solution based on brute-force integration of the exact FD spectral (plane wave) representation and inverse FFT (see Galdi and Felsen[12] for details).

We begin with the case of a sine-tapered linearly-phased aperture field distribution,

$$g(x) = \sin(\pi x/d), \quad \phi_{NL}(x) = 0, \tag{25}$$

which has a null at the center and also vanishes at the aperture edges, thus exhibiting two lobes. For time-harmonic excitation, the deviation between the free-space and transmitted nulls (boresight error) is typically used in radome applications to define the effect of the layer. For pulsed excitation, where the radiated and transmitted fields may have several nulls/minima which depend on the excitation waveform, identifying a TD boresight error is not straightforward and will not be considered further here. Instead, we limit ourselves to showing that the NW-PB synthesis remains accurate in the presence of multiple lobes in the aperture distribution. Typical results, for the near-zone layer-transmitted field, are shown in Fig. 2. Specifically, Figs. 2a and 2b show the comparison between the PB synthesis and the reference solution for the temporal behavior at a fixed observation point, and for a transverse (x) cut at a fixed time, respectively, in the near zone. Very good agreement is observed. Note that the layer thickness has been selected so as to allow time resolution of the various roundtrip reflections. The convergence, in terms of the normalized r.m.s. error Δe, with respect to Q and N_q is illustrated in Figs. 2c and 2d, respectively, for a fixed observation point in the near zone. One notes that accurate syntheses ($\Delta e \leq -30$ dB) are achieved with values of $Q < 0.1$ and $N_q \geq 3$.

As a further example, we consider an aperture distribution with

$$g(x) = 1, \quad \phi_{NL}(x) = 0, \tag{26}$$

which is discontinuous at the aperture edges, thus producing ray shadow boundary transitions regions. Typical results are shown in Figs. 3a-3d. It is observed that the NW-PB synthesis is actually able to capture the edge diffraction effects, and exhibits the same convergence characteristics as in the previous example.

The reader is referred to Galdi and Felsen[12] for further examples, details of the calibration and assessment of the range of applicability, as well as for discussion of computational costs. As a general conclusion, accurate and robust predictions were observed within the range Q<0.1 and with N_q chosen such that the magnitude of the first omitted term is less than 0.1% of that of the leading term.

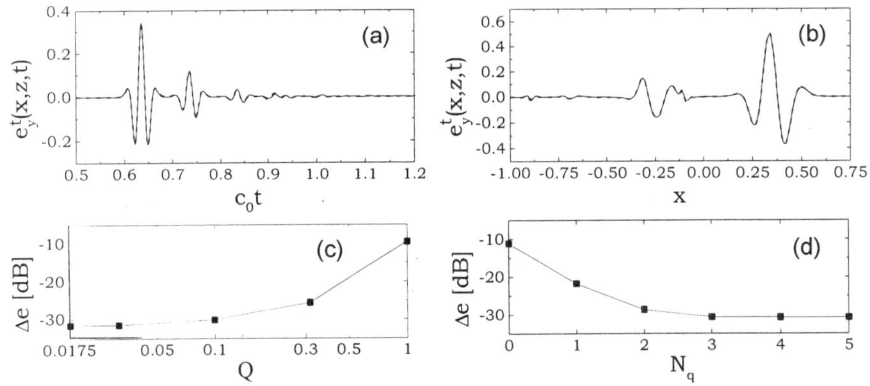

Figure 2. TD transmitted field: Sine-tapered linear-delay aperture field distribution in Eq. (25), with $\theta_A = 10°$. Geometry as in Fig. 1a, with $d = 1$ (arbitrary units), $c_0T = 0.1d$, $z_L = 0$, $d_1 = 0.5c_1T$, $\varepsilon_{r1} = 10$, $\sigma_1 = 0.01$ S/m. Pulsed excitation as in Eq. (13), with $j = 4$, $P_0 = T^4/30000$, $\varsigma = 1/\sqrt{50}$. Comparison between PB synthesis and reference solution for layer-transmitted field in the near zone. (a): Temporal profile at $(x = 0.25, z = d_1 + 0.5)$. (b): Spatial transverse profile at $z = d_1 + 0.5$ and $c_0t = 0.65$. —— Reference solution; - - - PB synthesis with $N_b = 120$ beams and $N_q = 3$ internal reflections. (c), (d): Normalized r.m.s. error vs. Q (at $N_q = 3$) and vs. N_q (at Q=0.1), respectively, for observation point at $(x = 0.25, z = d_1 + 0.5)$.

The normalized r.m.s. error is defined as $\Delta e = \int \left(e^t_{yb} - e^t_{yr}\right)^2 dt \Big/ \left[\int \left(e^t_{yb}\right)^2 dt \int \left(e^t_{yr}\right)^2 dt\right]^{1/2}$, with the subscripts "b" and "r" denoting the PB and the reference solution, respectively.

5. CONCLUSIONS

A *direct* TD formulation of a previously developed FD Gabor-based narrow-waisted quasi-ray Gaussian beam algorithm[8] has been presented for short-pulse 2-D wavefields radiated by 1-D aperture field distributions and propagated through planar slightly lossy dielectric layers. Numerical examples involving abruptly-truncated and multilobe aperture field distributions have been presented. Validations and calibrations against a numerical independently generated rigorous reference solution have yielded encouraging results. The TD extension of previously developed FD beam algorithms to propagation through *curved* layers[9-10] is under consideration, as is the extension to multiple layer configurations.

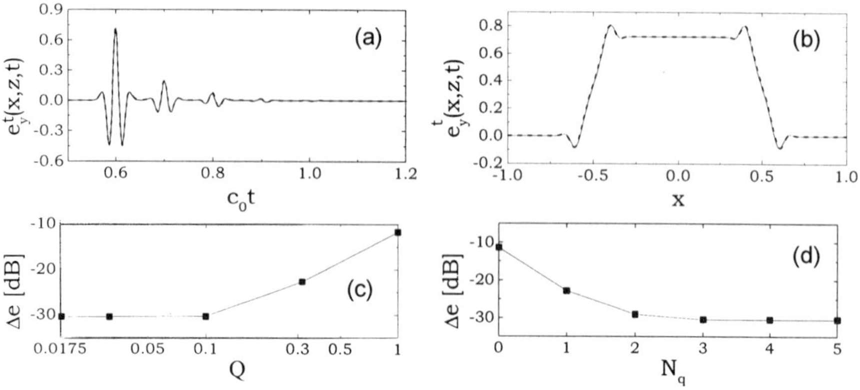

Figure 3. As in Fig. 2, but for the rectangular aperture field distribution in Eq. (26), with $\theta_A = 0$. (a): Temporal profile at $(x = 0, z = d_1 + 0.5)$. (b): Spatial transverse profile at $z = d_1 + 0.5$ and $c_0 t = 0.6$. —— Reference solution; - - - PB synthesis with $N_b = 120$ and $N_q = 3$. (c), (d): Normalized r.m.s. error vs. Q (at $N_q = 3$) and vs. N_q (at Q=0.1), respectively, for observation point at $(x = 0, z = d_1 + 0.5)$.

ACKNOWLEDGEMENTS

The work of V. Galdi was supported by a European Union postdoctoral fellowship through the University of Sannio, Benevento, Italy. L.B. Felsen acknowledges partial support from Grant No. 9900448 by the US-Israel Binational Science Foundation, Jerusalem, Israel, and from Polytechnic University, Brooklyn, NY 11201 USA.

REFERENCES

1. L.B. Felsen and N. Marcuvitz, *Radiation and Scattering of Waves* (Prentice Hall, Englewood Cliffs, 1973; Classic reissue, IEEE Press, Piscataway, 1994).
2. L.M. Brekhovskikh, *Waves in Layered Media* (Academic Press, New York, 1980).
3. L.B. Felsen and F. Niu, Spectral analysis and synthesis options for short pulse radiation from a point dipole in a grounded dielectric layer, *IEEE Trans. Antennas Propagat.* **41**(6), 747-754 (1993).
4. F. Niu and L.B. Felsen, Time-domain leaky modes on layered media: dispersion characteristics and synthesis of pulsed radiation, *IEEE Trans. Antennas Propagat.* **41**(6), 755-761 (1993).
5. F. Niu and L.B. Felsen, Asymptotic analysis and numerical evaluation of short pulse radiation from a point dipole in a grounded dielectric layer, *IEEE Trans. Antennas Propagat.* **41**(6), 762-769 (1993).
6. R.W.P. King and S.S. Sandler, The electromagnetic field of a vertical electric dipole in the presence of a three-layered region, *Radio Sci.* **29**(1), 97-113 (1994).

7. A.G. Tijhuis and A.R. Bretones, Transient excitation of a layered dielectric medium by a pulsed electric dipole, *IEEE Trans. Antennas Propagat.* **48**(10), 1673-1684 (2000).
8. J.J. Maciel and L.B. Felsen, Gaussian beam analysis of propagation from an extended aperture distribution through dielectric layers, Part I - plane layer, *IEEE Trans. Antennas Propagat.* **38**(10), 1607-1617 (1990).
9. J.J. Maciel and L.B. Felsen, Gaussian beam analysis of propagation from an extended aperture distribution through dielectric layers, Part II - circular cylindrical layer, *IEEE Trans. Antennas Propagat.* **38**(10), 1618-1624 (1990).
10. J.J. Maciel and L.B. Felsen, Gabor-based narrow-waisted Gaussian beam algorithm for transmission of aperture-excited 3D vector fields through arbitrarily shaped 3D dielectric layers, to be published in *Radio Sci.*, **38**(2) (2002).
11. V. Galdi, L.B. Felsen, and D.A. Castañon, Narrow-waisted Gaussian beam discretization for two-dimensional time-dependent radiation from large apertures, *IEEE Trans. Antennas Propagat.* **49**(9), 1322-1332 (2001).
12. V. Galdi and L.B. Felsen, Two-dimensional pulsed propagation from extended aperture field distributions through planar dielectric layers via quasi-ray Gaussian beams, to be published in *IEEE Trans. Antennas and Propagat.*, **51**(6) (2003).
13. W.D. Wang and G.A. Deschamps, Application of complex ray tracing to scattering problems, *Proc. IEEE* **62**(11), 7541-7551 (1974).
14. X.J. Gao and L.B. Felsen, Complex ray analysis of beam transmission through two-dimensional radomes, *IEEE Trans. Antennas Propagat.* **33**(9), 963-975 (1985).
15. M. Abramowitz and I.A. Stegun, *Handbook of Mathematical Functions* (Dover, New York, 1964).

AXIAL SCATTERING FROM THIN CONES

Carl E. Baum*

1. INTRODUCTION

A recent paper[4] considers the physical-optics (PO) approximation for early-time scattering. For axial backscattering examples of perfectly conducting bodies of revolution (BOR), including paraboloids and circular cones, seem to have good PO results as compared to exact and asymptotic results. However, the perfectly conducting angular sector is not approximated at all by PO (giving zero axial backscattering) for the case of the incident electric field parallel to the plane of the angular sector.

It would then seem useful to understand scattering from cones of various cross sections and compare the results to the PO approximation.[11] The general problem with arbitrary interior cone angle has not been solved. However, we know the exact form the solution must take and that it factors the angular dependence from the temporal/frequency dependence. This applies to various types of cones (e.g., dielectric), but our examples here are for perfectly conducting cones.

The incident field is given by

$$\vec{E}^{(inc)}(\vec{r},t) = E_0 f\left(t - \frac{\vec{1}_i \cdot \vec{r}}{c}\right)\vec{1}_e, \qquad \tilde{\vec{E}}^{(inc)}(\vec{r},t) = E_0 \tilde{f}(s) e^{-\gamma \vec{1}_i \cdot \vec{r}}\,\vec{1}_e,$$

$$\vec{H}^{(inc)}(\vec{r},t) = \frac{E_0}{Z_0} f\left(t - \frac{\vec{1}_i \cdot \vec{r}}{c}\right)\vec{1}_i \times \vec{1}_e, \qquad \tilde{\vec{H}}^{(inc)}(\vec{r},s) = \frac{E_0}{Z_0} \tilde{f}(s) e^{-\gamma \vec{1}_i \cdot \vec{r}}\,\vec{1}_i \times \vec{1}_e, \tag{1.1}$$

* Carl Baum, Air Force Research Laboratory, Directed Energy Directorate, Kirtland Air Force Base, NM 87117-5776.

where $\sim$ denotes the two-sided Laplace transform over time t, $c = 1/(\mu_0\varepsilon_0)^{1/2}$ is the speed of light, $Z_0 = (\mu_0/\varepsilon_0)^{1/2}$ is the wave impedance, $s = \Omega + j\omega$ is the Laplace transform variable or complex frequency, $\gamma = s/c$ is the propagation constant, $f(t)$ is the waveform,

$$\vec{1}_e \cdot \vec{1}_e = 0, \quad \vec{1}_i \equiv \text{direction of incidence}, \quad \vec{1}_e \equiv \text{polarization}.$$

The scatterer (target) being a cone, the cone tip is taken as the coordinate origin ($\vec{r} = \vec{0}$).

The scattered far field is given by (general case)

$$\vec{E}_f^{(sc)}(\vec{r},t) = \frac{1}{4\pi r}\overleftrightarrow{\Lambda}(\vec{1}_0, \vec{1}_i;t) \circ \vec{E}^{(inc)}\left(\vec{0}, t - \frac{r}{c}\right),$$

$$\tilde{\vec{E}}_f^{(sc)}(\vec{r},s) = \frac{e^{-\gamma r}}{r}\tilde{\overleftrightarrow{\Lambda}}(\vec{1}_0, \vec{1}_i;s) \cdot \vec{E}^{(inc)}\left(\vec{0}, s\right), \tag{1.2}$$

where $r = |\vec{r}|$, $\circ \equiv$ is convolution with respect to time,

$\tilde{\overleftrightarrow{\Lambda}}^{(sc)}(\vec{1}_0, \vec{1}_i;s)$ is the scattering dyadic,

$\tilde{\overleftrightarrow{\Lambda}}^{(sc)}(\vec{1}_0, \vec{1}_i;t) \circ$ is the scattering-dyadic operator,

$\overleftrightarrow{\Lambda}(\vec{1}_0, \vec{1}_i;t)$ is the scattering-dyadic impulse response,

$\vec{1}_o$ is the scattering direction (direction to observer), and

$\vec{E}_f^{(sc)}(\vec{r},t) \cdot \vec{1}_o = 0.$

For backscattering we have

$$\vec{1}_0 = -\vec{1}_i, \qquad \overleftrightarrow{\Lambda}_b(\vec{1}_i;t) \equiv \overleftrightarrow{\Lambda}(-\vec{1}_i, \vec{1}_i;t). \tag{1.3}$$

Reciprocity implies

$$\overleftrightarrow{\Lambda}(\vec{1}_0, \vec{1}_i;t) \equiv \overleftrightarrow{\Lambda}^T(-\vec{1}_i, -\vec{1}_0;t), \qquad \overleftrightarrow{\Lambda}_b(\vec{1}_i;t) \equiv \overleftrightarrow{\Lambda}_b^T(\vec{1}_i;t). \tag{1.4}$$

These relations are general and not specific to the cone. We also have the transverse dyadics

$$\overrightarrow{1}_i \equiv \overleftrightarrow{1} - \overrightarrow{1}_i \overrightarrow{1}_i, \quad \overrightarrow{1}_o \equiv \overleftrightarrow{1} - \overrightarrow{1}_o \overrightarrow{1}_o, \quad \overrightarrow{1}_r \equiv \overleftrightarrow{1} - \overrightarrow{1}_r \overrightarrow{1}_r,$$

$$\overrightarrow{1}_z \equiv \overleftrightarrow{1} - \overrightarrow{1}_z \overrightarrow{1}_z, \quad \overleftrightarrow{1} - \overrightarrow{1}_x \overrightarrow{1}_x + \overrightarrow{1}_y \overrightarrow{1}_y + \overrightarrow{1}_z \overrightarrow{1}_z \equiv \text{three dimensional identity,} \tag{1.5}$$

for later use.

In Ref. 4 we introduced some notation for multiple time integrals of the scattering-dyadic impulse response as

$$\int_{0_-}^{t}\int_{0_-}^{t_1}\cdots\int_{-\infty}^{t_{n-1}} \overleftrightarrow{\Lambda}(\overrightarrow{1}_o, \overrightarrow{1}_i; t_n)dt_n dt_n \cdots dt_1 = \int_{0_-}^{t}\int_{0_-}^{t_1}\cdots\int_{-\infty}^{t_{n-1}} \overleftrightarrow{\Lambda}(\overrightarrow{1}_o, \overrightarrow{1}_i; t_n)dt_n dt_n \cdots dt_1 \tag{1.6}$$

$$\equiv \overleftrightarrow{\Lambda}^{(n)}(\overrightarrow{1}_o, \overrightarrow{1}_i; t),$$

where

$$\overleftrightarrow{\Lambda}^{(0)}(\overrightarrow{1}_o, \overrightarrow{1}_i; t) \equiv \overleftrightarrow{\Lambda}(\overrightarrow{1}_o, \overrightarrow{1}_i; t), \qquad \overleftrightarrow{\Lambda}^{(n)}(\overrightarrow{1}_o, \overrightarrow{1}_i; s) = s^{-n}\, \tilde{\overleftrightarrow{\Lambda}}(\overrightarrow{1}_o, \overrightarrow{1}_i; s),$$

and similarly for backscattering.

For the cone we have the general form of the solution (general cone with dilation symmetry)[10]

$$\overleftrightarrow{\Lambda}(\overrightarrow{1}_o, \overrightarrow{1}_i; t) = cu(t)\overleftrightarrow{K}(\overrightarrow{1}_o, \overrightarrow{1}_i), \qquad \overleftrightarrow{\Lambda}(\overrightarrow{1}_o, \overrightarrow{1}_i; s) = \frac{c}{s}\overleftrightarrow{K}(\overrightarrow{1}_o, \overrightarrow{1}_i),$$

$$\overleftrightarrow{\Lambda}^{(1)}(\overrightarrow{1}_o, \overrightarrow{1}_i; t) = ct\,u(t)\overleftrightarrow{K}(\overrightarrow{1}_o, \overrightarrow{1}_i), \qquad \overleftrightarrow{\Lambda}^{(2)}(\overrightarrow{1}_o, \overrightarrow{1}_i; t) = \frac{1}{2}ct^2u(t)\,\overleftrightarrow{K}(\overrightarrow{1}_o, \overrightarrow{1}_i), \tag{1.7}$$

$$\overleftrightarrow{K}(\overrightarrow{1}_o, \overrightarrow{1}_i) = \overleftrightarrow{K}^T(-\overrightarrow{1}_i, -\overrightarrow{1}_o), \qquad \overleftrightarrow{K}_b(\overrightarrow{1}_i) = \overleftrightarrow{K}(-\overrightarrow{1}_i, \overrightarrow{1}_i) = \overleftrightarrow{K}_b^T(\overrightarrow{1}_i).$$

In the present paper we apply this result to thin cones for the special case of axial incidence. This will give some general results and allow us to calculate the axial backscatter from a perfectly conducting thin angular sector.

2. LOCAL TWO-DIMENSIONAL APPROXIMATION FOR THIN CONE

Let, as in Fig. 2.1, there be a thin cone centered on the +z axis with

$$\overrightarrow{1}_o = \overrightarrow{1}_z \tag{2.1}$$

Let this cone be characterized by an interior angle ψ which is related to some characteristic dimension of $z_S \tan(\psi)$ on a cross section of constant $z = z_S$. This might be the maximum distance from the z axis or any other convenient reference.

Where then should we place the z axis? It clearly should pass through the cone tip. If the cone cross section has a symmetry plane it should lie on this. If it has a rotation axis this should be the z axis. Such cases include the circular cone, elliptic cone and angular sector (perfectly conducting) treated in the appendices.

Going deeper into the matter, imagine for a moment that the direction of incidence is not parallel to this axis. Then there is a net current and a net charge per unit length (for conductors and dielectrics) and corresponding magnetic parameters if permeable magnetic materials are used. So let us choose the z axis so that incidence parallel to this produces no net current (and hence no net charge per unit length) and similarly for magnetic current (magnetic polarization current). This of course assumes that the electric and magnetic axes so defined coincide. For cross sections with rotation axes (C_N for $N \geq 2$ including both shape and constitutive parameters[9]) this happens automatically. One can consider off-axis incidence for cases such as perfectly conducting cones with the arbitrary cross section given an effective radius for use of the thin-circular-cone results.

With the direction of incidence along the z axis then the net charge per unit length (and similar magnetic parameter) are zero. We can then go on to the induced transverse electric- and magnetic-dipole moments per unit length. With ψ small we can locally think of the cross section in Fig. 2.1B as part of a two-dimensional structure extended in the $\pm \vec{1}_z$ directions. In the presence of the incident fields we can look at the total field distribution around and near the cone cross section as quasi static. What we need to calculate are the electric and magnetic polarizabilities per unit length so as to obtain the electric- and magnetic-dipole moments per unit length via

$$\tilde{\vec{p}}'(z_s,s) = \varepsilon_0 \tilde{\overleftrightarrow{p}}'(z_s,s) \cdot \tilde{\vec{E}}^{(inc)}(z_s \vec{1}_z, s), \qquad \tilde{\vec{m}}'(z_s,s) = \tilde{\overleftrightarrow{M}}'(z_s,s) \cdot \tilde{\vec{H}}^{(inc)}(z_s \vec{1}_z, s). \quad (2.2)$$

At this point it is useful to note certain properties (symmetries) of these polarizabilities per unit length. The very forms in (2.2) exhibit an invariance to reversal of the signs of the incident fields, equivalent to a rotation by π in the x, y plane which is C_2 symmetry.[9] This is in addition to any additional two-dimensional point symmetries of the electric and/or magnetic properties of the scatterer cross section on a constant z plane.

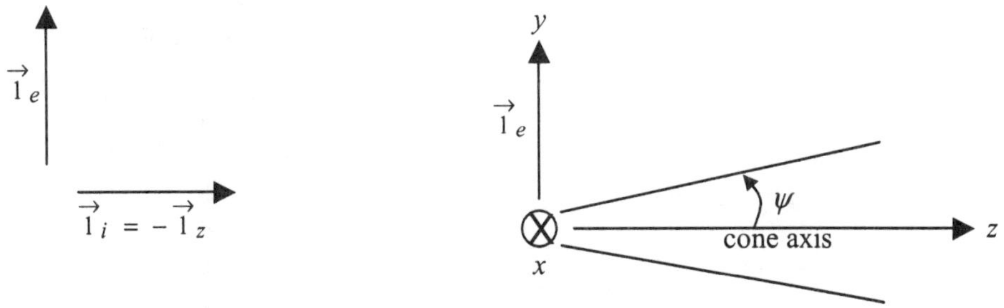

A. Side view

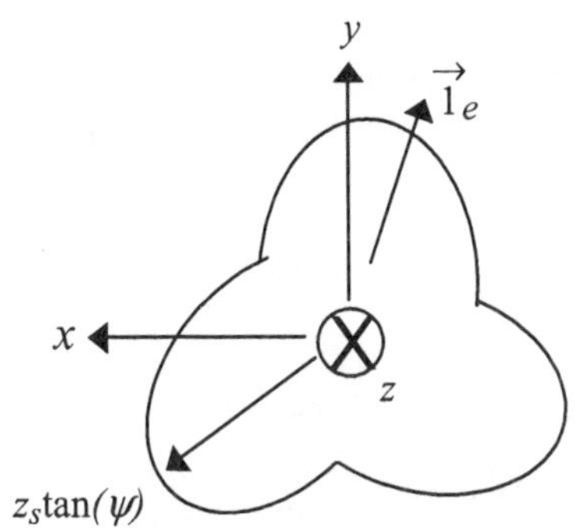

B. Cross-section view (plane of constant *z*)

Figure 2.1 Axial incidence on thin cone of general cross section.

For perfectly conducting scatterers one can solve for various cross-section geometries by conformal transformation (as in the appendices). There is also a symmetry on interchange of the transverse electric and magnetic fields (similar to self duality[9]). Considering a cross section as in Fig. 2.1B and noting that both fields are derived from the same complex potential we note that an incident electric field in the x direction has the same effect as an incident magnetic field in the *y* direction except for a change in sign. (The two-dimensional perfectly conducting body concentrates electric flux (positive polarizability) and excludes magnetic flux (negative polarizability.) In two-dimensional form we define

$$\overleftrightarrow{\tau}_z = \begin{pmatrix} 0 & -1 \\ 1 & 0 \end{pmatrix}, \qquad \overleftrightarrow{\tau}_z^{-1} = -\overleftrightarrow{\tau}_z = \overleftrightarrow{\tau}_z^{T}, \tag{2.3}$$

so that $\overleftrightarrow{\tau}_z \cdot$ is equivalent to $\vec{1}_z \times$, a positive $\pi/2$ rotation in the *x, y* plane. In three dimensions one can use

$$\overleftrightarrow{\tau}_z = \begin{pmatrix} 0 & -1 & 0 \\ 1 & 0 & 0 \\ 0 & 0 & 0 \end{pmatrix}. \tag{2.4}$$

Note also that $\cdot\ \overleftrightarrow{\tau}_z$ is equivalent to $\times\ \overrightarrow{1}_z\ \cdot$; so we have

$$\begin{aligned} \overleftrightarrow{M}' &= \overleftrightarrow{\tau}_z \cdot \overleftrightarrow{P}' \cdot \overleftrightarrow{\tau}_z = \overrightarrow{1}_z \times \overleftrightarrow{P}' \times \overrightarrow{1}_z, \\ \overleftrightarrow{P}' &= \overleftrightarrow{\tau}_z \cdot \overleftrightarrow{M}' \cdot \overleftrightarrow{\tau}_z = \overrightarrow{1}_z \times \overleftrightarrow{M}' \times \overrightarrow{1}_z, \end{aligned} \tag{2.5}$$

which for bodies of revolution or others with polarizabilities proportional to the transverse identity (due to special two-dimensional symmetries) gives

$$\overleftrightarrow{P} = P\begin{pmatrix} 1 & 0 \\ 0 & 1 \end{pmatrix} = P\,\overrightarrow{1}_z, \qquad \overleftrightarrow{M} = M\begin{pmatrix} 1 & 0 \\ 0 & 1 \end{pmatrix} = M\,\overrightarrow{1}_z, \qquad M = -P. \tag{2.6}$$

for perfectly conducting two-dimensional scatterers.

For present purposes we will assume that the polarizabilities per unit length $\overleftrightarrow{P}'$ and $\overleftrightarrow{M}'$ are independent of frequency. (Special frequency dependences are allowed within the constraint of dilation symmetry[10].) Given that the cross-section area is proportional to z_s^2 the polarizabilities per unit length can be written as

$$\overleftrightarrow{P}' = \overleftrightarrow{P}'_0\, z_s^2 \tan^2(\psi), \qquad \overleftrightarrow{M}' = \overleftrightarrow{M}'_0\, z_s^2 \tan^2(\psi), \tag{2.7}$$

where $\overleftrightarrow{P}'_0 = \overleftrightarrow{M}'_0$ are dimensionless. It is these quantities which we will need for any given cross section to include our general scattering results. The appendices consider these for circular, strip and elliptic cross sections.

3. SCATTERING DYADIC FOR AXIAL INCIDENCE ON THIN CONES

Now consider the scattering from elementary dipoles on the z axis (coordinate z_s). We have

$$\overrightarrow{R}(z_s) = \overrightarrow{r} - z_s\,\overrightarrow{1}_z, \qquad z_s\,\overrightarrow{1}_z \equiv \text{source position}, \qquad \overrightarrow{r} = r\,\overrightarrow{1}_0 \equiv \text{observer position},$$

$$R(z_s) = \left|\vec{R}(z_s)\right| = r\left|\vec{1}_0 - \frac{z_s}{r}\vec{1}_z\right| = r - \vec{1}_0 \cdot \vec{1}_z z_s + O\left(r^{-1}\right) \text{ as } r \to \infty,$$

$$e^{-\gamma R} = e^{-\gamma r} e^{\gamma \vec{1}_0 \cdot \vec{1}_z}\left[1 + O\left(r^{-1}\right)\right] \text{ as } r \to \infty \;, \quad \vec{1}_0 \cdot \vec{1}_z \equiv \cos(\theta), \tag{3.1}$$

which can be inserted in the formulae for elementary dipoles at the source position. For electric dipoles we have the far fields[1,3]

$$\tilde{\vec{E}}_f^{(p)}(\vec{r},s) = -e^{-\gamma r} e^{\gamma\cos(\theta) z_s} \frac{\mu_0 s^2}{4\pi r} \overset{\leftrightarrow}{1}_0 \cdot \tilde{\vec{p}}(z_s,s),$$

$$\tilde{\vec{H}}_f^{(p)}(\vec{r},s) = +e^{-\gamma r} e^{\gamma\cos(\theta) z_s} \frac{s^2}{4\pi r} \frac{1}{c^2} \vec{1}_0 \cdot \tilde{\vec{p}}(z_s,s). \tag{3.2}$$

For magnetic dipoles we similarly have

$$\tilde{\vec{E}}_f^{(m)}(\vec{r},s) = -e^{-\gamma r} e^{\gamma\cos(\theta) z_s} \frac{\mu_0 s^2}{4\pi r} \frac{1}{c} \vec{1}_0 \times \tilde{\vec{m}}(z_s,s),$$

$$\tilde{\vec{H}}_f^{(m)}(\vec{r},s) = -e^{-\gamma r} e^{\gamma\cos(\theta) z_s} \frac{s^2}{4\pi r} \frac{1}{c^2} \vec{1}_0 \cdot \tilde{\vec{m}}(z_s,s). \tag{3.3}$$

Integrating over z_s we have

$$d\tilde{\vec{p}}(z_s,s) = \tilde{\vec{p}}'(z_s,s)\, d z_s, \qquad d\tilde{\vec{m}}(z_s,s) = \tilde{\vec{m}}'(z_s,s)\, d z_s,$$

$$\tilde{\vec{E}}_f^{(sc)}(\vec{r},s) = -\frac{e^{-\gamma r}}{4\pi r}\mu_0 s^2 \int_0^\infty e^{\gamma\cos(\theta) z_s}\left[\overset{\leftrightarrow}{1}_0 \cdot \tilde{\vec{p}}'(z_s,s) - \frac{1}{c}\vec{1}_0 \times \tilde{\vec{m}}'(z_s,s)\right] d z_s \theta. \tag{3.4}$$

From (2.2) and (1.1) we have

$$\tilde{\vec{p}}'(z_s,s) = \varepsilon_o \overset{\leftrightarrow}{P}'(z_s,s) \cdot \tilde{\vec{E}}^{(inc)}(z_s \vec{1}_z,s) = \varepsilon_o E_0 \tilde{f}(s) e^{-\gamma z_s} \overset{\leftrightarrow}{p}'(z_s,s) \cdot \vec{1}_e,$$

$$\tilde{\vec{m}}'(z_s,s) = \overset{\leftrightarrow}{M}'(z_s,s) \cdot \tilde{\vec{H}}^{(inc)}(z_s \vec{1}_z,s) = \frac{1}{Z_0} E_0 \tilde{f}(s) e^{-\gamma z_s} \overset{\leftrightarrow}{M}'(z_s,s) \cdot \left[\vec{1}_z \times \vec{1}_e\right], \tag{3.5}$$

for axial incidence. The far scattered field then is

$$\vec{\tilde{E}}_f^{(sc)}(\vec{r},s) \tag{3.6}$$
$$= \frac{-e^{-\gamma r}}{4\pi r} E_o \tilde{f}(s) \gamma^2 \int_0^\infty e^{-[1-\cos(\theta)]\gamma z_s} \left[\vec{1}_o \cdot \overset{\leftrightarrow}{\tilde{P}}{}'(z_s,s) \cdot \vec{1}_e - \vec{1}_o \times \overset{\leftrightarrow}{\tilde{M}}{}'(z_s,s) \cdot \left[\vec{1}_z \times \vec{1}_e \right] \right]$$

Using the identity[7]

$$\overset{\leftrightarrow}{\tilde{M}}{}'(z_s,s) \cdot \left[\vec{1}_z \times \vec{1}_e \right] = \left[\overset{\leftrightarrow}{\tilde{M}}{}' \times \vec{1}_z \right] \cdot \vec{1}_e, \tag{3.7}$$

we have the scattering dyadic

$$\overset{\leftrightarrow}{\tilde{\Lambda}}{}^{(sc)}(\vec{1}_o, \vec{1}_z; s) = \overset{\leftrightarrow}{\tilde{\Lambda}}{}^{(sc)T}(-\vec{1}_z, -\vec{1}_o; s)$$
$$= -\gamma^2 \int_0^\infty e^{-[1-\cos(\theta)]\gamma z_s} \left[\overset{\leftrightarrow}{1}_o \cdot \overset{\leftrightarrow}{\tilde{P}}{}'(z_s,s) - \overset{\leftrightarrow}{1}_o \times \overset{\leftrightarrow}{\tilde{M}}{}'(z_s,s) \times \overset{\leftrightarrow}{1}_z \right] dz_s. \tag{3.8}$$

Note that the integration over z_s from 0 to ∞ is, in time domain, over 0 to $ct/[1-\cos(\theta)]$ (including any singularity at this point) based on the causal properties of the polarizabilities. Note that θ should not be too near the cone ($\theta >> \psi$). The polarizability dyadics have non-zero elements only for the transverse (x, y) coordinates, being zero for anything to do with the z coordinate (5 elements).

At this point let us note that (3.8) is quite general, allowing various forms of polarizabilities. These can be then converted to time domain via inverse Laplace Transform, or via convolution with $\overset{\leftrightarrow}{P}{}'(z_s,t)$ and $\overset{\leftrightarrow}{M}{}'(z_s,t)$.

4. SCATTERING DYADIC FOR AXIAL INCIDENCE ON PERFECTLY CONDUCTING CONES

For perfectly conducting cones we have

$$\overset{\leftrightarrow}{P}{}' = \overset{\leftrightarrow}{P}{}'_0 \tan^2(\psi) z_s^2, \quad \overset{\leftrightarrow}{M}{}' = \overset{\leftrightarrow}{M}{}'_0 \tan^2(\psi) z_s^2,$$
$$\overset{\leftrightarrow}{M}{}' = \overset{\leftrightarrow}{\tau}_z \cdot \overset{\leftrightarrow}{P}{}'_0 \cdot \overset{\leftrightarrow}{\tau}_z = \vec{1}_z \times \overset{\leftrightarrow}{P}{}'_0 \times \vec{1}_z, \tag{4.1}$$
$$\overset{\leftrightarrow}{M}{}'_0 \times \vec{1}_z = -\vec{1}_z \times \overset{\leftrightarrow}{P}{}'_0 = -\overset{\leftrightarrow}{\tau}_z \cdot \overset{\leftrightarrow}{P}{}'_0.$$

The scattering dyadic is then

$$\begin{aligned}
&\overset{\leftrightarrow}{\tilde{\Lambda}}^{(sc)}(\vec{1}_o, \vec{1}_z; s) \\
&= -\gamma^2 \tan^2(\psi)\left[\overset{\leftrightarrow}{1}_o \cdot \overset{\leftrightarrow}{P}'_0 - \vec{1}_o \times \overset{\leftrightarrow}{M}'_0 \times \vec{1}_z\right]\int_0^\infty e^{-[1-\cos(\theta)]\gamma z_s} z_s^2 dz_s \\
&= -\gamma^2 \tan^2(\psi)\left[\overset{\leftrightarrow}{1}_o + \vec{1}_o \times \overset{\leftrightarrow}{\tau}_z\right] \cdot \overset{\leftrightarrow}{P}'_0 \int_0^\infty e^{-[1-\cos(\theta)]\gamma z_s} z_s^2 dz_s \\
&= -\frac{2}{\gamma}\frac{\tan^2(\psi)}{[1-\cos(\theta)]^3}\left[\overset{\leftrightarrow}{1}_o + \vec{1}_o \times \overset{\leftrightarrow}{\tau}_z\right] \cdot \overset{\leftrightarrow}{P}'_0 .
\end{aligned} \tag{4.2}$$

Noting that the perfectly conducting cone has an exact form for the solution as[9]

$$\overset{\leftrightarrow}{\tilde{\Lambda}}(\vec{1}_o, \vec{1}_i; s) = \gamma^{-1}\overset{\leftrightarrow}{K}(\vec{1}_o, \vec{1}_i), \tag{4.3}$$

we then have

$$\overset{\leftrightarrow}{K}(\vec{1}_o, \vec{1}_z) = -2\frac{\tan^2(\psi)}{[1-\cos(\theta)]^3}\left[\vec{1}_o + \vec{1}_o \times \overset{\leftrightarrow}{\tau}_z)\right] \cdot \overset{\leftrightarrow}{P}'_0 . \tag{4.4}$$

The result of (4.2) can also be derived directly in time domain as

$$\begin{aligned}
\overset{\leftrightarrow}{\Lambda}^{(sc)}(\vec{1}_o, \vec{1}_z; t) &= -\frac{\tan^2(\psi)}{c^2}\left[\overset{\leftrightarrow}{1}_o + \vec{1}_o \times \overset{\leftrightarrow}{\tau}_z\right] \cdot \overset{\leftrightarrow}{P}'_0 \frac{\partial^2}{\partial t^2}\int_0^\infty \delta\left(t - \frac{1-\cos(\theta)}{c} z_s\right) z_s^2 dz_s \\
&= -\frac{\tan^2(\psi)}{c^2}\left[\overset{\leftrightarrow}{1}_o + \vec{1}_o \times \overset{\leftrightarrow}{\tau}_z\right] \cdot \overset{\leftrightarrow}{P}'_0 \left[\frac{c}{1-\cos(\theta)}\right]^3 \frac{\partial^2}{\partial t^2} \\
&\quad \int_0^\infty \delta\left(t - \frac{1-\cos(\theta)}{c} z_s - t\right)\left[\frac{1-\cos(\theta)}{c} z_s\right]^2 d\left[\frac{1-\cos(\theta)}{c} z_s\right] \\
&= -\frac{\tan^2(\psi)c}{[1-\cos(\theta)]^3}\left[\overset{\leftrightarrow}{1}_o + \vec{1}_o \times \overset{\leftrightarrow}{\tau}_z\right] \cdot \overset{\leftrightarrow}{P}'^{3}_0 \frac{\partial^2}{\partial t^2}\left[t^2 u(t)\right] \\
&= -2c\frac{\tan^2(\psi)}{[1-\cos(\theta)]^3}\left[\overset{\leftrightarrow}{1}_o + \vec{1}_o \times \overset{\leftrightarrow}{\tau}_z\right] \cdot \overset{\leftrightarrow}{P}'_0 u(t).
\end{aligned} \tag{4.5}$$

Specializing to axial backscattering we have

$$\vec{1}_o = -\vec{1}_z, \qquad \overleftrightarrow{1}_o = \overleftrightarrow{1}_z, \qquad \theta = \pi,$$

$$\left[\overleftrightarrow{1}_z - \vec{1}_z \times \overleftrightarrow{\tau}_z\right] \cdot \overleftrightarrow{P}'_0 = \left[\overleftrightarrow{1}_z - \overleftrightarrow{\tau}_z^{\,2}\right] \cdot \overleftrightarrow{P}'_0 = 2\,\overleftrightarrow{1}_z \cdot \overleftrightarrow{P}'_0 = 2\,\overleftrightarrow{P}'_0 = 2\,\overleftrightarrow{P}'_0,$$

$$\overleftrightarrow{K}_b(\overleftrightarrow{1}_z) = \overleftrightarrow{K}(-\overleftrightarrow{1}_z, \overleftrightarrow{1}_z) = -\frac{1}{2}\tan^2(\psi)\,\overleftrightarrow{P}'_0, \tag{4.6}$$

which is a rather simple result.

Appendix A considers the polarizability of a perfectly conducting circular cylinder of radius a, giving

$$\overleftrightarrow{P}' = 2\pi a^2\,\overleftrightarrow{1}_z. \tag{4.7}$$

Applying this to the thin perfectly conducting circular cone we have

$$\begin{aligned} a(z_s) &= z_s\tan(\psi), & \overleftrightarrow{P}'(z_s) &= \overleftrightarrow{P}'_0\, z_s^2\tan^2(\psi), \\ \overleftrightarrow{P}'_0 &= 2\pi\,\overleftrightarrow{1}_z, & \overleftrightarrow{K}_b &= -\pi\tan^2(\psi)\,\overleftrightarrow{1}_z = -\pi\psi^2\,\overleftrightarrow{1}_z, \end{aligned} \tag{4.8}$$

which is in agreement with both the physical-optics and small-ψ-asymptotic solution for axial backscattering.[4,8]

With this agreement for the circular cone, let us turn to the strip of width $2a$ in Appendix B. In this case, letting the plane of the strip lie on the $y = 0$ plane so that only $P'_{0_{x,x}}$ is non zero as

$$\overleftrightarrow{P}'_0 = \pi a^2\begin{pmatrix} 1 & 0 \\ 0 & 0 \end{pmatrix} \tag{4.9}$$

in 2 x 2 form. This then gives

$$a = z_s\tan(\psi), \quad \overrightarrow{K}_b = -\frac{\pi}{2}\tan^2(\psi)\begin{pmatrix} 1 & 0 \\ 0 & 0 \end{pmatrix} \simeq -\frac{\pi}{2}\psi^2\begin{pmatrix} 1 & 0 \\ 0 & 0 \end{pmatrix}. \tag{4.10}$$

So now we have a solution for axial backscattering from a thin angular sector. As pointed out in Ref. 4 this is in disagreement with PO, which gives zero since such a scatterer has zero cross-section area. Note, furthermore, that the nonzero element has exactly half the value of each element for the circular cone with the angular sector and circular cone both having the same value of $a = z_S\tan(\psi)$ describing the maximum extent from the z axis.

For more insight into the deviation of these results for thin cones from the physical-optics approximation consider the perfectly conducting elliptic cone with semiaxes a_x and a_y in Appendix B. This has

$$\overset{\leftrightarrow}{P}{}'_o = \pi\left[a_x + a_y\right]\begin{pmatrix} a_x & 0 \\ 0 & a_y \end{pmatrix}. \tag{4.11}$$

With a as the larger of a_x and a_y we have

$$a = z_s \tan(\psi), \qquad \overset{\leftrightarrow}{K}_b(\overrightarrow{1}_z) = -\frac{\pi}{2}\tan^2(\psi)\left[\frac{a_x}{a} + \frac{a_y}{a}\right]\begin{pmatrix} \frac{a_x}{a} & 0 \\ 0 & \frac{a_y}{a} \end{pmatrix}. \tag{4.12}$$

As $a_y \to 0$ this has both elements bounded by those for the minimum circumscribing circular cone. As $a_y \to 0$ this goes smoothly to the result for the angular sector. As $a_y \to a_x$ it smoothly goes to the result for the circular cone.

The cross-section area of the circular cone is

$$A(z_s) = \pi z_s^2 \tan^2(\psi)\frac{a_x}{a}\frac{a_y}{a}. \tag{4.13}$$

The physical-optics approximation then gives

$$\overset{\leftrightarrow}{K}{}_b^{(po)}(\overrightarrow{1}_z) = -\pi\tan^2(\psi)\frac{a_x}{a}\frac{a_y}{a}\overset{\leftrightarrow}{1}_z. \tag{4.14}$$

As we can see this goes to zero smoothly for a strip in either orientation. Normalizing the result in (4.12) to the physical optics case gives

$$\left[-\pi\tan^2(\psi)\frac{a_x}{a}\frac{a_y}{a}\right]^{-1}\overset{\leftrightarrow}{K}_b(\overrightarrow{1}_z) = \frac{1}{2}\left[\frac{a}{a_x} + \frac{a}{a_y}\right]\begin{pmatrix} \frac{a}{a_y} & 0 \\ 0 & \frac{a}{a_x} \end{pmatrix}, \tag{4.15}$$

showing the discrepancy. The PO result is smaller for both polarizations. The ratio of the matrix elements is not 1, but rather a_x / a_y showing the polarization dependence.

5. CONCLUDING REMARKS

Now we have a procedure for calculating the scattering from narrow cones for the case of axial incidence (or scattering by reciprocity). A few examples have been considered here for comparison to the PO approximation for axial backscattering. While the PO approximation seems adequate for perfectly conducting bodies of resolution (as, for example, the paraboloid and circular cone[4]), it clearly breaks down for the case of the elliptic cone. The elliptic cone is important in that a shadow boundary does not appear (until one truncates the cone). The limit of a circular cone is an angular sector, and the scattering dyadic for the elliptic cone smoothly goes to that of the angular sector (for which the PO approximation gives zero).

Noting that the present model is appropriate only for thin cones (small ψ), we still need accurate calculations for fat cones (large ψ but $<\pi/2$) for further comparison to the PO approximation.[11] In its own right, however, this small-ψ model can now be applied to various thin structures, including ones that are not conical.

APPENDIX A. POLARIZABILITY PER UNIT LENGTH OF PERFECTLY CONDUCTING CIRCULAR CYLINDER

For computing the fields in the vicinity of a perfectly conducting object one can conveniently use conformal transformations in the complex-coordinate plane.[6] As developed in Ref. 2 we have E_0 is a constant in V/m, $\zeta = x + jy$ is a complex coordinate, $w(\zeta) = u(\zeta) + jv(\zeta)$ is the normalized complex potential,

$$e_0(\zeta) = e_{0_x}(\zeta) - je_{0_y}(\zeta) = \frac{dw(\zeta)}{d\zeta} = \text{normalized complex electric field, and}$$

$$\vec{E}(x,y) = -\nabla\Phi(x,y) = E_0\left[e_{0_x}(\zeta)\,\vec{1}_x + e_{0_y}(\zeta)\,\vec{1}_y\right] = \text{electric field.} \tag{A.1}$$

Here we deal with the electric field, but as is well known the results apply to the magnetic field as well.

As in Fig. A.1 we map a strip on $u = 0$ in the w plane to a circle of radius a in the ζ plane via

$$\frac{w}{a} = \frac{\zeta}{a} - \frac{a}{\zeta}. \tag{A.3}$$

It is convenient to use cylindrical (Ψ,ϕ) coordinates via

$$\zeta = \Psi e^{j\phi}, \qquad \frac{w}{a} = \frac{\Psi}{a}e^{j\phi} - \frac{a}{\Psi}e^{-j\phi}. \tag{A.4}$$

For large $|\zeta|$ we have

$$\vec{E}(x,y) \to E_o\,\vec{1}_x \text{ as } |\zeta| \to \infty. \tag{A.5}$$

The surface charge density on the cylinder is

$$\rho_s = \varepsilon_0 E_\Psi = \varepsilon_0 \vec{E}(x,y) \cdot \vec{1}_\Psi \Big|_{|\zeta|=a} = \varepsilon_0 E_0 \left[e_{0_x} \cos(\phi) + e_{0_y} \sin(\phi) \right] \Big|_{|\zeta|=a}$$

$$= \varepsilon_0 E_0 \left[\mathrm{Re}\left[\frac{dw}{d\zeta}\right] \cos(\phi) - \mathrm{Im}\left[\frac{dw}{d\zeta}\right] \sin(\phi) \right] \Big|_{|\zeta|=a} . \tag{A.6}$$

Using, on the cylinder,

$$\frac{w}{a} = e^{j\phi} - e^{-j\phi} = j2\sin(\phi), \qquad d\zeta = ad\left[e^{j\phi}\right] = jae^{j\phi}d\phi, \qquad \frac{dw}{d\zeta} = 2e^{-j\phi}\cos(\phi), \tag{A.7}$$

we have

$$\rho_s(\phi) = 2\cos(\phi)\varepsilon_{0E_0}. \tag{A.8}$$

This exhibits the well-known enhancement of the electric field by a factor of two on the circular cylinder.

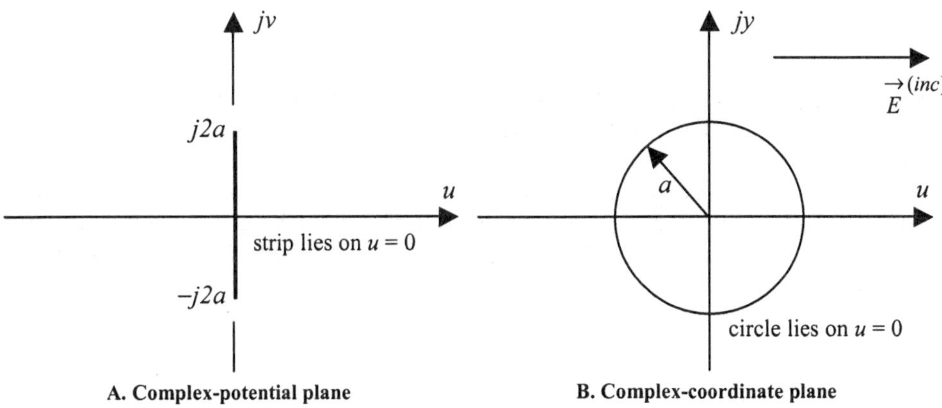

Figure A.1. Conformal transformation for circular cylinder.

We can now calculate the induced electric-dipole moment per unit length as

$$\vec{p}' = \vec{1}_x \int_0^{2\pi} \rho_s(\phi) x d[a\phi] = \vec{1}_x 2a^2 \varepsilon_0 E_0 \int_0^{2\pi} \cos^2(\phi) d\phi = \vec{1}_x 2\pi a^2 \varepsilon_0 E_0 = \varepsilon_0 \overleftrightarrow{P}' \cdot \vec{1}_x E_0. \quad (A.9)$$

Noting the rotation symmetry we have

$$\overleftrightarrow{P}' = 2\pi a^2 \vec{1}_z = 2\pi a^2 \begin{pmatrix} 1 & 0 & 0 \\ 0 & 1 & 0 \\ 0 & 0 & 0 \end{pmatrix} \quad (A.10)$$

or in two-dimensional form

$$\overleftrightarrow{P}' = 2\pi a^2 \begin{pmatrix} 1 & 0 \\ 0 & 1 \end{pmatrix} \quad (A.11)$$

for the electric-polarizability dyadic.

Since we are dealing with a perfectly conducting two-dimensional structure we also have

$$\overleftrightarrow{M}' = -\overleftrightarrow{P}' = -2\pi a^2 \begin{pmatrix} 1 & 0 \\ 0 & 1 \end{pmatrix} \quad (A.12)$$

is consistent with (2.6).

APPENDIX B. POLARIZABILITY PER UNIT LENGTH OF PERFECTLY CONDUCTING STRIP

Following the same procedure as in Appendix A we construct an appropriate conformal transformation in two steps. As indicated in Fig. B.1 we first transform a strip of width $2a$ on the $u = 0$ axis to a circle of radius $a/2$ with intermediate complex coordinates ζ_1 as

$$\frac{2w}{a} = \frac{2\zeta_1}{a} - \frac{a}{2\zeta_1}. \quad (B.1)$$

In turn this is transformed to a strip of width $2a$ on the x axis of the ζ plane as

$$\frac{2\zeta}{a} = \frac{2\zeta_1}{a} + \frac{a}{2\zeta_1}. \quad (B.2)$$

Combining these transformations we have

$$\frac{2\zeta_1}{a} = \frac{\zeta}{a} + \left[\left[\frac{\zeta}{a}\right]^2 - 1\right]^{\frac{1}{2}} , \quad \frac{2w}{a} = \frac{\zeta}{a} + \left[\left[\frac{\zeta}{a}\right]^2 - 1\right]^{\frac{1}{2}} - \left[\frac{\zeta}{a} + \left[\left[\frac{\zeta}{a}\right]^2 - 1\right]^{\frac{1}{2}}\right]^{1} \tag{B.3}$$

$$= \frac{\zeta}{a} + \left[\left[\frac{\zeta}{a}\right]^2 - 1\right]^{\frac{1}{2}} - \frac{\zeta}{a} + \left[\left[\frac{\zeta}{a}\right]^2 - 1\right]^{\frac{1}{2}} = 2\left[\left[\frac{\zeta}{a}\right]^2 - 1\right]^{\frac{1}{2}} .$$

The branch has been chosen such that for ζ/a real and greater than unity the square root is positive. On the strip (the branch cut) on the top side ($y = 0_+$) we have

$$\frac{w}{a} = j\frac{v}{a} = j\left[1-[x/a]^2\right]^{1/2} \tag{B.4}$$

with opposite sign on the bottom side ($y = 0_-$). We can account for both sides by multiplying the expressions by two and integrating on the top side.

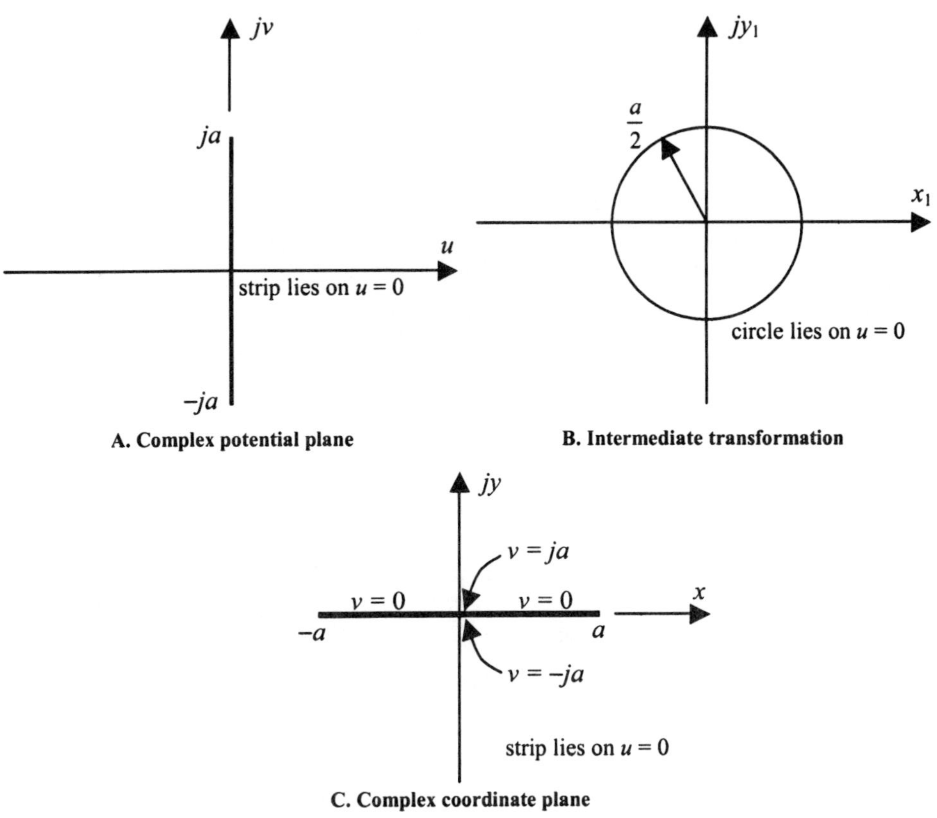

Figure B.1 Conformal transformation for strip.

The surface charge density, accounting for both sides, is

$$\rho_s = 2\varepsilon_0 E_y, \qquad y=0_+, \qquad E_y = -E_0\frac{dv}{dx}. \tag{B.5}$$

The electric-dipole moment per unit length is

$$\vec{p}\,' = \vec{1}_x \int_{-a}^{a} \rho_s(x)x\,dx = -\vec{1}_x 2\varepsilon_0 E_0 \int_{-a}^{a} x\frac{dv}{dx}dx. \tag{B.6}$$

Integrating by parts gives

$$\vec{p}\,' = -\vec{1}_x 2\varepsilon_0 E_0\left[xv\Big|_{-a}^{a} - \int_{-a}^{a} v\,dx\right] = \vec{1}_x 2\varepsilon_0 E_0 \int_{-a}^{a} v\,dx = \vec{1}_x 2\varepsilon_0 E_0 a \int_{-a}^{a}\left[1-\left[\frac{x}{a}\right]^2\right]^{\frac{1}{2}} dx,$$

with

$$\frac{x}{a} \equiv \sin(\xi), \qquad dx = a\cos(\xi)d\xi = \vec{1}_x \pi^2 \varepsilon_0 E_0, \qquad \vec{p}\,' = \vec{1}_x 2a^2\varepsilon_0 E_0 \int_{-\frac{\pi}{2}}^{\frac{\pi}{2}} \cos^2(\xi)d\xi. \tag{B.7}$$

This is for an incident electric field in the x direction. An incident electric field in the y direction, being perpendicular to the flat strip produces no induced electric-dipole moment per unit length. The electric polarizability per unit length is then

$$\vec{p}\,'_o = \pi a^2 \begin{pmatrix} 1 & 0 \\ 0 & 0 \end{pmatrix} \tag{B.8}$$

in two-dimensional form.

The magnetic polarizability per unit length from (2.5) is

$$\vec{M}\,' = \overleftrightarrow{\tau}_z \bullet \overleftrightarrow{p}\,' \bullet \overleftrightarrow{\tau}_z = \pi a^2 \begin{pmatrix} 0 & -1 \\ 1 & 0 \end{pmatrix} \bullet \begin{pmatrix} 1 & 0 \\ 0 & 0 \end{pmatrix} \bullet \begin{pmatrix} 0 & -1 \\ 1 & 0 \end{pmatrix} = -\pi a^2 \begin{pmatrix} 0 & 0 \\ 0 & 1 \end{pmatrix}. \tag{B.9}$$

This then scatters a y-directed magnetic field (only) as we expect.

APPENDIX C. POLARIZABILITY PER UNIT LENGTH OF PERFECTLY CONDUCTING ELLIPTIC CYLINDER

Electric and magnetic polarizabilities of various perfectly conducting objects including elliptic cylinders are tabulated in Refs. 5 and 12. With two symmetry planes the axis is well defined. The semiaxes are a_x and a_y for the subscripted coordinate directions. While in Fig. C.1 we have shown $a_x > a_y$, it can be $a_x < a_y$ as well. The results are

$$\overset{\leftrightarrow}{P}{}' = \pi\left[a_x + a_y\right]\begin{pmatrix} a_x & 0 \\ 0 & a_y \end{pmatrix}, \qquad \overset{\leftrightarrow}{M}{}' = -\pi\left[a_x + a_y\right]\begin{pmatrix} a_y & 0 \\ 0 & a_x \end{pmatrix}. \tag{C.1}$$

Note that for $a_x = a_y = a$ this gives the result for the circular cylinder in Appendix A. For $a_x = a$ and $a_y = 0$, this gives the result for the strip in Appendix B.

The ratio of the two elements is just a_x/a_y. For comparison to the case of the circular cylinder define

$$a \equiv \max\{a_x, a_y\}, \tag{C.2}$$

so that the elliptic cylinder has its maximum cross-section dimension equal to that of the circular cylinder. Further define

$$\alpha \equiv \frac{a_x}{a}, \qquad \beta \equiv \frac{a_y}{a}, \qquad a \le 1, \qquad \beta \le 1, \qquad \frac{\alpha + \beta}{2} \le 1, \tag{C.3}$$

giving

$$\overset{\leftrightarrow}{P}{}' = 2\pi a^2 \begin{pmatrix} \alpha\frac{\alpha + \beta}{2} & 0 \\ 0 & \beta\frac{a + \beta}{2} \end{pmatrix}. \tag{C.4}$$

Each matrix element above is less than or equal to 1, showing the comparison to that for the minimum circumscribing circular cylinder.

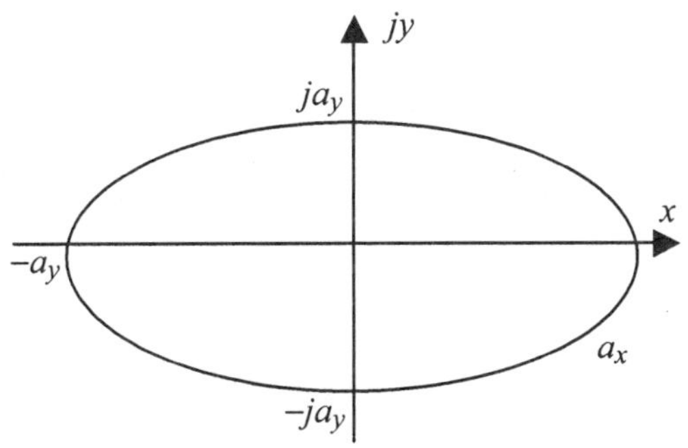

Fig. C.1 Elliptic Cylinder

ACKNOWLEDGEMENTS

This work was sponsored in part by the Air Force Office of Scientific Research and in part by the Air Force Research Laboratory, Directed Energy Directorate.

REFERENCES

1. C.E. Baum, Some characteristics of electric and magnetic dipole antennas for radiating transient pulses, *Sensor and Simulation Note 125*, Air Force Research Laboratory, Kirtland AFB NM (January 1971).
2. C.E. Baum, Aperture efficiencies for IRAs, *Sensor and Simulation Note 328*, Air Force Research Laboratory, Kirtland AFB NM (June 1991).
3. C.E. Baum, Location and orientation of electrically small transmitting and receiving antenna pairs with common linear polarization and beam direction for minimal mutual coupling, *Sensor and Simulation Note 446*, Air Force Research Laboratory, Kirtland AFB NM (March 2000).
4. C.E. Baum, The physical-optics approximation and early-time scattering, *Interaction Note 563*, Air Force Research Laboratory, Kirtland AFB NM (October 2000).
5. S.A. Schelkunoff and H.T. Friis, *Antennas: Theory and Practice* (Wiley, New York, 1952).
6. R.V. Churchill, *Complex Variables and Applications* (McGraw-Hill, New York, 1960).
7. J. Van Bladel, *Electromagnetic Fields* (Hemisphere, New York, 1985).
8. J.J. Bowman, The Cone, in: *Electromagnetic and Acoustic Scattering by Simple Shapes*, edited by J.J. Bowman, T.B.A. Senior, and P.L.E. Uslenghi (Hemisphere, New York, 1987), Ch. 18, pp. 637–701.
9. C.E. Baum and H.N. Kritikos, Symmetry in Electromagnetics, in: *Electromagnetic Symmetry*, edited by C.E. Baum and H.N. Kritikos (Taylor & Francis, New York, 1995), Ch. 1, pp. 1–90.
10. C.E. Baum, Continuous Dilation Symmetry in Electromagnetic Scattering, in: *Electromagnetic Symmetry*, edited by C.E. Baum and H.N. Kritikos (Taylor & Francis, New York, 1995), Ch. 3, pp. 143–183.
11. T.B.A. Senior, private communication.
12. K.S.H. Lee, private communication.

AN OBSERVATION CONCERNING THE ENTIRE FUNCTION IN SEM SCATTERING

Carl E. Baum*

1. INTRODUCTION

In the singularity expansion method (SEM) one of the hardest things to pin down has been the entire function (in the $s = \Omega + j\omega$ plane). In the time domain, this is an early-time contribution not representable by complex exponentials (poles). This is discussed in some detail in Ref. 1, which has extensive references. In the present paper we revisit this matter and gain additional insight into this entire function.

As usual ~ denotes the two-sided Laplace transform over time t, and s is the Laplace-transform variable or complex frequency for going back and forth between time and frequency.

2. CLASS-1 FORM OF INTERACTION REPRESENTATION

In Ref. 1 the currents on a body illuminated by a plane wave (an interaction problem) were considered. Summarizing we have an incident plane wave (electric field)

$$\vec{E}^{(inc)}(\vec{r},t) = E_0 \, f\left(t - \frac{\vec{1}_i \cdot \vec{r}}{c}\right)\vec{1}_e, \quad \vec{E}^{(inc)}(\vec{r},t) = E_0 \, e^{-\gamma \vec{1}_i \cdot \vec{r}} \tilde{f}(s)\, \vec{1}_e. \tag{2.1}$$

Referring to Fig. 2.1 we have $c = 1/(\mu_0\varepsilon_0)^{1/2}$ is the speed of light, $\gamma = s/c$ is the propagation constant, $f(t)$ is the waveform,

* Air Force Research Laboratory, Directed Energy Directorate, Kirtland Air Force Base, NM 87117-5776.

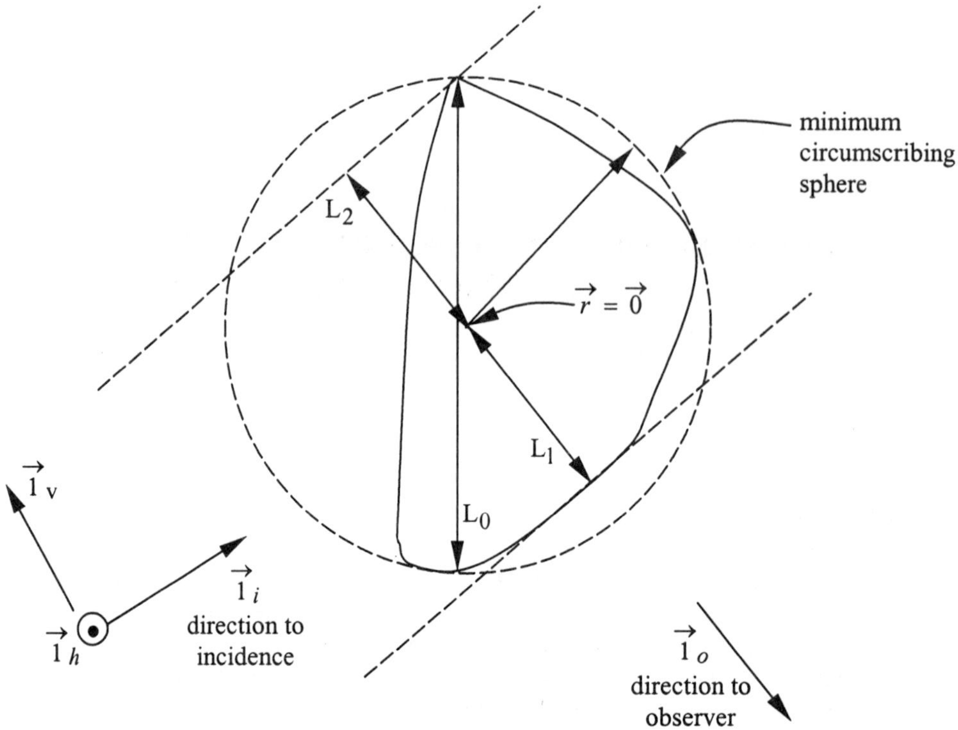

Figure 2.1 Finite-size scatterer in free space illuminated by a plane wave

$$\vec{1}_e \equiv \text{polarization (some weighted combination of radar coordinates } \vec{1}_h \text{ and } \vec{1}_v\text{)},$$

$$\vec{1}_e \equiv \text{direction of incidence},$$

$$\vec{1}_e \cdot \vec{1}_e = 0. \tag{2.2}$$

The interaction of this incident wave with the body is described by an integral equation of the form

$$\tilde{\vec{E}}_t^{(inc)}(\vec{r}_s,s) = \left\langle \tilde{\vec{Z}}_t(\vec{r}_s,\vec{r}\,'_s;S); \tilde{\vec{J}}_s(\vec{r}\,'_s,S) \right\rangle = \overleftrightarrow{1}_S(\vec{r}_s) \cdot \tilde{\vec{E}}^{(inc)}(\vec{r}_s,s),$$

where

$$\overset{\leftrightarrow}{1}_S(\vec{r}_s) = \overset{\leftrightarrow}{1} - \vec{1}_S(\vec{r}_s)\vec{1}_S(\vec{r}_s) \equiv \text{transverse dyadic to } \overset{\leftrightarrow}{1}_S(\vec{r}_s),$$

$$\overset{\leftrightarrow}{1}_S(\vec{r}_s) \equiv \text{outward pointing normal to surface } S \text{ at surface coordinate } \vec{r}_s . \tag{2.3}$$

In this form the scatterer is treated as perfectly conducting so that the integration is over S (coordinates $\vec{r}\,'_s$). However, this is not essential to the development. The above is also in the form of the E-field or impedance integral equation, but other forms can be used as well.

The formal solution to (2.3) is

$$\vec{J}_s(\vec{r}_s,S) = \left\langle \tilde{\overset{\leftrightarrow}{Z}}_t^{-1}(\vec{r}_s\,\vec{r}\,'_s;s);\ \tilde{\vec{E}}_t^{(inc)}(\vec{r}\,'_s,s) \right\rangle . \tag{2.4}$$

In class-1 SEM form the solution is

$$\tilde{\vec{J}}_s(\vec{r}_s,s) = E_0\tilde{f}(s)\left[\sum_\alpha \eta_\alpha(\vec{1}_i,\vec{1}_e)\vec{j}_{s_\alpha}(\vec{r}_s)\frac{e^{-[s-s_\alpha]t_i}}{s-s_\alpha} + \text{possible entire function}\right],$$

$$\tilde{\vec{J}}_s(\vec{r}_s,t) = E_0 f(t)\circ\left[\sum_\alpha \eta_\alpha(\vec{1}_i,\vec{1}_e)\vec{j}_{s_\alpha}(\vec{r}_s)e^{s_\alpha t}u(t-t_i) + \text{possible entire function (temporal form)}\right],$$

where t_i is the initial or turn-on time, typically taken as t_f, the time the incident wave first touches the body, $\circ$ ≡ convolution with respect to time,

$$\left\langle \tilde{\overset{\leftrightarrow}{Z}}_t(\vec{r}_s\,\vec{r}\,'_s;s_\alpha);\ \vec{j}_{s_\alpha}(\vec{r}\,'_s) \right\rangle = \vec{0} \text{ (for natural frequencies and modes)},$$

$$\mathsf{U}_\alpha = \left\langle \vec{j}_{s_\alpha}(\vec{r}_s); \frac{d}{ds}\tilde{\overset{\leftrightarrow}{Z}}_t(\vec{r}_s,\vec{r}\,'_s;s)\Big|_{s=s_\alpha};\ \vec{j}_{s_\alpha}(\vec{r}\,'_s) \right\rangle^{-1},$$

$$\eta_\alpha(\vec{1}_i,\vec{1}_e) = \mathsf{U}_\alpha\,\vec{1}_i\cdot\vec{C}_\alpha(\vec{1}_i) \equiv \text{coupling coefficient}, \tag{2.5}$$

$$\vec{C}_\alpha(\vec{1}_i) = \left\langle e^{-\gamma_\alpha\vec{1}_i\cdot\vec{r}\,'_s},\ \vec{j}_{s_\alpha}(\vec{r}\,'_s) \right\rangle, \qquad \gamma_\alpha \equiv \frac{s_\alpha}{c}.$$

Here first order poles have been used, appropriate to perfectly conducting bodies. Special cases of loaded bodies can have some higher-order poles, but this does not impact the argument.

As discussed in Ref. 1 a critical role is played by U_α . As one proceeds to sum over terms with more negative $\mathrm{Re}[s_\alpha]$ there is the asymptotic behavior

$$U_\alpha = O_e\left(\gamma_\alpha L_0\right) = O_e\left(s_\alpha t_0\right) \text{ as } \mathrm{Re}\left[s_\alpha\right] \to -\infty,$$

$$L_0 \equiv \text{maximum linear dimension of object } (t_0 \equiv L_0/c), \tag{2.6}$$

where the exponential order means a bound of the form $e^{s_\alpha t_0}$ times a function, which grows more slowly than any exponential. We also have

$$\vec{C}_\alpha(\vec{1}_i) = O_e(-s_\alpha t_b), \tag{2.7}$$

where t_b is the time for incident wave to reach all of the body (t back, positive). Ignoring $E_0\tilde{f}(s)$, the residues then take the bound

$$O_e\left(s_\alpha t_0\right)O_e\left(-s_\alpha t_b\right)O_e\left(s_\alpha t_f\right) = O_e\left(s_\alpha\left[t_0 + t_f - t_b\right]\right). \tag{2.8}$$

Note that for all $\vec{1}_i$ we have

$$t_0 \geq t_b - t_f . \tag{2.9}$$

Then with a bound on the number of poles smaller than an exponential as s_α sweeps to the left in the s plane the pole series converges. In time domain this implies convergence for all times greater than t_f. In order to avoid problems with convergence to a δ function (although this is often handled in the case of Fourier series) one can treat the step (or ramp) response by choice of $\tilde{f}(s)$ as s^{-1} etc., and including this in the pole residues.

While convergence of the class-1 series for all times greater than t_f is a necessary condition for the series to be an accurate representation of the current, this does not preclude ipso facto the absence of an entire function to complete the representation. However, for this purpose we can appeal to the numerous calculations that have been made to see how many poles (damped sinusoids) are required to approximate the currents to some degree of accuracy. (See Refs. 3-5 and the references therein.) For the various bodies (usually, but not always, perfectly conducting) it does not take many (say 10 or so) to reasonably agree with the numerically computed step response (often calculated by solution in the frequency domain followed by inverse Fourier transformation). Based on this let us define:

Simple scatterer (object) $\equiv$ one for which the currents can be represented by class-1 poles (damped sinusoids) without an additional entire function.

Higher order poles can also be allowed. This recognizes that there may be exceptional cases of finite-size bodies in free space for which this does not apply.

3. CLASS-1 FORM OF SCATTERING REPRESENTATION

In Ref. 1 the far fields scattered from a body were also considered. In that paper the convergence of the class-1 pole series for these fields was also considered, with different results from those for the currents.

The far scattering takes the form

$$\tilde{\vec{E}}_f(\vec{r},s) = \frac{e^{-\gamma r}}{4\pi}\tilde{\overleftrightarrow{\Lambda}}(\vec{1}_o,\vec{1}_i;s)\cdot\tilde{\vec{E}}^{(inc)}(\vec{0},s) = E_0\tilde{f}(s)\frac{e^{-\gamma r}}{4\pi}\tilde{\overleftrightarrow{\Lambda}}(\vec{1}_o,\vec{1}_i;s)\cdot\vec{1}_e.$$

$$\vec{1}_o \equiv \text{direction to observer at } \vec{r}. \tag{3.1}$$

The scattering can also be described by

$$\tilde{\vec{E}}_f(\vec{r},s) = -\frac{s\mu_0 e^{-\gamma r}}{4\pi r}\left\langle \overleftrightarrow{1}_o e^{\gamma \vec{1}_o\cdot\vec{r}_s};\tilde{\vec{J}}_s(\vec{r}_s,s)\right\rangle,$$

$$\overleftrightarrow{1}_o = \overleftrightarrow{1} - \vec{1}_o\vec{1}_o \equiv \text{dyadic transverse to } \vec{1}_o, \tag{3.2}$$

$$\overleftrightarrow{1}_i = \overleftrightarrow{1} - \vec{1}_i\vec{1}_i \equiv \text{dyadic transverse to } \vec{1}_i,$$

$$\overleftrightarrow{1} = \vec{1}_x\vec{1}_x + \vec{1}_y\overleftrightarrow{1}_y + \vec{1}_z\vec{1}_z \equiv \text{three-dimensional identity.}$$

From the integral-equation solution (2.4) we then have

$$\tilde{\overleftrightarrow{\Lambda}}_f(\vec{1}_o,\vec{1}_i;s) = -s\mu_0\left\langle \overleftrightarrow{1}_o e^{\gamma \vec{1}_o\cdot\vec{r}_s};\tilde{\overleftrightarrow{Z}}_t^{-1}(\vec{r}_s,\vec{r}'_s;s)\overleftrightarrow{1}_i e^{-\gamma\vec{1}_i\cdot\vec{r}'_s}\right\rangle \tag{3.3}$$

$$= \tilde{\overleftrightarrow{\Lambda}}^T(-\vec{1}_i-\vec{1}_o;s) \quad \text{(reciprocity)},$$

which for backscattering (monostatic) reduces to

$$\begin{aligned}\tilde{\overleftrightarrow{\Lambda}}_b(\vec{1}_i,s) &\equiv \tilde{\overleftrightarrow{\Lambda}}(-\vec{1}_i,\vec{1}_i;s)\\ &= -s\mu_0\left\langle \vec{1}_i e^{-\gamma \vec{1}_i\cdot\vec{r}_s}; \tilde{\overleftrightarrow{Z}}_t(\vec{r}_s,\vec{r}\,'_s;s); \vec{1}_i e^{-\gamma\vec{1}_i\cdot\vec{r}_s}\right\rangle \\ &= \tilde{\overleftrightarrow{\Lambda}}{}_b^T(\vec{1}_i,s), \quad \text{(symmetric)}\end{aligned} \tag{3.4}$$

where $\vec{1}_o = -\vec{1}_i$ and $\overleftrightarrow{1}_o = \vec{1}_i$.

Writing the scattering dyadic in class-1 form as

$$\tilde{\overleftrightarrow{\Lambda}}(\vec{1}_o,\vec{1}_i;s) = \sum \frac{e^{-[s-s_\alpha]t_p}}{s-s_\alpha}\vec{c}_\alpha(\vec{1}_o)\,\vec{c}_\alpha(\vec{1}_i) + \text{entire function},$$

$$\tilde{\overleftrightarrow{\Lambda}}_b(\vec{1}_i,s) = \sum \frac{e^{-[s-s_\alpha]t_p}}{s-s_\alpha}\vec{c}_\alpha(\vec{1}_i)\,\vec{c}_\alpha(\vec{1}_i) + \text{entire function}, \tag{3.5}$$

$$\vec{c}_\alpha(\vec{1}_i) = w_\alpha \vec{C}_\alpha(\vec{1}_i)\ , \quad W_\alpha = w_a^2 = -s_\alpha\mu_0 \mathrm{U}_\alpha .$$

In Ref. 1 it is shown that the turn-on time t_p for the poles is, in typical cases, after the first signal to reach the observer at a retarded time $2t_f$ (negative). Otherwise the pole series diverges as one sums over poles with more negative $\mathrm{Re}[s_\alpha]$.

Specialize to the worst case for this divergence problem, namely backscattering. Referring to Fig. 2.1 we have

$$t_1 = -\frac{L_1}{c} = \text{retarded time corresponding to the closest position on the scatterer to the observer (negative)},$$

$$t_2 = \frac{L_2}{c} = \text{retarded time corresponding to the farthest position on the scatterer from the observer (positive)},$$

$$t_3 = t_2 - t_1, \qquad 0 \le t_3 \le t_0 = L_0/c. \tag{3.6}$$

Comparing to (2.7) the case of backscattering has

$$t_1 = t_f, \qquad t_2 = t_b. \tag{3.7}$$

From (2.6), (2.7), and (3.5) we have a bound for the residues as

$$\mathrm{O}_e\!\left(s_\alpha t_p\right)\mathrm{O}_e\!\left(s_\alpha t_0\right)\mathrm{O}_e\!\left(-s_\alpha t_2\right)\mathrm{O}_e\!\left(-s_\alpha t_2\right) = \mathrm{O}_e\!\left(s_\alpha\left[t_p + t_0 - -2t_2\right]\right). \tag{3.8}$$

This gives convergence with

$$t_p > 2t_2 - t_0. \tag{3.9}$$

Now the first signal from the target to reach the observer is at $2t_1$ (negative) in retarded time. So we have a time

$$t_e = t_p - 2t_2 > 2[t_2 - t_1] - t_0 \tag{3.10}$$

as a time window when the pole series does not converge and gives a bound on how large t_p should be. Note for this bodies (compared to L_0) that t_e can be negative. However, in a worst case sense $\overrightarrow{1}_i$ is aligned with L_0 giving $t_e = t_0$. So we have

$$0 \leq t_e \leq t_0 \tag{3.11}$$

including a limiting case in the case of worst-case alignment. The point is:

> For a time window of width t_e an entire function (temporal form) is required to adequately represent the far scattering.

Now let us approach the entire-function question from a different (new) direction. In Section 2 we defined a simple scatterer as one for which the currents can be represented by class-1 poles without an entire function, and noted that this has found to be the case by many examples. From (2.5) we then write in time domain:

$$\overrightarrow{J}_s(\overrightarrow{r},t) = E_0 f(t) \circ \left[\sum_{\alpha} \eta_\alpha(\overrightarrow{1}_i, \overrightarrow{1}_e)\, \overrightarrow{j}_{s_\alpha}(\overrightarrow{r}_s) e^{s_\alpha t} u\left(t - t_f\right) \right]. \tag{3.12}$$

An alternate form is the step response from $\tilde{f}(s) = 1/s$ as

$$\begin{aligned} \tilde{\overrightarrow{J}}_s(\overrightarrow{r}_s,t) &= E_0 \sum_{\alpha} \eta_\alpha(\overrightarrow{1}_i, \overrightarrow{1}_e)\, \overrightarrow{j}_{s_\alpha}(\overrightarrow{r}_s) \frac{e^{-[s-s_\alpha]t_f}}{s_\alpha [s - s_\alpha]} + \text{pole at } s = 0, \\ \overrightarrow{J}_s(\overrightarrow{r}_s,t) &= E_0 \sum_{\alpha} \eta_\alpha(\overrightarrow{1}_i, \overrightarrow{1}_e)\, \overrightarrow{j}_{s_\alpha}(\overrightarrow{r}_s) \frac{1}{s_\alpha} e^{s_\alpha t} u\left(t - t_f\right) + \text{step function.} \end{aligned} \tag{3.13}$$

It is this form that has been used for many of the numerical-verification examples. Note[2] that the low-frequency far scattering is proportional to s^2, making the step response in far scattering proportional to s. This insures that the late-time step-induced far scattering goes to zero at late time, even with a pole at $s = 0$.

From the formula for the far scattered fields (3.2) then we have

$$\vec{\tilde{E}}_f\left(\vec{r}, t + \frac{r}{c}\right) = \frac{\mu_0}{4\pi r} \overleftrightarrow{1}_0 \cdot \left[\frac{\partial}{\partial t} \int_S \vec{J}_s\left(\vec{r}_s, t + \frac{\vec{1}_e \cdot \vec{r}_c}{c}\right) dS\right]. \tag{3.14}$$

Applying this termwise to (3.12) or (3.13) gives terms of the form

$$\begin{aligned} &\int_S e^{s_\alpha\left[t + \frac{\vec{1}_o \cdot \vec{r}_s}{c}\right]} u\left(t - t_f + \frac{\vec{1}_o \cdot \vec{r}_s}{c}\right) dS \\ &= e^{s_\alpha t} \int_S e^{s_\alpha \frac{\vec{1}_o \cdot \vec{r}_s}{c}} u\left(t - t_f + \frac{\vec{1}_o \cdot \vec{r}_s}{c}\right) dS \\ &= e^{s_\alpha t} \int_S e^{s_\alpha \frac{\vec{1}_o \cdot \vec{r}_s}{c}} dS \ \text{ for } t \geq t_f + t_2, \\ t_2 &= \frac{L_2}{c} = \sup_S \left[-\frac{\vec{1}_o \cdot \vec{r}_s}{c}\right]. \end{aligned} \tag{3.15}$$

So for times (retarded at the observer) greater than $t_f + t_2$ the scattered field (for $f(t)$ a delta- or step-function) can be represented by *damped sines only*. The temporal form of the entire function exists only for times less than this and after the first scattered signal reaches the observer.

Applying this to backscattering we have

$$\vec{1}_o = -\vec{1}_i, \qquad t_f = t_1, \tag{3.16}$$

with pure damped sines for

$$t \geq t_1 + t_2, \tag{3.17}$$

and the first signal reaching the observer (retarded)

$$t = 2t_1 . \tag{3.18}$$

This gives an upper bound on the temporal width of the entire function as

$$t_1 + t_2 - 2t_1 = t_2 - t_1 = t_3 , \tag{3.19}$$

which is one transit through the body in the direction of the observer. Noting that for thin bodies

$$0 \le t_3 \le t_0 , \tag{3.20}$$

we have a result comparable to (3.11). Now we have an upper bound on the entire-function width, whereas previously we had a lower bound.

Summarizing we have for backscattering:

> The entire-function (temporal form) of $\overleftrightarrow{\Lambda}_b(\vec{1}_i,t)$ for simple scatterers has a temporal width of no greater than t_3, one transit through the body. This window begins at the time $2t_1$, the time of the first scattered signal.

4. CONCLUDING REMARKS

In Ref. 1 a lower bound was found for the temporal width of the entire function in scattering. Noting the lack of an entire function in interaction (currents on the body) for simple scatterers, we now have an upper bound of t_3 (one transit of the body) for the temporal width of the entire function. These results are complementary and more tightly pin down the scattering entire function as an early-time phenomenon.

There are related issues discussed in Ref. 1. In particular there is the *identification time* t_d, defined as the retarded time for the observer to receive a scattered signal from all points on S (or more generally V for a volume type of scatterer). This occurs at a time of $2t_3$ in backscattering after first scattered signal in the simplest cases. Accounting for t_3 (at most) for the entire function this gives a time of t_3 during which a *unique* pole series does not apply, since more than one scatterer may give exactly the same scattering during this time, the differences between the two scatterers having not yet been "seen". Afterwards the pole series can correspond to only one target, unless we include various parts (e.g., inside), which experience no fields. For more general targets t_d can be longer, for example., for perfectly conducting targets for which waves have to travel around the scatterer by non-straight geodesic paths (fat targets).

5. ACKNOWLEDGEMENT

This work was sponsored in part by the Air Force Office of Scientific Research and in part by the Air Force Research Laboratory, Directed Energy Directorate.

6. REFERENCES

1. C.E. Baum, Representation of surface current density and far scattering in EEM and SEM with entire functions, *Interaction Note 486*, Air Force Research Laboratory, Kirtland AFB NM (February 1992); also in: *New Perspectives on Problems in Classical and Quantum Physics, Part II, Acoustic Propagation and Scattering, Electromagnetic Scattering*, edited by P.P. Delsanto and A.W. Saenz (Gordon and Breach, 1998), Ch. 13, pp. 273–316.
2. C.E. Baum, J.E. Mooney, and L.S. Riggs, Use of modified pole series in the singularity expansion method, *Interaction Note 525*, Air Force Research Laboratory, Kirtland AFB NM (April 1997).
3. K.A. Michalski, Bibliography of the singularity expansion method and related topics, *Electromagnetics*, 493–511 (1981).
4. C.E. Baum, The singularity expansion method: background and developments, *IEEE Antennas Propagat. Soc. Newsletter* **28**(4), 15–23 (August 1986).
5. C.E. Baum, The singularity expansion method, in: *Transient Electromagnetic Fields*, edited by L.B. Felsen (Springer-Verlag, New York, 1976), Ch. 3, pp. 129–179.

TWO METHODS FOR PROFILE RECONSTRUCTION IN SCATTERING THEORY

F. Seydou & T. Seppañen*

1. INTRODUCTION

In this paper we discuss the problem of finding a profile for the scattering of electromagnetic waves with fixed frequencies. This type of problem is one of the most important and interesting problems arising in mathematical physics and has applications in geophysics, technology, medicine and nondestructive testing (cf. Rfs. 1–5). The difficulty in solving such problems is due to two unpleasant facts: They are nonlinear and, more seriously, they fall in a group of problems called ill-posed that is the solution-if it exists at all-does not depend on the data. Many authors suggested methods for overcoming these difficulties. The methods can be broadly divided into two classes: Optimization and iterative methods. In the previous works most algorithms were for TM wave illuminations whereas much less was reported for the most complicated TE case. The TE incident field is complicated compared to the TM case because of the strong nonlinearity and singularity of the kernel of the integral equation. In our work we consider one optimization (the dual space) method and one iterative (the simplified Newton) method. We shall consider the two scattering problems: the TM and the TE polarizations. It is well known that a combination of the two cases improves the quality of reconstruction. Our objective is to implement numerically the two methods to both problems and discuss the advantages and disadvantages in each case. We shall emphasize on the frequency and the type of profiles for the reconstructions. The simplified Newton method we shall develop has many advantages over other iterative methods: It is computationally inexpensive and the forward problem does not need to be solved at each iteration step. We shall compare the results with the implementation of the dual space method for different profiles and frequencies.

*University of Oulu, Department of Electrical Engineering P.O. Box 3000, 90401 Oulu Finland.

Ultra-Wideband Short-Pulse Electromagnetics 6
Edited by Mokole *et al.*, Kluwer Academic/Plenum Publishers, 2003

2. STATEMENT OF THE PROBLEM

Consider the scattering of electromagnetic waves in an inhomogeneous medium. The governing Maxwell equations can be reduced to the following two sets of problems :

Problem 1

$$\Delta u + k^2 s u = 0, \text{ in } \mathbf{R}^2, \tag{1}$$

$$u = u^i + u^s, \tag{2}$$

$$\lim_{r\to\infty} \sqrt{r}(\frac{\partial}{\partial r} - ik)u^s = 0, \qquad r = |\mathbf{x}|, \tag{3}$$

Problem 2

$$\nabla \cdot (q\nabla u) + k^2 u = 0, \text{ in } D, \qquad \Delta u_0 + k^2 u_0 = 0, \text{ in } \mathbf{R}^2\backslash\overline{D}, \tag{4}$$

$$u_0 = u, \quad \frac{\partial u}{\partial \nu} = \frac{\partial u_0}{\partial \nu}, \quad \text{on } \partial D \tag{5}$$

$$u_0 = u^i + u^s, \tag{6}$$

$$\lim_{r\to\infty} \sqrt{r}(\frac{\partial}{\partial r} - ik)u^s = 0, \qquad r = |\mathbf{x}|. \tag{7}$$

In the two problems $u^i(\mathbf{x}) = \exp(ik\mathbf{x}\cdot\mathbf{d})$ is the incident wave with incident direction $\mathbf{d}$, with $|\mathbf{d}| = 1$, the functions s and q are the refractive indices, $k = \omega/c_0$ is the positive wave number, ω is the frequency and c_0 the sound speed.

We shall assume that $m := 1-q$ and $m' := 1-s$ are real valued (no absorption) and vanish outside the domain D (describing the scatterer) with a C^2 boundary ∂D.

Problem 1 is referred to as the TM-mode and Problem 2 is called TE-mode. The direct problem corresponding to Problem 1 (resp. Problem 2) is to find u (resp. u and u_0) for given s (resp. q), D, k, and $\mathbf{d}$.

We are interested in the inverse problem which is to determine s (resp. q) from the knowledge of the scattered field u^s at long distances from inhomogeinities, i.e. the far field pattern u_∞ solution of Problem 1 (resp. Problem 2).

It can be shown (cf. Ref. 1, Thm. 2.5) that the scattered field u^s has the asymptotic behavior

$$u^s(\mathbf{x}) = \frac{\exp(ik|\mathbf{x}|)}{\sqrt{|\mathbf{x}|}} \left\{ u_\infty(\hat{\mathbf{x}};\mathbf{d}) + O(\frac{1}{|\mathbf{x}|}) \right\}, \quad |\mathbf{x}| \to \infty, \quad \hat{\mathbf{x}} = \frac{\mathbf{x}}{|\mathbf{x}|}. \tag{8}$$

In the sequel we shall use $\hat{\mathbf{x}} := (\cos(\theta), \sin(\theta))$, $\mathbf{d} := (\cos(\alpha), \sin(\alpha))$ and Ω denotes the unit

3. THE DUAL SPACE METHOD

The DSM is to solve the inverse problem as follows (cf. Ref. 2) and the references therein for details). We assume that we have approximate measurements of the far field pattern for given incident waves and wave directions. In addition, we assume that a number b is a priori known such that the domain $\overline{D}$ is contained in the interior of $B := \{\mathbf{x} \in \mathbf{R}^2 : |\mathbf{x}| < b\}$. The method is then divided into two steps. In the first step, from the measured data (u_∞), we need to find $g_p(\theta) \in L^2[-\pi, \pi]$ solutions to the first kind integral equation

$$\int_{-\pi}^{\pi} [u_\infty(\theta;\alpha) - i\gamma \exp(ika\sin(\theta - \alpha))]\, g_p(\theta)\, d\theta = \gamma \exp(ip\alpha - i\frac{p\pi}{2}), \tag{9}$$

for $-P \leq p \leq P$, $\quad P \in \mathbf{N}$, where

$$\gamma = \sqrt{2/(\pi k)} \exp(-i\pi/4),$$

a is a positive real number assumed to satisfy the conditions: $b < a$ and $J_l(ka) \neq 0$ for every l, where J_l is the Bessel function of order l.

Since, in (9), we have expressions that represent far field patterns, we can use Rellich Lemma to show that the scattered field corresponding to the far field on the left hand side is equal to the scattered field of the expression on the right hand side. In particular, for u_∞ being the far field pattern corresponding to Problem 1, setting $h_p(\theta) := g_p(\theta - \pi)$, it is easy to show that equation (9) is satisfied if and only if there exists a total field $w_{p,1} := w_p^i + w_p^s$ that satisfies

$$\Delta w_{p,1} + k^2 s w_{p,1} = 0 \quad \text{in} \quad B \tag{10}$$

$$w_{p,1} - w_p^i - v_p - u_p = 0 \text{ and } \frac{\partial}{\partial r}(w_{p,1} - w_p^i - v_p - u_p) = 0 \qquad \text{on} \quad \partial B, \tag{11}$$

where

$$w_p^i(\mathbf{x}) = \int_{-\pi}^{\pi} u^i(\mathbf{x};\alpha) h_p(\alpha)\, d\alpha,$$

$$w_p^s(\mathbf{x}) = \int_{-\pi}^{\pi} u^s(\mathbf{x};\alpha)h_p(\alpha)\,d\alpha,$$

$$u_p(r;\theta) = H_p^{(1)}(kr)\exp(ip\theta),$$

and

$$v_p(\mathbf{x}) = \frac{i}{a}\int_{|\mathbf{y}|=a} H_0^{(1)}(k|\mathbf{x}-\mathbf{y}|)(Rh_p)\left(\frac{1}{a}\mathbf{y}\right)\,ds(\mathbf{y}), \tag{12}$$

where $(Rh)(\phi) := \sum_{-\infty}^{\infty} i^l h_l \exp(il\phi)$, h_l is the $l$the Fourier coefficient of h and $H_p^{(1)}$ is the Hankel function of first kind and order p.

Similarly, for u_∞ being the far field pattern corresponding to Problem 2, we have that equation (9) is satisfied if and only if there exist $w_{p,2} := w_p^i + w_p^s$ and $w_{p,0}$ that satisfy

$$\nabla\cdot(q\nabla w_{p,2}) + k^2 w_{p,2} = 0, \text{ in } D, \qquad \Delta w_{p,0} + k^2 w_{p,0} = 0, \text{ in } B\backslash\overline{D}, \tag{13}$$

$$w_{p,2} = w_{p,0}, \quad \frac{\partial w_{p,2}}{\partial\nu} = \frac{\partial w_{p,0}}{\partial\nu}, \quad \text{on } \partial D \tag{14}$$

$$w_{p,2} - w_p^i - v_p - u_p = 0 \text{ and } \frac{\partial}{\partial r}(w_{p,2} - w_p^i - v_p - u_p) = 0 \qquad \text{on} \quad \partial B. \tag{15}$$

Now we have the second step. Given the approximated solution of (9), we determine s (resp. q) by minimizing the functional

$$J(q) = \|w_p - w_p^i - v_p - u_p\| - \|\frac{\partial}{\partial r}(w_p - w_p^i - v_p - u_p)\|$$

on ∂B where w_p represents $w_{p,1}$ (resp. $w_{p,2}$).

4. THE SIMPLIFIED NEWTON METHOD

Using Green identities, we can show that Problem 1 is equivalent to the Lippmann-Schwinger equation

$$u(\mathbf{x}) = u^i(\mathbf{x}) + k^2\int_Q \Phi(\mathbf{x},\mathbf{y})m'(\mathbf{y})u(\mathbf{y})\,d\mathbf{y}. \tag{16}$$

Similarly Problem 2 is equivalent to

$$u(\mathbf{x}) = u^i(\mathbf{x}) + \int_Q \nabla_{\mathbf{y}} \Phi(\mathbf{x}, \mathbf{y}) \cdot m(\mathbf{y}) \nabla u(\mathbf{y})\, d\mathbf{y}, \tag{17}$$

where $Q = [-\pi, \pi]^2$ is a square that contains $\overline{D}$ in its interior and $\Phi(\mathbf{x}, \mathbf{y}) = i/4 H_0^{(1)}(k|\mathbf{x} - \mathbf{y}|)$ is the fundamental solution of the Helmholtz equation.

To fix the ideas let us consider Problem 2. The extension of the method to Problem 1 is trivial. Since $u = u^i + u^s$, we get

$$u^s(\mathbf{x}) = \int_Q \nabla_{\mathbf{y}} \Phi(\mathbf{x}, \mathbf{y}) \cdot m(\mathbf{y}) \nabla u(\mathbf{y})\, d\mathbf{y}. \tag{18}$$

Using (8) we obtain

$$u_\infty(\hat{\mathbf{x}}) = -\frac{ik\beta_k \hat{\mathbf{x}}}{4} \cdot \int_Q \exp(-ik\hat{\mathbf{x}} \cdot \mathbf{y}) m(\mathbf{y}) \nabla u(\mathbf{y})\, d\mathbf{y}, \quad \beta_k = \frac{\exp(-i\pi/4)}{\sqrt{8\pi k}}. \tag{19}$$

Define $V : L^2(Q) \to H^1(Q)$ and $W : L^2(Q) \to L^2(\Omega)$, such that

$$V(\phi)(\mathbf{x}) = \int_Q \nabla_{\mathbf{y}} \Phi(\mathbf{x}, \mathbf{y}) \cdot \phi(\mathbf{y})\, d\mathbf{y},$$

and

$$W(\psi)(\hat{\mathbf{x}}) = -\frac{ik\beta_k \hat{\mathbf{x}}}{4} \cdot \int_Q \exp(-ik\hat{\mathbf{x}} \cdot \mathbf{y}) \psi(\mathbf{y})\, d\mathbf{y},$$

for some vector functions ϕ and ψ. From (17), (19) and the definitions of V and W we can write the inverse problem as: Given a (measured) far field pattern f, find m and u such that $T(z) = R$, where $z = (m, u)$, $R = (u^i, f)$, and $T : C(Q) \times H^1(Q \times \Omega) \to H^1(Q \times \Omega) \times C(\Omega \times \Omega)$ is given by

$$T(z) = (u - V(m\nabla u), W(m\nabla u)).$$

The simplified Newton method (SNM) is given by :

$$z_{l+1} = z_l - \left[T'(\tilde{z})\right]^{-1} \left[T(z_l) - R\right],$$

for some fixed $\tilde{z}$ and $z_l := (m_l, u_l)$. By T' we mean the Fréchet derivative of T.

Since the components of T are bilinear it is easy to see that the Fréchet derivative of T at a point $\mu := (p, v) \in C(Q) \times H^1(Q \times \Omega)$ is given by

$$T'(m,u)\mu = (v - V(p\nabla u) - V(m\nabla v), W(p\nabla u) + W(m\nabla v)).$$

Choosing $\tilde{z} = (0, u^i)$ we get $T'(\tilde{z})(\mu) = (v - V(p\nabla u^i), W(p\nabla u^i))$. Thus the algorithm SNM can be rewritten as $z_{l+1} = z_l - \mu$, where μ is the solution of $T'(\tilde{z})(\mu) = T(z_l) - R$ which we rewrite as

$$\begin{cases} v = u_l + V(p\nabla u^i) - V(m_l \nabla u_l) - u^i \\ W(p\nabla u^i) = W(m_l \nabla u_l) - f. \end{cases} \tag{20}$$

Let us write the second equation in the system as a problem of finding p from $W(p\nabla u^i) = g$ for a known g.

$$\int_Q \exp(-ik(\hat{\mathbf{x}} - \mathbf{d}) \cdot \mathbf{y}) p(\mathbf{y})\, d\mathbf{y} = \frac{4}{\beta_k k^2 \hat{\mathbf{x}} \cdot \mathbf{d}} g(\hat{\mathbf{x}}, \mathbf{d})$$

To approximate p, we define (the regularization parameter) $N \in \mathbf{N}$ such that $N \leq \sqrt{2}k$, the set

$$\mathbf{Z}_N := \{j := (j_1, j_2) \in \mathbf{Z}^2, \quad |j_s| \leq N, s = 1, 2\}$$

and

$$L_N := \left\{ \sum_{j \in \mathbf{Z}_N} a_j \exp(ij \cdot \mathbf{x}), \quad a_j \in \mathbf{C} \right\}.$$

Then for every $j \in \mathbf{Z}_N$, we can fix the unit vectors $\hat{\mathbf{x}}_j$ and $\mathbf{d}_j$, such that $j = k(\hat{\mathbf{x}}_j - \mathbf{d}_j)$.

Hence $p = L_1 g$ where

$$L_1 g(\mathbf{x}) = \sum_{j \in \mathbf{Z}_N} \frac{4}{(2\pi)^2 \beta_k k^2 \hat{\mathbf{x}}_j \cdot \mathbf{d}_j} g(\hat{\mathbf{x}}_j, \mathbf{d}_j) \exp(ij \cdot \mathbf{x}).$$

The algorithm is now
Step 1: Set $z_0 = (0, u^i)$ and $l = 0$.
Step 2: Set $\mu := (p, v)$, where

$$p = L_1 \left(W(m_l \nabla u_l) - f \right)$$

and

$$v = u_l + V(p\nabla u^i) - V(m_l \nabla u_l) - u^i$$

Step 3 Set $l = l + 1$, $z_{l+1} = z_l - \mu$ and go to *Step 2*.

5. NUMERICAL RESULTS

In this section we shall present our numerical results. We shall use synthetic data as follows. For a given wave number k, refractive index s or q, direction $\mathbf{d}$ and scatterer D, we solve the direct problem to obtain an approximation of the far field pattern u_∞. The approximated far field can then be used to try and recover the index of refraction using the algorithms presented in the previous sections. We have used FEM-BEM method for solving the direct problems and FEM for the approximation of the functional in the DSM. For approximating m and m' for the SNM algorithm we used an $M \times M$ pixels to compute the integrals, where M is a fixed integer. Our numerical results show that we have a convergence problem

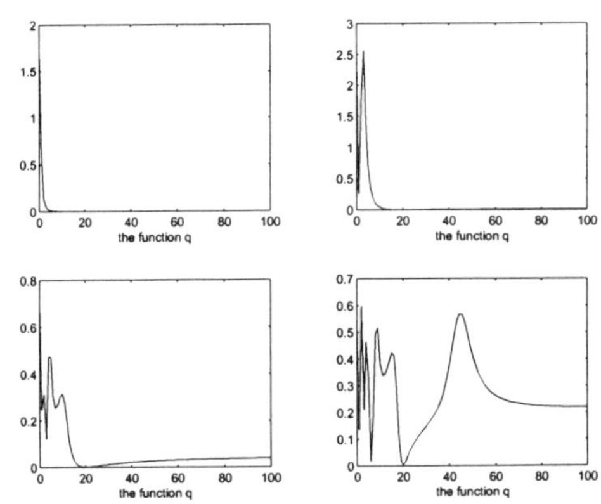

Figure 1. The functional J against the function q for different values of k for the TE-mode. Here $k = 2, 4, 8, 16$. The bigger we take k the more local minima we have. For small k it is hard to locate the minimum. The value $q = 20$ in the scatterer was used for computing the far field.

for SNM when the index is big, especially for the TE case. However SNM can give accurate results for small indices. The wave number does effect the reconstruction since the regularization parameter N depends on k. We have also seen that the TE case needs more iteration than the TM case. This may be due to the nonlinearity of the TE mode.

For small indices the first iteration in SNM (Born approximation) should give a reasonably good approximation. We tried to reconstruct a function s or q which is equal to 0.8 for $|x| < \pi/2$ and one otherwise. Using TM polarization, for $k = 5.2$ we get a relative error of 25% for $N = 1$, 18% for $N = 2$ and 5% for $N = 7$. When

$k = 10.2$ we get an error of 9% for $N = 5$. The latter result is plotted in Figure 5. When we use TE polarization the best result is achieved for $N = 3$ when $k = 5.2$ (Figure 6) . For $N > 3$, the bigger we take N the bigger is the error. This suggests that the results are more accurate for the TM case compared to the TE mode.

Let us now consider the DSM. It also depends on the wave number as we shall see later. Here, instead of recovering the index we analyze the behavior of the functional J in the second Step. This is very important since the functional J consists of the necessary information for finding the index. A method for reconstruction is discussed in Ref. 2. It is very slow compared to the SNM algorithm. Our approach here is as follows. For given index (q or s), scatterer D, wave number k we compute the far field pattern. The approximated far field f is then used to compute an approximation of g_p by solving the integral equation in the first step of DSM. Given the approximated g_ps we compute the functional J in the second step for different qs or ss. So, when we plot J against q or s the minimum should be achieved at the value of q or s used for computing the far field f.

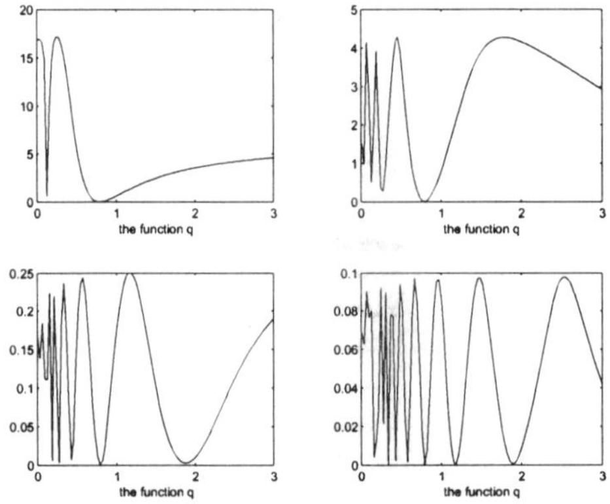

Figure 2. The functional J against the function q for different values of k for TE-mode. Here $k = 2, 4, 8, 16$. The bigger we take k the more local minima we have. Small k gives a good minimum. The value $q = 0.8$ in the scatterer was used for computing the far field.

For our numerical computation we have always assumed that D is a circle of radius $R = 2$ for the computation of the far field pattern. In everyone of Figures 1-4 we plotted the functional J against the index for different values of k, $k = 2$ (top, left), $k = 4$ (top, right), $k = 8$ (bottom, left) and $k = 16$ (bottom, right). Now, let us discuss the results for the two polarizations in details. First consider the TE-mode, i.e. the inverse problem related to Problem 2. In Figure 1 we plotted the functional J against the index q, where $q = 20$ for $|x| < 2$ and one otherwise. We see that The bigger we take k the more local minima we have. For small k it is hard to locate the minimum. This suggests that small wave numbers and very big wave numbers may not give a good approximation of the index q. When trying to plot the functional against a smaller index, $q = 0.8$ for $|x| < 2$ and one otherwise (Figure 2), we see that we have the same problem with very big wave numbers.

But small wave numbers give a good minimum. For the TM-mode i.e. the inverse problem related to Problem 1, we have roughly the same scenario like the TE-mode when the index is small (Figure 3. Here $s = 0.5$ for $|x| < 2$ and one otherwise). But when it is plotted against a big s, $s = 20$ for $|x| < 2$ and one otherwise, we have local minima for every range of k (Figure 4).

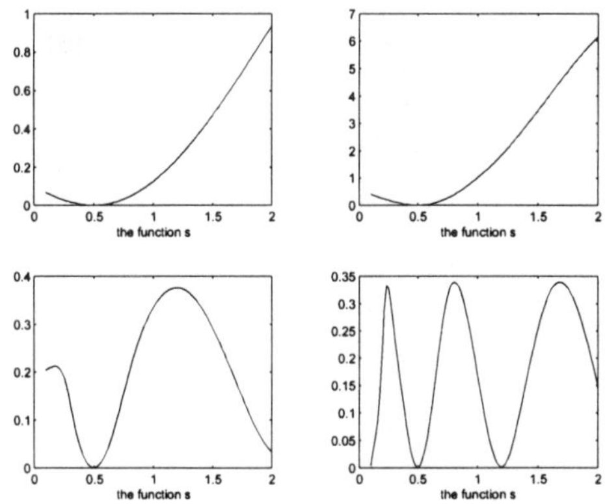

Figure 3. The functional J against the function s for different values of k for the TM-mode. Here $k = 2, 4, 8, 16$. The bigger we take k the more local minima we have. Small k gives a good minimum. The value $s = 0.5$ in the scatterer was used for computing the far field.

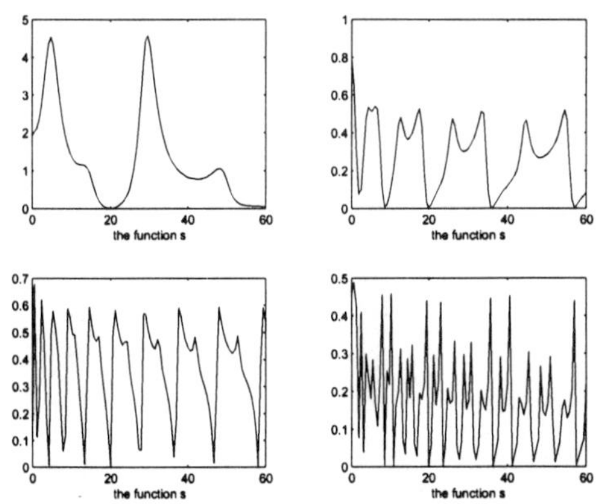

Figure 4. The functional J against the function s for different values of k for the TM-mode. Here $k = 2, 4, 8, 16$. For all ks we have local minima. The value $s = 20$ in the scatterer was used for computing the far field.

Figure 5. The Coefficient on left and its reconstruction for the TM case using SNM for k=10.2 and $N = 4$.

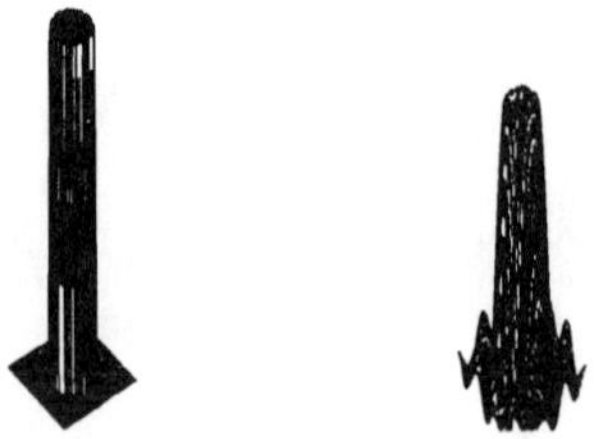

Figure 6. The Coefficient on left and its reconstruction for the TE case using SNM for k=5.2 and $N = 3$.

6. CONCLUSION

In this paper we have analyzed two methods (DSM and SNM)for the reconstruction of the refractive index for TE and TM scattering problems. Both methods are sensitive to the choice of the wave number as well as the size of the index to be reconstructed. Our results suggest that, while SNM is a faster algorithm it may lead to unstable solutions for big indices and works better for the TM case than the TE case. DSM works for all sizes of indices in the TE case provided the wave number is appropriately chosen. For the TM case the best results are for small indices.

REFERENCES

1. D. Colton and R. Kress, Inverse Acoustic and *Electromagnetic Scattering Theory* (Berlin, Springer, 1992).

2. F. Seydou. Profile inversion in scattering theory: the TE case. *Journal of Comp. and Appl. Math.*

3. S. Gutman and M. Klibanov, Regularized quasi-Newton method for inverse scattering problems. *Math. Comput. Modelling* **18** 5-31 (1993).

4. A. Kirsch, *An Introduction to the Mathematical Theory of Inverse Problems* (Springer, Berlin, 1996).

5. H. Moritz, General considerations regarding inverse and related problems, in *Inverse problems: Principles and Applications in Geophysics, Technology, and Medicine* Edited by G. Anger, D. Anger, R. Gorentflo, H. Jochmann, H. Moritz, W. Webers. (Springer Verlag, Berlin, 1993) 11-23.

OBJECT IMPULSE RESPONSE EVALUATION FOR ULTRAWIDEBAND APPLICATION

Vladimir I. Koshelev, Valeriy T. Sarychev, and Sergey E. Shipilov*

1. INTRODUCTION

Active methods for investigation of different objects suggest sounding by pulses of different waveforms with subsequent receiving and processing of the passed or reflected signal. Linear systems in which the input and output signal relation is determined by the convolution $Y(t) = h(t) \otimes X(t)$ are considered in this paper. Here, $Y(t)$ is the output signal, $h(t)$ is the impulse response (IR) of the investigated object, $X(t)$ is the sounding pulse. The IR completely reflects the properties inherent to this object in the sounding pulse frequency band and can be used for its recognition.

Difficulties in IR estimation arise mainly from two reasons. First, in practice, an investigator has a set of data in the observation interval that is limited both by time and by frequency. Second, any observation supposes noise availability. A consequence of these limitations of the initial data set is the deterioration of the estimation precision of an unknown function in the best case. In the worse case, the solution becomes unstable and there is no method to guarantee its correctness.

The instability of estimation is determined by the availability of nulls in the estimation of complex spectra (CS) obtained by limited sets of initial data. One method of null elimination is regularization.[1] Although regularization is a mathematical method for eliminating the consequences of having nulls, it does not provide a rationale for their existence. The reason for the appearance of nulls in the spectrum can be found in the assumption that beyond the observation window the signal is equal to zero. This assumption can be removed by selecting an approximation of the initial data within the observation window and by allowing the signal to continue outside the window. The most valid approach, from mathematical and physical points of view, should consider signal approximation in the time domain by exponentially damped oscillatory functions like

* Institute of High Current Electronics, RAS, 4 Akademichesky Avenue, 634055 Tomsk, Russia.

$$Q_m(t) = \theta(t)\left[a_m \exp(-iq_m t) - a_m^* \exp(iq_m^* t)\right], \; q_m = \Omega_m + i\gamma_m.$$

Here, $\theta(t)$ is the Heaviside unit-step function. The asterisk denotes the operation of complex conjugation. The corresponding frequency-domain function $Q_m(\omega)$ of $Q_m(t)$ is a very simple meromorphic function with two first-order poles at q_m and $-q_m^*$:

$$Q_m(\omega) = \frac{a_m^*}{\omega + q_m^*} - \frac{a_m}{\omega - q_m}. \tag{1}$$

Associated with the complex amplitude a_m are the frequency Ω_m and the decrement γ_m.

A similar approximation was suggested by Prony.[2] The same approximation is the basis for the singularity expansion method[3] that was suggested for investigating non-stationary ultrawideband (UWB) signals. As shown in this work, the CS of signals reflected from metal objects has typical sets of poles. This fact was used to solve the recognition problem of the objects probed by pulses having spatial lengths comparable to the object dimensions. The signal model presented in this way is hereinafter called a Pole Model (PM), and the functions approximating such signals are called the pole functions.

However, all existing methods of digital signal processing that use some form of PM suggest the absence of time delay between the functions describing the signals. This circumstance mandates using a large number of pole functions for signal approximation, which can result in discrepancies between the calculated poles and the actual poles that characterize an object and can decrease the stability of solving the IR-estimation problem on the whole. Besides, it is interesting to consider the more common case when the approximating model parameters are distributed continuously (not discretely) and are described by a certain distribution density function (DDF).

The main aim of the investigations presented in this work is the development of a mathematical model for approximating the investigated signals by an ensemble of exponentially damped oscillations with random and discrete values of the PM parameters. The results suggest that a search for analytical models of the DDFs of these parameters, based on the principle of maximum entropy, may prove to be fruitful. Further, on the basis of the obtained approximations of the input and output signals, evaluations of the impulse responses of the UWB system are made, and the possibility of solving the recognition problem is discussed.

2. CONTINUOUS POLE MODEL OF SIGNALS

As a rule, UWB signals are localized in time. This fact should be taken into account in the development of a mathematical model of this type of signal. The spectra of UWB signals from the null through the switching moment should satisfy a definite dispersion relation resulting from the causality principle.[4] Pole models of spectra automatically satisfy this principle. A Discrete Pole Model of a Signal (DPMS) and a Continuous Pole Model of a Signal (CPMS) should be distinguished. A DPMS looks like

$$S(t) = \sum_m Q_m(t - \tau_m), \tag{2}$$

where τ_m is the time delay.

In general, one must consider a CPMS when all oscillation parameters are random (frequency, decrement, amplitude, time delay). The multi-variate distribution of the parameters is specified by the DDF $f(\Omega, \gamma, a, \tau)$ and requires estimation. An expression for the time sequence $S(t_i)$ corresponding to the function $f(\Omega, \gamma, a, \tau)$ is determined by the multi-dimensional integral

$$< S(t_i) >= \int_{-\infty}^{\infty} d\omega \int_{-\pi}^{\pi} d\Omega \int_{0}^{\infty} d\gamma \int_{-\infty}^{\infty} da \int_{0}^{\infty} d\tau f(\Omega, \gamma, a, \tau) \frac{a \exp(i\omega(t_i - \tau))}{(\omega - \Omega - i\gamma)}. \tag{3}$$

The quantities Ω, γ, a, and τ are dimensionless. Transition to dimensional quantities is made by means of multiplication or division by the discrete time value Δt.

The problem of estimating the function $f(\Omega, \gamma, a, \tau)$ by the limited time sequence $S(t_i)$ is ill-posed, which may be removed by imposing an additional limitation on $f(\Omega, \gamma, a, \tau)$. The most acceptable estimation method, both from the mathematical perspective and from the physical perspective, is the principle of maximum entropy. According to Boltzmann, the entropy for this problem is

$$Entropy = -\int_{-\pi}^{\pi} d\Omega \int_{0}^{\infty} d\gamma \int_{-\infty}^{\infty} da \int_{0}^{\infty} d\tau f(\Omega, \gamma, a, \tau) \ln(f(\Omega, \gamma, a, \tau)). \tag{4}$$

To ensure the integrability of $f(\Omega, \gamma, a, \tau)$ with respect to the amplitude a, Eq. (3) should be supplemented with the condition

$$\int_{-\infty}^{\infty} d\omega \int_{-\pi}^{\pi} d\Omega \int_{0}^{\infty} d\gamma \int_{-\infty}^{\infty} da \int_{0}^{\infty} d\tau f(\Omega, \gamma, a, \tau) a^2 = \min. \tag{5}$$

Maximizing the entropy in Eq. (4) subject to the constraints of Eq. (3) and Eq. (5) yields the following parametric form of f:

$$\begin{aligned} f(\Omega, \gamma, a, \tau) &= C \exp\left(\frac{< a >^2 - (a - < a >)^2}{2\sigma^2} \right), \\ < a > &= \sum_k \lambda_k \exp(-\gamma(t_k - \tau)) \sin(\Omega(t_k - \tau)). \end{aligned} \tag{6}$$

Here, C is the normalizing parameter, λ_k are the Lagrange coefficients providing fulfillment of the relation (3), and σ is the mean square root of amplitude providing fulfillment of the condition (5). In order to estimate the values of λ_k and σ, one should substitute Eq. (6) into Eq. (3) and perform the indicated integration. By analytically integrating Eq. (5) with respect to the frequency ω and the amplitude a, one obtains

$$< S(t_i) >= C\int_0^{\pi} d\Omega \int_0^{\infty} d\gamma \int_0^{t_i} d\tau < a > \exp\left(-\gamma(t_i - \tau) + \frac{< a >^2}{2\sigma^2}\right)\sin(\Omega(t_i - \tau)). \quad (7)$$

By accounting for the symmetry of the subintegral involving the sign change of the frequency Ω, one may integrate with respect to the variable Ω over the interval $[0, \pi]$. Unfortunately, owing to the complex dependence of the integrand on the integration variables, the remaining integrals are difficult to evaluate analytically. However, the expression in Eq. (7) admits a linearization relative to the Lagrange coefficients. As soon as this linearization is applied, the integrals are calculated analytically. The normalizion parameter C in Eq. (7) is proportional to the number of oscillations from which the signal $<S(t_i)>$ is synthesized. Results of the signal measurement make no limitations on the choice of this parameter value. For any essentially large value of C, the ratio $|<a>|/\sigma$ becomes much less than unity and Eq. (7) assumes the more simple form

$$\begin{aligned} < S(t_i) > &= C\int_0^{\pi} d\Omega \int_0^{\infty} d\gamma \int_0^{t_i} d\tau < a > \exp(-\gamma(t_i - \tau))\sin(\Omega(t_i - \tau)) \\ &= \sum_k \lambda_k \int_0^{\pi} d\Omega \int_0^{\infty} d\gamma \int_0^{t_k} d\tau \exp(-\gamma(t_i + t_k - 2\tau)) \\ &\qquad \times\left[\cos(\Omega(t_i - t_k)) - \cos(\Omega(t_i + t_k - 2\tau))\right]. \end{aligned} \quad (8)$$

As shown above, the discretization interval is taken as a time unit. In this case, one can let $t_i = i$. After evaluating the temporal integration with $t_i = i$, Eq. (8) reduces to

$$S_i = \sum_k \lambda_k B_{ik},\ B_{ik} = \begin{cases} ci(\pi \mid i - k \mid) - ci(\pi(i+k)), & i \neq k; \\ Ce + \ln(\pi(i+k)) - ci(\pi(i+k)) - 1, & i = k, \end{cases} \quad (9)$$

where $ci(x)$ is the integral cosine and Ce (= 0.577215664) is Euler's constant.

Solving the linear system (9) yields a CPMS on the basis of maximizing the entropy of the signal S_i ($i = 1,\ldots,N$). A CPMS permits interpolating by discrete counts of the signal S_i, that is, by calculating the values of $S(t)$ for any $t \in [0, N]$. The interpolation expression is determined as a result of the calculation in Eq. (8) for an arbitrary (not just integer) value of t. This expression has the following form

$$S(t) = \sum_{i=1}^{N} \lambda_i \psi_i(t),$$

$$\psi_i(t) = \frac{\sin(\pi(i-t))}{\pi(i-t)}\left(\ln\left(\frac{i+t}{|i-t|}\right) - 1\right) + \frac{\sin(\pi(i+t))}{\pi(i+t)} + ci(\pi \mid i-t \mid) - ci(\pi \mid i+t \mid). \quad (10)$$

In its turn, the availability of a continuous signal model in the form of (10) allows the convolution operation

$$Y(t) = \int_0^t h(\tau)X(t-\tau)d\tau \qquad (11)$$

that couples the input $X(t)$ and the output $Y(t)$ of a linear system. Equation (11) can be used for IR evaluation when the input and output signals are known. The parametric forms of $X(t)$ and the IR are taken as the same as those in Eq. (10), that is,

$$X(t) = \sum_{i=1}^{N} \lambda_i \psi_i(t), \quad h(t) = \sum_{k=1}^{K} \mu_k \psi_k(t). \qquad (12)$$

Hence, in order to evaluate the convolution in Eq. (11), it is necessary to determine the convolutions of the functions $\psi_k \otimes \psi_i$. In this work, the convolutions were determined numerically, and function tables for

$$\varphi_{k,i}(t) = \psi_k(t) \otimes \psi_i(t). \qquad (13)$$

were obtained. Joint use of the expressions (11)–(13) leads to the following linear system of algebraic equations for estimating the coefficients μ_k of the IR:

$$Y_m = \sum_{k=1}^{K} \mu_k \sum_{i=1}^{N} \lambda_i \varphi_{k,i}(m); \qquad m = 1, ..., K+N. \qquad (14)$$

The set of μ_k values is a compact form of IR recording and can be used in the problem of object recognition by the IR.

3. DISCRETE POLE MODEL OF SIGNALS

In the majority of cases, the signal approximation problem can be simplified by imposing a series of restrictions on the DDF $f(\Omega, \gamma, a, \tau)$. In particular, the authors assume that Ω, γ, a are discrete and that τ is described by the distribution density function of the pole-function delay $f_m(\tau)$. In this case, to approximate the CS of the corresponding signal $\tilde{X}(t)$ and the IR $\tilde{h}(t)$, use

$$S(\omega) = \sum_{m=1}^{M} \int_{-\infty}^{\infty} f_m(\tau) Q_m(\omega) \exp(i\omega\tau) d\tau\,. \qquad (15)$$

This approach reduces the number of parameters needed for signal description and increases the stability of IR estimation. The locations of the IR poles in the complex plane and the form of $f_m(\tau)$ are the identification features of the investigated object and can be used for its recognition.

At non-simultaneous "switching-on" of the damped oscillations, the functions $f_m(\tau)$ must be estimated. Each time-delay distribution $f_m(\tau)$ is estimated with one of the following two variants:

1) Equi-probable distribution in a certain semi-bounded time interval:

$$f_m(\tau) = \begin{cases} 1, & \tau \geq \tau_m; \\ 0, & \tau < \tau_m. \end{cases} \tag{16}$$

2) Equi-probable distribution in a certain bounded time interval:

$$f_m(\tau) = \begin{cases} 1/T_m, & \tau \in [\tau_m, T_m + \tau_m]; \\ 0, & \tau \notin [\tau_m, T_m + \tau_m]. \end{cases} \tag{17}$$

Next, an analytical expression for $\tilde{Y}(t)$ is obtained by applying convolution

$$\tilde{Y}(t) = \sum_{k=1}^{K} \int_{-\infty}^{\infty} d\tau f_k^h(\tau) Q_k^h(\tau - \tau_k^h) \otimes \sum_{m=1}^{M} \int_{-\infty}^{\infty} d\tau f_m^X(\tau) Q_m^X(\tau - \tau_m^X). \tag{18}$$

The expression (18) allows estimating the values of the IR's PM parameters at the known PM values of the sounding pulse $X(t)$ and the known output signal $Y(t)$. A valid criterion for quantifying the accuracy of the estimate of the pole model parameters is whether the discrepancy Φ between the output signal $Y(t)$ and the model signal $\tilde{Y}(t)$ is minimized. The discrepancy is calculated by substituting Eq. (18) into

$$\Phi = \left[Y(t) - \tilde{Y}(t)\right]^2 / [Y(t)]^2. \tag{19}$$

For a model description, the minimum number of poles is sought that yields the discrepancy threshold value determined by the noise level in signals. In case the noise level is unknown a priori, the number of poles N_p is determined from the condition,

$$\Phi_Y(N_p) - \Phi_Y(N_p + 1) << \Phi_Y(N_p - 1) - \Phi_Y(N_p).$$

This condition implies that further increasing the number of poles results in a non-essential decrease in the discrepancy value.

4. POLE MODEL APPLICATION FOR CABLE SYSTEMS

To verify the efficiency of the suggested approach, experimental data for UWB pulse propagation through three coaxial cables were collected: (1) RC 50-2-11 of length 2 m; (2) RC 50-4-11 of length 8 m; (3) RC 50-2-11 of length 20 m. A detailed description of the experimental setup for obtaining the time realizations of UWB pulses was presented earlier.[5] To record the voltage $X(t)$ directly at the generator output, a short piece of nearly distortionless cable was used.

Figure 1 displays the monopolar pulses $Y1(t)$, $Y2(t)$, and $Y3(t)$ after the generated pulse $X(t)$ has passed through cables No. 1, 2, and 3, respectively. In addition, three bipolar pulses are obtained from the same three cables by combining two 1-ns duration monopolar pulses of opposite polarity, shifted relative to each other by 1 ns (Fig. 2). The choice of two sounding pulses for the experiment allows demonstrating the stability of the suggested approach of IR evaluation.

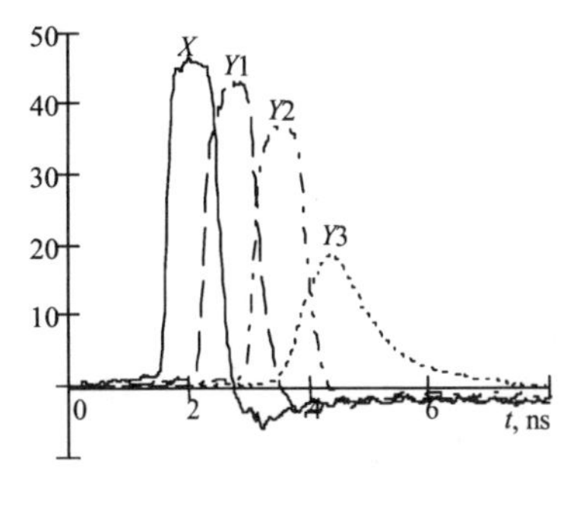

Figure 1. Monopolar signals from the generator after propagation through the cables.

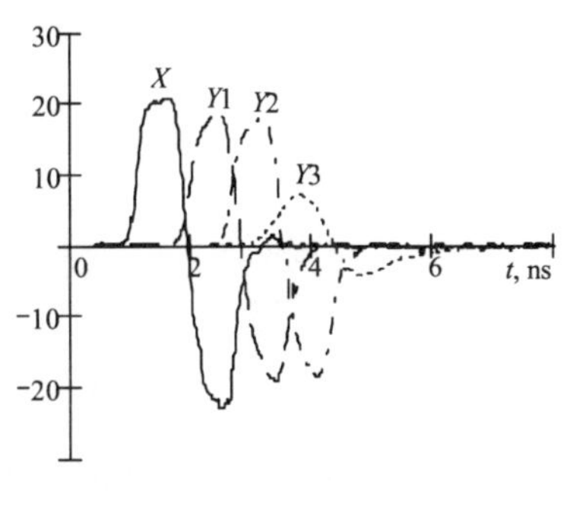

Figure 2. Bipolar signals from the generator after propagation through the cables.

By using CPMS and two variants of DPMS, the impulse responses (Figs. 3, 4) of cable No. 3 for the sounding monopolar and bipolar pulses were estimated. In the case of CPMS, the choice $h1$ of the discretization parameter of the IR was made on the following assumption for the output signal frequency: $\Delta t = 1/\Delta f$, where Δf is the frequency band containing 97% of the output signal energy. For cable No. 3, Δt was equal to 0.51 ns. In DPMS, two poles for $h2$ in the model (16) and 1 pole for $h3$ in the model (17) were estimated. This set of the PM parameters was sufficient so that the discrepancy (19) was not higher than 1%. The nonlinear parameters Ω_m, γ_m, and τ_m of the poles in DPMS were estimated by the alternating-variable descent method; initial approximations were calculated by means of specially developed algorithms. Comparison of Figs. 3 and 4 results in conclusions about the similarity of the obtained IRs for different PMs, as well as about the stability of the IR estimation with respect to the sounding pulse waveforms. Special attention was given to investigating the value of estimation stability of the IR model parameters, calculated by means of CPMS and DPMS, relative to measurement noise. Stability of estimations of parameter values allows using them in the problem of recognition.

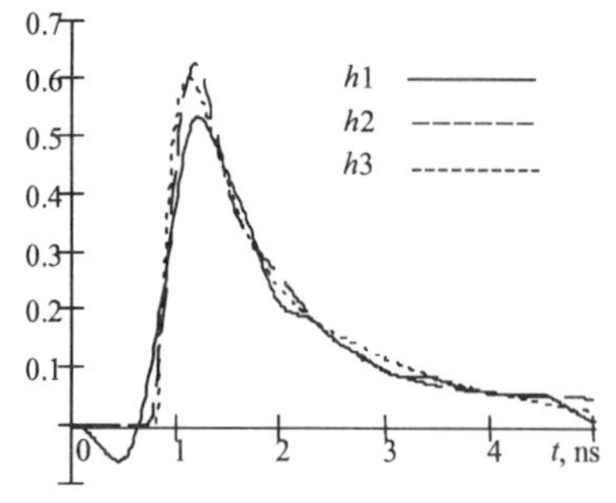

Figure 3. Impulse responses of cable No. 3 on sounding by a monopolar pulse.

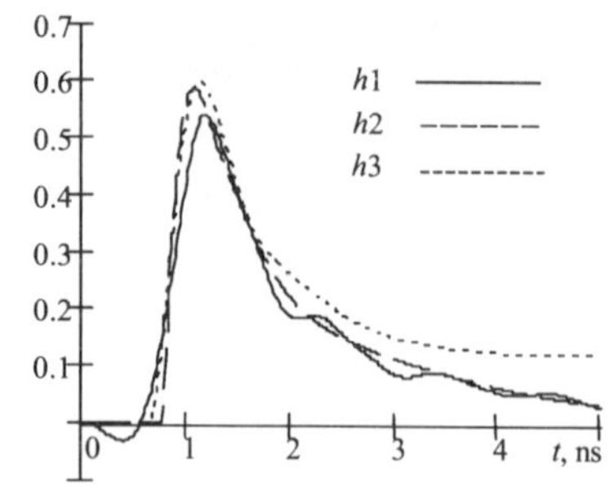

Figure 4. Impulse responses of cable No. 3 on sounding by a bipolar pulse.

Numerical simulation for verifying the parameter-value stability of the suggested models in the presence of white noise that was additively added to the output signal $Y(t)$ was made. Monopolar signals $Y1$ and $Y2$ were chosen for the simulation. The difference between their time waveforms after normalizing by the amplitude was no higher than 0.7%. The white noise level was 5% of the maximum signal Y. The IR was determined by averaging 100 realizations of the noise and looked like the IR of cable No. 3 in Fig. 3. The maximum and the width at the half-power level were 3.64 and 0.24 ns for cable No. 1 and 3.62 and 0.22 ns for cable No. 2, respectively. Figure 5 presents the results of the average values μ_k of the CPMS, determined from 100 realizations for these two cables. The mean square roots of $\mu 1$ and $\mu 2$ were 4.2% and 3.4%, respectively. Because the confidence intervals for estimating $\mu 1$ and $\mu 2$ have a 50% overlap, one cannot draw reliable conclusions about the IRs of the two cables.

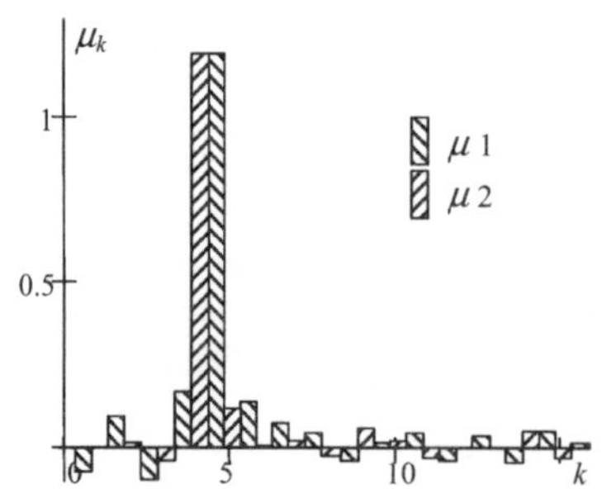

Figure 5. Lagrange coefficients of the impulse responses for cables No. 1 and No. 2.

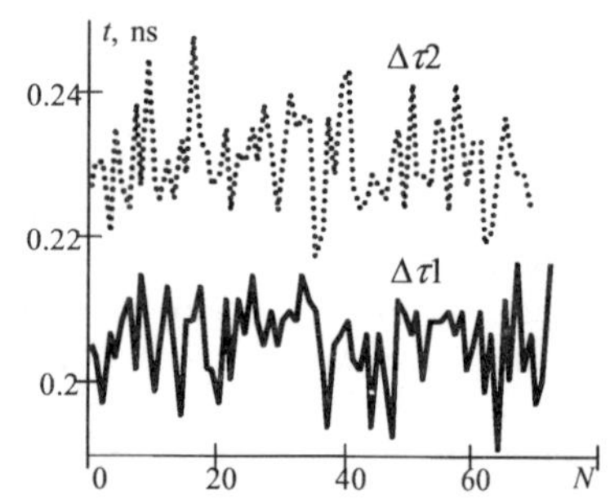

Figure 6. Values of the DPMS parameter $\Delta\tau$ of model (16) for cables No. 1 and No. 2.

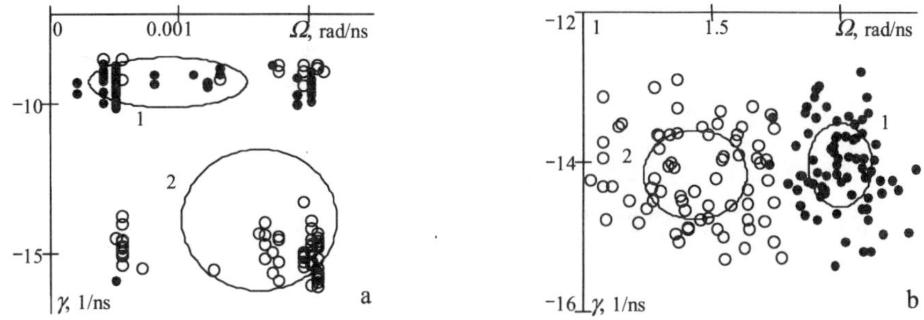

Figure 7. Position of the IR poles of the cables determined by model (16): a. No.1; and b. No. 2.

For analogous calculations using DPMS with $f_m(\tau)$ in the form of (16), the IRs of both cables had a time waveform similar to that presented in Fig. 3. The maximum value and half-power width are 3.55 and 0.25 ns for cable No. 1 and 3.55 and 0.21 ns for cable No. 2, respectively. The differential time delay $\Delta\tau$ of the IR pole model for each cable is calculated with $\Delta\tau = \tau_2 - \tau_1$ for 75 noise realizations and the results are displayed in Fig. 6. The positions of the first (Fig. 7a) and second (Fig. 7b) poles in the complex plane (Ω,γ), with corresponding confidence intervals in the form of ellipses, are presented for cables No. 1 and No. 2, respectively. From Figs. 6 and 7 one can discern that the spreads of the PM parameters (Ω_m, γ_m, and τ_m) at the 5% noise level for two similar IRs are not overlapped for most realizations. This result permits one to entertain the possibility of using these types of parameters for solving the recognition problem.

Application of the DPMS with $f_m(\tau)$ in the form of (17) for approximation of the cable IR allows reducing the number of poles to one, while preserving the discrepancy value (19). The IR time waveform differs from the IR calculated with application of expression (16) by less than 5%. The pole positions calculated by this model, with the corresponding confidence intervals in the form of ellipses, are displayed in the complex plane (Ω,γ) for cables No. 1 and No. 2 (Fig. 8). Spreads of other parameters (T and τ) of the models for the 5% noise level are presented in Fig. 9. Note that the overlapping for all model parameters is absent. A small number of parameters for signal approximation and the absence of overlapping of the parameter values make this model promising for solving the recognition problem. At the same time, the DPMS of (16) is more general.

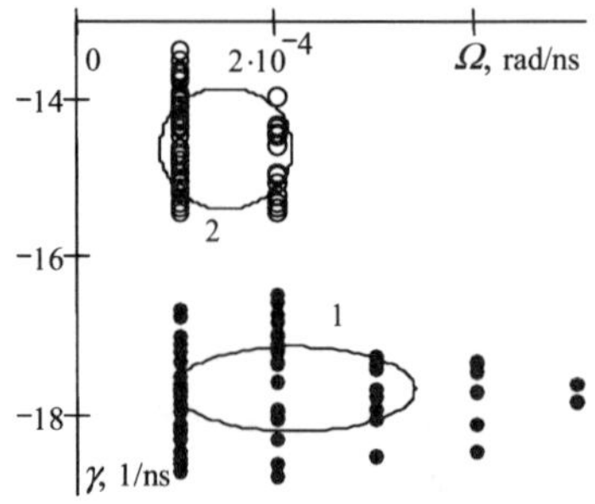

Figure 8. Position of the IR poles of cables No. 1 and No. 2 as determined by model (17).

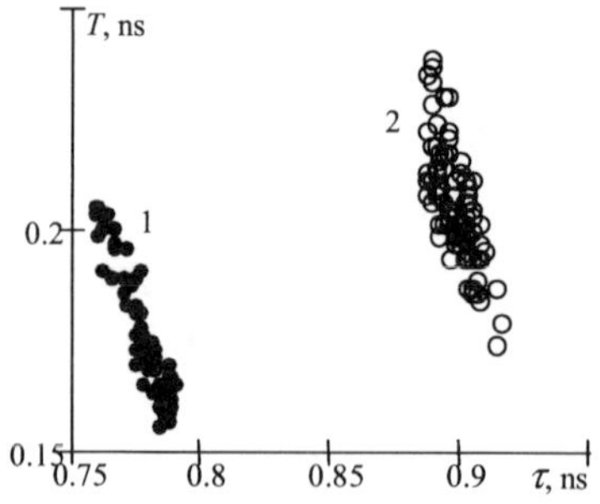

Figure 9. Values of DPMS parameters (T, τ) of model (17) for cables No. 1 and No. 2.

5. POLE MODELS APPLICATION FOR OBJECTS IN FREE SPACE

The PMs were used as well for evaluating the IR of an object in free space. Previously published data[6] were used as the experimental results. Measurements of scattered radiation from different objects were carried out in an anechoic chamber. TEM antennas were used to radiate and receive signals.

The pulse Y_{sph} scattered from the sphere of radius $R = 10$ cm (Fig. 10) and the pole model of its IR h_{sph} (Fig. 11) obtained from the data[7] were used. The waveform of the sounding pulse $\tilde{X}$ was reconstructed on this basis (Fig. 10), taking into account distortions inserted by the receiving system. Six pole functions of the DPMS of (16) were used to calculate $\tilde{X}$. The PM of the sounding pulse $\tilde{X}$ (Fig. 10) was also used for the IR evaluation of a 62-cm long and 25-cm diameter metal cylinder h_{cyl} (Fig. 11). The incident wave propagated along the cylinder axis.

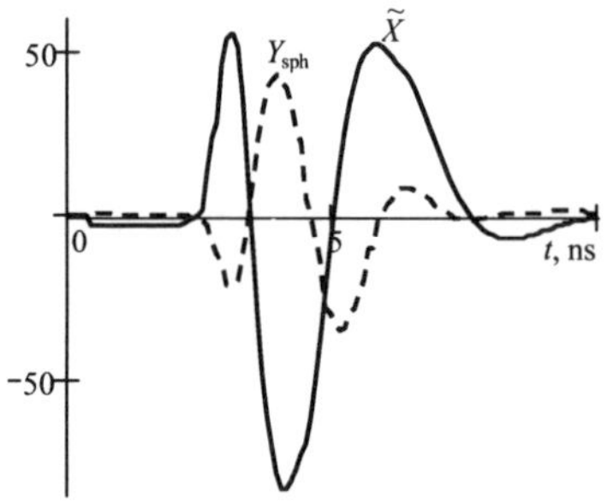

Figure 10. Pulse Y_{sph} scattered from the sphere of radius $R = 10$ cm and the reconstructed sounding pulse $\tilde{X}$.

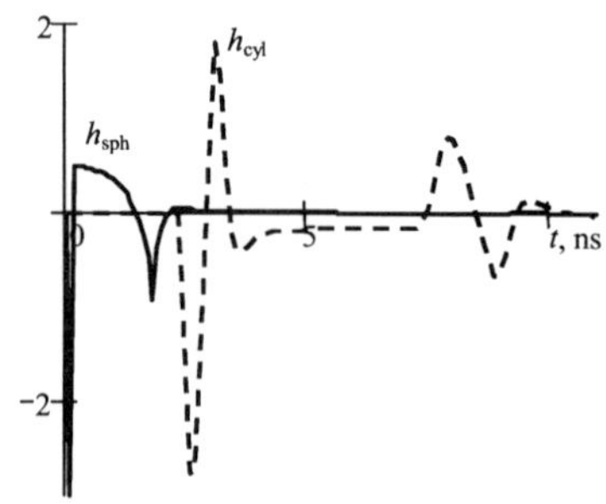

Figure 11. Impulse responses of the sphere h_{sph} and cylinder h_{cyl}.

6. CONCLUSION

Three pole models of signals have been developed that show similar and stable results at evaluation of the impulse response. A discrete pole-model application allowed stable values of the model parameters to be obtained. The regions of the discrete pole-model parameter values for the impulse responses of two similar temporal waveforms (which differ by 0.7%) do not overlap. This result suggests the possibility of using this model for solving the recognition problem.

REFERENCES

1. A.N. Tikhonov, A.V. Goncharsky, V.V. Stepanov, and A.G. Yagola, *Numerical Methods for Solving Ill-Posed Problems* (Nauka, Moscow, 1990).
2. S.L. Marple, Jr., *Digital Spectral Analysis with Applications* (Prentice-Hall, Englewood Cliffs NJ, 1987).
3. C.E. Baum, E.J Rothwell, K. Chen, and D.P. Nyquist, The singularity expansion method and its application to target identification, *Proc. IEEE* **79**(10), 1481–1492 (1991).
4. H.M. Nussenzveig, *Causality and Dispersion Relations* (Academic Press, New York, 1972).
5. V.I. Koshelev, V.T. Sarychev, and S.E. Shipilov, Using the Kramers-Kronig relation for estimation of the impulse responses of extra-wide-band system, *Radiophysics and Quantum Electronics*, **43**(5), 390–395 (2000).
6. M. Le Goff, P. Pouliguen, Y. Chevalier, Y. Imbs, B. Beillard, J. Andrieu, B. Jacko, G. Bouillon, and B. Juhel, UWB short pulse sensor for target electromagnetic backscattering characterization, in: *Ultra-Wideband, Short-Pulse Electromagnetics 4,* edited by E. Heyman, B. Mandelbaum, and J. Shiloh (Kluwer Academic/Plenum Publishers, New York, 1999), pp. 195–202.
7. E.M. Kennaugh and D.L. Moffatt, Transient and impulse response approximations, *IEEE Trans. Antennas Propogat.*, **53**(8), 893–901 (1965).

THE LIFE AND TIMES OF A SEA SPIKE: A STUDY OF WIDEBAND SEA CLUTTER AT HIGH RANGE RESOLUTIONS AND VERY LOW GRAZING ANGLES

Fred Posner and Karl Gerlach*

1. INTRODUCTION

Because the wideband spiky sea clutter that is frequently observed at low grazing angles and high resolutions plays a critical role in the performance of modern naval radars, it is necessary to analyze two important quantities, the mean spike duration and the mean interval between spikes. These analyses need to be conducted upon representative examples of such wideband clutter taken from a suitable database. Microwave backscatter (X-band, 9.5—10.0 GHz) observed at near grazing incidence (0.2 degrees) from the sea off the coast of Kauai, Hawaii, was measured with a wideband radar characterized by a high spatial resolution in range (0.3 meters) and a high temporal resolution (2000 Hz pulse repetition frequency). Extensive amounts (over 20 minutes per measurement) of both vertically and horizontally polarized wideband sea clutter data were taken with upwind and crosswind transmit geometries during the data collection campaign. The analyses of the dependence of the mean spike duration and the mean interval between spikes upon three parameters, the spike amplitude threshold, the minimum spike width, and the minimum interval between spikes, were carried out using all of the extensive data within a number of selected individual range cells for both vertically and horizontally polarized clutter, measured with an upwind transmit geometry.

The performance of high resolution radars in the detection and tracking of targets with low radar cross sections, such as submarine periscopes, may be seriously affected by the strong, targetlike returns from the sea, often referred to as sea spikes, that can occur at low grazing angles. The performance of lower resolution radars, such as those carried by sea skimming missiles and used in the detection and tracking of relatively low cross section ship targets, may be similarly affected by the spiky sea backscatter frequently encountered at low grazing angles.

To add to our understanding of sea clutter, various analyses will be presented here of

* Radar Division, Naval Research Laboratory, 4555 Overlook Avenue S.W., Washington DC 20375.

specific but representative examples of wideband sea clutter taken from an extensive data base which contains multiple transmit geometries and polarizations. It has been observed that the complex behavior of sea clutter at low grazing angles and high range resolutions is dependent upon transmit geometry and polarization, as well as upon the scale of observation, in both space and time. More extensive discussions of such material, with numerous references to the relevant literature, can be found in Posner.[1, 2]

This paper is organized as follows. Section 2 describes the sea backscatter data collection and the wideband radar system used in the campaign. Section 3 discusses certain aspects of the phenomenology of spiky sea clutter. Section 4 presents analyses of the dependence of the mean spike duration and the mean interval between spikes upon the spike amplitude threshold, the minimum spike width, and the minimum interval between spikes. Section 5 is a short conclusion and summary.

2. EXPERIMENTAL MEASUREMENTS

Between August and December of 1994, the Naval Research Laboratory (NRL), with the Naval Air Warfare Center/China Lake, the Johns Hopkins University Applied Physics Laboratory, and Texas Instruments/Raytheon, carried out extensive sea clutter measurements from various sites in Kauai, Hawaii, as part of the Advanced Radar Periscope Detection and Discrimination program, which was sponsored by the Office of Naval Research. The system used to gather the data was NRL's Advanced Profile Testbed Radar, an enhanced version of the APS-137 radar. An itemized description of the measurement radar appears in Table 1. To go from one polarization, either vertical transmit, vertical receive (VV) or horizontal transmit, horizontal receive (HH), to the other required a feedhorn change; as a consequence, the data are of only one polarization.

There are three major descriptors of the measured backscatter data. The first is grazing angle, which is determined by the radar height and the radar range. These clutter measurements were taken at the Lihue Airport site, where the antenna was 23 m above mean sea level. The center of the 156 m range swath was 5.74 km (6.11 km) downrange, well out in deep water, for the VV (HH) data collection run, and the VV (HH) grazing angle was 0.23 deg (0.22 deg). The azimuthal beamwidth of 2.4 deg gives a footprint width of 240 m (255 m) at the center of the VV (HH) data collection swath. The second is transmit geometry, which is determined by the radar bearing and the wind direction. Broadly speaking, two different transmit geometries, upwind (UP) and crosswind (CR), were available at the Lihue Airport site, which, situated on the windward side of Kauai, faces east into the prevailing trade winds. The third is polarization, which is either VV or HH. Sea backscatter measurements characterized by these descriptors were made in data collection runs that typically lasted over 1200 s, for which period of time the wind speed and the significant wave height would be determined. The above characterization of the measured sea backscatter data is summarized in Table 2. In this table, the 8 digit data tag consists of the year, the month, and the day when the measurement was made, followed by the number of the data collection run from which the specific examples of data were taken for subsequent analysis.

The vertically (horizontally) polarized, upwind transmit geometry (VV/UP (HH/UP)) data collection run, data tag 94110807 (94110816), was taken between 12:00 noon (6:13PM) and 12:26 PM (6:36 PM) local time. Although both of these measurements were made the same day, the UP runs in the two different polarizations were separated in time by 6 hours. For the VV/UP run, the bearing of the radar was 86 deg, while the wind direction (true) was 75°; this is clearly an UP alignment. For the HH/UP run, the bearing of the radar was 78°, which is very close to that for the VV/UP run. If, during the HH/UP run, the unmeasured wind direction was also close to that during the VV/UP run which occurred 6 hours earlier (a reasonable assumption at this site with its prevailing trade winds), then, once again, this would be an UP alignment. During the VV/UP run, the measured wind speed was 9.0 m/s and the measured significant wave height was 2.9 m. These conditions are typically associated with a hydrographic sea state of 4.

Table 1. Radar Descriptors

Operating Frequency	X-band (9.5—10.0 GHz)
Peak Power	500 KW
Pulse Length	500 ns compressed to 2.5 ns
Range Resolution	0.3 m
Signal Processing	synchronous demodulation for I and Q channels, 8 bit analog-to-digitial conversion at 500 MHz
Pulse Repetition Frequency	2000 Hz
Data Collection Extent	156 m range swath, 512 individual range cells
Data Collection Mode	spotlight due to fixed antenna pointing
Antenna Beamwidths	2.4° in azimuth and 4.0° in elevation
Polarization	vertical transmit, vertical receive (VV) or horizontal transmit, horizontal receive (HH)

Table 2. Data Descriptors

Data Tag	Start Time	Transmit/Receive Polarization	Radar Height	Radar Range	Grazing Angle
94110807	2200 GMT	vertical/vertical	23 m	5.74 km	0.23°
94110816	0413 GMT	horizontal/horizontal	23 m	6.11 km	0.22°
Data Tag	**Radar Bearing**	**Wind Direction**	**Transmit Geometry**	**Wind Speed**	**Significant Wave Height**
94110807	086°	075°	upwind	9.0 m/s	2.9 m
94110816	078°	—	upwind	—	—

3. PHENOMENOLOGY OF SPIKY SEA CLUTTER

To get a quick overview of spiky sea clutter, we will observe data that have been temporally averaged by noncoherently integrating 100 pulses at a time. This use of noncoherent integration effectively reduces, or spoils, the full temporal resolution from 2000 Hz down to 20 Hz. Figure 1 presents plots of radar cross section (RCS), in square meters (sm), for UP sea clutter from range cell one occurring over a twenty minute time scale, with the data at the top (bottom) being VV (HH). The maximum value of the RCS

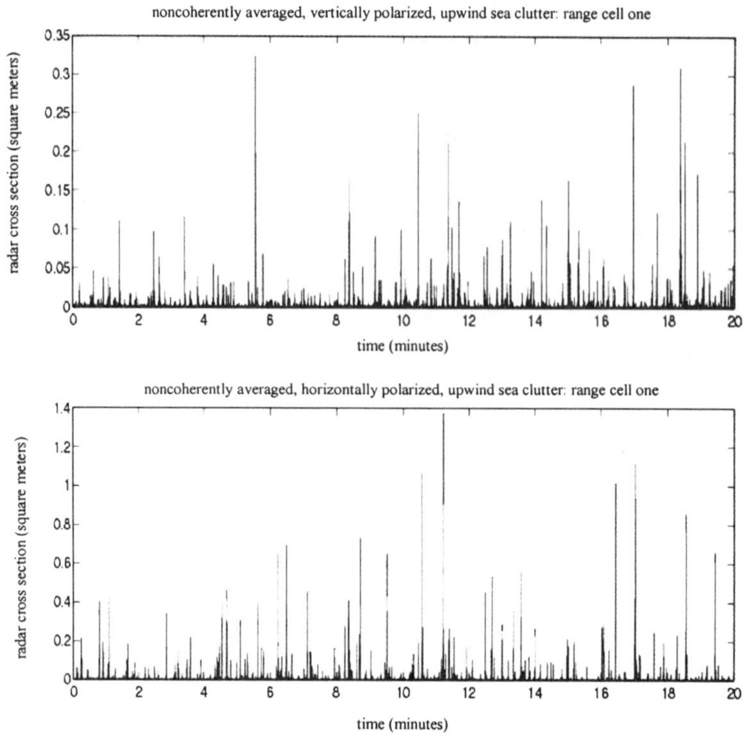

Figure 1. Radar cross section, in square meters, for VV/UP range cell one (above) and HH/UP range cell one (below), after noncoherent temporal averaging.

For the rest of this paper, all of the data to be presented and discussed will be at the full temporal resolution of 2000 Hz. In Figure 2 there are plots, on Weibull axes, of the cumulative distribution function (CDF) of all of the RCS data in range cell one, with the data at the top (bottom) being VV/UP (HH/UP). For range cell one, there are 1528 s (1335 s) of data for VV/UP (HH/UP). For VV/UP (HH/UP), the RCS axis extends from –40 dBsm (–35 dBsm) to 0 dBsm (+5 dBsm); these dynamic ranges cover both the clutter-dominated (the upper 25 dBsm) and the noise-dominated (the lower 15 dBsm) domains of the data. For subsequent clutter analyses in this paper, we will consider VV/UP (HH/UP) data above –25 dBsm (–20 dBsm).

Two observations that apply to both VV/UP and HH/UP: (1) For the very largest values of the RCS, the conspicuous change in the behavior of the CDFs due to receiver saturation produces the clipping of the returns. (2) It can be seen from the closeness to linearity of the clutter data, in the region below the clipping and above the noise, that the Weibull distribution does provide good fits to the data. But it can also be seen that the CDFs do have some slight curvature in this region, and thus the clutter data do depart, to some degree, from being Weibull-distributed.

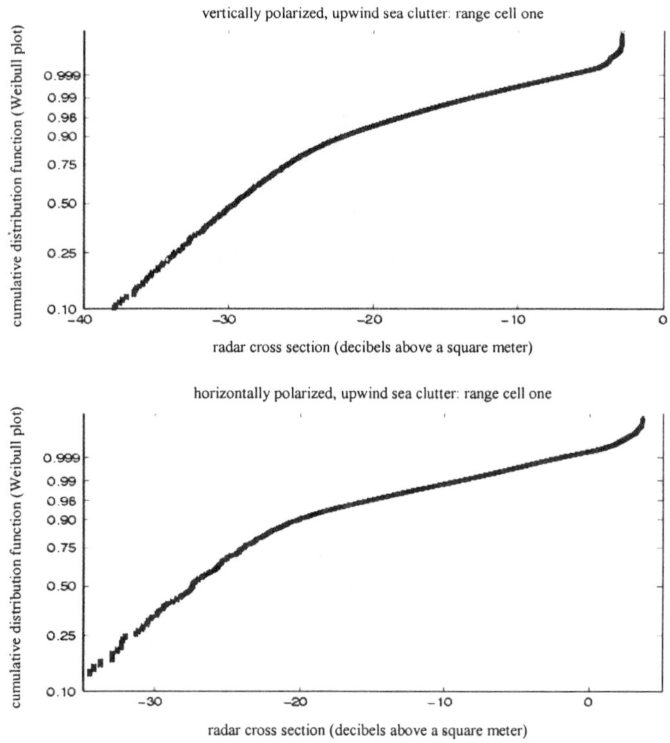

Figure 2. Weibull plots of cumulative distribution functions for VV/UP range cell one (above) and HH/UP range cell one (below).

Figure 3 presents plots of RCS, in square meters, for examples of UP sea clutter from range cell one occurring over a two second time scale, consisting of all four thousand pulses, with the data at the top (bottom) being VV (HH). Note that the maximum value of the RCS axis is 0.6 sm (2.5 sm) for VV/UP (HH/UP). Because both of these examples were selected to display conspicuously spiky behavior, clipping effects are certainly present. Compared to the VV/UP clutter, the HH/UP clutter is sharper, spikier, and more intermittent, with larger values of the RCS.

4. CHARACTERIZATION OF SPIKY SEA CLUTTER

The examples in Figure 3 show that a series of radar returns from a single range cell must meet criteria determined by three quantitative parameters in order to be considered a sea spike. The first parameter is the spike amplitude threshold, in square meters, defined

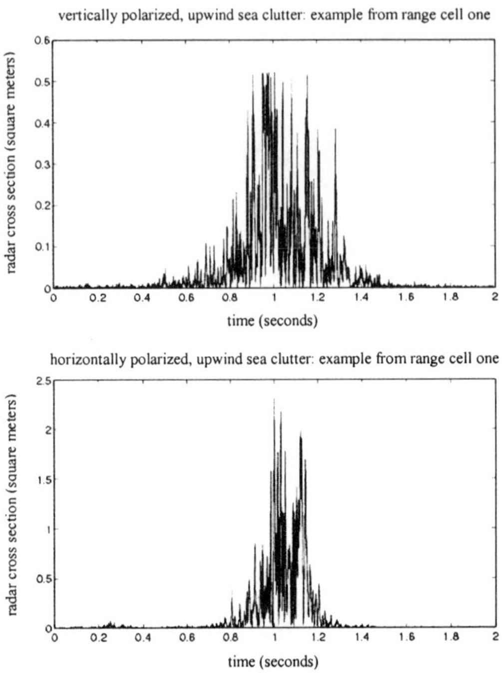

Figure 3. Radar cross section, in square meters, for examples of spiky sea clutter from VV/UP range cell one (above) and HH/UP range cell one (below).

so that the amplitudes of the returns must exceed a specified radar cross section. The second parameter is the minimum spike width, in seconds, defined so that the amplitudes of the returns must remain above the spike amplitude threshold for at least as long as the specified minimum spike width. The third parameter is the minimum interval, in seconds, between spikes, defined so that if the amplitudes of the returns do fall below the spike amplitude threshold, they may not then remain below the spike amplitude threshold longer than the specified minimum interval between spikes; if they do remain below longer, then this set of returns would be considered as consisting of separate sea spike candidates, each of which must meet the three criteria. Using the above formal definition of a sea spike, one can then consider two important statistical measures of sea spike demographics, the first being the mean spike duration, in seconds, and the second being the mean interval, in seconds, between spikes, and examine their functional dependence upon the three sea spike defining parameters. Related research can be found in Olin.[3-5]

At the top of Figure 4 is a plot of the mean spike duration, in seconds, for all of the RCS data in VV/UP range cell one. In this plot the maximum value of the mean spike duration axis is 0.35 s. At the bottom of Figure 4 is a plot of the mean interval between spikes, in deciBels above a second (dBsec), for all of the RCS data in VV/UP range cell one. In this plot the mean interval between spikes axis extends from 0 dBsec to 30 dBsec. For both plots in Figure 4 the minimum spike width is fixed at 0.01 s and the minimum interval between spikes is fixed at 0.1 s. And finally, for both plots in Figure 4, as well as for all subsequent figures, the spike amplitude threshold axis extends from –25 dBsm to 0 dBsm for VV/UP, this being the clutter-dominated domain of the data. Similar analyses were carried out for all of the RCS data in HH/UP range cell one, with the only difference being that the clutter-dominated domain of the data extends from –20 dBsm to +5 dBsm for HH/UP.

Regarding dependence on spike amplitude threshold: (1) With the other two parameters held fixed, the mean spike duration is a decreasing function and the mean interval between spikes is an increasing function. (2) Over the range of the smaller values of the spike amplitude threshold, mean spike durations are longer and mean intervals between spikes are shorter for VV/UP than for HH/UP. (3) Mean spike durations and mean intervals between spikes both show greater variability for VV/UP than for HH/UP.

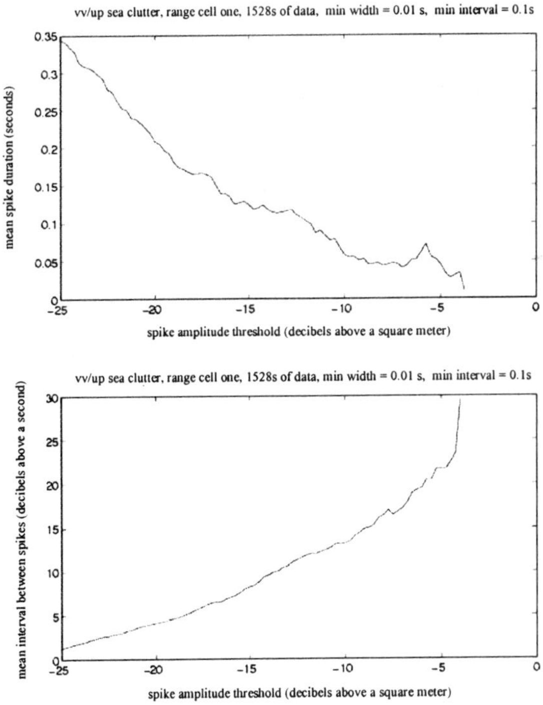

Figure 4. Mean spike duration (above) and mean interval between spikes (below) as functions of spike amplitude threshold for VV/UP range cell one.

In Figure 5 there are multiple plots of the mean spike duration, in seconds, for all of the RCS data in VV/UP range cell one. For these plots the maximum value of the mean spike duration axis is slightly larger than 0.6 s. In Figure 6 there are multiple plots of the mean interval between spikes, in deciBels above a second, for all of the RCS data in VV/UP range cell one. For these plots the mean interval between spikes axis extends from −5 dBsec to 30 dBsec. In Figures 5 and 6 there are five plots corresponding to five different minimum spike widths of 0.005, 0.010, 0.020, 0.050, and 0.100 s, while the minimum interval between spikes is held fixed at 0.1 s.

Regarding dependence on minimum spike width: (1) With the other two parameters held fixed, the mean spike duration is a decreasing function and the mean interval between spikes is an increasing function. (2) Mean spike durations and mean intervals between spikes both show greater variability for VV/UP than for HH/UP.

In Figure 7 there are multiple plots of the mean spike duration, in seconds, for all of the RCS data in VV/UP range cell one. For these plots the maximum value of the mean spike duration axis is slightly larger than 0.6 s. In Figure 8 there are multiple plots of the mean interval between spikes, in deciBels above a second, for all of the RCS data in VV/UP range cell one. For these plots the mean interval between spikes axis extends from −5 dBsec to 30 dBsec. In Figures 7 and 8 there are three plots corresponding to three different minimum intervals between spikes of 0.05, 0.10, and 0.20 s, while the minimum spike width is held fixed at 0.01 s.

Regarding dependence on minimum interval between spikes: (1) With the other two parameters held fixed, the mean spike duration and the mean interval between spikes are both increasing functions. (2) Mean spike durations and mean intervals between spikes both show greater variability for VV/UP than for HH/UP.

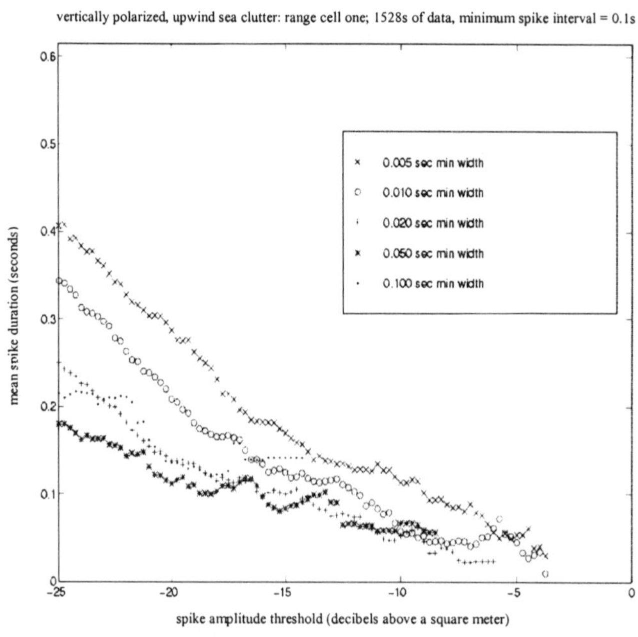

Figure 5. Mean spike duration as a function of spike amplitude threshold for selected values of minimum spike width for VV/UP range cell one.

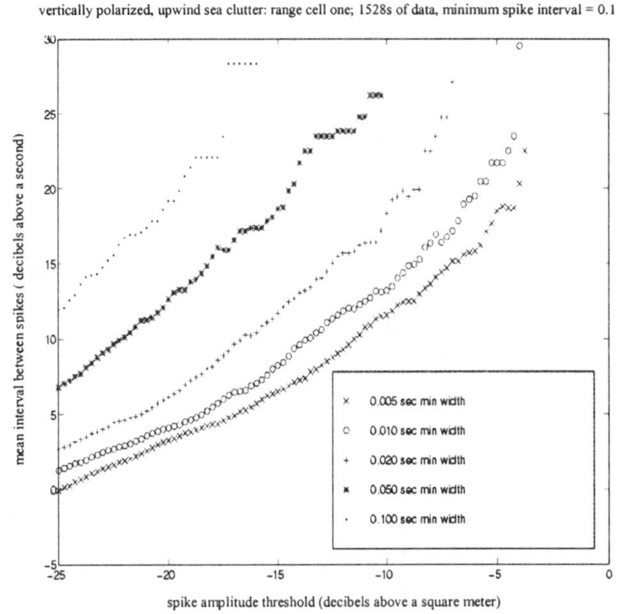

Figure 6. Mean interval between spikes as a function of spike amplitude threshold for selected values of minimum spike width for VV/UP range cell one.

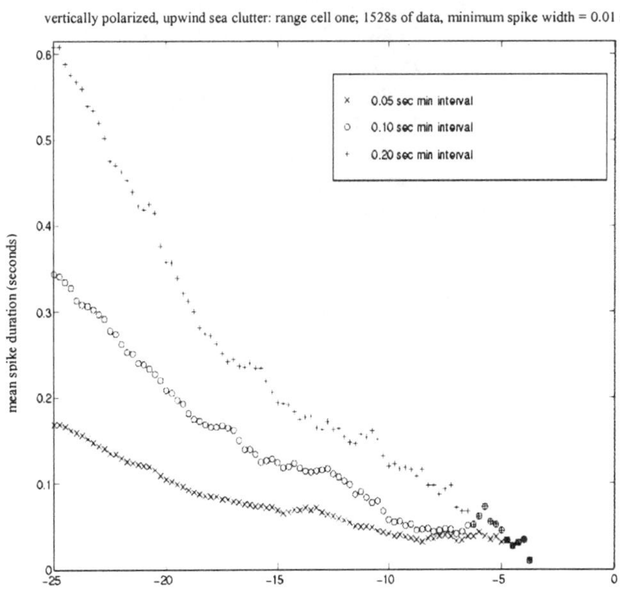

Figure 7. Mean spike duration as a function of spike amplitude threshold for selected values of minimum interval between spikes for VV/UP range cell one.

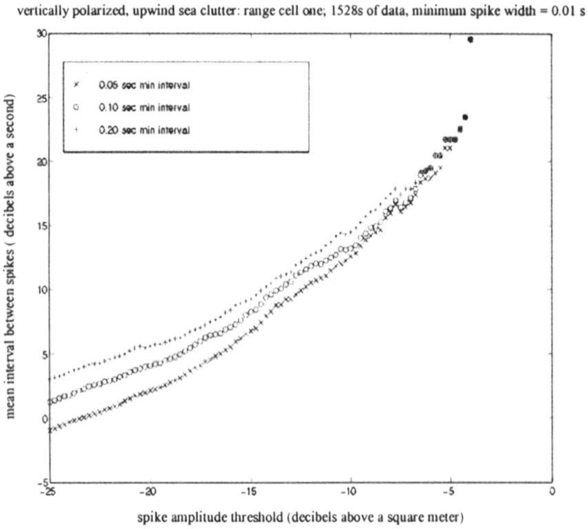

Figure 8. Mean interval between spikes as a function of spike amplitude threshold for selected values of minimum interval between spikes for VV/UP range cell one.

5. CONCLUSIONS

By characterizing spiky sea clutter in terms of three parameters, the spike amplitude threshold, the minimum spike width, and the minimum interval between spikes, individual sea spike events can be singled out and identified. The dependence of two important statistical measures of sea spike demographics, the mean spike duration and the mean interval between spikes, on these three parameters has been analyzed by processing extensive amounts of wideband microwave backscatter data. From these analyses, the dependence of the two statistical quantities on radar polarization has also been determined for the case of an upwind transmit geometry.

6. ACKNOWLEDGEMENTS

This work has been sponsored by the Office of Naval Research, Code ONR 31.

REFERENCES

1. F.L. Posner, Experimental Observations at Very Low Grazing Angles of High Range Resolution Microwave Backscatter from the Sea, Naval Research Laboratory Report NRL/MR/5310--98--8326, 31 December 1998.
2. F.L. Posner, Spiky Sea clutter at high range resolutions and very low grazing angles, *IEEE Transactions on Aerospace and Electronic Systems* **38**(1), 58–73 (2002).
3. I.D. Olin, Amplitude and temporal statistics of sea spike clutter, *Radar 82*, IEE Conference Publication 216, pp. 198–202, October 1982.
4. I.D. Olin, Characterization of spiky sea clutter for target detection, *Proceedings of the 1984 IEEE National Radar Conference*, pp. 27–31, August 1984.
5. A.T. Jessup, W.K. Melville, and W.C. Keller, Breaking waves affecting microwave backscatter – 1. detection and verification, *J.Geophysical Res.* **96**(C11), 20547–20559 (1991).

A FULL WAVE APPROACH TO ELECTROMAGNETIC, SCATTERING BY NANOSTRUCTURES WITH SMALL AND LARGE SCALE PERIODIC AND RANDOM ROUGH SURFACES AND LATERAL FLUCTUATIONS IN THE PERMITTIVITY AND PERMEABILITY

Ezekiel Bahar*

1. INTRODUCTION

The methodology employed in analyzing electromagnetic scattering by irregular layered structures, in which the heights of the interfaces as well as the medium parameters fluctuate laterally, is such that all the simplifying assumptions introduced in order to make the rigorous solutions to the problems more tractable, are made a posteriori rather than a priori. In this way the same analysis can be used to obtain the high frequency physical optics solutions that can be applied to structures with scales of roughness that are much larger than the wavelength as well as to obtain the low frequency small perturbation type solutions for structures with scales of roughness that are comparable to the wavelength. Thus this analysis can be applied to structures with multiple scales of roughness without introducing an artificial scale parameter that dictates the solution to the problem. In addition the same analysis can be used to obtain the far field approximations suitable for structures with scales of roughness comparable and larger than the wavelength, as well as to obtain the near field approximations that are suitable for structures with sub-wavelength scales of roughness. By imposing reciprocity it can be shown that near field measurements can be implied from far field measurements, provided that objects in the vicinity of the interfaces excite the near fields. Thus the analysis should account for evanescent as well as propagating waves, the lateral waves and the guided, surface waves of the irregular structures.

To this end, a full wave approach, based on the complete expansion of the fields as well as on the imposition of exact boundary conditions at the rough interfaces, is used in

* University of Nebraska, Lincoln, NE 68588-0511, 402-472-1966 (telephone); 402-472-4732 (facsimile); ebahar1@unl.edu.

Ultra-Wideband Short-Pulse Electromagnetics 6
Edited by Mokole *et al.*, Kluwer Academic/Plenum Publishers, 2003

the analysis. Since these expansions of the complete fields do not necessarily converge uniformly at the irregular interfaces, careful mathematical procedures must be followed such that any interchange in the order of summation and differentiation of the complete field expansions is strictly avoided. Thus for example it is shown that using the far field approximations, the solutions for the scattered fields are expressed as integrals over the spatial variables. On the other hand when the near field approximations are used, the scattered fields are expressed as integrals over the wave vector variables. This full wave analysis can also be applied to anisotropic media such as chiral materials.

A brief outline of the principal elements of the full wave approach (complete field expansions and the imposition of exact boundary coalitions) used to analyze electromagnetic wave scattering from irregular layered structures, in which the surface heights of interfaces as well as the electric permittivity and magnetic permeability of the medium are functions of the lateral spatial variables, is presented in section 2. The details of the approach can be found in a recent review article in the Wiley Encyclopedia of Electrical and Electronics Engineering,[1] which contains extensive references to the published scientific/technical literature as well as to engineering applications in radar remote sensing. The principal properties of these solutions (reciprocity, duality and realizability) are also presented in this section.

The principal results of the full wave approach, leading to the conversion of Maxwell's Equations into sets of Generalized Telegraphists' equations similar to those derived by Schelkunoff for waveguides with irregular cross sections,[2] are presented in this section. For the open structures considered in this work, the complete expansions of the fields are associated with two branch cut integrals for the radiation term and the lateral wave term as well as the guided, surface waves, of the layered structure associated with the residues at the poles where the reciprocal of the composite reflection coefficients for the layered structure vanish.

Iterative, single scatter solutions for the telegraphists equations that provide the like and cross polarized scattered fields from rough surfaces with small slopes are presented in Section 3. Thus the scattering mechanism is represented by a 2 x 2 surface element scattering matrix and the scattered fields are expressed as integrals over the wave vector variables as well as the lateral spatial variables. These original, small slope, full wave solutions are extended to surfaces with larger slopes and the fields, multiply scattered by rough surfaces, are also considered. This approach provides the basis for the full wave solutions for the enhanced backscatter cross sections for random rough surfaces. The expressions for the scatter cross sections are initially eight dimensional integrals (four spatial variables and four wave vector variables. By using coordinate transformations and introducing physically based approximations associated with Fresnal zones between the transmitter and receiver,[3] these integrals are reduced to four-dimensional integrals (two spatial and two wave vector variables).

In section 4, low frequency approximations are introduced to obtain the small perturbation solutions and high frequency approximations are introduced to obtain the associated physical optics solutions. When the observer is in the far field, the scatter cross sections are expressed as two-dimensional integrals over the spatial variables. In this case it is assumed that the scales of the structure are of the order of the wavelength or larger. To probe features of the structure that are smaller than the wavelength, it is necessary to derive the near fields scattered by the structure. The expressions for the scattered near fields are expressed as two-dimensional integrals over the wave vector variables. For structures that are irregular in one-dimensional only, the corresponding integrals are over one spatial or one wave vector variable only. Both high frequency and low frequency approximations are obtained for the near fields. Section 5 deals with scattering by periodic or quasi-periodic rough surfaces.

Section 6 contains a discussion of the full wave solutions for the coupling between the radiation fields, the lateral waves as well as the surface waves of the irregular layered structures. It is shown that while the introduction of approximations a posteriori, rather than a priori involves more stringent analytical procedures, the benefits of such procedures more than make up for the additional effort invested in the initial analysis. Moreover some of the limitations associated with the a priori approximations, such as the Kirchhoff approximations for the surface fields, that result in nonreciprocal solutions, can be resolved when high frequency approximations are introduced a posteriori.

2. PRINCIPLE ELEMENTS OF THE FULL WAVE APPROACH; THE GENERALIZED TELEGRAPHIST'S EQUATIONS

The full wave approach to scattering from irregular stratified structures with rough interfaces and laterally fluctuating medium parameters is shown schematically in Fig. 1. Thus the surface height h between media i and $i+1$ is given by

$$f(x,y,z) = y - h_{i,i+1}(x,z) = 0. \tag{2.1}$$

Note that the complex permittivity $\varepsilon(x,z)$ and permeability $\mu(x,z)$ of the medium are also functions of the lateral variables.

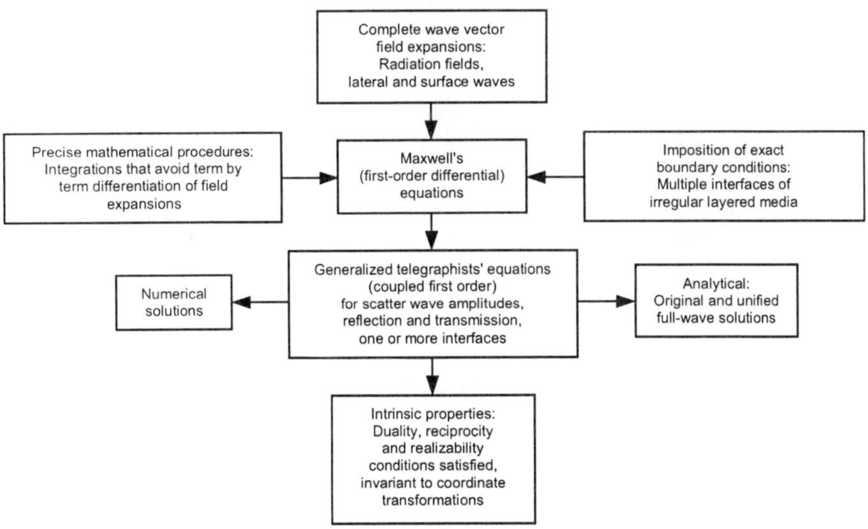

Figure 1. Schematic of procedures for deriving the generalized telegraphists' equations.

Maxwell's equation for the transverse (*y*,*z*) components (denoted by subscript *T*) of the electric $\bar{E}$ and magnetic $\bar{H}$ fields can be expressed as follows:[4-6]

$$-\frac{\partial \bar{E}_T}{\partial x} = i\omega\mu\left(\bar{H}_T \times \bar{a}_x\right) - \frac{1}{i\omega}\nabla_T\left(\frac{1}{\varepsilon}\nabla_T \cdot \left(\bar{H}_T \times \bar{a}_x\right)\right) + \bar{M}_T \times \bar{a}_x + \frac{1}{i\omega}\nabla_T \frac{J_x}{\varepsilon}, \tag{2.2}$$

$$-\frac{\partial \bar{H}_T}{\partial x} = i\omega\varepsilon\left(\bar{a}_x \times \bar{E}_T\right) - \frac{1}{i\omega}\nabla_T\left(\frac{1}{\mu}\nabla_T \cdot \left(\bar{a}_x \times \bar{E}_T\right)\right) + \bar{a}_x \times \bar{J}_T + \frac{1}{i\omega}\nabla_T \frac{M_x}{\mu}, \tag{2.3}$$

where the operator ∇_T is

$$\nabla_T = \bar{a}_y \frac{\partial}{\partial y} + \bar{a}_z \frac{\partial}{\partial z}, \tag{2.4}$$

and the transverse vectors are

$$\bar{A}_T = \bar{a}_y A_y + \bar{a}_z A_z \,. \tag{2.5}$$

and $\bar{J}\left(\mathrm{A/m^2}\right)$ and $\bar{M}\left(V/m^2\right)$ are the electric and (dual) magnetic current densities. The exact boundary conditions imposed at each of the interfaces of the irregular layered structure are the continuity of the tangential components of the electric and magnetic fields,

$$\left[\bar{n}\times\bar{E}\right]_{h_{i,i+1}^-}^{h_{i,i+1}^+} = 0,\ \left[\bar{n}\times\bar{H}\right]_{h_{i,i+1}^-}^{h_{i,i+1}^+} = 0,\ i = 0,1,\ldots m\,, \tag{2.6}$$

in which $\bar{n}$ is the unit vector normal to the interfaces,

$$\bar{n}_{i,i+1} = \left(-\frac{\partial h_{i,i+1}}{\partial x}, 1, -\frac{\partial h_{i,i+1}}{\partial z}\right) \Big/ \left[\left(\frac{\partial h_{i,i+1}}{\partial x}\right)^2 + 1 + \left(\frac{\partial h_{i,i+1}}{\partial z}\right)^2\right]^{1/2}, \tag{2.7}$$

and $y = h_{i,i+1}$ is the interface between medium *i* and medium *i* + 1. The full wave, complete expansions for the vertically (*V*) an the horizontally (*H*) polarized electric and magnetic fields are given in terms of the transverse basis functions $\bar{e}_T^V, \bar{h}_T^V, \bar{e}_T^H$ and $\bar{h}_T^H$ as

$$\bar{E}_T(x,y,z) = \sum_v \int_{-\infty}^{\infty}\left[E^V(x,v,w)\bar{e}_T^V + E^H(x,v,w)\bar{e}_T^H\right]dw\,, \tag{2.8}$$

$$\bar{H}_T(x,y,z) = \sum_v \int_{-\infty}^{\infty}\left[H^V(x,v,w)\bar{h}_T^V + H^H(x,v,w)\bar{h}_T^H\right]dw\,. \tag{2.9}$$

The summation symbol with index v in (2.8) and (2.9) denotes summation (integration) over the complete wave vector spectrum, consisting of the radiation term and lateral waves (associated with branch cut integrals) and the waveguide modes (or surface waves) of the layered structure (associated with the residues at the poles of the reflection coefficients).[4-6] In (2.8) and (2.9) the scalar field transforms for the vertically ($P = V$) and horizontally ($P = H$) polarized electric and magnetic fields are

$$E^P(x,v,w) = \int_{-\infty}^{\infty} \bar{E}_T(x,y,z)\cdot\left(\bar{h}_P^T \times \bar{a}_x\right) dydz\,, \tag{2.10}$$

$$H^P(x,v,w) = \int_{-\infty}^{\infty} \bar{H}_T(x,y,z)\cdot\left(\bar{a}_x \times \bar{e}_P^T\right) dydz\,, \tag{2.11}$$

where the complementary (reciprocal) basis functions are $\bar{e}_V^T, \bar{h}_V^T, \bar{e}_H^T$ and $\bar{h}_H^T$.

The basis functions and the reciprocal basis functions satisfy the biorthogonal relationships

$$\left.\begin{aligned} &\int_{-\infty}^{\infty} \bar{e}_T^P \cdot \left(\bar{h}_Q^T \times \bar{a}_x\right)' dydz \\ &\int_{-\infty}^{\infty} \bar{h}_T^P \cdot \left(\bar{a}_x \times \bar{e}_P^T\right)' dydz \end{aligned}\right\} = \delta_{P,Q}\Delta(v-v')\delta(w-w')\,. \tag{2.12}$$

In (2.12) the Kronecker delta δ_{PQ} implies that the vertically (P, $Q = V$) and horizontally (P, $Q = H$) polarized basis functions are orthogonal in uniform layered structures. Furthermore, the Dirac delta $\delta(w-w')$ appearing in (2.12) is a result of the completeness and orthogonality relationships associated with the z dependent functions.

The symbol $\Delta(v,v')$ in (2.13) is the product of the Kronecker delta $\delta_{q,r}$ (subscripts q and r denote the radiation fields, the later waves, or the surface waves) and the Dirac delta function $\delta(v,v')$ for the radiation and lateral wave terms or the Kronecker delta $\delta_{v,v'}$ for the guided (surface) waves of the layered structure. Thus the radiation fields, the lateral waves and the guided varies of the full wave (spatial) spectrum are mutually orthogonal. The radiation fields and the lateral waves are associated with branch cut integrals in the complex wave number plane (with branch points at $k = k_0$, uppermost medium and $k = k_m$, lowermost medium). The guided waves of the layered structure are associated with the residues at the poles of the composite reflection coefficient seen from above or below the layered structure.

In this work, it is convenient to express the vertically and horizontally polarized scalar field transforms (2.10) and (2.11) in terms of the vertically and horizontally polarized forward wave amplitude a^P and backward wave amplitude b^P as follows:

$$H^P = a^P \pm b^P \text{ and } E^P = a^P \mp b^P \text{ for } P = \begin{cases} V(upper\ sign) \\ H(lower\ sign) \end{cases}. \tag{2.13}$$

Substituting the complete field transforms for the transverse components into Maxwell's equations ((2.2)–(2.5)), making use of the biorthogonal relationships (2.12) and imposing the exact boundary conditions at each interface of the irregular layered structure (2.6) yield the following generalized telegraphists' equations:

$$-\frac{da^P}{dx}-iua^P=\sum_Q\sum_{v'}\int_{-\infty}^{\infty}\left[s_{PQ}^{BA}a^Q+S_{PQ}^{BB}b^Q\right]dw'-A^P, \tag{2.14}$$

$$-\frac{db^P}{dx}+iub^P=\sum_Q\sum_{v'}\int_{-\infty}^{\infty}\left[s_{PQ}^{AA}a^Q+S_{PQ}^{AB}b^Q\right]dw'+B^P, \tag{2.15}$$

in which A^P and B^P are associated with the source terms $\bar{J}$ and $\bar{M}$ in (2.2)–(2.5). Furthermore, S_{PQ}^{BA} and S_{PQ}^{AB} are transmission scattering coefficients, while S_{PQ}^{AA} and S_{PQ}^{BB} are reflection scattering coefficients. These scattering coefficients vanish when the layered medium is horizontally stratified with homogeneous medium in each layer. In this case, the forward and backward wave amplitudes for the vertically and horizontally polarized waves are decoupled and analytical closed form solutions are readily obtained. However, if the rough surface height or the complex permittivities and permeabilities are functions of x and z, the lateral (horizontal) variables, the wave amplitudes are coupled. In the general case, the basis functions do not individually satisfy the irregular boundary conditions and the complete field expansions do not uniformally converge at the boundaries. Thus, on following precise mathematical procedures, the orders of integration (summation) and differentiation are not interchanged.[2]

The intrinsic properties of the full wave solutions are, duality, reciprocity, realizability and invariance to coordinate transformations. All the above properties follow directly from Maxwell's equations and they are not a result of any additional constraints imposed on the results. When the lowermost and/or uppermost half space is perfectly conducting or a good conducting medium, the two boundary conditions (2.6) at the lowermost (and/or uppermost) interface can be replaced by a single (approximate) surface impedance boundary condition.

$$\bar{n}\times\bar{E}=Z_s\bar{n}\times\left(\bar{H}\times\bar{n}\right). \tag{2.16}$$

In (2.16) the unit vector $\bar{n}$ is normal to the interface and points into the conducting half space. For the isotropic conducting half space the surface impedance Z_s is a scalar. In general, the surface impedance can be represented by a dyad. Contributions from integrals associated with one (or two) branch cuts are eliminated when impedance boundary conditions are used.

The scattering coefficients $S_{PQ}^{\alpha\beta}$ appearing in the generalized telegraphist's equation ((2.14) and (2.15)) can be expressed in closed form upon employing the orthogonal relationships between the basis function. For example, if the rough surface height parameters h and medium parameters ε,μ are functions of x only, the basis functions associated with the vertical (y) variable are also implicitly dependent on the variable through the function $h_{r-1,r}$ (r = 1, ...m) and the complex permittivities $\varepsilon_r(x)$ and permeabilities $\mu_r(x)$ in the medium denoted by the subscript r.

The coupling coefficients can be expressed explicitly in terms of the derivatives of the scalar basis functions

$$\frac{\partial \Psi_p(u,y)}{\partial x} \quad \text{and} \quad \frac{\partial^2 \Psi_P(u,y)}{\partial x \partial y}.$$

Thus in this case $\partial \Psi_p / \partial x$ can be expressed as

$$\frac{\partial}{\partial x}\Psi_P(u,y) = \left[\sum_{r=0}^{m}\left(\varepsilon_r' \frac{\partial}{\partial \varepsilon_r} + \mu_r' \frac{\partial}{\partial \mu_r}\right) + \sum_{r=1}^{m} h_{r-1,r}' \frac{d}{dh_{r-1,r}}\right]\Psi_P(u,y) \tag{2.17}$$

and a similar expression can be written for $\partial^2 \Psi_P(u,y)/\partial x \partial y$. In (17) the primes in ε_r', μ_r' and $h_{r-1,r}'$ denote derivatives with respect to x.

3. PRINCIPAL PROPERTIES OF THE ORIGINAL AND UNIFIED FULL WAVE SOLUTIONS

The principal properties of the full wave solutions and their relationships to the low frequency small perturbation solutions and the high frequency physical-geometrical/optics solutions are shown schematically in Fig. 2.

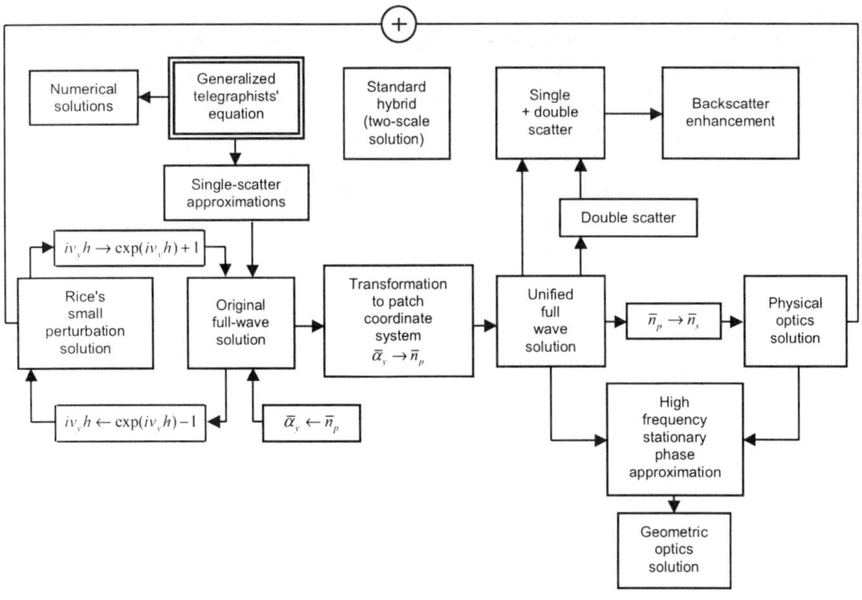

Figure 2. Principal properties of the original and unified full-wave solutions.

Interative analytical procedures, as well as numerical techniques, are used to solve the Generalized Telegraphists' equations ((2.14) and (2.15)) for the forward and backward wave amplitude scattered by two dimensionally rough surfaces. The analytical procedures are dealt with in this section. To obtain the single scatter approximations for the wave amplitudes, the expressions for the primary fields impressed upon the rough interface, due to the sources, are first derived from (2.14) and (2.15) upon neglecting all the coupling terms manifested by the scattering coefficients S_{PQ}^{BA}. When the sources are in the far field, the primary incident fields impressed upon the rough surface are vertically and horizontally polarized plane waves propagating in the direction of the (free space) wave vector

$$\bar{k}_o^i = k_{ox}^i \bar{a}_x + k_{oy}^i \bar{a}_y + k_{oz}^i \bar{a}_z = k_o \bar{n}^i, \tag{3.1}$$

where $\bar{n}^i$ is a unit vector and $k_o = \omega\sqrt{\mu_o \varepsilon_o}$. Thus, the primary electric fields impressed upon the rough surfaces between medium 0 and medium 1 are

$$\bar{E}^i = \bar{a}^P E_o^i \exp\left[-i\left(k_{ox}^i x + k_{oz}^i z\right)\right]\left[\exp\left(ik_{oy}^i y\right) + R_o^P \exp\left\{ik_{oy}^i (2h - y)\right\}\right], \tag{3.2}$$

in which R_o^P is the $P = V$, H polarized Fresnel reflection coefficient for waves incident from medium 0 (free space) upon medium 1, and $\bar{a}^P$ is the unit vector parallel ($P = V$, vertical/parallel) or perpendicular ($P = H$, horizontal) to the plane of incidence. The primary fields are proportional to the local basis functions. The corresponding vertically, or horizontally polarized field transforms and waves amplitudes are obtained using (2.10), (2.11), and (2.13). In view of the biorthogonality relationships (2.12), the primary wave amplitudes are proportional to the delta functions corresponding to the polarization (Q = V,H) and direction $\bar{k}_o^i\left(u_o^i, v_o^i, w_o^i\right)$ of the incident waves. When these expressions for the primary wave amplitudes a^Q and b^Q are substituted on the right hand side of (2.13) and (2.14) (with the source terms A^P and B^P suppressed) the (iterative) differential equations for single, diffusely scattered wave amplitudes are obtained.

For plane wave excitations, the full wave solution for the diffusely scattered propagating and evanescent waves from two dimensionally rough surfaces are expressed in matrix form as follows:

$$G_s = \left(\frac{k}{2\pi i}\right)^2 \int \frac{D\left(\bar{k}',\bar{k}^i\right)}{v_y'}\left[\exp\left(-\bar{v}'\cdot\bar{r}_s\right) - \exp\left(i\bar{v}'\cdot\bar{r}_t\right)\right]\frac{dx_s dz_s dk_y' dk_z'}{k_x'} G_i = G^f - G_d^f, \tag{3.3}$$

in which G_s and G_i are 2 x 1 column matrices whose elements are the vertically and horizontally polarized complex wave amplitudes for the scattered and incident waves respectively. The elements of the 2 x 2 matrix $D\left(\bar{k}',\bar{k}^i\right)$ are the surface element scattering coefficients $D^{PQ}\left(\bar{k}',\bar{k}^i\right)$ in which P and Q denote V (vertical) or H (horizontal) polarizations. They are given by

$$D\left(\bar{k}',\bar{k}'\right)=2\cos\theta'\cos\theta^{i}R\left(\bar{k}',\bar{k}^{i}\right). \tag{3.4}$$

The wave vectors for the scattered and incident waves are respectively

$$\bar{k}'=k_{o}\left(\sin\theta'\cos\phi'\bar{a}_{y}+\cos\theta'\bar{a}_{y}+\sin\theta'\sin\phi'\bar{a}_{z}\right)=k'_{ox}\bar{a}_{x}+k'_{oy}\bar{a}_{y}+k'_{oz}\bar{a}_{z}, \tag{3.5a}$$

$$\bar{k}^{i}=k_{o}\left(\sin\theta^{i}\cos\phi^{i}\bar{a}_{x}-\cos\theta^{i}\bar{a}_{y}+\sin\theta^{i}\bar{a}_{z}\right)=k^{i}_{ox}\bar{a}_{x}+k^{i}_{oy}\bar{a}_{y}+k^{i}_{oz}\bar{a}_{z}, \tag{3.5b}$$

in which $k_{o}=\omega\left(\mu_{o}\varepsilon_{o}\right)^{1/2}$ is the wave number in the medium above the rough interface (free space) and the incident wave impinges upon the medium with wave number $k_{1}=\omega\left(\mu_{1}\varepsilon_{1}\right)^{1/2}$. The complex electric permittivity and magnetic permeability are ε and μ respectively and ω is the frequency in radians per second. The mean reference surface is given by y = 0. The position vectors from the origin to the observation point and to points on the mean surface and to the rough surface respectively are

$$\bar{r}=x\bar{a}_{x}+y\bar{a}_{y}+z\bar{a}_{z},\ \bar{r}_{t}=x_{s}\bar{a}_{x}+z_{s}\bar{a}_{z},\ \bar{r}_{s}=\bar{r}_{t}+h\left(x_{s},z_{s}\right)\bar{a}_{y}, \tag{3.6}$$

in which $\bar{a}_{x},\bar{a}_{y}$ and $\bar{a}_{z}$ are unit vectors.

The expressions for $R^{PQ}\left(\bar{k}',\bar{k}^{i}\right)$ the elements of the matrix $R\left(\bar{k}',\bar{k}^{i}\right)$ depend on the polarizations of the incident waves Q = V, H and scattered waves P = V, H, the medium parameters on both sides of the interface and the incident and scatter (spherical coordinate) angles θ^{i},ϕ^{i} and θ',ϕ' respectively.[7] For flat surfaces R^{PP} are the Fresnel reflection coefficients and $R^{PQ}=0$ for $P\neq Q$. The vector $\bar{v}'$ is

$$\bar{v}'=\bar{k}'-\bar{k}^{i}=v'_{x}\bar{a}_{x}+v'_{y}\bar{a}_{y}+v'_{z}\bar{a}_{z}. \tag{3.7}$$

In (3.3) the integrations are over the rough surface (transverse) variables $x_{s}\left(-L_{x},L_{x}\right)$ and $z_{s}\left(-L_{z},L_{z}\right)$ as well as the wave vector variables k'_{y} and k'_{z} $\left(-\infty,\infty\right)$. The first term G^{f} in (3.3) contains the exponent $\exp\left(iv_{y}h\right)$ while the second term G^{f}_{d} does not. On integrating the second term with respect to x_{s} and z_{s} it is shown that G^{f}_{d} is the specularly reflected wave from flat surfaces.

It is readily shown that the full wave solution reduces to the small height-small slope perturbation solution of Rice[8] provided that is assumed that $k_{o}h\ll 1$. On retaining the first two terms of the Taylor series expansion of exp $\left(iv_{y}h\right)$ it follows that

$$\exp\left(i\bar{v}'\cdot\bar{r}_{s}\right)-\exp\left(i\bar{v}'\cdot\bar{r}_{t}\right)\approx iv'_{y}h\left(x,z\right)\exp\left(i\bar{v}'\cdot\bar{r}_{t}\right). \tag{3.8}$$

In this small height-small slope limit, the original full wave solution[9] is indistinguishable from the small perturbation solution for the far fields scattered by slightly rough surfaces (See Fig. 2). These limiting perturbation forms of the full wave solutions are however, no longer invariant to coordinate transformations. Furthermore, it is shown that they are valid only if the height and slopes are of the same order of smallness.

Turning to the high frequency limit, it is assumed that the radii of curvature of the larger scale rough surfaces are very large compared to wavelength. The unit vector $\bar{n}$ normal of these larger scale patches of rough surface is assumed to have arbitrary orientation. Thus the planes of the incidence and scatter with respect to the reference coordinate systems (normal to $\bar{n}^i \times \bar{a}_y$ and $\bar{n}^f \times \bar{a}_y$ respectively) are not the same as the local planes of incidence and scatter with respect to the local coordinate systems (normal to $\bar{n}^i \times \bar{n}$ and $\bar{n}^f \times \bar{n}$ respectively). Furthermore the sines and cosines of the angles of incidence and scatter appearing in the scattering coefficients (3.4) are replaced by the sines and cosines of the local angles of incidence and scatter. In order to account for the arbitrary slope of the larger scale surface, the surface element scattering matrix in (3.4) is replaced by[10]

$$D\left(\bar{k}^f, \bar{k}^i\right) \rightarrow T^f D\left(\bar{k}^f, \bar{k}^i\right) T^i . \tag{3.9}$$

In (3.9) the matrix operator T^i decomposes the waves, which are vertically and horizontally polarized with respect to the reference plane of incidence (normal to $\bar{n}^i \times \bar{a}_y$), into vertically and horizontally polarized waves with respect to the local plane of incidence (normal to $\bar{n}^i \times \bar{n}$). Similarly, the matrix operator T^f decomposes the waves which are vertically and horizontally polarized with respect to the local plane of scatter (normal $\bar{n}^f \times \bar{n}_y$) back into vertically and horizontally polarized waves with respect to the reference plane of scatter (normal to $\bar{n}^f \times \bar{a}_y$).

The changes represented by (3.9) constitute the transformation into the larger scale (patch) coordinate system.[11] It is readily shown that at high frequencies, the major contributions come from the vicinity of the stationary phase, specular points on the rough surface where $\bar{n}$ is along the bisector between $\bar{n}^f$ and $-\bar{n}^i$. Pursuant to the transformation (3.9), it can be shown that at these stationary phase points R^{VV} and R^{HH} reduce to the familiar Fresnel reflection coefficients at the specular points (where $\bar{n} \rightarrow \bar{v}'/v$ and $\bar{v} = \bar{k}' - \bar{k}^i$) while the cross polarized terms R^{VH} and R^{HV} vanish at the specular points. Thus in these limits, the full wave solution reduces to the Physical Optics solution for the diffuse scattered fields.[12] If in addition (3.3) is evaluated analytically using stationary phase approximations, the full wave solution reduces to the geometric optics solution (See Fig. 2).[12] However, in order to account for multiple scatter at the same rough surface, it is necessary to return to the original form (3.3) even at high frequencies.[13] Moreover if the scales of roughness or medium fluctuations are smaller than the wavelength and near fields must be evaluated, the original form of the full wave solution (3.3) must be used.

The mean intensities of the diffusely scattered like and cross polarized fields are

$$\begin{aligned}\left\langle G_S^P G_S^{P*}\right\rangle = \left(\frac{k}{2\pi}\right)^4 \Bigg\langle \int \frac{D^{PQ}\left(\bar{k}', \bar{k}^i\right) D^{PQ*}\left(\bar{k}'', \bar{k}^i\right)}{v_y' v_y''} \exp\left(-i\left(\bar{k}' - \bar{k}''\right)\cdot \bar{r}\right) \\ \exp\left(\bar{i v}'\cdot \bar{r}_t'\right)\left[\exp\left(i v_y' h'\right) - 1\right] \exp\left(-\bar{i v}''\cdot \bar{r}_t''\right)\left[\exp\left(-i v_y'' h''\right) - 1\right] \\ \frac{dx_s' dz_s' dk_y' dk_z' dx_s'' dz_s'' dk_y'' dk_z''}{k_x' k_x''} \Bigg\rangle .\end{aligned} \tag{3.10}$$

The symbol $\langle\cdot\rangle$ denotes the statistical average. To facilitate the integration, the following changes of variables are made

$$\bar{r}_t' - \bar{r}_t'' = \bar{r}_d, (\bar{r}_t' + \bar{r}_t'')/2 = \bar{r}_c . \tag{3.11}$$

Thus (3.10) is expressed as

$$\left\langle G_S^P G_S^{P*} \right\rangle = \left(\frac{k}{2\pi}\right)^4 \Bigg\langle \int \frac{D^{PQ}(\bar{k}', \bar{k}^i)}{v_y'} \frac{D^{PQ*}(\bar{k}'', \bar{k}^i)}{v_y''} \exp\left[-i(\bar{k}' - \bar{k}'') \cdot \bar{r}\right]$$
$$\exp\left[i(\bar{v}' + \bar{v}'')\bar{r}_d / 2\right] \exp\left[i(\bar{v}' - \bar{v}'') \cdot \bar{r}_c\right] \left[\exp(iv_y'h') - 1\right]\left[\exp(iv_y''h'') - l\right]$$
$$d\bar{r}_c d\bar{r}_d \frac{dk_y' dk_z' dk_y'' dk_z''}{k_x' k_x''} > . \tag{3.12}$$

For points on the surface, the significant contributions to the fields come from the small Fresnel zone[3] around the points at $\bar{r}_t'$ and $\bar{r}_t''$. On integrating (3.12) with respect to $\bar{r}_c$ therefore, the limits of integration can be assumed to be infinite. Furthermore

$$\frac{1}{(2\pi)^2} \int_{-\infty}^{\infty} \exp\left[i(\bar{v}' - \bar{v}'') \cdot \bar{r}_c\right] dx_c dz_c = \delta(v_x' - v_x'')\delta(v_z' - v_z'') . \tag{3.13}$$

Thus integrating the results with respect to $dk_y'' dk_z''$ and noting that $dk_y / k_x = -dk_x / k_y$ (3.12) reduces to

$$\left\langle G_S^P G_S^{P*} \right\rangle = \frac{k^4}{(2\pi)^2} \Bigg\langle \int \left| \frac{D^{PQ}(\bar{k}', \bar{k}^i)}{v_y'} \right|^2 \exp(i\bar{v}' \cdot \bar{r}_d)$$
$$\left[\exp(iv_y'h') - 1\right]\left[\exp(-iv_y'h'') - 1\right] d\bar{r}_d \frac{dk_x' dk_z'}{(k_y')^2} \Bigg\rangle . \tag{3.14}$$

4. LOW AND HIGH FREQUENCY APPLICATIONS FOR THE NEAR FIELDS

For nanostructures with scales of roughness much smaller than the wavelength of the illuminating source it is assumed that $kh << 1$. Thus approximate $[\exp(iv_y h) - 1]$ with $(iv_y h)$ etc. and note that the Fourier transform of the surface height autocorrelation function $\langle h'h'' \rangle$ is the surface height spectral density function W:

$$\frac{1}{(2\pi)^2}\int_{-\infty}^{\infty}\langle h'h''\rangle \exp(i\bar{v}'\cdot\bar{r}_d)\,d\bar{r}_d = W(v'_x,v'_z). \tag{4.1}$$

Thus (3.14) reduces to

$$\langle G_S^P G_S^{P*}\rangle = k^4\int_{-\infty}^{\infty}\left|\frac{D^{PQ}(\bar{k}',\bar{k}^i)}{k'_y}\right|^2 W(v'_x,v'_z)\,dk'_x dk'_z. \tag{4.2}$$

Note that the result in (4.2) accounts for both propagating waves, with $k_y=\sqrt{1-k_x^2-k_z^2}$ real and evanescent waves with k_y imaginary. Both must be accounted for when fields very near the surface are evaluated.

Since the surface height spectral density function $W(v'_x,v'_z)$, which is related to the surface height autocorrelation function, appears in the expression for the near fields scattered by nanoscale surface roughness, this work can be applied to the synthesis and modification of nanostructure surfaces.

It is interesting to compare the above results with those for the high frequency limit ($kh >> 1$). For this case the following substitutions are made in (3.14):

$$h' = h_c + (x_d h_x + z_d h_z)/2, \qquad h'' = h_c - (x_d h_x + z_d h_z)/2, \tag{4.3}$$

$$h_c = (h'+h'')/2, \quad h_x = \frac{\partial h}{\partial x}, \quad h_z = \frac{\partial h}{\partial z}. \tag{4.4}$$

Thus for the high frequency limit,

$$\begin{aligned}\langle G_S^P G_S^{P*}\rangle &= \frac{k^4}{(2\pi)^2}\left\langle \int\left|\frac{D^{PQ}(\bar{k}',\bar{k}^i)}{v'_y}\right|^2 \exp\left[i\left(v'_x x_d + v'_y(x_d h_x + z_d h_z) + v'_z z_d\right)\right]dx_d dz_d \frac{dk'_x dk'_z}{(k'_y)^2}\right\rangle \\ &= k^4\left\langle \int\left|\frac{D^{PQ}(\bar{k}',\bar{k}^i)}{v'_y}\right|^2 \delta(v'_x + v'_y h_x)\,\delta(v'_z + v'_y h_z)\frac{dk'_x dk'_z}{(k'_y)^2}\right\rangle \\ &= k^4\int\left|D^{PQ}(\bar{k}',k^i)\right|_S^2 p(h_{xs},h_{zs})\frac{dk'_x dk'_z}{(v'_y)^4 (k'_y)^2},\end{aligned} \tag{4.5}$$

in which D^{PQ} and the slope probability density function $p(h_x,h_z)$ are evaluated at the specular points where

$$h_{xs} = -\frac{v'_x}{v'_y} = -\left(\frac{k'_x - k^i_x}{k'_y - k^i_y}\right) \quad h_{zs} = -\frac{v'_z}{v'_y} = -\left(\frac{k'_z - k^i_z}{k'_y - k^i_y}\right) \tag{4.6a}$$

and the unit vector normal to the surface at the specular points is

$$\bar{n}_s = \left(-h_{xs}\bar{a}_x + \bar{a}_y - h_{zs}\bar{a}_z\right) / \left(1 + h_{xs}^2 + h_{zs}^2\right)^{1/2} = \bar{v}' / v . \tag{4.6b}$$

Since the surface slope probability density function appears in the expression for the near fields scattered by larger scale rough surface, this work can be extended to the synthesis and modification of nanostructure surfaces with undulating slopes. The expression for the scattered power density (average Poynting vectors) can be evaluated in manner similar to (3.14). The ratio between the horizontal components of the electric and magnetic fields is given by the wave impedance $\omega\mu / k_y$ and $k_y / \omega\varepsilon$ for horizontally and vertically polarized waves respectively. For evanescent waves k_y is imaginary, thus the horizontal components of electric and magnetic fields are $\pi / 2$ radians out of phase, corresponding to no average scattered power flow in the vertical direction associated with evanescent waves.

5. SCATTERING FROM PERIODIC AND QUASI-PERIODIC-ROUGH SURFACES

The fullwave analysis can be readily applied to near and/or far field scattering from high density quantum dot arrays used in optics, data storage and other electronic applications. These structures are usually (quasi) periodic. For two dimensionally periodic structures, the full wave solutions for the scattered fields corresponding to (3.3) is

$$G_S = \left(\frac{k}{2\pi i}\right)^2 \int A_x\left(N_x v_x' L_x\right) A_z\left(N_z v_z' L_z\right) \frac{D\left(\bar{k}', \bar{k}^i\right)}{v_y'} \exp\left(-i\bar{k}' \cdot \bar{r}\right)$$
$$\left[\exp\left(i\bar{v}' \cdot \bar{r}_s\right) - \exp\left(i\bar{v}' \cdot \bar{r}_t\right)\right] \frac{dx_s dz_s dk_y' dk_z'}{k_{x'}} G_i , \tag{5.1}$$

in which the $\left(N_x \cdot N_z\right)$ element array factors are

$$A_x\left(N_x v_x' L_x\right) = \frac{\sin\left(N_x v_x' L_x / 2\right)}{\sin\left(v_x' L_x / 2\right)} \exp\left(-i v_x' L_x / 2\right), \tag{5.1a}$$

$$A_z\left(N_z v_z' L_z\right) = \frac{\sin\left(N_z v_z' L_z / 2\right)}{\sin\left(v_z' L_z / 2\right)} \exp\left(-i v_z' L_z / 2\right). \tag{5.1b}$$

The integrations with respect to $dx_s dz_s$ are over the area $L_x L_z$ of one element of the periodic structure. The integrations with respect to $dk_y' dk_z'$ are as in (3.3) over the propagating and evanescent spectra of scattered waves. A small random rough component can also be added to the (quasi) periodic surface height. When the quasi-periodic surface height has a small random component $h = h_p + h_r$ (subscripts p and r denote periodic and random respectively), the statistical average (over random heights) of

(5.1) is first obtained, resulting in the multiplication of $\exp(i\bar{v}' \cdot \bar{r}_s)$ in (5.1) by the random rough surface height characterization functions $\chi_r(v_y)$.

For observation points near the surface, the full wave solutions account for coupling between the propagating and evanescent waves, the lateral waves and the surface waves associated with rough surface scattering. When the rough surface is assumed to be perfectly conducting the contribution from the branch cut integral associated with the lateral waves vanishes and there are no residue contributions (associated with surface waves) from the singularities of the reflection coefficients. When the approximate impedance boundary condition is used, the lateral wave contribution is eliminated. Since the perturbation and physical optics solution are far field approximations, the surface waves and the lateral waves are not accounted for in these solutions.

The full wave method can also be used to determine the fields scattered upon transmission across rough surfaces.[14] When scattering from more than one rough interface in irregular stratified media is considered, in general, it becomes necessary, to account for scattering upon reflection and transmission across rough interfaces. The analysis presented here to obtain the scattered intensities can be readily extended to the derivation of the full wave solutions for all sixteen elements of the Mueller Matrix that relates the incident to the scattered Stokes vectors that fully characterize scattering form irregular structures.

6. COUPLING BETWEEN THE RADIATION FIELDS, THE LATERAL WAVES, AND THE SURFACE WAVES

For simplicity, in this section, the rough surface height and the medium parameters are assumed to be independent of the z variable and the incident wave vector $\bar{k}'$ is restricted to the xy plane. The scattered wave vectors $\bar{k}'$ are also restricted to the xy plane and the scattered waves are not depolarized. The uppercase superscripts or subscripts P, $Q = V, H$ are dropped from the notation for the generalized telegraphists' equation. Instead the low case subscripts p and q are inserted to identify the particular contribution to the complete wave spectrum. Thus, $p, q = 0$ identifies the radiation term; $p, q = 1$ identifies the lateral wave term (associated with the branch cuts that end on the branch points at k_o and k_m, the wave numbers for the uppermost and lowermost media respectively) and $p, q = s$ identifies the surface wave terms associated with the residues at the poles of the total reflection coefficients seen from above (or below) the structure. The generalized telegraphists' equations are expressed as

$$\frac{-da_p}{dx} - iua_p = \sum_q S_{pq}^{BA} a_q + S_{pq}^{BB} b_q - A^P, \tag{6.1}$$

$$\frac{-db_p}{dx} - iub_p = \sum_q S_{pq}^{AA} a_q + S_{pq}^{AB} b_q + B^P. \tag{6.2}$$

Thus in (6.1) and (6.2) the scattering coefficients $S_{10}^{\alpha\beta}$ and $S_{s0}^{\alpha\beta}$ $(\alpha, \beta = A, B)$ account for coupling between the incident radiation fields and the scattered lateral waves and surface waves respectively. Explicit expressions for these rough surface scattering coefficients and the reciprocal expressions $S_{01}^{\alpha\beta}$, $S_{0s}^{\alpha\beta}$ have been published[15,16] for both vertically and horizontally polarized excitations. Interactive, single scatter solutions to (6.1) and (6.2) are derived using the procedures used to derive the scattered radiation fields. Thus the primary, source dependent impressed fields are obtained from (6.1) and (6.2) on neglecting the coupling terms. The first order scattering terms are obtained on substituting the expressions for the impressed fields on the right hand side of (6.1) and (6.2) and on suppressing the source terms A^P and B^P. The resulting expressions for the lateral wave or surface wave amplitudes are substituted into (2.8) and (2.9) to obtain the electric and magnetic field expressions associated with the lateral waves and the surface waves that are excited by the radiation fields incident upon the rough surface.[15,16]

7. CONCLUDING REMARKS

A full wave approach based on the complete expansions of the scattered electromagnetic fields into the radiation term, lateral waves and surface waves associated with layered media have been used to convert Maxwell's equations into generalized telegraphists' equations. Exact boundary conditions are imposed at the rough interface between laterally fluctuating media. The coupling between the wave amplitudes is due to surface height and medium fluctuations. Both random as well as (quasi) periodic rough surfaces are considered in the analysis of the scattered near and far fields. Both large scale and small-scale (including subwavelength) features of nanostructures can be characterized. The procedures followed in this paper can be used to derive all sixteen elements of the Mueller Matrix that completely characterize the scattered fields. This approach can also be used to analyze, irregular structures with anisotropic media such as chiral materials. This work is applicable to the synthesis of nanostructures in order to modify its mechanical and/or electrical characteristics.

ACKNOWLEDGMENTS

The author wishes to thank Professor Vasundara V. Veredan for her constructive suggestions regarding the presentation of this material, especially in pointing out the benefits of introducing the high and low frequency approximations as well as far and near field approximations a posteriori rather than a priori. This manuscript was prepared by Jane Craig.

REFERENCES

1. E. Bahar, Full wave solutions for electromagnetic fields scattered by irregular stratified media with rough interfaces: theory and experiment, *Encyclopedia of Electrical and Electronic Engineering, Vol. 17*, edited by J.G. Webster (John Wiley & Sons, New York, 1999), pp. 635–654.
2. S.A. Schelkunoff, Conversion of Maxwell's equations into generalized telegraphists' equations, *Bell System Technical J.* **34**, 995–1045 (1955).
3. P. Beckmann and A. Spizzichino, *The Scattering of Electromagnetic Waves from Rough Surfaces* (Macmillan, New York, 1963).

4. E. Bahar, Depolarization of electromagnetic waves excited by distribution of electric and magnetic sources in inhomogeneous multilayered structures of arbitrarily varying thickness, generalized field transform, *J. Mathematical Phys.* **14** (11), 1502–1509 (1973).
5. E. Bahar, Depolarization of electromagnetic waves excited by distribution of electric and magnetic sources in inhomogeneous multilayered structures of arbitrarily varying thickness full wave solutions, *J. Mathematical Phys.* **14** (11), 1510–1515 (1973).
6. E. Bahar, Depolarization in nonuniform multilayered structures - full wave solutions, *J. Mathematical Phys.* **15**(2), 202–208 (1974).
7. E. Bahar, Full wave solutions for the depolarization of the scattered radiation fields by rough surface of arbitrary slope, *IEEE Trans. Antennas Propagat.* **AP-29**(3), 443–454 (1981).
8. S.O. Rice, Reflection of electromagnetic waves from a slightly rough surface, *Commun. Pure Applied* Math **4**, 351–378 (1951).
9. E. Bahar and G. Govindarajan, Radiation by a line source over nonuniform stratified earth, *J. Geophysical Research* **78**(2), 393–406 (1973).
10. E. Bahar, C.L. Rufenach, P.E. Barrick, and M.A. Fitzwater, Scattering cross section modulation for arbitrarily oriented composite rough surfaces - full wave solutions, *Radio Sci.* **18**(5), 675–690 (1983).
11. E. Bahar and Y. Zhang, A new unified full wave approach to evaluate the scatter cross sections of composite random rough surfaces, *IEEE Trans. Geoscience Remote Sensing* **34**(4), 973–980 (1996).
12. E. Bahar and B.S. Lee, Full wave solutions for rough surface bistatic radar cross sections: comparison with small perturbation, physical optics, numerical and experimental results, *Radio Sci.* **29**(2), 407–429 (1994).
13. E. Bahar and M. El-Shenawee, Enhanced backscatter from one-dimensional random rough surfaces: stationary-phase approximations to full wave solutions, *J. Optical Soc. Am.* **12**(1), 151–161 (1995).
14. E. Bahar and B.S. Lee, Transmission scatter cross sections across two-dimensional random rough surfaces - full wave solutions and comparisons with numerical results, *Waves in Random* Media **6**, 25–48 (1966).
15. E. Bahar, Excitation of lateral waves and scattered radiation fields by rough surfaces of arbitrary slope, *Radio Sci.* **15**(6), 1095–1104 (1980).
16. E. Bahar, Excitation of surface waves and the scattered radiation fields by rough surfaces of arbitrary slope, *IEEE Trans. Microwave Theory Techniques* **MTT-28**(9), 999–1006 (1980).

FRAME-BASED BEAM-SUMMATION ALGORITHMS FOR ULTRA WIDEBAND RADIATION FROM EXTENDED APERTURES.

PART I: FORMULATIONS IN THE MULTI-FREQUENCY DOMAIN

Amir Shlivinski, Ehud Heyman, Amir Boag, and Christine Letrou*

1. INTRODUCTION

Beam based phase-space formulations are an important tool in wave theory since they provide a systematic framework for ray-based construction of spectrally uniform local solutions in complex configurations. An important property of these formulations is that the spectrum of beam propagators is overcomplete and thus may be a priori discretized as, for example, in the Gabor series. The latter, however, is *critically complete* and hence suffers from two *inherent difficulties*: (a) Non-locality and instability of the expansion coefficients, and (b) the phase space beam lattice (origins and directions) varies with frequency. Property (b) makes the conventional Gabor-based beam formulation *inapplicable* for wideband applications since it requires the tracking of different beams for each frequency. Property (a) makes it inconvenient even for monochromatic fields.

Recently we introduced a novel wideband frame-based beam representation that accommodates the difficulties noted above.[1,2,3] Using an overcomplete basis where the overcompleteness is inversely proportional to the frequency, we were able to obtain a beam series representation that has the *same beam lattice for all frequencies.* The overcompleteness also localizes and stabilizes the expansion

*Amir Shlivinski, Ehud Heyman, and Amir Boag are with the Faculty of Engineering, Department of Physical Electronics, Tel-Aviv University, Tel-Aviv, 69978 Israel. Christine Letrou is with the I.N.T., 9 rue Charles Fourier, F-91011 EVRY Cedex, France.

Ultra-Wideband Short-Pulse Electromagnetics 6
Edited by Mokole *et al.*, Kluwer Academic/Plenum Publishers, 2003

coefficients.[4] Furthermore, using properly chosen set of *iso-diffracting* Gaussian-Beam (ID-GB) window functions, we obtained the "snuggest" frame representation for *all frequencies* within the excitation spectrum. Another important property of these frame functions is that they can be transformed in closed form into the Time Domain (TD), where they yield the so-called iso-diffracting pulsed beams (ID-PB) ,[5] leading to a new phase space formulation for short-pulse fields directly in the TD.[6,7]

The new formulation is presented here in the context of wideband radiation from an extended aperture. Emphasis is placed on analytical study of the expansion and beam parameters that render the beam representation most effective and localized (Sec. 4). The frequency bandwidth for which the formulation is most effective is of the order of one octave. For wider excitation bands, we introduce in Sec. 5 a special hierarchy of frequency bands with a related set of spatial and directional discretizations that leads to a self consistent representation wherein the beam sets at the lower bands are decimated sub-sets of those at the highest band. Consequently, only the set of GBs at the highest frequency band needs to be traced, while for the lower bands one may use properly decimated subsets.

2. LOCAL SPECTRUM AND BEAM SUMMATION: A REVIEW

The formulation is presented in the context of radiation from an extended aperture in the $z=0$ plane with a prescribed field $\widehat{u}_0(\boldsymbol{x},\omega)$. The 3D coordinates convention is $\boldsymbol{r}=(\boldsymbol{x},z)$ with $\boldsymbol{x}=(x_1,x_2)$ being the coordinates transverse to z. Henceforth, here, $\widehat{\ }$ denotes a frequency domain (FD) constituent with an $e^{-i\omega t}$ time-dependence. The field is constrained to the band $\Omega=(\omega_{\min},\omega_{\max})$ that may extend over several octaves. In the expressions bellow we shall not mention the frequency variable explicitly, yet the reader should recall that all expressions are functions of frequency over a wide frequency band.

2.1. Continuous Formulation: Plane Wave and Local Spectrum Formulations

Since the theory addresses wide band fields, we use the frequency-normalized spatial spectrum coordinate $\boldsymbol{\xi}=(\xi_1,\xi_2)=\boldsymbol{k}_x/k$ where $\boldsymbol{k}_x=(k_{x_1},k_{x_2})$ is the spectral wave number and $k=\frac{\omega}{c}$, c being the light speed. This leads to the transform pair

$$\widehat{\widetilde{u}}_0(\boldsymbol{\xi})=\int d^2x\,\widehat{u}_0(\boldsymbol{x})e^{-ik\boldsymbol{\xi}\cdot\boldsymbol{x}},\qquad \widehat{u}_0(\boldsymbol{x})=\left(\frac{k}{2\pi}\right)^2\int d^2\xi\,\widehat{\widetilde{u}}_0(\boldsymbol{\xi})e^{ik\boldsymbol{\xi}\cdot\boldsymbol{x}},\tag{1}$$

where $\widetilde{\ }$ denotes spatial spectrum constituents. Since the spectral variable $\boldsymbol{\xi}$ is normalized with respect to k, the spectral constituents propagate in *frequency independent* directions (θ,ϕ) defined via $\boldsymbol{\xi}=\sin\theta(\cos\phi,\sin\phi)$ for $|\boldsymbol{\xi}|<1$.

Next we define the local transform pair as

$$\widehat{U}_0(\boldsymbol{X})=\int d^2x\,\widehat{u}_0(\boldsymbol{x})\widehat{\varphi}^*_{\boldsymbol{X}}(\boldsymbol{x}),\qquad \widehat{u}_0(\boldsymbol{x})=\left(\frac{k}{2\pi\|\widehat{\psi}\|}\right)^2\int d^4X\,\widehat{U}_0(\boldsymbol{X})\widehat{\psi}_{\boldsymbol{X}}(\boldsymbol{x}).\tag{2}$$

Here $\widehat{\varphi}_{\boldsymbol{X}}(\boldsymbol{x})=\widehat{\varphi}(\boldsymbol{x}-\boldsymbol{x}_0)e^{ik\boldsymbol{\xi}_0\cdot(\boldsymbol{x}-\boldsymbol{x}_0)}$ and $\widehat{\psi}_{\boldsymbol{X}}(\boldsymbol{x})=\widehat{\psi}(\boldsymbol{x}-\boldsymbol{x}_0)e^{ik\boldsymbol{\xi}_0\cdot(\boldsymbol{x}-\boldsymbol{x}_0)}$ are local Fourier kernels that are associated with the "analysis" window $\widehat{\varphi}(\boldsymbol{x})$ and its dual "synthesis" window $\widehat{\psi}(\boldsymbol{x})$ (in the continuous phase space formulation of (2) we have $\widehat{\varphi}=\widehat{\psi}$; cf. the large oversampling approximation in (8)). If the windows are localized around $\boldsymbol{x}=0$, then $\widehat{U}_0$ is the local spectrum of $\widehat{u}_0$ near $\boldsymbol{x}_0$. The phase space coordinates are denoted conveniently as $\boldsymbol{X}\equiv(\boldsymbol{x}_0,\boldsymbol{\xi}_0)$.

The phase space representation of $\widehat{u}_0$ in the second expression of (2) can be propagated into $z>0$ by propagating the functions $\widehat{\psi}_{\mathbf{X}}(\boldsymbol{x})$, giving the beam summation representation of the radiating field

$$\widehat{u}(\boldsymbol{r})=\left(\frac{k}{2\pi\|\widehat{\psi}\|}\right)^2\int d^4X\,\widehat{U}_0(\boldsymbol{X})\widehat{B}_{\mathbf{X}}(\boldsymbol{r}),\quad \widehat{B}_{\mathbf{X}}(\boldsymbol{r})=\left(\frac{k}{2\pi}\right)^2\int d^2\xi\,\widehat{\widetilde{\psi}}_{\mathbf{X}}(\boldsymbol{\xi})e^{ik(\boldsymbol{\xi}\cdot\boldsymbol{x}+\zeta z)},\quad (3)$$

where $\zeta=\sqrt{1-\boldsymbol{\xi}\cdot\boldsymbol{\xi}}$. If $\widehat{\psi}$ is wide on a wavelength scale, the spatial and spectral distributions of $\widehat{\psi}_{\mathbf{X}}(\boldsymbol{x})$ are localized around $(\boldsymbol{x},\boldsymbol{\xi})=(\boldsymbol{x}_0,\boldsymbol{\xi}_0)$ and the propagator $\widehat{B}_{\mathbf{X}}(\boldsymbol{r})$ is a collimated beam whose axis emerges from the point $\boldsymbol{x}_0$ in the $z=0$ plane in the direction (θ_0,ϕ_0) defined by $\boldsymbol{\xi}_0$. The 4-fold $\boldsymbol{X}$ integration in (3) expresses the field as a superposition of beams that emerge from all points $\boldsymbol{x}_0$ and in all directions $\boldsymbol{\xi}_0$. The amplitude of the beams in (2) is given by the local spectrum $\widehat{U}_0$ that matches this spectrum of beams to the source distribution $\widehat{u}_0$.

2.2. Discretized Formulation: The Difficulties of the Gabor Series Approach

The representation (2) is overdetermined since there are many 4-variable kernels $\widehat{U}_0$ that can be used in (2) and yield $\widehat{u}_0$. Of particular interest is the Gabor representation, which is based on sampling of $\widehat{U}_0$ on a grid in the phase space $\boldsymbol{X}$ defined by

$$\boldsymbol{X}_{\boldsymbol{\mu}}=(\boldsymbol{x}_{\boldsymbol{m}},\boldsymbol{\xi}_{\boldsymbol{n}})=(\boldsymbol{m}\bar{x},\boldsymbol{n}\bar{\xi})\qquad\text{with}\quad k\bar{x}\bar{\xi}=2\pi,\quad (4)$$

where $(\bar{x},\bar{\xi})$ are the dimensions of the phase space unit cell and we use the 4-index $\boldsymbol{\mu}=(\boldsymbol{m},\boldsymbol{n})=(m_1,m_2,n_1,n_2)$. In the Gabor formulation, the functions $\widehat{\psi}_{\mathbf{X}}(\boldsymbol{x})$ and $\widehat{\varphi}_{\mathbf{X}}(\boldsymbol{x})$ are biorthogonal.[8,9] (see Fig. 1(f))

When applied to radiation problems, this formulation suffers from two fundamental difficulties: (*a*) The analysis function $\widehat{\varphi}(\boldsymbol{x})$ that extracts the phase-space information from $\widehat{u}_0$ is neither smooth nor localized, and is even not L_2 integrable (see Fig. 1(f)); (*b*) The unit cell dimensions $(\bar{x},\bar{\xi})$ in (4) change with frequency yielding different beam trajectories for each frequency, thereby making this Gabor-based formulation inconvenient for wideband applications[10].

Both difficulties are solved by the frame-based wideband beam representation introduced originally in Ref. 1, presented in Sec. 4, and further extended in Sec. 5. We start, however, in Sec. 3 by summarizing some relevant properties of the Gabor frames.

3. ELEMENTS OF GABOR FRAME FORMULATION

In $L^2(\mathbb{R})$, two types of frames are mostly used: the wavelet frames and the Weyl-Heisenberg frames (also termed the windowed Fourier transform frames or Gabor frames). In this paper, we shall be interested in the latter. Extensive treatment of frame theory can be found in Ref. 11.

The windowed Fourier (Gabor) frames are constructed from a single window function which is translated both in the spatial and spectral domains. Choosing a proper window function $\psi(x) \in L^2(\mathbb{R})$, the frame elements ψ_μ are

$$\psi_\mu(x) = \psi(x - m\bar{x})e^{in\bar{k}_x(x-m\bar{x})}, \quad \mu = (m,n) \in \mathbb{Z}^2, \quad x \in \mathbb{R}, \tag{5}$$

where $(\bar{x}, \bar{k}_x)$ are the spatial and spectral dimensions of the phase-space unit cell. For the set $\{\psi_\mu\}$ to be a frame, it is necessary that the unit-cell area $\bar{x}\bar{k}_x$ satisfies

$$\bar{x}\bar{k}_x = 2\pi\nu, \quad \text{with} \quad \nu \leq 1, \tag{6}$$

where ν describes the frame overcompleteness (or the redundancy; ν^{-1} is the over-sampling factor). The frame is overcomplete for $\nu < 1$ and is critically complete in the Gabor limit $\nu \uparrow 1$ of (4). The effect of ν on the frame properties is studied in Fig. 1.

The frame properties depend also on how well the spatial and spectral distributions of ψ fit into the phase-space grid. This property is quantified by the "match" (or aspect ratio) parameter

$$\rho = \frac{\Delta_x/\bar{x}}{\Delta_{k_x}/\bar{k}_x}, \tag{7}$$

where $\Delta_x = \|x\psi(x)\|/\|\psi(x)\|$ and $\Delta_{k_x} = \|k_x\widetilde{\psi}(k_x)\|/\|\widetilde{\psi}(k_x)\|$ are the spatial and spectral widths of ψ, respectively, and an over tilde denotes Fourier transform. When $\rho = 1$, the spatial and spectral window coverage of the unit cell are balanced, but if $\rho < 1$ or $\rho > 1$, the window is spatially narrow or wide, respectively, (i.e., spectrally wide or narrow) with respect to the grid dimensions $(\bar{x}, \bar{k}_x)$. This property is schematized in Fig. 2(a) which depicts the phase space coverage by a Gaussian window for $\nu = 0.5$ and for three values of ρ. As demonstrated in Fig. 2(b), the snuggest frame for a given ν, and, thus, the most localized dual function φ, are obtained when $\rho = 1$.

The dual frame set $\{\varphi_\mu\}$ corresponding to the set $\{\psi_\mu\}$ is given by

$$\varphi_\mu(x) = \varphi(x - m\bar{x})e^{in\bar{k}_x(x-m\bar{x})}, \qquad \varphi(x) \approx [\nu/\|\psi\|^2]\psi(x). \tag{8}$$

The exact dual window $\varphi(x)$ should be calculated by numerical inversion of the frame operator. This can be done numerically by various techniques, for example using a Neumann series approach.[11] Keeping only the first term in this series yields the approximation on the right equation in (8), which is valid for $\nu \leq \frac{1}{2}$ (see Fig. 1(a–c)). The limit $\nu \downarrow 0$ recovers the continuous transform used in (2) wherein $\varphi = \psi$.

Finally, the frame representation for a function $f(x)$ is given by[11]

$$f(x) = \sum_\mu a_\mu \psi_\mu(x), \quad a_\mu = \langle f, \varphi_\mu \rangle = \int dx f(x)\varphi_\mu^*(x). \tag{9}$$

It should be recalled that the frame representation (9) is not unique, yet it minimizes the ℓ_2 norm of the coefficient seriesx $\{a_\mu\}$.

3.1. Example: Gaussian Windows

We consider the Gaussian window

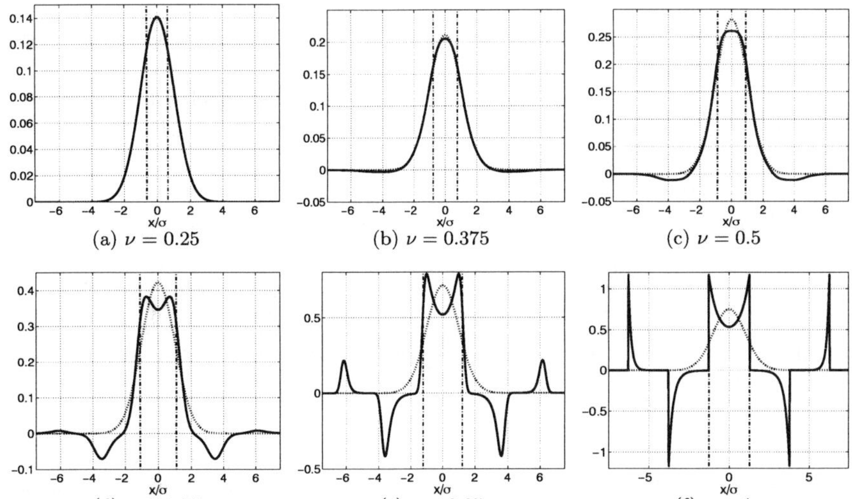

(a) $\nu = 0.25$ (b) $\nu = 0.375$ (c) $\nu = 0.5$

(d) $\nu = 0.75$ (e) $\nu = 0.95$ (f) $\nu = 1$

Figure 1: The exact and approximate dual functions (solid and dotted lines) corresponding to the Gaussian window (10) for 6 different values of ν. In all cases, the phase-space grid is matched to the window as noted after (10), and the unit cell boundaries at $\pm\bar{x}/2$ are depicted as two vertical lines.

$$\psi(x) = e^{-x^2/2\sigma^2}, \quad \sigma^2 \text{ real.} \tag{10}$$

Using in (7) $\Delta_x = \sigma 2^{-1/2}$ and $\Delta_{k_x} = \sigma^{-1}2^{-1/2}$ yields $\rho = \sigma^2\bar{k}_x/\bar{x}$. The window is matched to the grid if $\rho=1$, giving $\bar{x}=\sigma\sqrt{2\pi\nu}$ and $\bar{k}_x=\sigma^{-1}\sqrt{2\pi\nu}$.

The dual function corresponding to (10) is depicted in Fig. 1 for several values of ν. For each case, the cell dimensions are chosen to match the window as discussed after (10). The x axis is normalized to x/σ and the cell's boundaries at $\pm\bar{x}/2$ are depicted by vertical lines. The full lines denote the numerically calculated dual functions and the dotted lines are the approximation on the right equation of (8) that replicates ψ up to a proportionality constant. One observes that this approximation is valid for $\nu \lesssim \frac{1}{2}$, while for $\nu \gtrsim \frac{1}{2}$, φ is increasingly less localized and in the limit $\nu \uparrow 1$ it tends to the highly non-local and irregular Gabor biorthogonal function (Fig. 1(f)), yielding non-local and unstable coefficients a_μ in (9).

The role of the matching parameter ρ of (7) is explored in Fig. 2(b) which depicts the numerically calculated φ for $\nu=\frac{1}{2}$ and three values of ρ: $\rho=0.5$, 1, and 2 (i.e., ψ is spatially narrow, matched and wide with respect to the grid $(\bar{x}, \bar{k}_x)$ as schematized in Fig. 2(a)). One observes that the smoothest and most localized φ is obtained for the balanced case $\rho=1$.

4. WIDEBAND BEAM SUMMATION: THE BASIC ALGORITHM

Returning to the problem of radiation from a wideband aperture distribution $\widehat{u}_0(\boldsymbol{x})$ stated at the beginning of Sec. 2, we start by choosing a reference frequency $\bar{\omega}$ such that $\bar{\omega} > \omega_{\max}$ (typically $\bar{\omega} = P\omega_{\max}$ where $2 \leq P \leq 3$; see discussion after

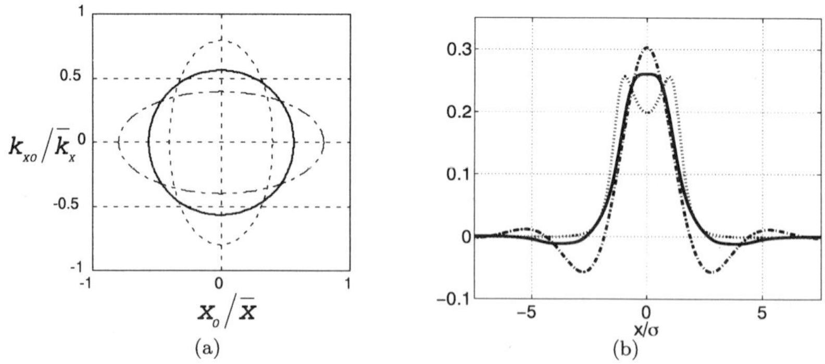

Figure 2: (a) The phase space coverage by the Gaussian window (10) with $\nu = 0.5$ and for three values of ρ: $\rho = 0.5, 1$ and 2 (dotted, solid and dash-dotted lines, respectively). (b) The corresponding dual function φ.

(17)), and define the phase-space grid $(\bar{x}, \bar{\xi})$ to be critically complete at $\bar{\omega}$ (see (4))

$$\bar{k}\bar{x}\bar{\xi} = 2\pi, \qquad \text{where } \bar{k} = \bar{\omega}/c. \tag{11}$$

We use the same $(\bar{x}, \bar{\xi})$ for all $\omega < \bar{\omega}$, hence for all $\omega < \bar{\omega}$ the grid is overcomplete with

$$\bar{k}_x \bar{x} = k\bar{\xi}\bar{x} = 2\pi\nu \qquad \text{where } \nu = \omega/\bar{\omega} < 1. \tag{12}$$

Thus, the beam lattice is *frequency independent* for $\omega < \bar{\omega}$ if $\nu = \omega/\bar{\omega}$.

Next, for a proper window function $\widehat{\psi}(\boldsymbol{x};\omega)$ and its dual $\widehat{\varphi}(\boldsymbol{x};\omega)$, we introduce the Gabor-type frames in $L_2(\mathbb{R}^2)$ for all $\omega < \bar{\omega}$ (cf. (5) and (8))

$$\widehat{\psi}_{\boldsymbol{\mu}}(\boldsymbol{x}) = \widehat{\psi}(\boldsymbol{x} - \boldsymbol{x_m}) e^{ik\boldsymbol{\xi_n}\cdot(\boldsymbol{x}-\boldsymbol{x_m})}, \quad \widehat{\varphi}_{\boldsymbol{\mu}}(\boldsymbol{x}) = \widehat{\varphi}(\boldsymbol{x} - \boldsymbol{x_m}) e^{ik\boldsymbol{\xi_n}\cdot(\boldsymbol{x}-\boldsymbol{x_m})}, \tag{13}$$

where the phase-space grid points are denoted conveniently $\boldsymbol{X_\mu} = (\boldsymbol{x_m}, \boldsymbol{\xi_n})$ and we use the vector indices $\boldsymbol{\mu} = (\boldsymbol{m}, \boldsymbol{n})$ introduced in (4). For all $\omega < \bar{\omega}$, the frame representations of the initial field $\widehat{u}_0(\boldsymbol{x})$ and of the propagated field $\widehat{u}(\boldsymbol{r})$ for $z > 0$ are (see (9))

$$\widehat{u}_0(\boldsymbol{x}) = \sum_{\boldsymbol{\mu}} \widehat{a}_{\boldsymbol{\mu}} \widehat{\psi}_{\boldsymbol{\mu}}(\boldsymbol{x}), \qquad \widehat{u}(\boldsymbol{r}) = \sum_{\boldsymbol{\mu}} \widehat{a}_{\boldsymbol{\mu}} \widehat{B}_{\boldsymbol{\mu}}(\boldsymbol{r}), \tag{14a}$$

$$\text{with} \qquad \widehat{a}_{\boldsymbol{\mu}} = \langle \widehat{u}_0(\boldsymbol{x}), \widehat{\varphi}_{\boldsymbol{\mu}}(\boldsymbol{x}) \rangle \equiv \int d^2x \, \widehat{u}_0(\boldsymbol{x}) \widehat{\varphi}^*(\boldsymbol{x} - \boldsymbol{x_m}) e^{-ik\boldsymbol{\xi_n}\cdot(\boldsymbol{x}-\boldsymbol{x_m})}, \tag{14b}$$

$\widehat{a}_{\boldsymbol{\mu}}$ are recognized as the local spectrum of $\widehat{u}_0$ around $\boldsymbol{X_\mu}$ with respect to the "analysis" window $\widehat{\varphi}$. The beams propagators $\widehat{B}_{\boldsymbol{\mu}}(\boldsymbol{r})$ are obtained by propagating the "synthesis" window $\widehat{\psi}_{\boldsymbol{\mu}}(\boldsymbol{x})$ as in (3). They emerge from the points $\boldsymbol{x_m}$ in the $z{=}0$ plane, in the directions $\boldsymbol{\xi_n} = \sin\theta_n(\cos\phi_n, \sin\phi_n)$. Actually, the expansion for $\widehat{u}(\boldsymbol{r},\omega)$ involves only the propagating beams with $|\boldsymbol{\xi_n}| < 1$ since all other decay away from $z{=}0$.

4.1. Special Case: Iso-diffracting Gaussian Windows

The Iso-diffracting (ID) Gaussian windows are our favorable windows for field processing and propagation for the following reasons: (*a*) they yield GB propagators in the FD or iso-diffracting pulsed beams (ID-PB) in the TD;[6,7] both are analytically tractable[5]; (*b*) their width is scaled with the frequency such that they yield the *snuggest frames* for all ω (see (16)).

The simplest ID window has the form

$$\widehat{\psi}_{\mathrm{ID}}(\boldsymbol{x}) = \exp(-k|\boldsymbol{x}|^2/2b), \quad \text{where } b > 0. \tag{15}$$

The *frequency independent* parameter b is identified as the collimation (Rayleigh) distance of the resulting GB, hence the designation ID since all frequency components diffract at the same z. For collimated beams we must choose $kb \gg 1$ for all ω in the band.

The condition for matching the window to the grid is obtained by combining the grid condition (6) with the matching condition $\rho = 1$ in (7) using $\Delta_x = (b/k)^{1/2}$ and $\Delta_\xi = (kb)^{-1/2}$, yielding

$$b = \bar{x}/\bar{\xi} = \bar{k}\bar{x}^2/2\pi, \qquad \Rightarrow \quad \text{snuggest frame } \forall\, \omega < \bar{\omega}. \tag{16}$$

As noted in (11), the reference is chosen to be $\bar{\omega} = P\omega_{\max}$ where typically $2 \leq P \leq 3$. Consequently $\nu < P^{-1}$ for all ω in the band and $\widehat{\varphi}_{\mathrm{ID}}$ can be approximated analytically by (cf. (8) and Figs. 1(a)–(c))

$$\widehat{\varphi}_{\mathrm{ID}} \approx (\nu^2/\|\widehat{\psi}_{\mathrm{ID}}\|^2)\widehat{\psi}_{\mathrm{ID}} = (\omega/\bar{\omega})^3(2/\bar{x}^2)\widehat{\psi}_{\mathrm{ID}}, \tag{17}$$

In this ν range, $\widehat{\varphi}$ is local and it gives rise to localized and stable $\widehat{a}_{\boldsymbol{\mu}}$. This choice also simplifies the TD algorithm since in this case both $\widehat{\psi}$ and $\widehat{\varphi}$ have closed form TD counterparts[7].

For $\omega \ll \frac{1}{2}P^{-1}\bar{\omega}$, on the other hand, the oversampling becomes too large making the representation inefficient. Consequently, this formulation is stable and efficient for a frequency band of the order of one octave $\omega_{\min} > \frac{1}{2}\omega_{\max}$, so that $\frac{1}{2}P^{-1} \leq \nu \leq P^{-1}$ and the oversampling ratio ν^{-1} is $P \div 2P$. For wider bandwidth, we utilize the multiscale formulation outlined in the following section.

5. MULTISCALE FORMULATION FOR UWB FIELDS

To overcome the difficulties noted in the previous paragraph for signal bandwidth larger than one octave, we divide the bandwidth into J bands of one octave each (Fig. 3):

$$\Omega^{(j)} = (\omega_{\max}2^{-j}, \omega_{\max}2^{-j+1}), \quad j = 1, \ldots J, \qquad J = \left[\log_2 \frac{\omega_{\max}}{\omega_{\min}}\right] + 1, \tag{18}$$

where [ ] denotes *the integer part of*, and then construct a wide-band beam summation formulation for each band as follows. For $\Omega^{(j)}$ we choose the reference frequency $\bar{\omega}^{(j)}$ to be P times the highest frequency of $\Omega^{(j)}$, i.e.,

$$\bar{\omega}^{(j)} = \bar{\omega}2^{-j+1}, \quad \text{where} \quad \bar{\omega} = P\omega_{\max}, \tag{19}$$

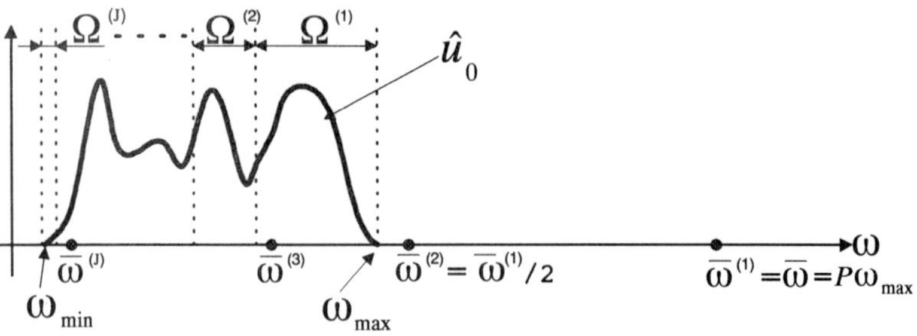

Figure 3: Frequency domain construction of the filter bands. Each band $\Omega^{(j)}$ has a one-octave bandwidth.

so that in each band we have $\frac{1}{2}P^{-1} \leq \nu \leq P^{-1}$. Thus choosing $2 \leq P \leq 3$ yields ν in the favorable range discussed in the last paragraph of Sec. 4.1. Next we choose the phase space grid $(\bar{\xi}^{(j)}, \bar{x}^{(j)})$ to be complete at $\bar{\omega}^{(j)}$, i.e.,

$$\bar{k}^{(j)}\bar{\xi}^{(j)}\bar{x}^{(j)} = 2\pi, \quad \text{where } \bar{k}^{(j)} = \bar{\omega}^{(j)}/c, \tag{20}$$

leading to

$$\bar{\xi}^{(j)}\bar{x}^{(j)} = \bar{\xi}^{(1)}\bar{x}^{(1)}2^{j-1}. \tag{21}$$

Eq. (21) implies that $\boldsymbol{X}_{\boldsymbol{\mu}}^{(j)}$ and the beam lattices for $j > 1$ can be constructed by decimation of the lattice at the highest band $j = 1$ (see Fig. 4); hence the beams need to be traced only once. Two of the preferable decimation schemes are:

a. **Interlaced decimation, first in x and then in ξ**: the lattice is decimated spatially for even j and spectrally for odd j (see Fig. 4(a)), i.e.

$$\bar{x}^{(j)} = \bar{x}^{(1)}2^{\lfloor j/2 \rfloor}, \qquad \bar{\xi}^{(j)} = \bar{\xi}^{(1)}2^{\lfloor (j-1)/2 \rfloor}, \qquad b^{(j)} = \begin{cases} b^{(1)} & j \text{ odd}, \\ 2b^{(1)} & j \text{ even}. \end{cases} \tag{22a}$$

In each band, we used a matched ID window $\widehat{\psi}_{\text{ID}}^{(j)} = \exp(-k|\boldsymbol{x}|^2/2b^{(j)})$ as in (15) with $b^{(j)}$ obtained from (16), giving $b^{(j)} = \bar{x}^{(j)}/\bar{\xi}^{(j)}$.

b. **Interlaced decimation, first in ξ and then in x**: the lattice is decimated spectrally for even j and spatially for odd j (see Fig. 4(b)), i.e.,

$$\bar{x}^{(j)} = \bar{x}^{(1)}2^{\lfloor (j-1)/2 \rfloor}, \qquad \bar{\xi}^{(j)} = \bar{\xi}^{(1)}2^{\lfloor j/2 \rfloor}, \qquad b^{(j)} = \begin{cases} b^{(1)} & j \text{ odd}, \\ \frac{1}{2}b^{(1)} & j \text{ even}, \end{cases} \tag{22b}$$

where in each band, we used, again, a matched ID window with $b^{(j)} = \bar{x}^{(j)}/\bar{\xi}^{(j)}$. The dual window $\widehat{\varphi}_{\text{ID}}^{(j)}$ corresponding to $\widehat{\psi}_{\text{ID}}^{(j)}$ may now be approximate using (17) yielding $\widehat{\varphi}_{\text{ID}}^{(j)} = (\omega/\bar{\omega}^{(j)})^3(2/(\bar{x}^{(j)})^2)\widehat{\psi}_{\text{ID}}^{(j)}$.

Finally, using, the frame parameters corresponding to the j'th frequency band, the field for $\omega \in \Omega^{(j)}$ may be expand as in (14))

$$\widehat{u}_0(\boldsymbol{x}) = \sum_{\boldsymbol{\mu}} \widehat{a}_{\boldsymbol{\mu}}^{(j)}\widehat{\psi}_{\boldsymbol{\mu}}^{(j)}(\boldsymbol{x}), \quad \widehat{u}(\boldsymbol{r}) = \sum_{\boldsymbol{\mu}} \widehat{a}_{\boldsymbol{\mu}}^{(j)}\widehat{B}_{\boldsymbol{\mu}}^{(j)}(\boldsymbol{r}), \tag{23a}$$

$$\widehat{a}_{\boldsymbol{\mu}}^{(j)} = \langle \widehat{u}_0(\boldsymbol{x}), \widehat{\varphi}_{\boldsymbol{\mu}}^{(j)}(\boldsymbol{x}) \rangle, \tag{23b}$$

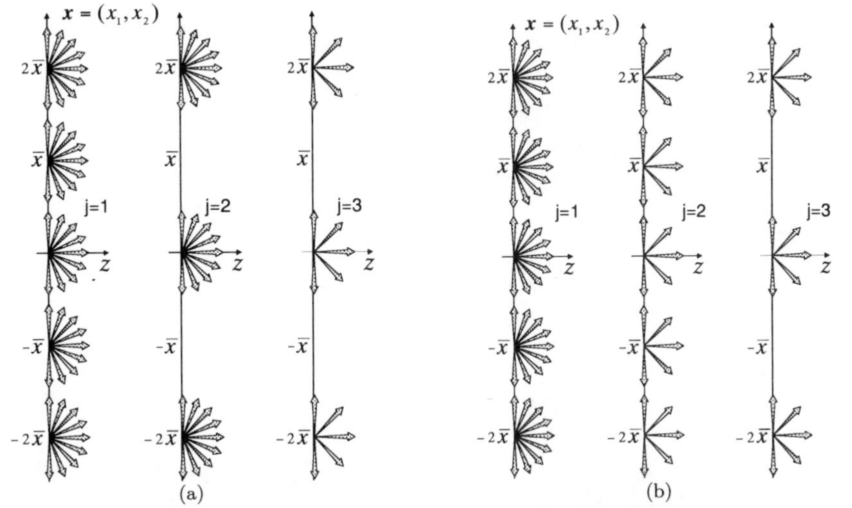

Figure 4: The beam lattices for the $j = 1, 2, 3$ bands (left, center, right axis respectively). (a) interlaced decimation, first in x and then in ξ,(b) interlaced decimation, first in ξ and then in x.

where $\{\widehat{\psi}_{\boldsymbol{\mu}}^{(j)}\}$ and $\{\widehat{\varphi}_{\boldsymbol{\mu}}^{(j)}\}$ are the ID dual frame sets and $\widehat{B}_{\boldsymbol{\mu}}^{(j)}$ is the beam propagator corresponding to $\widehat{\psi}_{\boldsymbol{\mu}}^{(j)}$ as obtained via (3)

5.1. Example

As an example we consider the expansion coefficients for a wideband focusing aperture distribution

$$\widehat{u}_0(\boldsymbol{x}) = e^{ik|\boldsymbol{x}|^2/2R}, \qquad \Omega = (\omega_{\min}, \omega_{\max}), \qquad \omega_{\max}/\omega_{\min} > 2, \tag{24}$$

within the square aperture $|x_{1,2}| \leq L/2$, and zero otherwise, where we choose $L = 400$. The parameter R is the radius of curvature of wavefront in the aperture and it is taken here to be $R = -400$, so that the field focuses to a cusp at $z = 400$, see Fig. 5(a). All units are normalized to $\bar{k} = 1$, i.e., $k_{\max} = P^{-1}$, etc., where we choose $P = 2$.

To demonstrate and contrast the decimation schemes discussed above, we calculated via (23) the expansion coefficients $\widehat{a}_{\boldsymbol{\mu}}^{(j)}$ for several representative frequencies. The resulting coefficients are depicted in Fig. 5 in the discretized (x, ξ) domain (note the different unit cell sizes in the sub-figures).

Fig. 5(b) depicts $\widehat{a}_{\boldsymbol{\mu}}^{(j)}$ at the upper frequency $\omega_{\max}$. Here $j = 1$ and, since $\bar{\omega} = 2\omega_{\max}$ we have $\nu = \frac{1}{2}$. The window parameter b in (15) has been chosen to be $b^{(1)} = |R| = 400$ to provide collimation at the focus, and the cell dimension $(\bar{x}^{(1)}, \bar{\xi}^{(1)})$ were then calculated via (16) for *snuggest* phase space coverage. Next in Figs. 5(c), 5(d) and 5(e) we show the coefficients at $\omega = \frac{1}{2}\omega_{\max}$ calculated via three different alternatives. In Fig. 5(c) this point is taken as the lowest frequency of the $j = 1$

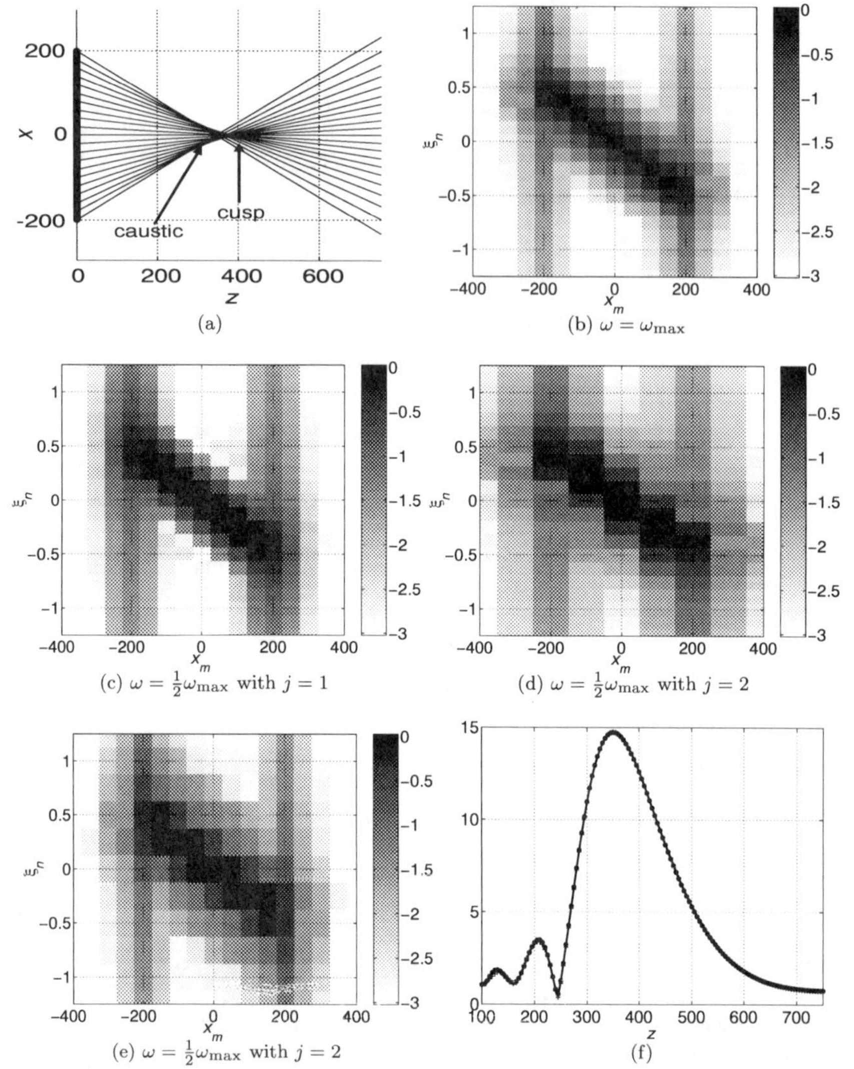

Figure 5: Phase space representation of radiation from the UWB focused aperture (24). (a) Ray trajectories. (b) The phase space coefficients $\widehat{a}_{\mu}^{(j)}$ at $\omega = \omega_{\max}$ (log scale). (c) The coefficients at $\omega = \frac{1}{2}\omega_{\max}$ obtained by regarding it as the lowest frequency in $\Omega^{(1)}$. (d) and (e) The coefficients at $\omega = \frac{1}{2}\omega_{\max}$ obtained by regarding it as the highest frequency in $\Omega^{(2)}$ and using the interlaced decimation scheme in (22a) and (22b), respectively. (f) The field along the z axis at $\omega = \frac{1}{2}\omega_{\max}$ calculated via the schemes in Figs. 5(c), 5(d), and 5(e) (full, dotted, and + + + lines, respectively).

band, and the coefficients $\widehat{a}_{\mu}^{(1)}$ were calculated via (23) with the parameters noted above. In Figs. 5(d) and 5(e), on the other hand, $\omega = \frac{1}{2}\omega_{\max}$ is taken as the highest frequency in the $j = 2$ band. In Fig. 5(d), we chose the decimated lattice $\boldsymbol{X}_{\mu}^{(2)}$

according to the scheme in (22a) (see Fig. 4(a)) giving the parameters $b^{(2)}=2b^{(1)}$, $\bar{x}^{(2)}=2\bar{x}^{(1)}$, and $\bar{\xi}^{(2)}=\bar{\xi}^{(1)}$. In Fig. 5(e), on the other hand, we used the decimation scheme (22b) (see Fig. 4(b)) giving the parameters $b^{(2)}=\frac{1}{2}b^{(1)}$, $\bar{x}^{(2)}=\bar{x}^{(1)}$, and $\bar{\xi}^{(2)}=2\bar{\xi}^{(1)}$. Finally in Fig. 5(f), we compared the radiated field along the z axis calculated for $\omega=\frac{1}{2}\omega_{\max}$ via (23) using the three expansions in Figs. 5(c), 5(d) and 5(e) (full line, dotted line and $+++$ line), with no noticeable difference between the results (even not near the focus at $z=400$).

It should be noted that in all cases the dominant (or relevant) coefficients are concentrated near the *discretized Lagrange manifold* $\boldsymbol{\xi}=\boldsymbol{x}/R$, which provides the ray skeleton of the geometrical optics (see ray trajectories in Fig. 5(a)), and near the *diffraction manifold*[3] $\boldsymbol{x}=\pm L/2$. Note also that in the decimated lattices for $j=2$ in Figs. 5(d) and 5(e) there are fewer dominant coefficients than in the $j=1$ lattice in Figs. 5(c) (due to a larger cell area in the former case fewer elements are needed to cover the phase space), leading to a more efficient representation at that frequency.

6. CONCLUSIONS

A discrete frame based beam summation scheme for ultra-wideband radiation from apertures was introduce. The formulation utilizes a frequency independent beam lattice (unlike the conventional Gabor scheme), emerging from a discrete set of points and orientations in the aperture. The propagators are ID-GB that guarantee snuggest frames for all ω. The expansion coefficients are windowed Fourier transforms of the source distribution sampled at the frequency independent grid points.

The key feature in this novel scheme is the observation that a frequency independent beam lattice is obtained if the oversampling parameter ν is proportional to the frequency and is given by $\nu=\omega/\bar{\omega}$ where $\bar{\omega}$ is a properly chosen reference frequency (see (12)). The expansion bandwidth posses a tradeoff between analytical simplicity and numerical efficiency. For $\nu\leq\frac{1}{2}$ the dual function which is used to calculate the expansion coefficients (the "analysis" function) is known analytically to be approximately a Gaussian (see (17)), thereby providing a local and stable expansion. For $\nu\leq\frac{1}{4}$, on the other hand, the oversampling is too large (greater than 4) and the representation becomes inefficient. This limits the bandwidth to be of the order of one octave so that $\nu=\frac{1}{2}$ at $\omega_{\max}$ and $\nu=\frac{1}{4}$ at $\omega_{\min}$.

For sources with bandwidth greater than one octave, efficient representation is obtained using an hierarchy of one-octave bands, such that the beam lattices at the lower frequency bands are obtained by decimation of the lattice at the highest band. Consequently, only the beam set at the highest frequency band needs to be traced.

Acknowledgement

E. Heyman acknowledges partial support by the Israel Science Foundation (ISF), Grant No. 216/02, and by the AFOSR, Grant No. F49620-01-C-0018. A. Boag acknowledges partial support by the ISF, Grant No. 577/00.

REFERENCES

1. A. Shlivinski, E. Heyman, A. Boag, D. Lugara and C. Letrou, "Gabor-frame phase space beam summation formulation for wideband radiation from extended apertures," in *Proc. of URSI International Symposium on Electromagnetic Theory*, Victoria, Canada, 2001, pp. 56–58.
2. D. Lugara, C. Letrou, A. Shlivinski, E. Heyman and Amir Boag, "The frame based Gaussian beam summation method: theory and application," *Radio Sci.*, 2002 (accepted).
3. A. Shlivinski, E. Heyman, A. Boag, D. Lugara and C. Letrou, "A frame-based phase-space beam summation formulation for wideband radiation," *IEEE Trans. Antennas Propagat.*, 2001 (submitted).
4. D. Lugara and C. Letrou, "Alternative to Gabor's representation of plane aperture radiation," *Electron. Lett.* **34**, 2286–2287, 1998.
5. E. Heyman and L.B. Felsen, "Gaussian beam and pulsed beam dynamics: Complex source and spectrum formulations within and beyond paraxial asymptotics," *J. Opt. Soc. Am. A*, **18**, 1588–1611, 2001.
6. A. Shlivinski, E. Heyman, A. Boag, D. Lugara and C. Letrou, "A discretized-phase-space pulsed beam representation for time dependent radiation," in *Proc. of URSI International Symposium on Electromagnetic Theory*, Victoria, Canada, 2001, pp. 71–73.
7. A. Shlivinski, E. Heyman, A. Boag and C. Letrou, "Frame-based beam-summation algorithms for ultra wideband radiation from extended apertures. Part II: Time domain formulations," *this issue*.
8. M. Bastiaans, "The expansion of an optical signal into a discrete set of Gaussian beams," *Optik*, **57**, 95–102, 1980.
9. P. Einziger, S. Raz, and M. Shapira, "Gabor representation and aperture theory," *J. Opt. Soc. Am. A*, **3**, 508–522, 1986.
10. B. Steinberg and E. Heyman, "Phase space beam summation for time dependent radiation from large apertures: Discretized parameterization," *J. Opt. Soc. Am. A*, **8**, 959–966, 1991.
11. I. Daubechies, *Ten Lectures on Wavelets*, vol. 61 of *CBMS-NSF Regional Conference Series in Applied Mathematics*. Philadelphia: SIAM Press, 1992.

FRAME-BASED BEAM-SUMMATION ALGORITHMS FOR ULTRA WIDEBAND RADIATION FROM EXTENDED APERTURES. PART II: TIME DOMAIN FORMULATION

Amir Shlivinski, Ehud Heyman, Amir Boag, and Christine Letrou*

1. INTRODUCTION

The wideband frame-based beam summation algorithms introduced in the first part of this two parts sequence[1] (see also Ref. 2) have several attractive features that make them amenable for an extension into the Time Domain (TD): (*a*) The same beam lattice is used for all frequencies; (*b*) The iso-diffracting Gaussian beam (ID-GB) basis provides the "snuggest" frame representation for all frequencies; (*c*) The propagation parameters of the resulting beams are frequency-independent and need to be calculated only once at a reference frequency. These properties were utilized in Ref. 3 to derive new TD representations for radiation from extended apertures, wherein the field is expanded in a discrete phase space lattice of shifted and tilted iso-diffracting pulsed beam propagators (ID-PB). The excitation coefficients of these propagators are extracted from the aperture source distribution by using the ID-PBs as "analysis windows" in a "discrete local slant stack transform" (DL-SST). From a broader perspective, this transform involves a new class of Radon-frames and can be regarded as a "discrete local Radon transform" in $\mathbb{R}^3$ that may be relevant in other disciplines, such as image processing.

This new TD formulation is further extended here, emphasizing an effective treatment of the time dependence, which may comprise many scales of frequencies

*Amir Shlivinski, Ehud Heyman, and Amir Boag are with the Faculty of Engineering, Dept. of Physical Electronics, Tel–Aviv University, Tel–Aviv, 69978 Israel. Christine Letrou is with the I.N.T., 9 rue Charles Fourier, F-91011 EVRY Cedex, France.

Ultra-Wideband Short-Pulse Electromagnetics 6
Edited by Mokole *et al.*, Kluwer Academic/Plenum Publishers, 2003

and time. This is achieved by utilizing a *multi-scale* (filter-bank) hierarchy of space-time Radon frames within the setting of the DL-SST, wherein each scale accounts for a frequency band of the order of one octave. The small scale windows sample a spatio-temporal source distribution at a dense phase-space grid, while the grids corresponding to the large scale windows are decimated sets of this basic grid. They give rise to ID-PB propagators with narrow space-time support that emerge from the aperture in a dense phase space lattice of beam-axes, while lattice associated with the large scale windows and propagators are decimated sets of this basic lattice. Consequently, only the basic set of ID-PB needs to be calculated, while for the larger scales, one uses properly decimiation subsets. Closed form expressions for the "analysis" and "synthesis" Radon frames in terms of the standard ID-PB are provided together with guidelines for choosing all the expansion parameters.

2. PLANE WAVE SPECTRUM AND LOCAL SPECTRUM IN THE TIME DOMAIN: A REVIEW

Following Ref. 1, the formulation is presented in the context of radiation into the half space $z>0$ from to an extended aperture in the $z=0$ plane with a prescribed time-dependent scalar field $u_0(\boldsymbol{x},t)$. The 3D coordinate notation is $\boldsymbol{r}=(\boldsymbol{x},z)$ with $\boldsymbol{x}=(x_1,x_2)$ being the coordinates transverse to z. The signal is constrained to the band $\Omega=(\omega_{\min},\omega_{\max})$ that may extend over several octaves.

One way to express the field is by a superposition of TD plane-waves as formalized by the *slant stack transform* (SST)

$$\widetilde{u}_0(\boldsymbol{\xi},\tau)=\int d^2x\, u_0[\boldsymbol{x},\tau+c^{-1}\boldsymbol{\xi}\cdot\boldsymbol{x}], \tag{1a}$$

$$u(\boldsymbol{r},t)=-(2\pi c)^{-2}\int d^2\xi\, \partial_t^2\widetilde{u}_0[\boldsymbol{\xi},t-c^{-1}(\boldsymbol{\xi}\cdot\boldsymbol{x}+\zeta z)]. \tag{1b}$$

where $\widetilde{}$ denotes spatial spectrum constituents. Eq. (1a) is a Radon transform of u_0 in the 3D $(\boldsymbol{x},t)$ data space, consisting of projections along planes of linear delay $t-c^{-1}\boldsymbol{\xi}\cdot\boldsymbol{x}=\tau=$ const. (Fig. 1(a)). The transform extracts from $u_0(\boldsymbol{x},t)$ the TD plane wave $\widetilde{u}_0(\boldsymbol{\xi},\tau)$ that propagates in the direction (θ,ϕ) defined via $\boldsymbol{\xi}=\sin\theta(\cos\phi,\sin\phi)$. The projections, whose normal lies within a $\frac{\pi}{4}$-cone about the ct-axis, constitute the propagating spectrum with $|\boldsymbol{\xi}|<1$, while the evanescent spectrum consists of those projections that make larger angles. For $z=0$, Eq. (1b) is the inverse SST: It also represents propagation to $z>0$ by adding the spectral delay ζz, where $\zeta=\sqrt{1-\boldsymbol{\xi}\cdot\boldsymbol{\xi}}$, thus expressing the field as an angular superposition of TD plane waves $\widetilde{u}_0$ (Fig. 1(b)). Note that in order to accommodate the complex

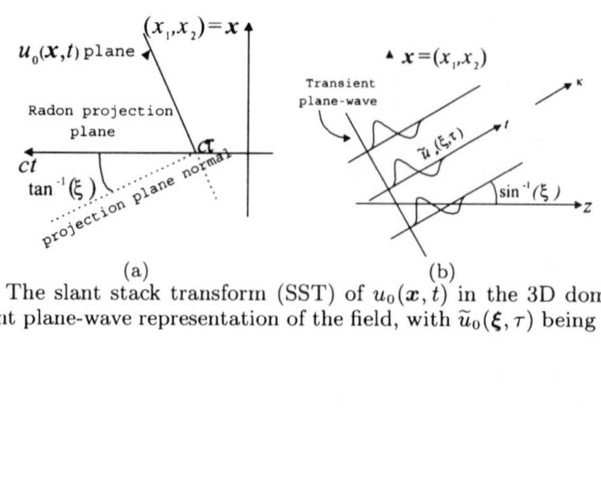

Figure 1: (a) The slant stack transform (SST) of $u_0(\boldsymbol{x},t)$ in the 3D domain $(\boldsymbol{x},t)$. (b) Time-dependent plane-wave representation of the field, with $\widetilde{u}_0(\boldsymbol{\xi},\tau)$ being the plane-wave waveform.

time-delay in the evanescent spectrum where $\zeta = i|\zeta|$, the expression for $z>0$ should be understood in an analytic signal sense.[4,5]

The *local time-dependent spectrum* of the field is obtained by windowing the SST in Eq. (1a). Given a space-time window function $\psi(\boldsymbol{x},t)$ localized around $(\boldsymbol{x},t)=(\mathbf{0},0)$ and its dual $\varphi(\boldsymbol{x},t)$ (in the continuous formulation considered in this section $\varphi=\psi$ as in Ref. 1; other conditions will be discussed later), we construct a continuous set of shifted and rotated windows

$$\psi_{\boldsymbol{Y}}(\boldsymbol{x},t)=\psi[\boldsymbol{x}-\boldsymbol{x}_0, t-t_0-c^{-1}\boldsymbol{\xi}_0\cdot(\boldsymbol{x}-\boldsymbol{x}_0)] \quad \text{and} \quad \varphi_{\boldsymbol{Y}}=\varphi[\boldsymbol{x}-\boldsymbol{x}_0, t-t_0-c^{-1}\boldsymbol{\xi}_0\cdot(\boldsymbol{x}-\boldsymbol{x}_0)] \quad (2)$$

which are localized about $(\boldsymbol{x},t)=(\boldsymbol{x}_0,t_0)$ with spectral tilt $\boldsymbol{\xi}_0$ (Fig. 2(a)). For brevity we denote the 5-dimensional phase-space as $\boldsymbol{Y}\equiv(\boldsymbol{X},t_0)=(\boldsymbol{x}_0,\boldsymbol{\xi}_0,t_0)$. The local SST pair is therefore given by[6]

$$U_0(\boldsymbol{Y})=\int d^2x\int dt\, u_0(\boldsymbol{x},t)\,\varphi_{\boldsymbol{Y}}(\boldsymbol{x},t)=\langle u_0(\boldsymbol{x},t),\varphi_{\boldsymbol{Y}}(\boldsymbol{x},t)\rangle, \quad (3a)$$

$$u_0(\boldsymbol{x},t)=-(2\pi c)^{-2}\int d^5Y\, U_0(\boldsymbol{Y})\, N(t)\overset{t}{\otimes}\psi_{\boldsymbol{Y}}(\boldsymbol{x},t), \quad (3b)$$

where $N(t)=\frac{1}{2\pi}\int d\omega\,(-i\omega)^2\|\widehat{\psi}(\boldsymbol{x};\omega)\|^{-2}e^{-i\omega t}$ and $\overset{t}{\otimes}$ is a temporal convolution. This transform can be derived by inverting Eq. (2) of Ref. 1 into the TD, with ψ, φ, $\psi_{\boldsymbol{Y}}$ and $\varphi_{\boldsymbol{Y}}$ being the TD counterparts of the FD windows $\widehat{\psi}$, $\widehat{\varphi}$, $\widehat{\psi}_{\boldsymbol{X}}$ and $\widehat{\varphi}_{\boldsymbol{X}}$, respectively.

$U_0(\boldsymbol{Y})$ in (3a) is the projection of $u_0(\boldsymbol{x},t)$ onto the Radon-type window $\varphi_{\boldsymbol{Y}}(\boldsymbol{x},t)$. In view of the space-time and shifts and the spectral tilt in $\varphi_{\boldsymbol{Y}}$ (see (2)), $U_0(\boldsymbol{Y})$ is recognized as a *Local (windowed) Slant Stack Transform* (L-SST) that extracts from $u_0(\boldsymbol{x},t)$ the local directional information near $(\boldsymbol{x}_0,t_0)$.

The phase-space representation (3b) can be propagated into $z>0$ by propagating the phase space windows $\psi_{\boldsymbol{Y}}$ via (1b), giving[6,5]

$$u(\boldsymbol{r},t)=-(2\pi c)^{-2}\int d^5Y\, U_0(\boldsymbol{Y})N(t)\overset{t}{\otimes}B_{\boldsymbol{Y}}(\boldsymbol{r},t), \quad (4a)$$

where $B_{\boldsymbol{Y}}(\boldsymbol{r},t)$ are the propagators corresponding to $\psi_{\boldsymbol{Y}}(\boldsymbol{x};t)$ and are given by

$$B_{\boldsymbol{Y}}(\boldsymbol{r},t)=-(2\pi c)^{-2}\int d^2\xi\,\partial_t^2\widetilde{\psi}_{\boldsymbol{Y}}[\boldsymbol{\xi},t-c^{-1}(\boldsymbol{\xi}\cdot\boldsymbol{x}+\zeta z)], \quad (4b)$$

with $\widetilde{\psi}_{\boldsymbol{Y}}(\boldsymbol{\xi},\tau)=\widetilde{\psi}(\boldsymbol{\xi}-\boldsymbol{\xi}_0,\tau-t_0+c^{-1}\boldsymbol{\xi}\cdot\boldsymbol{x}_0)$ being the SST (1a) of $\psi_{\boldsymbol{Y}}(\boldsymbol{x},t)$; it is expressed here in term of $\widetilde{\psi}(\boldsymbol{\xi},\tau)$ which is the STT of the "mother window" $\psi(\boldsymbol{x},t)$. Noting that $\psi_{\boldsymbol{Y}}$ is localized about $(\boldsymbol{x},\boldsymbol{\xi},t)=(\boldsymbol{x}_0,\boldsymbol{\xi}_0,t_0)$ it follows that $B_{\boldsymbol{Y}}$ is a localized space-time wavepacket (Pulsed Beam, PB) that emerges from the $z=0$

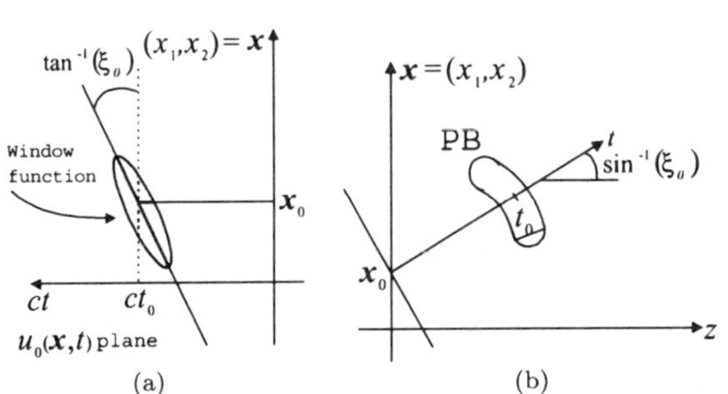

Figure 2: (a) The local slant-stack transform (L-SST) (3a). (b) Field reconstruction via phase-space PB summation (4b).

plane at $(\boldsymbol{x},t)=(\boldsymbol{x}_0,t_0)$ in the direction (θ_0,ϕ_0) defined via $\boldsymbol{\xi}_0=\sin\theta_0(\cos\phi_0,\sin\phi_0)$ (Fig. 2(b)). Further properties of the PB propagators $B_{\boldsymbol{Y}}$ will be discussed in Sec. 3.1 in connection with the frame formulation. Explicit space-time expressions for $B_{\boldsymbol{Y}}(\boldsymbol{r},t)$ are given in Ref. 5 and Ref. 7.

3. TIME DOMAIN RADON-FRAME FORMULATION

Our starting point is the frequency-independent phase-space grid $\boldsymbol{X}_{\boldsymbol{\mu}}$ introduced in Eq. (11) of Ref. 1, whose unit cell dimensions $(\bar{x},\bar{\xi})$ satisfy the completeness condition $\bar{k}\bar{\xi}\bar{x}=2\pi$ at the reference frequency $\bar{\omega}$, which is chosen to be $\bar{\omega}>\omega_{\max}$; typically $\bar{\omega}=P\omega_{\max}$ where $2\leq P\leq 3$ as was discussed in Eq. (17) of Ref. 1. This grid provides an overcomplete phase-space coverage for all $\omega<\omega_{\max}$. We then define the 5D discretized-space / continuous-time phase space coordinates

$$\boldsymbol{Y}_{\boldsymbol{\mu},\tau}=(\boldsymbol{x}_{\boldsymbol{m}},\boldsymbol{\xi}_{\boldsymbol{n}},\tau)=(\boldsymbol{m}\bar{x},\boldsymbol{n}\bar{\xi},\tau)\quad\text{with}\quad\boldsymbol{\mu}=(\boldsymbol{m},\boldsymbol{n})=(m_1,m_2,n_1,n_2,). \tag{5}$$

Next, we introduce the synthesis and analysis window functions $\psi(\boldsymbol{x},t)$ and $\varphi(\boldsymbol{x},t)$ to be the TD counterparts of $\widehat{\psi}(\boldsymbol{x},\omega)$ and $\widehat{\varphi}(\boldsymbol{x},\omega)$ in Eq. (5) and (8) of Ref. 1 and define the Radon-type frame and its dual as the TD counterparts of the Gabor frames in Eq, (13) of Ref. 1

$$\psi_{\boldsymbol{\mu},\tau}(\boldsymbol{x},t)=\psi[\boldsymbol{x}-\boldsymbol{x}_{\boldsymbol{m}},t-\tau-c^{-1}\boldsymbol{\xi}_{\boldsymbol{n}}\cdot(\boldsymbol{x}-\boldsymbol{x}_{\boldsymbol{m}})], \tag{6a}$$

$$\varphi_{\boldsymbol{\mu},\tau}(\boldsymbol{x},t)=\varphi[\boldsymbol{x}-\boldsymbol{x}_{\boldsymbol{m}},t-\tau-c^{-1}\boldsymbol{\xi}_{\boldsymbol{n}}\cdot(\boldsymbol{x}-\boldsymbol{x}_{\boldsymbol{m}})]. \tag{6b}$$

One may readily verify that these function sets indeed form frames in $L_2(\boldsymbol{x},t)$. Examples for these window functions will be given in Sec. 3.1.

The TD frame expansion of the field $u_0(\boldsymbol{x},t)$, the TD counterparts of Eq. (14) of Ref. 1, is given by

$$u_0(\boldsymbol{x},t)=\sum_{\boldsymbol{\mu}}\int d\tau\, a_{\boldsymbol{\mu}}(\tau)\psi_{\boldsymbol{\mu},\tau}(\boldsymbol{x},t), \tag{7a}$$

$$a_{\boldsymbol{\mu}}(\tau)=\langle u_0(\boldsymbol{x},t),\varphi_{\boldsymbol{\mu},\tau}(\boldsymbol{x},t)\rangle,\qquad\text{where }\langle\,,\,\rangle\equiv\int d^2x\int dt. \tag{7b}$$

The frame expansion coefficients $a_{\boldsymbol{\mu}}(\tau)$ in (7b), the TD counterparts of $\widehat{a}_{\boldsymbol{\mu}}(\omega)$ in Eq. (14b) of Ref. 1, are calculated here *directly* from the TD data by projecting $u_0(\boldsymbol{x},t)$ onto $\varphi_{\boldsymbol{\mu},\tau}(\boldsymbol{x},t)$. From (3a) they are readily identified as the L-SST of $u_0(\boldsymbol{x},t)$ sampled at the discrete points $\boldsymbol{x}_{\boldsymbol{m}}$ with spectral tilts $\boldsymbol{\xi}_{\boldsymbol{n}}$ (see Fig. 2(a)). The field reconstruction formula (7a) involves a discrete summation over all locations $\boldsymbol{x}_{\boldsymbol{m}}$ and spectral tilts $\boldsymbol{\xi}_{\boldsymbol{n}}$, and an integration over the window center τ, which in view of (6a) is recognized as a convolution.

The field reconstruction formula (7a) may readily be propagated away from the $z=0$ plane, giving the following expression for the propagating field

$$u(\boldsymbol{r},t)=\sum_{\boldsymbol{\mu}}\int d\tau\, a_{\boldsymbol{\mu}}(\tau)B_{\boldsymbol{\mu},\tau}(\boldsymbol{r},t) \tag{8a}$$

where the PB propagators $B_{\boldsymbol{\mu},\tau}(\boldsymbol{r},t)$ in (8a) are the fields radiated by the fields $\psi_{\boldsymbol{\mu},\tau}(\boldsymbol{x},t)$ in the $z=0$ plane, and the summation involves only the propagating PBs with $|\boldsymbol{\xi}_{\boldsymbol{n}}|<1$ (see Fig. 2(b)). $B_{\boldsymbol{\mu},\tau}$ are given by (4b) with $\boldsymbol{Y}\to(\boldsymbol{x}_{\boldsymbol{m}},\boldsymbol{\xi}_{\boldsymbol{n}},\tau)$, which is rewritten here as

$$B_{\boldsymbol{\mu},\tau}(\boldsymbol{r},t)=-(2\pi c)^{-2}\int d^2\xi\,\partial_t^2\widetilde{\psi}[\boldsymbol{\xi}-\boldsymbol{\xi}_{\boldsymbol{n}},\,t-\tau-c^{-1}(\boldsymbol{\xi}\cdot(4x-\boldsymbol{x}_{\boldsymbol{m}})+\zeta z)], \tag{8b}$$

where $\widetilde{\psi}(\boldsymbol{\xi},\tau)$ is the SST (1a) of $\psi(\boldsymbol{x},t)$. For isodiffracting (ID) windows (Sec. 3.1), one obtains explicit expressions for these PBs that are similar to those derived in Ref. 5 and Ref. 7.

3.1. Special Case: Isodiffracting Pulsed Beam (ID-PB) Windows

The ID-PB windows are the TD counterparts of the ID-GB windows used in Sec. 4.1 of Ref. 1. They have the following favorable properties: (*a*) they provide the *snuggest frames* for all the frequencies in the excitation band; (*b*) they yield ID-PB propagators that can be tracked analytically in inhomogeneous media or through interaction with boundaries. They are also closely related to the so called "complex source PB." A a recent review of this class of functions can be found in Ref. 7.

The ID window and its dual have been defined in Eq. (15) of Ref. 1 as

$$\widehat{\psi}_{\rm ID}(\boldsymbol{x},\omega)=\widehat{f}(\omega)e^{-k|\boldsymbol{x}|^2/2b},\quad \widehat{\varphi}_{\rm ID}(\boldsymbol{x},\omega)\approx\widehat{g}(\omega)(\omega/\bar{\omega})^3(2/\bar{x}^2)\widehat{\psi}_{\rm ID}(\boldsymbol{x},\omega),\quad b=\bar{x}/\bar{\xi}, \tag{9}$$

where in the approximation for $\widehat{\varphi}_{\rm ID}$ applies for[1] $\omega \lesssim \frac{1}{2}\bar{\omega}$, while the condition on b guarantees snugness for all ω. Note that in order to transform the windows of Ref. 1 to the TD they have been multiplied by the "filters" functions $\widehat{f}(\omega)$ and $\widehat{g}(\omega)$. These filters should satisfy the condition

$$\widehat{f}(\omega)\widehat{g}^*(\omega) = 1 \qquad \text{for } \omega \in \Omega = [\omega_{\min}, \omega_{\max}] \tag{10}$$

that guarantees a faithful reconstruction within the frequency band Ω of the excitation. Outside Ω, these functions may be chosen arbitrarily, hence they will be chosen so they decay there smoothly in order to yield smooth and localized spatio-temporal windows. Other conditions on $\widehat{f}$ and $\widehat{g}$ will be discussed in Sec. 4.1.

The inversion to the TD is greatly simplified by utilizing the analytic (one sided) inverse Fourier transform

$$\overset{+}{f}(t) = \frac{1}{\pi}\int_0^\infty d\omega\, \widehat{f}(\omega)e^{-i\omega t}, \quad \operatorname{Im} t \leq 0, \tag{11a}$$

$$f(t) = \operatorname{Re} \overset{+}{f}(t) \qquad \text{for } t \text{ real}. \tag{11b}$$

Clearly, the integral in (11a) defines the analytic function in $\mathbb{C}^-$, the lower half of the complex t-plane. This function is the so called analytic signal $\overset{+}{f}(t)$ associated with the frequency spectrum $\widehat{f}(\omega)$. The limit of this function on the real t axis then defines the real signal via (11b) in terms of the limit of $\overset{+}{f}(t)$ there (other properties of the analytic signal representation can be found, for example, in Ref. 4–Ref. 7).

Transforming (9) to the TD via (11), the *real* spatio-temporal windows are found to be

$$\psi_{\rm ID}(\boldsymbol{x};t)=\operatorname{Re}\{\overset{+}{f}(t-i|\boldsymbol{x}|^2/2bc)\},\quad \varphi_{\rm ID}(\boldsymbol{x};t)\approx(2/\bar{\omega}^3\bar{x}^2)\operatorname{Im}\{\overset{+}{g}{}'''(t-i|\boldsymbol{x}|^2/2bc)\}, \tag{12}$$

where $\overset{+}{f}$ and $\overset{+}{g}$ are the analytic signals corresponding to frequency filters $\widehat{f}$ and $\widehat{g}$. As discussed above, the windows in (12) and the associated propagators are termed ID-PB. The temporal windowing in (12) is due to the pulse shape of $\overset{+}{f}(t)$ and $\overset{+}{g}(t)$, while the $\boldsymbol{x}$-windowing is due to the negative imaginary part of the time-argument in the analytic signals that increases quadratically with $|\boldsymbol{x}|$ causing a decay of the signal. As discussed after (10), the functions $\overset{+}{f}(t)$ and $\overset{+}{g}(t)$ should be chosen such that these windows will converge well for numerical processing of the L-SST in (7).

Specific examples for the filters $\widehat{f}$ and $\widehat{g}$ and for the corresponding windows $\psi_{\rm ID}$ and $\varphi_{\rm ID}$ will be given in Fig. 6, below, where further conditions on these functions will be defined. We therefore defer any further discussion of the properties of (12) to Sec. 4.1.

Finally, using the ID windows ψ_{ID} to construct the Radon-frame $\psi_{\boldsymbol{\mu},\tau}$ (6b) leads to explicit space-time expressions for the propagators[5,7] $B_{\boldsymbol{\gamma}}(\boldsymbol{r},t)$.

4. A MULTISCALE RADON FRAME SCHEME

As discussed in Eq. (17) of Ref. 1, the formulation in Sec. 3 is localized and computationally efficient if the excitation bandwidth Ω is smaller than one octave: In this case $\frac{1}{2}P^{-1} \leq \nu \leq P^{-1}$ so that on one hand ν is sufficiently small and $\widehat{\varphi}$ can be approximated by $\widehat{\psi}$ as in (9), while on the other hand, the oversampling is not too large. For wider bandwidth we introduced in Sec. 5 of Ref. 1 a scheme wherein the frequency band Ω of the signal is divided into a hierarchy of one-octave sub-bands $\Omega^{(j)}$.

In this section this scheme is extended to the TD Radon frame formulation. We note, however, that the sharp division of Ω into distinct $\Omega^{(j)}$ as in Eq. 18 of Ref. 1 yields poorly converging sinc-type windows in the TD. To overcome this difficulty, we separate the signal spectrum to the sub-bands by using two filter-bank sets

$$\widehat{g}^{(j)}(\omega) = \widehat{g}(2^{j-1}\omega), \quad \widehat{f}^{(j)}(\omega) = \widehat{f}(2^{j-1}\omega), \quad j = 1 \dots J, \tag{13}$$

where $\widehat{g}(\omega)$ and $\widehat{f}(\omega)$ are the "mother filters" at the $j = 1$ band. In (17) and (20), they will be associated with the analysis and synthesis windows respectively.

The data $\widehat{u}_0(\boldsymbol{x},\omega)$ can now be separated into sub-bands via (see Fig. 3)

$$\widehat{u}_0(\boldsymbol{x};\omega) = \sum_{j=1}^{J} [\widehat{f}^{(j)}(\omega)\widehat{g}^{(j)*}(\omega)]\, \widehat{u}_0(\boldsymbol{x};\omega), \tag{14}$$

The filters $\widehat{f}$ and $\widehat{g}$ are chosen such that $[\widehat{f}^{(j)}\widehat{g}^{(j)*}]$ covers $\Omega^{(j)}$ but its support is wider and therefore overlap with neighboring bands as schematized in Fig. 3. These overlapping filters should provide a uniform coverage of Ω, i.e.,

$$\sum_{j=1}^{J} \widehat{f}^{(j)}(\omega)\widehat{g}^{(j)*}(\omega) = 1, \qquad \forall\, \omega \in \Omega. \tag{15}$$

In order to obtain well behaved TD windows, it is also required that $\widehat{g}$ and $\widehat{f}$ will be tapered smoothly and that $\overset{+}{g}(t)$ and $\overset{+}{f}(t)$ will be known analytically. There are many possibilities to construct such filters; a simple example is given in Fig. 5 and Eq. (23).

Next, in each band $j = 1, \dots J$, we apply the Gabor-frame formulation of Sec. 3 as in Sec. 5 of Ref. 1: The reference frequencies $\bar{\omega}^{(j)}$ and the phase space grids $(\bar{\xi}^{(j)}, \bar{x}^{(j)})$ are given by (see Eqs. (19), (21) of Ref. 1)

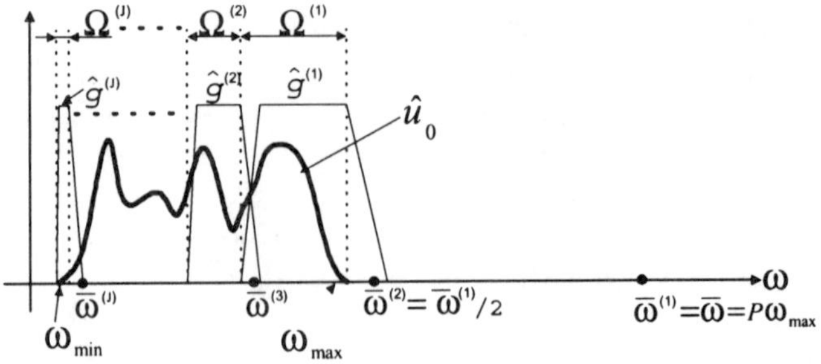

Figure 3: Frequency domain construction of the filter bands.

$$\bar{\omega}^{(j)} = \bar{\omega}2^{-j+1}, \qquad \text{with} \quad \bar{\omega} = P\omega_{\max}, \tag{16a}$$
$$\bar{\xi}^{(j)}\bar{x}^{(j)} = \bar{\xi}^{(1)}\bar{x}^{(1)}2^{j-1}, \qquad \text{with} \quad \bar{k}^{(1)}\bar{\xi}^{(1)}\bar{x}^{(1)} = 2\pi, \tag{16b}$$

where (16b) represents a decimation of the $j=1$ grid (see Eqs. (22) and Fig. 4 of Ref. 1).

For each j we then construct a dual frame sets $\{\widehat{\psi}^{(j)}_{\boldsymbol{\mu}}, \widehat{\varphi}^{(j)}_{\boldsymbol{\mu}}\}$ and, utilizing (14), we expand the FD field as

$$\widehat{u}_0(\boldsymbol{x};\omega) = \sum_j \sum_{\boldsymbol{\mu}} \widehat{a}^{(j)}_{\boldsymbol{\mu}}(\omega)\widehat{f}^{(j)}(\omega)\widehat{\psi}^{(j)}_{\boldsymbol{\mu}}(\boldsymbol{x};\omega), \tag{17a}$$
$$\widehat{a}^{(j)}_{\boldsymbol{\mu}}(\omega) = \langle \widehat{u}_0(\boldsymbol{x};\omega), \widehat{g}^{(j)}(\omega)\widehat{\varphi}^{(j)}_{\boldsymbol{\mu}}(\boldsymbol{x};\omega)\rangle. \tag{17b}$$

Transforming to the TD yields for each j a formulation that is similar to (7), hence

$$u_0(\boldsymbol{x},t) = \sum_j \sum_{\boldsymbol{\mu}} \int d\tau\, a^{(j)}_{\boldsymbol{\mu}}(\tau)\psi^{(j)}_{\boldsymbol{\mu},\tau}(\boldsymbol{x},t), \tag{18a}$$
$$a^{(j)}_{\boldsymbol{\mu}}(\tau) = \langle u_0(\boldsymbol{x},t), \varphi^{(j)}_{\boldsymbol{\mu},\tau}(\boldsymbol{x},t)\rangle, \tag{18b}$$

where the Radon-frame functions are now given by (cf. (6))

$$\psi^{(j)}_{\boldsymbol{\mu},\tau}(\boldsymbol{x},t) = \psi^{(j)}[\boldsymbol{x}-\boldsymbol{x}^{(j)}_{\boldsymbol{m}}, t-\tau-c^{-1}\boldsymbol{\xi}^{(j)}_{\boldsymbol{n}}\cdot(\boldsymbol{x}-\boldsymbol{x}^{(j)}_{\boldsymbol{m}})], \tag{19a}$$
$$\varphi^{(j)}_{\boldsymbol{\mu},\tau}(\boldsymbol{x},t) = \varphi^{(j)}[\boldsymbol{x}-\boldsymbol{x}^{(j)}_{\boldsymbol{m}}, t-\tau-c^{-1}\boldsymbol{\xi}^{(j)}_{\boldsymbol{n}}\cdot(\boldsymbol{x}-\boldsymbol{x}^{(j)}_{\boldsymbol{m}})], \tag{19b}$$

with the mother window functions

$$\psi^{(j)}(\boldsymbol{x},t) = \mathcal{F}^{-1}\{\widehat{f}(2^{j-1}\omega)\widehat{\psi}^{(j)}(\boldsymbol{x};\omega)\}, \qquad \varphi^{(j)}(\boldsymbol{x},t) = \mathcal{F}^{-1}\{\widehat{g}(2^{j-1}\omega)\widehat{\varphi}^{(j)}(\boldsymbol{x};\omega)\}, \tag{20}$$

where $\mathcal{F}^{-1}$ is the inverse Fourier transform to the TD.

Finally, the radiated field $u(\boldsymbol{r};t)$ is given by (18a) with the synthesis windows $\psi^{(j)}_{\boldsymbol{\mu},\tau}(\boldsymbol{x},t)$ replaced by the corresponding PB propagators $B^{(j)}_{\boldsymbol{\mu},\tau}(\boldsymbol{r},t)$ (see (8b) and Fig. 4(b)).

Eq. (18a) expresses the data $u_0(\boldsymbol{x},t)$ using a set of Radon frames tagged by the index j. Since the phase space grid $(\bar{x},\bar{\xi})$ at higher j is formed by decimation of the grids at the lower j, it follows that in order to cover the phase space without "gaps" the frame functions $\psi^{(j)}$ and $\varphi^{(j)}$ at the higher j are wider (both temporarily and spatially) than those at the lower j (see Fig. 4(a) and also explicit expressions in (22)). Thus, the sets $\{\psi^{(j)}_{\boldsymbol{\mu},\tau}, \varphi^{(j)}_{\boldsymbol{\mu},\tau}\}_1^J$ are recognized as a *multiscale* hierarchy of space-time Radon frames, and the coefficients $a^{(j)}_{\boldsymbol{\mu}}(\tau)$ in (18) are obtained as a L-SST projections using this multiscale Radon frames.

4.1. Special Case: Multiscale ID-PB Windows

As discussed in Sec. 3.1, the ID-PB are the preferable space-time windows, hence we shall derive closed form expressions for their utilization in the multiscale expansion. These expressions also demonstrate the general properties of this expansion discussed following (20).

The j-band ID window has the form[1]

$$\widehat{\psi}^{(j)}_{\rm ID}(\boldsymbol{x},\omega) = \widehat{f}(2^{j-1}\omega)e^{-k|\boldsymbol{x}|^2/2b^{(j)}}, \qquad \widehat{\varphi}^{(j)}_{\rm ID}(\boldsymbol{x},\omega) \approx \widehat{g}(2^{j-1}\omega)(\omega/\bar{\omega}^{(j)})^3(2/(\bar{x}^{(j)})^2)\widehat{\psi}^{(j)}_{\rm ID}, \tag{21}$$

where $(\bar{x}^{(j)},\bar{\xi}^{(j)})$ are defined in (15) and $b^{(j)} = \bar{x}^j/\bar{\xi}^{(j)}$ is taken to provide matched windows.

Using these windows in (20), we obtain (cf. (12)

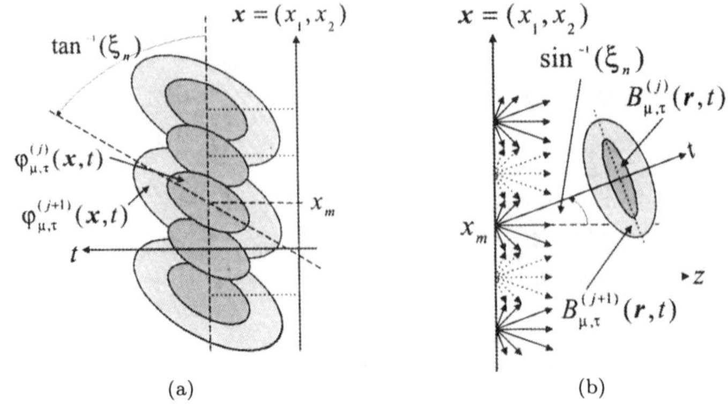

Figure 4: The multiscale discretized local slant stack transform (L-SST) (a) Schematization of the transform for two scale level j and $j+1$: The phase space lattice (locations/orientations/times) at the $j+1$ level is a decimation of the j-level lattice and the space-time windows of the former ($\varphi_{\mu,\tau}^{(j+1)}$; light grey) are wider than those of the latter (dark grey). (b) Field reconstruction. All the arrows are used at the j level, but only the full-line arrows are used at the $j+1$ level. The propagators $B_{\mu,\tau}^{(j)}$ at the $j+1$ level (light grey) are wider than those of the j level (dark grey).

$$\psi_{\mathrm{ID}}^{(j)}(\boldsymbol{x},t) = 2^{1-j}\mathrm{Re}\left\{\overset{+}{f}\left[2^{1-j}\left(t - i|\boldsymbol{x}|^2/2b^{(j)}c\right)\right]\right\} \tag{22a}$$

$$\varphi_{\mathrm{ID}}^{(j)}(\boldsymbol{x},t) \approx \frac{2^{5-4j}}{(\bar{\omega}^{(j)})^3(\bar{x}^{(j)})^2}\mathrm{Im}\left\{\overset{+}{g}'''\left[2^{1-j}\left(t - i|\boldsymbol{x}|^2/2b^{(j)}c\right)\right]\right\}. \tag{22b}$$

Note the spatio-temporal scale of these window functions as implied by the 2^{-j+1} term in the argument of $\overset{+}{f}$ and $\overset{+}{g}$. These windows are now used in (19) to construct the multiscale Radon frames. The propagators $B_{\mu,\tau}^{(j)}(\boldsymbol{r},t)$ associated with $\psi_{\mathrm{ID}}^{(j)}(\boldsymbol{x},t)$ are the ID-PB discussed in connection with (12), where as schematized in Fig. 4(b), the higher j corresponds to large-scale PBs (light grey).

One example for the mother filters $\widehat{g}(\omega)$ and $\widehat{f}(\omega)$ is depicted in Fig. 5, where the upper and lower transition zones are tapered linearly. They belong to a general class of windows with smoother tapers that satisfy the partition of unity condition (15).

For a given reference frequency $\bar{\omega}$, $\widehat{g}(\omega)$ is characterized by the frequencies $\bar{\omega}/F$ and $\frac{1}{2}\bar{\omega}/F$, with transition zones whose widths are 2α and α, respectively. The

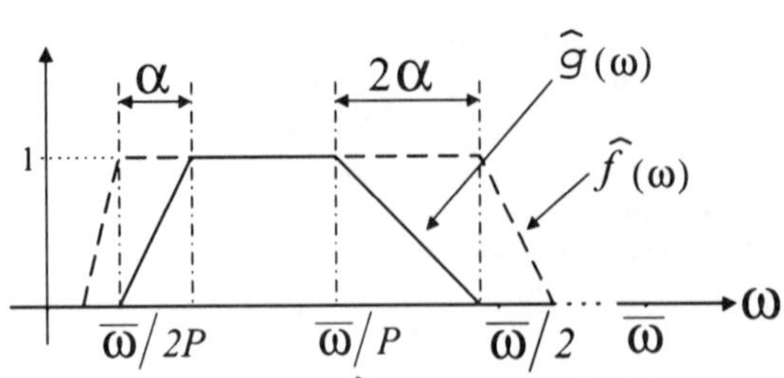

Figure 5: The mother filters $\widehat{g}(\omega)$ and $\widehat{f}(\omega)$ (full and dashed lines, respectively). $\widehat{g}$ is defined by two parameters: $\bar{\omega}/F$ that defines the upper and lower frequencies and α which defines the transition zone.

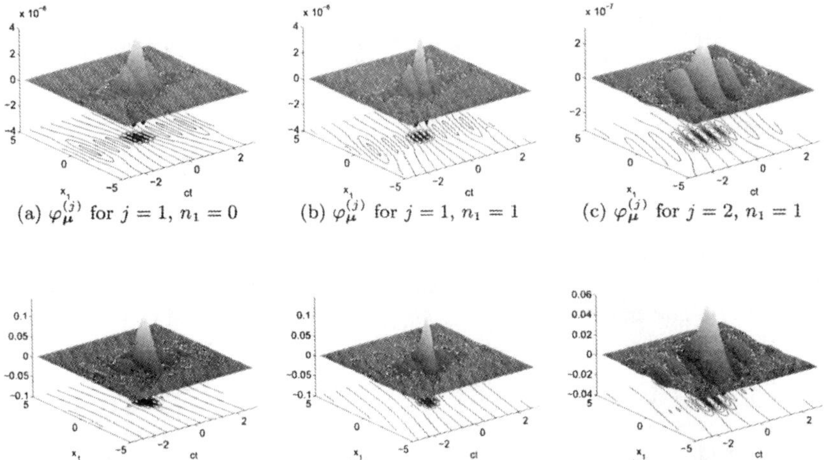

(a) $\varphi_{\boldsymbol{\mu}}^{(j)}$ for $j=1$, $n_1=0$ (b) $\varphi_{\boldsymbol{\mu}}^{(j)}$ for $j=1$, $n_1=1$ (c) $\varphi_{\boldsymbol{\mu}}^{(j)}$ for $j=2$, $n_1=1$

(d) $\psi_{\boldsymbol{\mu}}^{(j)}$ for $j=1$, $n_1=0$ (e) $\psi_{\boldsymbol{\mu}}^{(j)}$ for $j=1$, $n_1=1$ (f) $\psi_{\boldsymbol{\mu}}^{(j)}$ for $j=2$, $n_1=1$

Figure 6: The analysis and synthesis frame functions $\varphi_{\boldsymbol{\mu}}^{(j)}$ and $\psi_{\boldsymbol{\mu}}^{(j)}$. They are plotted in the (x_1, ct) plane for $x_2=0$, where all axes are normalized with respect to the $j=1$ unit cell $\bar{x}^{(1)}$, and for several representative indices j and $\boldsymbol{\mu}$.

slope of $\widehat{g}$ in the transition zones determines the temporal decay rate of the analytic signals $\overset{+}{g}(t)$ and, thereby, of the window $\varphi_{\mathrm{ID}}^{(j)}$ in (22b). Using a more moderate slope (or preferably a smoother transition, e.g., a higher order polynomial) yields smoother spatio-temporal windows. This slope cannot be too large though, since the frequency band of the window should be limited by $\nu \le \frac{1}{2}$, leading to $0 \le \alpha < \frac{1}{2}(\frac{1}{2} - P^{-1})\bar{\omega}$ (e.g., for $P=3$, α should be $\le \frac{1}{12}$).

The analytic signal $\overset{+}{g}(t)$ associated with the mother filter $\widehat{g}$ in Fig. 5 is given by

$$\overset{+}{g}(t) = \frac{e^{i\gamma}}{i\pi\alpha t^2}\left[2e^{-i(\bar{\omega}/2P+\alpha/2)t}\sin(\alpha t/2) - e^{-i(\bar{\omega}/P+\alpha)t}\sin(\alpha t)\right], \tag{23}$$

where the spectrum of $\widehat{g}$ has been multiplied by $e^{i\gamma}$. This phase parameter allows the user to control the balance between the real and imaginary parts of the analytic signal when taking the real part of $\overset{+}{g}$ via (22b). One may readily verify that (23) is analytic in $\mathbb{C}^-$, and that its decay to zero as $t\to\infty$ is dominated by $\mathcal{O}(t^{-2}e^{-it\bar{\omega}/2P})$ while for $t\to 0$, $\overset{+}{g} \approx e^{i\gamma}(\bar{\omega}/P + \alpha)/2\pi$.

The mother filter $\widehat{f}(\omega)$ is chosen next to be $e^{i\gamma}$ within the support of $\widehat{g}(\omega)$ as discussed after (17) (see dashed line in Fig. 5). Outside this range, $\widehat{f}(\omega)$ can be selected quite arbitrarily, and it should be chosen such that the synthesis window $\psi^{(j)}$ in (22a) will give rise to smooth and trackable PB propagators. Clearly, the filter banks $\{\widehat{g}^{(j)}, \widehat{f}^{(j)}\}_1^J$ satisfy the partition of unity condition (15).

Figures 6(a)-6(c) depict the analysis Radon frame $\varphi_{\boldsymbol{\mu},\tau}^{(j)}$ (19b) corresponding to the ID-PB windows (22b) with $g(t)$ of (23) (for the parameters $\bar{\omega}=1$, $P=3$, $\alpha=\frac{1}{12}$ and $\gamma=0$). The windows are shown in the (x_1, ct) plane for $x_2=0$ (the axes are normalized with respect to the $j=1$ unit cell $\bar{x}^{(1)}$), and for the following indices: Fig. 6(a) depicts the "mother window" (22b) (i.e. $j=1$ and $\boldsymbol{\mu}=\mathbf{0}$). Fig. 6(b) is for the same j but for $\boldsymbol{\mu}=(m_1, m_2, n_1, n_2)=(0,0,1,0)$, leading to a rotated window in the (x_1, ct) plane as schematized in Fig. 4(a). Finally Fig. 6(c) is for the same μ as in Fig. 6(b), but for $j=2$, where we used the decimation scheme of Eq.22a and Fig. 4(a) of Ref. 1 hence $\bar{x}^{(2)} = 2\bar{x}^{(1)}$ and $\bar{\xi}^{(2)} = \bar{\xi}^{(1)}$. Note the wider spatial-temporal distribution. Finally, Figs. 6(d)-6(f) depict the synthesis frame functions $\psi_{\mu}^{(j)}$ corresponding to the analysis frame functions in Figs. 6(a)-6(c).

5. SUMMARY AND CONCLUSIONS

A discrete Pulsed Beam (PB) summation scheme for short-pulse radiation from aperture sources directly in the time domain was introduced. The scheme utilizes a new class of spatio-temporal Radon frames (Eq. (6)). The frame elements give rise to ID-PB propagators that emerge from a discrete set of points and orientations in the aperture plane. The expansion coefficients are obtained via a Discrete Local Slant Stack Transform (DL-SST) of the time-dependent source distribution. This is a discrete local Radon transform structured around a discrete grid in space, time and orientations (Eq. (7) and Fig. 2(a)).

For sources with bandwidth greater than one octave, an efficient representation is obtained by using a binary multiscale hierarchy of Radon frames (Eqs. (19) with (22), and Fig. 4). The basic level frame $(j=1)$ has the finest space-time scale and it samples the spatio-temporal source distribution at the densest phase space grid while the grids corresponding to higher j-levels (larger scales) are decimated sets of this basic grid (Eq. (16b) and Fig. 4(a)). The basic scale level frame gives rise to narrow PB propagators that emerge from the aperture along a dense beam lattice: the higher j propagators have larger space-time support and more decimated beam lattices (Fig. 4(b)). Consequently, only the basic set of ID-PBs needs to be traced through the medium, while those of higher j are found by proper decimation of the lattice and scaling of the ID-PB functions (as in (22)). Closed form expressions for the "analysis" and "synthesis" Radon frames were provided (Eq. (22)), together with guidelines for choosing the expansion parameters (Eqs. (16), (21) and (23)).

Acknowledgement

E. Heyman acknowledges partial support by the Israel Science Foundation (ISF), Grant No. 216/02, and by the AFOSR, Grant No. F49620-01-C-0018. A. Boag acknowledges partial support by the ISF, Grant No. 577/00.

REFERENCES

1. A. Shlivinski, E. Heyman, A. Boag and C. Letrou, "Frame-based beam-summation algorithms for ultra wideband radiation from extended apertures. Part I: Formulations in the multi frequency domain," *this issue*, 2002.
2. A. Shlivinski, E. Heyman, A. Boag, D. Lugara and C. Letrou, "Gabor-frame phase space beam summation formulation for wideband radiation from extended apertures," in *Proc. of URSI International Symposium on Electromagnetic Theory*, Victoria, Canada, 2001, pp. 56–58.
3. A. Shlivinski, E. Heyman, A. Boag, D. Lugara and C. Letrou, "A discretized-phase-space pulsed beam representation for time dependent radiation," in *Proc. of URSI International Symposium on Electromagnetic Theory*, Victoria, Canada, 2001, pp. 71–73.
4. E. Heyman, "Transient plane wave spectrum representation for radiation from volume source distribution," *J. Math. Phys.*, **37**, 658–681, 1996.
5. E. Heyman and T. Melamed, "Space-time representation of ultra-wideband signals," in *Advances in Imaging and Electron Physics* (P. Hawkes, Ed.), **103**, 1–66, Boston: Academic Press, 1998.
6. B. Steinberg, E. Heyman, and L. Felsen, "Phase space beam summation for time dependent radiation from large apertures: Discretized parameterization," *J. Opt. Soc. Am. A*, **8**, 959–966, 1991.
7. E. Heyman and L. Felsen, "Gaussian beam and pulsed beam dynamics: complex source and spectrum formulations within and beyond paraxial asymptotics," *J. Opt. Soc. Am. A*, **18**, 1588–1610, 2001.

PULSE CENTROVELOCITY: ASYMPTOTIC AND FFT RESULTS

Natalie A. Cartwright and Kurt E. Oughstun*

1. INTRODUCTION

Over the past 100 years a variety of definitions have been introduced in order to describe the velocity of an electromagnetic pulse as it travels through a dispersive material. Although these velocity measures provide comparable results in those frequency regions of the material dispersion where the loss is small, they disagree wherever the material loss is large. In fact, some definitions of the pulse velocity yield nonphysical results (superluminal or negative velocities), while others apply only to certain pulse characteristics. In 1970, Smith[1] addressed these issues while proposing a new definition for the velocity of an electromagnetic pulse which he called the centrovelocity. This centrovelocity is defined by the quantity

$$\left|\nabla\left(\int_{-\infty}^{\infty} tE^2(\mathbf{r},t)dt \Big/ \int_{-\infty}^{\infty} E^2(\mathbf{r},t)dt\right)\right|^{-1}$$

where $\mathbf{E}(\mathbf{r},t)$ is the *real* electric intensity vector. Notice that the ratio of the two integrals is analogous to a center of mass calculation since it tracks the temporal center of gravity of the intensity of the pulse. Recently, Peatross, Glasgow and Ware[2] introduced a variant of Smith's centrovelocity which tracks the temporal center of gravity of the *real* Poynting vector $\mathbf{S}(\mathbf{r},t)$ rather than that of the intensity of the pulse, where the Poynting vector is defined (in MKS units) as

$$\mathbf{S}(\mathbf{r},t) = \mathbf{E}(\mathbf{r},t) \times \mathbf{H}(\mathbf{r},t). \tag{1}$$

In this paper, the centrovelocity of a rectangular-modulated plane-wave electromagnetic pulse traveling in the positive z -direction in a Lorentz model[3] dielectric material is calculated in two independent ways: the first uses an asymptotic approximation and the second uses a numerical code based on the fast Fourier transform algorithm. The results show that, in the asymptotic limit as $z \to \infty$, the centrovelocity approaches the speed at which the Brillouin precursor travels through the causally dispersive material.

* Natalie A. Cartwright, Department of Mathematics and Statistics, Univerisity of Vermont, Burlington, Vermont, 05405. Kurt E. Oughstun, College of Engineering and Mathematics, University of Vermont, Burlington, Vermont, 05405.

2. FORMULATION

It is assumed that the dielectric material is nonmagnetic with magnetic susceptibility $\mu = \mu_0$ and that it occupies all of space. The field source lies in a bounded region in the negative half-space $z < 0$. A single-resonance Lorentz model[3] is used to characterize the dielectric material with the complex index of refraction given by

$$n(\omega) = \left(1 - \frac{b^2}{\omega^2 - \omega_0^2 - 2i\delta\omega}\right)^{\frac{1}{2}}, \tag{2}$$

where ω_0 is the resonance frequency of the material, b is the plasma frequency, and δ is the phenomenological damping constant. The same material parameters as used by Brillouin in his original work[4] are employed here, viz. $\omega_0 = 4.0 \times 10^{16}$ rad/s, $b = \sqrt{20.0 \times 10^{32}}$ rad/s, $\delta = 0.28 \times 10^{16}$ rad/s. These material values are characteristic of a highly attenuative medium.

The electromagnetic field is chosen as a transverse plane-wave traveling in the positive z-direction. Without loss of generality, the field is assumed to be linearly polarized with the electric field in the y-direction and the magnetic field in the x-direction. For concreteness, we choose a rectangular-modulated sine wave as the initial pulse at the plane $z = 0$. This pulse is ultrawideband for all initial pulse durations $T > 0$. For a plane wave pulse traveling in the positive z -direction with initial value specified at $z = 0$, the centrovelocity v_c is determined by the equation

$$v_c = \frac{z}{< t_z > - < t_0 >} \tag{3}$$

where $< t_z >$ is the arrival time of the temporal centroid of the field's Poynting vector at the plane $z > 0$ given by

$$< t_z > = \frac{\hat{z} \cdot \int_{-\infty}^{\infty} t\mathbf{S}(z,t)\, dt}{\hat{z} \cdot \int_{-\infty}^{\infty} \mathbf{S}(z,t)\, dt}. \tag{4}$$

Notice that calculation of the centrovelocity requires knowledge of the propagated Poynting vector $\mathbf{S}(z,t)$, which, in turn, requires expressions for $\mathbf{E}(z,t)$ and $\mathbf{B}(z,t) = \mu_0\mathbf{H}(z,t)$. Unfortunately, there are no closed form solutions for the propagated field vectors $\mathbf{E}(z,t)$ and $\mathbf{B}(z,t)$ for this canonical problem. Instead, these quantities are determined here using two different approaches, an asymptotic approximation and a numerical code utilizing the fast Fourier transform algorithm.

3. ASYMPTOTIC ANALYSIS

By use of the angular spectrum of plane waves representation, the electromagnetic field on the plane $z > 0$ is found to be given by[4,5]

$$\mathbf{E}(z,t) = \hat{\mathbf{y}}\frac{1}{2\pi}\Re\left\{i\int_{ia-\infty}^{ia+\infty} \tilde{u}(\omega - \omega_c)e^{i[\tilde{k}(\omega)z - \omega t]}d\omega\right\} \tag{5a}$$

$$\mathbf{B}(z,t) = -\hat{\mathbf{x}}\frac{1}{2\pi c}\Re\left\{i\int_{ia-\infty}^{ia+\infty} n(\omega)\tilde{u}(\omega - \omega_c)e^{i[\tilde{k}(\omega)z - \omega t]}d\omega\right\}, \tag{5b}$$

where a is greater than the abscissa of absolute convergence for the initial pulse envelope $u(t)$ with Fourier transform, $\tilde{u}(\omega)$, $\tilde{k}(w) = (\omega/c)n(\omega)$ is the complex wavenumber, and c is the speed of light in vacuum. Here, $\hat{\mathbf{x}}$ is a unit vector along the positive x -direction and $\hat{\mathbf{y}}$ is a unit vector along the positive y -direction. By constructing the rectangular envelope function as the difference between two unit step function pulses separated by the initial time duration T of the rectangular envelope, the integral representations in Eq.(5) become[5]

$$\begin{aligned}\mathbf{E}(z,t) =& -\hat{\mathbf{y}}\frac{1}{2\pi}\Re\left\{\int_{ia-\infty}^{ia+\infty}\tilde{u}(\omega-\omega_c)e^{\frac{z}{c}\phi(\omega,\theta)}d\omega\right.\\ &\left.-e^{-i\omega_c T}\int_{ia-\infty}^{ia+\infty}\tilde{u}(\omega-\omega_c)e^{\frac{z}{c}\phi(\omega,\theta)+i\omega T}d\omega\right\}\end{aligned}\tag{6a}$$

$$\begin{aligned}\mathbf{B}(z,t) =& \hat{\mathbf{x}}\frac{1}{2\pi c}\Re\left\{\int_{ia-\infty}^{ia+\infty}n(\omega)\tilde{u}(\omega-\omega_c)e^{\frac{z}{c}\phi(\omega,\theta)}d\omega\right.\\ &\left.-e^{-i\omega_c T}\int_{ia-\infty}^{ia+\infty}n(\omega)\tilde{u}(\omega-\omega_c)e^{\frac{z}{c}\phi(\omega,\theta)+i\omega T}d\omega\right\},\end{aligned}\tag{6b}$$

where $\tilde{u}(\omega)$ now represents the Fourier transform of the Heaviside unit step function, and where

$$\phi(\omega) = i\omega[n(\omega)-\theta] = i\frac{c}{z}\left[\tilde{k}(\omega)z-\omega t\right]\tag{7}$$

is the complex phase function, and where $\theta = ct/z$ is a dimensionless space-time parameter.

By defining $\phi_T(\omega,\theta) \equiv i\omega[n(\omega)-\theta_T]$, where $\theta_T = c(t-T)/z$, the integrals in Eq.(6) can be expressed as

$$\begin{aligned}\mathbf{E}(z,t) =& -\hat{\mathbf{y}}\frac{1}{2\pi}\Re\left\{\int_{ia-\infty}^{ia+\infty}\tilde{u}(\omega-\omega_c)e^{\frac{z}{c}\phi_0(\omega,\theta)}d\omega\right.\\ &\left.-e^{-i\omega_c T}\int_{ia-\infty}^{ia+\infty}\tilde{u}(\omega-\omega_c)e^{\frac{z}{c}\phi_T(\omega,\theta)}d\omega\right\}\end{aligned}\tag{8a}$$

$$\begin{aligned}\mathbf{B}(z,t) =& \hat{\mathbf{x}}\frac{1}{2\pi c}\Re\left\{\int_{ia-\infty}^{ia+\infty}n(\omega)\tilde{u}(\omega-\omega_c)e^{\frac{z}{c}\phi_0(\omega,\theta)}d\omega\right.\\ &\left.-e^{-i\omega_c T}\int_{ia-\infty}^{ia+\infty}n(\omega)\tilde{u}(\omega-\omega_c)e^{\frac{z}{c}\phi_T(\omega,\theta)}d\omega\right\}.\end{aligned}\tag{8b}$$

In order to facilitate the asymptotic analysis, the generic integral

$$A(z,t) = -\frac{1}{2\pi}\Re\left\{\int_{ia-\infty}^{ia+\infty}\tilde{u}(\omega-\omega_c)e^{\frac{z}{c}\phi(\omega,\theta)}d\omega\right\}\tag{9}$$

is introduced. With the determination of the asymptotic behavior of $A(z,t)$ as $z\to\infty$, the asymptotic behavior of the electric field vector $\mathbf{E}(z,t)$ as $z\to\infty$ is also determined since the first integral in Eq. (8a) is $A(z,t)$ and the second integral in Eq. (8a) is of the same form but retarded in time. In order to determine the asymptotic behavior of the magnetic field vector $\mathbf{B}(z,t)$ as $z\to\infty$, the same technique is applied with the multivalued complex function $n(\omega)$ introduced into the integrand. This additional factor poses no complications since the poles of $n(\omega)$ are the same as the branch cut locations for $\phi(\omega,\theta)$, which are naturally avoided in the asymptotic analysis[6].

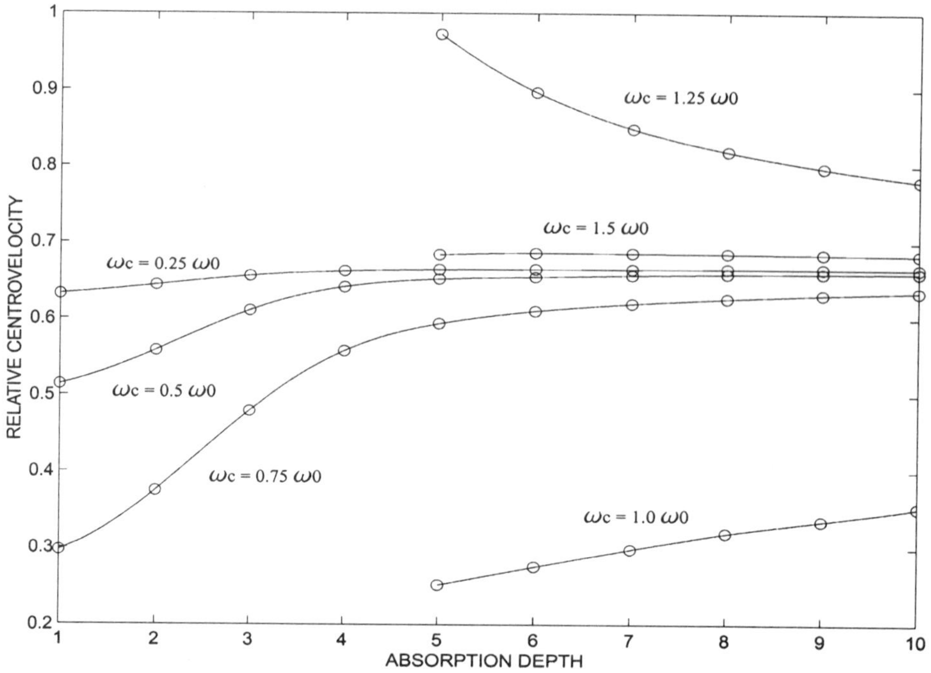

Figure 1. Relative centrovelocities in a single resonance Lorentz model dielectric as calculated using the asymptotic code for intermediate propagation distances.

As presented by Oughstun and Sherman[5-9], the asymptotic expansion of $A(z,t)$ consists of three contributions to the integral, where

$$A(z,t) = A_S(z,t) + A_B(z,t) + A_C(z,t). \tag{10}$$

Here $A_S(z,t,)$ is the contribution due to the distant saddle points of $\phi(\omega,\theta)$ and is referred to as the Sommerfeld precursor, $A_B(z,t)$ is the contribution due to the near saddle points of $\phi(\omega,\theta)$ and is referred to as the Brillouin precursor, and $A_C(z,t)$ is the contribution due to any poles of the integrand (caused by any singularities in the initial spectrum) and is referred to as the main signal. The asymptotic code employs the asymptotic approximation of these three contributions using numerically determined saddle point locations[10]. In this code, the appropriate uniform asymptotic approximation is used in the determination of the Sommerfeld and Brillouin precursors, while a nonuniform approximation is used to determine the main signal contribution.

It has been shown using these asymptotic techniques[5-7] that the Sommerfeld precursor is of the order $\mathcal{O}(z^{-1/2}e^{-\alpha_s z})$ as $z \to \infty$, the Brillouin precursor is of the order $\mathcal{O}(z^{-1/2})$ as $z \to \infty$, and the pole contribution is of the order $\mathcal{O}(e^{-\alpha_c z})$ as $z \to \infty$. Here, $\alpha_s = \Im\{\tilde{k}(\omega_{SP_D}^+\} > 0$ is the attenuation coefficient characteristic of the Sommerfeld precursor and $\alpha_c = \Im\{\tilde{k}(\omega_c\} > 0$ is the attenuation coefficient that is characteristic of the pole contribution, where $\omega_{SP_D}^+(\theta)$ denotes the location of the distant saddle point in the right half-plane and ω_c denotes the carrier frequency. Thus, with increasing absorption depth, the Brillouin precursor dominates all other contributions to the field behavior.

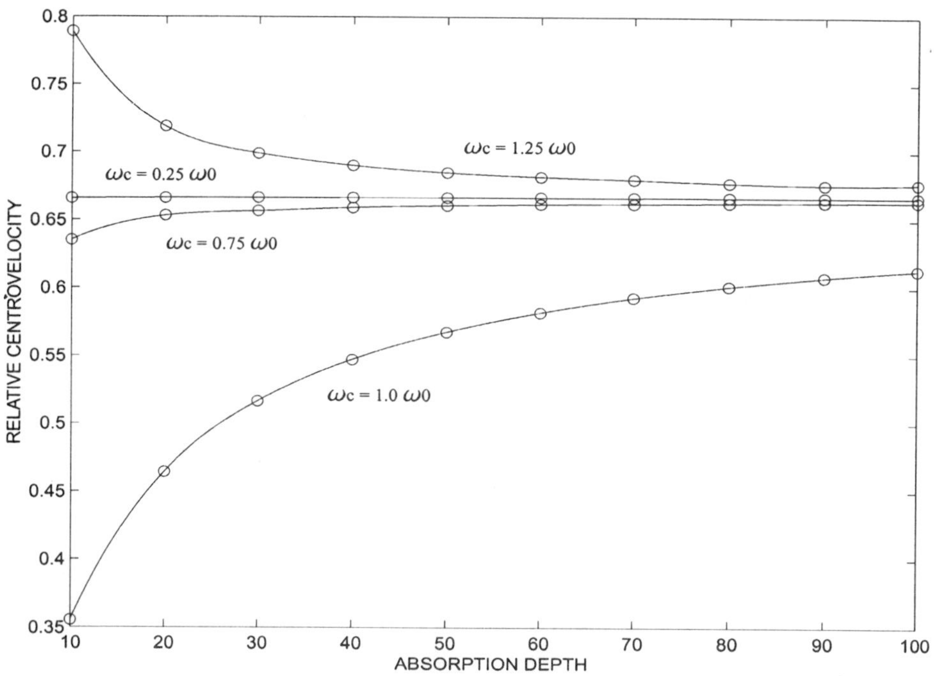

Figure 2. Relative centrovelocities in a single resonance Lorentz model dielectric as calculated using the asymptotic code for large propagation distances.

With the asymptotic approximation for both $\mathbf{E}(z,t)$ and $\mathbf{B}(z,t)$, the Poynting vector for the plane wave field is then easily calculated as $\mathbf{S}(z,t) = (1/\mu_0)|\mathbf{E}||\mathbf{B}|\hat{\mathbf{z}}$. Since the Brillouin precursor for the electric and magnetic fields occur during the same θ-domain then, as $z \to \infty$, the Poynting vector will be dominated by the product of the electric and magnetic field Brillouin precursors. It is known[5,6] that the peak of the Brillouin precursor travels through the material at the rate $c/n(0) \equiv c/\theta_0$ which, for the Lorentz model and the chosen material parameters, equals $(2/3)c$. Hence, it is expected that the centrovelocity of this pulse approaches the value $(2/3)c$ in the asymptotic limit $z \to \infty$.

This centrovelocity has been calculated for carrier frequencies from $\omega_c = 0.25\omega_0$ to $\omega_c = 1.5\omega_0$ in intervals of $0.25\omega_0$. For each carrier frequency considered the calculation has been performed for propagation distances from 1 to 100 absorption depths at the pulse carrier frequency. The relative centrovelocities for propagation distances from one to ten absorption depths are shown in Figure 1. The relative centrovelocities for propagation distances from 10 to 100 absorption depths are shown in Figure 2. In these figures, the circles represent the actual values of the centrovelocity calculated by asymptotic methods and the solid lines are a cubic spline fit to each set of data points. Evident in Figure 1 and Figure 2 is the asymptotic limit $1/\theta_0 = 2/3$ that the relative centrovelocity approaches as $z \to \infty$.

4. NUMERICAL METHOD

The second method used to calculate the Poynting vector at any plane $z > 0$ utilizes the fast Fourier transform (FFT) algorithm to determine the propagated

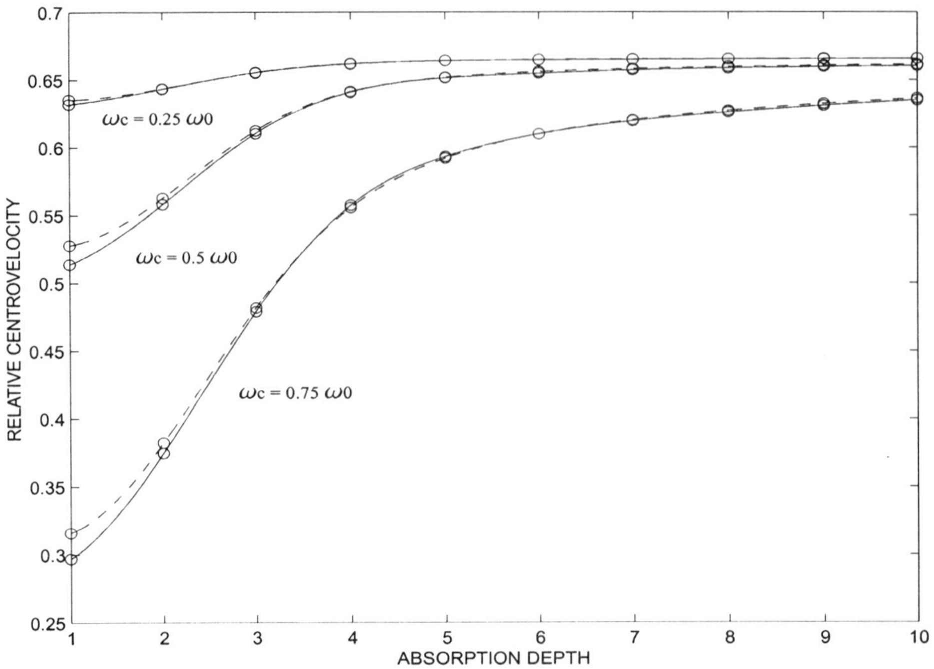

Figure 3. Relative centrovelocities for carrier frequencies $\omega_c = 0.25\omega_0$, $\omega_c = 0.5\omega_0$, and $\omega_c = 0.75\omega_0$. The circles represent data points, the solid curve is a cubic spine fit to the asymptotic code values and the dashed curve is a cubic spline fit to the numerical values.

electric and magnetic fields. In order to compute the propagated electric field vector $\mathbf{E}(z,t)$, this algorithm takes the Fourier transform of the initial electric field vector $\mathbf{E}(0,t)$, propagates each monochromatic component by multiplication with the propagation factor $\exp\,(i\tilde{k}(\omega)z)$, and then computes the inverse Fourier transform of the propagated spectrum, thereby constructing the propagated field. Likewise, the propagated magnetic field vector $\mathbf{B}(z,t)$ is computed by taking the Fourier transform of the initial electric field vector $\mathbf{E}(0,t)$, multiplying by $(-1/c)n(\omega)$, propagating each monochromatic component using the same propagation factor $\exp\,(i\tilde{k}(\omega)z)$, and then computing the inverse Fourier transform of the propagated spectrum.

The centrovelocity has been calculated for the same carrier frequencies as in the asymptotic approximation and, for each carrier frequency, the calculation has been performed for propagation depths from one to ten absorption depths. These numerical values are plotted along with the asymptotic centrovelocity values in Figure 3 and Figure 4. In these two figures the circles represent the computed data points, the solid curves are a cubic spline fit to the asymptotic results, and the dashed curves are a cubic spline fit to the numerical results. Figure 3 shows the results for the below resonance carrier frequencies $\omega_c = 0.25\omega_0$, $\omega_c = 0.5\omega_0$, and $\omega_c = 0.75\omega_0$ which lie in the normal dispersion regime below the absorption band. These results have a maximum relative error of less than four percent at one absorption depth and this relative error monotonically decreases as $z \to \infty$. Figure 4 shows the results for carrier frequencies $\omega_c = 1.0\omega_0$, $\omega_c = 1.25\omega_0$, and $\omega_c = 1.5\omega_0$ which lie in the anomalous dispersion region within the absorption

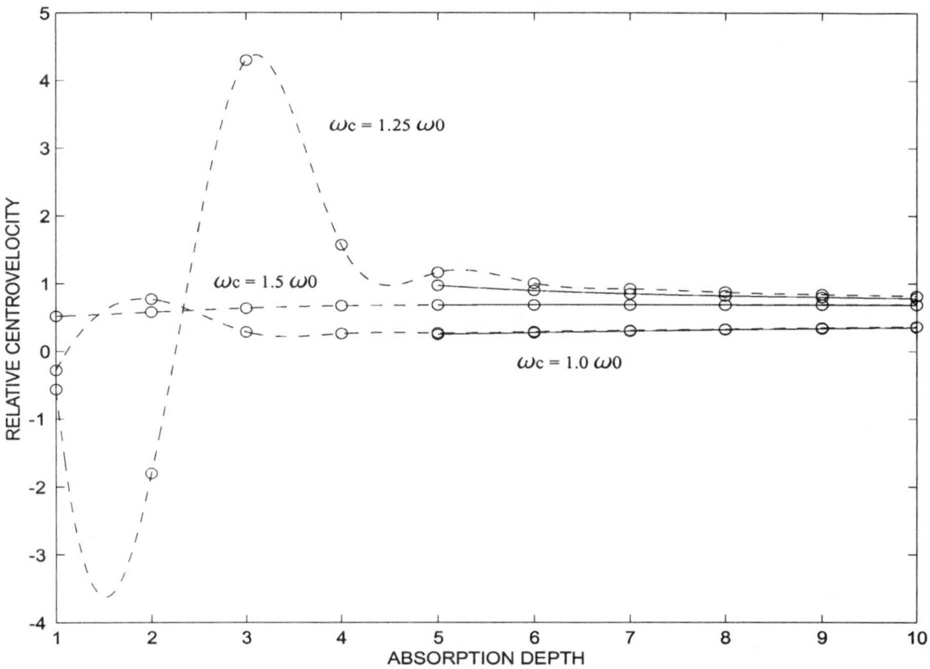

Figure 4. Relative centrovelocities for carrier frequencies $\omega_c = 1.0\omega_0$, $\omega_c = 1.25\omega_0$, and $\omega_c = 1.5\omega_0$. The circles represent data points, the solid curve is a cubic spline fit to the asymptotic code values and the dashed curve is a cubic spline fit to the numerical values.

band. These results have a maximum relative error of 13.4 percent at the smallest propagation distance considered for the asymptotic approximation and this relative error also decreases monotonically as $z \to \infty$, decreasing to less than four percent for propagation distances greater than eight absorption depths. Notice again that the relative centrovelocity approaches the asymptotic limit $1/\theta_0 = 2/3$ as $z \to \infty$.

5. CONCLUSIONS

For a plane wave with rectangular modulation traveling through a causally dispersive material asymptotic methods show that, as the propagation distance increases, the Brillouin precursor dominates the structure of the field. Using both an asymptotic and a numerical code, the centrovelocity of a plane wave with rectangular modulation traveling through a dispersive material has been calculated. The results show that, in all cases, the centrovelocity approaches the value $c/\theta_0 = c/n(0)$ set by the Brillouin precursor, in agreement with the asymptotic analysis.

6. ACKNOWLEDGMENTS

The research presented in this paper has been supported, in part, by the United States Air Force Office of Scientific Research under grant no 49620-01-1-0306. The first author would like to thank Dr. Richard Foote for his expert advice in typesetting this document.

7. REFERENCES

1. R. Smith, The Velocities of Light, *Am. J. Phys.* **38**(8), 978–984 (1970).
2. J. Peatross, S.A. Glasgow, M. Ware, Average energy flow of optical pulses in dispersive media, *Phys. Rev. Lett.* **84**(11), 2370–2373 (2000).
3. H.A. Lorentz, *The Theory of Electrons* (Leipzig, Teubner,1909), Ch. IV.
4. L. Brillouin, *Wave Propagation and Group Velocity* (Academic, New York, 1960).
5. K.E. Oughstun and G.C. Sherman, *Pulse Propagation in Causal Dielectrics* (Springer-Verlag, Berlin, 1994).
6. K.E. Oughstun and G.C. Sherman, Propagation of electromagnetic pulses in a linear dispersive medium with absorption (the Lorentz medium), *J. Opt. Soc. Am. B* **5**(4), 817–849 (1988).
7. K.E. Oughstun and G.C. Sherman, Uniform asymptotic description of electromagnetic pulse propagation in a linear dispersive medium with absorption (the Lorentz medium), *J. Opt. Soc. Am. A* **6**(9), 1394–1420 (1989).
8. K.E. Oughstun and G.C. Sherman, Uniform asymptotic description of ultrashort rectangular optical pulse propagation in a linear, causally dispersive medium, *Phys. Rev. A* **41**(11), 6090–6113 (1990).
9. K.E. Oughstun, Pulse Propagation in a Linear, Causally Dispersive Medium, *Proc. IEEE* **79**(10), 1394–1420(1991).
10. K.E. Oughstun and P.D. Smith, On the accuracy of asymptotic approximations in ultrawideband signal, short pulse, time-domain electromagnetics, *Proc. 2000 IEEE Ant. & Prop. Soc. Ann. Mtg.* 685–688(2000).

REVERSIBLE AND IRREVERSIBLE PROCESSES IN DISPERSIVE/DISSIPATIVE OPTICAL MEDIA

Electro-magnetic free energy and heat production

C. Broadbent, G. Hovhannisyan, M. Clayton, J. Peatross, S. A. Glasgow*

1. INTRODUCTION

Experiments involving slow[1] and fast[2-7] light have motivated various theoretical investigations[8,9] that interpret the phenomena arising from dispersion and dissipation in optical systems. Previously, we showed that the interpretation of these phenomena often referred to in the literature were intrinsically associated with incomplete energy accounting.[9] Specifically, we showed that for a dissipative dielectric, 1) there is an unambiguous notion of the total energy of the medium-field system, and 2) this total energy can in nowise demonstrate "fast" dynamics, neither locally nor globally.

This otherwise satisfying result is tempered by the realization that the total energy contains a component which is "dynamically unrecoverable," and should be viewed either as a static component of a closed system, or as part of a continual loss from an open system. In a closed system the "dynamically unrecoverable" energy is like a trail of bricks left behind a wagon; stationary pieces that are still considered part of the wagon-brick system. In the open system we think of the dropped bricks as having returned to the earth; once dropped they are no longer considered part of the wagon-brick system. Either viewpoint cheapens the result that total energy transport is luminal; one might suppose that luminality is only a consequence of a component which is either not moving at all or is completely lost from the system's energy accounting and whose transport properties should not be considered in the first place. In this work we do not address the question of luminality of the dynamically recoverable energy. Instead, we consider the heretofore-unsolved problem of unambiguously separating the recoverable from the unrecoverable energy in a general linear dielectric medium.

---

* S. A. Glasgow, G. Hovhannisyan, and M. Clayton, Department of Mathematics, J. Peatross and C. Broadbent, Department of Physics, Brigham Young University, Provo, Utah, 84602.

Ultra-Wideband Short-Pulse Electromagnetics 6
Edited by Mokole *et al.*, Kluwer Academic/Plenum Publishers, 2003

We use ideas from Landau and Lifshitz [LL][10] and Oughstun and Sherman [OS][11] by which the static component of energy is interpreted as the heat dissipated to the medium. [LL] explicitly make this interpretation, in the case of an arbitrary medium and arbitrary process, only in the limit of large time, while [OS] indicate that such a notion should exist at any time. In this sense [OS] consider a notion of heat production that is manifestly dynamical. [LL] establish a dynamical notion of *internal energy* (the sum of heat and free energy; the total energy), but they quickly restrict to *the non-dispersive case*, and proceed to define it as the difference between (some fixed notion of) the medium-field energy with and without the history of the instigating fields. They deem this type of definition as being thermodynamically significant. Of course, as they point out, the non-dispersive case is also the (macroscopically) conservative case. Consequently in the case they consider there can be no notion of heat dissipated, neither dynamically nor asymptotically. Therefore the total energy is just the free energy. Moreover [LL] continue by indicating that such a thermodynamically significant notion of (dynamical) internal energy probably cannot exist for a dissipative system. From the open system viewpoint this postulate is valid. Returning to the wagon-brick analogy, if one stops accounting for lost bricks, then the total amount of bricks in the wagon-brick system is constantly changing; bricks are no longer conserved. Consequently, with no notion of dynamical internal energy, they do not prescribe a meaningful notion of its components, free and heat energy.

In this paper we define the dynamical free and heat energy in a stationary, causal, and purely dissipative medium (a *simple* medium). We show how the definition 1) vindicates a certain component of the viewpoint of [OS], namely that there is a meaningful notion of dynamical heat production, 2) creates a thermodynamically significant notion of total energy, generalizing the [LL] notion of internal energy, and 3) makes connections with intuitive properties for the simple Lorentz model and validates the interpretation of those properties.

2. REVERSIBLE AND IRREVERSIBLE ENERGY

From Poynting's theorem, the well-known total energy density in the system is

$$u(t) := \frac{1}{2}\mathrm{E}^2(t) + \frac{1}{2}\mathrm{H}^2(t) + \int_{-\infty}^{t} \mathrm{E}(\tau)\dot{\mathrm{P}}(\tau)d\tau, \tag{1}$$

where we have restricted to the scalar case, in order to simplify the following discussion.

The first two terms in Eq. (1) give the electromagnetic field energy density, and the last term represents the energy density stored in the coupled medium-field system. We refer to this third component as the *interaction energy* and denote it as $u_{\text{int.}}(t)$.

The change in the free energy in a reversible isothermal process is defined as the difference between the change in the internal energy and the change in the heat energy during this process. An immediate consequence of this definition is that this change in free energy is equivalent to the work done on the system. Since the process is reversible, this work done by some external agent on the system is equivalent to the work the system can do on, or return to, the agent. We generalize this idea to irreversible/dissipative/non-equilibrium processes by defining the *dynamical free energy* in all cases to be the work the system can do on, or return to, an external agent. This generalized notion of free energy thus defined reduces to the classical notion in the reversible case. Though our generalization of the free energy is manifestly dynamical, in the following we will abbreviate and refer to this non-equilibrium object as, simply, the free energy.

It can be shown that the first two terms in Eq. (1) always constitute energy (density) that the medium-field system can return to a causal agent. Thus these first two terms always constitute a component of the free energy just defined. Consequently our search for the total free energy can be restricted to a discussion of the third term, the interaction energy $u_{\text{int.}}(t)$.

[OS] claim that $u_{\text{int.}}(t)$ should in principle be decomposed into two pieces, one of which represents energy stored in the medium "reactively" (their term), the other of which represents energy stored "latently" (our term). The free energy is then the sum of the field energy term and the reactive energy term. The heat energy corresponds to the latent term of the interaction energy. We will define and then find the reactive and latent energy terms.

Utilizing a theorem from our previous work,[9] we first rewrite the interaction energy

$$u_{\text{int.}}(t) = \int_{-\infty}^{+\infty} \omega \operatorname{Im}[\chi(\omega)] \left|\hat{\mathrm{E}}_t(\omega)\right|^2 d\omega, \tag{2}$$

where the *instantaneous* (or *causal*) *spectrum* $\hat{\mathrm{E}}_t(\omega)$ is defined by

$$\hat{\mathrm{E}}_t(\omega) := \frac{1}{\sqrt{2\pi}} \int_{-\infty}^{t} \mathrm{E}(\tau) e^{i\omega\tau} \, d\tau. \tag{3}$$

and $\chi(\omega)$ is the susceptibility of the medium: $\hat{\mathrm{P}}(\omega) = \chi(\omega)\hat{\mathrm{E}}(\omega)$.

Manifestly, $u_{\text{int.}}(-\infty) = 0$. We note that by the work of [LL], $u_{\text{int.}}(+\infty)$, the [LL] asymptotic heat, is non-negative for all fields interacting with any simple medium. Equation (2) generalizes this result and shows that a passive dielectric also has the property that the energy of interaction never runs a deficit at any finite time: $u_{\text{int.}}(t) \geq 0$ for any finite t. This is to say that, at any given time t, the field has done more work on the medium than the medium has done against the field. This property is important in order to make an unambiguous, dynamical separation of this term into reactive and latent energy pieces.

A physically relevant and unambiguous way to separate the interaction energy at each time t, for any field $\mathrm{E}(t)$ and any susceptibility (of a simple dielectric), into "reactive" and "latent" pieces, is to reinterpret "reactive," to mean *reversible*. We make this notion precise by extending a well-accepted interpretation. In the [LL] interpretation $u_{\text{int.}}[\mathrm{E}](+\infty) = u_{\text{int.}}[\mathrm{E}](+\infty) - 0 = u_{\text{int.}}[\mathrm{E}](+\infty) - u_{\text{int.}}[\mathrm{E}](-\infty)$ represents the energy dissipated or *lost* to the medium via the signal $\{E(\tau)|-\infty < \tau < +\infty\}$ since time $t = -\infty$. Consequently, $u_{\text{int.}}[E](+\infty) - u_{\text{int.}}[E](t)$ represents the energy lost to the medium due to the future signal $\{E(\tau)|t < \tau < +\infty\}$ since time t, given a preparation of the medium established by the past signal $\{E(\tau)|-\infty < \tau < t\}$ prior to time t. When this difference of energies is negative, we should refer to its opposite, $u_{\text{int.}}[E](t) - u_{\text{int.}}[E](+\infty)$, which is positive, as energy that, due to the future field, is returned to the field and so comprises a *reversible process* of borrowing energy from the field and returning it at a later time ($t = \infty$). Obviously, given any fixed past field interacting with a medium, some future fields yield larger energies in the borrow-return reversible process than others. Mathematically, the difference $u_{\text{int.}}[E](t) - u_{\text{int.}}[E](+\infty)$ is larger for some future fields than others. This idea suggests a definition.

Definition. The *reversible energy* $u_{rev.}[E](t)$ *of the signal* $\{E(\tau)|-\infty < \tau < t\}$ *at time t in the medium* is the supremum[†] of values $u_{\text{int.}}[E](t) - u_{\text{int.}}[E](+\infty)$ can obtain over alternative futures, i.e. over alternative future signals $\{E(\tau)|t < \tau < +\infty\}$.

We can rewrite this conceptual definition operationally as,

$$u_{rev.}[E](t) := \sup_{E_f}\left\{u_{\text{int.}}[EH_t^- + E_f](t) - u_{\text{int.}}[EH_t^- + E_f](+\infty)\right\}, \tag{4}$$

where

$$H_t^-(\tau) := \begin{Bmatrix} 1, & \tau < t \\ 0, & \text{otherwise} \end{Bmatrix}, \qquad E_f(\tau) = 0 \text{ for all } \tau < t, \tag{5}$$

and E_f denotes the possible future fields, $\{E(\tau)|t < \tau < +\infty\}$. Since the interaction energy is causal, i.e. since at any given time t it depends only on values of the field prior to t, the first term in Eq. (4) does not depend on the future field E_f. Therefore, we can simplify Eq. (4) to

$$u_{rev.}[E](t) := u_{\text{int.}}[E](t) - \inf_{E_f} u_{\text{int.}}[EH_t^- + E_f](+\infty). \tag{6}$$

Thus, from a computational point of view, the complement of the reversible energy is more fundamental. We will call this complement the irreversible energy:

$$u_{irrev.}[E](t) := \inf_{E_f} u_{\text{int.}}[EH_t^- + E_f](+\infty). \tag{7}$$

---

† The supremum (sup) and infimum (inf) are the least upper bound and greatest lower bound, respectively. For example, the sequence (1, .5, .25, .125, .0625, ...) has infimum equal to zero though no minimum exists. The supremum and infimum, however, are the maximum and minimum, respectively, of a series if the maximum or minimum exist.

Defined this way, $u_{irrev.}[\mathrm{E}](t)$ lends itself easily to the calculus of variations. Noting that any solution $\mathrm{E_f}(\tau)$ vanishes for all times $\tau < t$ (Eq. 5), the Fourier transform of the concatenated field $(\mathrm{E}H_t^- + \mathrm{E_f})(\tau)$

$$\mathcal{F}[\mathrm{E}H_t^- + \mathrm{E_f}](\omega) = \frac{1}{\sqrt{2\pi}} \int_{-\infty}^{t} \mathrm{E}(\tau) e^{i\omega\tau}\, d\tau + \frac{1}{\sqrt{2\pi}} \int_{t}^{\infty} \mathrm{E_f}(\tau) e^{i\omega\tau}\, d\tau \tag{8}$$

is given by $\hat{\mathrm{E}}_t(\omega) + \hat{\mathrm{E}}_f(\omega)$. Then, by using Eq. (2), and setting the variational derivative over $\mathrm{E_f}(\tau)$ to be zero, we have

$$\delta_{\mathrm{E_f}} u_{\mathrm{int.}}[\mathrm{E}H_t^- + \mathrm{E_f}](\infty) = \delta_{\hat{\mathrm{E}}_f} \int_{-\infty}^{+\infty} \omega \, \mathrm{Im}[\chi(\omega)] \left| \hat{\mathrm{E}}_t(\omega) + \hat{\mathrm{E}}_f(\omega) \right|^2 d\omega = 0. \tag{9}$$

Introducing new functions $\mathrm{E}_+(\tau) = \mathrm{E_f}(\tau + t)$, and $\mathrm{E}_-(\tau) = (\mathrm{E}H_t^-)(\tau + t)$, shifts the transforms according to $\hat{\mathrm{E}}_f(\omega) = e^{i\omega t} \hat{\mathrm{E}}_+(\omega)$ and $\hat{\mathrm{E}}_t(\omega) = e^{i\omega t} \hat{\mathrm{E}}_-(\omega)$. Since the extra phase factors $e^{i\omega t}$ can be neglected because of the modulus in Eq. (9), we can replace $\hat{\mathrm{E}}_f(\omega)$ and $\hat{\mathrm{E}}_t(\omega)$ with $\hat{\mathrm{E}}_+(\omega)$ and $\hat{\mathrm{E}}_-(\omega)$, which, importantly, are functions analytic in the upper and lower half planes, respectively. Finally, we take the variation over the new quantity $\hat{\mathrm{E}}_+(\omega)$, and using the symmetries of the Fourier transform of a real-valued time-series, we get

$$2 \int_{-\infty}^{+\infty} \omega \, \mathrm{Im}[\chi(\omega)] \left(\hat{\mathrm{E}}_-(\omega) + \hat{\mathrm{E}}_+(\omega) \right) \delta \hat{\mathrm{E}}_+^*(\omega) d\omega = 0. \tag{10}$$

Since $\hat{\mathrm{E}}_+(\omega)$ is analytic in the upper half plane, so is its variation $\delta \hat{\mathrm{E}}_+(\omega)$. Consequently, $\delta \hat{\mathrm{E}}_+^*(\omega)$ is analytic in the lower half plane. A certain subset of these variations is *admissible*. However, we will briefly postpone discussion of this requirement.

As $\delta \hat{\mathrm{E}}_+^*(\omega)$ is analytic in the lower half plane, but is otherwise arbitrary, Eq. (10) is impossible unless the rest of the integrand is also analytic in the lower half plane. If the entire integrand is analytic and rapidly vanishing in the lower half plane, we can obtain the integral in Eq. (10) by extending the contour with an arbitrarily large half-circle in the lower half plane. From Cauchy's theorem, the integral over the whole contour will be zero. Because the integrand is rapidly vanishing at infinity, the integral over the half-circle will go to zero as we extend the radius to infinity. This ensures that, regardless of the form of an admissible variation, that the integral over the real axis (in Eq. (10)) is zero. By defining a new quantity $\hat{\mathrm{Z}}_-(\omega)$, we can rewrite this condition on $\hat{\mathrm{E}}_+(\omega)$ as

$$\hat{\mathrm{Z}}_-(\omega) = \omega \, \mathrm{Im}[\chi](\omega) \left(\hat{\mathrm{E}}_-(\omega) + \hat{\mathrm{E}}_+(\omega) \right), \tag{11}$$

where $\hat{\mathrm{Z}}_-(\omega)$ is required to be analytic and rapidly vanishing in the lower-half complex plane.

Equation (11) is the main result of this paper. Its solution allows for the separation of the interaction energy into reversible and irreversible components for *any* simple medium. Notably, Eq. (11) constitutes one equation with two unknowns, $\hat{Z}_-(\omega)$ and $\hat{E}_+(\omega)$, and is only solvable by utilizing important constraints inherent in the original problem. First, passive susceptibilities are analytic in the upper-half complex plane, have the asymptotics such that for large real frequencies ω,

$$\mathrm{Re}[\chi(\omega)] = O(\frac{1}{\omega^2}),\ \mathrm{Im}[\chi(\omega)] = O(\frac{1}{\omega^3}),\ \mathrm{Im}[\chi(\omega)] \neq o(\frac{1}{\omega^3}),\quad \omega \to \infty, \tag{12}$$

and for small real frequencies ω,

$$\mathrm{Im}[\chi(\omega)] = O(\omega),\ \mathrm{Im}[\chi(\omega)] \neq o(\omega),\quad \omega \to 0, \tag{13}$$

and have the property that $\omega\,\mathrm{Im}[\chi(\omega)] \geq 0$ for all real frequencies, with equality only when $\omega = 0$. These asymptotics put restrictions on admissible solutions to this problem via the requirement that the associated interaction energy be finite. Second, $\hat{E}_+(\omega)$ is analytic in the upper-half complex plane, and $\hat{E}_-(\omega)$ and $\hat{Z}_-(\omega)$ are analytic in the lower-half complex plane. The general solution to Eq. (11), together with the constraints just mentioned, is found by recasting the problem in terms of a classical one from complex analysis: the Riemann-Hilbert problem.[12] The solution is (the Fourier transform of) an optimal future field (which we will call the reversal field and denote as $\{E_{rev.}[E,t](\tau)|\ t \leq \tau < \infty\}$) that will give the infimum [LL] asymptotic heat associated with some past field $\{E(\tau)|-\infty < \tau < t\}$; it is a future field that uniquely specifies $u_{irrev.}[E](t)$, and, by it's complimentary nature, $u_{rev.}[E](t)$. In section four we show in general that the irreversible energy is consistent with thermodynamic notions of heat; the irreversible energy is shown to never decrease. We also show that the reversible energy (the difference between the interaction energy and the irreversible energy) is consistent with thermodynamic notions of free energy.

3. SOLUTION FOR A SINGLE LORENTZ OSCILLATOR

In this section we utilize the definition for reversible and irreversible energy to find the reversal field, and consequently, the reversible and irreversible energies for a single species Lorentz oscillator. While the quantities associated with free energy and heat have been known for a dissipative Lorentz oscillator model of a medium since, at least, the work of Loudon in 1970,[13] we pursue the trivial Lorentz model to reinforce Loudon's interpretation for a well-known model of a medium. However, we have found the intuitive method used by Loudon for the single Lorentz oscillator model *cannot* determine the free and heat energy in general. That method involves determining the kinetic and potential energies of the components of the system and calling the sum of those free energy, and ascribing all other forms to energy dissipation. We have found, for the next simplest case of a double species Lorentz oscillator, for all times t (except $t = \infty$) the kinetic and potential energies are always greater than the reversible energy. As will be examined closely in an upcoming publication,[14] in which we consider general (non-Lorentzian) media, the reason for this is the effective irreversibility of (microscopic) decoherence among system elements.

Consider the single Lorentz oscillator with susceptibility $\chi(\omega)$,[‡]

$$\chi(\omega) = \frac{-\omega_p^2}{(\omega-(\omega_0-i\gamma))(\omega-(-\omega_0-i\gamma))}. \tag{14}$$

In contrast to the general method of solution, for a single Lorentz oscillator it is easier to use a Wiener-Hopf factorization method than the theory of the Riemann-Hilbert problem to find the reversal field (the solution to Eq. (11) with associated constraints). Once factored, the theorem from complex analysis dictating that bounded entire functions are constants gives the solution. The entire derivation will not be presented in this paper. Instead of solving for $\hat{\mathrm{E}}_+[\mathrm{E},t](\omega)$ in Eq. (11) it is necessary to solve for $\omega\hat{\mathrm{E}}_+[\mathrm{E},t](\omega)$. After some work, we find that

$$\begin{aligned}\omega\hat{\mathrm{E}}_+[\mathrm{E},t](\omega) = &-\left(\mathrm{Re}\left[\hat{\mathrm{E}}_-(\omega_0-i\gamma;t)\right]+\frac{\gamma}{\omega_0}\mathrm{Im}\left[\hat{\mathrm{E}}_-(\omega_0-i\gamma;t)\right]\right)\omega - \\ &\frac{i(\omega_0^2+\gamma^2)}{\omega_0}\mathrm{Im}\left[\hat{\mathrm{E}}_-(\omega_0-i\gamma;t)\right],\end{aligned} \tag{15}$$

where

$$\hat{\mathrm{E}}_-(\omega_0-i\gamma;t) := \frac{1}{\sqrt{2\pi}}\int_{-\infty}^{0}\mathrm{E}(t+\tau)e^{i(\omega_0-i\gamma)\tau}\,d\tau. \tag{16}$$

Then

$$\omega\hat{\mathrm{E}}_+[\mathrm{E},t](\omega) = C_1\omega + C_0, \tag{17}$$

where, to simplify the following discussion, C_n is the coefficient of ω^n in Eq. (15).

Given that $\omega\hat{\mathrm{E}}_+[\mathrm{E},t](\omega)$ does not generically have a zero at $\omega=0$ (i.e. $C_0 \neq 0$), Eq. (17) does not define $\hat{\mathrm{E}}_+[\mathrm{E},t](\omega)$ as an analytic function in the upper-half ω plane. Consequently the latter's inverse transform, the function $\tau \mapsto \mathrm{E}_+[\mathrm{E},t](\tau)$, will not vanish for times $\tau<0$, and finally the reversal field $\tau \mapsto \mathrm{E}_{rev.}[\mathrm{E},t](\tau)$, will not vanish for times $\tau<t$. This problem is generated by the fact that the [LL] asymptotic interaction energy

‡ We note that the denominator of the susceptibility given in Eq. (14) differs from the more common use of $(\omega_0^2 - i\gamma\omega - \omega^2)$ which emphasizes resonance and damping terms. The denominator in Eq. (14) proves more useful to this work in that it explicitly denotes the placement of the poles of the susceptibility in the complex plane.

(from which the irreversible energy is computed using the past/reversal concatenated field as in $u_{int.}[EH_t^- + E_{rev.}](+\infty)$) relies solely on the derivatives of the incident field; it can be rewritten as

$$u_{\text{int.}}(+\infty) = \int_{-\infty}^{+\infty} \frac{\text{Im}[\chi(\omega)]}{\omega} \left|-i\omega\hat{E}(\omega)\right|^2 d\omega, \tag{18}$$

and thus, cannot distinguish between fields that vary by a constant. However, this problem is easily remedied. Instead of solving for $\hat{E}_+[E,t](\omega)$ in Eq. (17), we map $\omega \to \omega + i\varepsilon$, and then solve for $\hat{E}_+[E,t](\omega + i\varepsilon)$

$$\hat{E}_+[E,t](\omega + i\varepsilon) = C_1 + \frac{C_0}{\omega + i\varepsilon}, \tag{19}$$

where ε denotes a positive infinitesimal. In this way $\hat{E}_+[E,t](\omega + i\varepsilon)$ becomes distributional and, most importantly, becomes analytic in the upper-half plane. Since the distribution $1/(\omega + i\varepsilon)$ can be decomposed into two distributions, namely a delta function and a principal value distribution, $\hat{E}_+[E,t](\omega + i\varepsilon)$ can be organized into three types of terms: a constant, a delta function, and a principal value distribution. In time τ these three become, respectively, a delta function (supported at $\tau = 0$), a constant (i.e. DC field), and a "sign" function, i.e. a piece-wise constant function returning the sign of its argument times 1. The amplitudes of the latter two components, both arising from the distribution $1/(\omega + i\varepsilon)$, are such that $E_+[E,t;\varepsilon](\tau)$ vanishes for $\tau < 0$, and so

$$E_{rev.}[E,t;\varepsilon](\tau) = \sqrt{2\pi}\left(C_1\delta(\tau - t) - iC_0 e^{-\varepsilon(\tau - t)} H_\tau^-(t)\right). \tag{20}$$

Equation (20) defines a family, in ε, of reversal fields for each time t. With the reversal field for each positive ε, there is an associated [LL] asymptotic heat. The infimum of this family of [LL] asymptotic heats, as ε tends to zero, is the irreversible energy at time t. As ε approaches arbitrarily close to zero the [LL] asymptotic heat (for the past/quasi-reversal concatenated field) approaches arbitrarily close to the infimum that defines the irreversible energy. Figures 1(b) and 1(e) on the following page illustrate that reversal fields with smaller ε yield [LL] asymptotic heats that approach closer to the irreversible energy. However, as ε gets smaller, it takes longer to realize the [LL] asymptotic heat for the past/quasi-reversal concatenated field. Figures 1(c) and 1(f) show that the interaction energy for a past/quasi-reversal field with smaller ε takes longer to approach its final asymptotic value. The infimum of the family, in ε, of the [LL] asymptotic heats can be found by the application of two limits as shown here,

$$u_{irrev.}\left[E\right](t) = \lim_{\varepsilon \to 0} \lim_{T \to \infty} u_{\text{int.}}\left[EH_t^- + E_{rev.}[E,t;\varepsilon]\right](T). \tag{21}$$

Importantly, as per the previous discussion, the limits in ε and T do not commute.

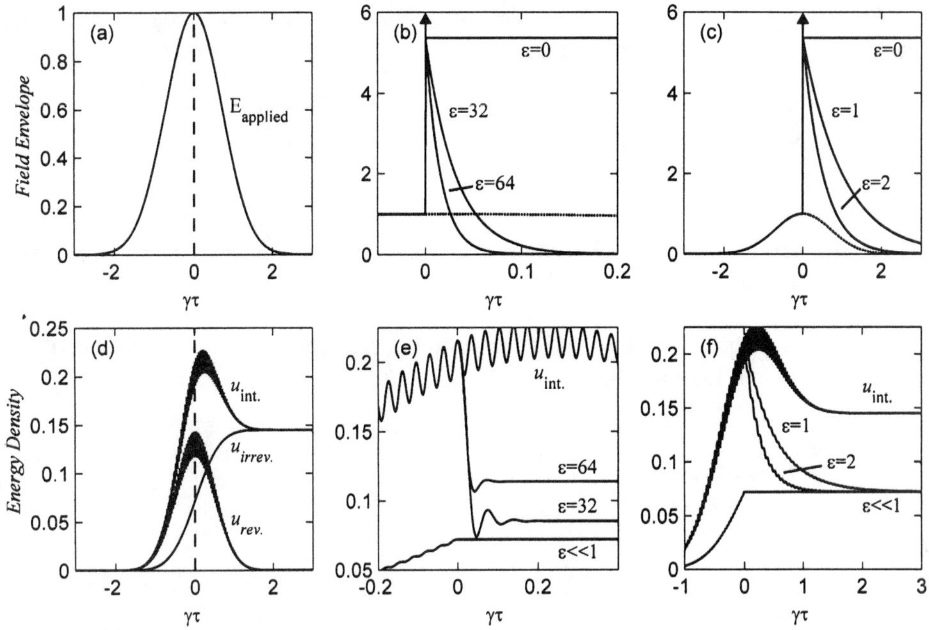

Figure 1. (a-c) Electric field envelopes in units of E_0. (a) An applied E-field. The dashed line in (a) and (d) denotes the time $\gamma t = 0.0005$. (b) Alternative future fields applied at $\gamma t = 0.0005$ taking the place of the applied E-field (dotted). Here $\varepsilon = 0, 32$ and 64. (c) Same as in (b) except $\varepsilon = 0, 1$, and 2. (d-f) Energy in units of $E_0^2/2$. (d) Interaction, reversible, and irreversible energy for the applied E-field. Here grey depicts the actual rapid oscillations; the black is time-averaged. (e) Interaction energy for alternative future fields. Here $\varepsilon = 32, 64$. The top line is the interaction energy for the applied E-field. The bottom line gives the irreversible energy as it would have evolved for $\varepsilon \ll 1$. (f) Same as (e) except $\varepsilon = 1, 2$.

Equations (21), (6), and (7) result in the following reversible and irreversible energies for the single Lorentz oscillator media:

$$u_{rev.}[\mathrm{E}](t) = \pi \frac{\omega_p^2}{\omega_0^2} \mathrm{Re}[(\omega_0 - i\gamma)\hat{\mathrm{E}}_-(\omega_0 - i\gamma; t)]^2 + \pi \frac{\omega_p^2}{\omega_0^2}\left(\omega_0^2 + \gamma^2\right)\mathrm{Im}[\hat{\mathrm{E}}_-(\omega_0 - i\gamma; t)]^2, \tag{22}$$

$$u_{irrev.}[\mathrm{E}](t) = \frac{4\pi\gamma\omega_p^2}{\omega_0^2}\int_{-\infty}^{t} \mathrm{Re}[(\omega_0 - i\gamma)\hat{\mathrm{E}}_-(\omega_0 - i\gamma; \tau)]^2 \, d\tau. \tag{23}$$

Manifestly, $u_{rev.}[\mathrm{E}](t)$, $u_{irrev.}[\mathrm{E}](t)$, and $\dot{u}_{irrev.}[\mathrm{E}](t) \geq 0$.

After a bit of work (involving the constitutive relation $\hat{P}(\omega) = \chi(\omega)\hat{E}(\omega)$ and Cauchy's theorem), it can be shown that

$$\mathrm{Im}[\hat{E}_{-}(\omega_0 - i\gamma; t)] = -\frac{\omega_0}{\sqrt{2\pi\omega_p^2}} P(t), \tag{24}$$

$$\mathrm{Re}[(\omega_0 - i\gamma)\hat{E}_{-}(\omega_0 - i\gamma; t)] = \frac{\omega_0}{\sqrt{2\pi\omega_p^2}} \dot{P}(t). \tag{25}$$

Substituting Eq. (24) and Eq. (25) back into Eq. (22) and Eq. (23), we find, verifying the interpretation of Loudon, that

$$u_{rev.}[E](t) = \frac{1}{2\omega_p^2} \dot{P}^2(t) + \frac{\left(\omega_0^2 + \gamma^2\right)}{2\omega_p^2} P^2(t), \tag{26}$$

$$u_{irrev.}[E](t) = \frac{2\gamma}{\omega_p^2} \int_{-\infty}^{t} \dot{P}^2(\tau) d\tau. \tag{27}$$

Equation (26) shows that the reversible energy for a single Lorentz oscillator is the sum of its kinetic and potential energy. However, in general (i.e. for an arbitrary susceptibility), the reversible energy is not equal to the sum of the kinetic and potential energies of (microscopic) system elements, but is found, generically, to be less than that sum. As mentioned previously, this is due to the effective irreversibility of decoherence among system elements.

4. FREE ENERGY AND HEAT

We will now demonstrate that the definitions we have made for reversible and irreversible (interaction) energy not only have the properties their names suggest, but also always have the properties classically defining free energy and heat. We have defined the reversible energy as the supremum on the amount of work the medium can do against the field. Since one defines (the change in) the free energy of a body thermodynamically as the work done on a body in a reversible isothermal process, the use of the term is justified from this semantic point of view. Better, however, is the fact that one can show that from the classic definition of free energy, in an irreversible isothermal process, volume also being held constant, free energy always decreases. The restrictions just stated amount to those that prohibit an introduction of total energy into the system. In the following, we will show that, after the instigating electric field has ceased subsidizing the reversible energy's existence, i.e. after the electric field ceases to introduce energy into the medium-field system, this component of the total energy can never increase and, moreover, decreases almost always.

The way that we will establish this result, however, is by showing first that the irreversible energy never decreases and increases almost always *regardless of the state of the field* $\mathrm{E}(t)$. In this way we establish this component of the energy as recording the irreversibility of the process defining the medium-field interaction, at least from an energetic point of view. This will be our justification for calling this component heat.

The proof that the irreversible energy cannot decrease follows directly from the definition of the irreversible energy, namely that, at a specific time t, it is the infimum of the interaction energy after this time: if these dynamically calculated infima were to ever decrease after time t, the original value for the infimum at time t would not have been an extrema after all. We are assured that such an infimum is guaranteed to exist by Eq. (2), which shows that the total interaction energy is positive definite, i.e. it can never obtain a value smaller than zero—see the paragraph after Eq. (2) alluding to this necessity. Having proved this fact regarding the dynamics of the irreversible energy, from now on we will call it the (dynamical) heat, and will denote it as $\mathcal{Q}[\mathrm{E}](t)$:

$$\mathcal{Q}[\mathrm{E}](t) := u_{irrev.}[\mathrm{E}](t). \tag{28}$$

It is now clear that the reversible energy can never increase *after the field* $\mathrm{E}(t)$ *has ceased*. Suppose that this field ceases at (i.e. is supported before) time t_0. Then, according to Eq. (6), (7), and now (28) for each time t greater than t_0,

$$u_{rev.}[\mathrm{E}](t) = u_{\mathrm{int.}}[\mathrm{E}](t_0) - \mathcal{Q}[\mathrm{E}](t), \tag{29}$$

an important feature of which is that the first term on the right is now a constant. From Eq. (29), it is now clear that the reversible energy decreases whenever the heat increases after time t_0. We will now call the reversible energy the free energy of the medium and will denote it as $F_{med.}[\mathrm{E}](t)$:

$$F_{med.}[\mathrm{E}](t) := u_{rev.}[\mathrm{E}](t) = u_{\mathrm{int.}}[\mathrm{E}](t) - \mathcal{Q}[\mathrm{E}](t). \tag{30}$$

5. SUMMARY

In this paper we have provided the essential definitions and concepts that allow for a general solution to the dynamical separation of free and heat energy first alluded to by [LL] in 1958, and later posed concretely by [OS] in 1994. We have vindicated a certain component of the viewpoint of [OS]: there is a meaningful notion of dynamical heat production. We have shown, contrary to the viewpoint of [LL], that there is a thermodynamically significant notion of internal (or total) energy for dissipative systems, and we have extended [LL]'s well known definition regarding evolved, asymptotic heat to pose a definition for the dynamical free energy and heat. This definition states that the amount of free energy in a medium-field system is the field energy plus the free energy of the medium energy. In turn, the free energy of the medium is defined as the supremum of all possible reductions in the internal energy of the medium. Consequently, heat is equivalent to the infimum of all possible future internal energies. We have applied this definition to the single Lorentz oscillator model of a dielectric and have shown that it verifies the interpretation of Loudon involving an intuitive decomposition of dynamical free energy and heat. Lastly, we have shown for an arbitrary simple dielectric, that the definitions posed for dynamical free energy and heat yield the properties that characterize them classically.

In forth-coming works we will present the development of the dynamical free energy and heat for arbitrary simple dielectrics,[12] demonstrate that the free energy is almost always less than the sum of the kinetic and potential energies of the microscopic elements comprising the system,[14] and discuss the application of these concepts to super/subluminal phenomenon.[15]

6. REFERENCES

1. L.V. Hau, S.E. Haris, Z. Dutton, and C.H. Behroozi, Light speed reduction to 17 metres per second in an ultracold atomic gas, *Nature* **397**, 594–598 (1999).
2. C.G.B. Garrett and D.E. McCumber, Propagation of a Gaussian light pulse through an anomalous dispersion medium, *Phys. Rev. A* **1**, 305–313 (1970).
3. S. Chu and S. Wong, Linear pulse propagation in an absorbing medium, *Phys. Rev. Lett.* **48**, 738–741 (1982).
4. R.Y. Chiao, Superluminal (but causal) propagation of wave packets in transparent media with inverted atomic populations, *Phys. Rev. A* **48**, R34–R37 (1993).
5. E.L. Bolda, J.C. Garrison, and R.Y. Chiao, Optical pulse propagation at negative group velocities due to a nearby gain line, *Phys. Rev. A* **49**, 2938–2947 (1994).
6. R.Y. Chiao and A.M Steinberg, Tunneling times and superluminality, *Progress in Optics* **37**, 347–406 (Emil Wolf ed., Elsevier, Amsterdam, 1997).
7. L.J. Wang, A. Kuzmmich, and A. Dogariu, Gain-assisted superluminal light propagation, *Nature* **406**, 277–279 (2000).
8. K.E. Oughstun and C.M. Balictsis, Gaussian pulse propagation in a dispersive, absorbing dielectric, *Phys. Rev. Lett.* **77**, 2210–2213 (1996).
9. S. Glasgow, M. Ware, and J. Peatross, Poynting's theorem and luminal total energy transport in passive dielectric media, *Phys. Rev. E* **64**, 446101–4461014 (2001).
10. L.D. Landau and E.M. Lifshitz, *Electrodynamics of Continuous Media*, 2nd ed. (Pergamon, New York, 1984), pp. 272–276.
11. K.E. Oughstun and G.C. Sherman, *Electromagnetic Pulse Propagation in Causal Dielectrics*, (Springer-Verlag, Berlin 1994), pp. 14–1.
12. M. Meilstrup, C. Broadbent, J. Peatross, G. Hovhannisyan, M. Clayton, and S. Glasgow are preparing the following publication to be submitted to *Phys. Rev. E*: "Real-time determination of free energy and loss in dispersive/dissipative optical media."
13. R. Loudon, The propagation of electromagnetic energy through an absorbing dielectric, *J. of Phys. A*, **3**, 233–245 (1970).
14. C. Broadbent, G. Hovhannisyan, J. Peatross, M. Clayton, and S. Glasgow, Dynamical free energy and loss in dispersive/dissipative dielectrics, to be submitted to *Phys. Rev. Lett.*
15. C. Broadbent, M. Meilstrup, J.Peatross, G. Hovhannisyan, and S. Glasgow, Luminality of free energy in optical absorbing media, to be submitted to *Phys. Rev. Lett.*

FAST TIME-DOMAIN INTEGRAL EQUATION APPROACH FOR WIDE-BAND PULSE PROPAGATION IN DISPERSIVE MEDIA

E. Bleszynski, M. Bleszynski, and T. Jaroszewicz *

1. INTRODUCTION

The availability of accurate and fast numerical tools capable of simulating electromagnetic wave propagation through ***dispersive*** media constitutes an important element of analysis of many aspects of wide-band pulse propagation (e.g., in the context of wide frequency band antennas or foliage penetration). Although there has been recently much interest in developing fast solution methods for time domain integral formulations of Helmholtz and Maxwell's equations (Refs. 1 – 4), relatively little attention has been given to the development of such tools for dispersive media, which pose a challenge from both analytic and numerical point of view.

We present here a novel time-domain integral equation formulation which takes advantage of the block-triangular and block Toeplitz structures of the impedance matrix in temporal indices, and the Toeplitz structure of the far-field component of the impedance matrix in spatial indices, hence allowing for the ***simultaneous application of Fast Fourier Transforms (FFTs) in space and time***. The algorithm relies only on the ***translational invariance*** of the Green's function. Its computational cost is ***independent of the degree of dispersion*** of the medium and scales as $O(N_t\, N_s \log N_t \log N_s)$ and $O(N_t\, N_s^{4/3} \log N_t \log N_s)$ for volume and surface problems, respectively, where N_t and N_s denote the number of temporal and spatial samples. An additional advantage of the method is its applicability to problems involving ***Green's functions given in tabulated as well as analytical forms***.

Another aspect of our approach, which we discuss in detail in this paper, is a new and ***accurate treatment*** of the ***causal behavior of the time-domain Green's function*** achieved through the utilization of a complete Galerkin procedure in space and time variables in the evaluation of the elements of the impedance matrix. It has been known in general, that the use of Galerkin's method constitutes a key element in obtaining numerical stability of the solution schemes for time domain integral equation approaches.

* Monopole Research, Thousand Oaks, CA 91360.

Although for one-dimensional wave propagation problems accounting for causality of the Green's function in the context of the Galerkin discretization is relatively straightforward, implementation in three dimensions presents a much more complicated task for the following two reasons:

(i) The Green's function depends on the relative time $t = t_1 - t_2$ and distance $r = |\mathbf{r}_1 - \mathbf{r}_2|$. Basis function supports have a non-zero spatial extent and some portions of them may not be, for a given t, causally connected. Hence implementation of causality requires careful evaluation of all relevant radial integrals over the spatial coordinates $\mathbf{r}_1$ and $\mathbf{r}_2$ associated with the basis functions entering the matrix element.

(ii) The three-dimensional Green's function has a singular term associated with $1/r = 1/|\mathbf{r}_1 - \mathbf{r}_2|$, which is in practice rather difficult to take into account in the context of three-dimensional integrals over vector variables $\mathbf{r}_1$ and $\mathbf{r}_2$.

The procedure we developed provides a simple solution to both problems ***(i)*** and ***(ii)***. Its essential element consists of a suitable transformation of the integral expressions for the matrix elements of the tensor and vector time-domain Green's functions into two-dimensional integrals over the relative (scalar) distance r and the relative time coordinate t. The expressions we derived have a simple form amenable to accurate and efficient implementation. Apart from the ability to treat rigorously causal dependence in matrix elements of all the Green's functions arising in the context of time-domain integral equations, the important advantage of our method is that it does not require any additional "singularity extraction" procedure; the singular term associated with the scalar Green's function is canceled by the r^2 factor arising from the Jacobian of the transformation.

The paper is organized as follows: in Sections 2 through 4 we present our accelerated solution scheme based on space-time FFT matrix compression, and in Sections 5 and 6 details of the Galerkin discretization of the relevant integral operators, in spatial and temporal variables, are described.

2. SOLUTION ACCELERATION BASED ON SPACE-TIME FFT MATRIX COMPRESSION METHOD

In this Section we describe a time-domain impedance matrix compression and solution method based on simultaneous application of FFTs in space and time; we refer to this method as FFT-TDIE (Fast Fourier Transform - Time Domain Integral Equation) method. Since the approach we propose relies only on the ***translational invariance*** of the Green's function, it is then sufficient to consider a generic form of a time-domain integral equation

$$\int_0^t \mathrm{d}t' \int_D \mathrm{d}r'\, G(t-t', \mathbf{r}-\mathbf{r}')\, x(t', \mathbf{r}') = b(t, \mathbf{r}) \quad \text{for } t > 0,\ \mathbf{r} \in D, \tag{1}$$

where x represents the unknowns (typically currents in a spatial domain D), and b is the external excitation.

The time-domain integral equation (1) can be discretized in a way analogous to the frequency-domain Method of Moments (MoM), except that here the basis functions depend also on time. Sources (currents) and fields are expanded in trial functions and probed with test functions. A typical parameterization of a space-time basis function $\Psi^{\mu}_{\alpha}(t, \mathbf{r})$ is

$$\Psi^{\mu}_{\alpha}(t, \mathbf{r}) = T^{\mu}(t)\, \Psi_{\alpha}(\mathbf{r}) \, , \tag{2}$$

where upper and lower indices refer to time and space coordinates, respectively. The function $\Psi_{\alpha}(\mathbf{r})$ is the usual spatial basis function with support localized near some point $\mathbf{r}_{\alpha}$, and $T^{\mu}(t)$ is a unit step function supported on an interval of length Δt (the time-step size), centered at some instant t_{μ}. We assume that all the functions T^{μ} are generated by time translation, $T^{\mu}(t) = T(t - t_{\mu})$, and that they do not overlap. We then expand currents as

$$x(t, \mathbf{r}) = \sum_{\mu} \sum_{\alpha} \Psi^{\mu}_{\alpha}(t, \mathbf{r})\, x^{\mu}_{\alpha} \, . \tag{3}$$

By using the ***Galerkin method*** we obtain a linear system of equations,

$$\sum_{\nu=1}^{\mu} \sum_{\beta=1}^{N_s} A^{\mu\nu}_{\alpha\beta}\, x^{\nu}_{\beta} = b^{\mu}_{\alpha} \qquad \text{for } \mu = 1, ..., N_t \, , \quad \alpha = 1, ..., N_s \, , \tag{4}$$

where

$$A^{\mu\nu}_{\alpha\beta} = \int \mathrm{d}t_1 \mathrm{d}r_1 \mathrm{d}t_2 \mathrm{d}r_2 \, \Psi^{\mu}_{\alpha}(t_1, \mathbf{r}_1)\, G(t_1 - t_2, \mathbf{r}_1 - \mathbf{r}_2)\, \Psi^{\nu}_{\beta}(t_2, \mathbf{r}_2) \tag{5}$$

has upper time indices and lower space indices, associated with temporal and spatial basis functions, and N_t and N_s denote the numbers of temporal and spatial samples.

Matrix $A = \{A^{\mu\nu}_{\alpha\beta}\}$ can be considered as consisting of blocks $A^{\mu\nu}$ labeled by time indices μ, ν, with elements in such blocks numbered by space indices α, β. In terms of this block structure:

(i) Matrix A is ***block lower triangular*** in time indices (as a consequence of causality and thus the use of the retarded Green's function).

(ii) Matrix A is ***block Toeplitz*** in time indices, $A^{\mu\nu} = A^{\mu-\nu}$ (as a consequence of invariance under time translations and uniform discretization in time).

(iii) For ***non-dispersive unbounded*** media, the individual blocks of matrix $A^{\mu-\nu}$ are ***partly sparse*** in spatial indices since the Green's function is nonzero only on the light cone; the number of non-zero elements in each row is then $O(N_s^{1/2})$. However, for dispersive media, the Green's function does not vanish inside the light cone. In the extreme case of a very strong dispersion, the blocks $A^{\mu-\nu}$ become dense, with $O(N_s)$ elements per row.

The conventional solution strategy of the linear system (4) is the marching-on-in-time (MOT) scheme.[1] This approach can be simply interpreted as the block forward-substitution solution method for the block-lower-triangular matrix A:

$$\sum_{\beta} A^{0}_{\alpha\beta}\, x^{\mu}_{\beta} = b^{\mu}_{\alpha} - \sum_{\nu=1}^{\mu-1} \sum_{\beta} A^{\mu-\nu}_{\alpha\beta}\, x^{\nu}_{\beta} \qquad \text{for all } \mu, \text{ all } \alpha \, . \tag{6}$$

If the MoM matrix is not compressed, the MOT operation count is

$$\mathcal{N}[Ax = b \,|\, \mathrm{MOT}] \sim \begin{cases} N_\mathrm{t}^2 N_\mathrm{s}^{3/2} & \text{for non-dispersive media,} \\ N_\mathrm{t}^2 N_\mathrm{s}^2 & \text{for strongly dispersive media.} \end{cases} \tag{7}$$

The factors $N_\mathrm{s}^{3/2}$ and N_s^2 come from the cost of vector multiplication by the matrix block $A^{\mu-\nu}$. (Equivalent complexity estimates are given sometimes, which utilize the relation $N_\mathrm{t} \sim N_\mathrm{s}^{1/2}$.)

We note that the MOT solution scheme takes advantage of the block-lower-triangularity of matrix A (through the block forward substitution algorithm), and totally ignores its block Toeplitz property. On the other hand, an iterative solution with the temporal FFT compression would utilize the block Toeplitz property of matrix A and ignore its block-lower-triangularity.

The scheme proposed below, similar to a problem analyzed in Ref. 5 in the context of block Hessenberg systems, takes advantage of both, the block-lower-triangularity and block Toeplitz structure of A.

We describe our solution algorithm first in the form which only utilizes the Toeplitz property in the ***time*** indices. Subsequently, we generalize it to a problem with full Toeplitz (FFT) compression in both space and time. The outline of this approach was first reported in Ref. 6.

3. COMPRESSION IN TEMPORAL COORDINATES

Our objective is to solve the equation $Ax = b$ (Eq.(4)) with a block-lower-triangular and block Toeplitz matrix

$$\begin{bmatrix} A^0 & 0 & \cdots & 0 \\ A^1 & A^0 & \cdots & 0 \\ \vdots & \vdots & & \vdots \\ A^{N_\mathrm{t}-1} & A^{N_\mathrm{t}-2} & \cdots & A^0 \end{bmatrix} \begin{bmatrix} x^1 \\ x^2 \\ \vdots \\ x^{N_\mathrm{t}} \end{bmatrix} = \begin{bmatrix} b^1 \\ b^2 \\ \vdots \\ b^{N_\mathrm{t}} \end{bmatrix} . \tag{8}$$

We organize now the forward substitution solution by means of a "divide-and-conquer" algorithm [5]. We assume, for simplicity, that N_t is a power of 2 (this condition can be easily relaxed), and split the matrix into equal-size blocks (north-west, south-west, etc.),

$$\begin{bmatrix} A_\mathrm{nw} & 0 \\ A_\mathrm{sw} & A_\mathrm{se} \end{bmatrix} \begin{bmatrix} x_\mathrm{n} \\ x_\mathrm{s} \end{bmatrix} = \begin{bmatrix} b_\mathrm{n} \\ b_\mathrm{s} \end{bmatrix} . \tag{9}$$

The solution of Eq.(8) is now obtained in two consecutive steps: $A_\mathrm{nw}\, x_\mathrm{n} = b_\mathrm{n}$ and $A_\mathrm{se}\, x_\mathrm{s} = b_\mathrm{s} - A_\mathrm{sw}\, x_\mathrm{n}$. Since the blocks A_nw and A_se have the same block-lower-triangular and block Toeplitz properties as the entire matrix, we continue splitting the matrix recursively, i.e., solving the problem in the first subinterval, and using the solution to solve in the second subinterval.

The crucial point of the procedure is that on each level of the recursion the computational work is dominated by the ***matrix block – vector multiplication*** $A_{\rm sw}\, x_{\rm n}$, and it is here where we utilize the block Toeplitz property of the matrix. To take advantage of this attribute we permute indices to convert the block Toeplitz matrix $A_{\rm sw}$ to a matrix whose blocks are Toeplitz. More specifically, on the k-th level of recursion ($k = 0$ being the top level), the product $y_{\rm s} \equiv A_{\rm sw}\, x_{\rm n}$ has the structure

$$\begin{bmatrix} y^{\mu+\nu} \\ \vdots \\ y^{\mu+2\nu-1} \end{bmatrix} = \begin{bmatrix} A^{\nu} & \cdots & A^{1} \\ \vdots & & \vdots \\ A^{2\nu-1} & \cdots & A^{\nu} \end{bmatrix} \begin{bmatrix} x^{\mu} \\ \vdots \\ x^{\mu+\nu-1} \end{bmatrix}, \tag{10}$$

with $\nu \times \nu$ matrix blocks each of size $N_{\rm s} \times N_{\rm s}$, where $\nu \equiv \nu_k = N_{\rm t}/2^k$. We represent it then in an equivalent form as

$$\begin{bmatrix} y_1 \\ \vdots \\ y_{N_{\rm s}} \end{bmatrix} = \begin{bmatrix} A_{11} & \cdots & A_{1N_{\rm s}} \\ \vdots & & \vdots \\ A_{N_{\rm s}1} & \cdots & A_{N_{\rm s}N_{\rm s}} \end{bmatrix} \begin{bmatrix} x_1 \\ \vdots \\ x_{N_{\rm s}} \end{bmatrix}, \tag{11}$$

where the matrix has $N_{\rm s} \times N_{\rm s}$ ***Toeplitz*** blocks

$$A_{\alpha\beta} = \begin{bmatrix} A^{\nu}_{\alpha\beta} & \cdots & A^{1}_{\alpha\beta} \\ \vdots & & \vdots \\ A^{2\nu-1}_{\alpha\beta} & \cdots & A^{\nu}_{\alpha\beta} \end{bmatrix} \tag{12}$$

of size $\nu \times \nu$. The matrix-vector product (11) can be therefore carried out in

$$\mathcal{N}[A_{\rm sw}\, x_{\rm n} \,|\, \text{level } k] \sim N_{\rm s}^2\, \nu_k \, \log_2 \nu_k = N_{\rm s}^2\, N_{\rm t}/2^k \, \log_2 N_{\rm t}/2^k \tag{13}$$

operations. There are 2^k such multiplications on each level k. By summing contributions from all levels, we find the total operation count

$$\mathcal{N}[Ax = b \,|\, \text{block Toeplitz}] \simeq \sum_{k=0}^{\log_2 N_{\rm t}} 2^k \mathcal{N}[A_{\rm sw} x_{\rm n} \,|\, \text{level } k] \sim \tfrac{1}{2} N_{\rm s}^2 N_{\rm t} \log_2^2 N_{\rm t}. \tag{14}$$

.We note that for $N_{\rm s} = 1$, i.e., for an ordinary lower triangular Toeplitz matrix, the algorithm has complexity $O(N_{\rm t} \log^2 N_{\rm t})$, and thus belongs to the class of "superfast" direct solutions, as defined, e.g., in Ref. 7.

When comparing the above result with that of Eq.(7), we see that, for non-dispersive media, the solution utilizing the block Toeplitz property in time achieves the same (up to logarithms) performance as the conventional MOT scheme. For strongly dispersive media, however, the operation count for the MOT scheme becomes $N_{\rm t}$ times larger (again, up to a logarithmic factor) than for the scheme utilizing the block Toeplitz compression in time.

In the next Section we describe how to utilize the Toeplitz property of the far-field component of the impedance matrix to further reduce the operation count.

4. DIRECT SOLUTION WITH SPATIAL AND TEMPORAL FFT COMPRESSION

The estimate (14) involves the logarithmic factor $\log N_{\rm t}$ due to the divide-and-conquer strategy, the factor $N_{\rm t} \log N_{\rm t}$ due to vector multiplications by Toeplitz matrix blocks, and the factor $N_{\rm s}^2$ coming from multiplication of a vector by a dense $N_{\rm s} \times N_{\rm s}$ matrix block, with elements indexed by the spatial elements α, β.

Clearly, the last factor can be reduced by applying the FFT based (AIM) matrix compression in spatial coordinates [8,9] to individual matrix blocks $A_{\alpha\beta}$. The matrix vector product speed-up and memory reduction is achieved through the use of far-field equivalent auxiliary basis functions

$$\mathbf{\Psi}_\alpha(\mathbf{r}) \simeq \widehat{\mathbf{\Psi}}_\alpha(\mathbf{r}) \equiv \sum_{\mathbf{u}} V_{\alpha\mathbf{u}}\, \delta^3(\mathbf{r}-\mathbf{u})\,, \tag{15}$$

given in terms of sets of point-like sources localized at nodes $\mathbf{u}$ of a regular Cartesian grid. The auxiliary basis functions are used only in the evaluation of matrix elements involving sufficiently large distances R between the source and the observation points (in practice, at least the size of the expansion box); at smaller distances the conventional MoM matrix elements are used. We can thus approximate, to any desired accuracy, a general matrix element as

$$A^{\mu-\nu}_{\alpha\beta} \simeq \overset{\rm N}{A}{}^{\mu-\nu}_{\alpha\beta} + \overset{\rm F}{A}{}^{\mu-\nu}_{\alpha\beta}\,, \tag{16}$$

where (for all distances)

$$\overset{\rm F}{A}{}^{\mu-\nu}_{\alpha\beta} = \int {\rm d}t_1 {\rm d}r_1 {\rm d}t_2 {\rm d}r_2\, T^\mu(t_1)\, \widehat{\mathbf{\Psi}}_\alpha(\mathbf{r}_1)\, G(t_1-t_2, \mathbf{r}_1-\mathbf{r}_2)\, T^\nu(t_2)\, \widehat{\mathbf{\Psi}}_\beta(\mathbf{r}_2)\,, \tag{17}$$

and the near-field matrix is defined as

$$\overset{\rm N}{A}{}^{\mu-\nu}_{\alpha\beta} = \begin{cases} A^{\mu-\nu}_{\alpha\beta} - \overset{\rm F}{A}{}^{\mu-\nu}_{\alpha\beta} & \text{for } {\rm dist}(\alpha,\beta) \le R\,, \\ 0 & \text{for } {\rm dist}(\alpha,\beta) > R\,, \end{cases} \tag{18}$$

i.e., for small distances as the MoM matrix with the subtracted far-field contribution, and for large distances simply as zero. The far-field zone matrix elements evaluated in terms of the above mentioned auxiliary basis functions are:

$$\overset{\rm F}{A}{}^{\mu-\nu}_{\alpha\beta} = \int {\rm d}t_1 {\rm d}t_2 \sum_{\mathbf{u},\mathbf{v}} T^\mu(t_1)\, V_{\alpha\mathbf{u}}\, G(t_1-t_2, \mathbf{u}-\mathbf{v})\, V^{\rm T}_{\mathbf{v}\beta}\, T^\nu(t_2) = \sum_{\mathbf{u},\mathbf{v}} V_{\alpha\mathbf{u}}\, G^{\mu-\nu}_{\mathbf{u}-\mathbf{v}}\, V^{\rm T}_{\mathbf{v}\beta}, \tag{19}$$

or

$$\overset{\rm F}{A}{}^{\mu-\nu} = V\, G^{\mu-\nu}\, V^{\rm T}. \tag{20}$$

By construction, the matrices $\overset{\mathrm{N}}{A}{}^{\mu-\nu}$ and V are sparse, and $G^{\mu-\nu}$ is a "block Toeplitz – Toeplitz block" matrix (i.e., a block Toeplitz matrix in temporal indices, composed of Toeplitz matrices in spatial indices) with

$$G_{\mathbf{u}-\mathbf{v}} = \begin{bmatrix} G^{\nu}_{\mathbf{u}-\mathbf{v}} & \cdots & G^{1}_{\mathbf{u}-\mathbf{v}} \\ \vdots & & \vdots \\ G^{2\nu-1}_{\mathbf{u}-\mathbf{v}} & \cdots & G^{\nu}_{\mathbf{u}-\mathbf{v}} \end{bmatrix}, \tag{21}$$

for all three-indices $\mathbf{u}$ and $\mathbf{v}$ corresponding to the N_{C} nodes of the Cartesian grid.

Now the matrix-vector multiplication

$$y^{\mu}_{\alpha} = \sum_{\nu}\sum_{\beta} A^{\mu-\nu}_{\alpha\beta}\, x^{\nu}_{\beta} = \overset{\mathrm{N}}{y}{}^{\mu}_{\alpha} + \overset{\mathrm{F}}{y}{}^{\mu}_{\alpha}, \tag{22}$$

where

$$\overset{\mathrm{N}}{y}{}^{\mu}_{\alpha} = \sum_{\nu}\sum_{\beta} \overset{\mathrm{N}}{A}{}^{\mu-\nu}\, x^{\nu}_{\beta}, \tag{23}$$

and

$$\overset{\mathrm{F}}{y}{}^{\mu}_{\alpha} = \sum_{\nu}\sum_{\beta} \overset{\mathrm{F}}{A}{}^{\mu-\nu}\, x^{\nu}_{\beta} = \sum_{\mathbf{uv}}\sum_{\nu\beta} V_{\alpha\mathbf{u}}\, G^{\mu-\nu}_{\mathbf{u}-\mathbf{v}}\, V^{\mathrm{T}}_{\mathbf{v}\beta}\, x^{\nu}_{\beta}, \tag{24}$$

can be carried out by means of a space-time FFTs of size $N_{\mathrm{C}}N_{\mathrm{t}}$,

$$y^{\mu}_{\alpha} = \sum_{m} \mathcal{F}^{-1\,\mu m}\left[\sum_{\beta} \widetilde{A}^{m}_{\alpha\beta}\, \widetilde{x}^{m}_{\beta} + \sum_{\mathbf{u}} V_{\alpha\mathbf{u}} \sum_{\mathbf{a}} \mathcal{F}^{-1}_{\mathbf{ua}}\, \widetilde{\widetilde{G}}{}^{m}_{\mathbf{a}} \sum_{\mathbf{v}} \mathcal{F}_{\mathbf{av}} \sum_{\beta} V^{\mathrm{T}}_{\mathbf{v}\beta}\, \widetilde{x}^{m}_{\beta}\right], \tag{25}$$

where $\mathcal{F}$ denotes a Fourier transform, and

$$\widetilde{A}^{m}_{\alpha\beta} = \sum_{\mu} \mathcal{F}^{m\mu}\, \overset{\mathrm{N}}{A}{}^{\mu}_{\alpha\beta}, \quad \widetilde{x}^{m}_{\beta} = \sum_{\mu} \mathcal{F}^{m\mu}\, x^{\mu}_{\beta}, \quad \widetilde{\widetilde{G}}{}^{m}_{\mathbf{a}} = \sum_{\mu}\sum_{\mathbf{u}} \mathcal{F}^{m\mu}\, \mathcal{F}_{\mathbf{au}}\, G^{\mu}_{\mathbf{u}}. \tag{26}$$

In Table 1 we list the consecutive operations required to compute near- and far-field matrix-vector multiplication resulting from the prescription (25), and the associated complexity at a given level of recursion k. Note that $N_{\mathrm{s}}^{\mathrm{Near}}$ is the number of current elements in the near-field range of a given element, independent of N_{s}.

We note that that the total operation count for the solution becomes $\sim N_{\mathrm{t}}N_{\mathrm{C}} \log N_{\mathrm{t}} \log(N_{\mathrm{C}})$. N_{C} is the total number of Cartesian grid nodes where the equivalent sources are located; $N_{\mathrm{C}} \sim N_{\mathrm{s}}^{4/3}$ for surface problems, and $N_{\mathrm{C}} \sim N_{\mathrm{s}}$ for volume problems. (The surface problem estimate assumes an alternative equivalent source representation located on a set of parallel planes.[10]) Hence, in terms of the original spatial and temporal unknowns, N_{t} and N_{s}, the total operation count for the solution is

$$\mathcal{N}[Ax=b\,|\,\text{FFT-time/space}] \sim \begin{cases} N_{\mathrm{t}}N_{\mathrm{s}} \log N_{\mathrm{t}} \log N_{\mathrm{s}} & \text{for volume problems}, \\ N_{\mathrm{t}}N_{\mathrm{s}}^{4/3} \log N_{\mathrm{t}} \log N_{\mathrm{s}} & \text{for surface problems}. \end{cases} \tag{27}$$

Table 1. Operation counts for near- and far-field matrix-vector products in FFT-TDIE.

	near-field	**complexity on level k**
1. FT of a matrix	$\widetilde{A}^{m}_{\alpha\beta} = \Sigma_{\mu}\, \mathcal{F}^{m\mu}\, \overset{\mathrm{N}}{A}{}^{\mu}_{\alpha\beta}$	$N_{\mathrm{t}}/2^k\, N_{\mathrm{s}}\, N_{\mathrm{s}}^{\mathrm{Near}} \log(N_{\mathrm{t}}/2)$
2. FT of a vector	$\widetilde{x}^{m}_{\beta} = \Sigma_{\mu}\, \mathcal{F}^{m\mu}\, x^{\mu}_{\beta}$	$N_{\mathrm{t}}/2^k\, N_{\mathrm{s}} \log(N_{\mathrm{t}}/2^k)$
3. product of FTs	$\widetilde{y}^{m}_{\alpha} = \Sigma_{\beta}\, \widetilde{A}^{m}_{\alpha\beta}\, \widetilde{x}^{m}_{\beta}$	$N_{\mathrm{t}}/2^k\, N_{\mathrm{s}}\, N_{\mathrm{s}}^{\mathrm{Near}}$
4. inverse FT of a vector	$y^{\mu}_{\alpha} = \Sigma_{\mu}\, \mathcal{F}^{-1\,\mu m}\, \widetilde{y}^{m}_{\alpha}$	$N_{\mathrm{t}}/2^k\, N_{\mathrm{s}} \log(N_{\mathrm{t}}/2^k)$
total complexity (all levels):		$N_{\mathrm{t}}\, N_{\mathrm{s}}\, N_{\mathrm{s}}^{\mathrm{Near}} \log^2(N_{\mathrm{t}})$

	far-field	**complexity on level k**
0. FT of the Green's function (precomputed)	$\widetilde{\widetilde{g}}{}^{m}_{\mathbf{a}} = \Sigma_{\mu}\Sigma_{\mathbf{u}}\mathcal{F}^{m\mu}\, \mathcal{F}_{\mathbf{au}}\, g^{\mu}_{\mathbf{u}}$	$N_{\mathrm{t}}/2^k\, N_{\mathrm{C}} \log(N_{\mathrm{t}}/2^k\, N_{\mathrm{C}})$
1. FT of a current vector	$\widetilde{x}^{m}_{\beta} = \Sigma_{\mu}\mathcal{F}^{m\mu}\, x^{\mu}_{\beta}$	$N_{\mathrm{t}}/2^k\, N_{\mathrm{s}} \log(N_{\mathrm{t}}/2^k)$
2. product of a sparse matrix and a vector	$\widehat{\widetilde{x}}{}^{m}_{\mathbf{v}} = \Sigma_{\beta} V^{\mathrm{T}}_{\mathbf{v}\beta}\, \widetilde{x}^{m}_{\beta}$	$N_{\mathrm{t}}/2^k\, N_{\mathrm{C}}\, N_{\mathrm{s}}^{\mathrm{Near}}$
3. FT of a vector	$\widetilde{\widetilde{x}}{}^{m}_{\mathbf{a}} = \Sigma_{\mathbf{v}}\mathcal{F}_{\mathbf{av}}\, \widehat{\widetilde{x}}{}^{m}_{\mathbf{v}}$	$N_{\mathrm{t}}/2^k\, N_{\mathrm{C}} \log(N_{\mathrm{C}})$
4. product of vectors	$\widetilde{\widetilde{y}}{}^{m}_{\mathbf{a}} = \widetilde{\widetilde{g}}{}^{m}_{\mathbf{a}}\, \widetilde{\widetilde{x}}{}^{m}_{\mathbf{a}}$	$N_{\mathrm{t}}/2^k\, N_{\mathrm{C}}$
5. inverse FT of a vector	$\widehat{\widetilde{y}}{}^{m}_{\mathbf{u}} = \Sigma_{\mathbf{a}}\mathcal{F}^{-1}_{\mathbf{ua}}\, \widetilde{\widetilde{y}}{}^{m}_{\mathbf{a}}$	$N_{\mathrm{t}}/2^k\, N_{\mathrm{C}} \log(N_{\mathrm{C}})$
6. product of a sparse matrix and a vector	$\widetilde{y}^{m}_{\alpha} = \Sigma_{\mathbf{u}} V_{\alpha\mathbf{u}}\, \widehat{\widetilde{y}}{}^{m}_{\mathbf{u}}$	$N_{\mathrm{t}}/2^k\, N_{\mathrm{C}}\, N_{\mathrm{s}}^{\mathrm{Near}}$
7. inverse FT of a vector	$y^{\mu}_{\alpha} = \Sigma_{m}\mathcal{F}^{-1\,\mu m}\, \widetilde{y}^{m}_{\alpha}$	$N_{\mathrm{t}}/2^k\, N_{\mathrm{s}} \log(N_{\mathrm{t}}/2^k)$
total complexity (all levels):		$N_{\mathrm{t}}\, N_{\mathrm{C}} \log(N_{\mathrm{t}}) \log(N_{\mathrm{C}})$

To summarize the main points of this Section, we stress that, since the proposed approach relies only on the ***translational invariance*** of the Green's function, it is directly applicable to all integral equation formulations utilizing Green's functions of ***unbounded and homogeneous media***. It applies, in particular, to (a) volumetric integral equation formulation for bulk, piecewise homogeneous dispersive and non-dispersive media utilizing free-space or infinite homogeneous medium Green's functions, or, (b) surface integral equation formulation for piecewise homogeneous dispersive and non-dispersive materials containing dispersive or non-dispersive interfaces characterized by surface resistivity parameters, R, S, and U, and utilizing the free-space Green's function.[11]

The complexity estimates (Eq.(27)) follow only from the Toeplitz structure of the matrices and do not depend on their degree of sparsity. Therefore, the performance of the proposed algorithm remains the same for the non-dispersive, as well as dispersive media, and does not deteriorate with the increasing strength of dispersion.

5. PROCEDURE FOR EVALUATION OF MATRIX ELEMENTS OF TIME DOMAIN INTEGRAL OPERATORS

In this Section, we will demonstrate that the causal behavior of the time domain (TD) Green's function can be accounted for rigorously through the utilization of the Galerkin procedure in space and time variables in the evaluation of the elements of the impedance matrix.

As an example we will consider electric field time domain integral equations (TDIEs) for geometries composed of dispersive penetrable and impenetrable surfaces embedded in free space,

$$\frac{1}{c}\int dt' \begin{bmatrix} R(t-t',\mathbf{r}) & U(t-t',\mathbf{r})\,\mathbf{n}(\mathbf{r})\times \\ U(t-t',\mathbf{r})\,\mathbf{n}(\mathbf{r})\times & S(t-t',\mathbf{r}) \end{bmatrix} \begin{bmatrix} \mathbf{J}(t',\mathbf{r}) \\ \mathbf{M}(t',\mathbf{r}) \end{bmatrix} +$$

$$\begin{bmatrix} \int\limits_{-\infty}^{t} d\tau \left(\frac{1}{c^2}\partial_\tau^2 - \nabla\nabla\cdot\right) \int dt' d^3r' g(\tau-t',\mathbf{r}-\mathbf{r}') & \frac{1}{c}\nabla\times\int dt' d^3r' g(t-t',\mathbf{r}-\mathbf{r}') \\ -\frac{1}{c}\nabla\times\int dt' d^3r' g(t-t',\mathbf{r}-\mathbf{r}') & \int\limits_{-\infty}^{t} d\tau\left(\frac{1}{c^2}\partial_\tau^2 - \nabla\nabla\cdot\right)\int dt' d^3r' g(\tau-t',\mathbf{r}-\mathbf{r}') \end{bmatrix}$$

$$\begin{bmatrix} \mathbf{J}(t',\mathbf{r}') \\ \mathbf{M}(t',\mathbf{r}') \end{bmatrix} = \begin{bmatrix} \mathbf{E}^{i}(t,\mathbf{r}) \\ \mathbf{H}^{i}(t,\mathbf{r}) \end{bmatrix}, \tag{28}$$

with the TD Green's function in free space given by

$$g(t,\mathbf{r}) = \int \frac{d\omega}{2\pi}\, e^{-i\omega t}\, \tilde{g}(\omega,\mathbf{r}) = \frac{1}{4\pi\,|\mathbf{r}|}\,\delta(t-|\mathbf{r}|/c)\,. \tag{29}$$

The quantities $\mathbf{E}^i(t,\mathbf{r})$, $\mathbf{H}^i(t,\mathbf{r})$, $\mathbf{J}(t',\mathbf{r}')$, $\mathbf{M}(t',\mathbf{r}')$, $R(t,\mathbf{r})$, $S(t,\mathbf{r})$, and $U(t,\mathbf{r})$ are the Fourier transforms of incident fields, electric and magnetic currents, and dimensionless resistivities; for example,

$$\mathbf{E}^i(t,\mathbf{r}) = \int \frac{d\omega}{2\pi}\, e^{-i\omega t}\, \tilde{\mathbf{E}}^i(\omega,\mathbf{r})\,. \tag{30}$$

By using the ***Galerkin discretization*** and the space-time basis functions of Eq.(2), Eqs. (28) give rise to three distinct types of matrix elements of the Green's function $(G_{\alpha\beta}^{\mu\nu(1)}, G_{\alpha\beta}^{\mu\nu(2)}, G_{\alpha\beta}^{\mu\nu(3)})$:

$$\begin{aligned} G_{\alpha\beta}^{\mu\nu(1)} &= c^{-2}\int dt_1\, dt_2\, d^3r_1\, d^3r_2 \\ &\quad T^\mu(t_1)\,\boldsymbol{\Psi}_\alpha(\mathbf{r}_1)\cdot \int_{-\infty}^{t_1} d\tau\, \partial_\tau^2\, g(\tau-t_2,\mathbf{r}_1-\mathbf{r}_2)\, T^\nu(t_2)\,\boldsymbol{\Psi}_\beta(\mathbf{r}_2)\,, \end{aligned} \tag{31a}$$

$$\begin{aligned} G_{\alpha\beta}^{\mu\nu(2)} &= -\int dt_1\, dt_2\, d^3r_1\, d^3r_2 \\ &\quad T^\mu(t_1)\,\boldsymbol{\Psi}_\alpha(\mathbf{r}_1)\cdot \int_{-\infty}^{t_1} d\tau\, \nabla\,\nabla\cdot g(\tau-t_2,\mathbf{r}_1-\mathbf{r}_2)\, T^\nu(t_2)\,\boldsymbol{\Psi}_\beta(\mathbf{r}_2)\,, \end{aligned} \tag{31b}$$

$$\begin{aligned} G_{\alpha\beta}^{\mu\nu(3)} &= c^{-1}\int dt_1\, dt_2\, d^3r_1\, d^3r_2 \\ &\quad T^\mu(t_1)\,\boldsymbol{\Psi}_\alpha(\mathbf{r}_1)\cdot \left(\nabla g(t_1-t_2,\mathbf{r}_1-\mathbf{r}_2)\times T^\nu(t_2)\,\boldsymbol{\Psi}_\beta(\mathbf{r}_2)\right)\,. \end{aligned} \tag{31c}$$

Evaluation of the above matrix elements constitutes a challenging analytical task and a critical element of the numerical implementation. The procedure we developed exploits the ***translational invariance in temporal and spatial coordinates*** of the time-domain integral operators. This property allows us to represent all the required matrix elements of Eqs. (31), as well as their analogues for dispersive media, in the following generic forms

$$G^{\mu\nu(m)}_{\alpha\beta} = \int \mathrm{d}t \int \mathrm{d}^3 r \, g(t,r)\, \eta^{\mu\nu(m)}(t)\, \rho^{(m)}_{\alpha\beta}(\mathbf{r}) \,, \qquad \text{for } m = 1, 2, \tag{32}$$

and

$$G^{\mu\nu(m)}_{\alpha\beta} = \int \mathrm{d}t \int \mathrm{d}^3 r \, [r \frac{\partial}{\partial r} g(t,r)]\, \eta^{\mu\nu(m)}(t)\, \rho^{(m)}_{\alpha\beta}(\mathbf{r}) \,, \text{ for } m = 3, \tag{33}$$

where $g(t,r) = g(t_1 - t_2, |\mathbf{r}_1 - \mathbf{r}_2|)$ is the scalar Green's function, and

$$\eta^{\mu\nu(m)}(t) = \eta^{\mu\nu(m)}(t_1 - t_2) \quad \text{and} \quad \rho^{(m)}_{\alpha\beta}(\mathbf{r}) = \rho^{(m)}_{\alpha\beta}(\mathbf{r}_1 - \mathbf{r}_2) \tag{34}$$

represent suitably constructed "temporal separation density functions" and "spatial separation density functions", dependent only on the relative temporal ($t = t_1 - t_2$) and spatial ($\mathbf{r} = \mathbf{r}_1 - \mathbf{r}_2$) coordinates. Further, by defining angular averages

$$\langle \rho^{(m)}_{\alpha\beta}(r) \rangle = (4\pi)^{-1} \int \mathrm{d}\Omega_{\mathbf{r}}\, \rho^{(m)}_{\alpha\beta}(\mathbf{r}) \,, \tag{35}$$

the matrix elements (32) can be expressed as double integrals

$$G^{\mu\nu(m)}_{\alpha\beta} = 4\pi \int \mathrm{d}t \int_0^\infty \mathrm{d}r\, r^2\, g(t,r)\, \eta^{\mu\nu(m)}(t)\, \langle \rho^{(m)}_{\alpha\beta}(r) \rangle \,, \tag{36}$$

where

$$\eta^{\mu\nu(1)}(t) = c^{-2} \int \mathrm{d}t'\, T^\mu(t+t')\, \partial_{t'} T^\nu(t') \,, \tag{37a}$$

$$\eta^{\mu\nu(2)}(t) = \int_t^\infty \mathrm{d}\tau \int \mathrm{d}t'\, T^\mu(\tau + t')\, T^\nu(t') \,, \tag{37b}$$

$$\eta^{\mu\nu(3)}(t) = c^{-1} \int \mathrm{d}t'\, T^\mu(t+t')\, T^\nu(t') \,, \tag{37c}$$

$$\begin{aligned}\langle \rho^{(1)}_{\alpha\beta}(r) \rangle &= (4\pi)^{-1} \int \mathrm{d}\Omega_{\mathbf{r}} \int \mathrm{d}^3 r_1 \mathrm{d}^3 r_2\, \delta^3(\mathbf{r} - \mathbf{r}_1 + \mathbf{r}_2)\, \mathbf{\Psi}_\alpha(\mathbf{r}_1) \mathbf{\Psi}_\beta(\mathbf{r}_2) \\ &= (4\pi)^{-1} \int \mathrm{d}\Omega_{\mathbf{r}} \int \mathrm{d}^3 r'\, \mathbf{\Psi}_\alpha(\mathbf{r} + \mathbf{r}') \cdot \mathbf{\Psi}_\beta(\mathbf{r}') \,,\end{aligned} \tag{37d}$$

$$\begin{aligned}\langle \rho^{(2)}_{\alpha\beta}(r) \rangle &= (4\pi)^{-1} \int \mathrm{d}\Omega_{\mathbf{r}} \int \mathrm{d}^3 r_1 \mathrm{d}^3 r_2\, \delta^3(\mathbf{r} - \mathbf{r}_1 + \mathbf{r}_2)\, \nabla \cdot \mathbf{\Psi}_\alpha(\mathbf{r}_1)\, \nabla \cdot \mathbf{\Psi}_\beta(\mathbf{r}_2) \\ &= (4\pi)^{-1} \int \mathrm{d}\Omega_{\mathbf{r}} \int \mathrm{d}^3 r'\, \nabla \cdot \mathbf{\Psi}_\alpha(\mathbf{r} + \mathbf{r}')\, \nabla \cdot \mathbf{\Psi}_\beta(\mathbf{r}') \,,\end{aligned} \tag{37e}$$

$$\begin{aligned}\langle \rho^{(3)}_{\alpha\beta}(r) \rangle &= (4\pi)^{-1} \int \mathrm{d}\Omega_{\mathbf{r}} \int \mathrm{d}^3 r_1 \mathrm{d}^3 r_2\, \delta^3(\mathbf{r} - \mathbf{r}_1 + \mathbf{r}_2) \mathbf{\Psi}_\alpha(\mathbf{r}') \cdot (\nabla \times \mathbf{\Psi}_\beta(\mathbf{r}_2)) \\ &= (4\pi)^{-1} \int \mathrm{d}\Omega_{\mathbf{r}} \int \mathrm{d}^3 r'\, \mathbf{\Psi}_\alpha(\mathbf{r} + \mathbf{r}') \cdot (\nabla \times \mathbf{\Psi}_\beta(\mathbf{r}')) \,.\end{aligned} \tag{37f}$$

We note that an alternative form of integral equations, obtained by taking the time derivative of Eq. (28), is also used in the literature. When discretized, this form would lead to matrix elements of the Green function expressible through the same spatial and temporal separation density functions (Eq. (37)), with the exception that an additional temporal separation density function, defined below, is needed:

$$\eta^{\mu\nu(4)}(t) = -c^{-2} \int \mathrm{d}t'\, \partial_t T^{\mu}(t+t')\, \partial_{t'} T^{\nu}(t') \ . \tag{38}$$

6. CONSTRUCTION OF TEMPORAL AND SPATIAL SEPARATION DENSITY FUNCTIONS

Our objective now is to provide explicit expressions for the spatial and temporal separation density functions, $\langle \rho^{(m)}_{\alpha\beta}(r) \rangle$ and $\eta^{\mu\nu(m)}(t)$, for a particular form of temporal and spatial basis functions. Such expressions can be arrived at by means of integration by parts and appropriate changes of integration variables.

As temporal basis functions, we use *non-overlapping* pulse basis functions in time,

$$T^{\mu}(t) = \frac{1}{\Delta t}\, \Theta(t - t_{\mu-1})\, \Theta(t_{\mu} - t) \ , \tag{39}$$

where $t_{\mu} = \mu\, \Delta t$. As spatial basis functions, we employ the RWG[12] rooftop-type basis functions,

$$\mathbf{\Psi}_{\alpha}(\mathbf{r}) = \begin{cases} (\mathbf{r} - \mathbf{v}^{+}_{\alpha})\, \chi^{+}_{\alpha}(\mathbf{r})/h^{+}_{\alpha} \equiv \mathbf{\Psi}^{+}_{\alpha}(\mathbf{r}) & \text{for } \mathbf{r} \in T^{+} \ , \\ (\mathbf{r} - \mathbf{v}^{-}_{\alpha})\, \chi^{-}_{\alpha}(\mathbf{r})/h^{-}_{\alpha} \equiv \mathbf{\Psi}^{-}_{\alpha}(\mathbf{r}) & \text{for } \mathbf{r} \in T^{-} \ , \\ 0 & \text{otherwise} \ , \end{cases} \tag{40}$$

where $T^{\pm}$ are two triangles adjacent to the edge α, $\chi^{\pm}_{\alpha}(\mathbf{r})$ are the characteristic functions of the triangles (equal to 1 on the triangle and 0 outside), $\mathbf{v}^{\pm}_{\alpha}$ denote vertices of the triangles $T^{\pm}$ opposite the edge α, and $h^{\pm}_{\alpha}$ are the heights of the triangles, measured from the common edge α.

We note that, in what follows, our formulae apply to "half-basis-functions" $\mathbf{\Psi}^{\pm}_{\alpha}$ defined on single triangles; we denote them generically $\mathbf{\Psi}_{\alpha}$.

Expressions for the temporal separation density functions are quite simple:

$$\eta^{\mu\nu(1)}(t) = T^{\mu-\nu+1}(t) - T^{\mu-\nu}(t) \ , \tag{41a}$$

$$\eta^{\mu\nu(2)}(t) = c \int_{t}^{\infty} \mathrm{d}\tau\, \eta^{\mu\nu(3)}(\tau) = H(\mu - \nu - t/\Delta t) \ , \tag{41b}$$

$$\eta^{\mu\nu(3)}(t) = \frac{1}{c\,\Delta t}\, h(\mu - \nu - t/\Delta t) \ , \tag{41c}$$

$$\eta^{\mu\nu(4)}(t) = -\frac{2}{c^2\,(\Delta t)^2} \left[\delta(t - (t_{\mu} - t_{\nu})) - \delta(t - (t_{\mu+1} - t_{\nu}))\right] \ , \tag{41d}$$

where

$$h(x) = \begin{cases} 1 - |x| & \text{for } 0 \le |x| \le 1 \ , \\ 0 & \text{otherwise} \ , \end{cases} \tag{42}$$

and

$$H(x) = \begin{cases} 0 & \text{for } |x| \le -1 \ , \\ \frac{1}{2}(1+x)^2 & \text{for } -1 \le x \le 0 \ , \\ 1 - \frac{1}{2}(1-x)^2 & \text{for } 0 \le x \le 1 \ , \\ 1 & \text{for } 1 \le x \ . \end{cases} \tag{43}$$

In the following we describe the procedure of evaluating the spatial density separation functions. Our technique eventually reduces the five-dimensional integrals of Eqs. (31) to two-dimensional nonsingular integrals. The main advantage of our method is that it does not require any "singularity extraction" procedure: the singular term associated with the scalar Green's function is canceled by the r^2 factor in the Jacobian, resulting in relatively simple two-dimensional integrals.

We define the triangular basis function supports, $\alpha : (\mathbf{v}_1, \mathbf{v}_2, \mathbf{v}_3)$ and $\beta : (\mathbf{w}_1, \mathbf{w}_2, \mathbf{w}_3)$, by the location of their vertices, numbered cyclically (so that $\mathbf{v}_4 = \mathbf{v}_1, \mathbf{w}_4 = \mathbf{w}_1$). In terms of the RWG basis functions, the spatial separation densities (Eq.(37)) acquire the form:

$$\langle \rho^{(1)}_{\alpha\beta}(r) \rangle = \frac{1}{4\pi h_\alpha h_\beta} \int d\Omega_{\mathbf{r}} \int d^3r' \, (\mathbf{r} + \mathbf{r}' - \mathbf{v}_\alpha)\, \chi_\alpha(\mathbf{r}+\mathbf{r}') \cdot (\mathbf{r}' - \mathbf{w}_\beta) \chi_\beta(\mathbf{r}'), \tag{44a}$$

$$\begin{aligned} \langle \rho^{(2)}_{\alpha\beta}(r) \rangle &= \frac{1}{4\pi h_\alpha h_\beta} \int d\Omega_{\mathbf{r}} \int d^3r' \\ &\quad \left(2\,\chi_\alpha(\mathbf{r}+\mathbf{r}') - (\mathbf{r}+\mathbf{r}'-\mathbf{v}_\alpha)\cdot \hat{\mathbf{u}}_{l_\alpha}\delta_{l_\alpha}\right)\left(2\,\chi_\beta(\mathbf{r}') - (\mathbf{r}'-\mathbf{w}_\beta)\cdot\hat{\mathbf{u}}_{l_\beta}\delta_{l_\beta}\right) \\ &\doteq \frac{1}{4\pi h_\alpha h_\beta} \int d\Omega_{\mathbf{r}} \int d^3r' \, 4\,\chi_\alpha(\mathbf{r}+\mathbf{r}')\,\chi_\beta(\mathbf{r}'), \end{aligned} \tag{44b}$$

$$\begin{aligned} \langle \rho^{(3)}_{\alpha\beta}(r) \rangle &= \frac{-1}{4\pi h_\alpha h_\beta r^2} \int d\Omega_{\mathbf{r}} \int d^3r' \\ &\quad [(\mathbf{v}_\alpha - \mathbf{w}_\beta)\cdot(\mathbf{r}\times\mathbf{r}') + (\mathbf{v}_\alpha \times \mathbf{w}_\beta)\cdot \mathbf{r}]\,\chi_\alpha(\mathbf{r}+\mathbf{r}')\,\chi_\beta(\mathbf{r}') \, . \end{aligned} \tag{44c}$$

To obtain the above formulas we used the relation $\nabla\chi_\alpha = -\delta_{\Gamma_\alpha}\hat{\mathbf{u}}_{\Gamma_\alpha}$, where δ_{Γ_α} is a δ-function supported on the boundary of the triangle α, and $\hat{\mathbf{u}}_{\Gamma_\alpha}$ is a unit vector perpendicular to the boundary of triangle α, located in the triangle plane and pointing outward the triangle. In (44b), we further used the fact that the term $(\mathbf{r}+\mathbf{r}'-\mathbf{v}_\alpha)\cdot\hat{\mathbf{u}}_\alpha\delta_{\Gamma_\alpha}$ is different from zero only on the edge l_α facing the vertex $\mathbf{v}_\alpha$ due to the vanishing of the scalar product, and similarly for the triangle β. We note however that, in (44b), terms $\sim \delta_{l_\alpha}$ and $\sim \delta_{l_\beta}$ do not contribute to the final result since they enter with opposite signs to full basis functions constructed out of two half-basis-functions adjacent to the same internal edge. If the edge is a boundary edge, these terms are not present at all. We will omit them in further derivations.

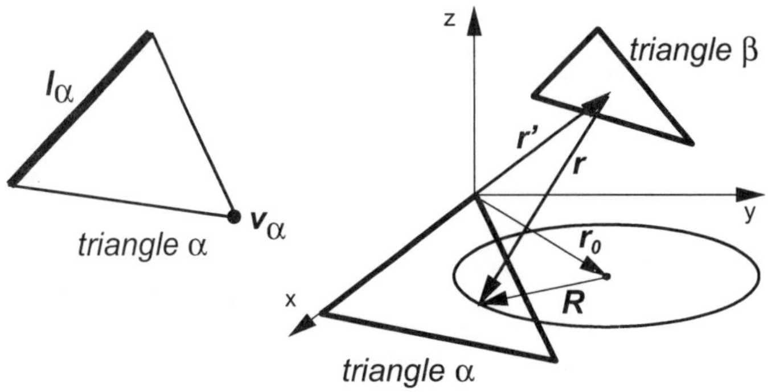

Figure 1: "Half basis function" support α associated with vertex $\mathbf{v}_\alpha$ and mutual orientation of half basis function supports α and β.

The basic steps of our integration procedure are:

1. Without losing generality, we position triangle the α in the (x, y) plane (Fig. 1). This allows us to explicitly extract the factor $\delta(z+z')$ from the characteristic function $\chi_\alpha(\mathbf{r}+\mathbf{r}') = \delta(z+z')\,\chi_\alpha(x+x', y+y')$.
2. We determine the range of r, outside which $\langle\rho_{\alpha\beta}^{(m)}\rangle$ vanish.
3. For each value of r we construct a sphere of radius r with the center at $\mathbf{r}'$ located on the triangle β.
4. We find the intersection of that sphere with the (x, y) plane coplanar with the triangle α. That intersection is a circle Γ of radius $R = \sqrt{r^2 - d^2}$, where $d = z'$ is the distance between the sphere center and the plane of triangle α. The circle is centered at $\mathbf{r}_0 = (x', y')$. The 2-dimensional integration over the solid angle $\mathrm{d}\Omega_{\mathbf{r}}$ reduces now to 1-dimensional integrals over arc segments Γ_α of Γ located inside the triangle α:

$$
\begin{aligned}
&\int \mathrm{d}\Omega_{\mathbf{r}}\delta(z+z')\chi_\alpha(x+x', y+y') \\
&\quad\to \int \mathrm{d}\cos\theta_{\mathbf{r}}\mathrm{d}\phi_{\mathbf{r}}\delta(rcos\theta_{\mathbf{r}}+z')\chi_\alpha(x+x', y+y') \\
&\quad\to \frac{1}{r}\int \mathrm{d}\cos\theta_{\mathbf{r}}\mathrm{d}\phi_{\mathbf{r}}\delta(\cos\theta_{\mathbf{r}}+\frac{z'}{r})\chi_\alpha(x+x', y+y') \to \frac{1}{r}\int_{\Gamma_\alpha} \mathrm{d}\phi_{\mathbf{r}}\,. \qquad (45)
\end{aligned}
$$

5. We carry out line integrals over arc segments of Γ lying inside the triangle α, with the integration limits determined by the circle intersection with the sides of the triangle. We observe that there are seven possible relative spatial configurations (Fig. 2) of triangles α and β, and equivalently, seven relative configurations of triangle α and circle Γ, for which integration limits should be found and integrals (44) evaluated. To find the crossing points of the sides of triangle α with circle Γ, we write the parametric equation of the circle Γ

$$
\Gamma(x', y', z') : \ \{\ \mathbf{r}_0(x', y') \ + \ R(z') \begin{pmatrix} \cos\phi \\ \sin\phi \\ 0 \end{pmatrix}, \quad 0 \le \phi < 2\pi\ \}, \qquad (46)
$$

where we indicate the dependence of Γ on the coordinates x', y', z' of the point $\mathbf{r}'$. Subsequently we write three parametric equations for the three sides of triangle α:

$$\text{side 1:} \quad \mathbf{r} = \mathbf{v}_2 + (\mathbf{v}_3 - \mathbf{v}_2)\,\eta\ , \tag{47a}$$

$$\text{side 2:} \quad \mathbf{r} = \mathbf{v}_3 + (\mathbf{v}_1 - \mathbf{v}_3)\,\eta\ , \tag{47b}$$

$$\text{side 3:} \quad \mathbf{r} = \mathbf{v}_1 + (\mathbf{v}_2 - \mathbf{v}_1)\,\eta\ , \tag{47c}$$

with $0 \le \eta \le 1$. Finding the crossing points amounts to solving a simple quadratic equation for the parameters ϕ and η. There may be, for a given side of triangle α, two ($\phi^{\pm}$), one (ϕ^{0}), or no solutions.

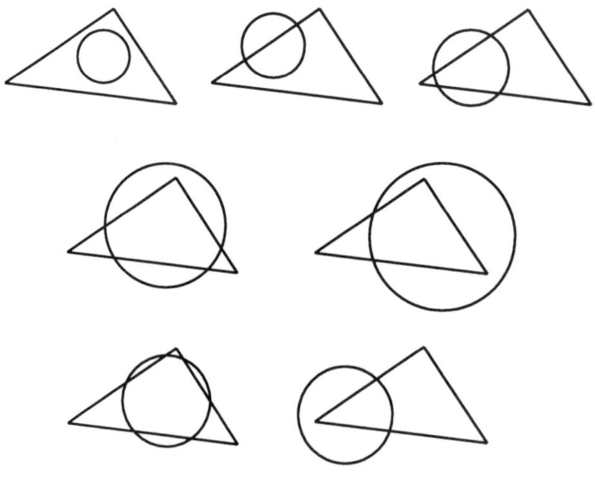

Figure 2: Seven spatial configurations for intersections of circle Γ with triangle α.

Now, the separation density $\langle \rho^{(m)}_{\alpha\beta}(r) \rangle$ can be written as a sum of contributions of circle arcs inside the triangle (see Fig. 3),

$$\langle \rho^{(m)}_{\alpha\beta}(r) \rangle = \sum_{n=1}^{n_{\max}} F^{(m)}_n(r)\ . \tag{48}$$

In evaluating the functions $F^{(m)}_n(r)$, we reduce Eqs. (44a) through (44c) to two-dimensional integrals over the surface of triangle β in barycentric coordinates with $\mathbf{r}'(\xi_1, \xi_2) = \mathbf{w}_1\,\xi_1 + \mathbf{w}_2\,\xi_2 + \mathbf{w}_3\,(1 - \xi_1 - \xi_2)$, with $0 \le \xi_1 \le 1,\ 0 \le \xi_2 \le 1$ and $0 \le \xi_1 + \xi_2 \le 1$. As the result we find the following explicit expressions for $F^{(m)}_n(r)$:

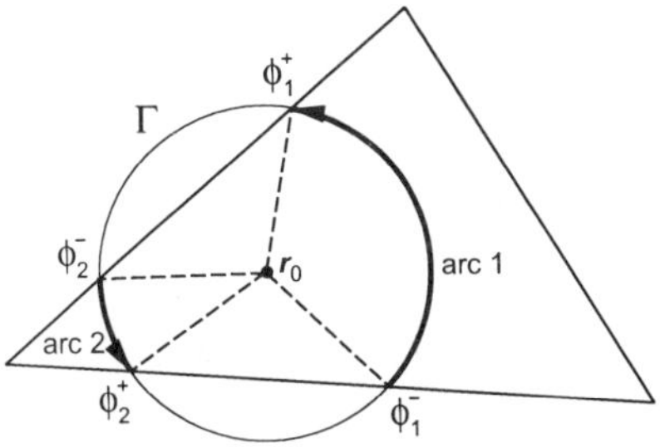

Figure 3: An example of contribution of two arcs to the integrals given in (36).

$$
\begin{aligned}
F_n^{(1)}(r) &= \frac{2A_\beta}{4\pi h_\alpha h_\beta r} \int \mathrm{d}\xi_1 \mathrm{d}\xi_2 \Big(\mathbf{r}'(\xi_1,\xi_2) - \mathbf{w}_\beta\Big) \cdot \int_{\phi_n^-(\xi_1,\xi_2)}^{\phi_n^+(\xi_1,\xi_2)} \mathrm{d}\phi \Big(\mathbf{C}(\xi_1,\xi_2) - \mathbf{v}_\alpha\Big), \\
F_n^{(2)}(r) &= \frac{2A_\beta}{\pi\, h_\alpha\, h_\beta r} \int \mathrm{d}\xi_1\, \mathrm{d}\xi_2 \int_{\phi_n^-(\xi_1,\xi_2)}^{\phi_n^+(\xi_1,\xi_2)} \mathrm{d}\phi, \\
F_n^{(3)}(r) &= \frac{1}{(8\pi)r^2} \frac{2A_\beta}{r h_\alpha h_\beta} (\mathbf{v}_\alpha - \mathbf{w}_\beta) \cdot \int d\xi_1 d\xi_2 \int_{\phi_n^-(\xi_1,\xi_2)}^{\phi_n^+(\xi_1,\xi_2)} \mathrm{d}\phi \\
&\quad [\mathbf{C}(\xi_1,\xi_2) - \mathbf{r}'(\xi_1,\xi_2)] \times [\mathbf{v}_\alpha + \mathbf{w}_\beta + \mathbf{C}(\xi_1,\xi_2) + \mathbf{r}'(\xi_1,\xi_2)], \qquad (49)
\end{aligned}
$$

where

$$
\mathbf{C}(\xi_1,\xi_2) = \mathbf{r}_0(\xi_1,\xi_2) + R(\xi_1,\xi_2) \begin{pmatrix} \cos\phi \\ \sin\phi \\ 0 \end{pmatrix}, \qquad (50)
$$

A_β is the area of triangle β, and the integration limits $\phi_n^\pm(\xi_1,\xi_2)$ correspond to the solutions of the above mentioned quadratic equations for the crossing points between the circle and the sides of triangle α.

The results of this Section demonstrate that, by proper transformation of variables, matrix elements of the full (tensor and vector) TD Green's function can be expressed as integrals of a product of the scalar Green's function and temporal and spatial separation density functions. The approach is, therefore, equally applicable to dispersive and non-dispersive media, and requires, in both cases, tabulation and computation of the Green's function as a function of t and r.

7. SUMMARY

We described elements of a new time-domain integral equation algorithm applicable to problems characterized by dispersive thin sheet and dispersive volumetric material properties. Two essential novel elements of our algorithm are (a) an impedance matrix compression and solution method based on ***simultaneous*** application of ***FFTs in space and time***, and (b) a rigorous treatment of the causal behavior of the time-domain Green's function executed in the context of ***complete Galerkin discretization in spatial and temporal variables***.

An important property of the proposed compression and solution method is that it relies only on the ***translational invariance*** of the Green's function, and not on its specific analytic form. Therefore, the method is directly applicable to all integral-equation formulations utilizing unbounded homogeneous medium Green's functions, either in the analytic or tabulated form. The method is equally applicable to non-dispersive and dispersive media, and its computational complexity is independent of the degree of dispersion.

Our Galerkin discretization procedure relies on a suitable ***transformation*** of the integral expressions for the elements of the impedance matrix into two-dimensional integrals over the ***relative (scalar) distance*** r and the ***relative time coordinate*** τ. Therefore, it allows for a rigorous and straightforward implementation in three dimensions. The procedure applies to the full time-domain Green's function, including tensor and vector components. Again, the procedure is equally applicable to dispersive and non-dispersive media.

Also, the proposed algorithm is based on the original causal form of the Green's function, and therefore does not require any additional procedures such as gating out anti-causal components of the signal, appearing, e.g., in Ref. 3.

ACKNOWLEDGMENT

This work is supported by the AFOSR grant F49620-01-1-0504.

REFERENCES

1. S.P. Walker, Developments in time-domain equation modeling at Imperial College, *IEEE Antennas and Propagation Magazine*, **39**, pp. 7-19 (1997).
2. D.A. Vechinski and S.M. Rao, A stable procedure to calculate the transient scattering by conducting surfaces of arbitrary shape, *IEEE Trans. on Antennas and Propagation*, **40**, pp. 661-665 (1992).
3. A.A. Egrin, B. Shanker, and E. Michielssen, The plane wave time domain algorithm for the fast analysis of transient wave phenomena, *IEEE Antennas and Propagation Magazine*, **41**, pp. 39-52 (1999).
4. B. Shanker, A.A. Egrin, K. Aygun, and E. Michielssen, Analysis of transient electromagnetic scattering phenomena using a two-level plane wave time domain algorithm, *IEEE Trans. on Antennas and Propagation*, **48**, pp. 510-523 (2000).
5. G.W. Stewart, On the solution of block Hessenberg systems, *Numer. Linear Alg. with Appl.*, **2**, pp. 287-296 (1995).
6. E. Bleszynski, M. Bleszynski, and T. Jaroszewicz, Fast time domain integral equations solver for electromagnetic scattering from dispersive media, *2001 IEEE Antennas and Propagation Society International Symposium Proceedings*, **4**, pp. 176-179 (2001).
7. T. Kailath and A. Sayed, Displacement structure: theory and applications, *SIAM Review*, **37**, pp. 297-386 (1995).
8. E. Bleszynski, M. Bleszynski, and T. Jaroszewicz, AIM: Adaptive Integral Method for solving large-scale electromagnetic scattering and radiation problems, *Radio Science*, **31**, pp. 1225-1251 (1996).
9. E. Bleszynski, M. Bleszynski, and T. Jaroszewicz, Least-squares based far-field expansion in the Adaptive Integral Method (AIM), *13th Annual Review of Progress in Applied Computational Electromagnetics Proceedings*, Monterey, CA, **2**,pp. 944-950 (1997).
10. O.P. Bruno and L.E. Kunyansky, Surface scattering in 3-D: an accelerated high order solver, *Journal of Computational Physics*, **169**, pp. 80-110 (2001).
11. E. Bleszynski, M. Bleszynski, and T. Jaroszewicz, Surface integral equations for electromagnetic scattering from impenetrable and penetrable sheets, *IEEE Antennas and Propagation Magazine*, **35**, pp. 14-25 (1993).
12. S.M. Rao, D.R. Wilton, and A.W. Glisson, Electromagnetic scattering by surfaces of arbitrary shape, *IEEE Trans. on Antennas and Propagation*, **30**, pp. 409-418 (1982).

EFFECT OF THE INVERSION SYMMETRY OF DISPERSION RELATIONS ON FOUR-WAVE INTERACTIONS IN PHOTONIC CRYSTAL

Anatoli Babin and Alexander Figotin*

1. INTRODUCTION

The effect of the spatial periodicity on nonlinear optical processes has been a subject of intensive studies in physics literature[1–27] and references therein. We study nonlinear wavepacket propagation assuming that its spatial range is much smaller than the dimension of the medium sample, therefore considering an infinite periodic medium. It is assumed that electromagnetic wave propagation is governed by the classical Maxwell equations

$$\nabla \times \mathbf{E}(\mathbf{r},t) = -\frac{1}{c}\partial_t \mathbf{B}(\mathbf{r},t) - \frac{4\pi}{c}\mathbf{J}_M(\mathbf{r},t), \ \nabla \cdot \mathbf{B}(\mathbf{r},t) = 0, \tag{1}$$

$$\nabla \times \mathbf{H}(\mathbf{r},t) = \frac{1}{c}\partial_t \mathbf{D}(\mathbf{r},t) + \frac{4\pi}{c}\mathbf{J}_E(\mathbf{r},t), \ \nabla \cdot \mathbf{D}(\mathbf{r},t) = 0, \tag{2}$$

where $\mathbf{H}$, $\mathbf{E}$, $\mathbf{B}$, and $\mathbf{D}$ are respectively the magnetic and electric fields, magnetic and electric inductions; $\mathbf{J}_E$ and $\mathbf{J}_M$ are impressed electric and so called impressed magnetic currents; $\mathbf{r} = (x_1, x_2, x_3) \in \mathbf{R}^3$ is the position vector. It is also assumed that there are no free electric and magnetic charges, and, consequently, the fields $\mathbf{B}$ and $\mathbf{D}$ are divergence free as indicated in the equations (1) and (2). The vectors $\mathbf{H}$, $\mathbf{E}$, $\mathbf{B}$ and $\mathbf{D}$ are naturally 3-component. We also consider two special cases: (i) the 2-dimensional case when the fields do not depend on x_3; (ii) the 1-dimensional case when the fields do not depend on x_2, x_3. Hence, we will consider three cases with the dimension $d = 1, 2, 3$. For simplicity, we consider nonmagnetic media, i.e.

$$\mathbf{B}(\mathbf{r},t) = \mu \mathbf{H}(\mathbf{r},t), \ \mu = 1. \tag{3}$$

The material relations between $\mathbf{D}$ and $\mathbf{E}$ are assumed to be of the standard form[28]

$$\mathbf{D} = \mathbf{E} + 4\pi \mathbf{P}(\mathbf{r}, t; \mathbf{E}), \tag{4}$$

* A. Babin and A. Figotin, Department of Mathematics, University of California at Irvine, CA 92697.

where the polarization $\mathbf{P}$ is a sum of its linear and the nonlinear parts

$$\mathbf{P}(\mathbf{r},t;\mathbf{E}(\cdot)) = \mathbf{P}^{(1)}(\mathbf{r},t;\mathbf{E}(\cdot)) + \mathbf{P}_{\mathrm{NL}}(\mathbf{r},t;\mathbf{E}(\cdot)), \tag{5}$$

with the nonlinear part $\mathbf{P}_{\mathrm{NL}}(\mathbf{r},t;\mathbf{E}(\cdot))$ being a cubic nonlinearity.

Following Ref. 29, we choose $\mathbf{D},\mathbf{B}$ as primary variables and rewrite the Maxwell equations (1), (2) in rescaled dimensionless variables as follows:

$$\partial_t \mathbf{U} = -i\mathbf{M}\mathbf{U} + \alpha\mathcal{F}_{\mathrm{NL}}(\mathbf{U}) - \mathbf{J};\ \mathbf{U}(t) = 0 \text{ for } t \le 0, \tag{6}$$

where

$$\mathbf{U} = \begin{bmatrix} \mathbf{D} \\ \mathbf{B} \end{bmatrix}, \ \mathbf{M}\mathbf{U} = i\begin{bmatrix} \nabla\times\mathbf{B} \\ -\nabla\times(\eta^{(1)}\mathbf{D}) \end{bmatrix}, \ \mathbf{J} = 4\pi\begin{bmatrix} \mathbf{J}_E \\ \mathbf{J}_M \end{bmatrix}, \tag{7}$$

$\eta^{(1)} = \eta^{(1)}(\mathbf{r}) = \varepsilon^{-1}(\mathbf{r})$ is the impermeability tensor, $\mathcal{F}_{\mathrm{NL}}(\mathbf{U})$ is a cubic nonlinearity. All the fields $\mathbf{D}, \mathbf{B}, \mathbf{J}_E$, and $\mathbf{J}_M$ are assumed to be divergence free. Both $\varepsilon(\mathbf{r})$ and $\mathbf{P}_{\mathrm{NL}}(\mathbf{r})$ are assumed to be periodic in $\mathbf{r}$ with the lattice of periods being a cubic lattice with the lattice constant 1. The parameter α in (6) evidently determines the relative significance of the nonlinearity. We assume that $\alpha \ll 1$, that is we study weakly nonlinear phenomena. More exactly, we study nonlinear regimes satisfying the following assumptions: (i) the amplitude of the wave component due to the nonlinearity does not exceed the amplitude of its linear component and (ii) the wave spatial range is much smaller than the dimension of the medium sample. Under these constraints the nonlinear processes are naturally described in terms of the Bloch modes and corresponding dispersion relations of the underlying linear periodic medium.[29]

1.1. Bloch Expansion

Basic spectral properties of the differential operator $\mathbf{M}$ with periodic coefficients are covered by the Floquet-Bloch theory.[30,31] Namely, all the eigenvalues and the eigenmodes of $\mathbf{M}$ are parametrized by two indices: zone (band) number $n = 1, 2, \ldots$, and the quasi-momentum $\mathbf{k}$ from the so called *Brillouin zone.* The Brillouin zone in our case is the cube $[-\pi,\pi]^d$. We assume that the positive eigenvalues are naturally ordered

$$0 \le \omega_1(\mathbf{k}) \le \omega_2(\mathbf{k}) \le \cdots, \quad \mathbf{k} \text{ in } [-\pi,\pi]^d. \tag{8}$$

Notice that the operator $\mathbf{M}$ has a property that if ω is an eigenfrequency then $-\omega$ is an eigenfrequency as well. To take into account the negative eigenfrequency we introduce pairs

$$\bar{n} = (\zeta, n) \text{ where } \zeta = \pm 1,\ n = 1, 2, \ldots, \tag{9}$$

and set

$$\omega_{\bar{n}}(\mathbf{k}) = \zeta\omega_n(\mathbf{k}), \text{ for } \bar{n} = (\zeta, n). \tag{10}$$

The corresponding Bloch eigenmodes $\tilde{\mathbf{G}}_{\bar{n}}(\mathbf{r},\mathbf{k})$ satisfy the following relations

$$\mathbf{M}\tilde{\mathbf{G}}_{\bar{n}}(\mathbf{r},\mathbf{k}) = \omega_{\bar{n}}(\mathbf{k})\,\tilde{\mathbf{G}}_{\bar{n}}(\mathbf{r},\mathbf{k}),\ \tilde{\mathbf{G}}_{\bar{n}}(\mathbf{r}+\mathbf{m},\mathbf{k}) = e^{i\mathbf{k}\cdot\mathbf{m}}\tilde{\mathbf{G}}_{\bar{n}}(\mathbf{r},\mathbf{k}),\ \mathbf{m}\text{ in }\mathbb{Z}^d, \quad (11)$$

and they are divergence-free. For every fixed quasimomentum $\mathbf{k}$ the eigenfunctions $\tilde{\mathbf{G}}_{\bar{n}}(\mathbf{r},\mathbf{k})$ with different $\bar{n}$ form an orthonormal basis. For any vector field $\mathbf{U}(\mathbf{r})$ we introduce its Bloch modal expansion

$$\mathbf{U}(\mathbf{r}) = \frac{1}{(2\pi)^d}\sum_{\bar{n}}\int_{[-\pi,\pi]^d}\tilde{U}_{\bar{n}}(\mathbf{k})\,\tilde{\mathbf{G}}_{\bar{n}}(\mathbf{r},\mathbf{k})\,d\mathbf{k}, \quad (12)$$

with $\tilde{U}_{\bar{n}}(\mathbf{k})$ being the Bloch amplitudes of $\mathbf{U}$.

1.2. First Nonlinear Response and Interacting Quads

If the excitation current $\mathbf{J}$ is chosen as an appropriate wavepacket, the solution $\mathbf{U}^{(0)}$ to the linear (with $\alpha = 0$) Maxwell equations becomes a zero order approximation to the solution of the nonlinear Maxwell equation (6).[29] It can be represented as

$$\mathbf{U}^{(0)}(t) = \sum_{\bar{n}}\frac{1}{(2\pi)^d}\int_{[-\pi,\pi]^d}\tilde{V}_{\bar{n}}^{(0)}(\mathbf{k},\tau)\,\mathbf{G}_{\bar{n}}(\mathbf{r},\mathbf{k})\,e^{-i\omega_{\bar{n}}(\mathbf{k})t}\,d\mathbf{k}, \quad \tau = \varrho t, \quad (13)$$

where τ is the "slow time" with $\rho \ll 1$, and $\tilde{V}_{\bar{n}}^{(0)}(\mathbf{k},\tau)$ are slowly varying in time modal amplitudes of the time harmonic carrier waves $\mathbf{G}_{\bar{n}}(\mathbf{r},\mathbf{k})\,e^{-i\omega_{\bar{n}}(\mathbf{k})t}$. The parameter $\varrho \sim \Delta\omega/\omega_0$ is the relative bandwidth of $\mathbf{J}$, where $\Delta\omega$ and ω_0 are respectively the bandwidth and the carrier frequency of $\mathbf{J}$.

Let us consider the first order approximation to $\mathbf{U}(t)$, namely

$$\mathbf{U}(t) = \mathbf{U}^{(0)}(t) + \alpha\mathbf{U}^{(1)}(t) + O\left(\alpha^2\right) \quad (14)$$

with $\mathbf{U}^{(1)}(t)$ being the first nonlinear response of the medium. Its Bloch expansion has the form,[29]

$$\mathbf{U}^{(1)}(t) = \sum_{\bar{n}}\frac{1}{(2\pi)^d}\int_{[-\pi,\pi]^d}\tilde{V}_{\bar{n}}^{(1)}(\mathbf{k},\tau)\,\mathbf{G}_{\bar{n}}(\mathbf{r},\mathbf{k})\,e^{-i\omega_{\bar{n}}(\mathbf{k})t}\,d\mathbf{k}, \quad \tau = \varrho t, \quad (15)$$

$$\tilde{V}_{\bar{n}}^{(1)}(\mathbf{k},\tau) = \frac{1}{\varrho}\sum_{\bar{n}',\bar{n}''}\int_0^{\tau}\int_{[-\pi,\pi]^d}\exp\left\{i\phi_{\bar{n}}(\mathbf{k},\mathbf{k}',\mathbf{k}'')\,\frac{\tau_1}{\varrho}\right\} \quad (16)$$

$$\breve{Q}_{\bar{n}}(\mathbf{k},\mathbf{k}',\mathbf{k}'')\,\tilde{V}_{\bar{n}'}^{(0)}(\mathbf{k}',\tau_1)\,\tilde{V}_{\bar{n}''}^{(0)}(\mathbf{k}'',\tau_1)\,\tilde{V}_{\bar{n}'''}^{(0)}\left(\mathbf{k}-\mathbf{k}'-\mathbf{k}'',\tau_1\right)\,d\mathbf{k}'d\mathbf{k}''d\tau_1,$$

$$\phi_{\bar{n}}(\mathbf{k},\mathbf{k}',\mathbf{k}'') = \omega_{\bar{n}}(\mathbf{k}) - \omega_{\bar{n}'}(\mathbf{k}') - \omega_{\bar{n}''}(\mathbf{k}'') - \omega_{\bar{n}'''}(\mathbf{k}-\mathbf{k}'-\mathbf{k}''), \quad (17)$$

where $\check{Q}_{\vec{n}}$ is a coefficient depending on the indices $\vec{n} = (\bar{n}, \bar{n}', \bar{n}'', \bar{n}''')$ and the quasimomenta $\mathbf{k}, \mathbf{k}', \mathbf{k}''$, and $\phi_{\vec{n}}$ is a 2π-periodic with respect to $\mathbf{k}, \mathbf{k}', \mathbf{k}''$ phase function.

The asymptotic behavior of $\tilde{V}_{\bar{n}}^{(1)}(\mathbf{k}, \tau)$ in (16) as $\varrho \to 0$ is determined primarily by the phase $\phi_{\vec{n}}$. Its form (17) signifies the well known fact that the nonlinear interactions for a cubic nonlinearity occur through quads of modes $(\bar{n}, \mathbf{k})$, $(\bar{n}', \mathbf{k}')$, $(\bar{n}'', \mathbf{k}'')$ and $(\bar{n}''', \mathbf{k}''')$ with the corresponding dispersion relations $\omega_{\bar{n}}(\mathbf{k})$, $\omega_{\bar{n}'}(\mathbf{k}')$, $\omega_{\bar{n}''}(\mathbf{k}'')$ and $\omega_{\bar{n}'''}(\mathbf{k}''')$. The representation (16), (17) explicitly takes into account the phase matching condition

$$\mathbf{k}''' = \mathbf{k} - \mathbf{k}' - \mathbf{k}'' \tag{18}$$

for the interacting modes following from the medium periodicity.

To analyze the interactions, we look at the impact of a triad of modes $(\bar{n}', \mathbf{k}')$, $(\bar{n}'', \mathbf{k}'')$ and $(\bar{n}''', \mathbf{k} - \mathbf{k}' - \mathbf{k}'')$ onto a mode $(\bar{n}, \mathbf{k})$, and observe that the amplitude of the first nonlinear response $\tilde{V}_{\bar{n}}^{(1)}(\mathbf{k}, \tau)$ depends on the amplitudes $\tilde{V}_{\bar{n}'}^{(0)}(\mathbf{k}', \tau_1)$, $\tilde{V}_{\bar{n}''}^{(0)}(\mathbf{k}'', \tau_1)$ and $\tilde{V}_{\bar{n}'''}^{(0)}(\mathbf{k} - \mathbf{k}' - \mathbf{k}'', \tau_1)$, $\tau_1 \leq \tau$, of the linear response. Let us fix a $\tau > 0$, and denote the contribution of amplitudes $\tilde{V}_{\bar{n}'}^{(0)}(\mathbf{k}')$, $\tilde{V}_{\bar{n}''}^{(0)}(\mathbf{k}'')$, and $\tilde{V}_{\bar{n}'''}^{(0)}(\mathbf{k} - \mathbf{k}' - \mathbf{k}'')$ to the amplitude $\tilde{V}_{\bar{n}}^{(1)}(\mathbf{k})$ by $\tilde{V}_{\bar{n}}^{(1)}(\bar{n}, \bar{n}', \bar{n}'', \bar{n}''', \mathbf{k}'', \mathbf{k}', \mathbf{k})$. It is shown in Ref. 29 that for $\varrho \to 0$ the first nonlinear response $\tilde{V}_{\bar{n}}^{(1)}$ vanishes as a power of ϱ, namely

$$\tilde{V}_{\bar{n}}^{(1)}(\bar{n}, \bar{n}', \bar{n}'', \bar{n}''', \mathbf{k}'', \mathbf{k}', \mathbf{k}) \sim \varrho^{q}, \ \varrho \to 0 \text{ where } 0 < q \leq \infty, \tag{19}$$

where the index q describes the intensity of the interaction. If, for instance, the modes $(\bar{n}, \mathbf{k})$, $(\bar{n}', \mathbf{k}')$, $(\bar{n}'', \mathbf{k}'')$, and $(\bar{n}''', \mathbf{k}''')$ are chosen "at random" then the above index q is infinite, and consequently the mode interaction is weaker than any power of ϱ as $\varrho \to 0$. But for special choices of quads the corresponding indices q can be finite with more appreciable nonlinear interactions. When q is finite, following Ref. 29 we introduce the *index of interaction* q_0 for the quad of modes by

$$q_0 = q_0(\vec{n}, \mathbf{k}'', \mathbf{k}', \mathbf{k}) = q + 1, \quad \vec{n} = (\bar{n}, \bar{n}', \bar{n}'', \bar{n}''') . \tag{20}$$

It turns out that the index q_0 takes on only a few universal values, most important of which are collected in tables of Section 3. Note that a smaller interaction index q_0 corresponds to a stronger interaction.

The formula (16) suggests that the integral

$$I_{\vec{n}}(\mathbf{k}, \tau) = \frac{1}{\varrho} \int_0^{\tau} \int_{[-\pi,\pi]^3} \exp\left\{ i\phi_{\vec{n}}(\mathbf{k}, \mathbf{k}', \mathbf{k}'') \frac{\tau_1}{\varrho} \right\} A(\mathbf{k}, \mathbf{k}', \mathbf{k}'', \tau_1)\, d\mathbf{k}' d\mathbf{k}'' d\tau_1, \tag{21}$$

$$A(\mathbf{k}, \mathbf{k}', \mathbf{k}'', \tau_1) = \check{Q}_{\vec{n}}(\mathbf{k})\, \tilde{V}_{\bar{n}'}^{(0)}(\mathbf{k}', \tau_1)\, \tilde{V}_{\bar{n}''}^{(0)}(\mathbf{k}'', \tau_1)\, \tilde{V}_{\bar{n}'''}^{(0)}\left(\mathbf{k} - \mathbf{k}' - \mathbf{k}'', \tau_1\right), \tag{22}$$

represents the nonlinear impact of the triad of modes $(\bar{n}', \mathbf{k}')$, $(\bar{n}'', \mathbf{k}'')$, $(\bar{n}''', \mathbf{k}''')$ onto the mode $(\bar{n}, \mathbf{k})$. We refer to integrals (21) as the *oscillatory interaction integrals*. Such interaction process is known in nonlinear optics as *four wave mixing*. We would like to emphasize that since the propagating wavepacket has a finite spatial range, its spectrum contains a continuum of modes all of which are involved in the four wave mixing. But for small ϱ almost all interactions are insignificant except for a few.[29]

1.3. Selection Rules

To single out and classify significant nonlinear interactions one has to study the asymptotic behavior of the integrals (21) as $\varrho \to 0$. An analysis of the interaction integrals $I_{\vec{n}}(\mathbf{k},\tau)$ of the form (21) carried out in Ref. 29 shows that a stronger nonlinear interaction can be found with the help of three selection rules. The first one, known as the *phase matching condition* (18), is already built in the very form of $I_{\vec{n}}(\mathbf{k},\tau)$. The second selection rule, known as the *frequency matching condition (FMC)*, is

$$\phi_{\vec{n}}\left(\mathbf{k},\mathbf{k}',\mathbf{k}''\right) = 0, \tag{23}$$

or, in view of (17),

$$\omega_{\bar{n}}(\mathbf{k}) - \omega_{\bar{n}'}(\mathbf{k}') - \omega_{\bar{n}''}(\mathbf{k}'') - \omega_{\bar{n}'''}\left(\mathbf{k}-\mathbf{k}'-\mathbf{k}''\right) = 0. \tag{24}$$

It is shown in Ref. 29 that if (24) is not fulfilled, the magnitude of the interaction integral $I_{\vec{n}}(\mathbf{k},\tau)$ would be of order ϱ^{q_0+1} compared with ϱ^{q_0} if (24) is satisfied. Therefore, the condition (24) must hold for stronger interactions.

The third selection rule, called the *group velocity matching condition*,[29] follows from the asymptotic analysis of the integral

$$I_{\vec{n}}(\mathbf{k}) = \frac{1}{\varrho}\int_{[-\pi,\pi]^d} \exp\left\{ i\phi_{\vec{n}}\left(\mathbf{k},\mathbf{k}',\mathbf{k}''\right)\frac{\tau_1}{\varrho}\right\} A\left(\mathbf{k},\mathbf{k}',\mathbf{k}'',\tau_1\right) d\mathbf{k}'d\mathbf{k}'', \tag{25}$$

which is an interior integral in (21). Evidently, the integral involves the large parameter $\frac{1}{\varrho}$ for $\varrho \ll 1$, and one can apply the stationary phase method. According to the method, for any fixed $\mathbf{k}$ the asymptotic behavior of $I_{\vec{n}}(\mathbf{k})$ as $\varrho \to 0$ is determined by critical points of the phase function $\phi_{\vec{n}}\left(\mathbf{k},\mathbf{k}',\mathbf{k}''\right)$, i.e. the points $(\mathbf{k}'_*,\mathbf{k}''_*)$ solving the following two d-component equations

$$\nabla_{k'}\phi_{\vec{n}}\left(\mathbf{k},\mathbf{k}',\mathbf{k}''\right) = 0,\ \nabla_{k''}\phi_{\vec{n}}\left(\mathbf{k},\mathbf{k}',\mathbf{k}''\right) = 0. \tag{26}$$

In view of (17), the equations (26) can be recast as

$$\nabla\omega_{\bar{n}'}(\mathbf{k}'_*) = \nabla\omega_{\bar{n}''}(\mathbf{k}''_*) = \nabla\omega_{\bar{n}'''}(\mathbf{k}'''_*),\ \mathbf{k}'''_* = \mathbf{k}-\mathbf{k}'_*-\mathbf{k}''_*, \tag{27}$$

justifying the name *group velocity matching condition (GVM)* since $\nabla\omega_{\bar{n}'}(\mathbf{k})$ is the group velocity of the $\mathbf{k}$-component of the linear wavepacket.

2. OUTLINE OF MAIN RESULTS

The stationary phase method analysis shows that the contribution of a critical point $(\mathbf{k}'_*, \mathbf{k}''_*)$ to the integral $I_{\vec{n}}(\mathbf{k})$ is of order ϱ^{q_0-1} where q_0 is determined by the properties of the phase function $\phi_{\vec{n}}$ at this point.[29] Notice that the critical point $(\mathbf{k}'_*, \mathbf{k}''_*)$ does not depend on the amplitude $A(\mathbf{k}', \mathbf{k}'')$ provided it is smooth, but the magnitude of the interaction clearly is proportional to $A(\mathbf{k}'_*, \mathbf{k}''_*)$. Such points must satisfy (27) whereas all other points and corresponding interactions are negligible. Observe that the vector equation (27) yields $2d$ scalar equations for $2d$ variables $\mathbf{k}', \mathbf{k}''$ and may have several solutions. The FMC condition (24) imposes one more constraint on $\mathbf{k}'_*, \mathbf{k}''_*$.

Note that for a fixed $\mathbf{k}$ the conditions (27) together with (24) impose $2d + 1$ equations on $2d$ variables, and *in a generic case one cannot expect these conditions to be satisfied.* Of course, there are cases when $2d + 1$ equations on $2d$ variables have a solution, but such cases are not typical. It turns out, though, that under the inversion (spectral) symmetry condition

$$\omega_n(\mathbf{k}) = \omega_n(-\mathbf{k}), \ \mathbf{k} \in [-\pi, \pi]^d, \ n = 1, 2, \ldots, \tag{28}$$

the equations have solutions. The *inversion symmetry* condition (28) on the dispersion relations is a special case of Wigner time-reversal symmetry[32]), and it always holds, for instance, for any nonmagnetic periodic medium. Note that (28) implies

$$\nabla\omega_n(\mathbf{k}) = -\nabla\omega_n(-\mathbf{k}). \tag{29}$$

In this paper we study the case when (28) holds and show that the overdetermined system (27), (23) *always has a solution.* In particular, for a given $(\vec{n}, \mathbf{k})$ the quad of strongly interacting modes solving the overdetermined system (27), (23) satisfies the following relations

$$n' = n'' = n''' = n. \tag{30}$$

In other words, the quad of modes belongs to the same band. In addition to that, the quasimomenta and the signs ζ in the index $\vec{n} = (\zeta, n)$ of $\omega_{\vec{n}}(\mathbf{k})$ are chosen as follows:

$$\mathbf{k}'''_* = -\mathbf{k}, \ \mathbf{k}'_* = \mathbf{k}''_* = \mathbf{k}, \quad \zeta''' = -\zeta, \ \zeta' = \zeta'' = \zeta. \tag{31}$$

If ζ are chosen as in (31) the phase function (17) takes the form

$$\phi_{\vec{n}}(\mathbf{k}, \mathbf{k}', \mathbf{k}'') = \pm[-\omega_n(\mathbf{k}) + \omega_{n'}(\mathbf{k}') + \omega_{n''}(\mathbf{k}'') - \omega_{n'''}(\mathbf{k} - \mathbf{k}' - \mathbf{k}'')],$$

and it is easy to examine that (27) and (24) are satisfied. There are two more quads similar to (31) with the same property, namely

$$\mathbf{k}'_* = -\mathbf{k}, \ \mathbf{k}''_* = \mathbf{k}'''_* = \mathbf{k}, \ \zeta'' = \zeta''' = \zeta, \zeta' = -\zeta, \tag{32}$$

and

$$\mathbf{k}''_* = -\mathbf{k}, \ \mathbf{k}'_* = \mathbf{k}'''_* = \mathbf{k}, \ \zeta''' = \zeta' = \zeta, \zeta'' = -\zeta. \tag{33}$$

In the nonlinear optics this type of interactions is known to occur in the process of a degenerate four-wave mixing, Ref. 28 p. 232.

Let us gauge now the intensity of nonlinear interactions using the interaction index q_0 and find the corresponding quads of modes. For simplicity, we consider $d = 1$ and $d = 2$. If we allow $\mathbf{k}$ in the system (24), (26) to be a variable together with $\mathbf{k}', \mathbf{k}''$ and if the dispersion relations do not have the inversion symmetry (28) and are generic, the set of all solutions is a $d-1$ dimensional manifold in the $\mathbf{k}, \mathbf{k}', \mathbf{k}''$-space. Let us denote this manifold by $\Xi = \Xi(\vec{n})$. Every point $(\mathbf{k}, \mathbf{k}', \mathbf{k}'')$ in $\Xi(\vec{n})$ corresponds to a strongly interacting quad.

Remarkably, in the presence of the inversion symmetry (28), if $n = n' = n'' = n'''$ and the signs ζ are as in (31), (32) or (33), the manifold $\Xi(\vec{n})$ is d-dimensional. In addition to that, the index $q_0(\vec{n}, \mathbf{k}, \mathbf{k}', \mathbf{k}'')$ of a strongly interacting quad depends on the rank of the Hessian $\nabla^2 \phi_{\vec{n}}(\mathbf{k}, \mathbf{k}', \mathbf{k}'')$ of the phase function at the point. For a typical quad from $\Xi(\vec{n})$ the Hessian at the point is not degenerate, its rank equals $2d$, and the interaction index $q_0 = d$. Let us denote by $\Xi' = \Xi'(\vec{n})$ a submanifold of $\Xi(\vec{n})$ on which the determinant of the Hessian of $\phi_{\vec{n}}(\mathbf{k}, \mathbf{k}', \mathbf{k}'')$ with respect to $\mathbf{k}', \mathbf{k}''$ vanishes, and by $\Xi'' = \Xi''(\vec{n})$ the submanifold of $\Xi'(\vec{n})$ on which the rank of the Hessian is $2d - 2$. As a rule, interactions corresponding to Ξ'' are stronger than the ones corresponding to Ξ', and interactions corresponding to Ξ' are stronger than the ones corresponding to Ξ. Our analysis shows that if the inversion symmetry (28) holds and $n = n' = n'' = n'''$ then the manifold $\Xi''(\vec{n})$ is *always* non-empty and its dimension is $d - 1$. For quads from Ξ'' generically the index $q_0 = d - \frac{1}{3}$, for $d = 1, 2$. When the inversion symmetry condition (28) does not hold, $\Xi''(\vec{n})$ may be empty. In fact, it is always empty for a generic one-dimensional medium, $d = 1$. For a generic two-dimensional medium, $d = 2$, without inversion symmetry it may be non-empty only for the quads related to the Third Harmonic Generation when

$$n' = n'' = n''', \ \mathbf{k}'_* = \mathbf{k}''_* = \mathbf{k}'''_* = \frac{1}{3}\mathbf{k}, \ \zeta = \zeta' = \zeta'' = \zeta'''. \tag{34}$$

In addition, if Ξ'' is non-empty, it may consist of only several points, that is the dimension of Ξ'' is zero, and the interaction index $q_0 = \frac{5}{3}$.

Therefore, the symmetry condition (28) dramatically changes the structure and the dimension of the manifold $\Xi(\vec{n})$ of solutions $(\mathbf{k}, \mathbf{k}', \mathbf{k}'')$ to the selections rules (23), (24) describing strongly interacting modes. Namely, for any $(\bar{n}, \mathbf{k})$ there always exist a quad of modes involving $(\bar{n}, \mathbf{k})$ and satisfying all the selection rules. In addition to that, the interactions are stronger than in the absence of the inversion symmetry (see tables in Section 4).

Recall that the optical Kerr effect (Ref. 28 p. 26) involves a quadruple of interacting modes with frequencies $-\omega_S$, ω_P, $-\omega_P$, ω_S. Similar interactions occur for a periodic medium with the inversion symmetry (28) for quads satisfying all the selection rules. Indeed, let us set

$$\omega_S = \omega_n(\mathbf{k}) = \omega_{n'}(\mathbf{k}'), \ \omega_P = \omega_{n'}(\mathbf{k}'') = \omega_{n'''}(\mathbf{k} - \mathbf{k}' - \mathbf{k}''),$$

$$n''' = n'', \; n = n', \mathbf{k} = \mathbf{k}', \; \mathbf{k}''' = -\mathbf{k}'', \; \zeta = \zeta' = \zeta'' = -\zeta'''. \tag{35}$$

Then the phase function takes the form

$$\phi_{\vec{n}}(, \mathbf{k}, \mathbf{k}', \mathbf{k}'')) = -\omega_n(\mathbf{k}) + \omega_{n'}(\mathbf{k}') + \omega_{n''}(\mathbf{k}'') - \omega_{n'''}(\mathbf{k} - \mathbf{k}' - \mathbf{k}'').$$

This function always satisfies FMC; obviously GVM holds for $\mathbf{k}'' = \mathbf{k}', n'' = n'$, but GVM may also hold for the case $n'' \neq n', \mathbf{k}'' \neq \mathbf{k}' = \mathbf{k}$ if the equation $\nabla\omega_{n'}(\mathbf{k}') = \nabla\omega_{n''}(\mathbf{k}'')$ is satisfied. The last condition yields d scalar equations on d variables $\mathbf{k}''$ and may be satisfied for several $\mathbf{k}'' = \mathbf{k}''(\mathbf{k})$.

3. OSCILLATORY INTEGRALS

For $\varrho \to 0$ the interaction integral of equation (25) involves a rapidly oscillating function $\exp\left\{i\phi_{\vec{n}}\left(\mathbf{k}, \mathbf{k}', \mathbf{k}''\right)\frac{\tau 1}{\varrho}\right\}$. To analyze its asymptotic behavior we consider the integral

$$I(\varrho, \mathbf{k}) = \int_{\mathbf{R}^d} e^{i\Phi(\mathbf{k},\mathbf{s})/\varrho} A(\mathbf{k}, \mathbf{s})\, ds, \; \mathbf{s} = \left(\mathbf{k}', \mathbf{k}''\right), \; \varrho \to 0, \tag{36}$$

where the the integration variable $\mathbf{s}$ has dimension $d_I = 2d$, $\mathbf{k}$ is a continuous parameter, and the phase function is

$$\Phi(\mathbf{k}, \mathbf{s}) = \phi_{\vec{n}}\left(\mathbf{k}, \mathbf{k}', \mathbf{k}''\right) \tag{37}$$

with $\phi_{\vec{n}}$ being defined by (17). The phase $\Phi(\mathbf{k}, \mathbf{s})$ depends on a d-dimensional parameter $\mathbf{k}$ and is assumed to satisfy the FMC equation $\Phi(\mathbf{k}, \mathbf{s}) = 0$ that imposes one constraint on d parameters. Hence, generally speaking, $\Phi(\mathbf{k}, \mathbf{s})$ is a $d - 1$ parametric family of phase functions. Note that the integral (25) has the factor ϱ^{-1} which is not included in the integral (36).

Let us review briefly the main points of the stationary phase method for oscillatory integrals of the form

$$I(\varrho) = \int_{\mathbf{R}^{d_I}} e^{i\Phi(\mathbf{s})/\varrho} A(\mathbf{s})\, ds, \; \varrho \to 0, \tag{38}$$

where $A(\mathbf{s})$ is assumed to be an infinitely smooth with a finite support. According to the stationary phase method, the main contribution to $I(\varrho)$ for $\varrho \to 0$ comes from small neighborhoods of critical points of the phase $\Phi(\mathbf{s})$, that is points $\mathbf{s}_*$ satisfying the equation

$$\nabla_{\mathbf{s}}\Phi(\mathbf{s}) = 0, \tag{39}$$

and those contributions are proportional to powers of ϱ. The corresponding coefficients before the powers are written in terms of values of $\Phi(\mathbf{s})$, $A(\mathbf{s})$ and its derivatives at critical points $\mathbf{s}_*$. Since (39) is a system of d_I equations for d_I variables, for a generic $\Phi(\mathbf{s})$ there is a finite number of such points.

3.1. Standard Asymptotic Approximations

In this subsection we list some standard asymptotic formulas used in our analysis. One of the most important characteristics of a critical point $\mathbf{s}_*$ is the rank of the Hessian $\Phi''(\mathbf{s}_*)$. The formulas provided below are for the cases when the rank of the Hessian is d_I or $d_I - 1$.

3.1.1. Non-Degenerate Points

When the rank of the Hessian is d_I we have a non-degenerate critical point $\mathbf{s}_*$, or a critical point of type A_1. All the eigenvalues $\mu_1, \ldots, \mu_{d_I}$ of the Hessian at this point are non-zero, and the following classical formula holds

$$I(\varrho) = b_0 \varrho^{\frac{d_I}{2}} b_{\mathcal{T}} A(\mathbf{s}_*) e^{i\Phi(\mathbf{s}_*)/\varrho} + O\left(\varrho^{\frac{d_I}{2}+1}\right), \ \varrho \to 0, \tag{40}$$

with

$$b_0 = \frac{(2\pi)^{\frac{d_I}{2}}}{\sqrt{|\det \Phi''(\mathbf{s}_*)|}} \exp\left\{i\eta\Phi(\mathbf{s}_*) + \frac{i\pi}{4} \operatorname{sign}[\Phi''(\mathbf{s}_*)]\right\}, \tag{41}$$

where sign $[\Phi''(\mathbf{s}_*)]$ is the signature of the matrix H (the algebraic sum of the signs of the matrix eigenvalues).

3.1.2. Simplest degenerate points

When the rank of the Hessian is $d_I - 1$ we have a critical point $\mathbf{s}_*$ of the type A_p, $p > 1$. One of the eigenvalues of the Hessian vanishes, i.e. $\mu_{d_I} = 0$. In this case along a curve tangent to the null-space of $\Phi''(\mathbf{s}_*)$ the function $\Phi(\mathbf{s})$ vanishes at $\mathbf{s}_*$ as $|\mathbf{s} - \mathbf{s}_*|^{p+1}$ with $p > 1$. For a critical point of the class $\mathcal{T} = A_p$ the leading term of the asymptotics takes the form (see, for example, Ref. 33, Section 6.1):

$$I(\varrho) = \varrho^{\frac{d_I-1}{2}+\frac{1}{p+1}} b_{\mathcal{T}} A(\mathbf{s}_*) e^{i\eta\Phi(\mathbf{s}_*)} + O\left(\varrho^{\frac{d_I-1}{2}+\frac{2}{p+1}}\right), \tag{42}$$

where the coefficient $b_{\mathcal{T}} = b_{A_p}$ is determined by the phase Φ at $\mathbf{s}_*$ by a formula similar to (41) and the oscillatory index is

$$q_{\mathcal{T}} = q_{A_p} = \frac{d_I - 1}{2} + \frac{1}{p+1}. \tag{43}$$

Formula (43) for the oscillatory index holds as well for a non-degenerate point with $p = 1$.

When the rank of the Hessian is $d_I - 2$ a more involved analysis of critical points $\mathbf{s}_*$ is needed. For the asymptotic expansions of oscillatory integrals with a general phase functions see Refs. 34,35, and references therein. Our analysis shows that to find the leading terms of asymptotics in generic situations when $\mathbf{k}$ is one-, two-, or three-dimensional variable it is sufficient to consider relatively simple singularities and the values of the index $q_{\mathcal{T}}$ can be found in these cases.

3.2. Oscillatory Integrals With Parameters

In this subsection we focus on the oscillatory integrals (36) that depend on the parameter $\mathbf{k}$. To make the results more transparent, we consider first a general phase $\Phi(\mathbf{k},\mathbf{s})$ and do not assume the special structure (37). Then we consider the case of (37) not assuming the inversion symmetry (28), and, finally, we will consider the case when (28) holds.

3.2.1. General Phase Function

Let us consider the integral (36) with a general phase function $\Phi(\mathbf{k},\mathbf{s})$ that does not have the special structure of (17). This example serves as a basis for a comparison and indicates the effect coming from the structure of (17).

Case 1. Let $d_I = 2d = 2$. In this case the GVM conditions (26) and the FMC condition (23) yield 3 equations on three one-dimensional parameters $\mathbf{k},\mathbf{k}',\mathbf{k}''$ selecting several values $(\mathbf{k}_*,\mathbf{s}_*)$ of the variables $\mathbf{s}$ and $\mathbf{k}$. Hence, the manifold Ξ has zero dimension, and for a fixed typical $\mathbf{k}\neq\mathbf{k}_*$ the system does not have a solution. In a generic case, the Hessian $\Phi''(\mathbf{k}_*,\mathbf{s}_*)$ with respect to $\mathbf{s}$ at these points is non-degenerate, $\det\Phi''(\mathbf{k}_*,\mathbf{s}_*)\neq 0$, therefore Ξ' is empty. The principal term of asymptotics of (36) by the classical stationary phase method (41) is of order ϱ.

Case 2. Let us take $d_I = 4$. The vector equations (26) and (23) yield 5 scalar equations on three two-dimensional parameters $\mathbf{k}$, $\mathbf{k}'$, $\mathbf{k}''$, that is 5 equations on 6 variables. Their solutions $\vec{\mathbf{k}} = (\mathbf{k},\mathbf{k}',\mathbf{k}'')$ form a manifold Ξ of the strongly interacting modes. In a generic case, Ξ is a one-dimensional manifold (a curve). This manifold may have singular points (self-intersections). When $\mathbf{k}$ from this curve is fixed, we have several critical points $\mathbf{s}_* = (\mathbf{k}'_*,\mathbf{k}''_*)$. When the Hessian $\Phi''(\mathbf{k},\mathbf{s}) = \nabla^2_{k'k''}\Phi(\mathbf{k},\mathbf{k}',\mathbf{k}'')$ of $\Phi(\mathbf{k},\mathbf{k}',\mathbf{k}'')$ with respect to $\mathbf{s}=(\mathbf{k}',\mathbf{k}'')$ is non-degenerate, i.e.

$$\det\nabla^2_{k'k''}\Phi\left(\mathbf{k},\mathbf{k}'_*,\mathbf{k}''_*\right)\neq 0,$$

we have non-degenerate critical points $\mathbf{s}_*=(\mathbf{k}'_*,\mathbf{k}''_*)$ of the phase function. According to (41) the contribution of such a point to the integral (36) is of order ϱ^2.

Since the interaction manifold Ξ is one-dimensional, there can be points where

$$\det\nabla^2_{k'k''}\Phi\left(\mathbf{k},\mathbf{k}'_*,\mathbf{k}''_*\right) = 0. \tag{44}$$

The solutions to (44) form the manifold Ξ'. If $\det\nabla^2_{k'k''}\Phi\left(\mathbf{k},\mathbf{k}',\mathbf{k}''\right)$ changes the sign along the curve at such a point (this is a generic case), such a point cannot be eliminated by small perturbations of the dispersion function $\Phi(\mathbf{k},\mathbf{s})$. Hence, the degeneration is robust and Ξ' may be non-empty. In a generic case, the rank of the Hessian is d_I-1, and a generic $\Phi(\mathbf{k},\mathbf{s})$ at $\mathbf{s}=\mathbf{s}^*$ vanishes in the direction of the null-space of the Hessian at cubic rate, resulting in a singularity of the type A_2. According to (42) the contribution of such point is of order $\varrho^{11/6}$.

3.2.2. Phase Function of Four-Wave Interactions

We consider now (36) with a phase function $\Phi(\mathbf{k},\mathbf{s})$ that has the special structure given by (17). Since $\omega_{\bar{n}}(\mathbf{k}) = \pm\omega_n(\mathbf{k}) = \zeta\omega_n(\mathbf{k})$, where the sign is determined by the ζ-component of $\bar{n} = (\zeta, n)$, we have by (17)

$$\begin{aligned}\Phi(\mathbf{k},\mathbf{s}) &= \phi_{\bar{n}}(\mathbf{k},\mathbf{k}',\mathbf{k}'') \\ &= \zeta\omega_n(\mathbf{k}) - \zeta'\omega_{n'}(\mathbf{k}') - \zeta''\omega_{n''}(\mathbf{k}'') - \zeta'''\omega_{n'''}(\mathbf{k}-\mathbf{k}'-\mathbf{k}''),\ \mathbf{s} = (\mathbf{k}',\mathbf{k}'').\end{aligned}$$

If $n''' \neq n''$ or $n''' \neq n'$, the analysis of (36) is similar to the analysis for a general phase function and it yields the same conclusions. If $d = 1$, $d_I = 2$, for a generic $\omega_n(\mathbf{k})$ even in the case $n''' = n'' = n'$ the situation is similar to the considered case of a general $\Phi(\mathbf{k},\mathbf{s})$. Namely, in the case $d = 1$ for a generic $\omega_n(\mathbf{k})$ there are only non-degenerate points, Ξ' is empty and the oscillatory integral (36) has the leading asymptotic term of order ϱ^1.

In the case $d = 2$, for a generic $\omega_n(\mathbf{k})$ the manifold Ξ is one-dimensional (if non-empty) and Ξ' is zero-dimensional (if non-empty). For a generic $\omega_n(\mathbf{k})$ points from Ξ' are of the type A_2, and we have a contribution from such a point of order $\varrho^{11/6}$ as for a general phase function. The special structure of the function $\Phi(\mathbf{k},\mathbf{s})$ implies certain restrictions on the Hessian. The Hessian $\Phi''(\mathbf{k},\mathbf{s})$ with respect to $(\mathbf{k}',\mathbf{k}'') = \mathbf{s}$ takes the block form

$$\Phi''(\mathbf{k}',\mathbf{k}'') = -\begin{pmatrix} \zeta'\omega''_{n'}(\mathbf{k}') + \zeta'''\omega''_{n'''}(\mathbf{k}''') & \zeta'''\omega''_{n'''}(\mathbf{k}''') \\ \zeta'''\omega''_{n'''}(\mathbf{k}''') & \zeta''\omega''_{n''}(\mathbf{k}'') + \zeta'''\omega''_{n'''}(\mathbf{k}''') \end{pmatrix} \tag{45}$$

where $\mathbf{k}''' = \mathbf{k} - \mathbf{k}' - \mathbf{k}''$.

The case $n''' = n'' = n'$ is special. In this case when $\zeta' = \zeta'' = \zeta'''$ the GVM condition (27) has a special family of solutions

$$\mathbf{k}'_* = \mathbf{k}''_* = \mathbf{k}'''_* = \frac{1}{3}\mathbf{k}. \tag{46}$$

If (23) is satisfied too, the special solutions correspond to the *Third Harmonic Generation.*

If $n''' = n'' = n'$, $\mathbf{k}'_* = \mathbf{k}''_* = \mathbf{k}'''_*$ (and if (23) is satisfied), an analysis shows that $\zeta' = \zeta'' = \zeta'''$ is the only case that makes an essential difference compared with the case of a general function. If the rank of $\omega''_{n'}(\mathbf{k}'_*)$ is d, the rank of $\Phi''(\mathbf{k}'_*,\mathbf{k}'_*)$ is $2d$. If the rank of $\omega''_{n'}(\mathbf{k}'_*)$ is $d-1$, then the rank of $\Phi''(\mathbf{k}'_*,\mathbf{k}'_*)$ in (45) is $2d-2$. Therefore on the curve Ξ at a generic point the Hessian is non-degenerate and the integral is of order ϱ. At several points where $\det\omega''_{n'}(\mathbf{k}'_*) = 0$, that is on Ξ'', the integral is of order $\varrho^{5/3}$. Note that the manifolds Ξ, Ξ' and Ξ'' may be empty.

3.2.3. Phase Function of Four-Wave Interactions With the Inversion (Spectral) Symmetry

In this case for a given $(\bar{n},\mathbf{k})$ the stronger interactions are attained for quads given by (30), that is $n''' = n'' = n' = n$, and (31), (32), or (33). The GVM conditions (26) and the FMC condition (23) are always satisfied. The manifold $\Xi(\vec{n})$ is obviously non-empty, and the index of an interacting quad is determined by the properties of the Hessian. The values of the indices are given below.

In the one-dimensional case, $d = 1$, $\omega_n''(\mathbf{k})$ is a scalar. For almost all $\mathbf{k}$ $\omega_n''(\mathbf{k}) \neq 0$, and, using (41), we get the integral of order ϱ. If $\omega_n''(\mathbf{k}_*) = 0$, by (45) we get a point from Ξ''. Generically, the oscillatory integral is of order $\varrho^{2/3}, q_0 = \frac{2}{3}$. Note that since FMC and GVM conditions are satisfied *for any* $\mathbf{k}$, we can find a point $\mathbf{k}_*$ such that $\omega_n''(\mathbf{k}_*) = 0$ (even in the one dimensional case, $d = 1$). Such $\mathbf{k}_*$ always exists since $\omega_n(\mathbf{k})$ is periodic and, hence, there always exists an inflection point. Therefore Ξ'' is non-empty and zero-dimensional.

In the two-dimensional case, $d = 2$, $d_I = 4$ and $\omega_n''(\mathbf{k})$ is a symmetric 2×2 quadratic matrix with 3 independent coefficients. When the matrix $\omega_n''(\mathbf{k})$ has a zero eigenvalue, the second eigenvalue generically is not zero. The rank of $\Phi''(\mathbf{k}', \mathbf{k}'')$ in (45) in this case equals 2, and we get a point of Ξ''. Note that the manifold Ξ'' is always non-empty. According to (31), (32), (33) the manifold Ξ'' consists of three curves in the $(\mathbf{k}, \mathbf{k}', \mathbf{k}'')$-plane, one of which is described by

$$\det \omega_n''(\mathbf{k}) = 0, \ \mathbf{k}' = \mathbf{k}'' = \mathbf{k}.$$

The function $\det \omega_n''(\mathbf{k})$ always has zeros thanks to the periodicity of $\omega_n(\mathbf{k})$. For modes from the curve Ξ'' we get the oscillatory integral of order $\varrho^{5/3}$, hence the interaction index $q_0 = \frac{5}{3}$. This is the minimal interaction index possible in a generic case. Indeed, it is impossible to make $\omega_1''(\mathbf{k}) = 0$ in a generic case, since it requires vanishing of all three independent entries of the matrix $\omega''(\mathbf{k})$. One also cannot eliminate the cubic dependence in the two null directions by one parameter variation.

One more special case is described by Kerr type condition (35). In this case ($n'' \neq n'$ or $\mathbf{k}'' \neq \mathbf{k}'$) generically Ξ'' is empty, and one may have a non-empty $(d-1)$-dimensional $\Xi', d = 1, 2$. Almost all points from Ξ' are of type A_2 with the index $q_0 = d - \frac{1}{6}$ and in the case $d = 2$ there may be several points of type A_3 with the index $q_0 = d - \frac{1}{4}$.

4. TABLES OF INTERACTION INDICES

In this section we provide tables of interaction indices q_0 involved in (19), (20). The tables compare the oscillatory interaction integrals for the following types of the phase functions.

(i) General phase function. This phase function is an abstract, nonstructured function that is given for comparison purposes.

(ii) 4-frequency phase function of the form (17) with dispersion relations $\omega_{\bar{n}}(\mathbf{k})$ without the inversion (spectral) symmetry.

(iii) Phase function of the form (17) when dispersion relations $\omega_{\bar{n}}(\mathbf{k})$ have the inversion symmetry.

Table 1. Indices of critical points for a generic one-dimensional periodic medium with a cubic nonlinearity.

One-dimensional case $d = 1$					
Phase function type	General	4-frequency without inversion symmetry	4-frequency with inversion symmetry		
Index q_0	1	1	$\frac{2}{3}$	$\frac{5}{6}$	1
Interaction manifold dimension	0	0	0	0	1

Table 2. Indices of critical points for a generic two-dimensional periodic medium with a cubic nonlinearity.

Two-dimensional case $d = 2$									
Phase function type	General		4-frequency without inversion symmetry			4-frequences with inversion symmetry			
Index q_0	$\frac{11}{6}$	2	$\frac{5}{3}$	$\frac{11}{6}$	2	$\frac{5}{3}$	$\frac{7}{4}$	$\frac{11}{6}$	2
Interaction manifold dimension	0	1	0	0	1	1	0	1	2

In all the three cases we assume that the phase function is generic, it does not have symmetries which are not explicitly imposed.

Every entry in the second row of the tables contains minimal values of q_0 for all possible interactions of strongly interacting quads. When several strongly interacting quads $(\bar{n}, \mathbf{k})$, $(\bar{n}', \mathbf{k}')$, $(\bar{n}'', \mathbf{k}'')$, $(\bar{n}''', \mathbf{k}''')$ with different q_0 occur for the same $(n, \mathbf{k})$, we choose only the minimal value of q_0 selecting the strongest interaction. Note that a smaller index q_0 corresponds to a higher intensity interaction. The third row in the tables gives the dimension of the interaction manifold in the $\mathbf{k}, \mathbf{k}', \mathbf{k}''$-space with the given value of q_0. Higher dimension means that more quads are involved in this type of interaction.

For example, let us look at the third column of the table describing the two-dimensional case. This column corresponds to media with the inversion symmetry and it has four subcolumns. The subcolumns correspond to 4 different values of $q_0 = \frac{5}{3}, \frac{7}{4}, \frac{11}{6}, 2$. The largest $q_0 = 2$ since there is a 2-dimensional family Ξ of quasimomenta $\mathbf{k}$ (that is almost all $\mathbf{k}$) with this index, consequently, the third row in this subcolumn has entry 2. Then, there may exist a 1-dimensional family Ξ' (a curve) with $q_0 = \frac{11}{6}$; there also may exist several points on Ξ' (a 0-dimensional set) with $q_0 = \frac{7}{4}$; and, finally, there always exists a special 1-dimensional family Ξ'' (a curve) of quasimomenta for which the interactions are the strongest, $\varrho^{q_0} = \varrho^{5/3}$ that is $q_0 = 5/3$.

5. ACKNOWLEDGMENT AND DISCLAIMER

The efforts of A. Babin and A. Figotin are sponsored by the Air Force Office of Scientific Research, Air Force Materials Command, USAF, under grant number F49620-01-1-0567. The US Government is authorized to reproduce and distribute reprints for governmental purposes notwithstanding any copyright notation thereon. The views and conclusions contained herein are those of the authors and should not be interpreted as necessarily representing the official policies or endorsements, either expressed or implied, of the Air Force Office of Scientific Research or the US Government.

REFERENCES

1. Optical gap solitons: Past, present, and future;theory and experiments, *Chaos*, **9**(3), 584-589 (2000).
2. G. Agrawal, *Nonlinear Fiber Optics* (Academic Press, 1995).
3. N. Aközbek and S. John, *Optical Solitary Waves in Two- and Three- Dimensional Nonlinear Photonic Band-Gap Structures*, Phys. Rev. E, **57**(2), 2287-2319 (1998).
4. J. Atai and B. A. Malomed, Bragg-grating solitons in a semilinear dual-core system, *Physical Review E*, **60**(6), 8713-8718 (2000).
5. A. Arraf and C. M. de Sterke, Coupled-mode equations for quadratically nonlinear deep gratings, *Physical Review E*, **58**(6), 7951-7958 (1998).
6. A. Arraf, H. He, and C. M. de Sterke, Deep Gratings with a $\chi^{(2)}$ Nonlinearity: Analysis and Solutions, *Optical Fiber Technology*, **5**, 223-234 (1999).
7. A. Arraf, C. M. de Sterke, and H. He, Bright stationary solitary waves in deep gratings with a quadratic nonlinearity, *Physical Review E*, **63**(2) (2001), 026611.
8. V. Belyakov, *Diffraction Optics of Complex-Structured Periodic Medium* (Springer-Verlag, 1992).
9. V. Berger, Nonlinear Photonic Crystals, *Phys. Rev. Letters*, **81**, 4136-4139 (1998).
10. N. G. R. Broderick, G. W. Ross, H. L. Offerhous, D. J. Richardson, and D. C. Hanna, Hexagonally Poled Lithium Niobate: A Two-Dimensional Nonlinear Photonic Crystal, *Physical Review Letters*, **84**(19), 4345-4348 (2000).
11. C. Conti, S. Trillo, and G. Assanto, Bloch function approach for parametric gap solitons, *Optics Letters*, **22**(7), 445-447 (1997).
12. B. J. Eggleton, C. M. de Sterke, and R. E. Slusher, Nonlinear pulse propagation in Bragg gratings, *J. Opt. Soc. Am. B*, **14**(11), 2980-2993 (1997).
13. Quasi-Phase-Matched Second Harmonic Generation: Tuning and Tolerances, *IEEE J. of Quantum Electronics*, **28**(11), 2631-2654 (1992).
14. M. Goul'kov, S. Odoulov, I. Naumova, F. Agullo-Lopez, G. Calvo, E. Podivilov, B. Sturman, and V. Pruneri, Degenerate parametric light scattering in periodically poled LiNbO3 : Y : Fe, *Physical Review Letters*, **86**(18), 4021-4024 (2001).
15. J. Haus, Viswanathan, M. Scalora., A. Kalocsai, J. Cole, and J. Theimer, Enhanced Second-Harmonic Generation in Media with a Weak Periodicity, *Phys. Rev. A*, **57**(3), 2120-2128 (1998).
16. H. He and P. D. Drummond, Theory of multidimensional parametric band-gap simultons, *Physical Review E*, **58**(4), 5025-5046 (1998).
17. T. Iizuka and C. M. de Sterke, Corrections to coupled mode theory for deep gratings, *Physical Review E*, **61**(4), 4491-4499 (2000).

18. J. Kumasako, M. Matsumoto, and S. Waiyapot, Linear stability analysis of dispersion-managed solitons controlled by filters, *J. of Lightwave Technology*, **18**(8), 1064-1068 (2000).
19. D. Mills, *Nonlinear Optics* (Springer-Verlag, 1991).
20. S. Mou, C.-F. Lin, H.-F. Chou, Quasi-3-D beam -propagation method for modeling nonlinear wavelength conversion, *J. of Lightwave Technology*, **19**(5), 772-779 (2001).
21. J. H. B. Nijhof, W. Forysiak, and N. J. Doran, The averaging method for finding exactly periodic dispersion-managed solitons, *IEEE J. of Selected Topics in Quantum Electronics*, **6**(2), 330-336 (2000).
22. M. Scalora, M. J. Bloemer, A. S. Manka, J. P. Dowling, C. M. Bowden, R. Viswanathan, and J. W. Haus, Pulsed second-harmonic generation in nonlinear, one-dimensional, periodic structures, *Physical Review A*, **56**(4), 3166-3174 (1997).
23. M. J. Steel and C. M. de Sterke, Parametric amplification of short pulses in optical fiber Bragg gratings, *Physical Review E*, **54**(4), 4271-4284 (1996).
24. C. M. de Sterke, D. G. Salinas, and J. E. Sipe, Coupled-mode theory for light propagation through deep nonlinear gratings, *Physical Review E*, **54**(2), 1969-1989 (1996).
25. C. M. de Sterke, B. J. Eggleton, and P. A. Krug, High -intensity pulse propagation in uniform gratings and grating superstructures, *J. of Lightwave Technology*, **15**(8), 1494-1502 (1997).
26. T. Yasui and M. Koshiba, Three-dimensional vector beam-propagation method for second harmonic generation analysis, *J. of Lightwave Technology*, **19**(5), 780-785 (2001).
27. Y. Zhu and N. Ming, *Dielectric Superlattices for Nonlinear Optical Effects, Optical and Quantum Electronics*, **31**, 1093-1128, (1999).
28. P. Butcher and D. Cotter, *The Elements of Nonlinear Optics* (Cambridge University Press, 1990).
29. A. Babin and A. Figotin, *Nonlinear photonic crystals: I. Quadratic nonlinearity*, Waves in Random Media, 2001.
30. N. Ashcroft and N. Mermin, *Solid State Physics* (Holt, Rinehart and Winston, New York-London, 1976).
31. M. Reed and B. Simon, *Analysis of Operators, Vol. 4* (Academic Press, 1978).
32. W. Jones and N. M. March, *Theoretical Solid State Physics Vols. 1,2* (Dover, New York, 1985).
33. N. Bleistein and R. A. Handelsman, *Asymptotic Expansions of Integrals* (Dover Publications, 1986).
34. V. I. Arnold, S. M. Gusein-Zade, and A. N. Varchenko, *Singularities of Differentiable mappings, Vol. 1*, Monographs in mathematics, **82** (Birkhauser, 1985).
35. V. I. Arnold, S. M. Gusein-Zade, and A. N. Varchenko, *Singularities of Differentiable mappings, Vol. 2*, Monographs in mathematics, **83** (Birkhauser, 1988).

A FINITE-DIFFERENCE-TIME-DOMAIN/ PERFECTLY-MATCHED-LAYER ALGORITHM APPLIED FOR NONLINEAR DISPERSIVE MEDIA

S. J. Yakura*

1. INTRODUCTION

With the advent of high power computers that provide fast execution times and great quantities of computer memory, we are at the stage where we can perform direct numerical calculations of Maxwell's equations in nonlinear dispersive materials. Out of many numerical techniques available in the computational electromagnetics community, one that has shown great promise in the time domain is the well-known finite-difference-time-domain (FDTD) method.[1] It is based on using a simple staggered differencing scheme in both time and space to calculate the transient behavior of electromagnetic field quantities. One of the greatest challenges of the FDTD methods has been the efficient and accurate formulation of electromagnetic wave interactions in unbounded regions. For those problems, an absorbing boundary condition must be introduced at the outer layer boundary to simulate the extension of the lattice to infinity. One approach that has given great promise in realizing such an absorbing outer boundary inside the finite volume computational domain is the well-known perfectly-matched-layer (PML) algorithm that was first introduced by J. P. Berenger[2] in 1994 for the free space absorbing boundary condition. Since that time Chew and Weedon[3] came up with the modified PML algorithm that is based on complex coordinate stretching, which is shown to be equivalent to the unsplit-field, uniaxial anisotropic PML medium approach introduced by Sacks *et al.*[4]

In this paper, we explore the formulation of a three-dimensional PML algorithm used to describe the behavior of electromagnetic quantities in outer absorbing boundary layers of a nonlinear dispersive medium that serves to absorb all outgoing waves within a finite

* Air Force Research Laboratory, Directed Energy Directorate, Kirtland AFB, NM 87117 USA.

Ultra-Wideband Short-Pulse Electromagnetics 6
Edited by Mokole *et al.*, Kluwer Academic/Plenum Publishers, 2003

computational volume. We consider the case where the electromagnetic wave propagates outwardly from a nonlinear dispersive medium to a nonlinear dispersive PML medium through a reflectionless nonlinear PML interface. We start the analysis based on the extension of the unsplit-field, uniaxial anisotropic PML formulation[4-8] of Maxwell's equations that are obtained in the frequency domain inside the nonlinear dispersive PML medium. We perform the inverse Fourier transform of these equations from the frequency domain to the time domain in order to obtain a set of first-order differential equations. Then, these equations are finite differenced in both time and space using the usual staggered Yee FDTD scheme,[1] while expanding the electric and magnetic field vectors in time using a Taylor series about the current time step to evaluate next time step values of the electromagnetic field quantities. Depending on the number of terms kept in the Taylor series expansion, we can numerically update the field values to any desired accuracy in time. In Section II, we use the piecewise-linear approximation, which is equivalent to keeping only the first-order, time-dependent term of a Taylor series, to show the process involved in obtaining a second-order accurate FDTD/PML algorithm. To obtain higher-order accurate FDTD/PML algorithms in time, we simply need to include higher-order, time-dependent terms of the Taylor series expansion and follow the same procedure outlined in Section 2. In the absence of the nonlinear PML interface, the FDTD/PML algorithm is shown to reduce to the known FDTD algorithm for nonlinear dispersive media.[9] In Section 3, we show the actual one-dimensional numerical demonstration of our FDTD/PML algorithm for an ultra-short electromagnetic pulse propagating inside a nonlinear Raman scattering medium with the nonlinear PML absorbing boundary condition to show the efficient absorption of outgoing waves. Finally, Section 4 gives concluding remarks.

2. PML FORMULATION FOR NONLINEAR DISPERSIVE MEDIA

For a wave propagating into anisotropic, uniaxial nonlinear dispersive PML media, the modified Maxwell equations under the unsplit-field, uniaxial anisotropic PML formulation[4] can be expressed in the frequency domain ($e^{j\omega t}$ convention) as

$$\underline{\nabla} \times \underline{\tilde{E}}(\omega;\underline{x}) = -j\omega\, \underline{\underline{\tilde{S}}}^{PML}(\omega) \bullet \mu_0 \mu_R\, \underline{\tilde{H}}(\omega;\underline{x}), \tag{2.1}$$

$$\underline{\nabla} \times \underline{\tilde{H}}(\omega;\underline{x}) = j\omega\, \underline{\underline{\tilde{S}}}^{PML}(\omega) \bullet \underline{\tilde{D}}(\omega;\underline{x}), \tag{2.2}$$

with

$$\underline{\widetilde{D}}(\omega;\underline{x}) = \varepsilon_0\varepsilon_R \underline{\widetilde{E}}(\omega;\underline{x}) + \varepsilon_0 \sum_{\rho=1}^{\rho_{max}} \underline{\widetilde{P}}_\rho^{L}(\omega;\underline{x}) + \varepsilon_0 \sum_{\rho=1}^{\rho_{max}} \underline{\widetilde{P}}_\rho^{NL}(\omega;\underline{x}) , \qquad (2.3)$$

$$\underline{\widetilde{P}}_\rho^{L}(\omega;\underline{x}) = \mathcal{F}\left\{\underline{P}_\rho^{L}(t;\underline{x})\right\} \equiv \mathcal{F}\left\{ \int_{-\infty}^{t} d\tau\, X_\rho^{L}(t-\tau)\, \underline{E}(\tau;\underline{x}) \right\} = \widetilde{X}_\rho^{L}(\omega)\, \underline{\widetilde{E}}(\omega;\underline{x}) , \qquad (2.4)$$

$$\underline{\widetilde{P}}_\rho^{NL}(\omega;\underline{x}) = \mathcal{F}\left\{ \underline{E}(t;\underline{x})\, R_\rho^{NL}(t;\underline{x}) \right\}$$

$$\equiv \mathcal{F}\left\{ \underline{E}(t;\underline{x}) \int_{-\infty}^{t} d\tau\, X_\rho^{NL}(t-\tau)\, [\underline{E}(\tau;\underline{x}) \bullet \underline{E}(\tau;\underline{x})] \right\} , \qquad (2.5)$$

where $\underline{\widetilde{E}}(\omega;\underline{x})$ is the complex electric field vector, $\underline{\widetilde{H}}(\omega;\underline{x})$ is the complex magnetic field vector, $\underline{\widetilde{D}}(\omega;\underline{x})$ is the complex displacement field vector, $\underline{\widetilde{P}}_\rho^{L}(\omega;\underline{x})$ is the complex linear (first-order) electric polarization vector, $\underline{\widetilde{P}}_\rho^{NL}(\omega;\underline{x})$ is the complex nonlinear (third-order) electric polarization vector, $\underline{\underline{\widetilde{S}}}^{PML}(\omega)$ is the complex uniaxial anisotropic PML matrix, ε_o is the free space electric permittivity, ε_R is the relative permittivity of a medium at infinite frequency, μ_0 is the free-space permeability, μ_R is the relative permeability, $X_\rho^{L}(t)$ and $X_\rho^{NL}(t)$ are the ρth terms of the collection consisting of ρ_{max} time-dependent linear and nonlinear electric susceptibility functions, where ρ_{max} is the maximum number of terms which we choose to consider for a particular formulation of (2.3). In (2.5) $R_\rho^{NL}(t;\underline{x})$ is introduced to isolate the part of $\underline{\widetilde{P}}_\rho^{NL}(\omega;\underline{x})$ that is represented by a convolution function. In the above equations, we use notations $\bullet$ and $\mathcal{F}\{\ \}$ to denote, respectively, a vector dot product and the Fourier transform operation. We also use symbol ~ above functions to denote ω-dependence in order to distinguish them from time-dependent functions. Elements of the uniaxial anisotropic PML matrix, $\underline{\underline{\widetilde{S}}}^{PML}(\omega)$, are given by[6]

$$\underline{\underline{\widetilde{S}}}^{PML}(\omega) = \text{diag}\left\{ \frac{\widetilde{S}_y(\omega)\,\widetilde{S}_z(\omega)}{\widetilde{S}_x(\omega)}, \frac{\widetilde{S}_z(\omega)\,\widetilde{S}_x(\omega)}{\widetilde{S}_y(\omega)}, \frac{\widetilde{S}_x(\omega)\,\widetilde{S}_y(\omega)}{\widetilde{S}_z(\omega)} \right\} , \qquad (2.6)$$

where "diag" is used to represent diagonal terms of a matrix, and $\widetilde{S}_x(\omega)$, $\widetilde{S}_y(\omega)$ and $\widetilde{S}_z(\omega)$ are ω-dependent functions that satisfy the impedance matching condition at the interface of the non-PML medium and the PML medium. It is a common practice in the computational electromagnetics community to choose $\widetilde{S}_x(\omega)$, $\widetilde{S}_y(\omega)$ and $\widetilde{S}_z(\omega)$ as

$$\widetilde{S}_x(\omega) = 1 + \frac{\sigma_x}{j\omega\varepsilon_0} \quad \text{with} \quad \frac{\sigma_x}{\varepsilon_0} = \frac{\sigma_x^*}{\mu_0} , \qquad (2.7\text{-}2.8)$$

$$\tilde{S}_y(\omega) = 1 + \frac{\sigma_y}{j\omega\varepsilon_0} \quad \text{with} \quad \frac{\sigma_y}{\varepsilon_0} = \frac{\sigma_y^*}{\mu_0}, \tag{2.9-2.10}$$

$$\tilde{S}_z(\omega) = 1 + \frac{\sigma_z}{j\omega\varepsilon_0} \quad \text{with} \quad \frac{\sigma_z}{\varepsilon_0} = \frac{\sigma_z^*}{\mu_0}, \tag{2.11-2.12}$$

where σ_x, σ_y and σ_z are PML electric conductivities, and σ_x^*, σ_y^* and σ_z^* are PML magnetic conductivities with subscripts x, y and z denoting the directions in which PML conductivities are assigned[2]. These PML conductivities are introduced for the purpose of properly handling the impedance matching condition at the nonPML/PML interface.

We first eliminate $\underline{\tilde{D}}(\omega;\underline{x})$ in favor of expressing Maxwell's equations in terms of $\underline{\tilde{E}}(\omega;\underline{x})$, $\underline{\tilde{P}}_\rho^L(\omega;\underline{x})$ and $\underline{\tilde{P}}_\rho^{NL}(\omega;\underline{x})$ by substituting (2.3) into (2.2). Then, taking the inverse Fourier transforms of (2.1), (2.2), (2.4) and (2.5) directly based on the delayed axial field convolution technique[10] with expressions shown in (2.7) through (2.12), we obtain the following time-dependent equations after some amount of manipulation:

$$\mu_0\mu_R \frac{\partial \underline{H}(t;\underline{x})}{\partial t} + \mu_0\mu_R \underline{\underline{\Xi_0}} \bullet \underline{H}(t;\underline{x}) + \mu_0\mu_R \underline{\underline{\Xi_1}} \bullet \underline{H}^{Delay}(t;\underline{x}) + \underline{\nabla} \times \underline{E}(t;\underline{x}) = 0, \tag{2.13}$$

$$\begin{aligned}
&\varepsilon_0\varepsilon_R \frac{\partial \underline{E}(t;\underline{x})}{\partial t} + \varepsilon_0 \sum_\rho \frac{\partial \underline{P}_\rho^L(t;\underline{x})}{\partial t} + \varepsilon_0 \sum_\rho \frac{\partial [\underline{E}(t;\underline{x})\, R_\rho^{NL}(t;\underline{x})]}{\partial t} \\
&+ \varepsilon_0 \underline{\underline{\Xi_0}} \bullet \Big[\varepsilon_R \underline{E}(t;\underline{x}) + \sum_\rho \underline{P}_\rho^L(t;\underline{x}) + \underline{E}(t;\underline{x}) \sum_\rho R_\rho^{NL}(t;\underline{x})\Big] \\
&+ \varepsilon_0 \underline{\underline{\Xi_1}} \bullet \Big[\varepsilon_R \underline{E}^{Delay}(t;\underline{x}) + \sum_\rho \underline{P}_\rho^{LDelay}(t;\underline{x}) + \sum_\rho \underline{P}_\rho^{NLDelay}(t;\underline{x})\Big] \\
&- \underline{\nabla} \times \underline{H}(t;\underline{x}) = 0,
\end{aligned} \tag{2.14}$$

with

$$\underline{P}_\rho^L(t;\underline{x}) = \int_{-\infty}^{t} d\tau\, X_\rho^L(t-\tau)\, \underline{E}(\tau;\underline{x}), \tag{2.15}$$

$$R_\rho^{NL}(t;\underline{x}) = \int_{-\infty}^{t} d\tau\, X_\rho^{NL}(t-\tau)\, [\underline{E}(\tau;\underline{x}) \bullet \underline{E}(\tau;\underline{x})], \tag{2.16}$$

$$\underline{H}^{Delay}(t;\underline{x}) = \int_{-\infty}^{t} d\tau\, \underline{\underline{\Phi}}(t-\tau) \bullet \underline{H}(\tau;\underline{x}), \tag{2.17}$$

$$\underline{E}^{\mathrm{Delay}}(t;\underline{x}) = \int_{-\infty}^{t} d\tau\, \underline{\underline{\Phi}}(t-\tau) \bullet \underline{E}(\tau;\underline{x}), \tag{2.18}$$

$$\underline{P}_{\rho}^{\mathrm{LDelay}}(t;\underline{x}) = \int_{-\infty}^{t} d\tau\, \underline{\underline{\Phi}}(t-\tau) \bullet \underline{P}_{\rho}^{L}(\tau;\underline{x}), \tag{2.19}$$

$$\underline{P}_{\rho}^{\mathrm{NLDelay}}(t;\underline{x}) = \int_{-\infty}^{t} d\tau\, \underline{\underline{\Phi}}(t-\tau) \bullet [\underline{E}(\tau;\underline{x})\, R_{\rho}^{NL}(\tau;\underline{x})], \tag{2.20}$$

where

$$\underline{\underline{\Xi}}_0 = \mathrm{diag}\,\{(\frac{\sigma_y}{\varepsilon_0}+\frac{\sigma_z}{\varepsilon_0}-\frac{\sigma_x}{\varepsilon_0}),(\frac{\sigma_x}{\varepsilon_0}+\frac{\sigma_z}{\varepsilon_0}-\frac{\sigma_y}{\varepsilon_0}),(\frac{\sigma_x}{\varepsilon_0}+\frac{\sigma_y}{\varepsilon_0}-\frac{\sigma_z}{\varepsilon_0})\}, \tag{2.21}$$

$$\underline{\underline{\Xi}}_1 = \mathrm{diag}\{(\frac{\sigma_y}{\varepsilon_0}-\frac{\sigma_x}{\varepsilon_0})(\frac{\sigma_z}{\varepsilon_0}-\frac{\sigma_x}{\varepsilon_0}),(\frac{\sigma_x}{\varepsilon_0}-\frac{\sigma_y}{\varepsilon_0})(\frac{\sigma_z}{\varepsilon_0}-\frac{\sigma_y}{\varepsilon_0}),(\frac{\sigma_x}{\varepsilon_0}-\frac{\sigma_z}{\varepsilon_0})(\frac{\sigma_y}{\varepsilon_0}-\frac{\sigma_z}{\varepsilon_0})\}, \tag{2.22}$$

$$\underline{\underline{\Phi}}(t-\tau) = \mathrm{diag}\{\exp[-(\frac{\sigma_x}{\varepsilon_0})(t-\tau)],\exp[-(\frac{\sigma_y}{\varepsilon_0})(t-\tau)],\exp[-(\frac{\sigma_z}{\varepsilon_0})(t-\tau)]\}. \tag{2.23}$$

In the preceding equations, $\underline{H}^{\mathrm{Delay}}(t;\underline{x})$, $\underline{E}^{\mathrm{Delay}}(t;\underline{x})$, $\underline{P}_{\rho}^{\mathrm{LDelay}}(t;\underline{x})$, and $\underline{P}_{\rho}^{\mathrm{NLDelay}}(t;\underline{x})$ are introduced to handle the delayed time-response behavior of $\underline{H}(t;\underline{x})$, $\underline{E}(t;\underline{x})$, $\underline{P}_{\rho}^{L}(t;\underline{x})$, and $[\underline{E}(t;\underline{x})R_{\rho}^{NL}(t;\underline{x})]$, respectively. These functions follow naturally from taking the inverse Fourier transforms of the convolution functions $[1/(j\omega\,\underline{\underline{I}}+\underline{\underline{A}})]\,\underline{\widetilde{H}}(\omega;\underline{x})$, $[1/(j\omega\,\underline{\underline{I}}+\underline{\underline{A}})]\,\underline{\widetilde{E}}(\omega;\underline{x})$, $[1/(j\omega\,\underline{\underline{I}}+\underline{\underline{A}})]\,\underline{\widetilde{P}}_{\rho}^{L}(\omega;\underline{x})$, and $[1/(j\omega\,\underline{\underline{I}}+\underline{\underline{A}})]\,\underline{\widetilde{P}}_{\rho}^{NL}(\omega;\underline{x})$ and by noting that the inverse Fourier transform of $[1/(j\omega\,\underline{\underline{I}}+\underline{\underline{A}})]$, which transforms back to the time domain, is simply given by $\exp(-\underline{\underline{A}}\,t)$, where $\underline{\underline{I}}$ is the identity matrix and $\underline{\underline{A}}$ is the time-independent diagonal matrix $\mathrm{diag}\{(\sigma_x/\varepsilon_o),(\sigma_y/\varepsilon_o),(\sigma_z/\varepsilon_o)\}$.

To solve (2.13) through (2.20), we need to specify expressions for linear and nonlinear electric susceptibility functions. In this paper we consider the general case in which both the linear and nonlinear electric susceptibility functions are expressed as complex functions that contain complex coefficients and exhibit exponential behavior in the time domain as follows:

$$\chi_{\rho}^{L}(t) = \mathrm{Re}\,\{\alpha_{\rho}^{L}\exp[-(\gamma_{\rho}^{L})t]\}\,U(t), \tag{2.24}$$

$$\chi_{\rho}^{NL}(t) = \mathrm{Re}\,\{\alpha_{\rho}^{NL}\exp[-(\gamma_{\rho}^{NL})t]\}\,U(t), \tag{2.25}$$

where Re{ } is used to represent the real part of a complex function, U(t) is the unit step function, and α_ρ^L, γ_ρ^L, α_ρ^{NL} and γ_ρ^{NL} are complex coefficients. We need to point out that by making the proper choices of complex coefficients and performing the Fourier transforms of (2.24) and (2.25), one can obtain the familiar forms of the complex permittivity in the frequency domain for the constant conductivity [i.e., by setting α_ρ^L to be a real number, $\gamma_\rho^L = 0$, $\alpha_\rho^{NL} = 0$ and $\gamma_\rho^{NL} = 0$], the Debye model [i.e., by setting α_ρ^L, γ_ρ^L, α_ρ^{NL} and γ_ρ^{NL} to be all real numbers] and the Lorentz model [i.e., by setting α_ρ^L and α_ρ^{NL} to be imaginary numbers, and γ_ρ^L and γ_ρ^{NL} to be complex numbers].

Upon substituting (2.24) and (2.25) into (2.15) and (2.16), we obtain the following expressions for $\underline{P}_\rho^L(t;\underline{x})$ and $R_\rho^{NL}(t;\underline{x})$:

$$\underline{P}_\rho^L(t;\underline{x}) \equiv \mathrm{Re}\,\{\underline{Q}_\rho^L(t;\underline{x})\} = \mathrm{Re}\,\{\alpha_\rho^L \int_0^t d\tau\, \exp[-(\gamma_\rho^L)(t-\tau)]\, \underline{E}(\tau;\underline{x})\}, \tag{2.26}$$

$$R_\rho^{NL}(t;\underline{x}) \equiv \mathrm{Re}\,\{Q_\rho^{NL}(t;\underline{x})\}$$
$$= \mathrm{Re}\,\{\alpha_\rho^{NL} \int_0^t d\tau\, \exp[-(\gamma_\rho^{NL})(t-\tau)]\, [\underline{E}(\tau;\underline{x}) \bullet \underline{E}(\tau;\underline{x})]\}, \tag{2.27}$$

where complex functions $\underline{Q}_\rho^L(t;\underline{x})$ and $Q_\rho^{NL}(t;\underline{x})$ are introduced in the above equations such that the real parts of these complex functions result in $\underline{P}_\rho^L(t;\underline{x})$ and $R_\rho^{NL}(t;\underline{x})$, respectively.

To derive FDTD expressions based on the staggered Yee scheme,[1] (2.13), (2.14), (2.25), (2,26) and (2.17) through (2.20) have to be solved numerically for $\underline{H}(t;\underline{x})$, $\underline{E}(t;\underline{x})$, $\underline{Q}_\rho^L(t;\underline{x})$, $Q_\rho^{NL}(t;\underline{x})$, $\underline{H}^{Delay}(t;\underline{x})$, $\underline{E}^{Delay}(t;\underline{x})$, $\underline{P}_\rho^{LDelay}(t;\underline{x})$ and $\underline{P}_\rho^{NLDelay}(t;\underline{x})$ at each time step by correctly carrying out the numerical integration of convolution integrals $\underline{Q}_\rho^L(t;\underline{x})$, $Q_\rho^{NL}(t;\underline{x})$, $\underline{H}^{Delay}(t;\underline{x})$, $\underline{E}^{Delay}(t;\underline{x})$, $\underline{P}_\rho^{LDelay}(t;\underline{x})$ and $\underline{P}_\rho^{NLDelay}(t;\underline{x})$. Therefore, the whole solution rests on the question of how to carry out the numerical integration of $\underline{Q}_\rho^L(t;\underline{x})$, $Q_\rho^{NL}(t;\underline{x})$, $\underline{H}^{Delay}(t;\underline{x})$, $\underline{E}^{Delay}(t;\underline{x})$, $\underline{P}_\rho^{LDelay}(t;\underline{x})$ and $\underline{P}_\rho^{NLDelay}(t;\underline{x})$ at each time step. For that reason, the rest of this section is devoted to the numerical formulation that treats $\underline{Q}_\rho^L(t;\underline{x})$, $Q_\rho^{NL}(t;\underline{x})$, $\underline{H}^{Delay}(t;\underline{x})$, $\underline{E}^{Delay}(t;\underline{x})$, $\underline{P}_\rho^{LDelay}(t;\underline{x})$ and $\underline{P}_\rho^{NLDelay}(t;\underline{x})$ into the overall FDTD scheme based on the recursive convolution approach.[9]

We first convert the convolution integrals $\underline{Q}_\rho^L(t;\underline{x})$, $Q_\rho^{NL}(t;\underline{x})$, $\underline{H}^{Delay}(t;\underline{x})$, $\underline{E}^{Delay}(t;\underline{x})$, $\underline{P}_\rho^{LDelay}(t;\underline{x})$ and $\underline{P}_\rho^{NLDelay}(t;\underline{x})$ into the following equivalent first-order differential equations:

$$\frac{\partial \underline{Q}_\rho^L(t;\underline{x})}{\partial t} + (\gamma_\rho^L)\, \underline{Q}_\rho^L(t;\underline{x}) = \alpha_\rho^L\, \underline{E}(t;\underline{x}), \tag{2.28}$$

$$\frac{\partial Q_\rho^{NL}(t;\underline{x})}{\partial t} + (\gamma_\rho^{NL})\, Q_\rho^{NL}(t;\underline{x}) = \alpha_\rho^{NL}[\underline{E}(t;\underline{x}) \bullet \underline{E}(t;\underline{x})], \tag{2.29}$$

$$\frac{\partial \underline{H}^{Delay}(t;\underline{x})}{\partial t} + \underline{\underline{\Phi}}(t) \bullet \underline{H}^{Delay}(t;\underline{x}) = \underline{H}(t;\underline{x}), \tag{2.30}$$

$$\frac{\partial \underline{E}^{Delay}(t;\underline{x})}{\partial t} + \underline{\underline{\Phi}}(t) \bullet \underline{E}^{Delay}(t;\underline{x}) = \underline{E}(t;\underline{x}), \tag{2.31}$$

$$\frac{\partial \underline{Q}_\rho^{LDelay}(t;\underline{x})}{\partial t} + \underline{\underline{\Phi}}(t) \bullet \underline{Q}_\rho^{LDelay}(t;\underline{x}) = \underline{Q}_\rho^{L}(t;\underline{x}), \tag{2.32}$$

$$\frac{\partial \underline{Q}_\rho^{NLDelay}(t;\underline{x})}{\partial t} + \underline{\underline{\Phi}}(t) \bullet \underline{Q}_\rho^{NLDelay}(t;\underline{x}) = \underline{E}(t;\underline{x})\, Q_\rho^{NL}(t;\underline{x}), \tag{2.33}$$

where complex functions $\underline{Q}_\rho^{LDelay}(t;\underline{x})$ and $\underline{Q}_\rho^{NLDelay}(t;\underline{x})$ are introduced in (2.32) and (2.33) such that the real parts of these complex functions result in $\underline{P}_\rho^{LDelay}(t;\underline{x})$ and $\underline{P}_\rho^{NLDelay}(t;\underline{x})$, respectively.

To show how we can use (2.13), (2.14) and (2.28) through (2.33) to come up with a three-dimensional FDTD/PML algorithm for nonlinear dispersive PML media, we integrate (2.13) and (2.30) from $t=(n-\frac{1}{2})\Delta t$ to $t=(n+\frac{1}{2})\Delta t$, and (2.14), (2.28), (2.29) and (2.31) through (2.33) from $t=n\Delta t$ to $t=(n+1)\Delta t$. Then (2.28) through (2.33) are solved exactly using the integrating factor technique for a given discrete time interval to go forward in time by Δt.

To obtain second-order accuracy in time from a finite differencing technique, $\underline{H}(t;\underline{x})$ and $\underline{E}(t;\underline{x})$ are taken to be piecewise-linear continuous functions over the entire temporal integration range such that $\underline{H}(t;\underline{x})$ and $\underline{E}(t;\underline{x})$ change linearly with respect to time over given discrete time step intervals. It is equivalent to saying that we use only the first-order, time-dependent term of the Taylor series expansion for $\underline{H}(t;\underline{x})$ and $\underline{E}(t;\underline{x})$, respectively, that are expanded in time about the current time step $t=(n-\frac{1}{2})\Delta t$ for $\underline{H}(t;\underline{x})$ and the current time step $t=n\Delta t$ for $\underline{E}(t;\underline{x})$. Mathematically, we can express $\underline{H}(t;\underline{x})$ and $\underline{E}(t;\underline{x})$ in the following forms in terms of $(\underline{H})_{ijk}^{n-\frac{1}{2}}$, $(\underline{H})_{ijk}^{n+\frac{1}{2}}$, $(\underline{E})_{ijk}^{n}$ and $(\underline{E})_{ijk}^{n+1}$ where superscripts $n-\frac{1}{2}$, n, $n+\frac{1}{2}$ and $n+1$ are used to denote discrete time steps at $t=(n-\frac{1}{2})\Delta t$, $t=n\Delta t$, $t=(n+\frac{1}{2})\Delta t$ and $t=(n+1)\Delta t$, respectively. Subscripts are used to denote discrete spatial locations, $\underline{x}=[i\Delta x, j\Delta y, k\Delta z]$ for $\underline{E}(t;\underline{x})$ and $\underline{x}=[(i-\frac{1}{2})\Delta x, (j-\frac{1}{2})\Delta y, (k-\frac{1}{2})\Delta z]$ for $\underline{H}(t;\underline{x})$ with Δx, Δy and Δz being the spatial grid sizes in the x, y and z directions, respectively.

$$\underline{H}(t;\underline{x}) = \begin{cases} (\underline{H})_{ijk}^{n-1/2} + \dfrac{[(\underline{H})_{ijk}^{n+1/2} - (\underline{H})_{ijk}^{n-1/2}]}{\Delta t}[t-(n-1/2)\Delta t] + \text{higher order terms}, \\ \qquad \text{for} \quad 0 \le (n-1/2)\Delta t \le t \le (n+1/2)\Delta t \\ 0, \qquad \text{for} \quad t<0 \end{cases} \tag{2.34}$$

$$\underline{E}(t;\underline{x}) = \begin{cases} (\underline{E})_{ijk}^{n} + \dfrac{[(\underline{E})_{ijk}^{n+1} - (\underline{E})_{ijk}^{n}]}{\Delta t}(t-n\Delta t) + \text{higher order terms}, \\ \qquad \text{for} \quad 0 \le n\Delta t \le t \le (n+1)\Delta t \\ 0, \qquad \text{for} \quad t<0 \end{cases} \tag{2.35}$$

Although we are not going to investigate higher than second-order accuracy in time in this paper, it is possible to obtain higher-order accurate FDTD algorithms by simply including more terms beyond the first-order, time-dependent term in the above Taylor series expansion.

Now, substituting (2.34) and (2.35) into (2.13), (2.14) and (2.28) through (2.33) by keeping only the first-order, time dependent term of the above Taylor series expansion, and then performing the time integration from $t=(n-1/2)\Delta t$ to $t=(n+1/2)\Delta t$ for field values that depend on the magnetic field [i.e., $\underline{H}(t;\underline{x})$ and $\underline{H}^{Delay}(t;\underline{x})$], and from $t=n\Delta t$ to $t=(n+1)\Delta t$ for field values that depend on the electric field [i.e., $\underline{E}(t;\underline{x})$, $\underline{Q}_\rho^{L}(t;\underline{x})$, $Q_\rho^{NL}(t;\underline{x})$, $\underline{E}^{Delay}(t;\underline{x})$, $\underline{Q}_\rho^{LDelay}(t;\underline{x})$ and $\underline{Q}_\rho^{NLDelay}(t;\underline{x})$], (2.13), (2.14) and (2.28) through (2.33) are manipulated and cast into the following cubic algebraic forms. These expressions are used to update $(\underline{H})_{ijk}^{n+1/2}$, $(\underline{E})_{ijk}^{n+1}$, $(\underline{Q}_\rho^{L})_{ijk}^{n+1}$, $(Q_\rho^{NL})_{ijk}^{n+1}$, $(\underline{E}^{Delay})_{ijk}^{n+1}$, $(\underline{H}^{Delay})_{ijk}^{n+1/2}$, $(\underline{Q}_\rho^{LDelay})_{ijk}^{n+1}$ and $(\underline{Q}_\rho^{NLDelay})_{ijk}^{n+1}$ at each time step:

$$\underline{\underline{\Omega}}_0 \bullet (\underline{H})_{ijk}^{n+1/2} + \underline{\underline{\Omega}}_1 \bullet (\underline{H})_{ijk}^{n-1/2} + \underline{\underline{\Omega}}_2 \bullet (\underline{H}^{Delay})_{ijk}^{n-1/2} + \underline{S}_E = 0, \tag{2.36}$$

$$\begin{aligned} &(\underline{\underline{\Lambda}}_0)_{ijk}^{n} \bullet (E)_{ijk}^{n+1} + (\underline{\underline{\Lambda}}_1)_{ijk}^{n} \bullet (\underline{E})_{ijk}^{n} + \mathrm{Re}\{\sum_\rho \underline{\underline{\Gamma}}_{\rho,0}^{L} \bullet (\underline{Q}_\rho^{L})_{ijk}^{n}\} + \underline{\underline{\Omega}}_2 \bullet (\underline{E}^{Delay})_{ijk}^{n} \\ &\quad + \underline{\underline{\Omega}}_3 \bullet \mathrm{Re}\{\sum_\rho (\underline{Q}_\rho^{LDelay})_{ijk}^{n}\} + \underline{\underline{\Omega}}_3 \bullet \mathrm{Re}\{\sum_\rho (\underline{Q}_\rho^{NLDelay})_{ijk}^{n}\} \\ &\quad + \mathrm{Re}\{\sum_\rho \underline{\underline{\Gamma}}_{\rho,0}^{NL}\} \bullet (\underline{E})_{ijk}^{n}\,[(\underline{E})_{ijk}^{n} \bullet (\underline{E})_{ijk}^{n}] \\ &\quad + \mathrm{Re}\{\sum_\rho \underline{\underline{\Gamma}}_{\rho,1}^{NL}\} \bullet (\underline{E})_{ijk}^{n}\,[(\underline{E})_{ijk}^{n} \bullet (\underline{E})_{ijk}^{n+1}] \\ &\quad + \mathrm{Re}\{\sum_\rho \underline{\underline{\Gamma}}_{\rho,2}^{NL}\} \bullet (\underline{E})_{ijk}^{n}\,[(\underline{E})_{ijk}^{n+1} \bullet (\underline{E})_{ijk}^{n+1}] \\ &\quad + \mathrm{Re}\{\sum_\rho \underline{\underline{\Gamma}}_{\rho,3}^{NL}\} \bullet (\underline{E})_{ijk}^{n+1}\,[(\underline{E})_{ijk}^{n} \bullet (\underline{E})_{ijk}^{n}] \end{aligned}$$

$$+\mathrm{Re}\{\sum_{\rho}\underline{\underline{\Gamma_{\rho,4}^{NL}}}\}\bullet(\underline{E})_{ijk}^{n+1}\,[\,(\underline{E})_{ijk}^{n}\bullet(\underline{E})_{ijk}^{n+1}\,]$$

$$+\mathrm{Re}\{\sum_{\rho}\underline{\underline{\Gamma_{\rho,5}^{NL}}}\}\bullet(\underline{E})_{ijk}^{n+1}\,[\,(\underline{E})_{ijk}^{n+1}\bullet(\underline{E})_{ijk}^{n+1}\,]+\underline{S_H}=0, \tag{2.37}$$

$$(\underline{Q}_{\rho}^{L})_{ijk}^{n+1}=\Theta_{\rho,0}^{L}\left[\,(\underline{Q}_{\rho}^{L})_{ijk}^{n}+\Theta_{\rho,1}^{L}\,(\underline{E})_{ijk}^{n}+\Theta_{\rho,2}^{L}\,(\underline{E})_{ijk}^{n+1}\right], \tag{2.38}$$

$$(\underline{Q}_{\rho}^{NL})_{ijk}^{n+1}=\Theta_{\rho,0}^{NL}\left\{(\underline{Q}_{\rho}^{NL})_{ijk}^{n}+\Theta_{\rho,1}^{NL}\,[\,(\underline{E})_{ijk}^{n}\bullet(\underline{E})_{ijk}^{n}\,]+\Theta_{\rho,2}^{NL}\,[\,(\underline{E})_{ijk}^{n}\bullet(\underline{E})_{ijk}^{n+1}\,]\right.$$
$$\left.+\Theta_{\rho,3}^{NL}\,[\,(\underline{E})_{ijk}^{n+1}\bullet(\underline{E})_{ijk}^{n+1}\,]\right\}, \tag{2.39}$$

$$(\underline{H}^{Delay})_{ijk}^{n+1/2}=\underline{\underline{\Omega_4}}\bullet\left[\,(\underline{H}^{Delay})_{ijk}^{n-1/2}+\underline{\underline{\Omega_5}}\bullet(\underline{H})_{ijk}^{n-1/2}+\underline{\underline{\Omega_6}}\bullet(\underline{H})_{ijk}^{n+1/2}\right], \tag{2.40}$$

$$(\underline{E}^{Delay})_{ijk}^{n+1}=\underline{\underline{\Omega_4}}\bullet\left[\,(\underline{E}^{Delay})_{ijk}^{n}+\underline{\underline{\Omega_5}}\bullet(\underline{E})_{ijk}^{n}+\underline{\underline{\Omega_6}}\bullet(\underline{E})_{ijk}^{n+1}\right], \tag{2.41}$$

$$(\underline{Q}_{\rho}^{LDelay})_{ijk}^{n+1}=\underline{\underline{\Omega_4}}\bullet\left[\,(\underline{Q}_{\rho}^{LDelay})_{ijk}^{n}+\underline{\underline{\Pi_{\rho,0}^{L}}}\bullet(\underline{Q}_{\rho}^{L})_{ijk}^{n}+\underline{\underline{\Pi_{\rho,1}^{L}}}\bullet(\underline{E})_{ijk}^{n}\right.$$
$$\left.+\underline{\underline{\Pi_{\rho,2}^{L}}}\bullet(\underline{E})_{ijk}^{n+1}\right], \tag{2.42}$$

$$(\underline{Q}_{\rho}^{NLDelay})_{ijk}^{n+1}=\underline{\underline{\Omega_4}}\bullet\left\{(\underline{Q}_{\rho}^{NLDelay})_{ijk}^{n}\right.$$
$$+(Q_{\rho}^{NL})_{ijk}^{n}\,\underline{\underline{\Pi_{\rho,0}^{NL}}}\bullet(\underline{E})_{ijk}^{n}+(Q_{\rho}^{NL})_{ijk}^{n}\,\underline{\underline{\Pi_{\rho,1}^{NL}}}\bullet(\underline{E})_{ijk}^{n+1}$$
$$+\underline{\underline{\Pi_{\rho,2}^{NL}}}\bullet(\underline{E})_{ijk}^{n}\,[\,(\underline{E})_{ijk}^{n}\bullet(\underline{E})_{ijk}^{n}\,]+\underline{\underline{\Pi_{\rho,3}^{NL}}}\bullet(\underline{E})_{ijk}^{n}\,[\,(\underline{E})_{ijk}^{n}\bullet(\underline{E})_{ijk}^{n+1}\,]$$
$$+\underline{\underline{\Pi_{\rho,4}^{NL}}}\bullet(\underline{E})_{ijk}^{n}\,[\,(\underline{E})_{ijk}^{n+1}\bullet(\underline{E})_{ijk}^{n+1}\,]+\underline{\underline{\Pi_{\rho,5}^{NL}}}\bullet(\underline{E})_{ijk}^{n+1}\,[\,(\underline{E})_{ijk}^{n}\bullet(\underline{E})_{ijk}^{n}\,]$$
$$\left.+\underline{\underline{\Pi_{\rho,6}^{NL}}}\bullet(\underline{E})_{ijk}^{n+1}\,[\,(\underline{E})_{ijk}^{n}\bullet(\underline{E})_{ijk}^{n+1}\,]+\underline{\underline{\Pi_{\rho,7}^{NL}}}\bullet(\underline{E})_{ijk}^{n+1}\,[\,(\underline{E})_{ijk}^{n+1}\bullet(\underline{E})_{ijk}^{n+1}\,]\right\}, \tag{2.43}$$

where $\underline{S_E}$ and $\underline{S_H}$ are given by

$$\underline{S_E}=\begin{pmatrix}\dfrac{\Delta t}{(\mu_0\mu_R)\Delta y}\left[(E_z)_{i(j+1/2)k}^{n}-(E_z)_{i(j-1/2)k}^{n}\right]-\dfrac{\Delta t}{(\mu_0\mu_R)\Delta z}\left[(E_y)_{ij(k+1/2)}^{n}-(E_y)_{ij(k-1/2)}^{n}\right]\\ \dfrac{\Delta t}{(\mu_0\mu_R)\Delta z}\left[(E_x)_{ij(k+1/2)}^{n}-(E_x)_{ij(k-1/2)}^{n}\right]-\dfrac{\Delta t}{(\mu_0\mu_R)\Delta x}\left[(E_z)_{(i+1/2)jk}^{n}-(E_z)_{(i-1/2)jk}^{n}\right]\\ \dfrac{\Delta t}{(\mu_0\mu_R)\Delta x}\left[(E_y)_{(i+1/2)jk}^{n}-(E_y)_{(i-1/2)jk}^{n}\right]-\dfrac{\Delta t}{(\mu_0\mu_R)\Delta y}\left[(E_x)_{i(j+1/2)k}^{n}-(E_x)_{i(j-1/2)k}^{n}\right]\end{pmatrix}, \tag{2.44}$$

$$\underline{S_H} = \begin{pmatrix} -\dfrac{\Delta t}{(\varepsilon_0\varepsilon_R)\Delta y}\left[(H_z)^{n+½}_{i(j+½)k} - (H_z)^{n+½}_{i(j-½)k}\right] + \dfrac{\Delta t}{(\varepsilon_0\varepsilon_R)\Delta z}\left[(H_y)^{n+½}_{ij(k+½)} - (H_y)^{n+½}_{ij(k-½)}\right] \\ -\dfrac{\Delta t}{(\varepsilon_0\varepsilon_R)\Delta z}\left[(H_x)^{n+½}_{ij(k+½)} - (H_x)^{n+½}_{ij(k-½)}\right] + \dfrac{\Delta t}{(\varepsilon_0\varepsilon_R)\Delta x}\left[(H_z)^{n+½}_{(i+½)jk} - (H_z)^{n+½}_{(i-½)jk}\right] \\ -\dfrac{\Delta t}{(\varepsilon_0\varepsilon_R)\Delta x}\left[(H_y)^{n+½}_{(i+½)jk} - (H_y)^{n+½}_{(i-½)jk}\right] + \dfrac{\Delta t}{(\varepsilon_0\varepsilon_R)\Delta y}\left[(H_x)^{n+½}_{i(j+½)k} - (H_x)^{n+½}_{i(j-½)k}\right] \end{pmatrix}. \tag{2.45}$$

The $(\underline{\underline{\Lambda}}_0)^n_{ijk}$ and $(\underline{\underline{\Lambda}}_1)^n_{ijk}$ that appear in the first and second terms of (2.37) are updated at each time step with the updated value of $(Q^{NL}_\rho)^n_{ijk}$ in order to include the effect of the nonlinear response. Other coefficients and matrices that appear in (2.36) through (2.43) are represented by $\Theta^L_{\rho,0}$, $\Theta^L_{\rho,1}$, $\Theta^L_{\rho,2}$, $\Theta^{NL}_{\rho,0}$, $\Theta^{NL}_{\rho,1}$, $\Theta^{NL}_{\rho,2}$, $\Theta^{NL}_{\rho,3}$, $\underline{\underline{\Omega}}_0$, $\underline{\underline{\Omega}}_1$, $\underline{\underline{\Omega}}_2$, $\underline{\underline{\Omega}}_3$, $\underline{\underline{\Omega}}_4$, $\underline{\underline{\Omega}}_5$, $\underline{\underline{\Omega}}_6$, $\underline{\underline{\Gamma}}^L_{\rho,0}$, $\underline{\underline{\Gamma}}^{NL}_{\rho,0}$, $\underline{\underline{\Gamma}}^{NL}_{\rho,1}$, $\underline{\underline{\Gamma}}^{NL}_{\rho,2}$, $\underline{\underline{\Gamma}}^{NL}_{\rho,3}$, $\underline{\underline{\Gamma}}^{NL}_{\rho,4}$, $\underline{\underline{\Gamma}}^{NL}_{\rho,5}$, $\underline{\underline{\Pi}}^L_{\rho,0}$, $\underline{\underline{\Pi}}^L_{\rho,1}$, $\underline{\underline{\Pi}}^L_{\rho,2}$, $\underline{\underline{\Pi}}^{NL}_{\rho,0}$, $\underline{\underline{\Pi}}^{NL}_{\rho,1}$, $\underline{\underline{\Pi}}^{NL}_{\rho,2}$, $\underline{\underline{\Pi}}^{NL}_{\rho,3}$, $\underline{\underline{\Pi}}^{NL}_{\rho,4}$, $\underline{\underline{\Pi}}^{NL}_{\rho,5}$, $\underline{\underline{\Pi}}^{NL}_{\rho,6}$, $\underline{\underline{\Pi}}^{NL}_{\rho,7}$. These coefficients and matrices can be evaluated at the beginning of the FDTD/PML simulation since they depend only on material properties and Δt. The material properties that are required for the evaluation of these coefficients and matrices are α^L_ρ, γ^L_ρ, α^{NL}_ρ, γ^{NL}_ρ, σ_x, σ_y and σ_z, which are known quantities since they are assigned for a specific FDTD/PML simulation problem. The explicit expressions of these coefficients and matrices found in Ref. 11.

Using the above FDTD/PML algorithm the computer simulation can be performed for electromagnetic waves that propagate inside nonlinear dispersive PML media by simply going through the following steps:

(1) First, as part of the initial condition, time-invariant coefficients $\Theta^L_{\rho,0}$, $\Theta^L_{\rho,1}$, $\Theta^L_{\rho,2}$, $\Theta^{NL}_{\rho,0}$, $\Theta^{NL}_{\rho,1}$, $\Theta^{NL}_{\rho,2}$ and $\Theta^{NL}_{\rho,3}$, and time-invariant matrices $\underline{\underline{\Omega}}_0$, $\underline{\underline{\Omega}}_1$, $\underline{\underline{\Omega}}_2$, $\underline{\underline{\Omega}}_3$, $\underline{\underline{\Omega}}_4$, $\underline{\underline{\Omega}}_5$, $\underline{\underline{\Omega}}_6$, $\underline{\underline{\Gamma}}^L_{\rho,0}$, $\underline{\underline{\Gamma}}^{NL}_{\rho,0}$, $\underline{\underline{\Gamma}}^{NL}_{\rho,1}$, $\underline{\underline{\Gamma}}^{NL}_{\rho,2}$, $\underline{\underline{\Gamma}}^{NL}_{\rho,3}$, $\underline{\underline{\Gamma}}^{NL}_{\rho,4}$, $\underline{\underline{\Gamma}}^{NL}_{\rho,5}$, $\underline{\underline{\Pi}}^L_{\rho,0}$, $\underline{\underline{\Pi}}^L_{\rho,1}$, $\underline{\underline{\Pi}}^L_{\rho,2}$, $\underline{\underline{\Pi}}^{NL}_{\rho,0}$, $\underline{\underline{\Pi}}^{NL}_{\rho,1}$, $\underline{\underline{\Pi}}^{NL}_{\rho,2}$, $\underline{\underline{\Pi}}^{NL}_{\rho,3}$, $\underline{\underline{\Pi}}^{NL}_{\rho,4}$, $\underline{\underline{\Pi}}^{NL}_{\rho,5}$, $\underline{\underline{\Pi}}^{NL}_{\rho,6}$ and $\underline{\underline{\Pi}}^{NL}_{\rho,7}$ are all calculated at the beginning of simulation for given values of α^L_ρ, γ^L_ρ, α^{NL}_ρ, γ^{NL}_ρ, α^L_ρ, σ_x, σ_y, σ_z and Δt. These values are stored in computer memory and used in calculating the updated field values at each time step.
(2) Using (2.36), $(\underline{H})_{ijk}{}^{n+½}$ is updated based on the known values of $(\underline{H})_{ijk}{}^{n-½}$ and $(\underline{H}^{Delay})_{ijk}{}^{n-½}$ and $(\underline{E})_{ijk}{}^{n}$.
(3) Using (2.40), $(\underline{H}^{Delay})_{ijk}{}^{n+½}$ is updated based on the known values of $(\underline{H}^{Delay})_{ijk}{}^{n-½}$, $(\underline{H})_{ijk}{}^{n+½}$ and $(\underline{H})_{ijk}{}^{n-½}$.
(4) $(\underline{\underline{\Lambda}}_0)^n_{ijk}$ and $(\underline{\underline{\Lambda}}_1)^n_{ijk}$ are updated based on the known values of $(Q_\rho{}^{NL})_{ijk}{}^{n}$.

(5) Using (2.37), $(\underline{E})_{ijk}^{n+1}$ is updated based on the known values of $(\underline{E})_{ijk}^{n}$, $(\underline{E}^{Delay})_{ijk}^{n}$, $(\underline{Q}_\rho^{L})_{ijk}^{n}$, $(\underline{Q}_\rho^{LDelay})_{ijk}^{n}$ and $(\underline{H})_{ijk}^{n+½}$. Because (2.37) represents the coupled, cubic, algebraic equation, the nonlinear Newton-Raphson method[12] is used to solve for $(\underline{E})_{ijk}^{n+1}$ by finding the zero of (2.37) with the previous time step value of $(\underline{E})_{ijk}^{n}$ as the initial guess.

(6) Using (2.38), (2.39), (2.41), (2.42) and (2.43), $(\underline{Q}_\rho^{L})_{ijk}^{n+1}$, $(Q_\rho^{NL})_{ijk}^{n+1}$, $(\underline{E}^{Delay})_{ijk}^{n+1}$, $(\underline{Q}_\rho^{LDelay})_{ijk}^{n+1}$ and $(\underline{Q}_\rho^{NLDelay})_{ijk}^{n+1}$ are updated based on the known values of $(\underline{E})_{ijk}^{n+1}$, $(\underline{E})_{ijk}^{n}$, $(\underline{Q}_\rho^{L})_{ijk}^{n}$, $(Q_\rho^{NL})_{ijk}^{n}$, $(\underline{E}^{Delay})_{ijk}^{n}$, $(\underline{Q}_\rho^{LDelay})_{ijk}^{n}$ and $(\underline{Q}_\rho^{NLDelay})_{ijk}^{n}$.

(7) Increment the time step by Δt. Go back to step (2) and repeat the whole process again until the end of simulation.

Shown in Figure 1 is the flow chart of numerical steps required to update field values as described above.

In the absence of the PML interface, σ_x, σ_y and σ_z can be set to zeroes. It results in reducing the nonlinear dispersive FDTD/PML algorithm to the usual nonlinear dispersive FDTD algorithm for the nonPML case.[9] Now, if we happen to set the material coefficient α_ρ^{NL} to zero, the nonlinear dispersive FDTD/PML algorithm reduces to the linear dispersive FDTD/PML algorithm reported before.[13] Furthermore, if both α_ρ^{L} and α_ρ^{NL} are set to zeroes, it reduces to the case of the simple free space FDTD/PML algorithm.[11]

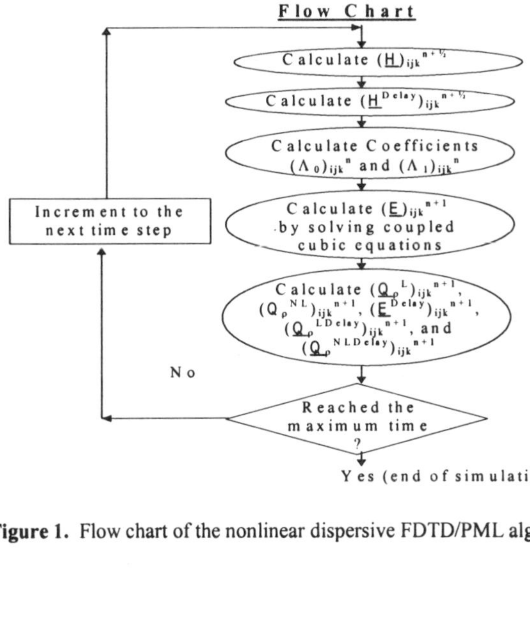

Figure 1. Flow chart of the nonlinear dispersive FDTD/PML algorithm.

3. ONE-DIMENSIONAL FDTD/PML NUMERICAL DEMONSTRATION FOR NONLINEAR RAMAN SCATTERING MEDIA

To demonstrate the validity of the FDTD/PML algorithm, we consider the formation of a temporal soliton in one-dimensional space inside a nonlinear dispersive medium. We simulate an ultra-short optical pulse propagating into a half-space, nonlinear Raman Scattering medium, which results in the nonlinear feedback of the delayed time response to the propagating wave.[9] First, we launch an optical pulse in free space and propagate it from left to right in the positive x-direction. After traveling a short distance, the optical pulse is incident on an infinite half-space, nonlinear Raman Scattering medium that is characterized by a single first-order (linear) susceptibility function, $X^{L}_{\rho=1}(t)$, and a single third-order (nonlinear) susceptibility function, $X^{NL}_{\rho=1}(t)$. We consider the case where $X^{L}_{\rho=1}(t)$ and $X^{NL}_{\rho=1}(t)$ are expressed in the following Lorentz forms[14,15] [i.e., set $\rho_{max}=1$ in (2.3) for a single species]:

$$X^{(1)}_{\rho=1}(t) = \frac{\omega_0^2(\varepsilon_S - \varepsilon_R)}{\sqrt{\omega_0^2 - \delta^2}} \exp(-\delta t)\, \sin(\sqrt{\omega_0^2 - \delta^2}\, t), \tag{2.46}$$

$$X^{NL}_{\rho=1}(t) = \chi^{NL}_{01} [(\tau_1^2 + \tau_2^2)/\tau_1\, \tau_2^2] \exp(-t/\tau_2) \sin(t/\tau_1), \tag{2.47}$$

where ω_0 is the resonant frequency, ε_S is relative permittivity of a medium at DC, ε_R is the relative permittivity of a medium at infinite frequency, δ is the first-order susceptibility damping constant, χ^{NL}_{01} is the Raman Scattering strength, $1/\tau_1$ is the optical phonon frequency, and τ_2 is the optical phonon lifetime.

Comparing (2.46) and (2.47) with (2.24) and (2.25), respectively, we can relate the above coefficients to the complex coefficients, α^{L}_{ρ}, γ^{L}_{ρ}, α^{NL}_{ρ} and γ^{NL}_{ρ}, as follows:

$$\alpha^{L}_{\rho=1} \Leftarrow j\frac{\omega_0^2(\varepsilon_S - \varepsilon_R)}{\sqrt{\omega_0^2 - \delta^2}} \quad \text{and} \quad \gamma^{L}_{\rho=1} \Leftarrow (\delta + j\sqrt{\omega_0^2 - \delta^2}), \tag{2.48-2.49}$$

$$\alpha^{NL}_{\rho=1} \Leftarrow j\chi^{NL}_{01} [(\tau_1^2 + \tau_2^2)/\tau_1\, \tau_2^2] \quad \text{and} \quad \gamma^{NL}_{\rho=1} \Leftarrow (\frac{1}{\tau_2} + j\frac{1}{\tau_1}), \tag{2.50-2.51}$$

We consider the incident optical pulse to have a sinusoidal-carrier electric field frequency ω_c of 8.61×10^{14} radians/sec enveloped inside a hyperbolic secant function that has a width constant T_w of 3.50 femtoseconds. The following mathematical form describes the time-dependent incident optical pulse that we use in our one-dimensional FDTD simulation:

$$\text{Incident Optical Pulse}(t) = A\ \cos[\omega_c(t - t_{delay})]\,\text{sech}[(t - t_{delay})/T_w], \qquad (2.52)$$

where A is the amplitude of the electric field and t_{delay} is the delay time for the incident optical pulse to reach its peak value. We arbitrarily assign A to take the value of 1.0 volt/meter and t_{delay} to take the value of 41.70 femtoseconds.

We select the total number of simulation cells to be 60,000, ranging from x =0 to x = 60,000, with the interface between free space and the nonlinear dispersive material located at x = 10,000. The right side of x = 10,000 is represented by free space and the left side of x = 10,000 by the nonlinear dispersive material. We launch an optical pulse into free space at 20 cells away from the leftmost edge (at x = 20) traveling in the positive x-direction. We use the free space PML absorbing boundary condition in the left and the nonlinear PML absorbing boundary condition in the right to terminate the computational space. At both outer absorbing boundary regions, we use 10 PML cells to absorb all waves that propagate out of the computational space. Listed below are important numerical parameters and material properties that we use to perform one-dimensional FDTD/PML simulation calculations.

- Basic FDTD parameters:
  - ❑ Uniform cell size = Δx = 5 nm;
  - ❑ Total simulation distance = Δx times the total number of cells = 0.1 cm;
  - ❑ Time step = $\Delta t = \Delta x/2c$ = 0.00834 fs, where c is the speed of light;
  - ❑ Total number of time steps = 400,000 (total simulation time = 3336 fs);
  - ❑ Quadratic increase in PML conductivities inside the PML cells.

- Linear dispersive material properties:
  - ❑ ε_s =5.25, ε_R =2.25, $\omega_o = 8.0\text{x}10^{14}$ radians/s, and $\delta = 4.0\text{x}10^{9}\ \text{s}^{-1}$.

- Nonlinear dispersive material properties:
  - ❑ τ_1 = 12.2 fs, τ_2 = 32.0 fs, and $\chi_{01}^{NL} = 2.0\ (\text{volts/meter})^{-2}$.

By choosing 5 nanometers for Δx, we estimate the free space numerical phase velocity error[16] to be around $5\text{x}10^{-6}$, which is about the same order of accuracy as the single precision calculation of our SPARC 60 UNIX workstation.

To update the electric field value for the next time step, we use the simple Newton-Raphson iterative method[12] by making use of the current-time-step electric field value as the initial guess to solve the cubic equation. For all calculations, the convergence criterion of 10^{-4} is satisfied with at most three iterations.

Figure 2 shows the time evolution of spatial electric field profiles of a delayed time response solitary wave inside the nonlinear Raman Scattering medium taken at five different time steps of 100,000Δt, 200,000Δt, 300,000Δt, 400,000Δt and 500,000Δt. As seen in this figure, at the beginning the solitary wave packet is imbedded inside the linear dispersive wave packet and it is lagged slightly behind the linear dispersive wave packet due to the nonlinear response of the Raman Scattering medium. As the wave propagates further into the medium, the solitary wave packet gets gradually isolated from the linear dispersive wave packet due to the slower speed of the moving solitary wave packet. By the time it reaches the time step of 300,000Δt, the solitary wave packet clearly forms its distinct solitary shape and propagates at constant amplitude while maintaining its shape. However, the shape of the solitary wave packet is uniquely different from the well-known secant function shaped wave packet that is obtained for the instantaneous time response of the Kerr-type case. Basically, the solitary wave packet inside the Raman Scattering medium appears retarded in time behind the linear dispersive wave packet as a consequence of the nonlinear delayed time response of the medium. Details of these differences for Kerr and Raman Scattering media are discussed in Yakura *et al.*[9] Also shown Figure 2 is the efficient absorption properties of the nonlinear PML located at the right-most edge of the computational domain. As seen in the spatial electric field profile at the time step of 500,000Δt, there is no observable reflection from the PML boundary.

To show the extent of reflection from the PML based on the nonlinear PML technique, we performed additional FDTD calculations using the Liao absorbing boundary condition.[17] These FDTD calculations are compared with the reference case where the interface of the absorbing boundary is extended to the right by an additional 60,000 grid cells so that reflection is not important around the region of the absorbing boundary for the simulation times of interest. Figure 3 shows the comparison of the electric field profiles obtained from the reference case, the PML absorbing boundary condition, and the Liao boundary condition. As seen in Figure 3, the PML absorbing boundary condition gives good agreement with the reference case and it reveals a much better absorption property as compared to the Liao absorbing boundary condition. The maximum reflection off the nonlinear PML absorbing boundary turns out to be –19 dB as compared to –9 dB for the Liao absorbing boundary case.

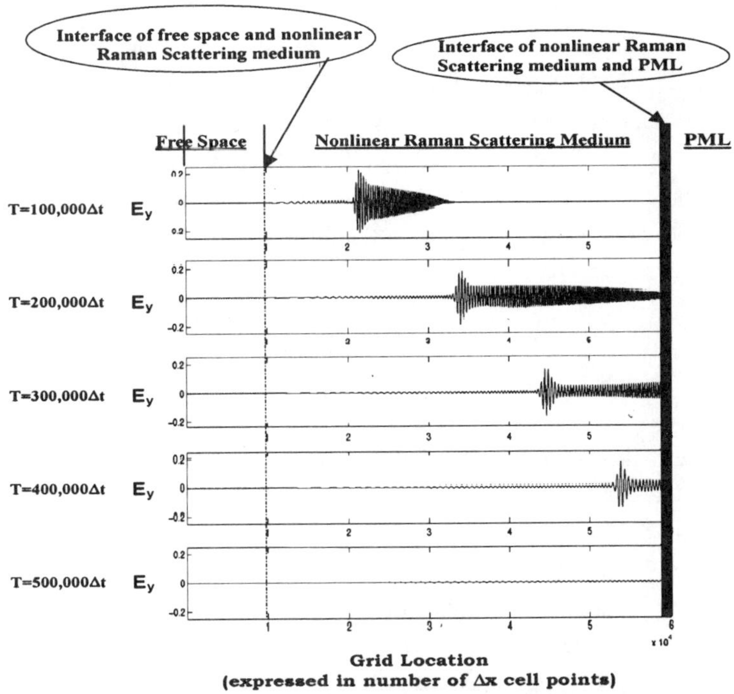

Figure 2. Comparison of the spatial electric field profiles that propagate inside the nonlinear Raman Scattering medium at various time steps. The right-side edge is terminated by 10 cells of the nonlinear PML absorbing boundary, with $\Delta t = 8.34 \times 10^{-18}$ s and $\Delta x = 5 \times 10^{-9}$ m. The grid location ranges from 0 to 60,000Δx.

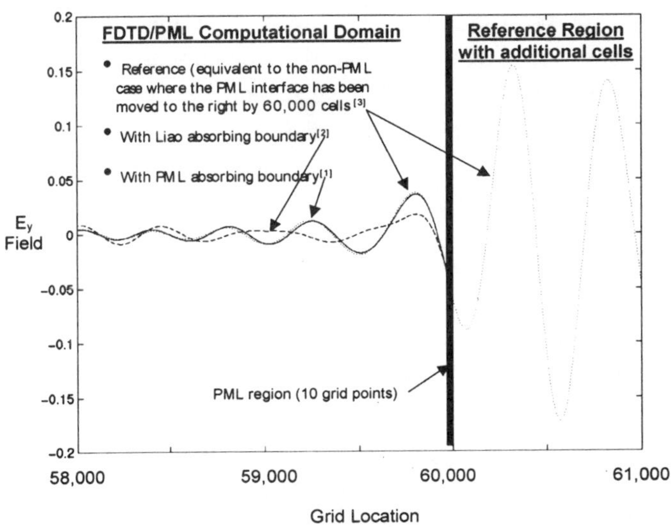

Figure 3. Comparing the spatial profiles of propagating waves at the instantaneous time step of T=450,000Δt for three different absorbing boundary conditions: (1) Terminated the right side by 10 cells of the nonlinear PML absorbing boundary; (2) Terminated the right side by the Liao absorbing boundary condition; and (3) Extended the absorbing boundary to the right by an additional 60,000 cells (The reference case with no effect of reflection).

To assure ourselves of the computational correctness of the results that we obtained from FDTD/PML algorithm, we repeated the same FDTD calculations using one-half the cell size and one-half the time step increment. These higher spatial and temporal resolution calculations resulted in a relative difference of less than 10^{-5} for the electric field value at the end of the simulation run. Since this relative difference is on the order of the single precision calculation error, we concluded that we are indeed calculating the correct electric field value. This procedure also served to validate the use of the piecewise-linear approximation.

4. CONCLUSIONS

We present in this paper the formulation of a three-dimensional FDTD/PML algorithm inside nonlinear dispersive PML media that is used to absorb outgoing electromagnetic waves within a finite simulation volume to create the notion of infinity at the outer layer boundary of the computational volume. Because of the use of the piecewise-linear approximation, the FDTD/PML algorithm provides second-order accuracy in time for the calculation of electromagnetic field quantities. The resulting forms of the FDTD/PML algorithm tell us that we need to solve coupled cubic equations for the three components of the electric field vector at each time step due to the nonlinear behavior of the third-order electric susceptibility function. In the absence of the nonlinear contribution, the electric field components are no longer coupled and each component of the electric field vector can be updated independently of the other two components.

For numerical validation of the FDTD/PML algorithm presented in this paper, we performed one-dimensional FDTD/PML calculations to demonstrate the efficient absorption of outgoing waves for an ultra-short optical pulse propagating inside a nonlinear Raman scattering medium with the nonlinear PML absorbing boundary.

REFERENCES

1. K.S. Yee, Numerical solution of initial boundary value problems involving Maxwell's equations in isotropic media, *IEEE Trans. Antennas Propagat.* **14**, 302–307 (1966).
2. J.P. Berenger, A perfectly matched layer for the absorption of electromagnetic waves, *J. Comput. Phys.* **114**, 185–200 (1994).
3. W.C. Chew and W. Weeden, A 3D perfectly matched medium from modified Maxwell's equations with stretched coordinates, *Microwave Opt. Tech. Lett.* **7**, 599–604 (1994).
4. Z.S. Sacks, D.M. Kingsland, R. Lee, and J.F. Lee, A perfectly matched anisotropic absorber for use as an absorbing boundary condition, *IEEE Trans. Antennas Propagat.* **4**, 1460–1463 (1995).
5. S.D. Gedney, An anisotropic PML absorbing media for FDTD simulation of fields in lossy dispersive media, *Electromagnetics* **16**, 399–415 (1996).
6. S.D. Gedney, An anisotropic perfectly matched layer absorbing media for the truncation of FDTD lattices, *IEEE Trans. Antennas Propagat.* **44**, 1630–1639 (1996).

7. A. Taflove, *Advances in Computational Electromagnetics: The Finite-Difference Time-Domain Method* (Artech House, Norwood MA, 1998), Chapter 5.
8. F.L. Teixeira, W.C. Chew, M.L. Oristaglio, and T. Wang, Perfectly matched layer and piecewise-linear recursive convolution for the FDTD solution of the 3D dispersive half-space problem, *IEEE Trans. Magnetics* **34**, 2247–2250 (1998).
9. S.J. Yakura, J.T. MacGillivray, and D. Dietz, A three-dimensional finite-difference time-domain algorithm based on the recursive convolution approach for propagation of electromagnetic waves in nonlinear dispersive media, *Appl. Comput. Electromag. Soc. (ACES) J.* **16**(1), 33–44 (March 2001).
10. S.J. Yakura, D. Dietz, and A.D. Greenwood, A dynamic stability analysis of the PML method, submitted to *J. Comput. Phys.*
11. S.J. Yakura, A Three-Dimensional Finite Difference Time Domain – Perfectly Matched Layer Algorithm for Nonlinear Dispersive Media, US Air Force Technical Report AFRL-DE-TR-2000-1031, Kirtland AFB NM, March 2000.
12. W.H. Press, S.A. Teukolsky, W.T. Vetterling, and B.P. Flannery, *Numerical Recipes in FORTRAN - The Art of Scientific Computing, Second Edition* (Cambridge University Press, Cambridge, 1992), pp. 372–375.
13. S.J. Yakura and D. Dietz, A Three-dimensional Finite Difference Time Domain – Perfectly Matched Layer Algorithm Based On Piecewise-linear Approximation For Linear Dispersive Media, US Air Force Technical Report AFRL-DE-TR-2000-1013, Kirtland AFB NM, February 2000.
14. P.M. Goorjian, A. Taflove, R.M. Joseph, and S.C. Hagness, Computational modeling of femtosecond optical solitons from Maxwell's equations, *IEEE J. Quantum Electr.* **28**, 2416–2422 (1992).
15. P.M. Goorjian and A. Taflove, Direct time integration of Maxwell's equations in nonlinear dispersive media for propagation of femtosecond electromagnetic solitons, *Opt. Lett.* **17**, 180–182 (1992).
16. A. Taflove, *Computational Electromagnetics: The Finite-Difference Time-Domain Method* (Artech House, Norwood MA, 1995), pp. 42–44.
17. Z.P. Liao, H.L. Wong, B.P. Yang, and Y.F. Yuan, A transmitting boundary for transient wave analysis, *Scientia Sinica* **XXVII**, 1063–1076 (1984).

FOCUSING OF ULTRA-WIDEBAND COMPOSITE X-WAVES

Amr M. Shaarawi, and Ioannis M. Besieris*

1. INTRODUCTION

The X-wave solutions introduced by Lu and Greenleaf[1-3] have attracted notable attention in the past few years.[4-7] These ultra-wideband slowly dispersing fields exhibit extended focus depth and can be very useful in applications ranging from high-resolution imaging, identification of buried objects, remote sensing and secure communication. There have been several studies of the generation of X-waves from finite apertures,[1,11,12] as well as studies of their scattering from objects and their transmission through dispersive media and multilayered structures.[13-19] Sources for acoustical, optical and microwave X-waves have been constructed and the theoretically predicted behavior of such pulses validated in a number of experiments.[2,5,6] The X-wave solution is an ultra-wideband superposition of spectral plane wave components traveling along wave vectors restricted to a conic surface characterized by the half-apex (axicon) angle ξ.[8-10] In the present work, we demonstrate that highly focused pulses can be shaped by exciting a finite aperture using an extended pulse train of distinct X-waves. The proposed scheme makes use of the fact that the peaks of X-waves, characterized by different axicon angles, travel at different velocities. Consequently, one can choose the initial excitation times of a sequence of several X-waves so that all their peaks meet at an *a priori* chosen focusing point.

An X-wave generated by a circular source of diameter D has two features that are essential to the focusing scheme introduced in this work. First, the focused depth of the generated X-wave is limited by the distance $z_d = D/(2\tan\xi)$, independently of the spectral bandwidth of the elements of the source.[1,11] The amplitude of the peak of the X-wave does not change until it reaches z_d. At farther distances, the peak of the X-wave

* Amr M. Shaarawi, the American University in Cairo, P.O.Box 2511, Cairo 11511, Egypt. Ioannis M. Besieris, Virginia Polytechnic Institute and State University, Blacksburg, Virginia, 24061.

initially falls at a very fast rate [much faster than $(1/z)$], before the decay rate adjusts to the usual $(1/z)$ rate. The second feature characterizing an X-wave generated by a finite aperture is that its peak travels initially at the superluminal speed $c/\cos\xi$ for distances $z << z_d$.[1,20] As it draws nearer to the z_d limit, the speed of the peak of the X-wave changes with distance. In particular, the X-wave slows down and the speed of its peak approaches asymptotically the velocity of light.[20]

There have been a number of synthesis methods suggesting the use of superpositions of X-waves for tailoring ultra-wideband pulses. These methods use superpositions over X-waves of different orders or over X-waves characterized by different ξ angles.[21-23] The second type has been discussed briefly in an earlier work by Lu *et al.* who introduced the term "Composite X-waves" to describe them.[21] In this work, we demonstrate that composite X-waves (CXW) can be used to produce highly focused pulses at a specific distance $z = z_f$ and time $t = t_f$. Our method uses the fact that the peak of an X-wave generated from a finite aperture travels at a speed equal approximately to $c/\cos\xi$, which depends on axicon angle ξ. Therefore, X-waves characterized by distinct spectral apex angles will travel without decay, but at different speeds, over distances of order z_d from their source. All pulses in a train of distinct X-waves with different ξ angles, each launched at an appropriately designated initial time, can be made to converge on the focusing point $z = z_f$ at time $t = t_f$.

The plan of this work is to describe in Sec. 2 the focusing scheme based on the use of CXW pulses. In Sec. 3, the advocated method is applied to the generation of CXW pulses from finite apertures. Numerical examples illustrating the focusing capabilities of our approach are provided and possible limitations on the focusing range are discussed. Conclusions and discussions of the prospects of this approach are included in Sec. 4.

2. FOCUSING USING COMPOSITE X-WAVES

A pulse train consisting of N_ℓ distinct first-order X-waves can be synthesized using a simple superposition, *viz.*,

$$\Psi(\rho, z, ct) = \sum_{\ell=1}^{N_\ell} \Phi_\ell(\rho, z, ct; z_\ell, \xi_\ell), \qquad (2.1.a)$$

where

$$\Phi_\ell(\rho, z, ct; z_\ell, \xi_\ell) = \frac{-A(z_\ell, \xi_\ell)\left[a_0 + i\left((z - z_\ell)\cos\xi_\ell - ct\right)\right]}{\left[(\rho\sin\xi_\ell)^2 + \left(a_0 + i\left((z - z_\ell)\cos\xi_\ell - ct\right)\right)^2\right]^{3/2}}. \qquad (2.1.b)$$

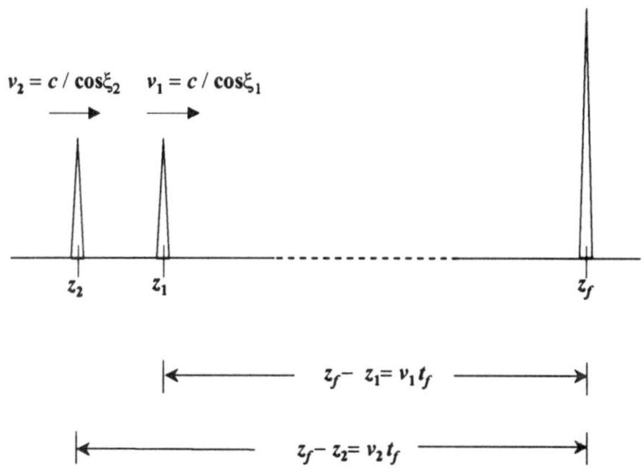

Figure 1. Schematic representation of the temporal focusing scheme. X-waves traveling at different speeds arrive simultaneously and add up at the focusing point z_f .

Here, the parameter ξ_ℓ is the spectral axicon angle characterizing the ℓth X-wave. The amplitude function $A(z_\ell, \xi_\ell)$ can be used to emphasize contributions from certain X-wave components. We choose the peaks of the individual X-waves to exist at the distinct points z_ℓ at time $t = 0$. Bearing in mind that these peaks move at different velocities $v_\ell = c/\cos\xi_\ell$, the z_ℓ and ξ_ℓ values can be selected so that all individual X-waves arrive simultaneously at a prescribed point z_f at the time $t = t_f$. Consequently, a highly focused CXW pulse is obtained at z_f. This concept is represented schematically in Fig. (1) and, accordingly, the following relationship should be satisfied by all individual X-waves:

$$(z_f - z_\ell)\cos\xi_\ell = ct_f . \tag{2.2}$$

One should note that z_f and t_f are the same for all individual X-waves. In designing the initial excitation of a finite source, we start by choosing the focusing point z_f. The simultaneous arrival time t_f is determined for specific choices of z_1 and ξ_1. Subsequently, all other values of ξ_ℓ and z_ℓ are determined for $\ell > 1$. Combining Eqs. (2.1) and (2.2), we obtain

$$\Psi(\rho, z, ct) = \sum_{\ell=1}^{N_\ell} \frac{A(z_f - (ct_f/\cos\xi_\ell), \xi_\ell)[a_0 + i((z - z_f)\cos\xi_\ell - c(t - t_f))]}{\left[(\rho\sin\xi_\ell)^2 + (a_0 + i((z - z_f)\cos\xi_\ell - c(t - t_f)))^2\right]^{3/2}} . \tag{2.3}$$

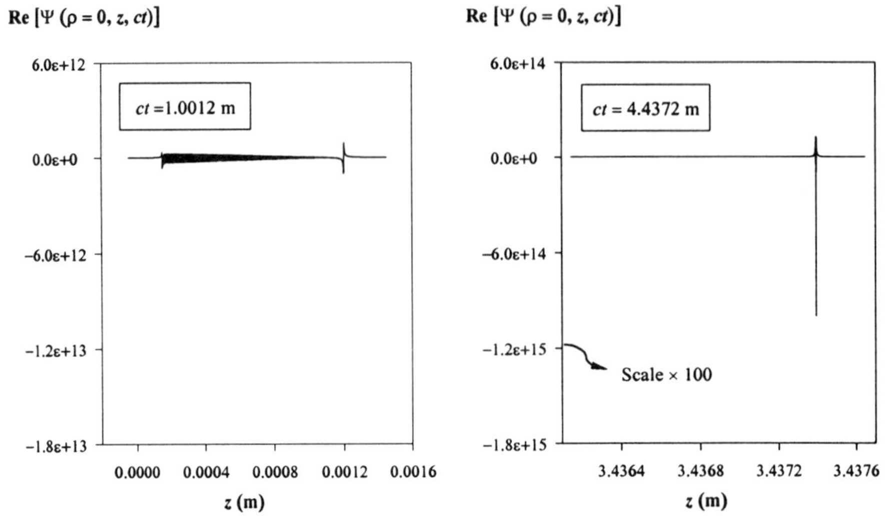

Figure 2. Focusing of a train of $N_\ell = 1001$ free-space X-waves designed to focus at $z_f = 3.4374$ m for $a_0 = 10^{-6}$ m, $z_1 = -1$ m and $0.5^\circ \le \xi_\ell \le 1.5^\circ$. The plots show the axial profile of the CXW pulse at $ct = 1.0012$ and 4.4372 m.

The effectiveness of the proposed focusing scheme can be clearly illustrated by considering the behavior of a source-free CXW. Consider a pulse train consisting of $N_\ell = 1001$ individual X-waves characterized by uniformly distributed angular parameters $\xi_\ell = 0.001(\ell - 1) + 0.5^\circ$. As ℓ acquires all integer values between 1 and 1001, the axicon angle attains values in the range $0.5^\circ \le \xi_\ell \le 1.5^\circ$. The parameter a_0 determines the waist of the individual X-waves as well as their spectral bandwidth.[11,12] These two factors balance each other in such a way that a_0 does not influence the diffraction range z_d.[1,11,12] For this reason, we choose $a_0 = 10^{-6}$ m to be the same for all the individual X-wave components. The parameter z_1 is chosen arbitrarily to equal -1 m. The focusing point $z_f = 3.4374$ m is chosen to be equal to $D/\tan\xi_0$, where $\xi_0 = 1^\circ$ is the central value of the chosen angular range. This value is very close to $z_{d\,\max} = D/(2\tan\xi_{\min})$ for $\xi_{\min} = 0.5^\circ$, which is the diffraction limit of the slowest decaying individual X-wave component. The corresponding value of t_f is calculated using Eq. (2.2). For all numerical examples used in this work, we choose $A(z_\ell, \xi_\ell) = 1$, so that the influence of this amplitude function is not investigated. In Fig. (2), we show plots of the axial profile of the CXW pulse at $ct = 1.0012$ and 4.4372 m. One should note the initial spread of the CXW excitation near $z = 0$. At the focusing point, $z = z_f = 3.4374$ m, the CXW pulse becomes very sharp and its peak amplitude equals 1000 times that of the initial excitation.

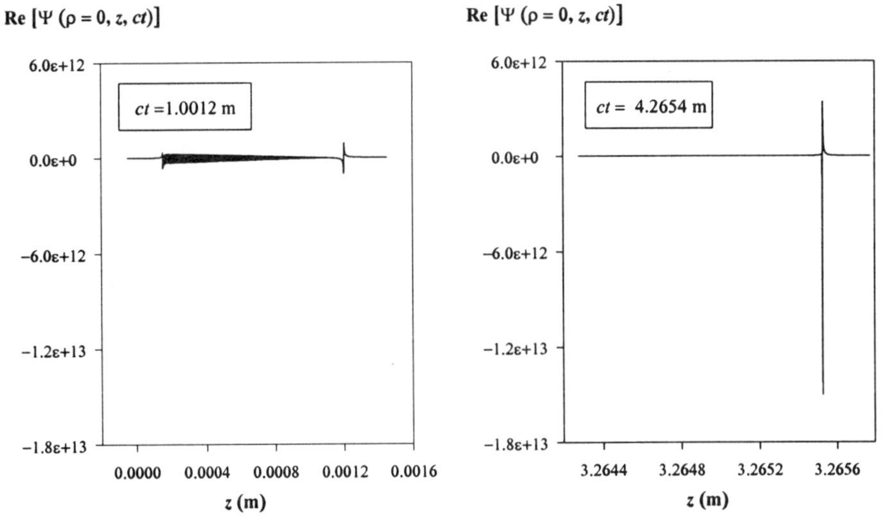

Figure 3. The axial profile of the CXW pulse radiated from a circular aperture having diameter $D = 6\,\text{cm}$. The pulse is designed to focus at $z_f = 3.4374\,\text{m}$. Plots show the axial profiles of the CXW pulse at $ct = 1.0012$ and 4.2654 m. The parameters characterizing the X-waves are the same as in Fig. 2.

3. COMPOSITE X-WAVES GENERATED BY A CIRCULAR APERTURE

Although the source-free CXW pulse illustrates clearly the focusing scheme proposed in this work, a physically generated CXW from a finite aperture behaves in a slightly different manner. In this section, we demonstrate that the focusing scheme discussed in Sec. 2 is also applicable to the CXW pulses generated by finite size sources. In the sequel, we assume that CXW pulses are launched from a finite circular aperture of diameter $D = 6\,\text{cm}$. The initial excitation of the aperture is a discretization of the waveform given in Eq. (2.3). The radiated field is calculated using the Rayleigh-Sommerfeld I formula,[24] *viz.*,

$$\Psi_{RSI}(0,z,t) = \sum_{n=1}^{N} \frac{A_n}{2\pi R_n}[-\partial_{z'}\Psi]. \tag{3.1}$$

Here, A_n is the area of the annular section labeled n, N is the number of annular sections, $\rho'_n = n\Delta\rho'$ is the radius of the nth annular section, and z is the point at which the field of the radiated pulse is observed. The distance from the source points to an axial $(\rho = 0)$ observation point is denoted by $R_n = \sqrt{(n\Delta\rho')^2 + (z - z_0)^2}$. The square brackets in Eq. (3.1) indicate that the quantities enclosed are evaluated at the retarded time $t' = t - R_n / c$.

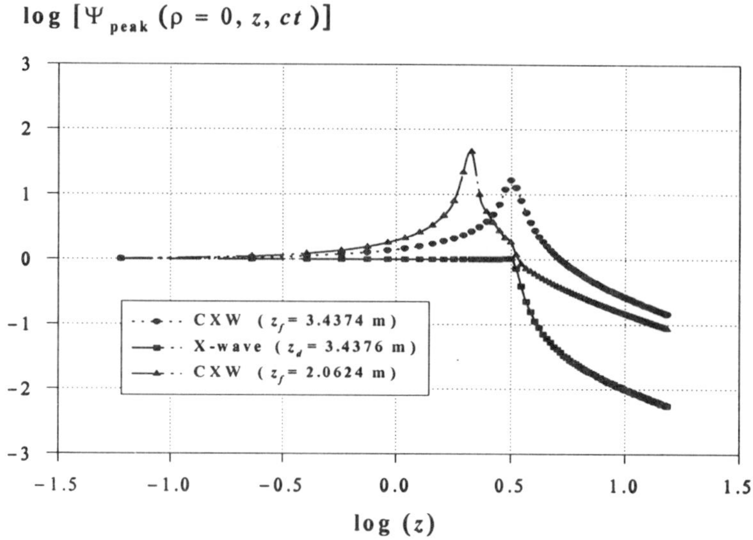

Figure 4. Logarithmic plots of the amplitudes of the peaks of the pulses, shown in Figs. 2 and 3 versus the axial distance from the source. The decay patterns of the peaks of the two CXW pulses are compared to that of an individual X-wave component exhibiting the longest diffraction range $z_{d\max} = 3.4376$ m corresponding to $\xi_1 = 0.5^\circ$.

Using the same parameters as in the case of the source-free X-wave, we calculate the radiated field from a circular aperture situated at $z_0 = 0$. The axial profile of the radiated CXW pulse is plotted in Fig. (3) for $z_f = 3.4374\,\mathrm{m}$ at $ct = 1.0012$, and $4.2654\,\mathrm{m}$. Notice that the pulse is focused at $z = 0.95 z_f$ and that the peak amplitude of the CXW pulse is only 16 times greater than that of the initial excitation. This is the case because the focusing point z_f is chosen to be approximately equal to $z_{d\max} = D/(2\tan\xi_{\min})$ $= 3.4376\,\mathrm{m}$. Recalling that the individual diffraction ranges are given by $z_{d\ell} = D/(2\,\tan\xi_\ell)$, all the individual X-wave components will be propagating beyond their diffraction limit as they approach the focusing point z_f. The fast roll-off in the amplitude of the individual X-waves as they travel beyond their diffraction lengths $z_{d\ell}$ is, therefore, one effect causing the decrease in the amplitude of the focused CXW pulse in comparison to that of the source-free CXW pulse. The second effect contributing to the relatively moderate focusing is that the peaks of the pulses generated from the finite aperture move with velocities that are close but not exactly equal to $c/\cos\zeta_\ell$.

To gain more insight into the decay behavior of CXW pulses, we provide in Fig. (4) logarithmic plots of the amplitudes of the peaks of such pulses versus the axial distance from the source. The plots illustrate the decay patterns of the peaks of the two CXW pulses designed to focus at $z_f = 2.0624\,\text{m}$ and 3.4374 m. As a reference to their performance, we have added a plot of the decay pattern of an individual X-wave exhibiting the longest diffraction range $z_{d\max} = 3.4376\,\text{m}$ (for $\xi = 0.5^\circ$). One should observe that the magnification in the focusing amplitude of the CXW pulses increases as we reduce z_f. The magnification is about 80-fold for the amplitude of the peak of the CXW pulse designed to focus at $z_f = 2.0624\,\text{m}$, in contrast to the 16-fold magnification in the amplitude of the pulse focusing at $z_f = 3.4374\,\text{m}$. This is the case because at $z_f = 2.0624\,\text{m}$ more of the individual X-wave components are contributing effectively to the focused peak of the CXW pulse. Furthermore, Fig. (4) illustrates that for a single X-wave the amplitude of the peak stays constant up to a distance $z = z_{d\max} = 3.4376\,\text{m}$. Beyond that distance, the amplitude of the pulse drops suddenly at a very fast rate before the decay slows down and starts to asymptotically approach the usual $1/z$ rate at farther distances. In contradistinction, the roll-off in amplitude for $z > z_{d\max}$ is slower for CXW pulses when compared to the single X-wave. The aforementioned sudden drop in amplitude is softened significantly for $z_f = 2.0624\,\text{m}$, while it is not observed for $z_f = 3.4374\,\text{m}$. Finally, we would like to point out that a CXW pulse exhibiting higher magnification in the focusing region has a faster decay profile in the far-field region. Beside the magnification of the amplitudes of the pulses in the focusing region, the two CXW pulses exhibit a much slower decay rate in the far-field region than the single X-wave. These advantages are achieved at the expense of using an intricate excitation scheme of a source plane that consumes much higher energies.

4. CONCLUDING REMARKS

It has been shown that strongly peaked CXW pulses can result from focusing a train of X-waves at a selected distance from the source. The focusing technique makes use of the fact that the peaks of X-waves characterized by different ξ_ℓ values travel at different speeds. We have described a scheme whereby one can coordinate the initial times of excitation of the different individual X-waves so that all of them arrive simultaneously at a selected focusing point. For CXW pulses generated from a finite aperture, two factors limit the extension of the focusing distance. First, individual X-waves decay at a very fast rate once they pass their respective diffraction limits $z_{d\ell} = D/(2\tan\xi_\ell)$. The second factor is that the peak of each X-wave component travels initially at the velocity $c/\cos\xi_\ell$ that gradually decreases to c as the $z_{d\ell}$ limit is approached. These two factors prevent some of the X-wave components from arriving at the focusing point at the designated time or, otherwise, arriving with negligible amplitudes.

The numerical examples examined in this work demonstrate that high focusing is attained at a point close to z_f. In general, high focusing is achieved for distances smaller than $z_{d\max}$. The magnification, however, is reduced as z_f is selected closer to or beyond $z_{d\max}$. Another important behavior of CXW pulses is that they enter the $1/r$ decay region at distances closer than the slowest decaying individual X-wave components. This means that in order to avoid the sudden drop in the amplitude of an individual X-wave, one can employ CXW pulses. For applications requiring that the peak amplitude of the received pulse be equal to that of the initial excitation, CXW pulses have major advantages over individual X-waves. This can be seen by comparing the decay behavior of the X-wave in Fig. (4) to that of the CXW pulse having $z_f = 3.4374\,\text{m}$. At the point where the peak of the CXW pulse has the same value as that of the initial excitation, the individual X-wave has fallen to about $1/20$ of the initial peak. Along similar lines, Song *et al.* have investigated the possibility of generating modified X-waves for which the angular parameter $\xi = \xi(\rho)$ is varied as a function of the transverse radial distance of the source.[25] They have shown that the resultant pulses have a larger focusing depth and exhibit better lateral profiles than the original X-waves. Song *et al.*'s approach, however, employs a single X-wave excitation characterized by a parameter ξ that varies with the radius of the source. On the other hand, our method uses a superposition of several X-waves each having a distinct ξ value.

The method described in this work is a pulse shaping technique that facilitates the tailoring of the initial pulsed excitation in order to generate pulses exhibiting a specific focusing pattern. The advocated pulse shaping method depends primarily on our understanding of the spatio-temporal properties of the X-wave solution. Nevertheless, the spectral content of the excitation can be easily deduced using methods introduced in other works.[11,12] Another advantage of using a superposition of X-waves to design the excitation of a focused CXW pulse is that each individual X-wave can be dealt with as an angular superposition over pulsed plane waves. This allows one to use available pulsed plane wave techniques for the analysis of scattering, diffraction and refraction problems involving X-waves.[13,19] For studies of diffraction, scattering or refraction of CXWs, the aforementioned approach applies to the X-wave components.[16-19] Consequently, the diffracted, scattered or refracted fields due to an incident CXW pulse can be calculated by summing up contributions from the constituent X-waves.

REFERENCES

1. J-Y. Lu and J.F. Greenleaf, Nondiffracting X waves - exact solutions to free-space scalar wave equation and their finite aperture realization, *IEEE Trans. on Ultrason., Ferroelect. Freq. Contr.* **39**, 19–31 (1992).
2. J-Y. Lu and J.F. Greenleaf, Experimental verification of nondiffracting X waves, *IEEE Trans. on Ultrason., Ferroelect., Freq. Contr.* **39**, 441–446 (1992).
3. I.M. Besieris, M. Abdel-Rahman, A.M. Shaarawi, and A.A. Chatzipetros, Two fundamental representations of localized pulse solutions of the scalar wave equation, *Progress in Electromagnetics Research* (PIER) **19**, 1–48 (1998).
4. R.W. Ziolkowski, I.M. Besieris, and A.M. Shaarawi, Aperture realizations of the exact solutions to homogeneous-wave equations, *J. Opt. Soc. Am. A* **10**, 75–87 (1993).
5. P. Saari and K. Reivelt, Evidence of X-shaped propagation-invariant localized light waves, *Phys. Rev. Lett.* **79**, 4135–4138 (1997).

6. E. Recami, On the localized 'X-shaped' superluminal solutions to maxwell's equations, *Physica A* **252**, 586–610 (1998).
7. D. Mugnai, A. Ranfagni, and R. Ruggeri, Observation of superluminal behaviors in wave propagation, *Phys. Rev. Lett.* **84**, 4830–4833 (2000).
8. J. Fagerholm, A.T. Friberg, J. Huttunen, D.P. Morgan, and M.M. Salomaa, Angular-spectrum representation of nondiffracting X waves, *Phys. Rev. E* **54**, 4347–4352 (1996).
9. J. Durnin, Exact solutions for nondiffracting beams. I. The scalar theory, *J. Opt. Soc. Am. A* **4**, 651–654 (1987).
10. J. Durnin, J.J. Miceli, Jr., And J.H. Eberly, Diffraction free beams, *Phys. Rev. Lett.* **58**, 1499–1501 (1987).
11. A.A. Chatzipetros, A.M. Shaarawi, I.M. Besieris, and M.A. Abdel-Rahman, Aperture synthesis of time-limited X-waves and analysis of their propagation characteristics, *J. Acoust. Soc. Am.* **103**, 2287–2295 (1998).
12. A.M. Attiya, E.A. El-Diwany, and A.M. Shaarawi, Generation of an approximate TE electromagnetic X-wave in free space using Huygen's aperture, *Proceedings of the 16th National Radio Science Conference*, NRSC'99, Ain Shams University, Cairo, Egypt, pp. B3.1–11 (1999).
13. A.M. Shaarawi, I.M. Besieris, A.M. Attiya, and E.A. El-Diwany, Acoustic X-wave reflection and transmission at a planar interface: spectral analysis, *J. Acoust. Soc. Am.* **107**, 70–86 (2000).
14. J-Y. Lu and J.F. Greenleaf, Diffraction-limited beams and their applications for ultrasonic imaging and tissue characterization, *Proc. SPIE* **1733**, 92–119 (1992).
15. J-Y. Lu, M. Fatemi, and J.F. Greenleaf, Pulsed-echo imaging with X wave, *Acoust. Imag.*, edited by P. Tortoli and L. Masotti, **22**, 191–196 (1996).
16. A.M. Attiya, E.A. El-Diwany, A.M. Shaarawi, and I.M. Besieris, Reflection and transmission of X-waves in the presence of planar layered media: the pulsed plane wave representation, *Progress in Electromagnetics Research* (PIER) **30**, 191–211 (2000).
17. A.M. Shaarawi, I.M. Besieris, A.M. Attiya, and E.A. El-Diwany, Reflection and transmission of an electromagnetic X-wave incident on a planar air-dielectric interface: spectral analysis, *Progress in Electromagnetics Research* (PIER) **30**, 213–250 (2000).
18. A.M. Attiya, E.A. El-Diwany, and A.M. Shaarawi, Transmission and reflection of TE electromagnetic X-wave normally incident on a lossy dispersive half-space, *Proceedings of the 17th National Radio Science Conference*, NRSC'2000, Minufiya University, Egypt, B11.1–12 (2000).
19. A.M. Attiya, *Transverse (TE) Electromagnetic X-waves: Propagation, Scattering, Diffraction and Generation Problems*, Ph. D. Thesis, Cairo University, May 2001.
20. A.M. Shaarawi and I.M. Besieris, On the superluminal propagation of X-shaped localized waves, *J. Phys. A: Math. Gen.* **33**, 7227–7254, (2000).
21. J.Y. Lu, An X-wave transform, *IEEE Trans. Ultrason. Ferroelec., Freq. Contr.* **47**, 1472–1481 (2000).
22. J. Salo, A.T. Friberg, and M. Salomaa, Orthogonal X-waves, *J. Phys. A: Math. Gen.* (accepted).
23. J. Salo, J. Fagerholm, A.T. Friberg, and M.M. Salomaa, Unified description of nondiffracting X and Y waves, *Phys. Rev. E* **62**, 4261–4275 (2000)
24. J.W. Goodman, *Introduction to Fourier Optics*, (McGraw-Hill, New York, 1968).
25. T.K. Song, J-Y. Lu, and J.F. Greenleaf, Modified X waves with improved field properties, *Ultrasonic Imag.* **15**, 36–47 (1993).

MULTI-RESOLUTION HOMOGENIZATION (MRH) FOR SHORT PULSE FIELDS AND FOR FAULT INTERROGATION IN COMPLEX LAMINATES

Vitaliy Lomakin, Ben Zion Steinberg and Ehud Heyman*

1. INTRODUCTION

The study of propagation and scattering of waves in multi-scale heterogeneity laminates is of fundamental importance in diverse areas of applications such as material synthesis, circuits design, fault interrogation, etc. Typically, these applications involve ultra wideband (UWB) short-pulse waves since they furnish the most complete data base for synthesis and interrogation. Yet, as the scales of the laminate heterogeneities ranges from the *micro* (a fraction of a wavelength) to the *macro* (wavelength and above), the analysis of the entire set of associated wave phenomena and their detailed structure pose a problem of overwhelming complexity.

In most applications, however, the micro-scale structure of the field is either weak or of no practical importance. This can be due to the field properties (e.g., near vs. far-zone fields) or due to the relevant measurement setup (e.g., the space-time resolution window of a detector). The role of the homogenization theory is to derive a physically cogent and simpler to solve *effective formulation* for the *macro-scale field* which smoothes out the micro-scale heterogeneities while retaining their effect on the macro scale observables.

This work extends the multiresolution homogenization theory (MRH) to UWB fields and explores several applications. The MRH is a *full-wave effective theory* for *source excited* electromagnetic (and other) fields in three dimensional (3D) finely layered structures.[1-4] Retaining the full-wave features enables one to construct rigorous spectral solutions for the field and to parameterize it physically in terms of wave mechanisms such as rays (including diffraction effects), modes (including leaky

*Faculty of Engineering, Tel–Aviv University, Tel–Aviv, 69978 Israel.

waves), traveling wavefronts or resonances, as well as hybrid combinations of the above.

The MRH is based on the decomposition of the field equations in a multiresolution basis, followed by a solution of the resulting equations in the scale domain at the desired scale (resolution) level while retaining self-consistently the effect of all the other scales. The theory utilizes a two-scale approach where the field and the medium are homogenized on different scales. This removes the need for a large gap between the micro and the macro scales in the medium heterogeneities, thus enables modeling of ultra wideband (multi-scale) sources and fields.

The objectives of this work are to extend the MRH to UWB fields and to explore its applications. Following the problem formulation (Sec. 2) and a review of the MRH (Sec. 3) we consider in Sec. 4 two alternative approaches to UWB homogenization: *dispersive* and *non-dispersive* approaches, and explore their effect on the time domain modeling of the *travelling-wavefronts* and *resonances* phenomena, including a bound on the resonance equivalence. In Sec. 5 we consider an application where the MRH is used to determine the conditions for fault interrogation and detection in multiscale heterogeneous laminate using transient fields.

2. TIME DOMAIN SPECTRAL REPRESENTATION OF THE FIELD (SLANT STACK TRANSFORM)

We consider 3D source-excited EM field in a complex laminate that resides in $0 < z < d$ surrounded by homogeneous domains at $z < 0$ and $z > d$. The medium has an unbounded cross section perpendicular to the stratification axis z, but other waveguide configurations with general cross sections may be considered in the same fashion. The constitutive parameters are assumed to be diagonal tensors whose components $\varepsilon_z, \varepsilon_t, \mu_z, \mu_t$ (the subscripts z and t denote the z and transverse components, respectively) are *multi-scale* functions of z comprising of both macro and micro scales. The field is excited by a time dependent current distribution $\mathbf{J}(\mathbf{r}, t)$ that can be either inside or outside the laminate.

The transient field in such configuration can be analyzed via the spectral theory of transients[5] (STT) (see also Ref. 6 for a free-space formulation) which is reformulated here in the context of EM fields in multilayered laminates. Referring to the problem described above, the field is expressed as a sum of E (or TM) and H (or TE) modes, denoted respectively, by the superscript $\alpha = e$ or h. The 3D vector field of each mode type can be synthesized from the scalar one dimensional amplitude functions $V^\alpha(z, t;\, \boldsymbol{\xi})$ and $I^\alpha(z, t;\, \boldsymbol{\xi})$ via the integral transform

$$\begin{Bmatrix} \mathbf{E}_t^\alpha(\mathbf{r}, t) \\ \mathbf{H}_t^\alpha(\mathbf{r}, t) \end{Bmatrix} = -\frac{1}{(2\pi c)^2}\frac{\partial^2}{\partial t^2}\iint_{-\infty}^{\infty} d\xi_x d\xi_y \begin{Bmatrix} \mathring{\boldsymbol{e}}^\alpha V^\alpha(z, t - c^{-1}\boldsymbol{\xi}\cdot\boldsymbol{\rho};\, \boldsymbol{\xi}) \\ \mathring{\boldsymbol{h}}^\alpha I^\alpha(z, t - c^{-1}\boldsymbol{\xi}\cdot\boldsymbol{\rho};\, \boldsymbol{\xi}) \end{Bmatrix}, \tag{1}$$

where $c = 1/\sqrt{\mu_0\varepsilon_0}$; $\eta = \sqrt{\mu_0/\varepsilon_0}$; $\mathbf{r} = (x, y, z) = (\boldsymbol{\rho}, z)$ denotes the 3D coordinate with $\boldsymbol{\rho} = (x, y)$ denoting the coordinates transverse to z; $\boldsymbol{\xi} = (\xi_x, \xi_y)$ denotes the transverse spectral coordinates; the unit vectors $\mathring{\boldsymbol{e}}^\alpha$ and $\mathring{\boldsymbol{h}}^\alpha$ (we denote unit-vectors by an over-circle) describe the spectral field polarization and are given by $\mathring{\boldsymbol{e}}^e = \mathring{\boldsymbol{\xi}}$, $\mathring{\boldsymbol{h}}^e = \mathring{\boldsymbol{z}} \times \mathring{\boldsymbol{e}}^e$ and $\mathring{\boldsymbol{h}}^h = \mathring{\boldsymbol{\xi}}$, $\mathring{\boldsymbol{e}}^h = \mathring{\boldsymbol{h}}^h \times \mathring{\boldsymbol{z}}$. The operation in (1) will be identified after (4) as an inverse *slant stack transform* (SST).

The spectral *amplitudes* $V^\alpha(z, t;\, \boldsymbol{\xi})$ and $I^\alpha(z, t;\, \boldsymbol{\xi})$ in (1) satisfy the time dependent *spectral transmission line-type* equations

$$\frac{\partial}{\partial z}V^\alpha = p\frac{1}{c}\frac{\partial}{\partial t}I^\alpha - v^\alpha\,, \qquad \frac{\partial}{\partial z}I^\alpha = q\frac{1}{c}\frac{\partial}{\partial t}V^\alpha - \imath^\alpha\,, \tag{2}$$

where the heterogeneity functions $p(z;\,\xi)$ and $q(z;\,\xi)$ are given by

$$\text{E (TM) modes :} \qquad p = \mu_t - \varepsilon_z^{-1}\xi^2, \qquad q = \varepsilon_t, \tag{3a}$$

$$\text{H (TE) modes :} \qquad p = \mu_t, \qquad q = \varepsilon_t - \mu_z^{-1}\xi^2\,, \tag{3b}$$

and $\xi \equiv |\boldsymbol{\xi}\cdot\boldsymbol{\xi}|^{1/2}$ and $(\varepsilon_{t,z}, \mu_{t,z})$ are the relative constitutive coefficients. The scalar sources in (2) are found from $\mathbf{J}(\mathbf{r}, t)$ via

$$v^e(z,t;\,\boldsymbol{\xi}) = \frac{\eta\xi}{\varepsilon_z}\iint_{-\infty}^{\infty} dx\,dy J_z(\mathbf{r}, t{+}c^{-1}\boldsymbol{\xi}\cdot\boldsymbol{\rho})\,,\ \imath^\alpha(z,t;\,\boldsymbol{\xi}) = \eta\iint_{-\infty}^{\infty} dx\,dy\,\mathring{\mathbf{e}}^\alpha\cdot\mathbf{J}(\mathbf{r}, t{+}c^{-1}\boldsymbol{\xi}\cdot\boldsymbol{\rho})\,.$$

and $v^h = 0$. (4)

The operation in (4) integrates the source excited fields along surfaces of linear delay $t{-}c^{-1}\boldsymbol{\xi}\cdot\boldsymbol{\rho}{=}$const., and is therefore termed the slant stack transform (SST). It is also identified as a 4D *Radon transform* in $(\mathbf{r}, t)$. Physically it extracts from $\mathbf{J}(\mathbf{r}, t)$ the time dependent plane wave that propagates in a direction specified by $\boldsymbol{\xi}$ and hence it is also termed the *time-dependent plane wave transform.* Its properties in a homogeneous 4D $(\mathbf{r}, t)$ domain have been explored in Ref. 7. Indeed one observes that in the homogeneous domain outside the slab region, the integrands in (1) describe, for a given $\boldsymbol{\xi}$, a source-excited time dependent plane wave with waveform $(V^\alpha(t), I^\alpha(t))$ that propagates along the unit vector direction $(\boldsymbol{\xi}, \sqrt{1-\xi^2})$, making an angle $\theta = \sin^{-1}\xi$ with the z axis. It is important though to realize that the STT-SST formulation above applies also within the heterogeneous laminate where the wavefronts are not planar.

Frequency domain (FD) interpretation: The STT representation in (1)–(4) can be obtained from its FD counterpart via the inverse Fourier transform

$$f(t) = (2\pi)^{-1}\int_{-\infty}^{\infty} \hat{f}(\omega)e^{-i\omega t}d\omega, \tag{5}$$

where the hat denotes FD constituents. The FD field can be represented via the conventional plane wave decomposition[8]

$$\begin{Bmatrix}\hat{\mathbf{E}}_t^\alpha(\mathbf{r},\omega)\\ \hat{\mathbf{H}}_t^\alpha(\mathbf{r},\omega)\end{Bmatrix} = \frac{k_0^2}{(2\pi)^2}\iint_{-\infty}^{\infty} d\xi_x d\xi_y \begin{Bmatrix}\mathring{\mathbf{e}}^\alpha\,\hat{V}^\alpha(z,\omega;\,\boldsymbol{\xi})\\ \mathring{\mathbf{h}}^\alpha\,\hat{I}^\alpha(z,\omega;\,\boldsymbol{\xi})\end{Bmatrix} e^{ik_0\boldsymbol{\xi}\cdot\boldsymbol{\rho}}, \tag{6}$$

where $\boldsymbol{\xi}$ is a frequency-normalized transverse wavenumber which is related to the conventional spectral wavenumber $\mathbf{k}_t$ via $\boldsymbol{\xi} = \mathbf{k}_t/k_0$ with $k_0 = \omega/c$. The spectral amplitudes $(\hat{V}^\alpha, \hat{I}^\alpha)$ in (6) satisfy the *spectral transmission line equations*[8]

$$\frac{d}{dz}\hat{V}^\alpha = ik_0 p\,\hat{I}^\alpha - \hat{v}^\alpha\,, \qquad \frac{d}{dz}\hat{I}^\alpha = i\,k_0\,q\,\hat{V}^\alpha - \hat{\imath}^\alpha, \tag{7}$$

where the heterogeneity functions (p, q) are given in (3) and the spectral sources $\hat{v}^\alpha(z,\omega;\,\boldsymbol{\xi})$ and $\hat{\imath}^\alpha(z,\omega;\,\boldsymbol{\xi})$ obtained via

$$\hat{v}^h = 0, \quad \hat{v}^e = \frac{\eta\xi}{\varepsilon_z}\iint_{-\infty}^{\infty} dx\,dy\,\hat{J}_z(\mathbf{r})e^{-ik_0\boldsymbol{\xi}\cdot\boldsymbol{\rho}}\,,\ \hat{\imath}^\alpha = \eta\iint_{-\infty}^{\infty} dx\,dy\,\mathring{\mathbf{e}}^\alpha\cdot\hat{\mathbf{J}}(\mathbf{r})\,e^{-ik_0\boldsymbol{\xi}\cdot\boldsymbol{\rho}}, \tag{8}$$

where $\hat{\mathbf{J}}(\mathbf{r})$ is the FD counterpart of $\mathbf{J}(\mathbf{r}, t)$. One further identifies $k_0\sqrt{pq}$ and $\sqrt{p/q}$ as the local spectral wavenumber and characteristic impedance along z.

Now, applying (5) to (6)–(8) we arrive at the STT representation in (1)–(4) where V, I, v and $\imath$ are the time domain counterparts of $\hat{V}$, $\hat{I}$, $\hat{v}$ and $\hat{\imath}$, respectively.

For brevity, we shall henceforth omit the superscript α in all the expressions.

3. EFFECTIVE FORMULATION IN THE FREQUENCY DOMAIN

3.1. Physical observables

Although the multi-scale structure of the medium induces multi-scale components in the three-dimensional (3D) electromagnetic field, the relevant field *observables* are described on a scale that is typically determined by the *wave physics* (e.g., the wavelength, near/far fields) or by the measurement arrangement (e.g., the spatio-temporal resolution of the detector, the dynamic range and the polarization, etc.). Accounting for these aspects permits simplification of the field calculations through an effective formulation that a priori retains only the large-scale components of the response, which thus define the *field homogenization scale* (F-HS). We shall choose the F-HS according to the wave physics. The most basic scale is characterized by the *local effective wavenumber* in the material; i.e. F-HS $= (k_0\sqrt{(\varepsilon\mu)_{\max}})^{-1}$.

Having chosen a relevant field homogenization scale, we invoke the theory of multiresolution analysis[9] to parameterize the F-HS macro and micro components in terms of the binary scale basis

$$\text{F-HS} = \bar{d}\,2^{-j}, \quad j = \text{integer}\,, \tag{9}$$

where $\bar{d}$ is an arbitrary reference scale. A given field function $f(z)$ is thereby decomposed into its "smooth" (macro-scale) and "detailed" (micro-scale) orthogonal components $f^s = \mathcal{P}_j f = \sum_n \langle f, \phi_{jn}\rangle\, \phi_{jn}$ and $f^d = \mathcal{D}_j f = \sum_{mn} \langle f, \psi_{mn}\rangle\, \psi_{mn}$, where the operators $\mathcal{P}_j$ projects f onto the linear spaces $\mathbb{V}_j \subset \mathbb{L}_2$ containing all scales $\geq \bar{d}\,2^{-j}$, while $\mathcal{D}_j$ projects f onto the orthogonal complementary space $\bigcup_{m=j}^{\infty} \mathbb{W}_m$ that contains all smaller scales; the wavelet space $\mathbb{W}_m$ consists only of the scale $\bar{d}\,2^{-m-1}$. The sets of scaling and wavelets functions $\{\phi_{jn}\}_{n\in\mathbb{Z}}$ and $\{\psi_{mn}\}_{n\in\mathbb{Z}}$ with $\mathbb{Z}$ being the set of all integers, constitutes a basis of the *homogenization space* (or *observables space*) $\mathbb{V}_j$ and of $\mathbb{W}_m$, respectively.

Having determined the F-HS, we are interested only in the smoothed field observables, and the corresponding "effective" medium that describes these observables. This is accomplished by projecting onto $\mathbb{V}_j$ the "actual" field $(\mathbf{E}, \mathbf{H})$ and the "effective" field $(\mathbf{E}^{\text{ef}}, \mathbf{H}^{\text{ef}})$ (the solution in the effective medium). It is required that these solutions will be identical to within a prescribed error, as formalized by the condition

$$\mathcal{P}_j(\mathbf{E}^{\text{ef}}, \mathbf{H}^{\text{ef}}) = \mathcal{P}_j(\mathbf{E}, \mathbf{H}). \tag{10}$$

The challenge to the modeler is to choose the effective medium consistent with this requirement as simple as possible so as to allow efficient numerical evaluation of the field and a cogent interpretation of the "effective" wave physics.

3.2. Effective medium

Referring to the spectral transmission line equations (7), we are interested in finding a simplified effective medium with smoother parameters $(p^{\text{ef}}, q^{\text{ef}})$. Such medium is obtained if one neglects all scales in the heterogeneity functions of (3) that are smaller than the *medium homogenization scale* (M-HS)

$$\text{M-HS} = \bar{d}\,2^{-\nu}, \quad \nu = \text{an integer to be determined.} \tag{11}$$

The M-HS is typically finer than the F-HS but it should be chosen as large as possible in order to simplify (by reducing the number of basis functions) the effective medium description. The effective and detail (neglected) parts of the medium are given by the respective projections

$$(p^{\mathrm{ef}}, q^{\mathrm{ef}}) = \mathcal{P}_\nu(p,q), \quad (p^d, q^d) = \mathcal{D}_\nu(p,q), \tag{12}$$

where $\mathcal{P}_\nu$ and $\mathcal{D}_\nu$ are defined after (9) with j replaced by ν. Applying (12) onto (3) one finds that the homogenized *effective constitutive parameters* are given by

$$\varepsilon_t^{\mathrm{ef}} = \mathcal{P}_\nu \varepsilon_t\,, \quad \varepsilon_z^{\mathrm{ef}} = [\mathcal{P}_\nu(\varepsilon_z^{-1})]^{-1}, \qquad \mu_t^{\mathrm{ef}} = \mathcal{P}_\nu \mu_t\,, \quad \mu_z^{\mathrm{ef}} = [\mathcal{P}_\nu(\mu_z^{-1})]^{-1}, \tag{13}$$

which implies an *effective anisotropy* (e.g., $\varepsilon_t^{\mathrm{ef}} \neq \varepsilon_z^{\mathrm{ef}}$ even if $\varepsilon_t = \varepsilon_z$).

3.3. Quality of the effective formulation

The quality of the effective formulation is described in (10) by examining the difference in $\mathbb{V}_j$ between the actual *source-excited* field solution and the solution of the effective formulation. It can be shown that this difference, denoted as the homogenization error, is bounded by[2]

$$\frac{\|\mathcal{P}_j \hat{V}^{\mathrm{ef}} - \mathcal{P}_j \hat{V}\|}{\|\mathcal{P}_j \hat{V}^{\mathrm{ef}}\|} \sim 2^j (d/\bar{d})\, N_\nu^2 M_\nu^2\,, \tag{14}$$

where $(\hat{V}^{\mathrm{ef}}, \hat{I}^{\mathrm{ef}})$ are the spectral amplitudes in the effective medium that satisfy the effective transmission line equations which are given by (7) but with (p,q) replaced by $(p^{\mathrm{ef}}, q^{\mathrm{ef}})$. The effective formulation error in (14) is controlled by the *dimensionless* scale parameter defined as

$$N_\nu \equiv \frac{\text{M-HS}}{\text{F-HS}} = 2^{-(\nu-j)} \ll 1\,. \tag{15}$$

The error also depends on the norm of the neglected detailed part q^d and/or p^d of the medium heterogeneity, which is expressed as a *non-dimensional* parameter

$$M_\nu^2 = \|q^d\|^2/\|q^{\mathrm{ef}}\|^2 + \|p^d\|^2/\|p^{\mathrm{ef}}\|^2, \tag{16}$$

where $\|q^{\mathrm{ef}}\|$ and $\|p^{\mathrm{ef}}\|$ refer only to the norm of the effective medium within the slab region (i.e. not to the entire z domain).

Note that the MRH is *not* a perturbation theory valid for weak micro heterogeneities (small M_ν) since the error in (14) is controlled by the small parameter N_ν^2 that can be made as small as required by increasing ν.

3.4. Outline of the error bound derivation

As discussed in the Introduction, the MRH is based on the decomposition of the field equations in a multiresolution basis, followed by a solution of the resulting equations in the scale domain at the desired scale (resolution) level. The scale decomposition is most conveniently performed if the transmission line equations (7) for the field amplitude (V, I) are replaced by an integral equation that express these amplitudes in terms of the solutions $(V^{\mathrm{ef}}, I^{\mathrm{ef}})$ in the effective medium. This integral equation is obtained by decomposing p, q into the *background* (smooth) and *foreground* (detail) components defined in (12), using the effective smooth components $p^{\mathrm{ef}}, q^{\mathrm{ef}} \in \mathbb{V}_\nu$ as the background. This leads to *two coupled* integral equations wherein the *foreground* components p^d, q^d act as *induced sources* in conjunction with the Green's function for the background. For simplicity, we assume here that there are only q-type micro-heterogeneities so that there is only one integral equation. Starting with the known background Green's function $g_{V_\imath}^{\mathrm{ef}}(z,z')$ which is the V-type response at z to a point $\imath$-type source at z' (i.e., $g_{V_\imath}^{\mathrm{ef}}(z,z')$ is the V-type solution of (7) with $\hat{v} = 0$ and $\hat{\imath} = \delta(z-z')/ik_0$ and with (p,q) replaced by the smooth backround $(p^{\mathrm{ef}}, q^{\mathrm{ef}})$), the integral equation for V in a medium with $p = p^{\text{ef}}$ but $q = q^{\text{ef}} + q^d$ is given by

$$\hat{V} = F + k_0^2 \mathcal{G}^{\text{ef}}_{V_\imath}\{q^d \hat{V}\}, \tag{17}$$

where the integral operator is defined by $\mathcal{G}^{\text{ef}}_{V_\imath}\{f(z)\} = \int_0^d \hat{g}^{\text{ef}}_{V_\imath}(z, z'')f(z'')dz''$, and the forcing term is given by $F = ik_0 \left(\mathcal{G}^{\text{ef}}_{I_\imath}\{\hat{\imath}\} + \mathcal{G}^{\text{ef}}_{I_v}\{\hat{v}\}\right)$, where $\mathcal{G}^{\text{ef}}_{I_\imath}$ is defined in a similar fashion but for the I-type response to an $\imath$-type source.

The *multiresolution decomposition* (MRD) of the integral equation is now performed by expressing $\hat{V} = \hat{V}^s + \hat{V}^d$ in terms of its smooth and detail components, and then take the inner product of (17) with the sets of scaling (ϕ_{jn}) and wavelet (ψ_{mn}) basis functions, thereby reducing (17) to the algebraic equation

$$\begin{pmatrix} \mathbf{I} - \mathbf{\Phi} & -\mathbf{C} \\ -\bar{\mathbf{C}} & \mathbf{I} - \mathbf{\Psi} \end{pmatrix} \begin{pmatrix} \boldsymbol{s} \\ \boldsymbol{d} \end{pmatrix} = \begin{pmatrix} \boldsymbol{s}^{\text{ef}} \\ \boldsymbol{d}^{\text{ef}} \end{pmatrix}, \tag{18}$$

where $\mathbf{I}$ is the identity matrix. Here, the unknown vectors $\boldsymbol{s}$ and $\boldsymbol{d}$ represent the smooth and detail *expansion coefficients* of V while the known vectors $\boldsymbol{s}^{\text{ef}}$ and $\boldsymbol{d}^{\text{ef}}$ are the corresponding expansion coefficients of F in (17). The elements of the matrices $\mathbf{\Phi}$, $\mathbf{C}$, $\bar{\mathbf{C}}$ and $\mathbf{\Psi}$, which describe the MRD of the integral operator, are defined by the projections

$$[\Phi]_{n,n'} = k_0^2 \left\langle \phi_{jn}, \mathcal{G}^{\text{ef}}_{V_\imath}\{q^d \phi_{jn'}\}\right\rangle, \qquad [C]_{n,m'n'} = k_0^2 \left\langle \phi_{jn}, \mathcal{G}^{\text{ef}}_{V_\imath}\{q^d \psi_{m'n'}\}\right\rangle,$$
$$[\bar{C}]_{mn,n'} = k_0^2 \left\langle \psi_{mn}, \mathcal{G}^{\text{ef}}_{V_\imath}\{q^d \phi_{jn'}\}\right\rangle, \qquad [\Psi]_{mn,m'n'} = k_0^2 \left\langle \psi_{mn}, \mathcal{G}^{\text{ef}}_{V_\imath}\{q^d \psi_{m'n'}\}\right\rangle, \tag{19}$$

where $\langle\ ,\ \rangle$ denotes the inner product in L_2. In these matrices, the indices n and n' span the z domain of the laminated slab, while m and m' extend from j to the smallest scale in the medium heterogeneities. The properties of the matrices in (19) are then explored in Ref. 2 using the asymptotic properties of the MRD projections as applied to the effective medium operator $\mathcal{G}^{\text{ef}}_{V_\imath}$.

The smooth/detail decomposition in (18) is exact and formally assigns equal importance to the "s" and "d" constituents. Since we are interested in an effective smooth-scale formulation, we eliminate $\boldsymbol{d}^q$ by substituting it from the second equation in (18) into the first equation to obtain

$$\Big[\mathbf{I} - \underbrace{\left(\mathbf{\Phi}^q + \mathbf{C}^q(\mathbf{I} - \mathbf{\Psi}^q)^{-1}\bar{\mathbf{C}}^q\right)}_{\mathbf{EMO}^q}\Big]\boldsymbol{s}^q = \boldsymbol{s}^{\text{ef}} + \mathbf{C}^q\left(\mathbf{I} - \mathbf{\Psi}^q\right)^{-1}\boldsymbol{d}^{\text{ef}}. \tag{20}$$

The "effective material operator" $\mathbf{EMO}^q$ in (20) incorporates the effects of the detail scales within the smooth-scale equation. The index q which indicates that it involves only q-type micro-heterogeneities will be omitted for brevity. The second term on the right hand side in (20) represents the coupling of the micro-scales in the forcing term $\boldsymbol{d}^{\text{ef}}$ into the macro-scale field $\boldsymbol{s}$ via the medium micro-heterogeneities; since the source functions $\hat{v}$ and $\hat{\imath}$ represent physical sources they are typically smooth. Consequently $\left\|\boldsymbol{d}^{\text{ef}}\right\| \approx 0$ hence this term will be neglected.

Solving the reduced equation (20) for $\boldsymbol{s}$ requires the inverse "operation" $(\mathbf{I} - \mathbf{EMO})^{-1}$. Since our goal is to approximate $\boldsymbol{s}$ by $\boldsymbol{s}^{\text{ef}}$, the norm $\left\|\mathbf{EMO}\right\|$ should be small. Choosing the problem parameters accordingly, the first order Neumann series solution of (20) becomes, using $(\mathbf{I}-\mathbf{EMO})^{-1} \sim \mathbf{I}+\mathbf{EMO}$, $\boldsymbol{s} \approx \boldsymbol{s}^{\text{ef}} + \mathbf{EMO}\,\boldsymbol{s}^{\text{ef}}$, leading to the error estimate $\left\|\boldsymbol{s} - \boldsymbol{s}^{\text{ef}}\right\| / \left\|\boldsymbol{s}^{\text{ef}}\right\| \sim \left\|\mathbf{EMO}\right\|$. Thus, the bound on the error may be calculated readily from the bound on the matrices that form the **EMO** in (20), leading to the error bound on the right hand of (14) (this bound applies, in fact, to the more general case where there are both q and p micro-heterogeneities, but it also applies to the present case by using only the q-part of M_ν in (16), i.e.,

$M_\nu^q = \|q^d\|/\|q^{\text{ef}}\|$). Using these bound we choose the homogenization parameters j and ν to make the bound on $\|\mathbf{EMO}\|$ sufficiently small so that it complies with the analysis above.

4. EFFECTIVE FORMULATION FOR UWB FIELDS

4.1. Time domain homogenization schemes

As discussed in connection with (14) the F-HS and M-HS are typically chosen relative to the wavelength. For ultra wideband excitations, the spectrum of the excitation pulse covers a wide range of wavelengths. This leads to two alternative homogenization strategies.

A dispersive homogenization approach:
Here the frequency spectrum of the source is divided into sub-bands and the M-HS is chosen for each band so that the error bound in (14) is sufficiently small for that band. The resulting effective medium is therefore frequency dependent, containing less detailed components at the lower frequencies. This approach provides the simplest (most economical) effective medium for each frequency within the source operation band. On the other hand, the field equations have to be solved on a frequency-by-frequency basis and, if time-domain solutions are required, the results should finally be transformed into the time domain.

A non-dispersive homogenization approach:
Here the M-HS is chosen to comply with (14) at the highest excitation frequency and then used for the entire operation band. This approach does not provide the most economical description of the medium at each frequency but, on the other hand, it allows for direct time domain solutions as the effective medium is frequency independent. For example, the pulse response in the effective medium may be calculated numerically via the finite difference time domain method (FDTD), or via a combined FDTD and multiresolution approach known as the multiresolution time domain method (MRTD).[10]

4.2. Example: Pulsed reflection from multiscale laminate

To demonstrate the effect of the homogenization on this representations, we consider a complex slab with $\mu = 1$ but complex (i.e., multiscale) ε. We synthesize this medium as $\varepsilon_t = 2\big(1+\sum_n \sum_{m=4}^{10} a_{mn}\psi_{mn}(z)\big)$, $\varepsilon_z = 2\big(1+\sum_n \sum_{m=4}^{10} b_{mn}\psi_{mn}(z)\big)^{-1}$ where $\psi_{mn}(z)=2^{m/2}\psi(2^m z-n)$ are wavelet functions. The coefficients a_{mn} and b_{mn} are generated randomly as $a_{mn} = 0.075\, f_a(m,n)2^{-m/2}$, $b_{mn} = 0.3\, f_b(m,n)2^{-m/2}$, where $f_a(m,n) = \pm 1$ and $f_b(m,n) = \pm 1$ are uniformly distributed random functions of m, n. The $\sim 2^{-m/2}$ behaviour of a_{mn} and b_{mn} is chosen so that the norms of $\varepsilon_{t,z}$ are evenly distributed in the scale domain m. The medium and the corresponding effective medium for $\nu = 4$ and $\nu = 6$, using $\bar{d}=1$ (see (11)) are depicted in Fig. 1.

As an example we consider a single spectral component in (1). Such a component may also be excited individually by a pulsed source distribution that diagonalize the SST operator in (4). Specifically, one may easily verify that a pulse source of the form $\mathbf{J}(\mathbf{r},t) = \mathring{x}\, j_x(z, t - x\,\xi_0/c)$ where $j_x(z,t)$ is a given function, excites only a single TM (i.e., E) mode with $(\xi_x, \xi_y) = (\xi_0, 0)$. From (1)–(4), the resulting

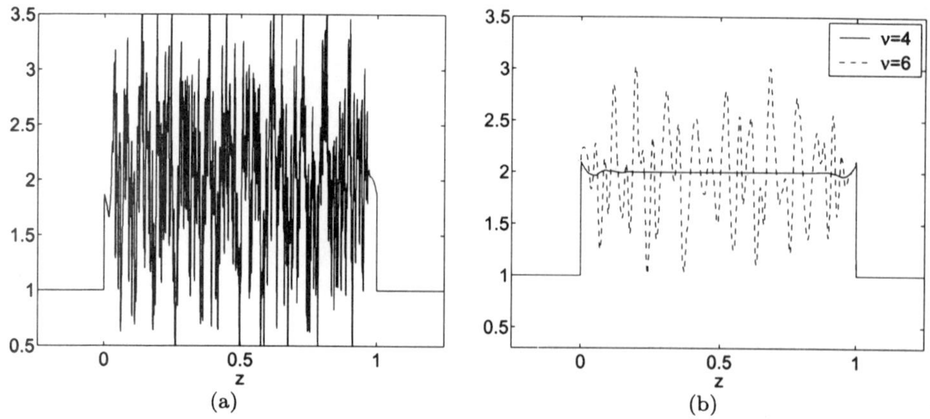

Figure 1: (a) The heterogeneous permittivity $\varepsilon_t(z)$; (b) The effective permittivity $\varepsilon_t^{\text{ef}}(z)$ using M-HS with $\nu = 4$ and (b) $\nu = 6$.

magnetic field has the form

$$\mathbf{H}(\mathbf{r},t) = \mathring{\boldsymbol{y}}\, I^e(z, t - x\,\xi_o/c), \tag{21}$$

where from the time dependent transmission line equation (2) with the q function from (3b) it follows that the function $I^e(z,t;\xi_o)$ satisfies the 1D wave equation

$$\left[\frac{\partial}{\partial z}\frac{1}{\varepsilon_t}\frac{\partial}{\partial z} - \left(\mu_t(z) - \frac{\xi_o^2}{\varepsilon_z}\right)\frac{1}{c^2}\frac{\partial^2}{\partial t^2}\right] I^e(z,t;\xi_o) = -\eta\varepsilon_t^{-1} j_x(z,t). \tag{22}$$

To obtain this result we first commuted the differentiation $-(2\pi c)^{-2}\partial_t^2$ in (1) to the spectral sources in (4). Then one can show the following identity $-(2\pi c)^{-2}\partial_t^2 \imath^e = (\mathring{\boldsymbol{\xi}}\cdot\mathring{\boldsymbol{x}}) j_x(t)\delta(\xi_x - \xi_o)\delta(\xi_y)$. (The reader may readily verify that a similar result is obtained for the time-harmonic excitation with $\hat{\mathbf{J}} = \mathring{\boldsymbol{x}}\hat{j}_x(z)e^{-i\omega\xi_o/c}$). Finally the field in (21) is obtain by substituting this expression into (1).

Specifically we assume henceforth that the source function $j_x(z,t)$ is located at $z<0$. It then follows from (22) that it excites an incident pulse of the form

$$I^e_{\text{inc}}(z,t;\xi_o) = f\left[t - \sqrt{1-\xi_o^2}\,z/c\right] \quad \Rightarrow \mathbf{H}_{\text{inc}}(\mathbf{r},t) = \mathring{\boldsymbol{y}}\, f\left[t - \left(\xi_o\,x + \sqrt{1-\xi_o^2}\,z\right)/c\right] \tag{23}$$

where $f(t)$ is a given waveform. The wave field $\mathbf{H}_{\text{inc}}$ in (23) is readily recognized as a pulsed plane wave impinging on the laminate from $z<0$ along the unit vector direction whose Cartezian components are given by $\mathring{\boldsymbol{\kappa}} = (\xi_o, 0, \sqrt{1-\xi_o^2})$.

In the numerical examples the incident pulse of length T is taken to be

$$f(t) = \pi^{-1}T[(t-4T)^2 + T^2]^{-1} \quad \text{with } cT = 2^{-3} \tag{24}$$

spanning frequencies from 0 up to $\omega/c \sim 2\pi/cT \sim 2^5$.

Figs. 2(a,b) compare the frequency spectrum of the exact reflected fields to that calculated in the effective medium with $\nu = 4$ and $\nu = 6$, respectively. We also show the spectrum of the incident pulse. One observes that choosing $\nu = 4$ or 6 provides a faithful representation up to $\omega/c \approx 2^3$ and 2^5, respectively. These results are in agreement with (14) noting that for the medium in Fig. 1 and with $\bar{d}=1$ we have there $2^{-j} \sim (\sqrt{2}\omega/c)^{-1}$.

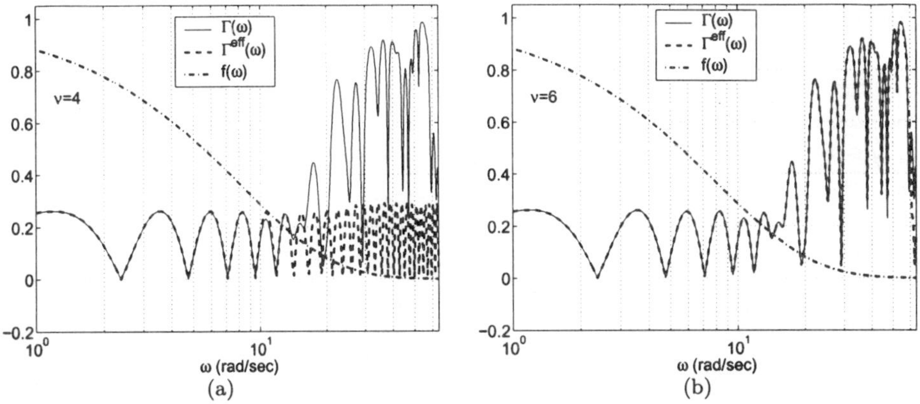

Figure 2: Frequency spectrum of the reflected fields calculated via the MRH with (a) $\nu = 4$ and (b) $\nu = 6$ (dashed lines), compared with the spectrum of the actual field (full lines).

Fig. 3 that depicts the transient reflected field calculated via the dispersive and non-dispersive homogenization approaches. One readily identifies the reflected signals from the front and back interfaces of the laminate at $ct = 0.5$ and 3.2, respectively. The dispersive approach provides an excellent agreement with the actual pulse (no noticeable difference) since the M-HS has been chosen for each frequency band to yield a negligible error there. The non-dispersive approach provides an excellent accuracy (no noticeable difference) for the finer M-HS $\nu = 6$ but small error (smoother waveform) for the coarse M-HS $\nu = 4$. This is due to the fact that the coarser M-HS is not accurate for frequencies greater than $\omega/c \approx 2^3$ whereas the finer M-HS is accurate over the entire excitation band (compare Figs. 2(a,b)). The signal in the coarser M-RH case misses the small ripples between $ct = 1$ and 2.5, yet depending on the application these ripples may be considered as non-observables, hence choosing $\nu = 4$ leads to the much simpler and easily interpretable effective medium in Fig. 1.

4.3. Singularity Expansion Method (SEM) representation

Instead of the traveling wave representation considered above, the field may be parameterized as in the SEM[11] using the oscillatory (standing wave) solutions that are associated with the complex resonance frequencies (poles) in the lower half of the complex frequency plane. The upper half of the frequency plane, on the other hand, is regular. Thus, prior to the wave arrival time, the integration contour in (5) may be closed in a large semi-circle about the upper half plane, and by utilizing Cauchy's theorem one obtains the null field there.[12] For $t > 0$, the integration contour may be closed about the lower half plane, thereby expressing the transient field as a sum of pole residue contributions $(V, I)(t) = \sum_{\omega_l} \text{Res}\{\hat{V}, \hat{I}\}|_{\omega_l} e^{-i\omega_l t}$ where ω_l are the SEM resonance frequencies. This description of the transient response as a series of decaying oscillations may serve as signatures for medium classification.

It has been established that the SEM resonances of the effective problem (denoted ω_l^{ef}) are equivalent to those of the actual problem with an error bound

$$\left|\omega_l - \omega_l^{\text{ef}}\right| \sim (\omega_l^{\text{ef}})^3[(\varepsilon^{\text{ef}}\mu^{\text{ef}})_{\max}/c^2]\, 2^{-2\nu} M_\nu^2, \tag{25}$$

where the homogenization parameters are defined and discussed in (14). This equivalence defines the conditions under which the effective formulation can be used to parameterize the actual (e.g., measured) resonance observables.

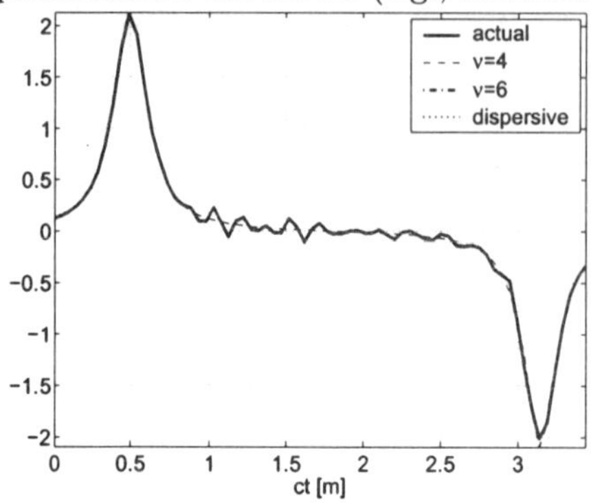

Figure 3: Transient reflected field for the non-dispersive approach for two values of the M-HS: $\nu = 4$ and 6 and for the dispersive approach

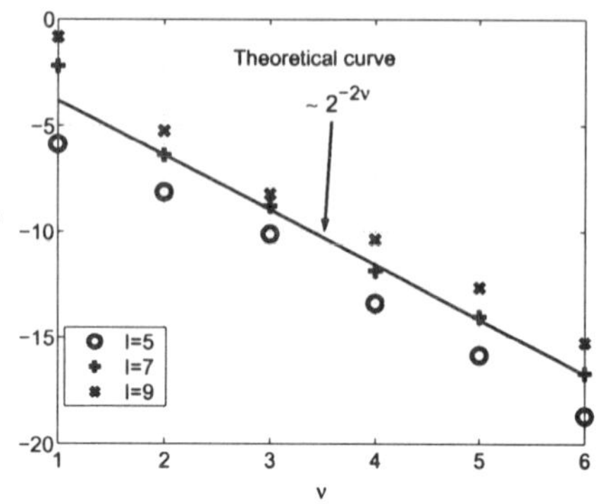

Figure 4: The error $\left|\omega_l - \omega_l^{\text{ef}}\right|$ in the SEM resonances, shown on a $\log_2$ scale as a function of the M-HS. The resonances plotted are $l = 5$, 7 and 9.

The equivalence between the actual and the effective resonances and the quality of the theoretical bound (25) are demonstrated in Fig. 4 for the medium used in the previous example. The error $\left|\omega_l - \omega_l^{\text{ef}}\right|$ is depicted on a $\log_2$ scale as a function of the M-HS index ν, showing that the error behaves like $2^{-2\nu}$ as predicted in (25).

5. FAULT INTERROGATION

Finally, we consider an application where the MRH is used to parameterize the UWB signature of faults in multiscale laminates. The class of faults considered is inclusions whose transversal size is large so that they can be described locally as layers in the laminate. This simplifies the analysis since the medium can still be considered as a 1D layered medium. The fault parameters are $\mu_f = 1$ while ε_f is some constant to be determined. It has a width d_f and is located at z_f within the heterogenous slab.

The role of the effective theory is to provide a physically cogent parameterization for the interrogating signal and for the fault signature in connection with identification. Unlike the multiscale laminate, the effective medium is relatively smooth and therefore leads to simple parameterization of the wave phenomena. As a simple example that clarify this point, consider the multiscale laminate in Fig. 1 where ε is rather complex but $\varepsilon^{\text{ef}} \approx 2$. The fault is detectable only if it has a sufficiently large contrast relative to the effective medium, i.e. if $\varepsilon_f \not\approx 2$.

5.1. Interrogation of a thick fault

If the width of the fault is larger than the pulse width (i.e., $d_f > cT\,(\mu^{\text{ef}}\varepsilon^{\text{ef}})^{-1/2}$), then its signature consists mainly of its width d_f and of the contrast between its spectral characteristic admittance $Y_f = \sqrt{p_f/q_f}$ and the admittance $Y^{\text{ef}} = \sqrt{p^{\text{ef}}/q^{\text{ef}}}$ of the ambient *effective* medium. Thus the fault is detectable if $\sqrt{p_f/q_f} \not\approx \sqrt{p^{\text{ef}}/q^{\text{ef}}}$.

In the numerical simulations we consider the multiscale laminate used in the previous examples (see Fig. 1, where $\varepsilon^{\text{ef}} \approx 2$ and $\mu^{\text{ef}}=1$). The fault has $\varepsilon_f = \mu_f = 1$ and is located between $z = 0.25$ and 0.75. For the pulse in (24) where $cT = 2^{-3}$ we have $cT(\varepsilon^{\text{ef}}\mu^{\text{ef}})^{-1/2} < d_f$ and $\varepsilon_f \neq \varepsilon^{\text{ef}}$ as required for resolving the fault. The reflected pulse is depicted in Fig. 5. One observe that the effective response agrees

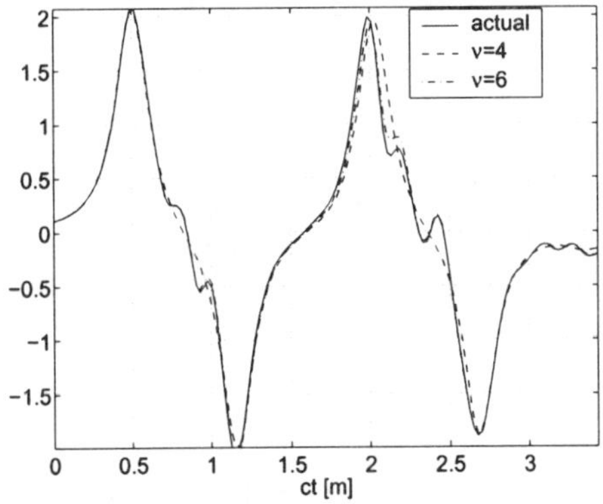

Figure 5: Pulse reflections from a heterogeneous laminate with a thick fault at $z_f = 0.25$ with $d_f = 0.5$ and $\varepsilon_f = 1$.

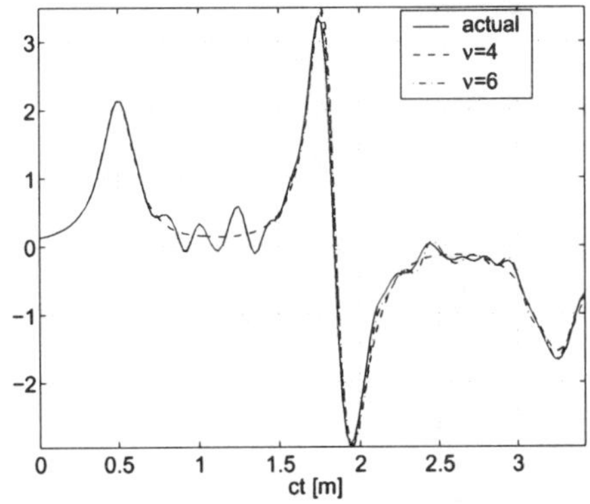

Figure 6: Pulse reflection from a heterogeneous laminate with a thin fault at $z_f = 0.5$ with $d_f = 2^{-7}$ and $\varepsilon_f = 20$.

with the actual one for $\nu = 6$ but there are minor discrepancies for $\nu = 4$. One also discerns the reflections from the front and back faces of the slab at $ct = 0.5$ and 3.2, as well as the doublet type reflection from the front and back faces of the fault at $ct = 1.1$ and 2 which is absent in the fault-free response of Fig. 3. Comparing the results in Figs. 3 and 5 it is then possible to assess the fault properties (location and electrical width).

5.2. Interrogation of a thin fault

If the fault is very thin on the scale of the interrogating pulse (i.e. $d_f \ll cT\,(\mu^{\text{ef}}\varepsilon^{\text{ef}})^{-1/2}$) then its scale is comparable to the micro-scale of the heterogeneities. In this case it cannot be observed unless it has a sufficient contrast with respect to the ambient *effective* medium. The precise condition is determined by noting that a thin fault with high ε_f can be replaced with an admittance sheet with *normalized surface admittance* (reactance)

$$Y_s = -i(\omega/c)\varepsilon_f d_f\,. \tag{26}$$

The reflected signal is therefore obtained via a simple transmission line model wherein the fault is described as the lumped admittance Y_s connected in parallel within an *effective* transmission line with *slowly varying* characteristic admittance $Y^{\text{ef}}(z) = \sqrt{p^{\text{ef}}(z)/q^{\text{ef}}(z)}$. The reflection coefficient is now estimated noting that the propagating short pulse is mainly affected by the local discontinuity due to Y_s and not by the slowly varying effective medium. The transmission line model for the local reflection coefficient is readily found to be

$$\Gamma(\omega) = \frac{-Y_s}{Y_s + 2Y^{\text{ef}}(z_f)}\,. \tag{27}$$

In the limiting cases of $Y_s \gtrless Y^{\text{ef}}(z_f)$ we find that $\Gamma \approx -1$ or $\Gamma \approx i(\omega/c)\varepsilon_f d_f/2Y^{\text{ef}}(z_f)$, respectively. Thus, for small Y_s the reflected pulse is a derivative of the incident pulse while for large Y_s the reflected pulse essentially replicates the incident one. Furthermore, the inclusion will be noticeable only if the reflection coefficient is non negligible leading to the condition $\varepsilon_f(d_f/cT) \gtrsim \sqrt{p^{\rm ef}/q^{\rm ef}}$.

The numerical simulations in Fig. 6 have been performed for a thin layer at $z_f = 0.5$ and $d_f = 2^{-7}$, with $\mu_f = 1$ and $\varepsilon_f = 20$ so that $Y_s \ll Y^{\rm ef}$. The differentiated reflected pulse around $ct = 1.8$ is well observed. Note the excellent agreement between the actual and the effective solutions for $\nu = 6$, with certain discrepancies for the $\nu = 4$.

Acknowledgement: This work has been supported in part by the US-Israel Binational Science Foundation, Jerusalem, Israel, under Grant No. 9900448, and in part by the Israel Ministry of Science. The authors would also like to thank Professors L.B. Felsen and J.J. McCoy for stimulating and constructive discussions.

References

1. B. Z. Steinberg, J. J. McCoy and M. Mirotznik, "A multiresolution approach to homogenization and effective modal analysis of complex boundary value problems," *SIAM J. Appl. Math,* **60**(3), 939-966, March 2000.
2. V. Lomakin, B. Z. Steinberg and E. Heyman, "Multiresolution homogenization for radiation and propagation in multiscale laminates: Part I – Effective formulation," submitted for publication.
3. V. Lomakin, B. Z. Steinberg and E. Heyman, "Multiresolution homogenization for radiation and propagation in multiscale laminates: Part II – Spectral representations of the field," submitted for publication.
4. V. Lomakin, B. Z. Steinberg and E. Heyman, "UWB analysis of EM fields in complex laminates: An MRA homogenization approach," in *Ultra-Wideband, Short-pulse Electromagnetics,* **5**, P.D. Smith ed., Plenum Press, NY, 2001, in print.
5. E. Heyman and L. B. Felsen, "Weakly dispersive spectral theory of transients (STT). Part I: Formulation and interpretation," "Part II: Evaluation of the spectral integral," and "Part III: Applications," *IEEE Trans. Antennas Propagat.,* **35**, 80–86, 574–580, and 1258–1266, 1987.
6. T. B. Hansen and A. D. Yaghjian, *Plane-Wave Theory of Time-Domain Fields,* IEEE Press, Piscataway NJ, 1999.
7. E. Heyman, "Transient plane wave spectrum representation for radiation from volume source distribution," *J. Math. Phys.*, **37**, 658–681, 1996.
8. L. B. Felsen, and N. Marcuvitz, *Radiation and Scattering of Waves,* Second Edition, IEEE Press, Piscataway NJ, 1994 (2nd printing of the 1973 original).
9. I. Daubechies, *Ten Lectures on Wavelets,* CBMS–NSF Series in Applied Mathematics, SIAM Publ., Philadelphia, 1992.
10. M. Krumpholz and L. P. B. Katehi, "MRTD: New time domain schemes based on multiresolution analysis," *IEEE Trans. Microwave Theory and Techniques,* **44**, 555–571, 1996.
11. C. E. Baum, "The Singularity Expansion Method," in *Transient electromagnetic fields.* L.B. Felsen, Ed. New York: Springer Berlag, 1976.
12. E. Heyman and L.B. Felsen, "A wavefront interpretation of the Singularity Expansion Method," *IEEE Trans. Antennas Propagat.*, **AP-33**, 706–718, 1985.

FOLIAGE PENETRATION STRIPMAP SAR IMAGING USING UWB RANDOM NOISE WAVEFORM

Ram M. Narayanan, Xiaojian Xu and Cihan Kumru*

Abstract: Ultra wideband (UWB) penetration imaging synthetic aperture radars (SAR) are among the latest generations of high resolution imaging radar systems. The University of Nebraska at Lincoln (UNL) has developed a new type of foliage penetration (FOPEN) SAR which uses random noise signal as the probing waveform. The system operates over the 250-500 MHz frequency band and uses two broadband dual-polarized log-periodic antennas. A dual-polarization heterodyne correlation receiver is used to preserve phase coherence and allow polarimetric SAR imaging. The ultra-wide bandwidth of the transmitted signal provides fine range resolution, while stripmap SAR image processing technique is used to obtain high cross-range resolution. In this paper, we present the random noise radar SAR imaging model, signal processing algorithms, and the 2-D imaging results obtained from field tests.

Key words: UWB, SAR, Random noise radar, Foliage penetration.

1. INTRODUCTION

Foliage penetration (FOPEN) and ground penetration (GPEN) ultra wideband (UWB) synthetic aperture radars (SAR) are among the latest generations of high resolution imaging radars. Such a radar system usually operates at a lower frequency band such as VHF or UHF bands, has exceptionally large fractional bandwidth, and antennas with wide angular beamwidths that span tens of degrees. One example of such a radar system is the Army Research Laboratory (ARL) Boom SAR which transmits an impulse waveform in the frequency band 20 MHz - 1.1 GHz [1], with a fractional bandwidth of 2 and antenna beamwidths larger than 90°. These types of imaging radars are grouped as ultra-wide band, wide angle (UWBWA) SARs [2]. In contrast to the signal processing of conventional SARs, different processing algorithms must be adopted for UWBWA SARs, because many of the simplifying assumptions made in conventional SAR imaging

* Ram M. Narayanan, Xiaojian Xu and Cihan Kumru, Department of Electrical Engineering, University of Nebraska-Lincoln, NE 68588-0511, USA. Tel: (402) 472-5141, Fax: (402) 472-4732, Email: rnarayanan@unl.edu.

are no longer valid [3]. Other UWB penetration imaging radar systems using linear frequency modulation (LFM) or step-frequency waveforms were also developed [4].

Under the support of the US Air Force Office of Scientific Research (AFOSR), the University of Nebraska at Lincoln (UNL) developed a new type of FOPEN SAR which uses a random noise signal as the transmit waveform. Some theoretical and simulation results were reported in the past [5,6]. In this paper, we first establish a model of the UWB coherent random noise radar system. Next, we point out that a UWB random noise radar can be modeled as a UWB short-pulse radar which transmits a pulse waveform and receives the returned pulses using an ideal sampler at specified time delays. We then develop the two-dimensional (2-D) SAR imaging model and image formation algorithm for a near-field bistatic UWBWA SAR imaging configuration using the random noise radar system.

2. RANDOM NOISE RADAR SYSTEM MODEL

We first consider the one-dimensional (1-D) imaging model of the random noise radar. Assume that the radar transmits a wide-sense stationary (WSS) band-limited stochastic waveform, $s(t)$, with zero mean. The signal received using a correlation receiver can be written as

$$r(t) = \int_{-L/2}^{L/2} \rho(\varsigma)s(t-\varsigma)d\varsigma + n(t) \tag{1}$$

where $\rho(t)$ is the 1-D target reflectivity function (or range profile), $n(t)$ represents any additive noise and/or interference, and L represents the time span of the target extension in down range.

The received signal by a coherent correlation receiver at a specified two-way delay $\tau = 2R/c$, can be expressed as

$$g_T(\tau) = \int_{-T/2}^{T/2} \int_{-L/2}^{L/2} \rho(\varsigma)s(t-\varsigma)s(t-\tau)d\varsigma dt + \int_{-T/2}^{T/2} n(t)s(t-\tau)dt \tag{2}$$

where R is the distance between the target and the radar, c is the wave propagation speed, and T is the correlation integration period. For a large integration period T, the second term in Eq.(2) tends to zero; thus Eq.(2) can be expressed as

$$g(\tau) = \lim_{T\to\infty} g_T(t) = \int_{-L/2}^{L/2} \rho(\varsigma)p(\tau-\varsigma)d\varsigma = \rho(\tau) \otimes p(\tau) \tag{3}$$

where $\otimes$ denotes convolution, and

$$p(\tau) = \lim_{T \to \infty} \int_{-T/2}^{T/2} s(t)s(t-\tau)dt \tag{4}$$

is the auto-correlation function of the transmit waveform. When the radar transmit waveform is an ideal band-limited Gaussian noise with central frequency f_0 and bandwidth B, we have

$$p(t) = A_0 \cos(2\pi f_0 t)\mathrm{sinc}(Bt)\,. \tag{5}$$

Equations (3) and (5) state that the UWB random noise radar can be modeled as a short-pulse radar transmitting a pulse $p(t)$ and receiving the returned pulse using an ideal sampler at specified time delays. For radars that use band-limited Gaussian noise as the probing waveform, $p(t)$ is a sinc-pulse modulated by a carrier frequency f_0. The pulse width is determined by the bandwidth B, which in turn, determines the down range resolution of the system.

3. 2-D STRIPMAP FOPEN SAR IMAGING MODEL

Consider a near-field bistatic strip-map SAR imaging scenario. The three-dimensional (3-D) and the equivalent 2-D data acquisition geometry for the FOPEN random noise stripmap SAR imaging are shown in Figure 1. The radar platform moves along a straight line path at a constant height, h, over the ground plane. In the equivalent 2-D slant-range/cross-range imaging plane, the radar coordinate system is u-v system with zero range reference at O. The distance between the transmit and the receive antennas is d. Thus, at time t, the position of the radar (which refers to the mid-point of the two antennas) can be denoted by $(u,0)$.

Consider a scattering point P at location (x,y). The raw radar imaging data acquired at a specific radar position (u,t) is the integration of the returned signals from each target point, *i.e.*,

$$d_p(u,t) = \int_{-L/2}^{L/2} \int_0^L \frac{1}{R_t(x,y,u)R_r(x,y,u)} g(x,y,u,t) \otimes p(t) \otimes \delta\left(t - \frac{R_v(x,y,u)}{c}\right) dydx \tag{6}$$

where the factor $\frac{1}{R_t R_r}$ takes account of the fact that both the incidence and the scattered fields propagate as spherical waves, $R_t(x,y,u) = \sqrt{(u-d/2-x)^2 + y^2 + h^2}$, $R_r(x,y,u) = \sqrt{(u+d/2-x)^2 + y^2 + h^2}$, and $R_v(x,y,u) = R_t(x,y,u) + R_r(x,y,u)$,

$$g(x,y,u,t) = a_t(x,y,u,t) \otimes \rho_f(x,y,u,t) \otimes a_r(x,y,u,t), \tag{7}$$

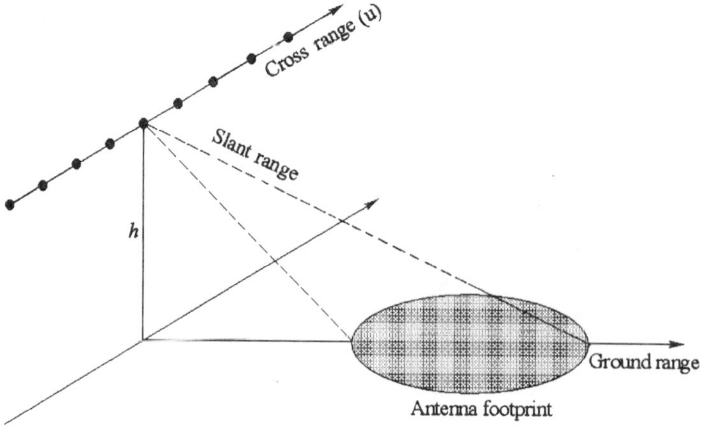

(a). 3-D data acquisition geometry.

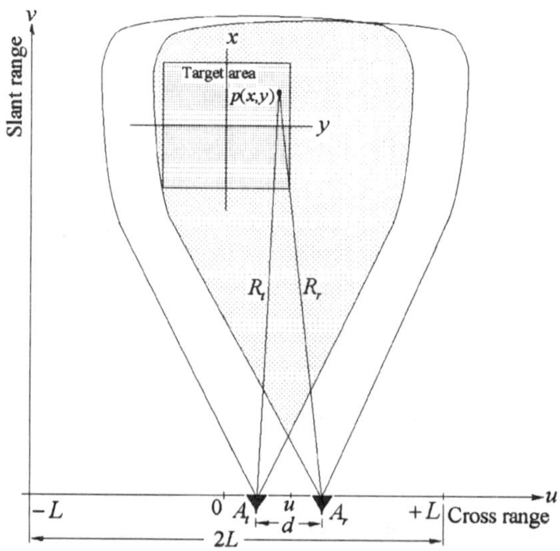

(b). Equivalent 2-D imaging plane.

Figure 1: Stripmap SAR imaging geometry.

and

$$\rho_f(x,y,u,t) = f_t(x,y,u,t) \otimes \rho(x,y,u,t) \otimes f_r(x,y,u,t)\,. \tag{8}$$

In Eq. (8), $\rho(x,y,u,t)$ is the target reflectivity function, which is a four-dimensional (4-D) function of target 2-D position, radar observation direction and time; $f_t(x,y,u,t)$ and

$f_r(x,y,u,t)$ represent the foliage transmission factors along transmit and receive channels, respectively, $a_t(x,y,u,t)$ and $a_r(x,y,u,t)$ represent the impact of the transmit and receive antennas.

4. 2-D IMAGE FORMATION

Eq.(6) can be rewritten as

$$d_p(u,t) = p(t) \otimes d(u,t) \tag{9}$$

where

$$d(u,t) = \int_{-L/2}^{L/2} \int_{0}^{L} \frac{1}{R_t(x,y,u)R_r(x,y,u)} g(x,y,u,t) \otimes \delta(t - \frac{R_v(x,y,u)}{c}) dydx \tag{10}$$

We consider the problem of image formation from "data" $d(u,t)$ first. In fact, $d(u,t)$ can be recovered from the blurred signal $d_p(u,t)$ by using an equalization filter, if the radar transmit waveform $p(t)$ is known. Instead of trying to obtain the 4-D target reflectivity function, we must make a tradeoff between the acceptable image quality and the available radar imaging data, *i.e.*, our purpose is to reconstruct the 2-D (foliage obscured) target function $\rho_f(x,y)$. Eqs.(7)-(10) suggest that a good estimate of the antenna characteristics blurred reflectivity function, $g(x,y)$ can be obtained from the radar data $d(u,t)$ by using a back-projection algorithm, *i.e.*,

$$g(x,y) = \int_{-\infty}^{\infty} \int_{-\infty}^{\infty} R_t(x,y,u)R_r(x,y,u)d(u,t)\delta(t - \frac{R_v(x,y,u)}{c})dudt, \tag{11}$$

where the factor R_tR_r is used for range compensation. When the characteristics of the antennas are known, the target function $\rho_f(x,y)$ can then be recovered using equalization filtering technique.

The 2-D image resolution obtained by using the above image formation procedure is limited by the radar bandwidth and the size of the synthesized aperture. Image enhancement processing algorithms, such as spatially variant apodization (SVA) [7], can be applied to further improve the 2-D resolution of the SAR images. SVA is a nonlinear filtering operation that significantly reduces the sidelobe levels without degrading mainlobe resolution of a sinc impulse response. A modified version of noninteger Nyquist SVA and an iterative super SVA procedure for SAR and ISAR image enhancement were developed and successfully applied to the FOPEN noise SAR images. More details of this image resolution enhancement technique can be found in [8].

5. EXPERIMENTAL RESULTS

Figure 2 illustrates the equivalent transmit short-pulse and the corresponding frequency spectrum for the 250-500 MHz experimental FOPEN random noise radar. To demonstrate the high resolution and foliage penetration capabilities of the random noise radar system, and to test the imaging algorithms, a variety of field tests were conducted. In the experiments, a boom van was used as the platform for the noise radar imaging system. The radar system was placed in the boom van. Two log-periodic antennas were installed on the boom which can rise to about 10 meters above the ground. An additional 50 m of low-loss cable was added to the time delay chain to provide a longer fixed delay. The primary objective of using this fixed length cable was to compensate for the two-way transmission length (by cables) between the radar and the antennas, and to set a reference center of the equivalent range gate. The fixed length cable, together with the variable delay line provided a range gate of 19.2 m centered at a distance of about 20 m from the radar antennas.

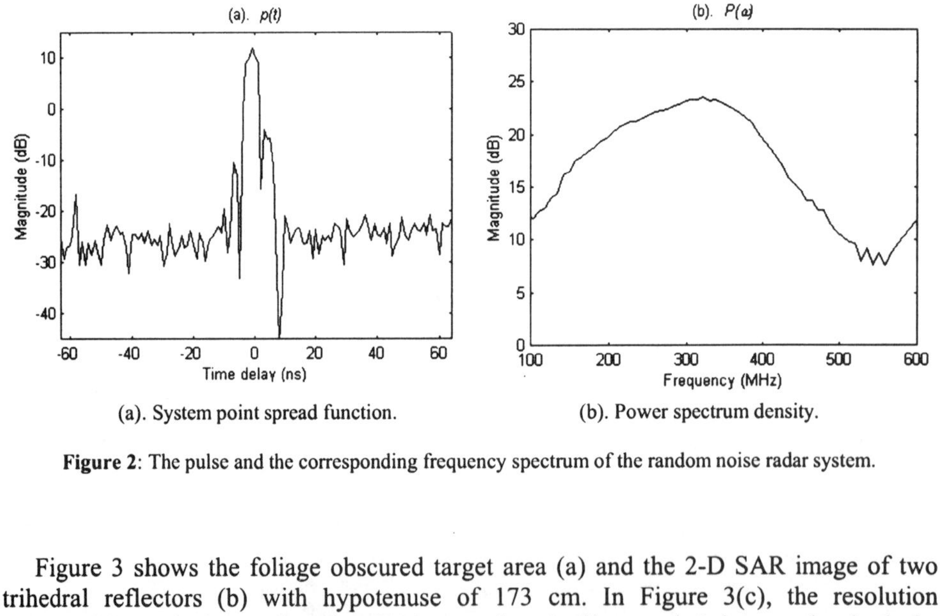

(a). System point spread function. (b). Power spectrum density.

Figure 2: The pulse and the corresponding frequency spectrum of the random noise radar system.

Figure 3 shows the foliage obscured target area (a) and the 2-D SAR image of two trihedral reflectors (b) with hypotenuse of 173 cm. In Figure 3(c), the resolution enhanced image using an iterative SVA procedure is illustrated. It is seen that both the trihedral reflectors and the separated trees in the target area scenario are all well resolved, thus showing the good foliage penetration and imaging capability for the UWB random noise radar system.

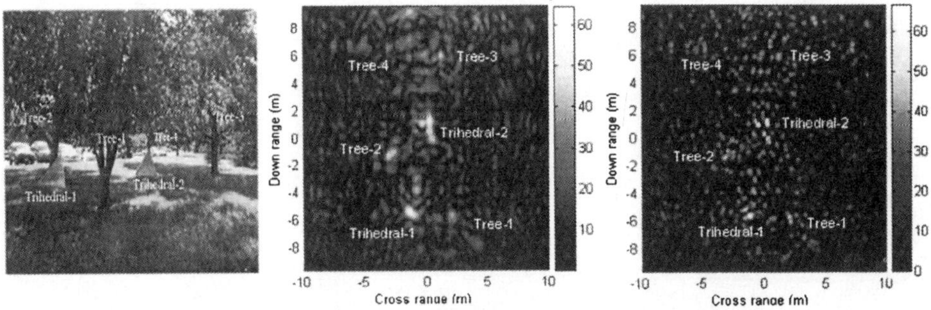

(a). Target scenario (b). 2-D Stripmap SAR image (c). Resolution enhanced image

Figure 3: FOPEN random noise SAR images of two trihedral reflectors obscured behind trees.

6. ACKNOWLEDGMENTS

This work was supported by US AFOSR under contract F49620-99-1-0209. Thanks are also due to Dr. Arje Nachman of AFOSR for helpful comments and support of the research.

7. REFERENCES

1. M. Ressler, L. Happ, L. Nguyen, T. Ton, and M. Bennett, The Army Research Laboratory ultra-wideband testbed radar," *Proc. of the IEEE 1995 Int. Radar Conf.*, Washington D.C., pp.686-691 (1995).
2. R. Rau and J. H. McClellan, Analytical models and postprocessing techniques for UWB SAR, *IEEE Trans. Aerospace and Electronic Systems*, Vol.12, No.4, pp.1058-1074 (2000).
3. M. Soumekh, Reconnaissance with ultra wideband UHF synthetic aperture radar, *IEEE Signal Processing Magazine*, Vo.12, No.4, pp.21-40 (1995).
4. M. Soumekh, D. A. Nobles, M. C. Wicks, and G. J. Genello, Signal processing of wide bandwidth and widebeamwidth P-3 SAR data, *IEEE Trans. on Aerospace and Electronic Systems*, Vol.37, No.4, pp.1122-1141 (2001).
5. X. Xu and R. M. Narayanan, FOPEN SAR imaging using ultra-wideband step-frequency and random noise waveforms," *IEEE Trans. on Aerospace and Electronic Systems*, Vol.37, No.4, pp.1287-1300 (2001).
6. X. Xu and R. M. Narayanan, Imaging performance analysis of a FOPEN UWB random noise radar," Digest *of 2001 IEEE AP-S Intl. Symp.*, Boston, MA, Vol.4, pp.273-276 (2001).
7. H. C. Stankwitz, R. J. Dallaire, and J. R. Fienup, Nonlinear apodization for sidelobe control in SAR imagery," *IEEE Trans. on Aerospace and Electronic Systems*, Vol.31, No.1, pp.267-279 (1995).
8. X. Xu and R. M. Narayanan, SAR image enhancement using noninteger Nyquist SVA technique," Digest *of 2002 IEEE AP-S Intl. Symp.*, San Antonio, TX, to appear (2002).

ITERATIVE SOLUTION OF FIELD PROBLEMS WITH A VARYING PHYSICAL PARAMETER

Anton G. Tijhuis, Martijn C. van Beurden, and A. Peter M. Zwamborn*

1. INTRODUCTION

In modern society different trends are recognized in the usage of the available electromagnetic spectrum. One can think of wireless communication or transport of (digital) information. The density of such applications is increasing rapidly. Obtaining electromagnetic compatibility and/or reducing electromagnetic interference sometimes seems to be an impossible task. Another trend is found in electromagnetic inverse scattering and profiling. For example, this development is used in the detection and classification of land mines and other unexploded ordnance. Regarding electromagnetic inversion, one can also think of medical applications such as tomography or the detection of defects in metallic heart valves. Finally, we would like to mention the problem of electromagnetic coupling into humans in the area of clinical hyperthermia or non-ionizing radiation hazards analysis. In these applications, a rigorous electromagnetic analysis is indispensable.

The focus of this chapter is found in computational tools in the field of electromagnetic analysis and design. One can identify the roadmap "going from engineering electromagnetics to electromagnetic engineering". One of the extensively used and most versatile methods is the integral equation technique. It takes into account that the irradiated object is present in free space and that it manifests itself through the presence of secondary sources or contrast sources. Integral equations can be solved by the method of moments[1]. This leads to a system of linear algebraic equations.

To solve this system, one can use direct discrete numerical solution methods, such as Gaussian Elimination or Singular Value Decomposition, or suitable iterative techniques such as a conjugate-gradient (CG) method. An overview of numerical solution methods for linear systems of equations can be found in the book by Golub and Van Loan[2]. In electromagnetic scattering and coupling problems, integral equations are often solved by using a Fast Fourier Transform to compute the spatial convolution of the integral operator

*Anton G. Tijhuis and Martijn C. van Beurden, Faculty of Electrical Engineering Eindhoven University of Technology P.O. Box 513, 5600 MB Eindhoven, the Netherlands. A. Peter M. Zwamborn, TNO Physics and Electronics Laboratory, P.O. Box 96864, 2509 JG 's-Gravenhage, the Netherlands

Ultra-Wideband Short-Pulse Electromagnetics 6
Edited by Mokole *et al.*, Kluwer Academic/Plenum Publishers, 2003

and a conjugate-gradient iterative scheme. This so-called CGFFT method has been used successfully for many electromagnetic scattering and coupling problems[3–9].

In analysis or design procedures, the engineer has the freedom to change one or more parameters to obtain an optimal design with respect to performance and costs. This means that he or she will need to consider the determination of electromagnetic fields for a (large) number of values of a physical parameter. In this chapter we present a strong approach for this type of problem, which utilizes the iterative schemes mentioned above. We restrict ourselves to the case where the linear system originates from one or more integral equations. We apply an iterative procedure based on the minimization of an integrated squared error, and start this procedure from an initial estimate that is a linear combination of the last few "final" results. When the coefficients in this extrapolation are determined by minimizing the integrated squared error for the actual value of the parameter, the built-in orthogonality in this type of scheme ensures that only a few iteration steps are required to obtain the solution. The success of this strategy has been demonstrated before[10–12]. However, it has not been applied to 3-D problems.

The outline of the chapter is as follows. In Section 2, the method of solution is discussed. Special attention is given to the iterative procedure and the implementation of a relevant initial estimate based on the previous solutions. In Section 3, a simple example is given to clarify the success of the present approach. Explicit examples are discussed in Section 4. Finally, Section 5 presents the conclusions.

2. METHOD OF SOLUTION

In the computational modeling of electromagnetic fields for practical applications, typically a large system of linear equations must be solved. This system originates from spatially discretizing Maxwell's differential equations (in "finite" or "local" techniques) or equivalent integral equations (in "global" techniques). In formal notation, such a system can be written as

$$L(p)\,u(p) = f(p), \tag{1}$$

where

$$\begin{aligned} L(p) &= \text{a linear operator,} \\ u(p) &= \text{the unknown field,} \\ f(p) &= \text{the forcing function,} \\ p &= \text{a physical parameter.} \end{aligned}$$

The operator $L(p)$ originates from discretizing its counterpart in the continuous equation, $u(p)$ is a discretized field and $f(p)$ corresponds to an impressed source or an incident field. We are interested in the situation where this problem must be solved for a large number of sampled values of the parameter p, e.g., $p = p_0 + m\Delta p$, with $m = 0, 1, \ldots, M$.

To solve the system of equations (1), we use the conjugate gradient method. This algorithm is described in detail by Van den Berg[3, 4].

Further, we will attempt to organize the space discretization such that the convolution structure of the continous equation is preserved. In that case, the matrix-vector products in the CG algorithm can be evaluated by FFT operations, which considerably improves the speed of the so-called CGFFT algorithm[3–9].

2.1. Initial estimate

In many applications of the conjugate-gradient method, a simple initial estimate is used. Typically, the scheme is started from an initial vector $u^{(0)} = 0$. Depending on the nature of the problem at hand, we can also start from an incident field or from the Kirchhoff approximation to an unknown surface current.

Our choice of the initial estimate is inspired by the fact that $u(p)$ depends in a well-behaved manner on the parameter p. Therefore, it should be possible to extrapolate, by choosing

$$u^{(0)}(p_m) = \sum_{k=1}^{K} \gamma_k \, u(p_{m-k}). \tag{2}$$

The interpretation of the conjugate-gradient scheme given above suggested that the values $\{\gamma_k \mid k = 1, \dots, K\}$ should be found by minimizing the squared error

$$< L(p_m)u^{(0)}(p_m) - f(p_m) \mid L(p_m)u^{(0)}(p_m) - f(p_m) > . \tag{3}$$

where we have defined the inner product as

$$< g \mid h >= \sum_j g_j^* h_j, \tag{4}$$

where g_j and h_j denote the components of g and h, and where the asterisk denotes complex conjugation.

Because of the built-in orthogonality of the conjugate-gradient method, we are then certain that this procedure must start its search for components of $f(p_m)$ outside the space spanned by the "previous" functions $\{Lu(p_{m-k}) \mid k = 1, \dots, K\}$.

The coefficients γ_k that minimize the squared error in Eq.(3) can be found from the system of linear equations

$$\sum_{k'=1}^{K} < L(p_m)u(p_{m-k}) \mid L(p_m)u(p_{m-k'}) > \gamma_{k'} =< L(p_m)u(p_{m-k}) \mid f(p_m) >, \tag{5}$$

with $k = 1, \dots, K$. Typically, we choose $K = 2$ (linear extrapolation) or $K = 3$ (quadratic extrapolation). For larger values of K, the basis vectors $L(p_m)u(p_{m-n})$ with $n = 1, \dots, K$ become almost linearly dependent, and therefore the coefficients $\{\gamma_k\}$ can no longer be resolved from Eq.(5).

3. A SIMPLE EXAMPLE

The authors are presently not aware of an exact analysis that provides an indication of the success of the extrapolation procedure outlined in Subsection 2.1. However, extensive numerical experimentation for a range of problems has at least given an indication what is going on, and how computational parameters like step size and desired accuracy should be chosen.

In the present section, we will try to explain the efficiency of the CG-extrapolation procedure. To illustrate this, we consider the simple system of three linear equations

$$\begin{bmatrix} 1 & 0 & 0 \\ 0 & 1 & 0 \\ 0 & 0 & 2 \end{bmatrix} \begin{bmatrix} u_1 \\ u_2 \\ u_3 \end{bmatrix} = \begin{bmatrix} 1 \\ 1 \\ 2\Delta \end{bmatrix}, \tag{6}$$

which has the solution $u = [1, 1, \Delta]^T$. The 3×3 matrix in the left-hand side is representative of a well-conditioned, discretized integral equation. Since this integral equation is supposed to originate from a passive, linear physical problem, all eigenvalues will have the same order of magnitude. The idealization is that the corresponding eigenvectors are supposed to be orthogonal with respect to the inner product Eq.(4). The real-valued parameter $\Delta \ll 1$ in the right-hand side represents the effect of a step Δp in the physical parameter p. We consider the situation where only u_1 and u_2 correspond to modes that were excited for the previous value of p. For the initial estimate we take $u^{(0)} = [1 + \delta, 1 + \delta, 0]^T$, where δ may be envisaged as the extrapolation error. For convenience, we assume that δ is real-valued as well.

Carrying out a single step of the CG algorithm then yields for the first correction vector u_{cor}:

$$u_{cor}^{(1)} = \frac{\delta^2 + 8\Delta^2}{\delta^2 + 32\Delta^2} \begin{bmatrix} -\delta \\ -\delta \\ 4\Delta \end{bmatrix}. \tag{7}$$

From Eq.(7), we observe that the first iteration step only leads to an effective correction of the error $u - u^{(0)} = [-\delta, -\delta, \Delta]$ when $|\delta| \ll |\Delta|$. Otherwise, we need to carry out all three iterations to obtain convergence.

In practice, the system of equations (1) is not always well conditioned. One possible cause is that the space discretization leading to such a system introduces spurious eigenvectors with a small eigenvalue. A second possibility is that the system is near resonance, and thus has a physical eigenmode with a small eigenvalue. In both cases, the requirements on the extrapolation error are even more severe. To illustrate this effect, we modify the system of equations (6) into

$$\begin{bmatrix} 1 & 0 & 0 \\ 0 & 1 & 0 \\ 0 & 0 & \Delta \end{bmatrix} \begin{bmatrix} u_1 \\ u_2 \\ u_3 \end{bmatrix} = \begin{bmatrix} 1 \\ 1 \\ \Delta \end{bmatrix}, \tag{8}$$

which has the solution $u = [1, 1, 1]^T$. We start from the same initial estimate $u^{(0)} = [1+\delta, 1+\delta, 0]^T$, which means that we have a 100% error in the direction of the eigenvector corresponding to the small eigenvalue. Now, the first correction is found as

$$u_{cor}^{(1)} = \frac{2\delta^2 + \Delta^4}{2\delta^2 + \Delta^6} \begin{bmatrix} -\delta \\ -\delta \\ \Delta^2 \end{bmatrix}, \tag{9}$$

while the actual error at the start of the iteration scheme is $u - u^{(0)} = [-\delta, -\delta, 1]$. This means that we now have an almost complete correction in the first step only if $|\delta| \ll |\Delta|^3$.

In the actual implementation, we organize the discretization such that the dimension of the linear system of Eq.(1) does not change, and that the elements of the system matrix $L(p)$ and the components of the vector $f(p)$ depend continuously on the physical parameter p. In that case, the changes represented by Δ in the above examples are proportional to the step Δp. The error in the initial estimate is represented by the small parameter δ. This means that in the realistic case represented by the second model even quadratic extrapolation with fixed coefficients, which leads to an error of $\delta = \mathcal{O}(\Delta p^3)$, may not suffice. The minimization procedure suggested in Subsection 2.1. completely cancels the error in the subspace spanned by the "previous" solutions, and therefore does lead to an acceptable convergence.

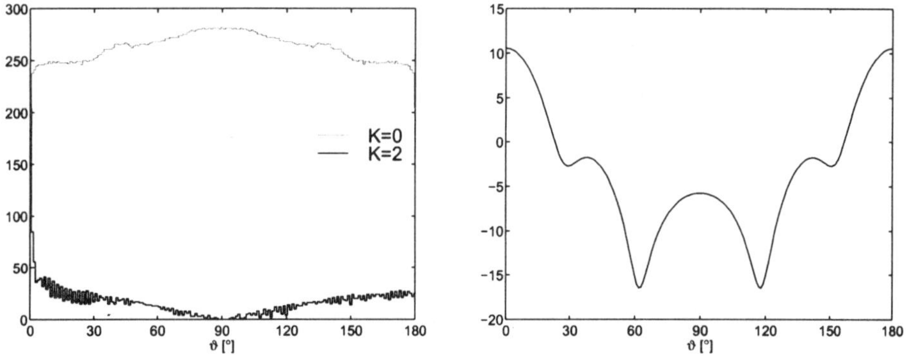

(a) (b)

Figure 1: Marching-on-in-angle version of the CGFFT method for a flat plate. (a) Number of iterations required to reach a relative error of 10^{-3} versus angle of incidence using zero (gray line) and two previous results (black line) as an initial estimate. (b) Monostatic RCS versus angle of incidence.

Finally, it should be mentioned that in these examples we have assumed that $\Delta \ll 1$. This suggests a restriction on the step size Δp. Numerical experimentation has shown that, for each problem, there appears to be a critical step size Δp beyond which the extrapolation procedure is no longer effective. One possible explanation can be derived from the second example, by assuming that the parameter Δ is governed by the space discretization and that δ depends on Δp.

4. SCATTERING BY THREE-DIMENSIONAL OBJECTS

To illustrate our approach, we have extended existing implementations of the CGFFT procedure for two three-dimensional objects that have become standards in the literature. In both cases, no special precautions were taken to enhance the discretization, which is first-order accurate as a function of the mesh size.

4.1. Scattering by a flat plate

The first example is a flat, rectangular plate in free space located at $0 < x < a$, $0 < y < b$ and $z = 0$. For this problem, we solve the well-known electric-field integral equation

$$\left[\nabla_T \nabla_T \cdot - \frac{s^2}{c_0^2}\right] \int_0^a dx' \int_0^b dy' \, \frac{\exp(-sR/c_0)}{4\pi R} \, \boldsymbol{J}_S(\boldsymbol{r}'_T, s) = -s\varepsilon_0 \, \boldsymbol{E}^i_T(\boldsymbol{r}_T, s), \quad (10)$$

where s is a complex frequency, $R = |\boldsymbol{r}_T - \boldsymbol{r}'_T|$, and where the subscript $_T$ stands for a transverse component. The unknown surface current $\boldsymbol{J}_S(\boldsymbol{r}_T, s)$ is approximated by rooftop functions, and we use a weak formulation of Eq.(10), weighted by the same rooftop functions[13]. In the resulting discretized form, the convolution symmetry is preserved, so that the matrix-vector products in the conjugate-gradient procedure can be evaluated with

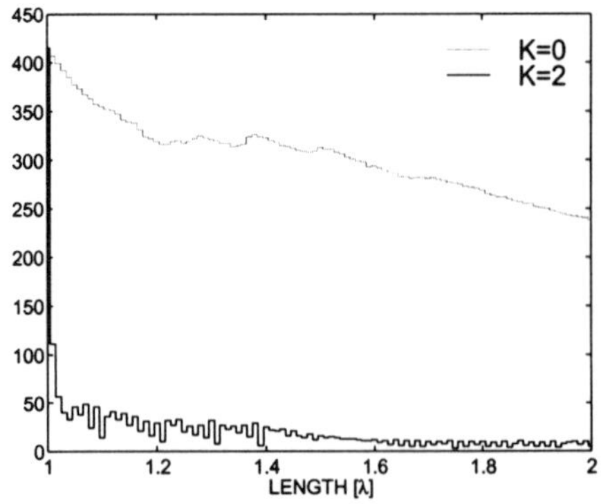

Figure 2: Number of iterations required to reach a relative error of 10^{-3} versus length of the plate for the marching-on-in-length version of the CGFFT method for a flat plate using zero (gray line) and two previous results (black line) as an initial estimate.

the aid of two-dimensional FFT operations.

In particular, we have computed the monostatic radar cross section of a $\lambda \times \lambda$ plate for the special case $s = j\omega$. A plane wave is incident on the plate at an angle θ with respect to the z-axis and an angle $\phi = 90°$ with respect to the x-axis. The incident plane wave is x-polarized. The discretized plate has a mesh of 31×31 points. Figure 1(a) shows the number of iterations for increasing θ. The gray line represents starting from a zero initial estimate, and the black line is for two previous results in the initial estimate, i.e. $K = 2$. Figure 1(b) presents the monostatic radar cross section of the $\lambda \times \lambda$ plate in the plane $\phi = 90°$.

Another result for the plate concerns marching on in length. The idea was inspired by the shape sensitivity analysis in[14, 15]. Here, we start from a $\lambda \times \lambda$ plate and we increase the length of the plate in 100 steps to a $2\lambda \times \lambda$ plate. We used a fixed spatial discretization of 62×31 mesh points. The number of iterations required to reach a relative error of 10^{-3} versus the length of the plate is shown in Figure 2. In the computations leading to Figures 1 and 2, it turned out that extrapolation with $K = 2$ was in fact more efficient than extrapolation with $K = 3$.

4.2. Scattering by an inhomogeneous dielectric cube

The second example is an inhomogeneous dielectric cube, again in free space. We formulate the scattering problem as a domain integral equation over the object domain $\mathcal{D}$ as

$$\boldsymbol{E}^i(\boldsymbol{r}, s) = \frac{\boldsymbol{D}(\boldsymbol{r}, s)}{\varepsilon(\boldsymbol{r}, s)} + \left(\frac{s^2}{c_0} - \nabla\nabla\cdot\right) \boldsymbol{A}(\boldsymbol{r}, s), \tag{11}$$

where s is a complex frequency and where the vector potential $\boldsymbol{A}(\boldsymbol{r}, s)$ is given by

$$\boldsymbol{A}(\boldsymbol{r}, s) = \frac{1}{\varepsilon_0} \iiint_{\mathcal{D}} \frac{\exp(-sR/c_0)}{4\pi R} \frac{\varepsilon(\boldsymbol{r}, s) - \varepsilon_0}{\varepsilon(\boldsymbol{r}, s)} \boldsymbol{D}(\boldsymbol{r}, s)\, dV', \tag{12}$$

where $R = |\boldsymbol{r} - \boldsymbol{r}'|$. We take the contrast function in Eq.(12) constant in each rectangular subdomain in the space discretization. Like the current in the plate problem, the dielectric displacement $\boldsymbol{D}(\boldsymbol{r}, s)$ is approximated by an expansion that is piecewise linear in the

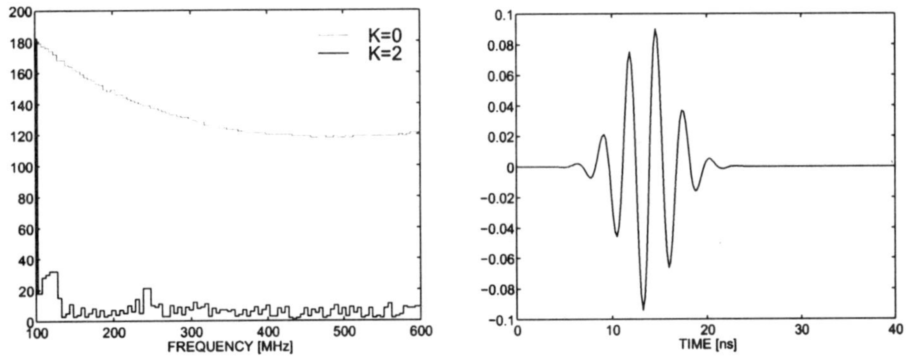

(a) (b)

Figure 3: Marching-on-in-frequency version of the CGFFT method for an inhomogeneous dielectric cube. (a) Number of iterations required to reach a relative error of 10^{-3} versus frequency for the using zero (gray line) and two previous results (black line) as an initial estimate. (b) Time domain signal at the center of the muscle cube for an incident x-polarized wave of 1 V/m.

longitudinal direction and constant in the transverse directions. The Green's function is replaced by a weak form, and the result is weighted by testing functions that are identical to the expansion functions. Again, the space discretization preserves the convolution symmetry of the continuous form of the integral equation given in Eq.(11) and Eq.(12). More details can be found in the papers by Zwamborn and Van den Berg[16, 17].

As an illustration, we have modeled a cube of muscle tissue centered inside a cube of fat tissue. The incident field is x-polarized with propagation vector parallel to the z-axis and a strength of 1 V/m. The dispersive tissues are modeled using a Debye model[18] and the dimensions of the inner and outer cubes are 0.14 m and 0.30 m, respectively. The discretized object has $30 \times 30 \times 30$ mesh points. The field is computed in the middle of the muscle cube for real-valued frequencies $f = \omega/2\pi = -js/2\pi$ of 100 to 600 MHz and then converted to a time domain signal. The number of iterations needed is shown in Figure 3(a), where the gray line is for a zero initial estimate, and the black line for minimization using two previous results. Again, using $K = 2$ in the extrapolation procedure led to the most rapid convergence. The time signal, shown in Figure 3(b), is computed by an FFT using the waveform $\exp\left[-(t-\tau)^2/(2T^2)\right]\sin(\omega_0 t)$, where $\tau = 14$ ns, $T = 2.75$ ns and $\omega_0/2\pi =$ 450 MHz.

5. CONCLUSIONS

In this chapter, we have extended the conventional conjugate-gradient method with a dedicated extrapolation procedure that considerably enhances the speed of convergence. Unfortunately, a rigorous analysis of the effectiveness of the combined procedure is not available. However, extensive numerical experimentation has at least given us an idea of what is going on. Further, we have illustrated this for two simple three-by-three systems. Although the procedure has already been demonstrated successfully for a range of applications, including transient scattering, RCS computations and inverse profiling, until now no applications to three-dimensional configurations have been reported. In the present chapter, this gap has been filled.

6. ACKNOWLEDGMENTS

The third author would like to thank the Deputy Director and the Integrated Topside Design Program Manager, both from the TNO Physics and Electronics Laboratory, for the financial support that enabled him to participate in this research project at the Eindhoven University of Technology.

7. REFERENCES

1. R.F. Harrington, *Field computation by Moment Methods*, (Macmillan, New York, 1968).
2. G.H. Golub and C.F. van Loan, *Matrix Computations Third Edition*, (The Johns Hopkins University Press, Baltimore, 1996).
3. P.M. van den Berg, Iterative computational techniques in scattering based upon the integrated square error criterion, *IEEE Trans. Antennas Propagat.* **32**, 1063–1071 (1984).
4. P.M. van den Berg, Iterative schemes based on the minimization of the error in field problems, *Electromagnetics* **5**, 237–262 (1985).
5. T.K. Sarkar, E. Arvas and S.M. Rao, Application of the fast Fourier transform and the conjugate gradient method for the solution of electromagnetic radiation from electrically large and small conducting bodies, *Electromagnetics* **5**, 99-122 (1985).
6. S.A. Bokhari and N.A. Balakrishnan, A method to extend the spectral iteration technique, *IEEE Trans. Antennas Propagat.* **34**, 51–57 (1986).
7. T.K. Sarkar, E. Arvas and S.M. Rao, Application of the FFT and the conjugate gradient method for the solution of electromagnetic radiation from electrically large and small conducting bodies, *IEEE Trans. Antennas Propagat.* **34**, 635–640 (1986).
8. Yuan Zhuang, Ke-Li Wu, Chen Wu, J. Litva, A combined full-wave CGFFT method for rigorous analysis of large microstrip antenna arrays, *IEEE Trans. Antennas Propagat.* **44**, 102–109 (1996).
9. J. Basterrechea and M.F. Catedra, Computatation of microstrip S-parameter using a CG-FFT scheme, *IEEE Trans. Microwave Theory Tech.* **42**, 234–240 (1994).
10. A.G. Tijhuis, Z.Q. Peng and A. Rubio Bretones, Transient excitation of a straight thin wire segment: a new look at an old problem, *IEEE Trans. Antennas Propagat.* **40**, 1132–1146 (1992).
11. A.G. Tijhuis and Z.Q. Peng, Marching-on-in-fequency method for solving integral equations in transient electromagnetic scattering, *IEE Proc. H* **138**, 347–355 (1991).
12. Z.Q. Peng and A.G. Tijhuis, "Transient scattering by a lossy dielectric cylinder: marching-on-in-frequency approach, *J. Electromagn. Waves Applicat.* **7**, 739–763 (1993).
13. A.P.M. Zwamborn and P.M. van den Berg, The weak form of the conjugate gradient method for plate problems, *IEEE Trans. Antennas Propagat.* **39**, 224-228 (1991).
14. J. Ureel and D. De Zutter, Shape sensitivities of capacitances of planar conducting surfaces using the method of moments, *IEEE Trans. Microwave Theory Tech.* **44**, 198–207 (1996).
15. J. Ureel and D. De Zutter, A new method for obtaining shape sensitivities of planar microstrip structures by a full-wave analysis, *IEEE Trans. Microwave Theory Tech.* **44** 249–260 (1996).
16. A.P.M. Zwamborn, and P.M. van den Berg, The three-dimensional weak form of the conjugate gradient FFT method for solving scattering problems, *IEEE Trans. Microwave Theory Tech.* **40**, 1757–1766 (1992).
17. A.P.M. Zwamborn, and P.M. van den Berg, Computation of electromagnetic fields inside strongly inhomogeneous objects by the weak conjugate gradient FFT method, *JOSA A* **11**, 1414–1421 (1994).
18. E.S.A.M. Lepelaars, Electromagnetic pulse distortion in living tissue, *Med. Biol. Eng. Comput.* **34**, 213–220 (1996).

A NEW FDTD SOLUTION METHOD WITHOUT A MARCHING-ON-IN-TIME SCHEME USING LAGUERRE POLYNOMIALS

Young-seek Chung, Tapan K. Sarkar, and Baek Ho Jung*

1. INTRODUCTION

The finite-difference time-domain (FDTD) method has been widely used for the numerical analysis of transient electromagnetic problems because it is conditionally stable and very easy to implement [1]. However, since the FDTD method is an explicit time-marching technique, its time step size should be limited by the well-known Courant-Friedich-Lecy (CFL) stability condition. Since the time step is dependent on the smallest length of the cell in a computational domain, this CFL condition may be too restrictive to solve problems with fine structures – thin material, slot, and via.

In recent times, to eliminate the CFL stability condition, the alternating-direction-implicit (ADI) method was proposed in order to formulate the implicit FDTD scheme [2]-[4]. The resulting ADI-FDTD method was found to be unconditionally stable. Therefore, one can use the larger value of the time step than dictated by the CFL limit. However, the larger value of the time step than the CFL limit results in larger dispersion error.

In this paper, we propose an unconditionally stable solution procedure for the FDTD method for the two-dimensional TE_z case using weighted Laguerre polynomials as temporal basis and testing function. Using the Laguerre polynomials and the weighting function, one can construct a set of orthogonal basis functions, which we call the weighted Laguerre polynomials. Physical quantities that are functions of time can be spanned in terms of these orthogonal basis functions – weighted Laguerre polynomials. Using the Galerkin's method, we introduce a temporal testing procedure, which results in an implicit FDTD formulation. When employing the conventional FDTD method, there is no matrix inversion involved with this computation procedure. However, the proposed method produces a banded sparse system matrix and is independent of the time step.

* Young-seek Chung and Tapan K. Sarkar, 121 Link Hall, Syracuse University, Syracuse, NY 13244. Baek Ho Jung, Depart. of ICE, Hoseo University, Asan, Chungnam 336-795, Korea.

However, this method uses also the same system matrix regardless of the order of basis functions to recursively solve for the unknowns. So one can assemble this sparse system matrix only once.

2. FDTD USING WEIGHTED LAGUERRE POLYNOMIALS

2.1. Finite Difference Time Domain For The TE_z case

Using the Yee algorithm [1], the TE_z equations are discretized as

$$E_x\big|_{i,j}^{n} = E_x\big|_{i,j}^{n-1} + C_y^E\big|_{i,j}\left(H_z\big|_{i,j}^{n-1/2} - H_z\big|_{i,j-1}^{n-1/2}\right) - \frac{\Delta t}{\varepsilon_{i,j}} J_x\big|_{i,j}^{n-1/2} \tag{1}$$

$$E_y\big|_{i,j}^{n} = E_y\big|_{i,j}^{n-1} - C_x^E\big|_{i,j}\left(H_z\big|_{i,j}^{n-1/2} - H_z\big|_{i-1,j}^{n-1/2}\right) - \frac{\Delta t}{\varepsilon_{i,j}} J_y\big|_{i,j}^{n-1/2} \tag{2}$$

$$H_z\big|_{i,j}^{n+1/2} = H_z\big|_{i,j}^{n-1/2} + C_y^H\big|_{i,j}\left(E_x\big|_{i,j+1}^{n} - E_x\big|_{i,j}^{n}\right) - C_x^H\big|_{i,j}\left(E_y\big|_{i+1,j}^{n} - E_y\big|_{i,j}^{n}\right) \tag{3}$$

where

$$C_x^E\big|_{i,j} = \frac{\Delta t}{\varepsilon_{i,j}\Delta\bar{x}_i},\quad C_y^E\big|_{i,j} = \frac{\Delta t}{\varepsilon_{i,j}\Delta\bar{y}_j},\quad C_x^H\big|_{i,j} = \frac{\Delta t}{\mu_{i,j}\Delta x_i},\quad C_y^H\big|_{i,j} = \frac{\Delta t}{\mu_{i,j}\Delta y_j}. \tag{4}$$

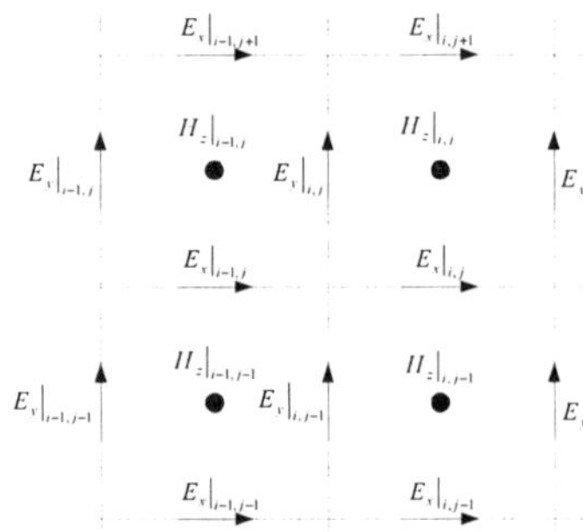

Figure 1. Position of the transient electric and magnetic field vector components on the TE_z plane.

For the above difference equations, *(i,j)* is not a real position but an array index of each field variable as shown in Fig. 1. Fig. 1 shows the position of the electric and magnetic field vector components over the two-dimensional cells. Δx_i and Δy_i are the lengths of the edge where the electric fields are located. $\Delta\bar{x}_i$ and $\Delta\bar{y}_j$ are the distances between the center nodes where the magnetic fields are located.

In this paper, we use the dispersive boundary condition (DBC) as an absorbing boundary condition [5]. Using the central difference scheme and the averaging technique, the 1st order DBC at $x=X$ (or $i=I$) can be written as

$$E_y\big|_{I,j}^{n} = E_y\big|_{I-1,j}^{n-1} - \gamma_1\left(E_y\big|_{I-1,j}^{n} - E_y\big|_{I,j}^{n-1}\right) \tag{5}$$

where

$$\gamma_1 = \frac{\Delta x - v_1 \Delta t}{\Delta x + v_1 \Delta t}. \tag{6}$$

Similarly, the difference relation at x=0 (or i=1) is

$$E_y\Big|_{1,j}^{n} = E_y\Big|_{2,j}^{n-1} - \gamma_1\left(E_y\Big|_{2,j}^{n} - E_y\Big|_{1,j}^{n-1}\right). \tag{7}$$

Since the conventional 2-D FDTD method is an explicit time domain technique, its time step is bounded by the CFL stability condition, which means that the maximum time step is bounded by the smallest length of the cell in a computational domain. This CFL condition may be too restrictive to solve problems with fine structures – thin material, slot, and via. So, in order to obtain an unconditionally stable solution regardless of the time step size, we propose a new FDTD algorithm in the following section, which uses weighted Laguerre polynomials as an entire domain temporal basis functions.

2.2. FDTD with Weighted Laguerre Polynomials

Consider the set of polynomials defined by

$$L_p(t) = \frac{e^t}{p!}\frac{d^p}{dt^p}\left(t^p e^{-t}\right), \text{ for } p \geq 0, \quad t \geq 0 \tag{8}$$

These are Laguerre polynomials of order p that are causal, which means that they exist for $t \geq 0$. These polynomials satisfy a recursive relationship given by

$$L_0(t) = 1, \quad L_1(t) = 1 - t \tag{9}$$

$$pL_p(t) = (2p - 1 - t)L_{p-1}(t) - (p-1)L_{p-2}(t), \text{ for } p \geq 2, \quad t \geq 0 \tag{10}$$

The Laguerre polynomials are orthogonal with respect to the weighting function e^{-t}, given by [6]

$$\int_0^\infty e^{-t} L_p(t) L_q(t)\, dt = \delta_{pq} \tag{11}$$

where δ_{pq} is a delta function for p=q and zero otherwise. Therefore, an orthonomal set of basis functions $\{\varphi_0, \varphi_1, \varphi_2, ...\}$ can be derived from Eq. (11) through the representation

$$\varphi_p(t,s) = e^{-s \cdot t/2} L_p(s \cdot t) \tag{12}$$

where $s > 0$ is a time scale factor. Note that these functions are absolutely convergent to zero as $t \to \infty$. Hence arbitrary functions spanned by these basis functions are also absolutely convergent to zero as $t \to \infty$.

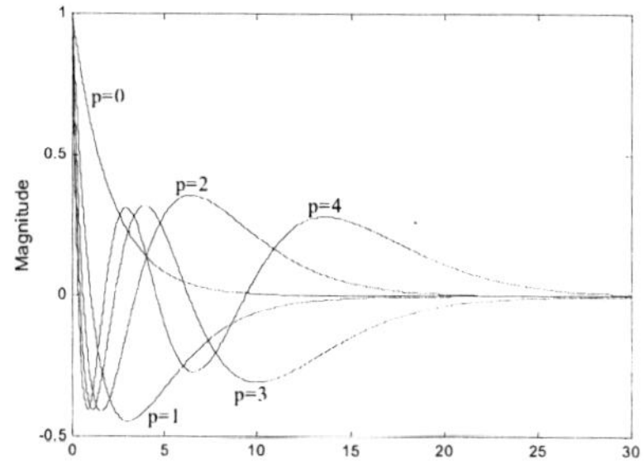

Figure 2. Weighted Laguerre Polynomials of different orders.

These basis functions are also orthogonal with respect to the scaled time variable $\bar{t}$ as

$$\int_0^\infty \varphi_u(\bar{t}) \cdot \varphi_v(\bar{t}) d\bar{t} = \delta_{uv} \tag{13}$$

where $\bar{t} = s \cdot t$ is the scaled time. Since the real time scale is quite small, in order to use the above basis functions properly, one should transform the real time scale using an appropriate scale factor s. These orthogonal functions can approximate causal electromagnetic responses quite well. Basis functions of order 0 to 4 are plotted in Fig. 2.

Using these basis functions, the electric and magnetic fields for the TE$_z$ case can be expanded as

$$E_x(\mathbf{r},t) = \sum_{p=0}^{\infty} E_x^p(\mathbf{r}) \varphi_p(\bar{t}), \quad E_y(\mathbf{r},t) = \sum_{p=0}^{\infty} E_y^p(\mathbf{r}) \varphi_p(\bar{t}), \quad H_z(\mathbf{r},t) = \sum_{p=0}^{\infty} H_z^p(\mathbf{r}) \varphi_p(\bar{t}) \tag{14}$$

Note that the time variable in the left hand side is different from the one on the right hand side. In the above equations, $\varphi_p(\bar{t})$ can be regarded as an entire domain temporal basis function. In [7], we can show that the 1st derivative of $U(\mathbf{r},t)$ with respect to time t is

$$\dot{U}(\mathbf{r},t) = s \sum_{p=0}^{\infty} \left(0.5 U_p(\mathbf{r}) + \sum_{k=0,\, p>0}^{p-1} U_k(\mathbf{r}) \right) \varphi_p(\bar{t}). \tag{15}$$

where $U(\mathbf{r},t)$ is a causal function. Inserting Eqs. (14)-(15) into the TE$_z$ equations, respectively, we have

$$s \sum_{p=0}^{\infty} \left(0.5 E_x^p(\mathbf{r}) + \sum_{k=0,\, p>0}^{p-1} E_x^k(\mathbf{r}) \right) \varphi_p(\bar{t}) = \frac{1}{\varepsilon(\mathbf{r})} \sum_{p=0}^{\infty} \frac{\partial}{\partial y} H_z^p(\mathbf{r}) \varphi_p(\bar{t}) - \frac{J_x(\mathbf{r},t)}{\varepsilon(\mathbf{r})} \tag{16}$$

$$s \sum_{p=0}^{\infty} \left(0.5 E_y^p(\mathbf{r}) + \sum_{k=0,\, p>0}^{p-1} E_y^k(\mathbf{r}) \right) \varphi_p(\bar{t}) = -\frac{1}{\varepsilon(\mathbf{r})} \sum_{p=0}^{\infty} \frac{\partial}{\partial x} H_z^p(\mathbf{r}) \varphi_p(\bar{t}) - \frac{J_y(\mathbf{r},t)}{\varepsilon(\mathbf{r})} \tag{17}$$

$$s \sum_{p=0}^{\infty} \left(0.5 H_z^p(\mathbf{r}) + \sum_{k=0,\, p>0}^{p-1} H_z^k(\mathbf{r}) \right) \varphi_p(\bar{t}) = \frac{1}{\mu(\mathbf{r})} \sum_{p=0}^{\infty} \left[\frac{\partial}{\partial y} E_x^p(\mathbf{r}) - \frac{\partial}{\partial x} E_y^p(\mathbf{r}) \right] \varphi_p(\bar{t}). \tag{18}$$

To eliminate the time dependent terms $\varphi_p(\bar{t})$, we introduce a temporal Galerkin testing procedure for Eqs. (16)-(18) by using the orthogonal property of the weighted Laguerre functions. We multiply both sides of Eqs. (16)-(18) by $\varphi_q(\bar{t})$ and integrate over $\bar{t} = [0, \infty)$. Then, we get

$$E_x^q\Big|_{i,j} = \bar{C}_y^E\Big|_{i,j} \left(H_z^q\Big|_{i,j} - H_z^q\Big|_{i,j-1} \right) - \frac{2}{s\varepsilon} J_x^q\Big|_{i,j} - 2 \sum_{k=0}^{q-1} E_x^k\Big|_{i,j} \tag{19}$$

$$E_y^q\Big|_{i,j} = -\bar{C}_x^E\Big|_{i,j} \left(H_z^q\Big|_{i,j} - H_z^q\Big|_{i-1,j} \right) - \frac{2}{s\varepsilon} J_y^q\Big|_{i,j} - 2 \sum_{k=0}^{q-1} E_y^k\Big|_{i,j} \tag{20}$$

$$H_z^q\Big|_{i,j} = -\bar{C}_x^H\Big|_{i,j} \left(E_y^q\Big|_{i+1,j} - E_y^q\Big|_{i,j} \right) + \bar{C}_y^H\Big|_{i,j} \left(E_x^q\Big|_{i,j+1} - E_x^q\Big|_{i,j} \right) - 2 \sum_{k=0}^{q-1} H_z^k\Big|_{i,j} \tag{21}$$

where

$$\bar{C}_x^E\Big|_{i,j} = \frac{2}{s\varepsilon_{i,j}\Delta x_i}, \quad \bar{C}_y^E\Big|_{i,j} = \frac{2}{s\varepsilon_{i,j}\Delta y_j}, \quad \bar{C}_x^H\Big|_{i,j} = \frac{2}{s\mu_{i,j}\Delta x_i}, \quad \bar{C}_y^H\Big|_{i,j} = \frac{2}{s\mu_{i,j}\Delta y_j} \tag{22}$$

$$J_x^q(\mathbf{r}) = \int_0^{T_f} J_x(\mathbf{r},t) \varphi_q(\bar{t}) d\bar{t}, \quad J_y^q(\mathbf{r}) = \int_0^{T_f} J_y(\mathbf{r},t) \varphi_q(\bar{t}) d\bar{t} \tag{23}$$

The upper limit of infinity can be replaced by a finite time interval T_f. This interval is chosen in such a way that the waveforms of interest have practically decayed to zero. This is the temporal testing procedure with respect to the basis function of order q, $\varphi_q(\bar{t})$.

Contrary to the conventional FDTD difference equations, Eqs. (1)-(3), Eqs. (19)-(21) have an implicit relation. Therefore, inserting Eq. (21) into Eq. (19) and Eq. (20), and rewriting Eq. (19) and Eq. (20), we have

$$\begin{aligned}
&-\bar{C}_y^H\Big|_{i,j-1} E_x^q\Big|_{i,j-1} + \left(1/\bar{C}_y^E\Big|_{i,j} + \bar{C}_y^H\Big|_{i,j} + \bar{C}_y^H\Big|_{i,j-1}\right) E_x^q\Big|_{i,j} - \bar{C}_y^H\Big|_{i,j-1} E_x^q\Big|_{i,j+1} \\
&\quad + \bar{C}_x^H\Big|_{i,j-1} E_y^q\Big|_{i-1,j} + \bar{C}_x^H\Big|_{i,j-1} E_y^q\Big|_{i-1,j-1} - \bar{C}_x^H\Big|_{i,j} E_y^q\Big|_{i,j} + \bar{C}_x^H\Big|_{i,j} E_y^q\Big|_{i+1,j} \\
&\quad = -\Delta y_j J_x^q\Big|_{i,j} - \frac{2}{\bar{C}_y^E\Big|_{i,j}} \sum_{k=0}^{q-1} E_x^k\Big|_{i,j} - 2\sum_{k=0}^{q-1}\left(H_z^k\Big|_{i,j} - H_z^k\Big|_{i,j-1}\right)
\end{aligned} \tag{24}$$

$$\begin{aligned}
&\bar{C}_y^H\Big|_{i,j} E_x^q\Big|_{i,j+1} - \bar{C}_y^H\Big|_{i-1,j} E_x^q\Big|_{i-1,j+1} - \bar{C}_y^H\Big|_{i,j} E_x^q\Big|_{i,j} + \bar{C}_y^H\Big|_{i,j} E_x^q\Big|_{i-1,j} \\
&\quad - \bar{C}_x^H\Big|_{i-1,j} E_y^q\Big|_{i-1,j} + \left(1/\bar{C}_x^E\Big|_{i,j} + \bar{C}_x^H\Big|_{i,j} + \bar{C}_x^H\Big|_{i-1,j}\right) E_y^q\Big|_{i,j} - \bar{C}_x^H\Big|_{i,j} E_y^q\Big|_{i+1,j} \\
&\quad = -\Delta x_i J_y^q\Big|_{i,j} - \frac{2}{\bar{C}_x^E\Big|_{i,j}} \sum_{k=0}^{q-1} E_y^k\Big|_{i,j} - 2\sum_{k=0}^{q-1}\left(H_z^k\Big|_{i,j} - H_z^k\Big|_{i-1,j}\right)
\end{aligned} \tag{25}$$

From Eq. (24) and Eq. (25), we can find that each electric field variable has the relationship with the adjacent six electric fields as shown in Fig. 1. Note that in Eq. (24) and Eq. (25), the magnetic fields are known because their orders are lower than those of the electric fields. Therefore, each row has seven non-zero terms. Rewriting Eq. (24) and Eq. (25) as a matrix equation, we have

$$[A]\{E^q\} = \{J^q\} + \{\beta^{q-1}\}, \; q = 0,1,2\ldots \tag{26}$$

In Eq. (26), $\{E^q\} = \{E_x^q, E_y^q\}^T$, and $\{J^q\} = \{J_x^q, J_y^q\}^T$ is the term due to an incident electric current expressed in Eq. (23). $\{\beta^{q-1}\}$ is the summation term from the order 0 to q–1. However, in the case of the boundary edges – perfect electric conductor (PEC) and absorbing boundary condition (ABC), we need a different procedure. First, in the case of the PEC boundary condition, all terms of its rows and columns except the diagonal term should be replaced with zero. Also the row term of the forcing vectors $\{J^q\}$ and $\{\beta^{q-1}\}$ should be zero.

In the case of the 1st order DBC, using Eq. (15), and applying the temporal testing procedure with $\varphi_q(\bar{t})$, we can eliminate the time derivative. Then, at $x=X$, we get

$$\frac{\partial}{\partial x} E_y^q(\mathbf{r}) + s v_1 \left(\frac{E_y^q(\mathbf{r})}{2} + \sum_{k=0}^{q-1} E_y^k(\mathbf{r}) \right) = 0 \tag{27}$$

Using the averaging technique and the central difference scheme at an auxiliary grid point (I-1/2, j), we can transform Eq. (27) into a difference equation.

$$\left(\frac{s}{4v_1} + \frac{1}{\Delta x}\right) E_y^q\Big|_{I,j} + \left(\frac{s}{4v_1} - \frac{1}{\Delta x}\right) E_y^q\Big|_{I-1,j} = -\frac{s}{2v_1} \sum_{k=0}^{q-1}\left(E_y^k\Big|_{I,j} + E_y^k\Big|_{I-1,j}\right) \tag{28}$$

Similarly, at $x=0$, we have the ABC difference equation as follows

$$\left(\frac{s}{4v_1}+\frac{1}{\Delta x}\right)E_y^q\Big|_{1,j}+\left(\frac{s}{4v_1}-\frac{1}{\Delta x}\right)E_y^q\Big|_{2,j}=-\frac{s}{2v_1}\sum_{k=0}^{q-1}\left(E_y^k\Big|_{2,j}+E_y^k\Big|_{1,j}\right) \tag{29}$$

Inserting Eq. (28), Eq. (29), and the PEC boundary condition into Eq. (26), we have a modified matrix equation as follows

$$\left[\tilde{A}\right]\left\{E^q\right\}=\left\{\tilde{J}^q\right\}+\left\{\tilde{\beta}^{q-1}\right\},\ q=0,1,2\ldots \tag{30}$$

Contrary to the conventional FDTD method, the proposed method has an implicit relationship between the field variables, which results in a sparse system matrix $\left[\tilde{A}\right]$. However, one can observe that the system matrix $\left[\tilde{A}\right]$ is independent of the order q of the temporal testing function, $\varphi_q(\bar{t})$. Note that the stability is no longer affected by the time step size. In Eq. (30), since $\left\{\tilde{J}^q\right\}$ is a known vector, therefore if we know the coefficients for the electric fields from order 0 to $q-1$, one can solve Eq. (30). Since at each recursion the proposed method uses the same system matrix first, one can perform the LU decomposition of $\left[\tilde{A}\right]$ only once at the beginning of the computation step. One can then solve Eq. (30) by using the back-substitution routine repeatedly. The magnetic fields can be obtained from Eq. (21).

In the next section, we explain upto what order of the temporal basis functions one should consider. This parameter is related to the accuracy of the solution regardless of the stability condition.

2.3. Choice of The Number of Temporal Basis Functions

It is assumed that the signal that we are interested in characterizing is practically bandlimited up to a frequency B. In addition, we are also interested in generating the same signal in the time domain up to the time duration T_f. Then, we represent the real time signal $P(t)$ by a Fourier series,

$$P(t)=\sum_u C_u e^{ju\omega_0 t} \tag{31}$$

where $\omega_0 = 2\pi/T_f$. Since $P(t)$ is real, $C_u^* = C_{-u}$ where * means conjugate transpose. If $P(t)$ is bandlimited to B Hertz, then the value of u can be fixed by

$$-B \le \frac{u}{T_f} \le B\,. \tag{32}$$

Therefore we have

$$P(t)=\sum_{u=-BT_f}^{BT_f} C_u e^{ju\omega_0 t}\,. \tag{33}$$

In Eq. (33), there are $2BT_f+1$ terms in the expansion of $P(t)$. Hence, the minimum number of temporal basis functions is

$$N_L = 2BT_f + 1\,. \tag{34}$$

In order to obtain an accurate solution, therefore, one should solve Eq. (30) recursively at least N_L times.

3. NUMERICAL EXAMPLE

In this section, two-dimensional parallel plate waveguide for the TE_z case is tested to validate our method. In this paper, we use the following sinusoidally modulated Gaussian pulse as an input electric current profile.

$$J(t)=\exp\left(-\left(\frac{t-T_c}{T_d}\right)^2\right)\sin(2\pi f_c(t-T_c)),\ T_d=\frac{1}{2f_c},\ T_c=3T_d\,. \tag{35}$$

In this paper, we use $f_c = 1$ GHz. And we choose $T_f = 11.71$ ns and $B = 5$ GHz. Inserting T_f and B into Eq. (34), we can evaluate the number of the weighted Laguerre polynomial functions, and we choose $N_L = 150$. And the time scale factor is $s = 6.07\times10^{10}$.

Consider a two-dimensional parallel plate waveguide with a thin PEC slot as shown in Fig. 3. In order to model this thin conductor plate of width 1.0 μm, we use the graded mesh technique.

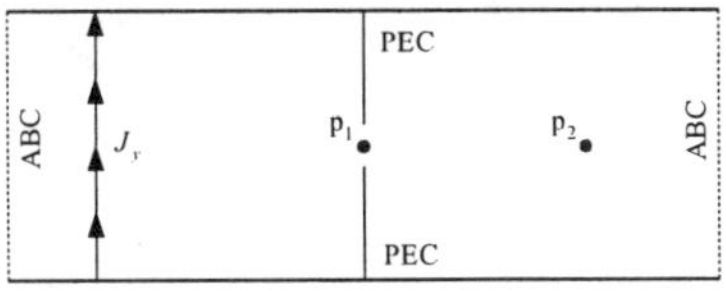

Figure 3. Two-dimensional parallel plate waveguide with the thin PEC slot of the thickness 1.0 μm and the distance 1.0 cm.

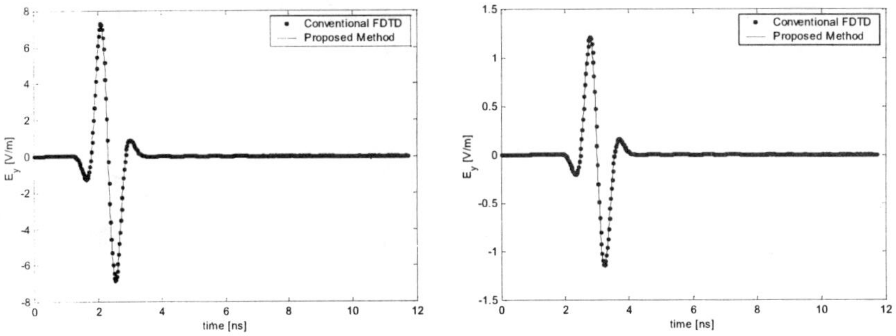

(a) at p_1 (b) at p_2

Figure 4. Transient electric fields of the y-component.

Table 1. Comparison of the computational efforts for the parallel plate waveguide with thin slot in a conductor.

	Δt	No. of iterations	CPU time
FDTD	3.33 fs	3570699	214 s
Our Method	10 ps	151	2.7 s

The PEC plate is divided into 2 cells, and the minimum cell size is 0.5 μm × 0.005 m. There are 120 and 12 subdivisions along the x- and y-directions, respectively. The CFL stability condition of this model is $\Delta t < 3.333$ fs. The time step size chosen for the proposed method is $\Delta t = 10.0$ ps to calculate the Laguerre coefficients of the excitation pulse, which is small enough to evaluate Eq. (23). Fig. 4 shows the y-components of the electric fields at p_1 and p_2, respectively. The agreement between the conventional FDTD method and the proposed method is very good. Table 1 represents the required computational resource and the computing time for the numerical simulations. Since the proposed method requires smaller number of iterations than those of the conventional FDTD method, the CPU time for the proposed method can be reduced to about 1.26% of the original FDTD method while maintaining the high accuracy. So by not using the CFL stability conditions and completely eliminating the time variable from the computations of the updating of the electric and the magnetic fields, the computation of the proposed method is less than the computation time of a conventional FDTD method by at least two orders of magnitude. All calculations in this paper have been performed on an Intel Pentium IV 2.2 GHz machine.

4. CONCLUSIONS

An unconditionally stable solution for the finite difference time domain algorithm has been proposed for the two-dimensional TE_z case with fine structures. To model a very thin slot in a conductor, graded discretization modeling has been employed to provide more flexibility. We utilize a marching-on-in-order method to solve the FDTD with weighted Laguerre polynomials. The advantages of using the weighted Laguerre polynomials as entire domain temporal basis and testing functions are: (1) It guarantees unconditional stability; (2) The solution is independent of the time discretization; (3) The temporal derivatives can be treated analytically, and (4) most importantly from a computational standpoint the spatial and the temporal variables can be separated resulting in an efficient and accurate solution. Transient fields obtained by the present method are unconditionally stable regardless of the time step size. Moreover the agreement between the results obtained using the proposed method and the conventional FDTD method is very good. Currently, it is being extended to 3D problems.

5. REFERENCES

1. A. Taflove and S. C. Hagness, *Computational Electrodynamics: the finite-difference time-domain method* (Artech House, Boston, 2000).
2. T. Namiki, A new FDTD Algorithm based on Alternating-Direction Implicit Method, *IEEE Trans. Microwave Theory Tech.*, **47**, 2003-2007, (1999).
3. T. Namiki, 3-D ADI-FDTD Method – Unconditionally Stable Time-Domain Algorithm for Solving Full Vector Maxwell's Equations, *IEEE Trans. Microwave Theory Tech.*, 48, 1743-1748, (2000).
4. T. Namiki and K. Ito, Numerical Simulation Using ADI-FDTD Method to Estimate Shielding Effectiveness of Thin Conductive Encloses, *IEEE Trans. Microwave Theory Tech.*, **49**, 1060-1066, (2001).
5. Z. Bi, K. Wu, C. Wu, and J. Litva, A Dispersive Boundary Condition for Microstrip Component Analysis Using The FD-TD Method, *IEEE Trans. Microwave Theory Tech.*, **40**, 774-777, (1992).
6. I. S. Gradshteyn and I. M. Ryzhik, *Table of Integrals, Series, and Products* (Academic Press, New York, 1980).
7. A. D. Poularikas, *The Transforms and Applications Handbook* (IEEE Press, 1996).
8. B. Z. Wang, Enhanced Thin-Slot Formalism for the FDTD Analysis of Thin-Slot Penetration, *IEEE Microwave Guided Wave Lett.*, **5**, 142-143,(1995).

TIME-DOMAIN INTEGRAL-EQUATION-BASED SOLVER FOR TRANSIENT AND BROADBAND PROBLEMS IN ELECTROMAGNETICS

Anatoliy O. Boryssenko* and Daniel H. Schaubert*

1. ABSTRACT

A time-domain (TD) numerical method has been developed based on the method-of moments (MoM) and marching-on-time (MoT) solution of the electrical-type integral equation with mixed potential formulation and free-space Green's functions. This time-domain integral-equation (TDIE) code is applicable for arbitrary 3-D antennas and arrays composed of conductors. It was developed particularly to simulate ultra-wideband (UWB), finite phased arrays of tapered slot antennas and some results from this study are included in the paper. The TDIE, which calculates wideband data in a single solution and utilizes basis functions defined only on the surfaces of conductors but not throughout the entire volume of the array, requires much less RAM and CPU time than frequency-domain and FDTD methods.

2. INTRODUCTION

Much recent research in computational electromagnetics has been devoted to solutions of integral equations through a MoM-MoT technique[1, 2, 3] called hereafter the TDIE. The main advantage of the TD MoM technique compared to FD is time savings for ultra-broadband simulations because the time-independent impedance matrix for the TDIE must be filled one time only. Also, the diagonal (explicit TDIE algorithm) or sparse diagonal-banded (implicit TDIE algorithm) portion of the impedance matrix must be inverted one time only. On the other hand, filling and inverting the dense frequency-dependent matrix must be repeated many times for the FD MoM solution. The electric-field integral equation (EFIE) is suitable for most geometries of conducting bodies while magnetic-field integral equation is applicable for conducting bodies with closed surfaces.

* Antenna Laboratory, University of Massachusetts, Amherst, MA, 01003, USA

Among EFIE formulations, we prefer a mixed-potential formalism through the equivalence principal with the free-space Green's functions[4]. In this way, the EFIE kernel is less singular, leading to easier evaluation of the required numerical integrals. The MoM-MoT formulation starts from the boundary condition on the PEC antenna surface, Eq. (1), that requires zero magnitude of the total tangential electrical field composed of incident *(i)* and scattered *(s)* fields, Eq. (2), and, in turn, the scattered field is expressed via the magnetic vector potential, Eq. (3), and electric scalar potential, Eq. (4),

$$\vec{E}^i + \vec{E}^s = 0 \tag{1}$$

$$\vec{E}^s(\vec{r},t) = -\frac{\partial}{\partial t}\vec{A}(\vec{r},t) - \nabla\Phi(\vec{r},t) \tag{2}$$

$$\vec{A}(\vec{r},t) = \mu_0 \cdot \int_S \vec{I}(\vec{r}',\tau)\big/(4\pi R)\cdot ds' \tag{3}$$

$$\Phi(\vec{r},t) = 1/\varepsilon_0 \cdot \int_S q(\vec{r}',\tau)\big/(4\pi R)\cdot ds' \tag{4}$$

The current, I, and the charge, q, densities are connected through the continuity equation:

$$\frac{\partial}{\partial t}q(\vec{r},t) + \nabla\cdot\vec{I}(\vec{r},t) = 0 \tag{5}$$

In Eqs. (2)-(5), $\tau = t - R/c$ is retarded time, $R = |\vec{r} - \vec{r}'|$ is the distance from the source to the observer; c is the free-space light velocity; ε_0, μ_0 are the fundamental electric and magnetic constants, respectively.

Eqs. (2)-(5) result in the EFIE, which is solved through the standard steps of MoM technique. First, the original problem is converted to a discrete one by employing a system of appropriate space-time basis functions[1, 4]. At this stage, potential integrals must be evaluated that can be done analytically[5] for the case of the free-space Green's function and subdomain basis functions in time and space[1]. Second, a testing procedure is applied to convert the discrete vector functional equation into a linear scalar system of equations to compute recursively the current through time stepping[1, 3, 6]

$$I^{(n+1)} = Z \cdot I^{(n)} + E^{(n)} \tag{6}$$

The above operator equation can be solved explicitly/implicitly in the given form, Eq. (6), where Z is a suitable matrix operator and E is a source vector[3]. Alternatively, a set of equations involving the potentials, the charge and the current can be treated[2, 7]. This method is utilized in the present study for reasons that are elucidated below. Specifically, the MoM testing must be done through a second integration over surface patches. The integration is applied to a dot product of a testing function and vector functions of the potentials. Usually, the same functions are used for weighting and testing to implement the Galerkin scheme. However, in practice, the integration is replaced usually by less accurate point matching[1, 3]. This simplification seems a main source of errors resulting in divergence of the computation and some noticeable anomalies[6, 8]. In this paper, we omit the well-known features of the TDIE formulation[1-3, 6, 7] and focus mainly on our specific contributions to support accurate and stable simulation of UWB phased array antennas of complex geometry with the TDIE code.

3. CHARGE IN THE TDIE INTERLEAVED COMPUTATIONAL SCHEME

Usually, the charge distribution is included implicitly into a discrete form of the EFIE through the continuity equation, Eq. (5). For the case of rooftop RWG basis functions, Figure 1a, the charge distribution is constant over triangular facets. However, the charge can be kept explicitly in the computational scheme to obtain some attractive benefits in computations[2, 7]. The TDIE algorithm, where the current is involved only, looks preferable due to lower computer storage/resources to simulate, for example, phased array antennas. Nevertheless, we used the algorithm similar to Refs. 2,7, which can be more stable than the single operator equation, Eq. (6), which is often used. From another perspective, if the charge is included directly into the TDIE computations, the updating recursive computational scheme can be treated like a digital filter where origins and intensity of digital noise can be traced, analyzed, and sometimes effectively suppressed.

For modeling the electric currents, we use the RWG spatial basis functions over triangular patches[4], Figure 1a. The same current mode models the delta-gap source, Figure 1b, necessary to feed antennas in transmit mode. We employ subdomain linear pyramid basis functions[9], Figure 1c, to express the charge surface distribution. The number of pyramid basis functions equals to the number of nodes in the grid partitioning the surface into a set of triangular patches. Each pyramid charge mode is maximum at the node and its total support is created by all triangles sharing the node. The pyramid basis functions do not adversely affect on computation time when carefully programmed because all manipulations with them are done at the pre-computing ("matrix-filling") stage to express the continuity equation (5) in accordance with topological relations, Figure 2b.

As seen in Figure 2b, the connection between the charge and the current, when the former is explicitly approximated by the pyramid functions, is more spatially spread, comparing to the case with triangle-wise constant charge, Figure 2a, over the same mesh. Thus, in fact, some spatial averaging is introduced automatically, which has been found to have a positive effect on stability[7]. Results in Refs. 2,7 and our observation reveal improvements in accuracy/stability of the TDIE when spatial interleaving for the charge and the current is introduced, Figure 2b. The same fruitful idea works in TD where sampling points in time for the charge and the current must be interleaved by a half of time step[2, 7, 10]. The both interleaving schemes in space and time are utilized in the TIDE cascaded code to maintain simplicity and accuracy and convergence in computations.

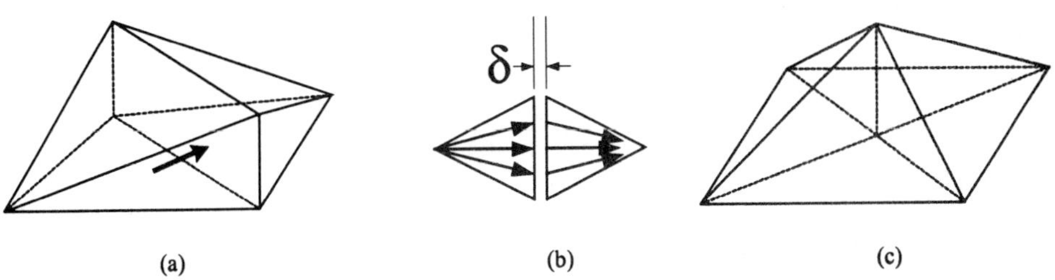

Figure 1. Basis functions: rooftop RWG mode for the current (a) including delta-gap source mode (b) and pyramid basis function (c) for the charge.

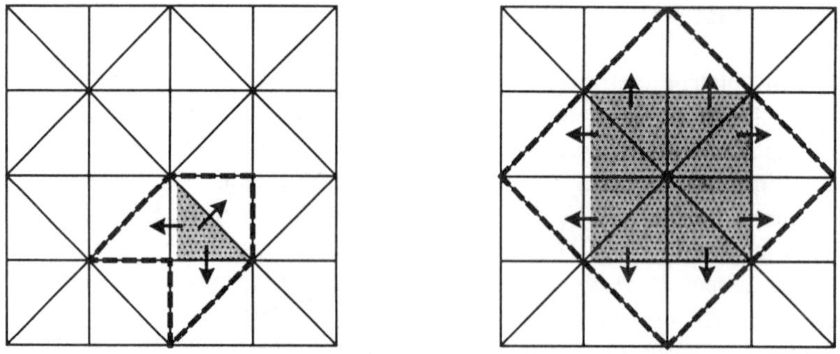

(a) (b)

Figure 2. Connection between current and charge for triangular-wise constant charge (a) and pyramid function (b). Shaded regions reflect support for the charge modes; bold broken lines specify the interaction regions between the current and the charge; arrows show RWG current modes involved in the continuity equation with specified charge modes.

4. SOME ISSUES ON CODE COMPUTATIONAL STABILITY IN LATE TIME

An important consideration for the TDIE solver, as any TD MoM-MoT scheme, is late-time instability caused by accumulated round-off errors [6, 7, 11-13]. To minimize this impact, besides more accurate modeling with explicit surface charge distribution discussed above, incorporating a stabilization procedure for the MoM-MoT solution is necessary. Known stabilizing procedures operate through averaging in time[6, 11, 14] and rarely in space [17]. For averaging in space or time, the underlying idea is "de-correlation" of accumulated computational noise and, of course, supplementary computer resources are required. In addition, irregularity of the mesh has an effect on accuracy/stability. Many real antennas, like the broadband tapered slot element[15] (TSA) shown in Figure 6, have complex shapes. The resulting mesh can be quite irregular in contrast to simple scattering scenarios with flat plates, etc[1-4]. Known approaches for stabilization do not work properly under such complexity/irregularity. A successful stabilization technique is essential to simulate TSA phased arrays where strong resonances lead to divergence and late-time instability.

In terms of signal processing, the instability means that spurious numerical poles appear in the right half of the complex frequency plane[3, 11, 16]. These poles are very close to the imaginary axis, but their small positive real part leads unavoidably to divergence and catastrophic instability in late time[11, 14]. Finite-impulse response (FIR) filtration[14] does not guarantee stability. The success of FIR filtration depends strongly on the geometry and mesh for specific problems[6, 14]. On the other hand, resistive loading introduced in the partial equivalent electrical circuit (PEEC) concept[13] seems helpful to suppress late-time instabilities in numerical EFIE solvers. We explored a similar idea but in a different mode. Our approach is relevant to the above formulation, where EM field quantities are utilized rather than circuit ones[13]. Doing so, we incorporate multiplication of the current computed at each time instant by an exponential coefficient with small damping factor multiplied by the current time.

In other words, we force all poles to leave the right half of the complex frequency plane. At the same time, the poles in the left half-plane will be shifted a little bit from their initial position to the left, Figure 3a. Finally, this heuristic stabilizing technique accomplished with low-order FIR filtration in the time-domain, spatial averaging through the charge mode with extended support, Figure 2b, and interleaving of charge/current in time/space is found to be very effective for all simulated cases for antennas and arrays of composite geometry where late-time instability is still a concern when existing stabilization techniques are used only.

The imposed numerical damping can be removed easy from a final transient solution by using the Prony extrapolation technique[17]. To this end, a correction needs to be done for all poles and zeros to compensate for the rotation in the complex frequency plane, Figure 3. However, this is not necessary in practice because the small damping that is needed to suppress amplifying growth of digital noise does not significantly influence the input impedance and pattern in the frequency domain. The damping acts like a windowing filtration, which precedes the Fourier processing. The effect of the damping on the magnitude of transient waveforms is nearly the same as the implicit MoM-MoT solvers cause compared with explicit solutions[3]. Furthermore, FIR filtration that is commonly used[6, 11, 14] has the nearly the same amplitude damping effect on transients waveforms as we estimated.

For practical computation with the TDIE, the magnitude of the damping factor σ in the term $exp(-\sigma t)$ can be estimated through some numerical experiments on the trial-and-error basis or, for example, by exploring an "impulse response" of the TDIE scheme[14] interpreting as a digital filter. Also, the pole-zero analysis may be utilized in the complex frequency plane[11, 16] for the same purpose. These approaches are based on an analysis over the whole mesh[16]. However, some our findings and results[13] corroborate that an analysis of local features in the mesh involving closely spaced, overlapping and non-planar modes only is sufficient.

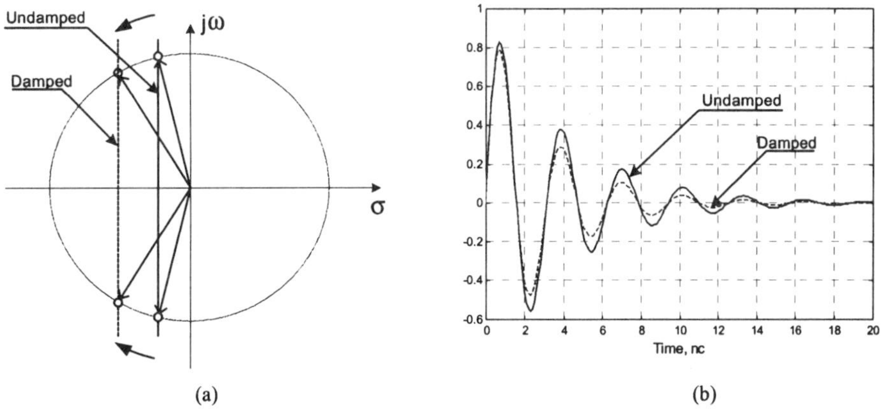

Figure 3. Damping effect on transient $exp(-\sigma t)\cdot sin(\omega t)$ in the complex frequency plane (a) and in time domain (b) by increasing the attenuation factor σ.

Indeed, these modes in the mesh are very sensitive to retardation with subsequent degradation in stability[13]. So, an analysis of such local parts of the mesh can be done to predict the quality of a simulation with the given mesh before running the complete TDIE simulation. The effect of the time domain basis function on stability and accuracy is a closely related to stability considerations. The simplest scheme, which is implemented in most references, is based on separation of temporal and spatial dependencies to avoid cumbersome integration[1, 2, 8, 14]. This is actually a point at issue because of the strong effect of the time-domain basis functions on late-time numerical stability[18].

5. RESULTS OF SIMULATIONS ON UWB ANTENNAS AND PHASED ARRAYS

Here we present results on transient behavior and equivalent FD properties for input impedance and VSWR of some UWB antennas and phased arrays. Radiation patterns and scattering properties, can be studied by the technique[19], but are not discussed here. All of the cases shown here and many others have been verified against published data and results of simulations with FD MoM and FDTD codes. Initially, the TDIE code was developed through combined Matlab-Fortran programming with a possibility to have a stand-alone version after compilation. After initial validation, key portions of the Matlab code have been converted to Fortran-90, resulting in an efficient code that was easy to develop. To reduce the computer RAM and CPU time, we have located an efficient representation of the sparse connection matrices.

Bow-tie antennas, Figure 4a, are frequently used for UWB applications. The waveforms in Figure 5a and b show stable results over long time intervals achieved by application of the Prony technique[17]. This technique uses a relatively short portion of transient waveform, Figure 5a, to compute a damped exponential representation of the current that is accurate to late time. Application of the Prony or other prediction method saves much CPU time for large problems and greatly reduces the chance of late-time divergence. The complex input impedance and VSWR for the bow-tie dipole are computed through the Fourier transform and shown in Figure 5c-d, correspondingly. A small low-frequency distortion in Figure 2c originates from spurious behavior of the RWG modes there[8].

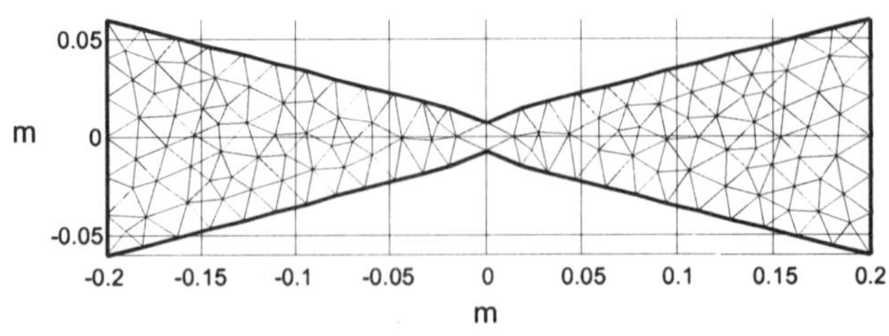

Figure 4. Geometry and mesh for bow-tie antenna with 242 triangles and 333 RWG modes.

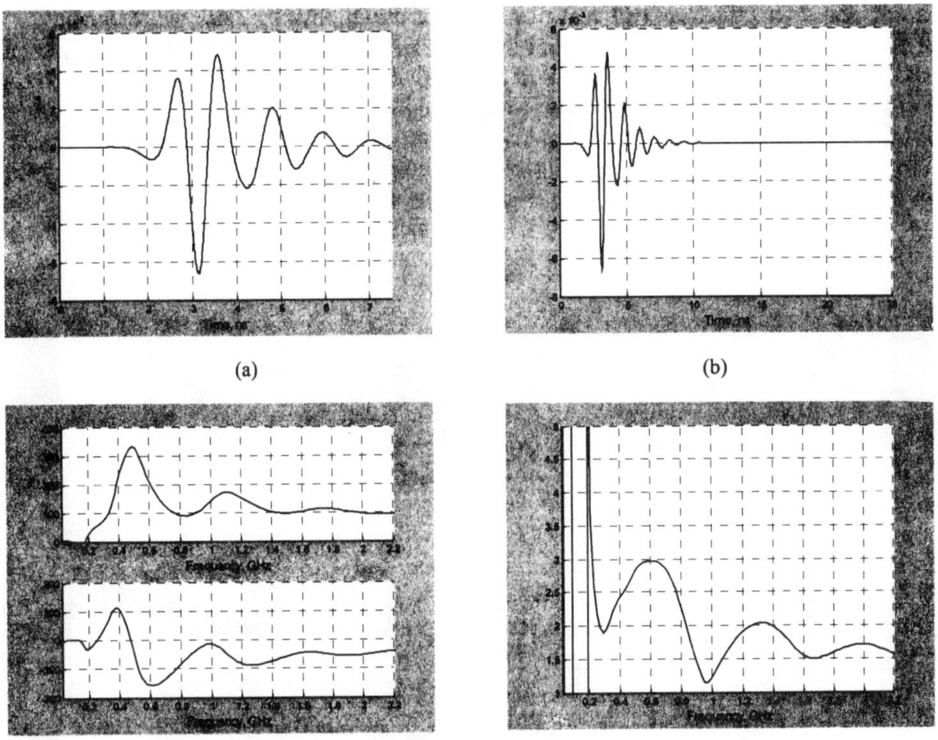

(a) (b)

(c) (d)

Figure 5. Transient simulated (a) and extrapolated (b) waveforms for antenna feed-point current. Antenna's FD features versus operating frequency: (c) input impedance and (d) VSWR for 50-Ohm reference.

Next simulated case is the TSA element[15] in Figure 6. The mesh has approximately 8 segments per wavelength at the highest frequency of interest. This coarse mesh was chosen to minimize the number of unknowns when a large array of such elements is simulated. The loss in accuracy is not great and the mesh is sufficient for most of the band of interest[19]. The broadband TSA element, as followed from Figure 7c-d, is the best-known candidate for UWB operation comparing to the matching data in Figure 5c-d for the bow-tie dipole of the almost same overall size. Only the first resonance is strong for the TSA while a couple of impressive resonances occurs for the bow-tie antenna within the same band.

An interesting effect in Figure 7a-b is associated with the Gaussian pulse that is commonly used as a source waveform. This pulse has a DC offset effect on the feed-point current in antenna, Figure 7b. The effect is observed for structures with a closed path for current in close proximity to the source region, like around the square resonator cavity of the TSA element, Figure 6. Although this effect due to low-frequency RWG anomalies[8] is annoying in the transient plot of the current, it does not alter antenna parameters such as input impedance except very near DC. To avoid this confusing time-domain behavior, we exploit the Rayleigh pulse, Figure 7a, which does not have DC spectral component.

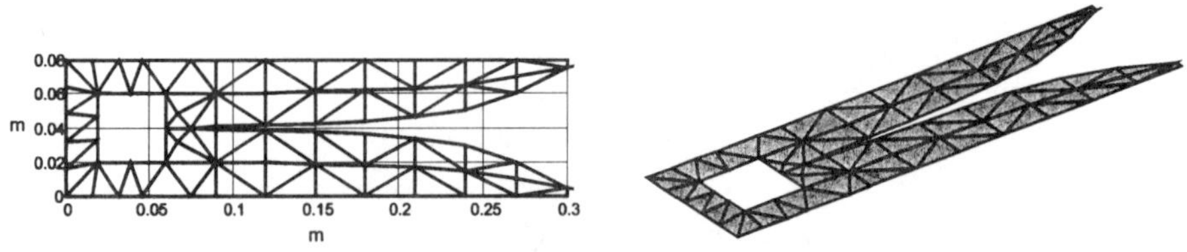

Figure 6. Geometry and triangular mesh for tapered slot antenna.

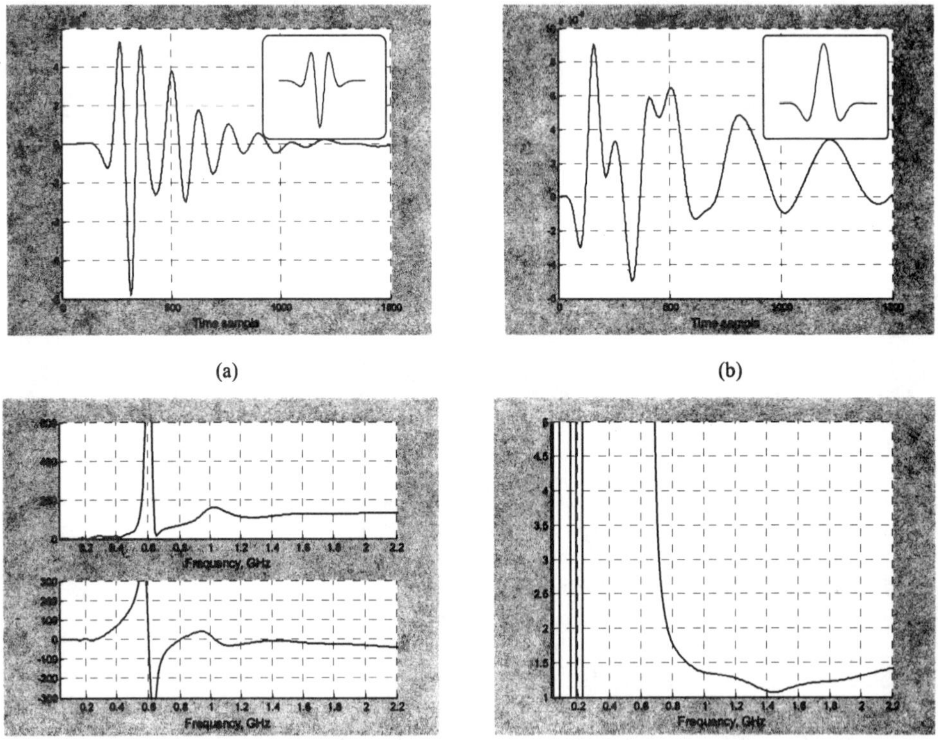

(a) (b)

(c) (d)

Figure 7. Transient current at the TSA feed-point for Rayleigh (a) and Gauss (b) pulses applied. TSA input impedance (c) and VSWR for 100-Ohm reference (d).

Next we show some results for an 8x8 phased array of TSA elements like that in Figure 6. The array is shown in Figure 8; the total number of unknowns is about 7,000. For maximum efficiency of this simulation with the TDIE, all existing symmetries and periodicity in the phased array were exploited. The impedance matrix resulting from Eq. (6), or its equivalent representation utilizing charge, has Toeplitz symmetries that are exploited to reduce storage and CPU time.

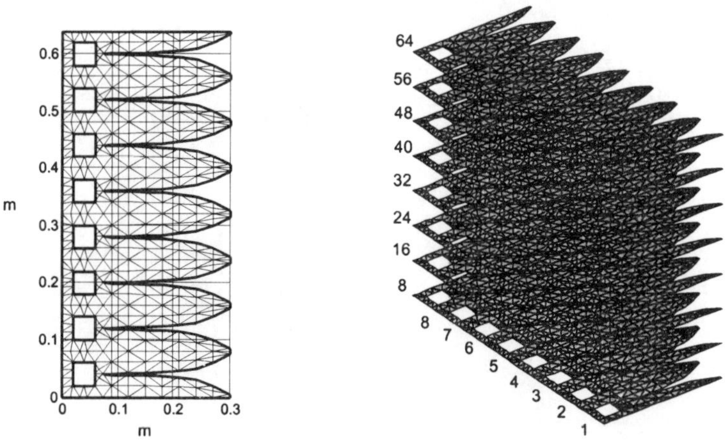

(a) (b)

Figure 8. Single sheet of array (a) comprised of eight tapered slot elements and meshed to 704 triangles and 916 RWG modes and geometry (b) of 8x8 phased array.

The sample of TDIE analysis results in Figure 9 reveal that UWB finite arrays are more susceptible to truncation effects than narrow band arrays[19]. There are usually a few frequency bands where some elements of the array are impacted strongly by a resonance, whereas others are well behaved. Narrow band arrays often do not encounter this problem because the resonance does not occur within their operating band. The exact physical origin and quantification of those resonances are still under investigation. It seems those resonances are associated with edge reflected waves. The resonances affect the central elements of the 3rd and 6th rows for broadside scanning direction and are slightly modified when the array scans to the 45-degree in the diagonal plane. Another interesting observation relates to the low-frequency limit of operation. The VSWR of the single element is very large below 0.7 GHz, Figure 7d. However, most elements of the array are well matched for frequencies as low as 0.4-0.5 GHz, a 30 per cent reduction in the minimum operating frequency or, equivalently, a 30 per cent increase in the ratio of maximum-to-minimum frequencies.

6. DISCUSSIONS AND DIRECTIONS OF FUTURE WORK

The TDIE code can analyze electrically large UWB problems like finite phased array antennas. For example, wideband impedance data for all elements of a 16x16 TSA phased array with about 25,000 unknowns is obtained in a few hours on a 1-GHz CPU computer platform taking 300-MB RAM. Currently we are working on a version of the code to handle phased arrays of size up to 100x100 TSA elements. For this purpose, different acceleration techniques are under testing. For instance, matrix multiplications that resemble convolutions are performed in the Fourier transform domain with considerable savings in computation time.

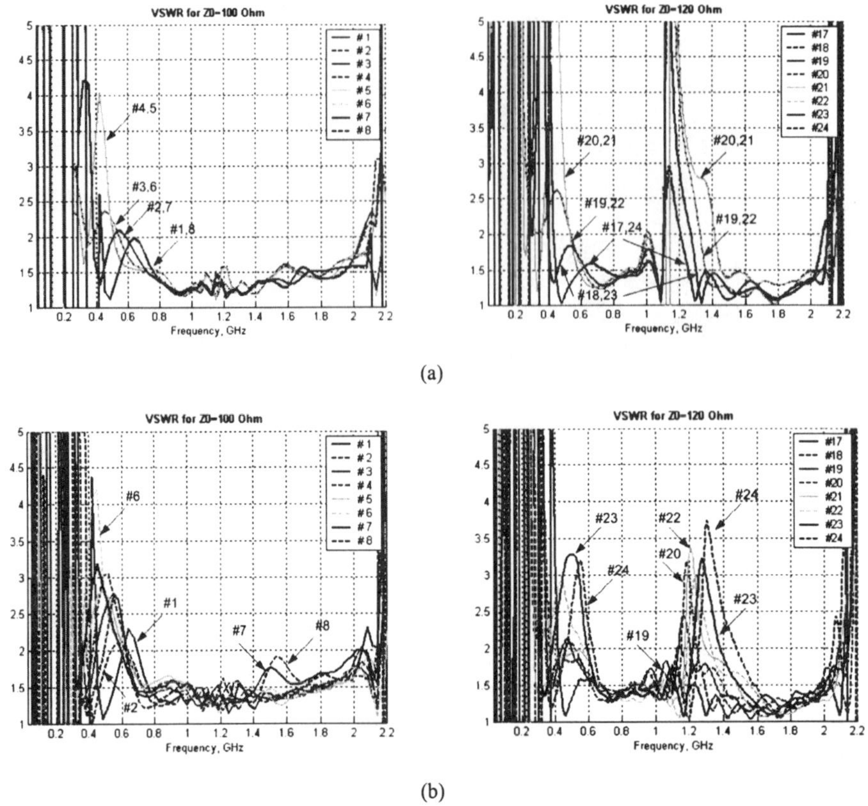

Figure 9. VSWR versus frequency in 10:1 band for broadside (a) and 45-degree diagonal plane scanning (b) for 1st and 3rd horizontal rows in 8x8 array.

Furthermore, we have found that some of our formulations for array analysis are amenable to either infinite array or finite array analysis through a combined TD-FD technique. As well, we are studying some asymptotic cases to simulate infinite and semi-infinite array configurations. Dielectric materials can be incorporated into the computational domain through the surface/volume equivalence principle, but at the price of additional computer resources. Also antennas built with real (lossy) conductors can be simulated through the TDIE but once more at the price of additional time because of the convolution nature of frequency dependent losses in TD representations.

In fact, the TSA phased arrays are extremely challenging for time marching computations because they are comprised of several parallel sheets of conductors, Figure 8, which support many high-Q resonances. These have a strong effect on late-time stability. We explored many stabilization approaches in the TDIE based on heuristic ideas to de-correlate computational noise and avoid its coherent accumulation to large amplitude in late time. These topics are still under study, but good results have been obtained as shown in this paper.

As a general rule, it seems that the best way to avoid late-time instability is accurate filling of the impedance matrix. For this we are exploiting an adaptive procedure whereby the filling for closely spaced elements is accomplished carefully with accurate MoM testing utilizing numerical integration and the filling for widely spaced elements is achieved with efficient point matching techniques.

7. ACKNOWLEDGEMENTS

This research was supported by ASTRON, Dwingeloo, the Netherlands, by DSO National Laboratories, Singapore, and by MURI grant DAAD19-01-1-0477.

8. REFERENCES

1. S. M. Rao, D. R. Witon, Transient Scattering Surfaces of Arbitrary Shape, *IEEE Trans. Ant. Propagat.* **39**(5), pp. 56-61 (1994).
2. B. P. Rynne, Time Domain Scattering from Arbitrary Surfaces Using the Electric Field Integral Equation, *J. Electromag. Waves Applicat.* **5**(1), pp. 93-112 (1991).
3. S. M. Rao, *Ed., Time Domain Electromagnetics (*Academic Press, London, 1999).
4. S. M. Rao, D. R. Wilton, A. W. Glisson, Electromagnetic Scattering by Surfaces of Arbitrary Shape, *IEEE Trans. Ant. Propagat.* **30**(5), pp. 409-418 (1982).
5. D. R. Wilton, S. M. Rao, *et al*, Potential Integrals for Uniform and Linear Source Distributions on Polygonal and Polyhedral Domains, *IEEE Trans. Ant. Propagat.* **32**(3), pp. 276-281 (1984).
6. S. Kashyap, M. Burton, A. Louie, A Stabilizing Scheme for the Explicit Time-Domain Integral-Equation Algorithm, *ACES J.* **13**(3), pp. 226-233 (1998).
7. P. J. Davies, On the Stability of Time-Marching Schemes for the General Surface Electric-Field Integral Equation, *IEEE Trans. Ant. Propagat.* **44**(11), pp. 1467-1473 (1996).
8. S. Kashyap, M. Burton, A Study of a New Moment-Method Algorithm that is Accurate to Very Low Frequencies, *ACES J.* **10**(3), pp. 58-68 (1995).
9. N. Ida, J. P. A. Bastos, *Electromagnetics and Calculation of Fields* (Springer, New York, 1997).
10. B. H. Jung, T. K. Sarkar, Time-Domain Electric-Field Integral Equation with Central Finite Difference, *Microwave Opt. Technol. Let.* **31**(6), pp. 429-435 (2001).
11. B. P. Rynne, P. D. Smith, Stability of Time Marching Algorithms of the Electric Field Integral Equation, *J. Electromag. Waves Applicat.* **4**(12), pp. 1181-205 (1990).
12. A. G. Tijhuis, Toward a Stable Marching-on-in-Time Method for Two-Dimensional Transient Electromagnetic Scattering Problems, *Radio Sciense* **19**(5), pp. 1311-1317 (1984).
13. J. E. Garrett, A. E. Ruehli, C. R. Paul, Accuracy and Stability Improvement of Integral Equation Model Using the Partial Element Equivalent Circuit (PEEC) Approach, *IEEE Trans. Ant. Propagat.* **46**(12), pp. 1824-1832 (1998).
14. A. Sadigh, E. Arvas, Treating the Instabilities in Marching-on-in-Time Method from a Different Perspective, *IEEE Trans. Ant. Propagat.* **41**(12), pp. 1695-1702 (1993).
15. J. Shin, D.H. Schaubert, A Parameter Study of Stripline-Fed Vivaldi Notch-Antenna Arrays, *IEEE Trans. Ant. Propagat.* **47**(5), pp. 879-886 (1999).
16. G. Manara, A. Monorchio, R. Reggiannini, A Space-Time Discretization Criterion for a Stable Time-Marching Solution of the EFIE, *IEEE Trans. Ant. Propagat.* **45**(3), pp. 527-532 (1997).
17. J. F. Callejon, A. R. Bretones, R. G. Martin, On Application of Parametric Models to the Transient Analysis of Resonant and Multiband Antennas, *IEEE Trans. Ant. Propagat.* **46**(3), pp. 312-317 (1998).
18. J.-L. Hu, C. H. Chan, Improved Temporal Basis Function for Time-Domain Electric Field Integral Equation method, *Electron. Let.* **35**(11), pp. 883-885 (1999).
19. A. O. Boryssenko, D. H. Schaubert, Predicted Performance of Small Arrays of Dielectric-Free Tapered Slot Antennas, in: *Antenna Application Symposium Digest* (Monticello, IL, 2001), pp. 250-279.
20. A. O. Boryssenko, D. H. Schaubert, Optimized Ultra-Wideband Radiation of Dipole Antennas with Triangle Driving Pulses, in: *This publication.*

EFFICIENT MIXED FINITE ELEMENTS FOR MAXWELL'S EQUATIONS IN TIME DOMAIN

Gary Cohen, Xavier Ferrières, Peter Monk, and Sébastien Pernet*

1. INTRODUCTION

Although efficient and easy to implement, FDTD meets its limits in the treatment of complex geometry. A first alternative to to this method is the finite volume methods which can be time consuming. For a long time, the use of high-order FETD for solving industrial problems seemed unrealistic since these methods introduced n-diagonal mass matrices which had to be inverted at each time-step. Even by using efficient solvers, the algorithms derived from this approach remained very expensive. A first step towards efficient FETD was realized by the use of Gauss-Lobatto quadrature rules to mass-lump the mass matrix[7,4]. This technique was then extended to the modeling of Maxwell's equations by an edge element coupled with a discontinuous element method (MFETD) which led to very efficient algorithms[5] both from the point of view of computational time and storage. In this paper, we extend this method to lossy Maxwell's equations in 2D and 3D. Then, we show the efficiency of this model in some numerical experiments for which we compare the efficiency of this method versus a FDTD code base on the Yee scheme[9].

2. CONSTRUCTION OF THE NUMERICAL MODEL

2.1. The Equations

We consider the following Maxwell's equations:

$$\underline{\underline{\varepsilon}}(x)\frac{\partial E}{\partial t}(x,t) - \nabla\times H(x,t) + \underline{\underline{\sigma}}(x)E(x,t) + J(x,t) = 0 \text{ in } \Omega, \tag{1a}$$

$$\underline{\underline{\mu}}(x)\frac{\partial H}{\partial t}(x,t) + \nabla\times E(x,t) = 0 \text{ in } \Omega, \tag{1b}$$

where $\Omega \subset \mathbb{R}^3$, J is a source term and the double underlined coefficients are matrices which can take

*Gary Cohen, INRIA, Domaine de Voluceau, Rocquencourt - BP 105, 78153 Le Chesnay cedex, France. Xavier Ferrières, Sébastien Pernet, ONERA, Département Electromagnétisme et Radar, unité CDE,2 av. Edouard Belin, BP 4025, 31055 Toulouse, France. Peter Monk, University of Delaware, Department of Mathematical Sciences, Newark DE 19716, USA.

into account anisotropic media. The support of J is a ball $B_R(x_0)$ of radius R and centered at x_0.

On the basis of these equations, we construct an appropriate PML approach[1] based on Chew and Weedon'approach to model unbounded domains[2,5].

The 2D model used in the numerical examples is the transverse-electric (TE) model.

2.2. The Functional Approximation

The electric field E is sought in a hexahedral or quadrilateral (even with curved edges) edge element space U_h^r and the magnetic field H, in a discontinuous element space V_h^r. With this notation, we can write the following approximation of (1a)-(1b), with a metallic condition on the boundary as follows:

$$\frac{\mathrm{d}}{\mathrm{d}t}\int_\Omega \underline{\underline{\varepsilon}} E_h \cdot \varphi_h \,\mathrm{d}x - \int_\Omega H_h \cdot \nabla \times \varphi_h \,\mathrm{d}x + \int_\Omega \underline{\underline{\sigma}} E_h \cdot \varphi_h \,\mathrm{d}x + \int_\Omega J \cdot \varphi_h \,\mathrm{d}x = 0, \ \forall \varphi_h \in U_h^r, \tag{2a}$$

$$\frac{\mathrm{d}}{\mathrm{d}t}\int_\Omega \underline{\underline{\mu}} H_h \cdot \psi_h \,\mathrm{d}x + \int_\Omega \nabla \times E_h \cdot \psi_h \,\mathrm{d}x = 0, \ \forall \psi_h \in V_h^r. \tag{2b}$$

Now, let us denote $\mathcal{Q}$ a mesh on Ω, $\hat{K} = [0,1]^3$, K_i an element of $\mathcal{Q}$ and F_i the mapping such that $F_i(\hat{K}) = K_i$. To F_i, we associate its Jacobian matrix DF_i and its Jacobian $J_i = \det DF_i$. Moreover, we denote

$$Q_r = \left\{ v(\hat{x}) = \sum_{\ell=0}^{r}\sum_{m=0}^{r}\sum_{n=0}^{r} a_{\ell,m,n}\hat{x}_1^{\ell}\hat{x}_2^{m}\hat{x}_3^{n}, \ a_{\ell,m,n} \in I\!R \right\}, \tag{3}$$

where $\hat{x} = (\hat{x}_1, \hat{x}_2, \hat{x}_3)^T \in \hat{K}$.

With this notation, the functions φ of U_h^r and ψ of V_h^r are such that $\forall K_i \in \mathcal{Q}$, $DF_i^* \varphi_{|K_i} \circ F_i \in [Q_r]^3$, $DF_i^* \psi_{|K_i} \circ F_i \in [Q_r]^3$, where DF_i^* is the transpose matrix of DF_i.

2.3. The Discretization

The functional frame of the approximation being defined, we now construct the discrete problem. For this purpose, we first define the 1D Gauss-Lobatto quadrature points $(\xi_i)_{i=0}^r$ on $[0,1]$. These points are such that $\xi_0 = 0$, $\xi_r = 1$ and $(\xi_i)_{i=1}^{r-1}$ are the $r-1$ roots of the derivative of the Legendre polynomial of order r.

On the other hand, we define the set of the $r+1$ Lagrange interpolation polynomials $(\theta_i)_{i=0}^r \subset Q_r$ on $[0,1]$ such that $\theta_i(\xi_j) = \delta_{ij}$, where δ_{ij} is the Kronecker symbol.

On the basis of these definitions, we can define the 3D Gauss-Lobatto quadrature points $\xi_{i,j,k} = (\xi_i, \xi_j, \xi_k)$, $0 \le i \le r$, $0 \le j \le r$, $0 \le k \le r$, and the $(r+1)^3$ Lagrange interpolation polynomials $\theta_{i,j,k} = \theta_i\theta_j\theta_k$ $(\in Q_r)$. Of course, these functions are such that $\theta_{i,j,k}(\xi_{\ell,m,n}) = \delta_{i\ell}\delta_{jm}\delta_{kn}$. Finally, we define the following set $\widehat{\mathcal{B}}$ of vector-valued functions: $\theta_{i,j,k}^{(1)} = (\theta_{i,j,k}, 0, 0)^T$, $\theta_{i,j,k}^{(2)} = (0, \theta_{i,j,k}, 0)^T$, $\theta_{i,j,k}^{(3)} = (0, 0, \theta_{i,j,k})^T$.

We can now construct the two sets of basis functions $\mathcal{B}_u$ of U_h^r and $\mathcal{B}_v$ of V_h^r as follows:

1. $\varphi_\ell \in \mathcal{B}_u \Longleftrightarrow \forall K_m \in \mathrm{supp}(\varphi_\ell), \exists \theta_{i,j,k}^{(n)} \in \widehat{\mathcal{B}}$ such that $\varphi_{\ell|K_m} \circ F_m = DF_m^{*-1}\theta_{i,j,k}^{(n)}$.

2. ψ_ℓ (such that $\mathrm{supp}(\psi_\ell) = K_m) \in \mathcal{B}_v \iff \exists \theta^{(n)}_{i,j,k} \in \widehat{\mathcal{B}}$ such that $\psi_\ell \circ F_m = DF_m^{*-1}\theta^{(n)}_{i,j,k}$.

Remark 1 *The support of a function of $\mathcal{B}_u$ can be composed of several hexahedra since U_h^r is a space of edge elements, i.e. whose tangential components are continuous. On the other hand, functions of $\mathcal{B}_v$ admit only one hexahedron as support since there is no continuity across the edges (Fig. 1).*

One of the main consequences of the above definitions is the following:

Let us decompose the stiffness term of (2a) on $\mathcal{B}_u$ and apply the change of variables defined by F_ℓ on each element K_ℓ. We get, for a basis function defined by $\varphi_{n|K_\ell} \circ F_\ell = DF_\ell^{*-1}\theta^{(m_0)}_{i_0,j_0,k_0}$:

$$\begin{aligned}
&\int_\Omega H_h \cdot \nabla \times \varphi_h \,\mathrm{d}x \\
&= \sum_{K_\ell \in Q} \sum_{i=0}^{r} \sum_{j=0}^{r} \sum_{k=0}^{r} \sum_{m=0}^{3} H^{(m)}_{\ell,i,j,k} \int_{\hat{K}} |J_\ell| DF_\ell^{*-1}\theta^{(m)}_{i,j,k} \cdot \frac{1}{J_\ell} DF_\ell \widehat{\nabla} \times \theta^{(m_0)}_{i_0,j_0,k_0} \,\mathrm{d}\hat{x} \\
&= \sum_{K_\ell \in Q} \sum_{i=0}^{r} \sum_{j=0}^{r} \sum_{k=0}^{r} \sum_{m=0}^{3} H^{(m)}_{\ell,i,j,k} \mathrm{sgn}(J_\ell) \int_{\hat{K}} \theta^{(m)}_{i,j,k} \cdot \widehat{\nabla} \times \theta^{(m_0)}_{i_0,j_0,k_0} \,\mathrm{d}\hat{x},
\end{aligned} \tag{4}$$

where $\widehat{\nabla} = (\partial/\partial\hat{x}_1, \partial/\partial\hat{x}_2, \partial/\partial\hat{x}_3)^T$.

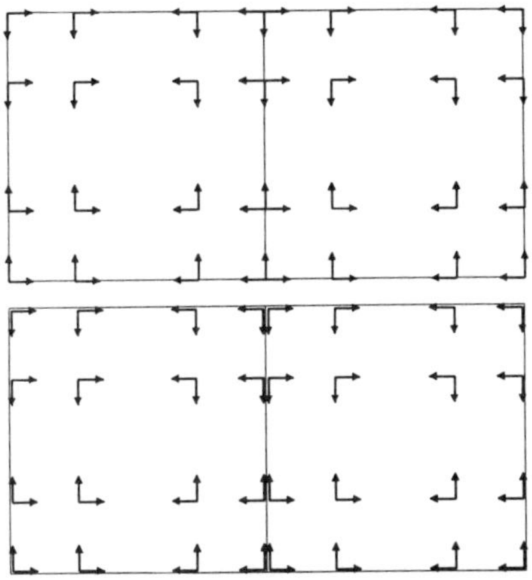

Figure 1. Degrees of freedom on two faces of regular hexahedra for U_h^r (left) and V_h^r (right), when $r = 3$ (which is the actual order of the elements used for the 2D numerical experiments).

Relation (4) shows that the knowledge of the stiffness terms on $\hat{K}$ is sufficient to compute the stiffness terms all over the mesh. In other words, we just have to store the elementary stiffness terms on $\hat{K}$ (and one sign of J_ℓ per element) to obtain the whole stiffness matrix, which induces a huge gain of storage! Moreover, since the other stiffness matrix is the transpose of this one, we avoid both storages.

After computing the other integrals in the same way and by applying the Gauss-Lobatto quadrature rule on each element, we get the following discrete problem:

$$B_\varepsilon \frac{\mathrm{d}\mathbf{E}}{\mathrm{d}t} + R_h\mathbf{H} + B_\sigma\mathbf{E} + \mathbf{J} = 0, \tag{5a}$$

$$B_\mu \frac{\mathrm{d}\mathbf{H}}{\mathrm{d}t} + R_h^*\mathbf{E} = 0, \tag{5b}$$

where B_ε and B_σ are $N \times N$ block-diagonal matrices such that N is the number of degrees of freedom in E around an interpolation point (generally, $3 \leq N \leq 6$) and B_μ is a 3×3 block-diagonal matrix. The stiffness matrices R_h and R_h^* are very sparse because of the orthogonal character of the basis functions on $\hat{K}$. The time discretization is made by a leapfrog scheme with a time-step small enough to keep the higher-order accuracy ($r = 3$ in our case) in space. The block-diagonal character of the mass matrices enables us to store their inverses, so that the resolution requires no matrix inversion at each time-step. All these features of the mass and the stiffness matrices lead to a very fast algorithm of resolution.

On the other hand, all the geometrical and physical information is contained in the mass-matrices whose block-diagonal character implies a very cheap storage (which is, as said above, the only one we need).

In the 2D case, everything can be written in the same way. In this case, the magnetic field which is scalar and provides a diagonal mass-matrix. A more detailed description of the method, based on Nédélec's work[8], can be found in Cohen's book[3].

3. NUMERICAL EXPERIMENTS

In the following numerical examples, we illustrate the different features of the MFETD approximation versus a FDTD code based on the Yee scheme. This comparison is made in the three following ways:

- Evaluation of the "higher-order effect" by comparing the solutions obtained for a square metallic cavity.
- Evaluation of the "geometrical effect" in an experiment of scattering by a metallic disk.
- Evaluation of the gain realized by using a mesh adapted to the wavelength in an experiment of damping by a forest.

3.1. The Metallic Cavity

In this 2D experiment, we consider a square of 10m × 10m with a metallic condition on its boundary. At the center of this square, we set a source which is a second derivative of a Gaussian function in time, distributed by a Gaussian function in space on a disk of radius 1. The center frequency of the pulse is at 0.3 GHz, which provides a wavelength λ equal to 1. In order to evaluate the accuracy of the solution, we draw three seismograms at the points (5.5,5.5) (near the center), (9,9) (at a corner) and (1,5) (on the boundary).

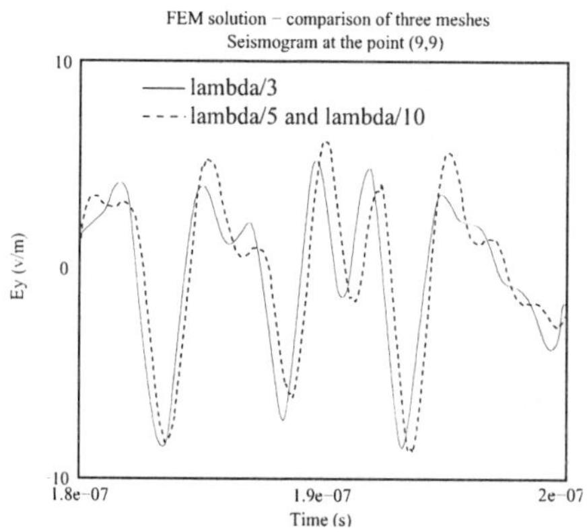

Figure 2. Seismograms at the point (9,9) for different MFETD approximations.

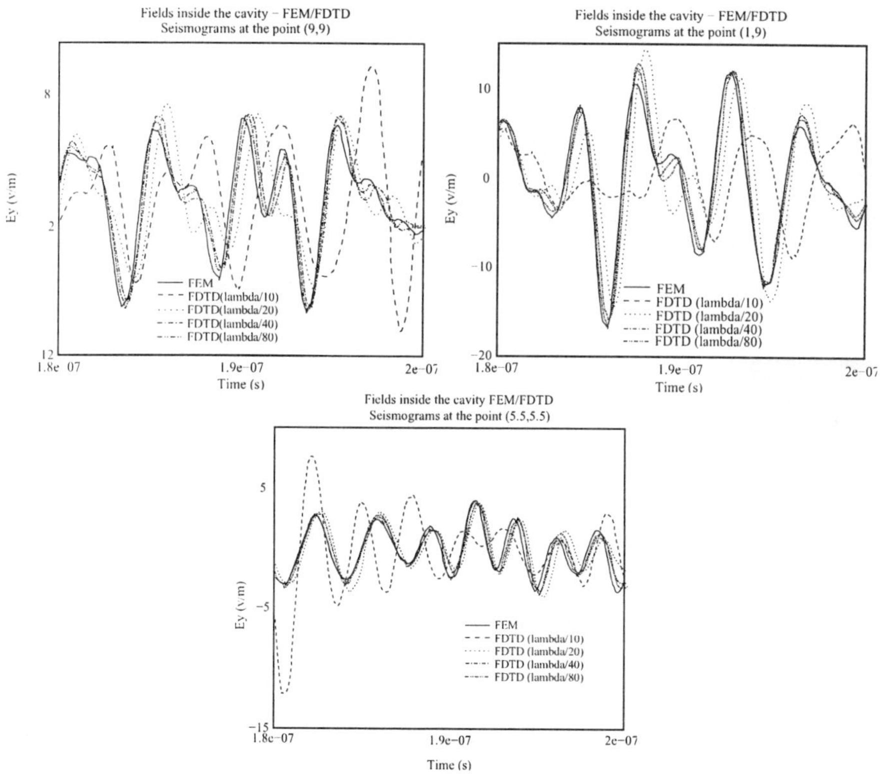

Figure 3. Seismograms at the points (9,9), (1,9) and (5.5,5.5) for different FDTD approximations compared to an accurate MFETD approximation.

In a first step, we compare the solutions given by the MFETD code (meshed by a regular orthogonal grid) with different space-steps equal to $\lambda/3$, $\lambda/5$ and $\lambda/10$. Fig. 2 shows that the optimal accuracy is already reached for $\lambda/5$.

In a second step, we compare the seismograms obtained by the MFETD approximation with a space-step equal to $\lambda/5$ with the FDTD code runned with different space steps ($\lambda/10$, $\lambda/20$, $\lambda/40$ and $\lambda/80$). In Fig 3, one can see the convergence of the FDTD approximations towards the MFETD one. In terms of CPU time, the MFETD approximation with $\lambda/5$ costs 3mn 38s whereas the FDTD approximation with $\lambda/40$ costs 5mn 9s and 50mn with $\lambda/80$. These results show the advantage of using higher-order approximations even in simple cases.

Remark 2 *The space-step for MFETD is the size of the element whereas the space-step for FDTD is the distance between 2 points. The number of points involved in one direction in the MFETD approximation is obtained by multiplying the number of elements by 3, so that, the average space-step between 2 points for MFETD (comparable to the space-step in FDTD) is given by dividing the size of the element by 3. For instance, $\lambda/5$ in MFETD corresponds to $\lambda/15$ in FDTD.*

3.2. Scattering by a Metallic Disk

This experiment is a classical one. We illuminate a perfectly metallic circle by a plane wave as shown in Fig. 4.

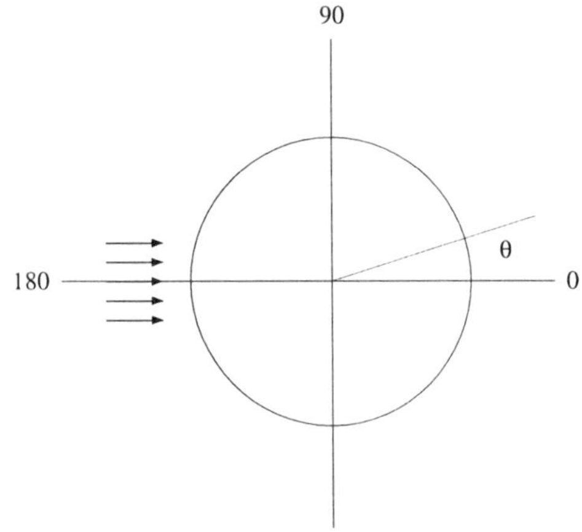

Figure 4. The illuminated metallic disk.

For this configuration, we compute the value of the RCS for different far points given around the circle for a angle range from 0° to 180°. The value of the RCS at a point r is given by $2\pi r|H_s(r)|/|H_{inc}(r)$, where $H_s(r)$ and $H_{inc}(r)$ are respectively the scattered and incident fields at the point r. The radius a of the circle, in this example, is taken such that $ka = 10$, where k is the wavenumber.

In Fig. 5, we show the RCS obtained by the MFETD approximation with average space-steps equal to $\lambda/6$ and $\lambda/12$. One can easily see that the two curves are almost the same, so that we can assume that an approximation with $\lambda/6$ provides a quasi-exact solution. The cost of this approximation is 3mn 15.

In a second step, we compute the RCS by using the FDTD code with different space-steps ($\lambda/15$, $\lambda/30$, $\lambda/50$). Fig. 5 shows that the smallest space-step is of course the closest to the MFETD solution but is not yet identical. In terms of CPU time, the time for $\lambda/30$ is comparable (but for a much less accurate result) to this of the MFETD with $\lambda/6$ (3mn 15) whereas the FDTD approximation with $\lambda/50$ costs 8mn 10.

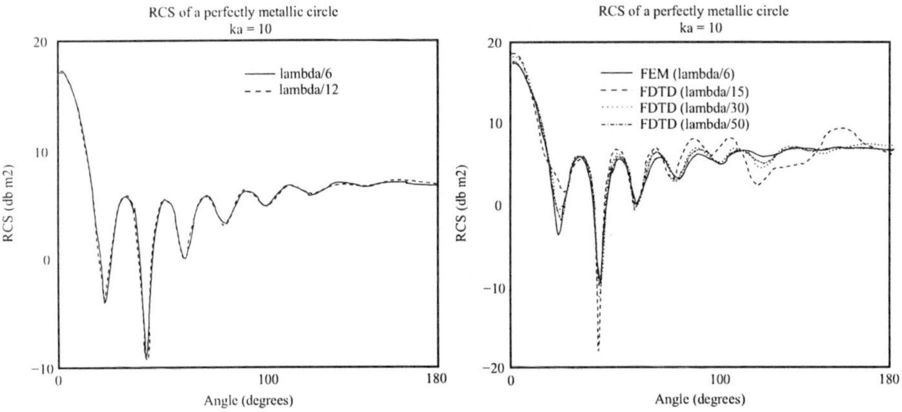

Figure 5. RCS for the MFETD approximation with two different space-steps (left) and for a FDTD approximation with different space-steps compared to a quasi-exact solution (right)

3.3. Damping by a Forest

In this experiment, we propagate a quasi-punctual pulse of frequency 100MHz, centered at the point (10,10), in a 300m $\times$ 45m domain composed of the soil at its bottom and the air and a forest as represented in Fig. 6. The dielectric characteristic for the soil and the forest are respectively ($\varepsilon_r = 5$, $\sigma = 1.e-3$) and ($\varepsilon_r = 1.03$, $\sigma = 5.e-5$). In this experiment, as in the previous one, the unbounded domain is taken into account by PML.

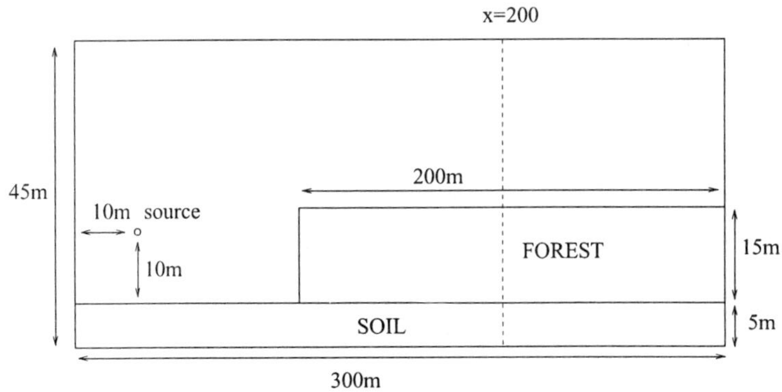

Figure 6. Configuration of the forest.

The purpose of this large scale experiment is to evaluate the amplification of a propagative wave inside the domain along a line located at $x = 200$m. The reference solution was computed by a parabolic equation model[6], which is a very accurate model. The solution was computed by the MFETD model by using a mesh adapted to the wavelength, i.e. using cells proportional to the wave-length in each medium connected by non-regular cells (Fig. 7), which induces a substantial gain of storage and CPU time. The whole mesh (with the PML) contains 44 000 elements and about 1 000 000 degrees of freedom in E.

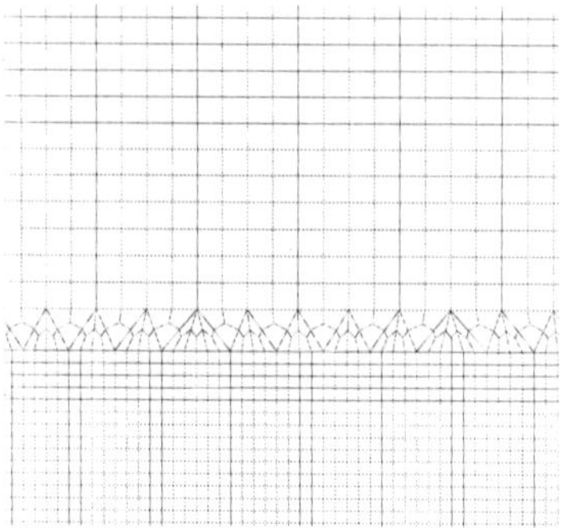

Figure 7. Detail of the mesh of the forest. Junction between the soil and the air.

The curve of amplification computed by using the MFETD model with the mesh defined above is close enough to the reference one. The CPU time for this experiment was 5677s. The FDTD code used with 3000×450 points ($\lambda/10$ versus the wavelength of the soil) leads to a bad approximation of the curve for a CPU time of 8046s. A 6000×900 point approximation gives an accurate enough definition of the curve but for cost is multiplied by about 11 versus the MFETD approach! All the curves are given in Fig. 8. Snapshots of the experiment can be seen in Fig. 9.

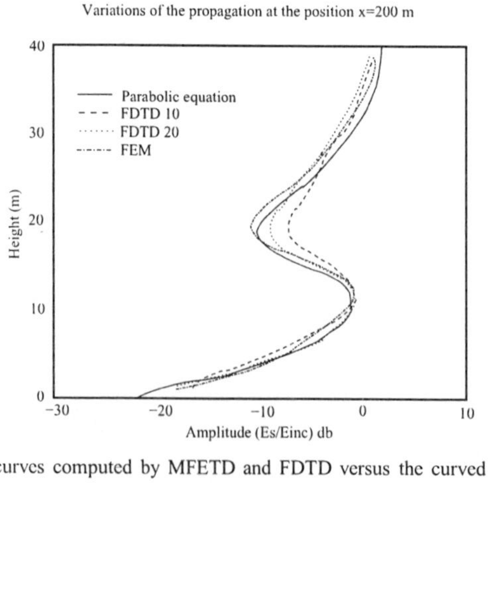

Figure 8. The different curves computed by MFETD and FDTD versus the curved computing by using the parabolic equation.

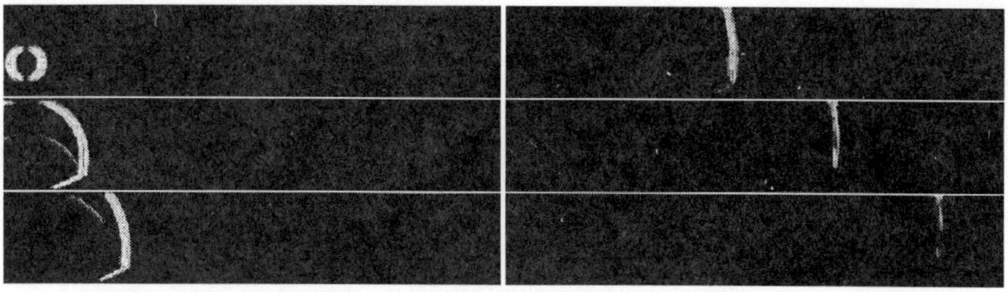

Figure 9. Snapshots of the experiment for the forest. The PML is included in the snapshots.

4. CONCLUSION

MFETD seems to be an efficient alternative to FDTD. It enables to take into account complex geometry and physics for a low cost both from storage and CPU time point of view. Its efficiency is particularly obvious in high contrast experiments. The 3D case is being studied.

REFERENCES

1. J.-P. Bérenger, A perfectly matched layer for the absorption of electromagnetic waves, *J. Comput. Phys.* **114**(2), pp. 185–200 (1994).
2. W. C. Chew and W. H. Weedon, A 3D perfectly matched medium from modified Maxwell equations with stretched coordinates, *IEEE Microwave and Optic. Tech. Letters*, **7**(13), pp. 599–604 (1999).
3. G. Cohen, *Higher-Order Numerical Methods for Transient Wave Equations*, (Springer Verlag, Heidelberg, 2002).
4. G. Cohen, P. Joly and N. Tordjman, Higher-order finite elements with mass lumping for the 1-D wave equation, *Finite Elements in Analysis and Design* **17**(3-4), pp. 329–336 (1994).
5. G. Cohen and P. Monk, Mur-Nédélec finite element schemes for Maxwell's equations, *Comp. Meth. in Appl. Mech. Eng.* **169**(3-4), pp. 197–217, (1999).
6. V. Fabbro, P.F Combes and N. Douchin, Three dimensional backscattering by a target above the sea surface, *Electromagnetics* **28**(6), pp. 451–466 (2001)
7. J.-P. Hennart, E. Sainz and M. Villegas, On the efficient use of the finite element method in static neutron diffusion calculations, *Computational Methods in Nuclear Engineering* **1**, pp. 3–87 (1979).
8. J.-C. Nédélec, A new family of mixed finite elements in $\mathbb{R}^3$, *Numer. Math.* **50**(1), pp. 57–81 (1986).
9. K. Yee, Numerical solutions of initial boundary value problems involving Maxwell's equations in isotropic media, *IEEE Trans. on Antennas and Propagation* **AP-16**, pp. 302–307 (1966).

SOLUTION AND STABILITY ANALYSIS OF A NEW INTEGRAL EQUATION FOR THE TRANSIENT SCATTERING BY A FLAT, RECTANGULAR CONDUCTING PLATE

Jos G.A. van Riswick and Anton G. Tijhuis*

1. INTRODUCTION

The electric-field integral equation (EFIE) is a common tool for determining scattered or radiating fields emanating from a perfectly conducting, open surface. In this type of approach the induced surface current and/or charge density are solved from the integral equations, and, subsequently, the scattered electromagnetic field is determined at any given point by evaluating the integrals in a conventional integral representation. Advantages of this approach are that the radiation condition is accounted for inherently, and that only two-dimensional surface currents and/or charges need to be computed over the area of the scattering surface, instead of three- dimensional electromagnetic fields over a volume surrounding that surface.

When the EFIE is solved directly in the time domain, instabilities are often observed in the computed results. Numerical experiments for a number of two-dimensional transient-scattering problems [1] as well as for perfectly conducting surfaces indicate that the occurrence of such unstable solutions is enhanced by the presence of space derivatives in these equations. In particular, the manner in which these derivatives are approximated in the space discretization has a considerable influence on the stability of the computational solutions. To investigate this observation in more detail, we consider the "canonical" problem of a rectangular, flat plate excited by a pulsed incident field. The role of the space derivatives is isolated from the complete computation by reformulating the procedure in three steps. First, we reduce the original EFIE to a Hallén-type integral equation, which consists of two coupled, scalar integral equations for the transverse currents $J_x(x,y,t)$ and $J_y(x,y,t)$. This step was inspired by the success of using Hallén's integral equation for the total current $I(z,t)$ along a straight conducting thin wire in time-domain computations

*Faculty of Electrical Engineering Eindhoven University of Technology P.O. Box 513, 5600 MB Eindhoven, the Netherlands.

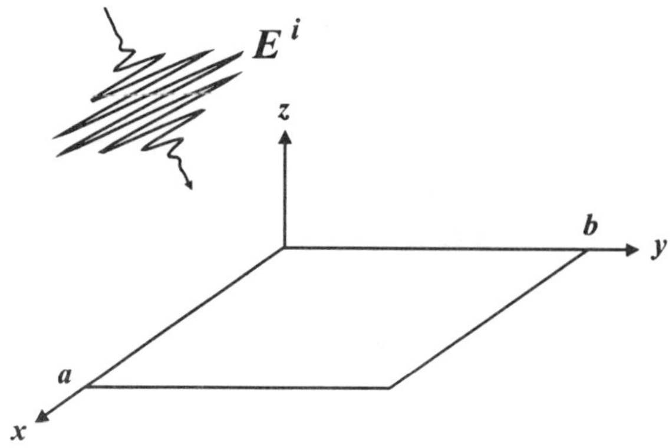

Figure 1: Pulsed plane wave incident on a flat rectangular plate.

[2]. Marching-on-in-time schemes based on our new integral equation are still unstable, but can be stabilized with the aid of time averaging.

Therefore, the second modification in our approach is to reformulate the integral equations in terms of two coupled differential equations that can be solved by a finite-difference time-domain scheme, and two uncoupled scalar integral equations, where the current densities $J_x(x, y, t)$ and $J_y(x, y, t)$ are resolved from the corresponding components of a vector potential $\boldsymbol{A}(x, y, t)$. Here, we found our inspiration in a similar treatment of the EFIE in [3]. A straightforward discretization of the differential equations results in a scheme that satisfies the well-known Courant stability criterion for two-dimensional wave propagation. The scalar integral equation is then solved by marching on in time, as described in [3]. For the case where the flat plate is infinite, both parts of the computation can be validated independently, by comparing numerical results with closed-form expressions for the field excited by a horizontally or vertically polarized, pulsed electric point dipole located above the plate. When the plate has finite dimensions, coupling occurs via the boundary conditions at its edges. This means that the two parts of the computation must be combined, and that the stability of the entire scheme depends on the implementation of these boundary conditions.

2. FORMULATION OF THE PROBLEM

A pulsed plane wave is incident on a two-dimensional, perfectly conducting rectangular plate as shown in Fig. 1. The Cartesian coordinate system is chosen such that the plate is located at $0 < x < a$, $0 < y < b$, and $z = 0$. The surrounding homogeneous, linearly and instantaneously reacting, isotropic medium has permittivity ε and permeability μ. We start from the standard form of the electric-field integral equation

$$\left(\nabla_T \nabla_T \cdot -\frac{1}{c^2}\partial_t^2\right) \boldsymbol{A}(\boldsymbol{r}_T, t) = -\varepsilon \partial_t \boldsymbol{E}_T^i(\boldsymbol{r}_T, t), \tag{1}$$

which holds for $\boldsymbol{r}_T$ on the plate. In (1), the subscript $_T$ stands for transverse, and the vector potential $\boldsymbol{A}(\boldsymbol{r}_T, t)$ is related to the unknown surface current density $\boldsymbol{J}_S(\boldsymbol{r}_T, t)$ in the plane $z = 0$ according to

$$\boldsymbol{A}(\boldsymbol{r}_T, t) = \int_0^a dx' \int_0^b dy' \, \frac{\boldsymbol{J}_S(\boldsymbol{r}'_T, t - R/c)}{4\pi R}, \tag{2}$$

with $R = |\boldsymbol{r}_T - \boldsymbol{r}'_T|$ and $c = 1/\sqrt{\varepsilon\mu}$. The aim of the computation is to determine $\boldsymbol{J}(\boldsymbol{r}_T, t)$ for $0 < x < a$ and $0 < y < b$ for a given behavior of the transverse incident field $\boldsymbol{E}^i_T(\boldsymbol{r}_T, t)$. Once this surface current is known, it can be treated as a "secondary" source that generates the electromagnetic field scattered by the plate.

3. HALLÉN-TYPE INTEGRAL EQUATION

As mentioned in the introduction, numerical differentations with respect to space and time are often the source of instabilities in time-domain computations. Therefore, they should be avoided as much as possible. To this end, we break up (1) into its x- and y-components. Since we restrict ourselves to the plane $z = 0$, we may replace $\boldsymbol{r}_T$ by the corresponding transverse coordinates x and y. This reduces (1) to a pair of coupled scalar electric-field integral equations:

$$\left(\partial_x^2 - \frac{1}{c^2}\partial_t^2\right) A_x(x, y, t) + \partial_x \partial_y A_y(x, y, t) = -\varepsilon \partial_t E_x^{\text{inc}}(x, y, t), \tag{3}$$

$$\left(\partial_y^2 - \frac{1}{c^2}\partial_t^2\right) A_y(x, y, t) + \partial_y \partial_x A_x(x, y, t) = -\varepsilon \partial_t E_y^{\text{inc}}(x, y, t). \tag{4}$$

In analogy with the case of a straight thin-wire segment, we identify the differential operators in the first terms on the left-hand sides as one-dimensional wave operators, that can be evaluated with the aid of a one-dimensional Green's function technique. This results in

$$\begin{aligned}
A_x(x, y, t) &= -\frac{1}{2}\int_0^a dx' \mathrm{sgn}(x - x')\, \partial_y \, A_y(x', y, t - |x - x'|/c) \\
&\quad + \frac{Y}{2}\int_0^a dx' E_x^{\text{inc}}(x', y, t - |x - x'|/c) \\
&\quad + F_0(t - x/c, y) + F_a(t - (a - x)/c, y), \\
A_y(x, y, t) &= -\frac{1}{2}\int_0^b dy' \mathrm{sgn}(y - y')\, \partial_x \, A_x(x, y', t - |y - y'|/c) \\
&\quad + \frac{Y}{2}\int_0^b dy' E_y^{\text{inc}}(x, y', t - |y - y'|/c) \\
&\quad + G_0(t - y/c, x) + G_b(t - (b - y)/c, x).
\end{aligned} \tag{5}$$

In (5), the time signals $F_0(t, y)$, $F_a(t, y)$, $G_0(t, x)$ and $G_b(t, x)$ correspond to homogeneous solutions of the relevant one-dimensional wave equations. To fix these signals, we need to augment the pair of integral equations in (5) with the boundary conditions

$$J_x(0, y, t) = 0, \quad J_x(a, y, t) = 0, \quad J_y(x, 0, t) = 0, \quad J_y(x, b, t) = 0, \tag{6}$$

which follow from the observation that $\boldsymbol{J}_S(x, y, t)$ is tangential at the edges of the plate.

Our first approach was to solve the system of equations formed by (5) and (6) by *marching on in time*. In the space discretization, the plate is subdivided into $L \times M$ cells with dimensions $\Delta x = a/L$ and $\Delta y = b/L$. The unknown currents are approximated by rooftop functions, and the integral equation is collocated on a staggered grid, at $x = \ell\Delta x$ and $y = (m + \frac{1}{2})\Delta y$, and $x = (\ell + \frac{1}{2})\Delta x$ and $y = m\Delta y$, respectively. This involves approximations of the form

$$J_x(x, y, t) = \sum_{\ell=1}^{L-1} \sum_{m=0}^{M-1} J_x[\ell, m; t] \, \Lambda(x - \ell\Delta x; \Delta x) \, \Pi(y - [m + \tfrac{1}{2}]\Delta y; \Delta y), \qquad (7)$$

where $\Delta x = a/L$, $\Delta y = b/M$, $\Lambda(\xi; h)$ is a triangle function of width $2h$ and $\Pi(\xi; h)$ is a rectangle function of width h. In (7), the first two boundary conditions of (6) are accounted for by leaving out the terms with $\ell = 0$ and $\ell = L$. Similarly, its counterpart for $J_y(x, y, t)$ satisfies the last two conditions of (6). In both cases, the relevant integral equation from (5) is collocated as well at these points to allow the determination of the unknown time signals. Finally, the time coordinate is discretized with time step Δt. The time retardation in the the delayed current $J_x(x, y, t - R/c)$ is handled by linear interpolation in time, and the differentiations with respect to y and x in (5) are evaluated via integration by parts. The resulting integrals over x' and y' in each cell can now be evaluated in closed form. The first integrals over x' and y' in the right-hand sides of (5) are approximated by a repeated midpoint rule, while the second ones are evaluated in closed form. Depending on the choice of the time step, the space-time discretization described above results in an explicit or an implicit scheme. In the latter case, a sparse matrix equation must be solved at each discrete time step.

As in the case of the original time-domain equation (1), the solution may exhibit instabilities. In our opinion, these instabilities can be explained from the fact that, by choosing a fixed space discretization, one inherently introduces an increasing error for increasing frequencies. Especially near the Nyquist frequency, this may lead to the introduction of spurious poles outside the unit circle in the discretized-frequency plane. The corresponding residual contributions are alternating, exponentially increasing time sequences. Based on this interpretation, several procedures seem available for stabilizing the solution. Until now, we have had partial success with smoothing the result after each time step, and with choosing a larger time step and using an implicit scheme. Smoothing does have a stabilizing effect, but generally affects the accuracy of the solution. Moreover, refining the discretization may bring back the instabilities. Using an implicit scheme is not as effective as we expected. The occurrence of instabilities depends critically on the condition of the system matrix of the linear system of equations that must be solved for each time step to obtain the field at that instant. This system matrix, in turn, depends directly on the manner in which the space-time integral equation is discretized.

4. FINITE-DIFFERENCE FORMULATION

A closed-form stability analysis of the procedure given in the previous chapter is, to our knowledge, not available. Even for the case of an infinite plate, generalizing the Von Neumann procedure that is customary for finite-difference calculations is too complicated. Therefore, we reformulate the integral equations in (5) in such a way that the vector potential can be solved from a system of differential equations with the aid of finite differences.

This part of the time-marching scheme can then be subjected to a Von Neumann stability analysis. This leaves the calculation of the the induced surface currents from the calculated potential. However, at least the consequences of the numerical differentiations in the scheme can be understood in this manner. For the infinite-plate problem, the analysis provides exact answers since we do not have to deal with boundary conditions. For a finite plate, the results are identical to those for the infinite plate until the induced currents reach the edge of the spatial grid. From that instant on, boundary conditions must be imposed and the two parts of the computation can no longer be carried out independently.

To arrive at the desired scheme, we introduce the auxiliary quantities

$$\begin{aligned} R_x(x,y,t) &= \frac{1}{2}\int_0^b dy' \operatorname{sgn}(y-y')\, A_x(x,y',t-|y-y'|/c), \\ R_y(x,y,t) &= \frac{1}{2}\int_0^a dx' \operatorname{sgn}(x-x')\, A_y(x',y,t-|x-x'|/c), \\ Q_x(x,y,t) &= \frac{1}{2}\int_0^b dy'\, A_x(x,y',t-|y-y'|/c), \\ Q_y(x,y,t) &= \frac{1}{2}\int_0^a dx'\, A_y(x',y,t-|x-x'|/c). \end{aligned} \tag{8}$$

Partial differentiation of the first and third lines of (8) with respect to y, and of the second and fourth lines of (8) with respect to x directly results in the first-order partial differential equations

$$\begin{aligned} \partial_y R_x(x,y,t) + \frac{1}{c}\,\partial_t Q_x(x,y,t) &= A_x(x,y,t), \\ \partial_x R_y(x,y,t) + \frac{1}{c}\,\partial_t Q_y(x,y,t) &= A_y(x,y,t), \\ \partial_y Q_x(x,y,t) + \frac{1}{c}\,\partial_t R_x(x,y,t) &= 0, \\ \partial_x Q_y(x,y,t) + \frac{1}{c}\,\partial_t R_y(x,y,t) &= 0. \end{aligned} \tag{9}$$

The two coupled Hallén-type integral equations given in (5) now reduce to

$$\begin{aligned} A_x(x,y,t) &= -\partial_y R_y(x,y,t) + P_x(x,y,t) \\ &\quad + F_0(t-x/c,y) + F_a(t-(a-x)/c,y), \\ A_y(x,y,t) &= -\partial_x R_x(x,y,t) + P_y(x,y,t) \\ &\quad + G_0(t-y/c,x) + G_b(t-(b-y)/c,x), \end{aligned} \tag{10}$$

where $P_x(x,y,t)$ and $P_y(x,y,t)$ are known functions that depend on the incident field:

$$\begin{aligned} P_x(x,y,t) &= \frac{Y}{2}\int_0^a dx' E_x^{\mathrm{inc}}(x',y,t-|x-x'|/c), \\ P_y(x,y,t) &= \frac{Y}{2}\int_0^b dy' E_y^{\mathrm{inc}}(x,y',t-|y-y'|/c). \end{aligned} \tag{11}$$

The system of partial differential equations can be consistently translated into a finite difference scheme by approximating the derivatives using the central difference rule, and using a grid staggered in space and time.

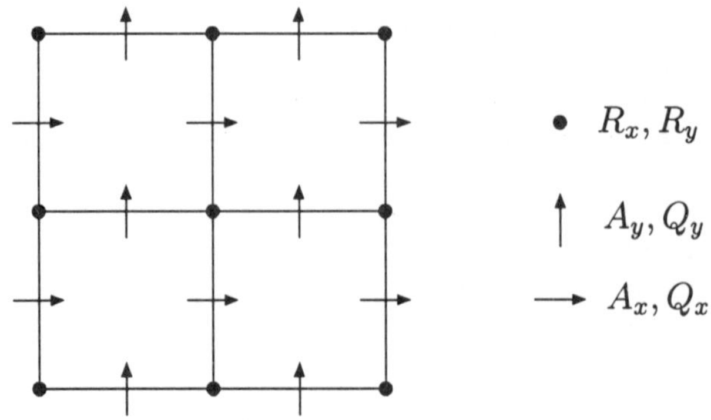

Figure 2: Four cells of the finite difference grid.

In formulating the difference equations, we use a short-hand notation to indicate the approximated version of the unknown quantities, e.g.,

$$A_x[\ell, m, n] = A_x(\ell\Delta x, m\Delta y, n\Delta t), \tag{12}$$

where ℓ, m and n denote integer numbers, and where Δx, Δy and Δt are the spatial cell sizes in the x- and y-directions, and the time step. To arrive at a consistent scheme, we evaluate the approximated quantities $A_x[\ell, m+\frac{1}{2}, n]$, $A_y[\ell+\frac{1}{2}, m, n]$, $R_x[\ell, m, n]$, $R_y[\ell, m, n]$, $Q_x[\ell, m+\frac{1}{2}, n+\frac{1}{2}]$, and $Q_y[\ell+\frac{1}{2}, m, n+\frac{1}{2}]$. By sampling in this manner, we arrive at an FDTD scheme on a Yee-type grid. The spatial part of the sampling is illustrated in Figure 2.

The resulting finite difference scheme can be subjected to a Von Neumann stability analysis to derive a stability condition for the scheme. We arrive at the usual Courant condition for two-dimensional propagation:

$$c\Delta t \leq 1/\sqrt{1/\Delta x^2 + 1/\Delta y^2}. \tag{13}$$

As can be observed from Figure 2, the sampling of A_x and A_y is consistent with the space discretization for the original Hallén-type equations (5). This means that we can use the same procedure to discretize the equations for obtaining J_x and J_y from the x- and y-components of (2). In the space discretization, we have introduced one refinement. To ensure that the retarded time $t' = t - R/c$ is as much as possible in the correct "time zone" $(n-k-1)\Delta t < t' < (n-k)\Delta t$, we carry out the time interpolation and the integration over x' and y' in (2) over a subgrid in each cell prior to the marching-on-in-time computation. In this manner, we allow each spatial cell to cover more than one "time zone", which has been demonstrated in the acoustic case to contribute to the stability of the solution [4].

Finally, it should be mentioned that each step of the procedure described above can be verified independently for the special case of excitation by a vertical electric dipole located above the center of the plate. Until the currents reach the end of the plate, the solution is identical to that for an infinite plate, for which $A_{x,y}$ and $J_{x,y}$ are available in closed form. This means that each step of the procedure can be validated independently.

5. REFERENCE SOLUTION: EXCITATION OF AN INFINITE PLATE BY A VERTICAL ELECTRIC DIPOLE

As mentioned above, we test our solution for the special case of an infinite plate excited by an electric dipole at $z = \boldsymbol{r}_0$. We first present an analytical expression for the vector potential for a dipole pointing in the $\hat{\boldsymbol{z}}$-direction, located at $\boldsymbol{r} = \boldsymbol{r}_0$ above an infinite perfectly conducting plane at $z = 0$. This reference solution will then be used to validate the numerical results of the procedure outlined in section 4..

With the aid of the method of images, the scattered electromagnetic field in the upper half-space of this configuration can directly be obtained from the the vector potential corresponding to a dipole located at the mirrored position $\boldsymbol{r}_0^- = r_{0_x}\hat{\boldsymbol{x}} + r_{0_y}\hat{\boldsymbol{y}} - r_{0_z}\hat{\boldsymbol{z}}$, also pointing in the z-direction:

$$\boldsymbol{A}^{\mathrm{s}}(\boldsymbol{r}, t) = \frac{I_0 L}{4\pi \tilde{R}} f(t - \tilde{R}/c)\hat{\boldsymbol{z}}, \tag{14}$$

where $\tilde{\boldsymbol{R}} = |\boldsymbol{r} - \boldsymbol{r}_0^-|$. However, the potential that is obtained from the Hallén equation is envisaged as being generated by induced surface currents on the plate, located in the horizontal plane $z = 0$. In the Lorentz gauge, surface currents in the horizontal plane produce a vector potential that has only components in the horizontal plane.

Thus to be able to compare the reference solution $\boldsymbol{A}^{\mathrm{s}}$ to that of the Hallén equation $\boldsymbol{A}^{\mathrm{h}}$, we need to introduce a gauge transformation of the form

$$A_x^{\mathrm{h}}(\boldsymbol{r}, t)\hat{\boldsymbol{x}} + A_y^{\mathrm{h}}(\boldsymbol{r}, t)\hat{\boldsymbol{y}} = A_z^{\mathrm{s}}(\boldsymbol{r}, t)\hat{\boldsymbol{z}} + \nabla\psi(\boldsymbol{r}, t), \tag{15}$$

in which ψ is an unknown scalar gauge function that satisfies the proper boundary conditions at infinity. This gauge function can be found by integration of the known z-component of (15):

$$\psi(\boldsymbol{r}, t) = \frac{I_0 L}{4\pi} \int_z^\infty dz' \frac{f(t - \tilde{R}/c)}{\tilde{R}}. \tag{16}$$

Now, the components of the reference vector potential that we are looking can be found by differentiation of 16 with respect to x and y, e.g.:

$$\begin{aligned} A_x^{\mathrm{h}}(\boldsymbol{r}, t) &= \partial_x \psi(\boldsymbol{r}, t) \\ &= \frac{I_0 L x}{4\pi} \int_0^\infty dz' \left(\frac{f(t - \tilde{R}/c)}{\tilde{R}^3} + \frac{1}{c} \frac{f'(t - \tilde{R}/c)}{\tilde{R}^2} \right). \end{aligned} \tag{17}$$

The integral over z' is again carried out numerically. In performing the numerical integration we exploit the fact that in the case of the dipole consisting of a pulsed current filament, only one or two small integration intervals need to be considered. If $f(t)$ is a Gaussian pulse with a maximum at t_0, then $f(t - \tilde{R}/c)$ has a maximum for $z_\pm = -r_{0_z} \pm R_1$ where $R_1 = \sqrt{(t - t_0)^2 - (x - r_{0x})^2 - (y - r_{0y})^2}$. In the numerical integration we therefore only take into account two intervals surrounding z_+ and z_-. When the two maxima are close together, we use a single integration interval.

Figure 3 shows a comparison of the solution of the FDTD scheme with the reference solution specified above. In the incident field, we chose f to be the derivative of a Gaussian pulse, i.e. the derivative of a Gaussian pulse $F(t)$ of width 2σ:

$$F(t) = \exp(-[t - t_0]^2/\sigma^2). \tag{18}$$

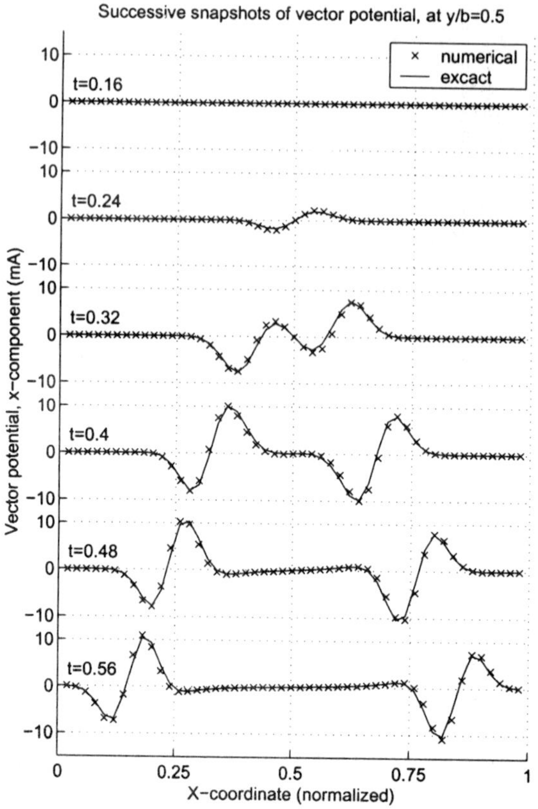

Figure 3: Snapshots of A_x for $y = b/2$ and varying x as specified in the text.

With this choice, we prevent the occurrence of a static electric incident field at late times. We consider a square plate, with $a = b$, and the space-time discretization was implemented for $L = M = 50$ and $c\Delta t = \Delta x/5$. Figure 3 presents snapshots of A_x for $y = b/2$ and varying x. The time coordinate is normalized with respect to the travel time a/c. The agreement between te numerical and analytical results is obvious. Unfortunately, results of subsequent steps in the calculation cannot be included here because of space limitations.

6. REFERENCES

1. A.G. Tijhuis, *Electromagnetic Inverse Profiling; Theory and Numerical Implementation*, VNU Science Press, Utrecht, the Netherlands, (1987), Sections 3.2 and 3.5.
2. A.G. Tijhuis, Z.Q. Peng and A. Rubio Bretones,"Transient excitation of a straight thin-wire segment: a new look at an old problem", *IEEE Trans. Antennas Propagat.* **40**, 1132-1146, (1992).
3. B.P. Rynne and P.D. Smith, "Stability of time marching algorithms for the electric field integral equation", *J. of Electromagnetic Waves and Applications* **4**, 1181-1205, (1990).
4. G.C. Herman, "Scattering of transient acoustic waves by an inhomogeneous obstacle", *J. Acoust. Soc. Am.* **69**, 909–915, (1981).

RESHAPING THE APERTURE FIELDS IN IMPULSE RADIATING ANTENNAS AND OSCILLATOR REFLECTOR ANTENNAS

J. Scott Tyo*

1. INTRODUCTION

Wideband, focused aperture antennas like impulse radiating antennas (IRAs) and oscillator reflector antennas (ORAs) can see improved aperture efficiency through control of the aperture fields and aperture shape. It has been suggested recently that the aperture efficiency can be further improved by reshaping the aperture fields through the use of uniaxially conducting reflectors, e.g. reflectors made of conducting wires rather than solid sheets.[1] In this paper, the benefit that can be obtained by using this strategy is explored analytically. Optimum filament orientation are shown, and the improvements possible in coplanar plate-fed IRAs and ORAs are presented. For common antenna configurations and impedances, the aperture height can be increased by amounts ranging from 7% to 67%. A method for extending the concept to lens IRAs is also proposed.

2. APERTURE EFFICIENCY IN IRAS AND ORAS

Impulse radiating antennas (IRAs)[2] and oscillator reflector antennas (ORAs)[3] belong to the class of focused aperture antennas. These antennas are particularly well suited for radiating energy generated by spark gap switches. The general configuration of an IRA or ORA is depicted in Fig. 1. The concept behind the antenna operation is that a spherical wave is launched by the source onto the non dispersive TEM transmission line feed structure of characteristic impedance Z_{line}. That spherical wave is converted into a plane wave in the geometric optics approximation by the focusing optic. The focusing optic can be a lens or a reflector (or even a timed array), but for many practical application, reflectors are

*Electrical and Computer Engineering Department, University of New Mexico, Albuquerque, NM 87131, USA, tyo@ieee.org.

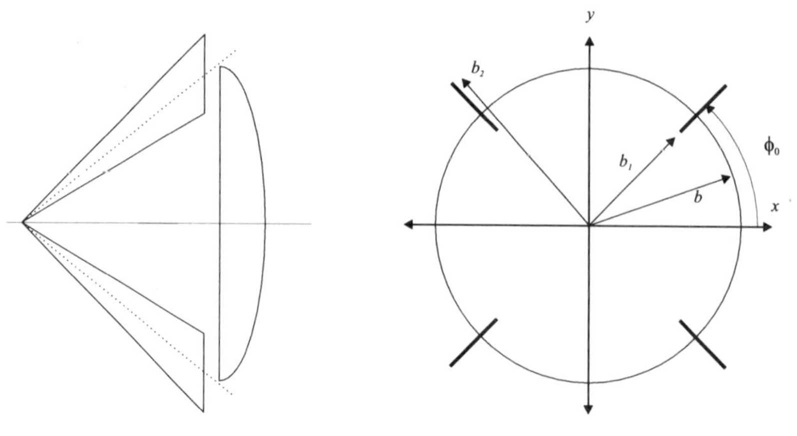

(a) (b)

Figure 1. Schematic diagram of IRAs and ORAs. The source is at the apex of the TEM feed line. This diagram is for reflector antennas, but the focusing optic could also be a lens.[2] A. Side view. B. Aperture plane (after stereographic projection).

preferable for their light weight and low volume.[4] Aperture theory (physical optics) is then used to compute the radiated far field. It is well known that the prompt field from an IRA in the direction of focus is given approximately as

$$E_{rad} = \frac{h_a}{2\pi r c f_g}\frac{dV}{dt}, \tag{1}$$

where V is the applied voltage waveform, r is the distance to the observation point, c is the speed of light, $f_g = Z_{line}/\eta_0$, and h_a is the aperture height given by

$$h_a = -\frac{f_g}{V_0}\iint_A E_y(x,y)dxdy. \tag{2}$$

The quantity V_0 in (2) is the peak applied voltage, and the surface integral in (2) is carried out over the extent of the focused aperture. In general h_a is a function of time, but for early times the voltage on the line and electric field in the aperture have the same temporal dependence. At later times, the fields in the aperture can be modified from the transmission line approximation. Strictly speaking, the aperture height is a vector quantity, with the x-component computed in the same fashion, but the symmetry of most practical IRAs forces the x-component of h_a to be zero.[2]

A concept of prompt aperture efficiency has been introduced for time-domain aperture antennas like IRAs and ORAs,[5] and it was shown that typical reflector IRAs with the feed arms oriented at angles of 45° with respect to the horizontal have aperture efficiencies on the order of 25%. The low aperture efficiency is due to two primary factors. First, a significant amount of power carried on the TEM feed line is propagating in fields that miss the focusing optic – an effect equivalent to spillover. Exactly 50% of the power misses the optic in IRAs fed by self reciprocal feed structures.[6] The second mechanism that reduces the aperture efficiency is field nonuniformity, which is inherent in all TEM modes on balanced transmission lines.

Several studies have explored mechanisms to increase aperture efficiency by addressing some of these issues. Buchenauer, *et al.*[5] showed that low-impedance, flat-plate TEM horns have aperture efficiencies that approach unity, but this type of feed is not suitable for use with reflectors. Tyo predicted numerically that moving the feed arms towards the vertical would increase field uniformity, providing a substantial increase in aperture efficiency.[7] These results were verified experimentally by Bowen, *et al.*, for reflector IRAs with feed arms at 60° from the horizontal.[8] Baum suggested that removal of portions of the aperture where E_y has the wrong sign from the integral in (2) could increase the aperture efficiency. This theory was verified numerically and experimentally by Baretela and Tyo.[9] Baretela and Tyo also showed that the aperture efficiency could be further improved by decreasing the size of the feed structure relative to the reflector and then trimming the resulting aperture accordingly.[9] All of these modification can improve the performance of an IRA or ORA without modifying the input impedance, and these changes do not affect the upstream components of the system.

3. REORIENTING THE APERTURE FIELDS

Baum has suggested recently that the aperture efficiency can be further increased by actually reorienting the fields in the aperture plane.[1] This can be accomplished by replacing the solid reflector with a mesh that conducts uniaxially. The solid reflector will carry current parallel to the local electric field, but the uniaxially conducting reflector can only carry currents along the direction of the wire. To maximize the response from the antenna, the direction of current flow at each point on the aperture should bisect the angle between the local tangential electric field and the vertical direction as shown in Fig. 2.

The early-time vector electric field of the TEM mode in the aperture plane can be derived from a complex scalar potential as follows. The complex coordinate is $z = x + jy$, and the complex potential is $w = u + jv$. The electric scalar potential is u and the magnetic scalar potential is v. The complex electric field is then[2]

$$E = E_x - jE_y = \frac{dw}{dz} = -\frac{du}{dx} + j\frac{du}{dy}. \tag{3}$$

Define the angle α from Fig. 2 as

$$\alpha = \tan^{-1}\frac{E_x}{-E_y}, \tag{4}$$

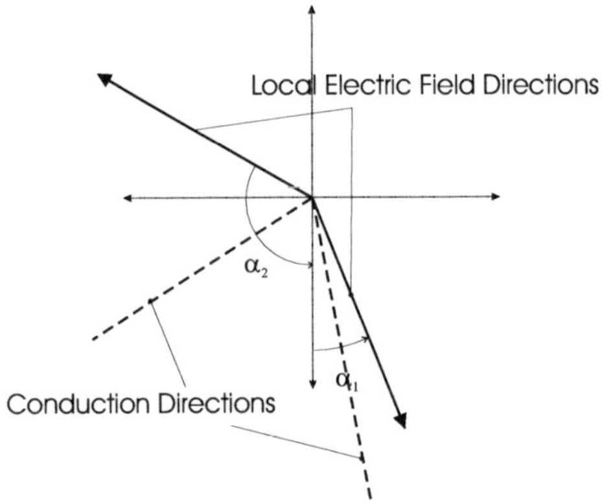

Figure 2. Two examples of the local conduction direction given the local electric field. The conduction direction bisects the angle α between the local electric field and the negative y-direction.

and the unit vector $\hat{\ell}$ that gives the desired direction of local current flow is

$$\hat{\ell} = \sin\frac{\alpha}{2}\hat{\mathbf{x}} - \cos\frac{\alpha}{2}\hat{\mathbf{y}}. \tag{5}$$

The resulting current flow is proportional to $(\mathbf{E}\cdot\hat{\ell})\hat{\ell}$, and the y-component is

$$J_y \propto -E_x \sin\frac{\alpha}{2}\cos\frac{\alpha}{2} + E_y\cos^2\frac{\alpha}{2} \tag{6}$$

$$\propto -\frac{1}{2}E_x \sin\alpha + E_y\frac{1+\cos\alpha}{2}. \tag{7}$$

Since $\sin\alpha = E_x/\sqrt{E_x{}^2+E_y{}^2}$ and $\cos\alpha = -E_y/\sqrt{E_x{}^2+E_y{}^2}$, (7) becomes

$$J_y \propto \frac{1}{2}\left[-\frac{E_x{}^2+E_y{}^2}{\sqrt{E_x{}^2+E_y{}^2}} + E_y\right] = \frac{1}{2}(E_y - |\mathbf{E}|). \tag{8}$$

The resulting quantity is always nonpositive, and is largest (most negative) when $\mathbf{E} = E_y\hat{\mathbf{y}}$ with E_y negative. The negative sign results from the fact that the top half of the TEM feed is usually taken as having positive voltage. The current flow on the wires is zero only at locations where $\mathbf{E} = E_y\hat{\mathbf{y}}$ with E_y positive. This strategy represents an improvement over the aperture trimming proposed by Baum[1] and Baretela and Tyo[9] because essentially all portions of the aperture can be made to contribute constructively to the surface integral in (2).

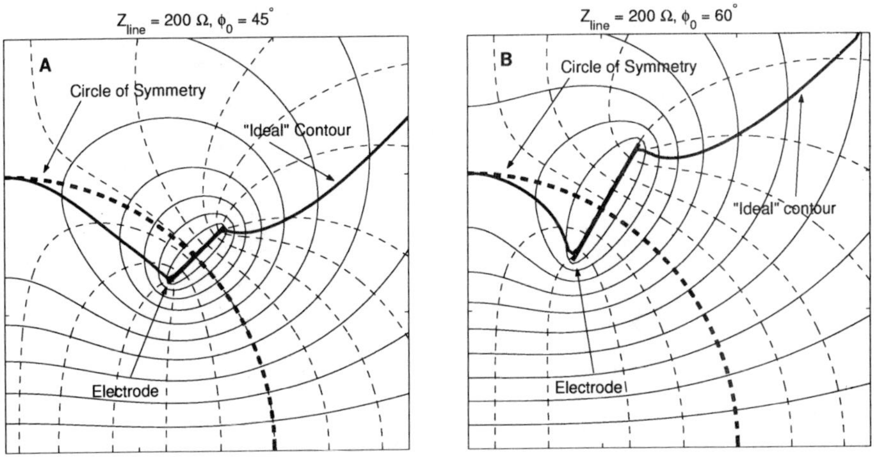

Figure 3. Potential (solid lines) and field distributions (dashed lines) for 200-Ω IRAs and ORAs with A. $\phi_0 = 45°$ and B. $\phi_0 = 60°$. Note that the circle of symmetry of the feed structure is also a field line.[6] The ideal contour indicated in the figures is the contour where $E_y = 0$. Trimming the portions of the aperture above this contour was shown to increase h_a.[9]

4. LOCAL CURRENT DIRECTIONS

The local direction of conduction is determined by computing the TEM mode distribution in the aperture plane and projecting that onto the paraboloidal surface through a stereographic projection. We will consider only the aperture plane here for simplicity (i.e. no stereographic projection has been performed). The potential and field distribution for one quadrant of 200-Ω-impedance reflector IRAs with feed arms at 45° and 60° from the horizontal are depicted in Fig. 3. The antenna with the 60° feed arms is significantly more aperture efficient than the antenna with the 45° feed arms when the circle of symmetry coincides exactly with the focused aperture. The increased efficiency is due to the fact that the fields are more uniform (and less of the aperture contributes destructively to the aperture integral).[9] The appropriate conducting paths are indicated for these two cases in Fig. 4.

5. IMPROVEMENT IN APERTURE HEIGHT

The aperture efficiency is given in terms of the aperture height as[5,7]

$$\eta_A = \frac{h_a^2}{f_g A}. \tag{9}$$

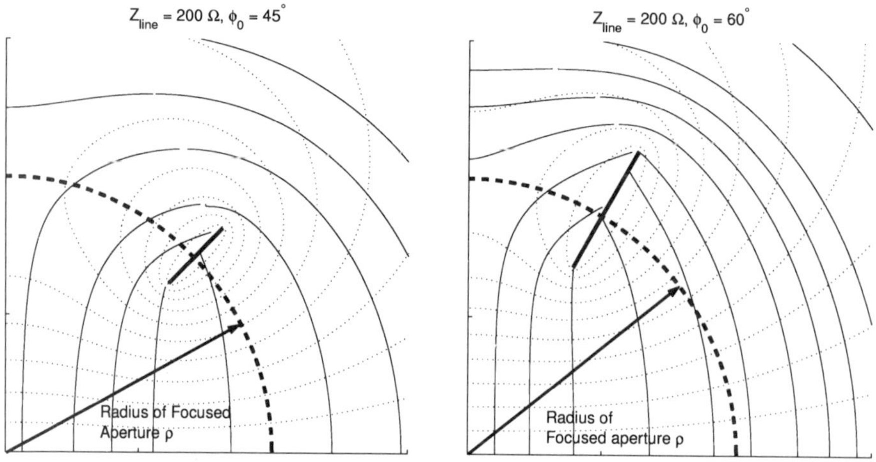

Figure 4. Local orientation of conductors for optimal aperture height. The solid lines bisect the angle between the local field direction displayed in Fig. 3 and the negative y-direction as shown in Fig. 2.

When the input impedance (f_g) is fixed and the aperture is circular with radius ρ, $A = \pi\rho^2$ and η_A is maximized when h_a/ρ is maximized. This quantity is termed here as the normalized aperture height and is defined as $\bar{h_a} \equiv h_a/\rho$. Fig. 5 shows the aperture height and normalized aperture height as a function of the radius of the circular aperture for 200-Ω IRAs with several feed arm angles.

Fig. 5A shows the aperture height for fixed-size feed arms as the radius of a circular aperture is increased. The solid lines show the aperture height with a solid reflector and the dashed lines show the aperture height using the uniaxially-conducting strategy outlined above. The aperture height for the solid reflector case increases until the circle intersects the outer tip of the electrode, at which point the aperture height remains constant. This result is consistent with results presented earlier.[9] The aperture height for the uniaxially conducting reflector continues to increase regardless of the size of the aperture. However, there is a clear optimum radius for a given feed arm angle as can be seen from Fig. 5B. This optimum radius is significantly higher than was found by using aperture trimming alone, as is detailed in table 1. As can be seen in the table, the increase realized by using a uniaxially conducting circular reflector is significant. The smallest improvement realized was 7.7% over the solid reflector with feed arms at 60° and $Z_{line} = 100\Omega$. For the near-optimal configuration of a 200Ω-antenna with $\phi_0 = 60°$,[7,8] aperture trimming provides little benefit because only a small part of the aperture is removed.[9] In that case, aperture trimming provides only a 2% increase in the aperture height. The inclusion of a uniaxially conducting reflector

Z_{line}	ϕ_0°	$\bar{h_a}$ - COS	$\bar{h_a}$ - trim	$\bar{h_a}$ - uni	r_{op}trim	r_{op} - uni
100	60	0.443	0.453	0.477	0.97	2.00
100	45	0.439	0.457	0.497	1.58	1.80
100	30	0.372	0.433	0.463	1.62	1.60
100	15	0.226	0.326	0.361	1.34	1.35
150	75	0.584	0.589	0.646	0.89	1.70
150	60	0.616	0.623	0.681	1.13	1.50
150	45	0.571	0.610	0.676	1.31	1.40
150	30	0.437	0.521	0.575	1.26	1.29
150	15	0.243	0.347	0.405	1.24	1.25
200	75	0.726	0.730	0.805	0.96	1.44
200	60	0.736	0.750	0.821	1.11	1.26
200	45	0.648	0.687	0.799	1.16	1.20
200	30	0.480	0.559	0.637	1.14	1.22
200	15	0.254	0.355	0.419	1.16	1.23
250	75	0.841	0.845	0.917	0.96	1.23
250	60	0.812	0.828	0.890	1.09	1.15
250	45	0.678	0.719	0.800	1.09	1.16

Table 1. Comparison between solid reflector, optimally trimmed reflector, and optimal uniaxially conducting reflector. The third column gives $\bar{h_a}$ for a circular aperture at the circle of symmetry of the feeds with no modifications to the reflector. The fourth column gives the optimum $\bar{h_a}$ for a trimmed aperture. The fifth column gives the optimum $\bar{h_a}$ for a uniaxially conducting aperture. The last two columns give the optimum radius for the trimmed and uniaxially conducting reflectors, respectively. Note that *substantial* performance improvements are possible with the uniaxially conducting reflectors. For common configurations, the improvement is on the order of 20%. Also note that the optimum radius for uniaxially conducting reflectors is usually much larger than the circle of symmetry or the optimum radius for trimmed apertures.

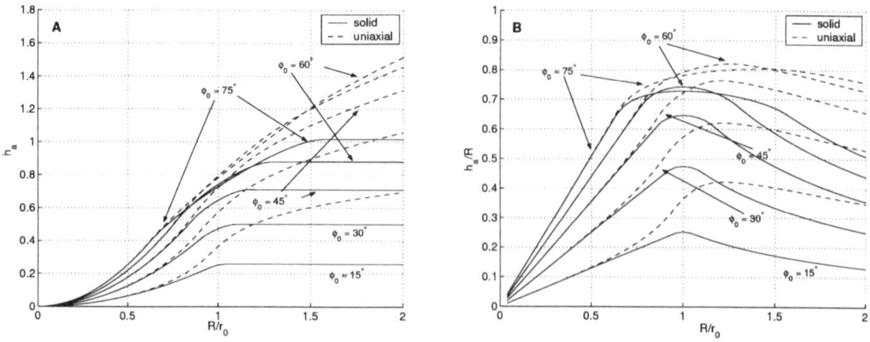

Figure 5. Aperture height and normalized aperture height for $Z_{line} = 200\Omega$ at a variety of feed angles ϕ_0. A. h_a as a function of radius for solid reflectors (solid lines) and uniaxially conducting reflectors (dashed lines). R is the radius of the circular aperture and r_0 is the radius of the circle of symmetry of the TEM feed structure. The aperture height flattens at the outer edge of the feed arms for the solid reflector, but continues to increase for the modified reflector. B. $\bar{h}_a$ as a function of radius. $\bar{h}_a$ has a clear optimum for all angles, and the improvement possible with the modified reflector is greater than 10% in all cases.

in this configuration provides a substantial 11% increase in the aperture height. At lower feed arm angles, the improvement is much more dramatic (as much as 66% at 15° and $Z_{line} = 200\Omega$). At these low feed arm angles, aperture trimming provides a significant benefit because large portions of the aperture are removed.[9] The uniaxially-conducting reflector can provide even greater benefits, since the portions of the aperture removed by trimming are made to contribute to the aperture height.

The detailed results for the $Z_{line} = 200\Omega$ are presented above because that is the most common impedance used in high power applications.[2] Similar calculations can also be performed for other impedances, and the results are summarized in Fig. 6. Fig. 6A presents the optimum value of $\bar{h}_a$ as a function of feed arm angle for several common impedances used for IRAs and ORAs. Fig. 6B presents the circular radius that yields that maximum value of $\bar{h}_a$ as a function of ϕ_0 for each impedance.

6. APPLICATION TO LENS IRAs and ORAs

The results above demonstrate that the strategy proposed in this paper can be very beneficial in virtually all reflector IRA and ORA configurations. A further question is whether or not similar results can be achieved for lens IRAs and ORAs by using the complementary strategy. With lens IRAs, we would like to transmit the y-polarized fields through the aper-

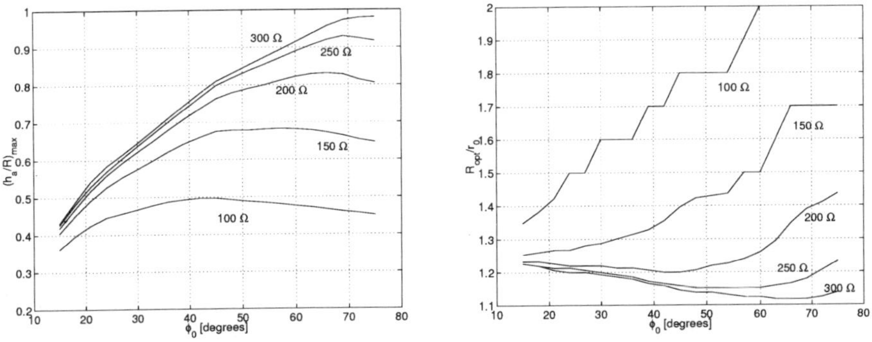

Figure 6. A. Optimum $\bar{h}_a$ as a function of feed arm angle ϕ_0 for common feed impedances. Note that the optimum angle for each feed impedance increases monotonically. B. The radius that the maximum $\bar{h}_a$ occurs at for the cases in panel A. The staircased nature of the data is due to coarse sampling. It is equivalent to either increase the size of the aperture for a fixed feed arm geometry or to decrease the size of the feed arms for a fixed maximum aperture size. versus phi

ture, not reflect them. In that case, the wires should be oriented at an angle that optimally reflects the unwanted fields. The angle of the wires should be $\beta/2$, where $\beta = \tan^{-1} E_y/E_x$ is the angle made by the field with respect to the positive y-direction as depicted in Fig. 7. . The effect of putting such a structure in place is to reflect the unwanted fields and allow the desired fields to pass. This concept has been used extensively to fabricate linear polarizers for use at infrared wavelengths.[10]

The feed geometries used for lens IRAs and ORAs are typically significantly more aperture efficient.[5] The reason for this can be seen in Fig. 8, which depicts the potential and field distribution for a circular, conical plate-fed lens IRA. This feed geometry is still self-reciprocal, meaning that 50% of the power misses the lens, but the aperture efficiency is 46% (as compared with 25% for 200-Ω reflector IRAs with 45° feed arms.[5]

Fig. 9 shows the computed values of $\bar{h}_a$ for a circular aperture that coincides with the circle of symmetry as a function of the electrode angle ϕ_0 depicted in Fig. 8. Fig. 9A shows the results for the optimum orientation of the wires.

There are some practical concerns associated with this strategy that may make it difficult to implement. First, the transmission through wire grid polarizers can be low for the desired polarization state if they are not properly fabricated. Second, the undesired fields are reflected by the polarizer and will propagate back down the TEM feed to the source. This can be highly detrimental to the source lifetime, especially in high power applications. Such an effect can be potentially mitigated by fabricating the wire grid from lossy wires, in which case they might act more like a typical dichroic polarizer used at optical

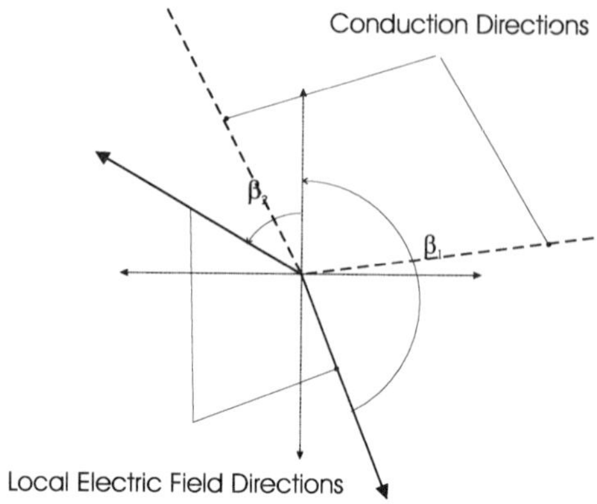

Figure 7. Conductor configuration for lens IRAs and ORAs. The local conduction direction is chosen now to reflect the *undesired* fields. This structure will act like a wire grid polarizer.[10]

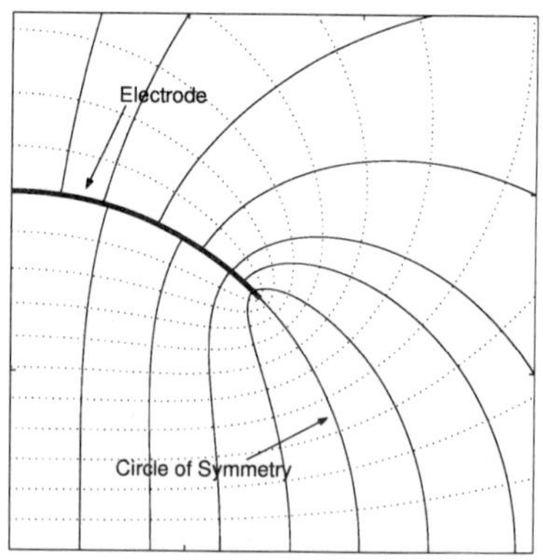

Figure 8. Field lines (solid) and potential distributions (dotted) for a circular, conical plate-fed lens IRA or ORA. This structure has $Z_{line} = 60\pi$ with the electrode tip at 45° from the vertical. The fields inside the circle of symmetry are far more uniform, and the majority of the loss of aperture efficiency is due to fields propagating outside the lens.

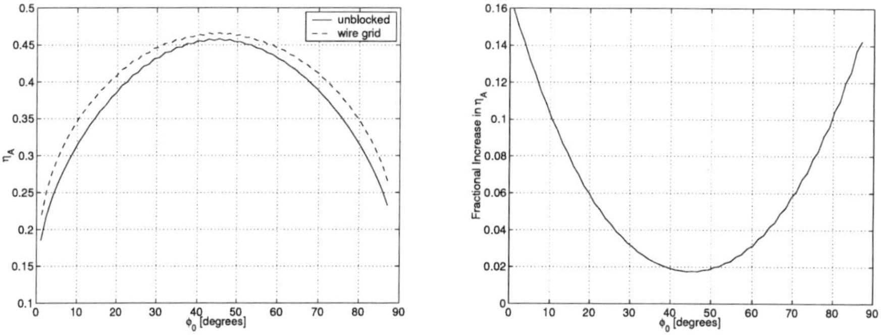

Figure 9. Aperture efficiency for lens IRAs and ORAs. A. η_A as a function of electrode angle for the wire grid (dashed) and unblocked (solid) apertures. B. Fractional improvement realized by using the polarizer.

wavelengths, i.e. the polarizer absorbs the unwanted energy instead of reflecting it. Such a strategy can be used in conjunction with the termination that is necessary to improve the late-time response of the antenna.[2]

The polarizer strategy does provide a small benefit for lens IRAs and ORAs, but the effect is nowhere near as dramatic as for reflector IRAs and ORAs. In lens IRAs, the dominant effect that reduces aperture efficiency is spillover.[5] The only method that has been proposed to reduce spillover involves the use of isorefractive materials to concentrate the power in the aperture.[11] A method has been proposed to realize such materials,[12] but to our knowledge none has been fabricated as yet.

7. DISCUSSION AND CONCLUSIONS

The results presented here clearly demonstrate that replacing the solid paraboloidal reflector with a mesh of uniaxially conducting wires has the potential to *substantially* increase the radiated fields from IRAs and ORAs. However, there are many additional factors that must be considered to determine how practical this solution is. In practice, the exact pattern necessary to achieve these ideal results can only be determined by numerical methods. In addition, a method must be devised that will allow these patterns to be synthesized while maintaining the structural integrity of the reflector.

For ORAs, the spacing of the wires can be chosen to be small compared with the center frequency as discussed by Baum.[3] The wires will appear inductive to the field, and will therefore need to be displaced by an appropriate distance from the actual paraboloidal surface in order to maintain the focusing properties of the optic. For IRAs, it is not as clear that these solutions can work. The broadband nature of the waveforms transmitted by

IRAs will force the spacing between conducting wires to be small, probably on the order of $ct_r/10$, where c is the speed of light and t_r is the rise time of the step applied to the IRA. The reflector still can be expected to ring in the late time, even with small spacing, and that will affect the transmitted response after the prompt signal.

The significant benefits predicted by this analysis seem to indicate that it is worth trying this strategy for reflector IRAs and ORAs. The late-time effects for IRAs are difficult to predict, but we hope to perform a set of experiments that will determine the feasibility of this strategy.

REFERENCES

[1] C. E. Baum, "Modification of TEM-fed reflector for increased efficiency," in: *Sensor and Simulation Notes #458*, edited by C. E. Baum (Air Force Research Laboratory, Albuquerque, NM, 2001).

[2] C. E. Baum, E. G. Farr, and D. V. Giri, "Review of impulse-radiating antennas," in: *Review of Radio Science*, edited by W. R. Stone (Oxford U. Press, New York, 1999), pp. 403-439.

[3] C. E. Baum, "Differential switched oscillators and associated antennas," in: *Sensor and Simulation Notes #457*, edited by C. E. Baum (Air Force Research Laboratory, Albuquerque, NM, 2001).

[4] D. V. Giri, H. Lackner, I. D. Smith, D. W. Morton, C. E. Baum, J. R. Marek, W. D. Prather, and D. W. Scholfield, "Design, fabrication, and testing of a paraboloidal reflector antenna and pulser system for impulse-like waveforms," *IEEE Trans. Plasma Sci.* **25**, 318-326 (1997).

[5] C. J. Buchenauer, J. S. Tyo, and J. S. H. Schoenberg, "Prompt aperture efficiencies of impulse radiating antennas with arrays as an application," *IEEE Trans. Antennas Propagat.* **49**, 1155-1165 (2001).

[6] E. G. Farr and C. E. Baum, "Radiation from self-reciprocal apertures," in: *Electromagnetic Symmetry*, edited by C. E. Baum and H. N. Kritikos (Plenum, New York, 1995), ch. 6, pp. 281-308.

[7] J. S. Tyo, "Optimization of the TEM feed structure for four-arm reflector impulse radiating antennas," *IEEE Trans. Antennas Propagat.* **49**, 607-614 (2001).

[8] L. H. Bowen, E. G. Farr, C. E. Baum, T. C. Tran, and W. D. Prather, "Experimental results of optimizing the location of feed arms in a collapsible IRA and a solid IRA," in: *Sensor and Simulation Notes #450*, edited by C. E. Baum (Air Force Research Laboratory, Albuquerque, NM, 2000).

[9] M. J. Baretela and J. S. Tyo, "Selective trimming of impulse radiating apertures to increase prompt radiated fields," in: *Sensor and Simulation Notes #461*, edited by C. E. Baum (Air Force Research Laboratory, Albuquerque, NM, 2001).

[10] R. M. A. Azzam and N. M. Bashara, *Ellipsometry and Polarized Light* (North-Holland, New York, 1977).

[11] J. S. Tyo, C. J. Buchenauer, and J. S. H. Schoenberg, "Use of isorefractive media to improve prompt aperture efficiency in a lens IRA," *IEEE Trans. Antennas Propagat.* **46**, 1114-1115 (1998).

[12] J. S. Tyo, "A class of longitudinally symmetric artificial material isorefractive with free space," *IEEE Trans. Antennas Propagat.* To be published in May 2003.

OPTIMIZATION OF IMPULSE RADIATING ANTENNAS

Leland H. Bowen, Everett G. Farr, Carl E. Baum, and Tyrone C. Tran*

1. INTRODUCTION

We describe here the development of an Impulse Radiating Antenna (IRA) through a series of modifications and improvements intended to achieve the following objectives:

1. Improve the Effective Gain
2. Reduce the Cross Polarization Coupling to the Antenna
3. Reduce Reflections (Improve Impedance Match)
4. Improve Mechanical Sturdiness (Make Flight Worthy)
5. Reduce the Side Lobes in the Antenna Pattern

In consideration of these objectives, we describe the theoretical basis for the modifications, the modifications themselves, and the experimental results.

All of the antennas developed during this series of experiments have a 46 cm (18-inch) diameter spun aluminum reflector with an *F/D* of 0.5. In this paper we describe improvements to the IRA-1 and IRA-2 over those previously described.[1, 2] We go on to develop new designs for the IRA-3 and IRA-4. We document how well a number of new ideas work to improve the performance of the IRA.[3]

We begin with the IRA-1, which has feed arms located in the traditional position of ±45° from the vertical. The modifications to the IRA-1 include improving the symmetry about the vertical E plane and tuning the resistors for low reflection.

Next, we consider the IRA-2. This antenna has the feed arms positioned at ±30° to vertical in order to increase gain and decrease crosspol. The design of the IRA-2 is based on theory developed by J. Scott Tyo[4] and Carl Baum.[5]

Next, we consider the IRA-3, which has a ground plane in the horizontal plane of symmetry, to reduce the crosspol and increase the sturdiness of the antenna. We performed a large number of experiments on the IRA-3 to improve the feed point design

* Leland H. Bowen, Everett G. Farr, Farr Research Inc.614 Paseo Del Mar NE, Albuquerque, NM, 87123, Carl E. Baum, and Tyrone C. Tran, Air Force Research Laboratory / Directed Energy Directorate, 3550 Aberdeen Ave. SE, Kirtland AFB, NM 87117-5776.

and the load resistor configuration. We also added a radome manufactured by Mission Research Corporation (MRC) to protect the antenna.

Finally, we repositioned the feed arms so their outer edge aligned with the outer edge of the reflector, resulting in the IRA-4. This was done to reduce the sidelobes, which might be enhanced by the high fields near the reflector rim. For the same reason, we also placed microwave absorber foam around the edge of the reflector.

The antenna measurements presented here were made using the outdoor time domain antenna range at Farr Research. The source antenna was a Farr Research Model FRI-TEM-01-50 sensor[6] and the IRA under test was the receiver. The distance between the apertures was 10 m. The technique for characterizing the antennas and processing the time domain measurements including calculation of the effective gain from the raw data are the same as used previously.[2, 6, 7]

On several versions of the IRA we also investigated the effects of improving the symmetry of the antenna, in order to reduce the crosspol coupling to the antenna. Carl Baum has shown that by carefully maintaining the symmetry of the antenna, the crosspol should be eliminated.[8]

A final improvement to the antenna design is due to an improved splitter. The splitter is the portion of the balun that splits the 50-ohm input into two 100-ohm transmission lines. Improvements in the splitter were incorporated into the IRA-4, resulting in much smaller reflections.

2. IRA-1

We investigate here an improved IRA-1, with feed arms at ±45°. We redesigned the original IRA-1[1] based on IRAs with a similar configuration with extremely flat TDRs.[2, 9] Such flat TDRs are useful in radar applications. Our improvements included modifications to the feed point and to the load resistors.

The feed arms for the IRA-1 have an included angle of 13.6°. For the original IRA-1, the load resistance at the end of each feed arm was made up of two 400-ohm High Voltage Resistors (HVRs) mounted in parallel on a small circuit card.[1, 3] For the improved IRA-1, shown in Figure 2.1, the high-voltage resistors have been replaced with a distributed resistance made up of five parallel strings with three resistors in each string. The RF characteristics of the improved IRA-1 are shown in Figures 2.2–2.3. In addition to the distributed load resistors, this version of the IRA-1 has a more symmetric Teflon support at the feed point. It also has a dummy cable on the upper left feed arm to improve the symmetry about the vertical plane of symmetry. Improving the symmetry in this way was intended to reduce the crosspol, as shown by Carl Baum.[8]

In Figure 2.2 we show the TDRs of both the original and improved versions of the IRA-1. The original IRA has a large dip in the impedance at the load resistors. The distributed load resistance reduces this dip and greatly improves the TDR of the antenna. The normalized impulse response has a FWHM of 35.7ps. The aperture height for this antenna is 124 mm, which is 12% better than the original IRA-1.

In Figure 2.3, we show the effective gain on boresight for both the copol and crosspol configurations of the original and modified versions of the IRA-1. There is considerable improvement in the crosspol rejection for the new IRA-1, as well as some improvement in the gain over most of the frequency range.

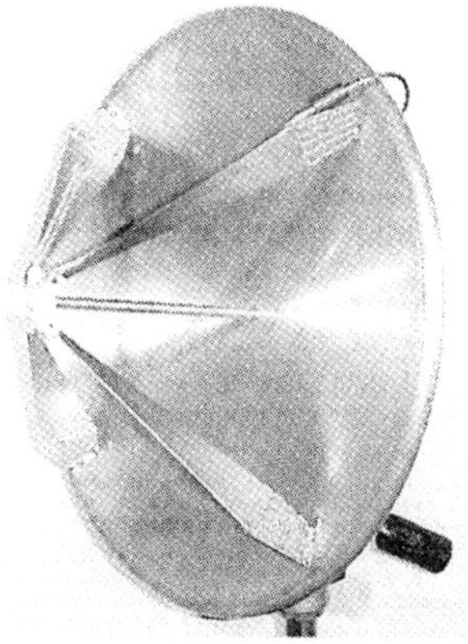

Figure 2.1. IRA-1 (Feed arms at ±45°).

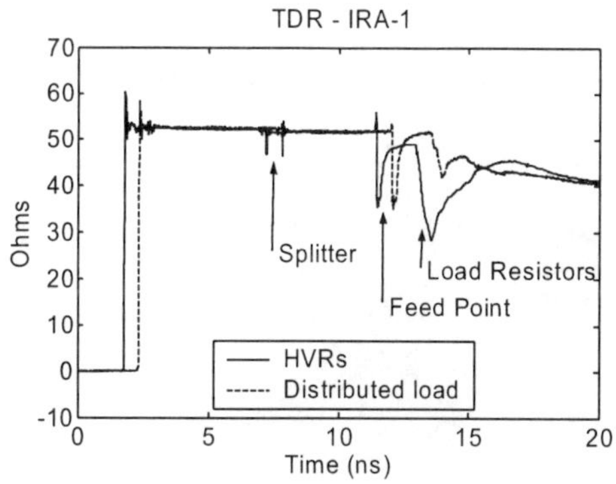

Figure 2.2. TDR of the IRA-1.

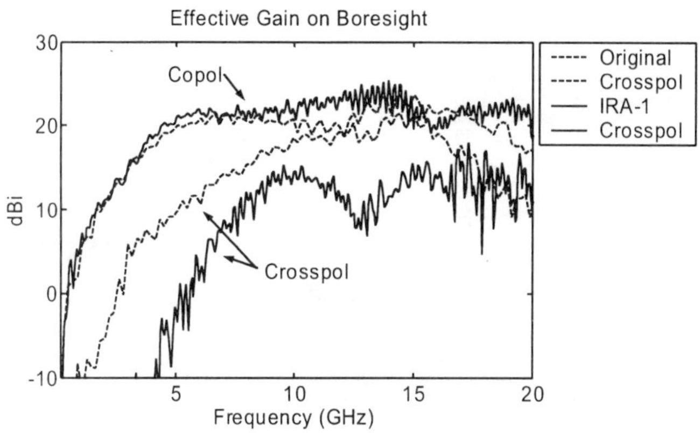

Figure 2.3. Comparison of the original and improved IRA-1.

3. IRA-2

We provide here corrected and expanded data on the version of the IRA-2 that was developed previously.[2] This version is based on Scott Tyo's work[4], with the feed arms positioned at ±30° to vertical. A photograph of the IRA-2 is shown in Figure 3.1. The feed arms for the IRA-2 have an included angle of 23.2°. The load resistor configuration is the same as for the original IRA-1, with the HVRs.

The TDR of the antenna is shown in Figure 3.2. As with the original IRA-1, we see a large dip in the impedance in the region of the resistors. Replacing the HVRs with an optimized load configuration would undoubtedly improve the TDR. The normalized impulse response on boresight has a FWHM of 34 ps and the aperture height is 133 mm. This is a 7% improvement over the IRA-1 presented above and 19% better than the original IRA-1. Since the IRA-2 is structurally closer to the original IRA-1, this improvement is even better than the expected 10%. This may be due in part to improvements in the feed point construction. The TDR at the feed point is somewhat better than that of the IRA-1. The effective gain is shown in Figure 3.3 where we overlay the gains of the IRA-1 and IRA-2. The boresight gain of the IRA-2 is higher than that for the IRA-1 especially from 14 to 18 GHz. The crosspol rejection for the IRA-2 is considerably better than the IRA-1, except at frequencies below 6 GHz.

Figure 3.1. IRA-2 (feed arms at ±30°).

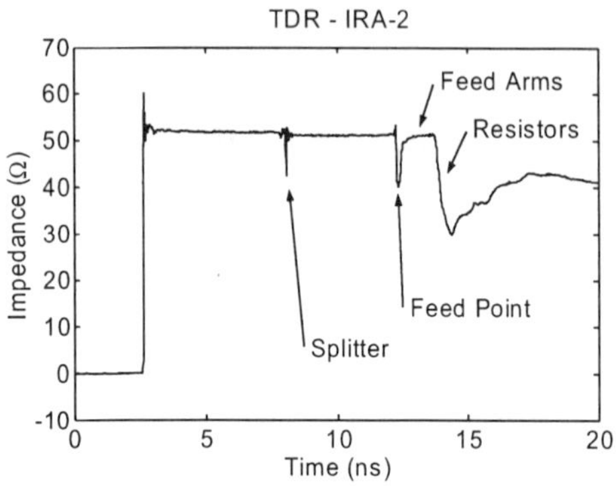

Figure 3.2. TDR of the IRA-2.

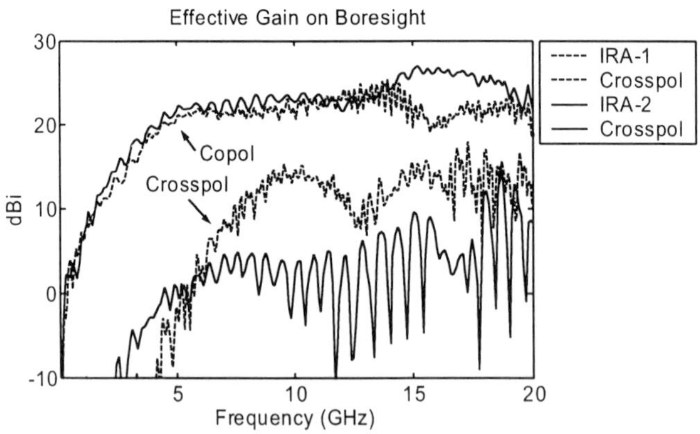

Figure 3.3. Comparison of IRA-1 and IRA-2 including crosspol.

4. IRA-3

We next added a ground plane to the IRA, resulting in the IRA-3. The ground plane should short out the horizontal component of the E field, thereby further reducing the crosspol coupling to the antenna. The ground plane along with the new feed arm support ring enhanced the mechanical strength or sturdiness of the antenna. This ring is made of UHMW, which has a relative dielectric constant of 2.3.

The initial experiments performed on this version of the IRA were intended to improve the TDR in the area of the load resistance. First we tried a distributed resistance in place of the two 400-ohm HVRs used in the past.[1] It was hoped that distributing the resistance over the end of the feed arm would improve the TDR, as it did for the IRA-1. Although several configurations were tested, we found that a distributed load was not optimal for the IRA-3 with feed arms positioned at ±30°. All of the configurations tested with a distributed load produced a large dip in the impedance in this region.

The dip in impedance at the load resistors is due in part to the electric field incident on the center of the parabolic reflector at time F/c where c is the speed of light in free space. The reflected field has a ramp-function time dependence, and it appears in the TDR as a negative dip at time $2F/c$.[10] We conducted a series of experiments to determine the effects of various load resistor, feed arm, and feed point configurations. Noting that the large dip in the TDR is a capacitive effect, we attempted to add some inductance at the right place to counteract it. The inductance was added by using small coils of various sizes and shapes wound in a figure 8 pattern to reduce the magnetic moment of the coils, thereby reducing emissions from the coil.[11] In addition to trying many different resistor and coil configurations, we tried various feed arm lengths and found that the optimum length for the feed arms is the focal length F. This length is such that the wave from the feed point reaches the end of the feed arm at the same time as it strikes the center of the reflector. For the final experiments we found that four 200 Ω metal film resistors in series/parallel have enough inductance that we were able to omit the coils.

Next, we moved the lower feed to the front of the feed point. Up to this point, the lower feed cable has always approached the feed point from the rear along the axis of symmetry similar to the IRA-1 and IRA-2. Moving the feed to the front launches the wave toward the reflector, which improves the TDR at the feed point and the overall antenna gain. In doing this, the crosspol rejection was also somewhat improved.

The next modification to the IRA-3 was the addition of dummy cables to improve the symmetry of the antenna near the feed point. C. E. Baum has shown that the crosspol coupling in an IRA can be greatly reduced by maintaining the symmetry about horizontal and/or vertical planes of symmetry.[8] In this case, we chose to maintain symmetry about the vertical plane. Thus, we place a short section of cable between the feed point and UHMW support ring, as shown in Figure 4.1 (left). On the bottom side of the ground plane we added another dummy cable, as shown in Figure 4.1 (right).

The overall characteristics of this antenna are quite good. The TDR of the IRA-3 is shown in Figure 4.2. The normalized impulse response as shown in Figure 4.3 has a FWHM of 35.5 ps. The aperture height is 139 mm. The effective gain as shown in Figure 4.4 has a peak of 26.5 dBi. The average crosspol rejection is 24 dB.

Next, we consider a radome that was attached to the IRA-3 by MRC Dayton. The radome is made from two glass/epoxy skins with a Nomex core. The IRA-3 with the radome is shown in Figure 4.5. The radome protects the feed elements of the IRA, and is required if the antenna is mounted to the outside skin of an aircraft. The radome only slightly changes the TDR and the peak of the normalized impulse response is reduced by about 10%. However, as we show in Figure 4.6, the gains both with and without the radome are almost identical. The radome reduces the peak gain slightly, but does not appreciably change the aperture height. We provide pattern plots in Figure 4.7 from both MRC (frequency domain) and FRI (time domain). The pattern plots using the two methods are in very good agreement.

Figure 4.1. IRA-3 with dummy cables added for symmetry.

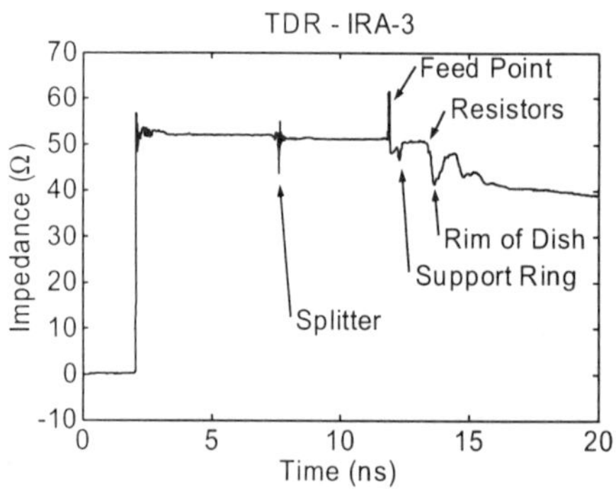

Figure 4.2. TDR of IRA-3 with dummy cables.

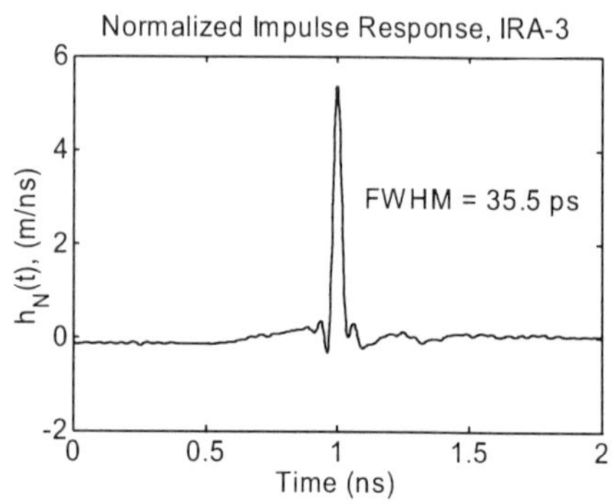

Figure 4.3. Normalized Impulse Response.

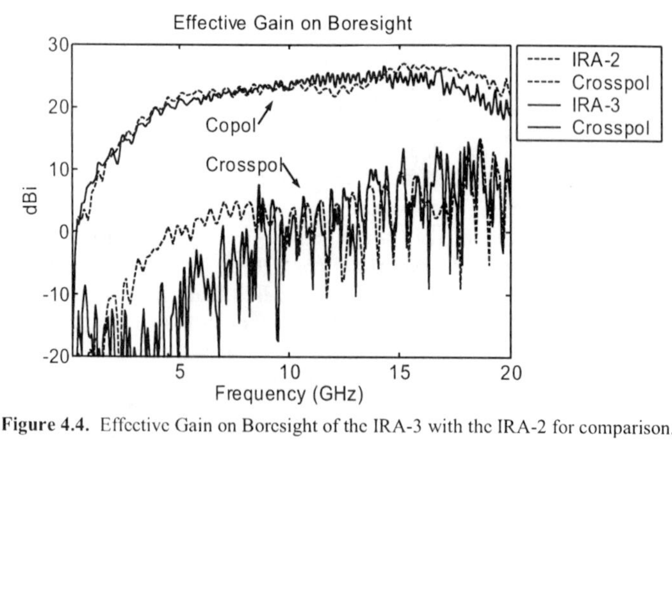

Figure 4.4. Effective Gain on Boresight of the IRA-3 with the IRA-2 for comparison.

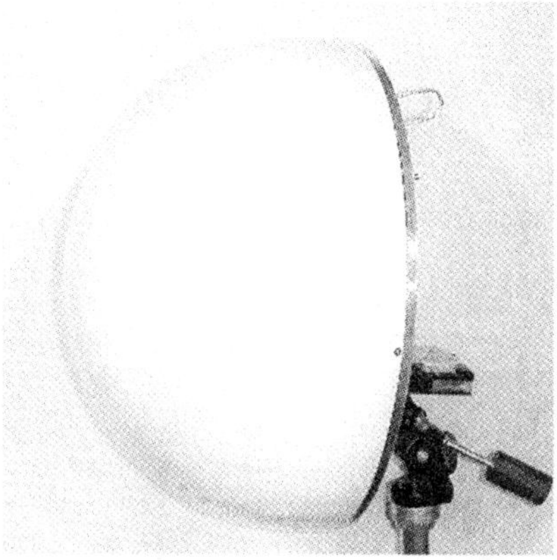

Figure 4.5. IRA-3 with radome (MRC).

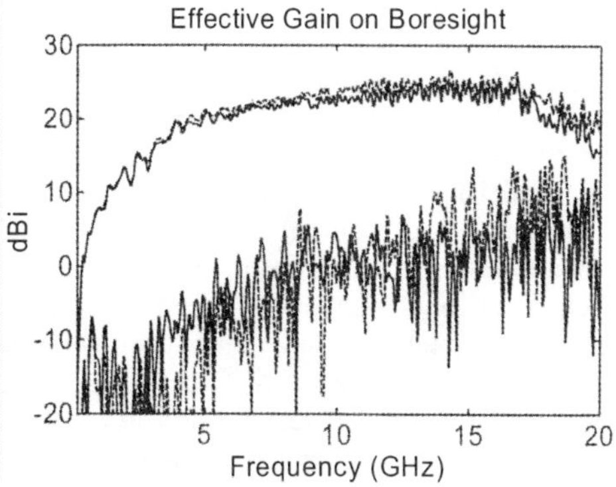

Figure 4.6. IRA-3 with and without radome.

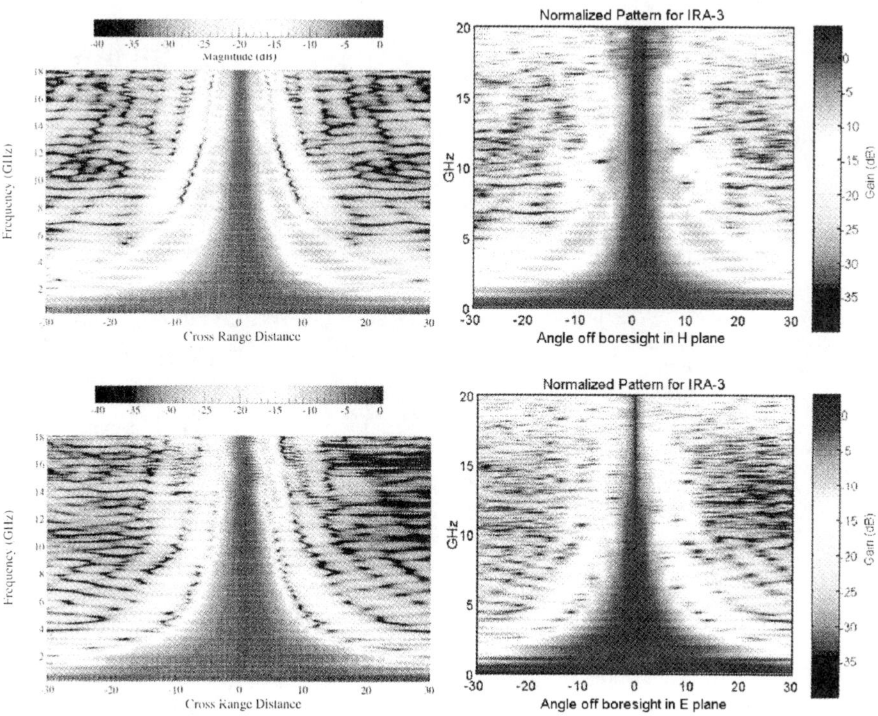

Figure 4.7. Pattern of the IRA-3, H plane (top), E plane (bottom), MRC(left), FRI (right).

5. IRA-4

Finally, we repositioned the feed arms so their outer edge is aligned with the outer edge of the reflector, resulting in the IRA-4. We did this to reduce the sidelobes by concentrating the field closer to the center of the reflector. The new feed arm location required recalculating the included angle of the feed arms to match the required 200-ohm impedance. The included angle of the feed arms for the IRA-4 is 20°.

Figure 5.1 is a photograph of the IRA-4. The RF characteristics of the IRA-4 were measured at various stages of its development. However, we will only show the final results. The final IRA-4 had dummy cables added to improve the symmetry about the vertical plane of symmetry and absorber foam around the rim of the reflector dish to reduce the fields in this area. The ¾-inch thick foam, shown in Figure 5.1 (right), is designed for use at frequencies above 3.5 GHz. Between the feed arms we extended the foam to cover the area where the electric field has the incorrect polarity. This is a larger area than that of the IRA-3. Covering this area should improve the gain of the antenna.

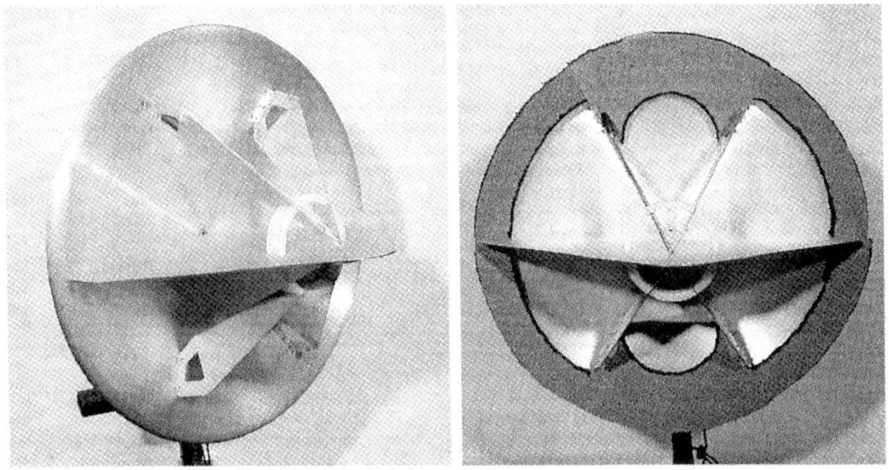

Figure 5.1. IRA-4 with and without absorber foam around rim.

A second modification we made in the IRA-4 was the development of an improved splitter. This device has a 50-ohm SMA connection that splits into two 100-ohm 0.141 in. diameter semi-rigid coax lines. The new splitter has an extremely flat TDR, as shown in Figure 5.2 as compared to the splitters shown in Figures 3.2 and 4.2.

The TDR of the final version of the IRA-4 is shown in Figure 5.2 and the normalized impulse response is shown in Figure 5.3. Next, the effective gain and crosspol gain are shown in Figure 5.4. While the gain of the IRA-4 is only slightly lower than that of the IRA-3, the crosspol has unfortunately increased considerably. Finally, the antenna pattern is shown in Figure 5.5, where we see a considerable increase in the sidelobes as compared to the IRA-3 in Figure 4.7.

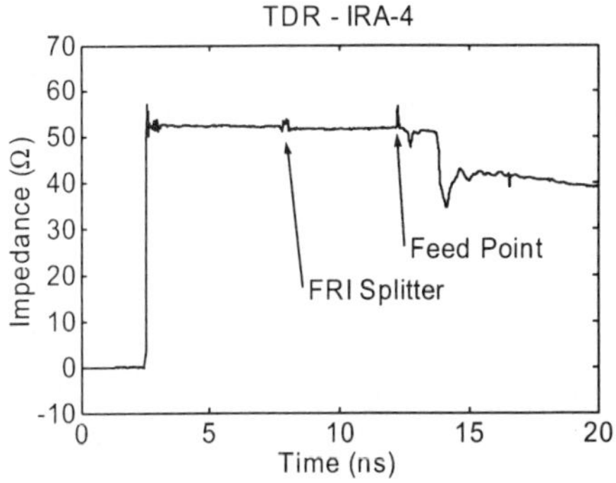

Figure 5.2. TDR of IRA-4.

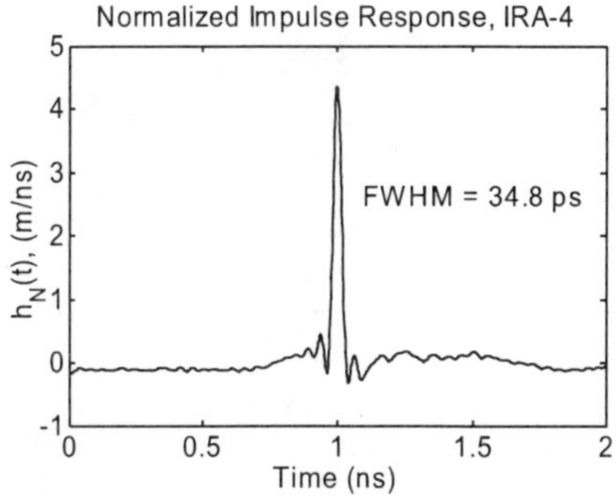

Figure 5.3. Normalized Impulse Response.

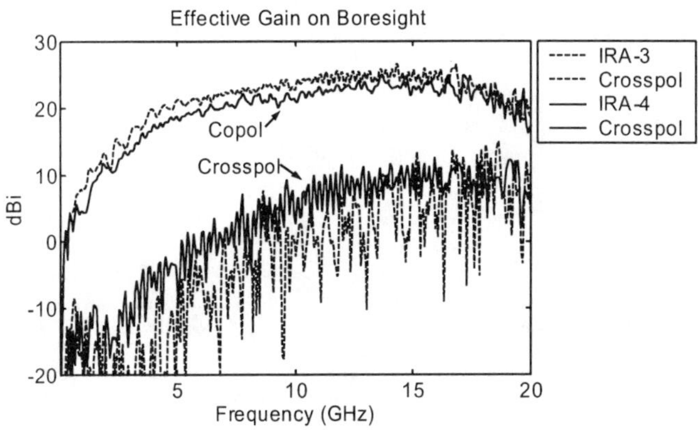

Figure 5.4. Comparison of the effective gain of the IRA-4 and the IRA-3.

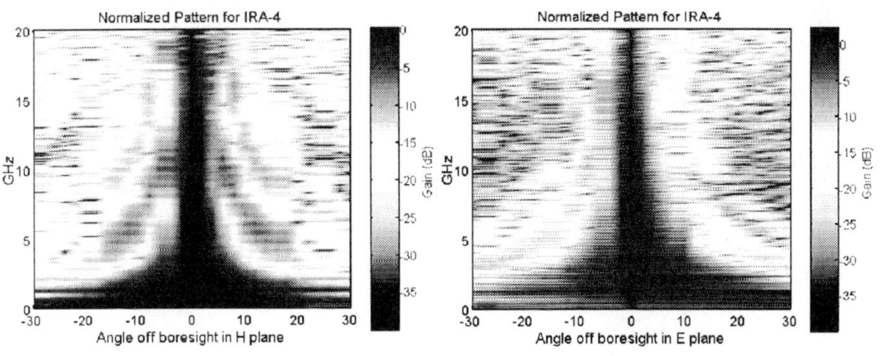

Figure 5.5. Pattern of the IRA-4.

6. Conclusions

Based on our experiments, we are able to make specific recommendations for improvements in the IRA design. By placing the feed arms at ±30° to vertical, we increased gain and reduced crosspol. By including a ground plane, we further reduced the crosspol and increased the mechanical strength of the antenna. By a variety of adjustments, we were able to reduce significantly the TDR reflections from three points on the antenna; the splitter, the feed point, and the resistors. This feature may be important in radar applications, which require low reflections. We have also tested a radome for the antenna built by MRC, which seems to provide good protection to the outside of the antenna with minimal transmission loss.

We also tried several adjustments that did not help as much as we had hoped. We found that dummy cables added to maintain symmetry do not help very much in reducing crosspol. In addition, we were unable to reduce the sidelobes by reducing the fields near the rim of the reflector. We tried this both by moving the feed arms closer to the center of the reflector, and by adding absorber foam. It is possible that a thicker foam might be effective, but this may not be practical.

Of the four antennas we studied, it appears that the IRA-3 is the preferred configuration for most situations. The IRA-1 has lower gain and higher crosspol than the IRA-3, with comparable reflection losses at the resistors. The IRA-2 has a higher crosspol than the IRA-3, and it is slightly less sturdy. The IRA-4 has slightly lower gain than the IRA-3, with higher crosspol and sidelobes.

7. References

1. E. G. Farr, L. H. Bowen, G. R. Salo, J. S. Gwynne, C. E. Baum, W. D. Prather, and T. C. Tran, Studies of an Impulse Radiating Antenna and a Pulse Radiating Antenna Element for SAR and Target Identification Applications, Sensor and Simulation Note 442, March 2000.
2. L. H. Bowen, E. G. Farr, C. E. Baum, T. C. Tran, and W. D. Prather, Experimental Results of Optimizing the Location of Feed Arms in a Collapsible IRA and a Solid IRA, Sensor and Simulation Note 450, November 2000.
3. L. H. Bowen, E. G. Farr, C. E. Baum, T. C. Tran, and W. D. Prather, Results of Optimization Experiments on a Solid Reflector IRA, Sensor and Simulation Note 463, January 2002.
4. J. S. Tyo, Optimization of the Feed Impedance for an Arbitrary Crossed-Feed-Arm Impulse Radiating Antenna, Sensor and Simulation Note 438, November 1999.
5. C. E. Baum, Selection of Angles Between Planes of TEM Feed Arms of an IRA, Sensor and Simulation Note 425, August 1998.
6. L. H. Bowen, E. G. Farr, W. D. Prather, An Improved Collapsible Impulse Radiating Antenna, Sensor and Simulation Note 444, April 2000.
7. L. H. Bowen, E. G. Farr, and W. D. Prather, "A Collapsible Impulse Radiating Antenna," in A. Stone (ed.) Ultra-Wideband, Short-Pulse Electromagnetics 5, Kluwer (in publication), (proceedings of the 2000 EUROEM conference held in Edinburgh, June 2000).
8. C. E. Baum, Symmetry in Single-Polarization Reflector Impulse Radiating Antennas, Sensor and Simulation Note 448, July 2000.
9. M. Abdallah, M. Skipper, D. V. Giri, *et al*, Evaluation of the Terminating Impedance in the Conical-Line Feed of the 6-foot IRA, Prototype IRA Memo 8, April 2001.
10. C. E. Baum, Some Topics Concerning Feed Arms of Reflector IRAs, Sensor and Simulation Note 414, October 1997.
11. D.V. Giri and C.E. Baum, Field Containing Solenoidal Inductors, Sensor and Simulation Note 368, July 1994.

ELECTROMAGNETIC LENS DESIGN

Alexander P. Stone and Carl E. Baum*

1. INTRODUCTION

In this paper we consider the design of transition regions (EM transient lenses) which serve to transmit waves, ideally with no reflection or distortion, from one transmission line to another. The design is accomplished by specifying the shape and medium of the lens region. The physical properties of these lenses, given by the permeability and permittivity, may vary from point to point, but we assume zero conductivity and frequency independence of the media parameters. The distortionless requirement dictates that wavelengths, at the high frequencies of interest, be small compared to the cross section dimensions. Since we may wish to change the direction of propagation of a wave being transmitted from one region to another, allowance must be made for distortion introduced at bends. Examples of EM transition regions include converging, diverging, and bending lens linking cylindrical and conical geometries. Solutions to design problems, which may be exact or approximate, are usually obtained by one of two general methods. One method is differential-impedance matching and transit-time conservation, while the other method is a differential geometric approach. An approach through approximations can also be employed. These transient lens have applications in the design of IRAs (Impulse Radiating Antennas).

2. DIFFERENTIAL GEOMETRIC SCALING

The basic idea is the creation of a class of electromagnetic problems, each having a complicated geometry and medium, which are equivalent under a differential-geometric scaling to an electromagnetic problem having a simpler geometry and medium. The latter problem we term the formal problem while the former problem is the lens (or real-world) problem. Solutions to Maxwell's equations can then be used to specify EM lenses for transitioning TEM waves, without distortion or reflection, between certain types of transmission lines. In this approach to lens design, it is assumed that the formal constitutive parameters, as expressed in some orthogonal curvilinear coordinate system, are of a particularly

* A. P. Stone, Department of Mathematics and Statistics, University of New Mexico, Albuquerque, NM 87131. C. E. Baum, Air Force Research Laboratory, Directed Energy Directorate, Albuquerque, NM 87117.

simple form, i.e., homogeneous, at least as they relate to the allowed field components. Furthermore, the constitutive parameters, before being transformed to the curvilinear system, are assumed to be those of an inhomogeneous but isotropic medium. From this many cases of isotropic inhomogeneous media have found for which certain types of electromagnetic wave propagation could be simply expressed.

In the approach described above, the medium is made inhomogeneous and perfectly conducting boundaries are geometrically arranged such that when they are transformed into the appropriate orthogonal curvilinear coordinates a simpler problem results which can be solved by more standard techniques. The approach can then be used to define geometries for perfectly conducting boundaries and distribution functions for inhomogeneous media such that devices built to such designs will transport electromagnetic waves in certain desirable ways. In particular, cases which in the curvilinear coordinate system corresponds to a problem of a TEM plane wave on a cylindrical transmission line is considered. In the reference Cartesian x, y, z coordinates the waves are still TEM, but not necessarily plane.

Examples of this technique include the design of converging, diverging, and bending lenses (based on bispherical, toroidal, and cylindrical coordinate systems, respectively), for transitioning TEM waves between conical and/or cylindrical transmission lines. These examples are described in Baum and Stone.[1] Special two dimensional lenses, based on conformal transformations, are also found in which one is required to vary only scalar ϵ or μ. Perfectly conducting boundaries are arranged to follow the appropriate coordinates in the orthogonal curvilinear system for which tangential $\vec{E}$ is zero.

3. GENERALIZED INHOMOGENEOUS TEM PLANE WAVES IN INHOMOGENEOUS BUT ISOTROPIC MEDIA

If we allow ϵ' and μ' to be inhomogeneous, we then seek formal fields, depending on the (u_1, u_2, u_3) coordinates, in the form

$$\begin{aligned} \vec{E}' &= \vec{E}_0'\,(u_1, u_2) f(t - \tau(u_3)),\ \vec{E}_0'\,(u_1, u_2)\cdot \vec{1}_3 = 0 \\ \vec{H}' &= \vec{H}_0'\,(u_1, u_2) f(t - \tau(u_3)),\ \vec{H}_0'\,(u_1, u_2)\cdot \vec{1}_3 = 0 \end{aligned} \tag{1}$$

If we restrict the fields to have no third components, then the formal parameters ϵ_3' and μ_3' play no role. The wave propagates in the μ_3 direction with possibly variable speed. The formal wave speed v', and formal impedance Z', may depend on coordinates and satisfy

$$v' = [\mu'\epsilon']^{-\frac{1}{2}}, \quad Z' = \left[\frac{\mu'}{\epsilon'}\right]^{\frac{1}{2}} \tag{2}$$

with $\mu' > 0,\ \epsilon' > 0$. We note that

$$\frac{\partial f}{\partial u_3} = \frac{\partial f}{\partial (t - \tau)}\left[-\frac{d\tau}{du_3}\right] = -\frac{\partial f}{\partial t}\frac{d\tau}{du_3}, \tag{3}$$

a result which is used later.

Consider one of the Maxwell equations

$$\nabla' \times \vec{E'} = -\frac{\partial \vec{B'}}{\partial t} = -\mu' \frac{\partial \vec{H'}}{\partial t} \tag{4}$$

where we think of the (u_1, u_2, u_3) formally as Cartesian (x, y, z) coordinates. This leads to the requirements

$$\frac{\partial}{\partial u_1} E'_{0_2} - \frac{\partial}{\partial u_2} E'_{0_1} = 0, \ \vec{1}_3 \times \vec{E'_0} \frac{d\tau}{du_3} = \mu' \vec{H'_0}, \quad \vec{E'_0} \frac{d\tau}{du_3} = \mu' \vec{1}_3 \times \vec{H'_0} \tag{5}$$

The second Maxwell equation yields a similar result, namely

$$\frac{\partial}{\partial u_1} H'_{0_2} - \frac{\partial}{\partial u_2} H'_{0_1} = 0, \ \vec{1}_3 \times \vec{H'_0} \frac{d\tau}{du_3} = -\epsilon' \vec{E'_0}, \ \vec{H'_0} \frac{d\tau}{du_3} = \epsilon' \vec{1}_3 \times \vec{E'_0} . \tag{6}$$

Thus these results lead to a constraint,

$$\frac{1}{\epsilon'} \frac{d\tau}{du_3} = \mu' \frac{du_3}{d\tau}, \quad \frac{d\tau}{du_3} = \pm[\mu' \epsilon']^{\frac{1}{2}} = \pm v'^{-1} \tag{7}$$

and hence, choosing the + sign for convenience, we obtain

$$\frac{d\tau}{du_3} = v'^{-1}(u_3), \quad \tau = \int_0^{u_3} \frac{du'_3}{v'(u'_3)} = \tau(u_3) \tag{8}$$

Thus the formal wave speed is

$$v' = [\mu' \epsilon']^{-\frac{1}{2}} = v'(u_3) \tag{9}$$

and is a function of u_3 only.

The requirements may now be summarized as

$$\begin{aligned} \frac{\partial}{\partial u_1} E'_{0_2} - \frac{\partial}{\partial u_2} E'_{0_1} &= 0, \vec{1}_3 \times \vec{E'_0} = Z' \vec{H'_0} \\ \frac{\partial}{\partial u_1} H'_{0_2} - \frac{\partial}{\partial u_2} H'_{0_1} &= 0, \vec{1}_3 \times \vec{H'_0} = -Z'^{-1} \vec{E'_0} \end{aligned} \tag{10}$$

Note the local TEM wave condition with the local field ratio given by the local wave impedance (Z' in general being coordinate dependent). There are three conditions now to be satisfied (the last two equations being equivalent). Note that since $\vec{E}'_0$ and $\vec{H}'_0$ are not functions of u_3 we have

$$Z' = \left[\frac{\mu'}{\epsilon'}\right]^{\frac{1}{2}} = Z'(u_1, u_2) \tag{11}$$

so that the ratio μ'/ϵ' is independent of u_3.

The forms of these equations suggest that we try solutions of the form

$$\vec{E_0'}(u_1,u_2) = \nabla_t'\Phi_e(u_1,u_2), \quad \vec{H_0'}(u_1,u_2) = \nabla_t'\Phi_h(u_1,u_2) \tag{12}$$

where the transverse gradients in the formal coordinates make this a two-dimensional potential problem. As usual, the electric potential is solved subject to specified potentials on the conductors passing through the lens, and the magnetic potential has to allow for a discontinuity in passing around a conductor (the discontinuity being equal to the current in the conductor). One finds that requiring Φ_e' to satisfy a Laplace equation for an inhomogeneous medium (variable ϵ') gives a two-dimensional potential which satisfies all of the necessary requirements. An analogous statement can be made about Φ_h'.

4. SCALING GENERALIZED INHOMOGENEOUS TEM PLANE WAVES

The generalized form of our TEM wave satisfies the following:

$$\begin{aligned}
&\vec{E'}=\vec{E_0'}(u_1,u_2)f(t-(u_3)) , \ \vec{E_0'}\cdot\vec{1}_3=0 \\
&\vec{H'}=\vec{H_0'}(u_1,u_2)f(t-(u_3)) , \ \vec{H_0'}\cdot\vec{1}_3=0 \\
&\vec{1}_3\times\vec{E_0'}=Z'\,\vec{H_0'}, \ \vec{H_0}=-Z'^{-1}\,\vec{1}_3\times\vec{E_0'} \\
&Z'=\left[\frac{\mu'}{\epsilon'}\right]^{\frac{1}{2}}=Z'(u_1,u_2) \text{ (a function of only } u_1 \text{ and } u_2) \\
&v'=[\mu'\epsilon']^{-\frac{1}{2}}=v'(u_3) \text{ (a function of only } u_3) \\
&\tau(u_3)=\int_0^{u_3}\frac{du_3'}{v'(u_3')} \\
&\vec{E_0'}(u_1,u_2)=-\nabla_t'\Phi_h(u_1,u_2) , \ \nabla_t'\cdot[\epsilon'\nabla_t'\Phi_e']=0 \\
&\vec{H_0'}(u_1,u_2)=-\nabla_t'\Phi_h(u_1,u_2) , \ \nabla_t'\cdot[u'\nabla_t'\Phi_h']=0
\end{aligned} \tag{13}$$

Here Φ_h' may be multiple valued (as in a conformal transformation) as one goes around a conductor, and one may verify that Maxwell's equations are satisfied.

We now regard the fields as formal and suppose that the (u_1,u_2,u_3) coordinates form an orthogonal curvilinear coordinate system. Then we have a distinction between ϵ' and ϵ and between μ' and μ and the general transformation equations will have the forms

$$\begin{aligned}
&\vec{E'}=(\alpha_{n,m})\cdot\vec{E} , \ \vec{H'}=(\alpha_{n,m})\cdot\vec{H} \\
&(\epsilon_{n,m}')=(\beta_{n,m})\cdot(\epsilon_{n,m}) , \quad (\mu_{n,m}')=(\beta_{n,m})\cdot(\mu_{n,m}) \\
&(\alpha_{n,m})=\begin{pmatrix} h_1 & 0 & 0 \\ 0 & h_2 & 0 \\ 0 & 0 & h_3 \end{pmatrix}, \quad (\beta_{n,m})=\begin{pmatrix} h_2h_3/h_1 & 0 & 0 \\ 0 & h_3h_1/h_2 & 0 \\ 0 & 0 & h_1h_2/h_3 \end{pmatrix}
\end{aligned} \tag{14}$$

This lets both the formal and real media be inhomogeneous but isotropic, noting that only 1, 1 and 2, 2 components are relevant since the 3-components of the fields are zero. Since the first two diagonal elements of the constitutive-parameter dyadics are equal we have

$$\mu' = \frac{h_2 h_3}{h_1}\mu = \frac{h_3 h_1}{h_2}\mu\,, \quad \epsilon' = \frac{h_2 h_3}{h_1}\epsilon = \frac{h_3 h_1}{h_2}\epsilon \tag{15}$$

which implies (taking the positive square root) $h_1 = h_2$. This in turn implies that *surfaces of constant u_3 can only be spheres or planes*. Furthermore we have

$$\mu' = h_3\mu\,,\ \epsilon' = h_3\epsilon\ \ \frac{\mu'}{\epsilon'} = \frac{\mu}{\epsilon}\ \text{(a function of } u_1 \text{ and } u_2 \text{ only)} \tag{16}$$

$$Z \equiv \left[\frac{\mu}{\epsilon}\right]^{\frac{1}{2}} = Z'\ \text{(a function of } u_1 \text{ and } u_2 \text{ only)},\quad v \equiv [\mu\epsilon]^{-\frac{1}{2}} = h_3[\mu'\epsilon']^{-\frac{1}{2}} = h_3 v'$$

While v' is a function of only u_3, v has the extra factor h_3 which can in general be a function of the various u_n.

5. UNIFORM PERMEABILITY LENSES

If we now impose the requirement that $\mu = \mu_0$, so that the permeability is uniform and that of free space, then a restriction on the scale factor h_3 results. In general, the permittivity is allowed to be variable. For physical realizability we also require in the lens region that $\epsilon \geq \epsilon_{\min} \geq \epsilon_0$ i.e., a permittivity bounded below by that of free space. Note that μ' and ϵ', being formal, do not have the same restrictions. For the wave speed we now have

$$v = [\mu_0\epsilon]^{-\frac{1}{2}} \leq [\mu_0\epsilon_0]^{-\frac{1}{2}} \equiv c \tag{17}$$

which is the basic causality restriction in the lens region. For the wave impedance we have

$$Z = \left[\frac{\mu_0}{\epsilon}\right]^{\frac{1}{2}} \leq \left[\frac{\mu_0}{\epsilon_0}\right]^{\frac{1}{2}} \equiv Z_0 \tag{18}$$

and consequently $\epsilon = \epsilon(u_1, u_2)$, so that the permittivity is a function of u_1, u_2 only. Similarly $v = [\mu_0\epsilon]^{-\frac{1}{2}}$ is a function of u_1, u_2 only, which restricts h_3 to the form

$$h_3 = \frac{v(u_1, u_2)}{v'(u_3)}. \tag{19}$$

If we now modify the usual cylindrical coordinates (Ψ, ϕ, z) so that our (u_1, u_2, u_3) coordinates become $u_1 = z$, $u_2 = \Psi$, and $u_3 = \Psi_{\max}\phi$, then a dielectric bending lens can be specified. Note that this is a case in which constant u_3 surfaces are planes. Here $0 < \Psi \leq \Psi_{\max}$ and $\Psi_{\max}$ is the maximum Ψ for the lens region. The scale factors then become

$$h_1 = h_2 = 1 \text{ and } h_3 = \frac{\Psi}{\Psi_{\max}} \tag{20}$$

and the constraints (17) and (19) are satisfied. If $v' = (\mu'\epsilon')^{-\frac{1}{2}} = (\mu_0\epsilon_0)^{-\frac{1}{2}} = c$, then

$$v = h_3 v' = h_3 c = \frac{\Psi}{\Psi_{\max}} c \leq c \tag{21}$$

and the formal medium has constitutive parameters

$$\mu' = h_3 \mu_0 = \frac{\Psi}{\Psi_{\max}} \mu_0, \quad \epsilon = \left(\frac{\mu_0}{\mu'}\right)^2 \epsilon_0 = h_3^{-2} \epsilon_0 = \left(\frac{\Psi_{\max}}{\Psi}\right)^2 \epsilon_0 \tag{22}$$

Hence when $0 < \Psi \leq \Psi_{max}$,

$$\epsilon = \left(\frac{\Psi_{\max}}{\Psi}\right)^2 \epsilon_0 \geq \epsilon_0, \quad Z = Z' = \left(\frac{\mu_0}{\epsilon}\right)^{\frac{1}{2}} = h_3 Z_0 = \frac{\Psi}{\Psi_{\max}} Z_0 \leq Z_0 \tag{23}$$

Note that since μ' is a formal parameter, it is of no consequence that $\mu' < \mu_0$. The formal fields can then be found via an electric potential Φ'_e from the equations

$$\vec{E}'_0 = -\nabla'_t \Phi'_e, \quad \vec{H}'_0 = \frac{1}{Z'} \vec{1}_0 \times \vec{E}'_0, \quad \nabla'_t \cdot [\epsilon' \nabla'_t \Phi'_0] = 0 \tag{24}$$

where $\nabla'_t = \vec{1}_z \frac{\partial}{\partial z} + \vec{1}_\Psi \frac{\partial}{\partial \Psi}$.

6. IMPEDANCE-MATCHING AND TRANSIT-TIME CONSERVATION

An alternative approach to exact transient lens design is one which might be termed a differential-impedance-matching and transit-time conservation method. We must first differentially match impedances at all waveguide and lens boundaries so that a TEM wave may be transmitted from one region to another without reflection. Secondly, in order that a wave be transmitted undistorted, a plane or spherical TEM wave in one region should go into one such wave in another region and consequently the travel times for waves following different paths should be conserved. These two conditions usually give rise to a system of nonlinear differential equations whose solutions will determine the lens geometry and medium.

This general approach to lens design has been used to construct a number of examples of exact anisotropic lenses. In these examples the permittivity is in general a function of position.

7. APPROXIMATE LENSES

If the equal time requirement is maintained, but the impedance matching condition relaxed from a differential one (along every ray) to a global one in which the transmission line impedance is kept constant on both sides of the lens (as well as inside the lens if possible), then one may develop a class of approximate lenses. What results is an average kind of impedance matching, allowing some small mismatch along ray paths, particularly at lens boundaries. By carefully choosing lens parameters the high-frequency properties can sometimes be made to have small degradation. A specific example of this approximate procedure is that of a lens feeding a circular coaxial line, specified so that a wave launched at an apex propagates through the lens region onto the coaxial line with minimum reflection and distortion. The equal-transit-time condition dictates that the lens shape is a prolate spheroid.

The examples of such approximate lenses can be expanded to include those for TEM waveguides more complicated than circular coaxes. Another interesting geometry is the balanced two-wire structure such as might be used for the feed structure launching a spherical TEM wave into a paraboloidal reflector. As one approaches the conical apex, the electric field strength can get quite large so that various dielectric insulation techniques are needed. Where the dielectric has a relative permittivity significantly greater than 1.0, the dielectric medium can be thought of as a lens which needs to be shaped (along with the conductors) to minimize the pulse distortion in this region.

The reasons for employing dielectric lenses in IRA applications may be summarized as follows. First of all, the ideal point source at the focal point is physically realized by a closing switch whose center is located at a point slightly inward (toward the reflector). The high voltage between the two electrodes of the switch requires the use of a dielectric medium surrounding the switch is required for high-voltage standoff, it can also be used to ensure a spherical TEM wave launch, as if an ideal point source was located at the focal point. All rays emanating from the switch center should exit the lens boundary and end up on a spherical wavefront annotated on the focal point at the same instant. This condition ensures an optimal spherical TEM wave launch onto the reflector. The feed and the radiating system (reflector in this case) should be able to propagate and radiate well into a couple of GHz. The frequency dependence of the dielectric constant of the lens medium is also a factor in the choice of the material.

8. CONCLUDING REMARKS

Thus concepts of differential geometry may be used to synthesize inhomogeneous, and also anisotropic media, in which electromagnetic waves propagate in a controlled way to make these special transition sections (lenses) for EM waves. This idea is built in part on the application made by A. Einstein of differential geometry for spacetime to general relativity; here, however, the lens problem is stated in a form related to 3-dimensional Euclidean space, in which 2-dimensional surfaces can be imbedded. These imbedded surfaces are in general non-Euclidean (i.e., Riemannian). The differential geometric approach to the design of these transition regions between certain types of transmission lines is a scaling, or transform, method. Such methods, which scale a problem and its solution into a class of equivalent problems, have long been of interest in fluid dynamics and mechanics, and are now finding application in electromagnetic theory.

The generalized form of a TEM plane wave involving inhomogeneous constitutive parameters can be used for the formal fields for generating lens designs based on differential-geometric scaling. Among other things this allows one to have purely dielectric lenses (free space permeability). The procedure described here has essentially given an extra degree of freedom on transient lens design via differential-geometric scaling.

While our theoretical analysis leads to lens materials with a specified inhomogeneous permittivity, the construction of these lenses is another matter. Bends in waveguides pose a problem for high-voltage ultra-wideband systems with low-loss and fast rise time requirements. The reason for this difficulty is that only straight sections of conventional transmission lines can support the pure TEM mode that is needed to preserve the rise time of a transmitted pulse. This problem is addressed in Refs. 1–4, for example, where it is shown that a waveguide bend can be embedded within a purely dielectric lens with an inhomogeneous, frequency independent, isotropic permittivity. Bigelow and Farr introduce the concept of a graded dielectric lens and consider several approaches to the implementation of a compensated waveguide bend.[6] Their designs represent a compromise between modeling fidelity and ease of construction. One concept discussed in Bigelow and Farr is a layered approximation of a graded dielectric strip line bend. The individual layers can be machined from sheets of low-loss, uniform permittivity plastic materials. The preliminary experimental results in Bigelow and Farr show that the rise time of a pulse propagated through a smooth bend through a total angle of 90° is reduced by the layered dielectric approximating Eq. (23).

This work was supported in part by the Air Force Research Laboratory (Kirtland AFB) and the Air Force Office of Scientific Research.

REFERENCES

1. C. E. Baum and A. P. Stone. *Transient lens synthesis: Differential geometry in electromagnetic theory* (Hemisphere, New York, 1991).
2. C. E. Baum, Two-Dimensional Inhomogeneous Dielectric Lenses for E-Plane Bends of TEM Waves Guided Between Perfectly Conducting Sheets, *Sensor and Simulation Note 388*, Air Force Research Lab, Kirtland AFB NM (October 1995).
3. C. E. Baum, Azimuthal TEM Waveguides in Dielectric Media, *Sensor and Simulation Note 397*, Air Force Research Lab, Kirtland AFB NM (March 31 1996).
4. C. E. Baum, Dielectric Body-of-Revolution Lenses With Azimuthal Propagation, *Sensor and Simulation Note 393*, Air Force Research Lab, Kirtland AFB NM (March 1996).
5. C. E. Baum, Use of Generalized Inhomogeneous TEM Plane Waves in Differential Geometric Lens Synthesis, *Sensor and Simulation Note 405*, Air Force Research Lab, Kirtland AFB NM (December 5 1998). Also in *Proceedings of 1998 International Symposium on Electromagnetic Theory* (The International Union of Radio Science, Thessaloniki, Greece, 1998), pp. 636-638.
6. W. S. Bigelow and E. G. Farr, Minimizing Dispersion in a TEM Waveguide Bend by a Layered Approximation of a Graded Dielectric Material, *Sensor and Simulation Note 416*, Air Force Research Lab, Kirtland AFB NM (January 1998).

LOCATION AND ORIENTATION OF ELECTRICALLY SMALL TRANSMITTING AND RECEIVING ANTENNA PAIRS WITH COMMON LINEAR POLARIZATION AND BEAM DIRECTION FOR MINIMAL MUTUAL COUPLING

C.E. Baum*

1. INTRODUCTION

In measurements of electromagnetic scattering (such as by a radar), one needs to transmit (from an antenna) an incident wave which in turn scatters from the object of interest (the target), the scattered field being sensed in turn via some receive antenna. The far-field amplitude decays with distance r proportional to r^{-1}. Similarly the scattered far field is proportional to r^{-1} for a total decay of the received signal (voltage) proportional to r^{-2}. If the same antenna is used for both transmission and reception (monostatic), there is then a problem in accurately measuring the scattered signal in the presence of the transmitted signal. One can use time gating to separate these if the time duration is shorter than the round-trip transit time to the target. So a short pulse in transmission offers some advantages. However, the transmitted signal should decay sufficiently rapidly in the source and antenna, noting the possibility of small reflections and resonances, which carry out in time. The decay should be sufficiently rapid such that at the round-trip time from the target the signal is negligible compared to the scattered signal in the antenna.

For a given (say linear) transmit polarization there are, in general, two scattering polarizations to measure. For two independent transmit polarizations (say horizontal h and vertical v) this gives a 2 x 2 scattering dyadic $\overset{\leftrightarrow}{\Lambda}$ which one can use to identify the

* Air Force Research Laboratory, Directed Energy Directorate, Kirtland Air Force Base, NM 87117-5776.

target if broadband pulses (or multiple frequencies) are used. One can use symmetry planes[6,7,9] to give both horizontal and vertical operation in the same antenna. For monostatic (backscatttering) crosspol measurements ($\Lambda_{b_{h,v}} = \Lambda_{b_{v,h}}$ (reciprocity)) the horizontal transmit does not couple (ideally) into the vertical receive (and conversely) as one has symmetric and the other antisymmetric fields with respect to a particular symmetry plane (there being possibly multiple symmetry planes). Practically speaking, geometric imperfections mean that the transmit signal (voltage) is reduced about two decades (depending on accuracy) in the receive channel. For copol measurements ($\Lambda_{b_{h,h}}$ and $\Lambda_{b_{v,v}}$) one does not have the advantage of the reduced coupling to the transmit signal.

One can improve the situation somewhat by having the transmit and receive channels not in the same antenna (including dual polarization), but by using physically separated transmit and receive antennas. In general, this gives a bistatic radar, but for present purposes, let us assume that the separation between the antennas is small compared to the distance to the target so that the orientations of the antennas with respect to the target are approximately the same, giving approximately a monostatic radar (quasi-monostatic, quasi-backscattering). One can easily correct for the different distances to the target (for staggered antennas) via the speed of light. We can then think of an approximate backscattering dyadic $\overleftrightarrow{\Lambda}_b$ which we can regard as 2 x 2 in the (h,v) coordinates.

With separated antennas we can still use a common symmetry plane for symmetric/antisymmetric measurements of crosspol.[5,6] This gives additional (perhaps decades) of isolation beyond that discussed above. More importantly, separated antennas give isolation between copol transmit/receive antennas. This suggests the question: What should be the relative positions of the two copol antennas for minimum coupling? This paper is primarily concerned with this question.

For simplicity we assume identical transmit and receive antennas. Furthermore, we assume that the antennas are electrically small for a definitive analysis. Such antennas are characterized by their electric and magnetic-dipole moments in transmission, and reciprocity takes care of reception. The canonical cases are electric dipoles, magnetic dipoles, and combined dipoles (balanced electric and magnetic moments at right angles). Furthermore, we assume that the two antennas are identical except for a spatial translation (that is, no rotation) so that they are both oriented the same with respect to the target.

2. FIELDS FROM DIPOLES

Fundamental to the following are the fields produced by electric and magnetic dipoles. For this discussion we have the illustration in Fig. 2.1. Dipoles subscripted 1 are located at the coordinate origin $\vec{r} = \vec{0}$. For convenience we take the dipole orientations as

$$\vec{p}_1 = p_1 \, \vec{1}_z, \qquad \vec{m}_1 = m_1 \, \vec{1}_x, \tag{2.1}$$

and think of these as characterizing the transmit antenna. The receive antenna is characterized by dipole moments (in transmission) subscripted 2. They are located at some arbitrary $\vec{r} \neq \vec{0}$ but are oriented as

$$\vec{p}_2 = p_2 \, \vec{1}_z, \qquad \vec{m}_2 = m_2 \, \vec{1}_x, \tag{2.2}$$

that is, parallel to the moments of antenna 1. Note that

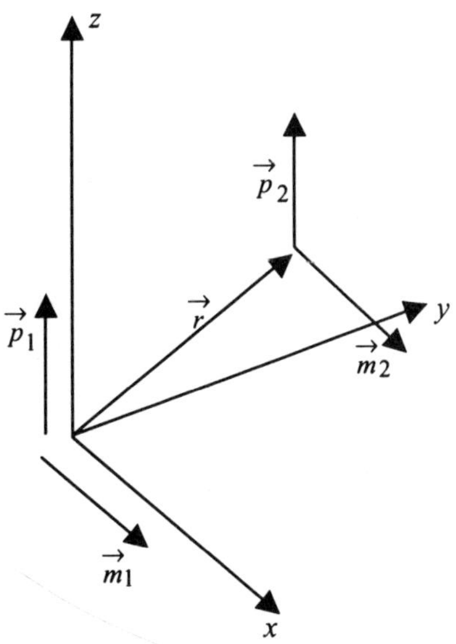

Figure 2.1. Dipole locations.

$$\vec{p}_n \bullet \vec{m}_n = 0 \ \text{ for } n = 1, 2, \tag{2.3}$$

so that when we come to combined dipoles the electric and magnetic moments are mutually perpendicular.[1] Note also that

$$\vec{1}_c \equiv \vec{1}_y = \vec{1}_z \times \vec{1}_x \tag{2.4}$$

is a particular direction of interest as the beam center for combined dipoles.

For the present analysis it is convenient to work in complex frequency domain, where ~ denotes the two-sided Laplace transform from time t to complex frequency s and

$$s = \Omega + j\omega \tag{2.5}$$

is the Laplace-transform variable or complex frequency. For convenience we also have

$\gamma = s/c$ is the propagation constant, $c = (\mu_0\varepsilon_0)^{-1/2}$ is the speed of light, $Z_0 = (\mu_0/\varepsilon_0)^{1/2}$ is the impedance of free space, μ_0 is the permeability of free space, and ε_0 is the permittivity of free space.

We have various vectors and dyadics as

$$\overleftrightarrow{1} \equiv \vec{1}_x \vec{1}_x + \vec{1}_y \vec{1}_y + \vec{1}_z \vec{1}_z = \vec{1}_r \vec{1}_r + \vec{1}_\theta \vec{1}_\theta + \vec{1}_\phi \vec{1}_\phi \equiv \text{identity},$$
$$\overleftrightarrow{1}_r \equiv \overleftrightarrow{1} - \vec{1}_r \vec{1}_r \equiv \text{transverse (to } r\text{) identity.} \tag{2.6}$$

Besides Cartesian coordinates (x,y,z), we have spherical coordinates (r,θ,ϕ). Spherical coordinates are normally defined with z as the preferred axis. Let us subscript these coordinates with z which is the preferred axis giving

$$\vec{r}_p = x\,\vec{1}_x + y\,\vec{1}_y + z\,\vec{1}_z$$
$$= r_p\left[\sin(\theta_z)\cos(\phi_z)\,\vec{1}_x + \sin(\theta_z)\sin(\phi_z)\,\vec{1}_y + \cos(\theta_z)\,\vec{1}_z\right]. \tag{2.7}$$

Similarly base a set of spherical coordinates on $\vec{m}_1$ (x axis) as

$$\vec{r}_m = r_m\left[-\sin(\theta_x)\cos(\phi_x)\,\vec{1}_z + \sin(\theta_x)\sin(\phi_x)\,\vec{1}_y + \cos(\theta_x)\,\vec{1}_x\right], \tag{2.8}$$

which is readily found by the substitutions $z \to x$ and $x \to -z$. Note the commonality of

$$\vec{r}_p = \vec{r}_m = \vec{r}, \quad \vec{1}_{r_p} = \vec{1}_{r_m} = \vec{1}_r, \quad \overleftrightarrow{1}_{r_p} = \overleftrightarrow{1}_{r_m} = \overleftrightarrow{1}_r. \tag{2.9}$$

2.1. Electric-Dipole Fields

Following Ref. 1 we have the fields from an electric dipole $\vec{p}_1$ as

$$\tilde{\vec{E}}_p = e^{-\gamma r}\left[-\left[\frac{1}{4\pi r^3}\frac{1}{\varepsilon_0} + \frac{sZ_o}{4\pi r^2}\right]\left[\overleftrightarrow{1} - 3\,\vec{1}_r\,\vec{1}_r\right] - \frac{s^2\mu_o}{4\pi r}\overleftrightarrow{1}_r\right]\cdot\tilde{\vec{p}}_1,$$
$$\tilde{\vec{H}}_p = e^{-\gamma r}\left[-\frac{s}{4\pi r^2} - \frac{s^2}{4\pi r}\frac{1}{c}\right]\vec{1}_r \times \tilde{\vec{p}}_1. \tag{2.10}$$

2.2. Magnetic-Dipole Fields

Likewise we have the fields from a magnetic dipole $\vec{\tilde{m}}_1$ as

$$\vec{\tilde{E}}_m = e^{-\gamma r}\left[\frac{s\mu_o}{4\pi r^2}+\frac{s^2\mu_o}{4\pi r}\frac{1}{c}\right]\vec{1}_r\times\vec{\tilde{m}}_1,$$

$$\vec{\tilde{H}}_m = e^{-\gamma r}\left[-\left[\frac{1}{4\pi r^3}+\frac{s}{4\pi r^2}\frac{1}{c}\right]\left[\overleftrightarrow{1}-3\vec{1}_r\vec{1}_r\right]-\frac{s^2}{4\pi r}\frac{1}{c^2}\overleftrightarrow{1}_r\right]\cdot\vec{\tilde{m}}_1. \tag{2.11}$$

2.3. Combined-Dipole Fields

For combined dipoles we use the special $\vec{p}\times\vec{m}$ form given by (2.2) together with[1]

$$p_n=\frac{m_n}{c}\ \text{for}\ n=1,2,\quad \vec{p}_n=-\frac{1}{c}\vec{1}_y\times\vec{m}_n,\quad \vec{m}_n=c\,\vec{1}_y\times\vec{p}_n. \tag{2.12}$$

The fields produced by antenna 1 are then

$$\begin{aligned}\vec{\tilde{E}} &= \vec{\tilde{E}}_p+\vec{\tilde{E}}_m = e^{-\gamma r}\Bigg[-\frac{1}{4\pi r^3}\frac{1}{\varepsilon_0}\left[\overleftrightarrow{1}-3\vec{1}_r\vec{1}_r\right]\cdot\vec{\tilde{p}}_1\\ &\quad-\frac{sZ_o}{4\pi r^2}\left[\overleftrightarrow{1}-3\vec{1}_r\vec{1}_r\right]\cdot\vec{\tilde{p}}_1-\vec{1}_r\times\frac{\vec{\tilde{m}}_1}{c}-\frac{s^2\mu_o}{4\pi r}\left[\vec{1}_r\cdot\vec{\tilde{p}}_1-\vec{1}_r\times\frac{\vec{\tilde{m}}_1}{c}\right]\Bigg]\\ &= e^{-\gamma r}\Bigg[-\frac{1}{4\pi r^3}\frac{1}{\varepsilon_0}\left[\overleftrightarrow{1}-3\vec{1}_r\vec{1}_r\right]\\ &\quad-\frac{sZ_o}{4\pi r^2}\left[\left[\overleftrightarrow{1}-3\vec{1}_r\vec{1}_r\right]-\vec{1}_y\vec{1}_r+\left[\vec{1}_r\cdot\vec{1}_y\right]\overleftrightarrow{1}\right]\\ &\quad-\frac{s^2\mu_o}{4\pi r}\left[\overleftrightarrow{1}_r-\vec{1}_y\vec{1}_r+\left[\vec{1}_r\cdot\vec{1}_y\right]\overleftrightarrow{1}\right]\Bigg]\cdot\vec{\tilde{p}}_1,\end{aligned} \tag{2.13}$$

$$\tilde{\overleftrightarrow{H}} = \tilde{\overleftrightarrow{H}}_m + \tilde{\overleftrightarrow{H}}_p = e^{-\gamma r}\left[-\frac{1}{4\pi r^3}\left[\overleftrightarrow{1} - 3\,\vec{1}_r\,\vec{1}_r\right]\bullet \tilde{\vec{m}}_1 \right.$$

$$- \frac{s}{4\pi r^2}\left[\left[\overleftrightarrow{1} - 3\,\vec{1}_r\,\vec{1}_r\right]\bullet \frac{\tilde{\vec{m}}_1}{c} + \vec{1}_r \times \tilde{\vec{p}}_1\right] - \frac{s^2}{4\pi r}\frac{1}{c}\left.\left[\overleftrightarrow{1}_r \bullet \frac{\tilde{\vec{m}}_1}{c} + \vec{1}_r \times \tilde{\vec{p}}_1\right]\right]$$

$$= e^{-\gamma r}\left[-\frac{1}{4\pi r^3}\left[\overleftrightarrow{1} - 3\,\vec{1}_r\,\vec{1}_r\right]\right.$$

$$- \frac{s}{4\pi r^2}\frac{1}{c}\left[\overleftrightarrow{1} - 3\,\vec{1}_r\,\vec{1}_r - \vec{1}_y\,\vec{1}_r + \left[\vec{1}_r \bullet \vec{1}_y\right]\overleftrightarrow{1}\right]$$

$$\left. - \frac{s^2}{4\pi r}\frac{1}{c^2}\left[\overleftrightarrow{1}_r - \vec{1}_y\,\vec{1}_r + \left[\vec{1}_r \bullet \vec{1}_y\right]\overleftrightarrow{1}\right]\right] \bullet \tilde{\vec{m}}_1. \tag{2.14}$$

See Ref. 8 for appropriate vector and dyadic identities. Here we see the symmetry in the expressions for $\tilde{\vec{E}}$ and $\tilde{\vec{H}}$, this being more transparent if we form $Z_0\tilde{\vec{H}}$ and use $\tilde{\vec{m}}_1/c$ to keep the units consistent with $\tilde{\vec{p}}_1$ (but not the orientation), and thereby make the coefficients of the various terms also agree. For example, we can see that the variation of $\tilde{\vec{E}}$ in the *yz* plane is the same as the variation of $\tilde{\vec{H}}$ in the *xy* plane.

In the $\vec{1}_y$ direction (beam center) the fields are

$$\tilde{\vec{E}} = e^{-\gamma r}\left[-\frac{1}{4\pi r^3}\frac{1}{\varepsilon_o} - \frac{sZ_0}{2\pi r^2} - \frac{s^2\mu_o}{2\pi r}\right]\vec{1}_z\,\tilde{p}_1,$$

$$\tilde{\vec{H}} = e^{-\gamma r}\left[-\frac{1}{4\pi r^3} - \frac{s}{2\pi r^2}\frac{1}{c} - \frac{s^2}{2\pi r}\frac{1}{c^2}\right]\vec{1}_x\,\tilde{m}_1 = \frac{1}{Z_0}\vec{1}_y \times \tilde{\vec{E}}, \tag{2.15}$$

showing the TEM character of the fields, including the near-field terms. In the $-\vec{1}_y$ (back) direction the fields are

$$\tilde{\vec{E}} = -\frac{e^{-\gamma r}}{4\pi r^3}\frac{\tilde{p}_1}{\varepsilon_o}\vec{1}_z, \qquad \tilde{\vec{H}} = -\frac{e^{-\gamma r}}{4\pi r^3}\tilde{m}_1\,\vec{1}_x = \frac{1}{Z_0}\vec{1}_y \times \tilde{\vec{E}}. \tag{2.16}$$

Note that the fields here are also TEM, but the Poynting vector is in the $+\vec{1}_y$ direction, that is, back toward the antenna.

Considering only the far field (the r^{-1} term) the radiated power has been shown[3,5] to be proportional to

$$\left[1+\cos(\theta_y)\right]^2 = \left[1 + \vec{1}_r \bullet \vec{1}_y\right]^2, \tag{2.17}$$

where θ_y is the angle of the observer from the y axis. This is often called a cardiod pattern.

3. FAR-FIELD COUPLING BETWEEN ANTENNAS

3.1. General Considerations

As discussed in Refs. 3 and 4, the far field radiated by an antenna (r^{-1} term) can be characterized by

$$\vec{\tilde{E}}_r = \frac{e^{-\gamma r}}{r} \vec{\tilde{F}}(\vec{1}_0,s)\tilde{V}_t, \qquad \vec{1}_0 \equiv \text{direction to the observer.} \tag{3.1}$$

Here we take the form of the equations where the voltage $\tilde{V}_t$ is the "transmitted" voltage and the source impedance is assumed matched to the antenna impedance (resistive and frequency-independent in some interesting cases).[10]

As a receiver such an antenna has a response characterized by an effective height as

$$\tilde{V}_r = \vec{\tilde{h}}(\vec{1}_i,s) \bullet \vec{\tilde{E}}^{(inc)}, \qquad \vec{1}_i \equiv \text{direction of incidence toward antenna,} \tag{3.2}$$

where the load impedance is also assumed matched to the antenna impedance. Reciprocity relates transmission and reception as

$$\vec{\tilde{F}}(\vec{1}_0,s) = \frac{s\mu_0}{2\pi R} \vec{\tilde{h}}(-\vec{1}_0,s), \tag{3.3}$$

R = source impedance = load impedance = antenna input impedance. Note that this constraint on the impedance is not essential and will not affect our choices of antenna locations.

3.2. Electric Dipoles

Considering both antennas are electric dipoles as in (2.11) we can identify from the r^{-1} term the transmission from antenna 1 via

$$\vec{1}_0 = \vec{1}_r, \qquad \tilde{f}_p \equiv \frac{\tilde{p}_1}{\tilde{V}_t},$$

$$\tilde{\vec{F}}_p(\vec{1}_r,s) = -\frac{s^2\mu_o}{4\pi}\overleftrightarrow{1}_r \cdot \frac{\tilde{\vec{p}}_1}{\tilde{V}_t} = -\frac{s^2\mu_o}{4\pi}\overleftrightarrow{1}_r \cdot \vec{1}_z\tilde{f}_p = \frac{s^2\mu_o}{4\pi}\vec{1}_{\theta_z}\sin(\theta_z)\tilde{f}_p. \tag{3.4}$$

Here $\tilde{f}_p$ merely characterizes the linear property of the antenna in terms of electric-dipole moment per unit driving voltage. The angle θ_z is the usual spherical polar angle (that is, θ) with respect to the electric-dipole orientation as in Fig. 2.1 and Eq. (2.8).

In reception an electric-dipole antenna at $\vec{r}$ with the same orientation as the first is characterized by

$$\vec{1}_i = -\vec{1}_r,$$

$$\tilde{\vec{h}}_p(-\vec{1}_r,s) = \frac{2\pi R}{s\mu_o}\tilde{\vec{F}}(\vec{1}_r,s) = -\frac{sR}{2}\overleftrightarrow{1}_r \cdot \vec{1}_z\tilde{f}_p = \frac{sR}{2}\vec{1}_{\theta_z}\sin(\theta_z)\tilde{f}_p. \tag{3.5}$$

Here the second antenna is assumed identical to the first giving the same $\tilde{f}_p$, but this is not necessary. Only the orientation is important for present purposes.

The response of antenna 2 to antenna 1 is then proportional to

$$\tilde{\vec{F}}_p(\vec{1}_r,s) \cdot \tilde{\vec{h}}_p(\vec{1}_r,s) = \frac{s^3R\mu_o}{8\pi}\sin^2(\theta_z)\tilde{f}_p^2, \tag{3.6}$$

which is zero on the z axis, that is, at

$$\theta_z = 0,\pi, \qquad \vec{r} = z\vec{1}_z, \qquad r \neq 0. \tag{3.7}$$

So parallel electric dipoles do not couple in the far-field approximation if they are *collinear*.

3.3. Magnetic Dipoles

The case of two parallel magnetic dipoles is dual to that of Section 3.2. Here we have in transmission (from (2.12))

$$\begin{aligned}
\vec{1}_0 &= \vec{1}_r, \qquad \tilde{f}_m \equiv \frac{\tilde{m}_1}{\tilde{V}_t}, \\
\vec{\tilde{F}}_m(\vec{1}_r,s) &= \frac{s^2\mu_0}{4\pi c}\vec{1}_r \times \vec{1}_x \tilde{f}_m \\
&= \frac{s^2\mu_o}{4\pi c}\left[-\sin(\theta_m)\cos(\theta_m)\vec{1}_y - \sin(\theta_m)\sin(\theta_m)\vec{1}_z\right]\tilde{f}_m \\
&= -\frac{s^2\mu_o}{4\pi c} f_m \sin(\theta_m) f_m \vec{1}_{\phi_m},
\end{aligned} \tag{3.8}$$

using the spherical coordinates in (2.9) based on the magnetic-dipole orientation.

In reception we have

$$\vec{\tilde{h}}_m(-\vec{1}_r,s) \bullet \frac{2\pi R}{s\mu_o}\vec{\tilde{F}}_m(\vec{1}_r,s) = -\frac{sR}{2c}\sin(\theta_m)\vec{1}_{\phi_m}\tilde{f}_m^2. \tag{3.9}$$

The response of antenna 2 to antenna 1 is then proportional to

$$\vec{\tilde{F}}_m(\vec{1}_r,s) = \vec{\tilde{h}}_m(\vec{1}_r,s) = \frac{s^3\mu_o R}{8\pi c^2}\sin^2(\theta_m)\tilde{f}_m, \tag{3.10}$$

which is zero on the x axis, that is, at

$$\theta_m = 0,\pi, \qquad \vec{r} = x\vec{1}_x, \qquad r \neq 0. \tag{3.11}$$

Parallel magnetic dipoles then do not couple in the far-field approximation if they are collinear. Note that this result is inconsistent with that for electric dipoles in (3.7) with electric dipoles on the z axis but magnetic dipoles on the x axis. With the given dipole orientations then we cannot simultaneously match both conditions.

3.4. Combined Dipoles

Now combine the electric and magnetic dipoles for both antennas 1 and 2 in the balanced form discussed in Section 2.3. From (2.14) we have from the r^{-1} term the antenna transmission as

$$\vec{1}_0 = \vec{1}_r, \qquad \tilde{f}_c = \frac{\tilde{p}_1}{\tilde{V}_t} = \frac{\tilde{m}_1}{c\tilde{V}_t},$$

$$\tilde{\vec{F}}_c(\vec{1}_r,s) = -\frac{s^2\mu_o}{4\pi}\left[\overleftrightarrow{1}_r - \vec{1}_y\vec{1}_r + \left[\vec{1}_r \bullet \vec{1}_y\right]\overleftrightarrow{1}\right] \bullet \vec{1}_z \tilde{f}_c \tag{3.12}$$

$$= -\frac{s^2\mu_o}{4\pi}\left[\overleftrightarrow{1}_r \bullet \vec{1}_z - \vec{1}_r \times \vec{1}_x\right]\tilde{f}_c.$$

From reciprocity an antenna at $\vec{r}$ with the same orientation as the first is characterized by

$$\vec{1}_i = \vec{1}_r,$$

$$\tilde{\vec{h}}_c(-\vec{1}_r,s) = \frac{2\pi R}{s\mu_o}\tilde{\vec{F}}_c(\vec{1}_r,s)$$

$$= -\frac{sR}{2}\left[\overleftrightarrow{1}_r - \vec{1}_y\vec{1}_r + \left[\vec{1}_r \bullet \vec{1}_y\right]\overleftrightarrow{1}\right] \bullet \vec{1}_z \tilde{f}_c \tag{3.13}$$

$$= -\frac{sR}{2}\left[\overleftrightarrow{1}_r \bullet \vec{1}_z - \vec{1}_r \times \vec{1}_x\right]\tilde{f}_c.$$

The response of antenna 2 to antenna 1 is then proportional to

$$
\begin{aligned}
&\tilde{\vec{F}}_c(\vec{1}_r,s) \bullet \tilde{\vec{h}}_c(\vec{1}_r,s) \\
&= \frac{s^3 R}{8\pi}\tilde{f}_c^2\left[\overset{\leftrightarrow}{1}_r \bullet \vec{1}_z - \vec{1}_r \times \vec{1}_x\right] \bullet \left[\overset{\leftrightarrow}{1}_r \bullet \vec{1}_z + \vec{1}_r \times \vec{1}_x\right] \\
&= \frac{s^3 R}{8\pi}\tilde{f}_c^2\left[\vec{1}_r \bullet \overset{\leftrightarrow}{1}_r \bullet \vec{1}_z - \left[\vec{1}_r \times \vec{1}_x\right] \bullet \left[\vec{1}_r \times \vec{1}_x\right]\right] \\
&= \frac{s^3 R}{8\pi}\tilde{f}_c^2\left[\vec{1}_z \bullet \overset{\leftrightarrow}{1}_r \bullet \vec{1}_z - \left[\vec{1}_r \bullet \vec{1}_r\right]\left[\vec{1}_x \bullet \vec{1}_x\right] + \left[\vec{1}_r \bullet \vec{1}_x\right]\left[\vec{1}_r \bullet \vec{1}_x\right]\right] \\
&= \frac{s^3 R}{8\pi}\tilde{f}_c^2\left[\vec{1}_z \bullet \overset{\leftrightarrow}{1}_r \bullet \vec{1}_z - 1 + \vec{1}_x \bullet \left[\vec{1}_r \vec{1}_r\right] \bullet \vec{1}_x\right] \\
&= \frac{s^3 R}{8\pi}\tilde{f}_c^2\left[-\vec{1}_z \bullet \left[\vec{1}_r \vec{1}_r\right] \bullet \vec{1}_z + \vec{1}_x \bullet \left[\vec{1}_r \vec{1}_r\right] \bullet \vec{1}_x\right],
\end{aligned}
\tag{3.14}
$$

which is zero, if and only if,

$$
\vec{1}_r \bullet \vec{1}_z = \pm \vec{1}_r \bullet \vec{1}_x \tag{3.15}
$$

This corresponds to two planes given by $x = \pm z$.

So one may position antenna 2 anywhere on these two planes as in Fig. 3.1, except at $\vec{r} = \vec{0}$ (or near here due to other moments associated with real antennas). One can, in principle, place antenna 2 on the $+y$ axis, but this may not be advisable due to high frequency blockage of antenna 1. At high frequencies an impulse radiating antenna[10] has a narrow beam and antenna 2 is best not located here since the electric and magnetic response is not in general perfectly balanced in a real antenna due to various imperfections in its construction. Using reciprocity it may be similarly inadvisable to locate antenna 2 on the $-y$ axis. This still leaves many locations on the $x = \pm z$ planes.

4. NEAR-FIELD COUPLING BETWEEN ANTENNAS

4.1 General Considerations

As one goes down in frequencies such that the distance r between the two antennas becomes comparable to a radian wavelength, the near field terms (r^{-2} and r^{-3}) in (2.10) and (2.11)) become important. Some of these have angular dependences different from those of the r^{-1} terms. Note, however, that the r^{-3} terms have the same angular dependence as the r^{-2} terms in both $\vec{E}_p$ and $\vec{H}_m$. Similarly r^{-2} and r^{-1} terms have the same angular dependence in both $\vec{H}_p$ and $\vec{E}_m$.

So let us consider the coupling of these additional terms between the two antennas. First, we have $\vec{p}_1$ to $\vec{p}_2$ and $\vec{m}_1$ to $\vec{m}_2$ coupling, followed by mixed dipole pairs, and finally by combined dipoles.

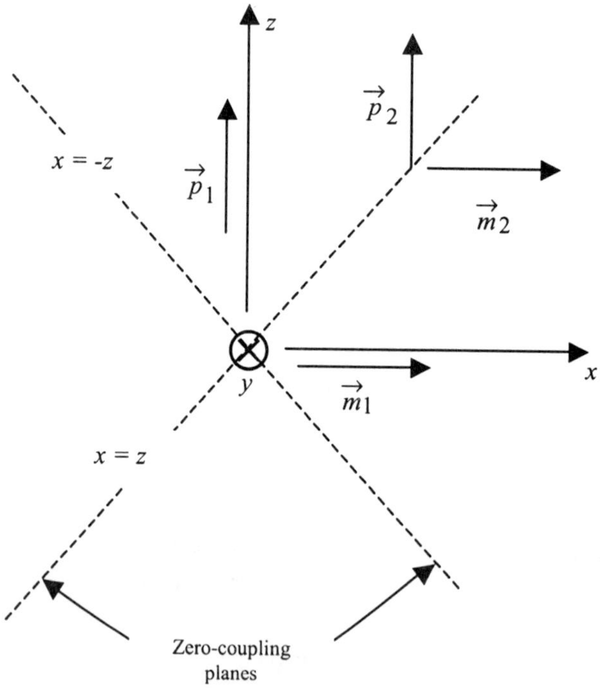

Figure 3.1. Planes of noncoupling combined dipoles in the far-field approximation.

4.2. Electric Dipoles

Consider first the coupling of r^{-2} and r^{-3} terms in $\vec{E}_p$, designated by superscript 2, to $\vec{p}_2$. This is proportional to

$$\vec{\tilde{E}}_p^{(2)} \bullet \vec{\tilde{p}}_2 = -e^{-\gamma r}\left[\frac{1}{4\pi r^3}\frac{1}{\varepsilon_o} + \frac{sZ_o}{4\pi r^2}\right]\vec{\tilde{p}}_1 \bullet \left[\overset{\leftrightarrow}{1} - 3\vec{1}_r\vec{1}_r\right] \bullet \vec{\tilde{p}}_2. \tag{4.1}$$

The angular part is

$$\vec{1}_z \bullet \left[\overset{\leftrightarrow}{1} - 3\vec{1}_r\vec{1}_r\right] \bullet \vec{1}_z = 1 - 3\cos^2(\theta_z), \tag{4.2}$$

which is zero at[2]

$$\cos(\theta_z) = \pm\frac{1}{\sqrt{3}}, \quad \theta_z \simeq 54.74°,\ 125.26°. \tag{4.3}$$

This solution describes two circular cones coaxial with the z axis. For a given r this describes two circles of radius $r\sqrt{2/3}$ on planes given by

$$z = r\cos(\theta_z) = \pm\frac{r}{\sqrt{3}}. \tag{4.4}$$

Note that this solution is in conflict with that for the far field for an electric-dipole pair in (3.7) for which $\theta_z = 0°, 180°$. However, it is *not* in conflict with the far-field solution for combined dipoles in (3.16) with $x = \pm z$. These two solutions intersect at

$$\begin{aligned} z^2 &= r^2\cos^2(\theta_z) = r^2\sin^2(\theta_z)\cos^2(\theta_z), \\ \cos(\phi_z) &= \cot(\theta_z) = \frac{1}{\sqrt{2}}, \quad \phi_z = 45°,\ 135°,\ 225°,\ 315°. \end{aligned} \tag{4.5}$$

Together with the two values of z (for fixed r) this gives eight points of common solution.

4.3 Magnetic Dipoles

Similarly the coupling of r^{-2} and r^{-3} terms in $\vec{\tilde{H}}_m$ (superscript 2) to $\vec{\tilde{m}}_2$ is proportional to

$$\vec{\tilde{H}}_m^{(2)} \bullet \vec{\tilde{m}}_2 = -e^{-\gamma r}\left[\frac{1}{4\pi r^3} + \frac{s}{4\pi r^2}\frac{1}{c}\right]\vec{\tilde{m}}_i \bullet \left[\overleftrightarrow{1} - 3\,\vec{1}_r\,\vec{1}_r\right] \bullet \vec{\tilde{m}}_2 . \tag{4.6}$$

The angular part is

$$\vec{1}_x \bullet \left[\overleftrightarrow{1} - 3\,\vec{1}_r\,\vec{1}_r\right] \bullet \vec{1}_x = 1 - 3\cos^2(\theta_x), \tag{4.7}$$

which is zero at[2]

$$\cos(\theta_x) = \pm\frac{1}{\sqrt{3}}, \qquad \theta_x \simeq 54.74^\circ, 125.26^\circ, \tag{4.8}$$

where now the spherical coordinates are defined with respect to the x axis as in (2.3).

This solution is just like that in Section 4.2 except that the solution circles are coaxial with the x axis. Merely replace the z subscripts with x subscripts giving

$$x = r\cos(\theta_x) = \pm\frac{r}{\sqrt{3}}, \qquad x^2 = r^2\cos^2(\theta_x) = r^2\sin^2(\theta_x)\cos^2(\theta_x),$$
$$\cos(\phi_x) = \cot(\theta_x) = \frac{1}{\sqrt{2}}, \qquad \phi_x = 45^\circ, 135^\circ, 225^\circ, 315^\circ . \tag{4.9}$$

Noting that $\theta_z = \theta_x$ for points of common solution, this says that the eight points of common solution are common between (4.5) and (4.9).

In Fig. 2.1 we see that looking parallel to the y axis both $\vec{\tilde{p}}_1$ to $\vec{\tilde{p}}_2$ coupling and $\vec{\tilde{m}}_1$ to $\vec{\tilde{m}}_2$ coupling in the near field are zero for $\vec{r}$ on the $x = \pm z$ planes, but not on the $y = 0$ plane. Note that the direction cosines have

$$\cos^2(\theta_x) + \cos^2(\theta_y) + \cos^2(\theta_z) = 1, \qquad \cos(\theta_y) = \pm\frac{1}{\sqrt{3}}. \tag{4.10}$$

So the common solution points lie on four lines equally spaced between the coordinate axes. Said another way, the solutions lie on eight directions with the three direction cosines taking all eight choices of combinations of $\pm 1/\sqrt{3}$.

4.4. Mixed Terms

Since electric dipoles produce magnetic fields, and conversely, we need to consider the response of the magnetic part of antenna 2 to $\vec{p}_1$ and the electric part of antenna 2 to $\vec{m}_1$. If we consider these couplings separately, we find that $\vec{m}_2$ needs to be on the $y = 0$ plane for zero coupling, and that $\vec{p}_2$ needs to be on the $y = 0$ plane for zero coupling. These two requirements are consistent with each other, but inconsistent with the results of Section 4.2 and 4.3 for the $\vec{p}_1$ to $\vec{p}_2$ and $\vec{m}_1$ to $\vec{m}_2$ near-field coupling.

Another approach consists of summing these mixed terms. For this purpose one needs to adjust them to allow for the balance between $\vec{p}_n$ and $\vec{m}_n/c$, and normalize the magnetic field as $Z_0 \vec{H}$. So we form

$$-Z_o \vec{H}_p^{(2)} \cdot \frac{\vec{m}_2}{c} + \vec{E}_m^{(2)} \cdot \vec{p}_2$$
$$= e^{-\gamma r} \frac{s}{4\pi r^2} \left[\vec{1}_x \cdot \left[\vec{1}_r \times \vec{1}_z \right] \tilde{p}_1 \frac{\tilde{m}_2}{c} Z_o + \vec{1}_z \cdot \left[\vec{1}_r \times \vec{1}_x \right] \tilde{p}_2 \tilde{m}_1 \mu_o \right]. \tag{4.11}$$

Notice the minus sign in the sum; this needs some explanation. This is related to the minus sign appearing in one of the two Maxwell equations. Depending on the convention used, there is a sign reversal on the electric- or magnetic-dipole term in transmission versus reception. This can be seen in some of the previous results. For example, the far-field coupling in Section 3 can be viewed from the r^{-1} terms in (2.11) and (2.12). Observe that for an observer on the y axis $\vec{E}_p \cdot \vec{p}_2$ has a minus sign and $\vec{H}_m \cdot \vec{m}_2$ also has a minus sign. However, these two terms *subtract* to give a zero far-field response of antenna 2 on the y axis (Section 3). The reader may wish to delve into this more deeply.

Returning to (4.11) we see that the angular part of the response is

$$\vec{1}_x \cdot \left[\vec{1}_r \times \vec{1}_z \right] + \vec{1}_z \cdot \left[\vec{1}_r \times \vec{1}_x \right]$$
$$= \vec{1}_r \cdot \left[\vec{1}_z \times \vec{1}_x \right] + \vec{1}_r \cdot \left[\vec{1}_x \times \vec{1}_z \right] = \vec{1}_x \cdot \vec{1}_y - \vec{1}_x \cdot \vec{1}_y = 0, \tag{4.12}$$

that is, the mixed terms cancel everywhere (except for $\vec{r} = \vec{0}$).

5. CONCLUSIONS

The pair of balanced combined dipoles with parallel $\vec{p}_n$ and parallel $\vec{m}_n$ does not couple from antenna 1 to antenna 2 (and conversely) under the conditions:

1. Far field terms: Zero coupling occurs if antenna 2 is located on the $x = \pm z$ planes (Section 3.4, Fig. 3.1).

2. Near field terms: Zero coupling occurs for both $\vec{p}_1$ to $\vec{p}_2$ and $\vec{m}_1$ and $\vec{m}_2$ separately (Sections 4.2 and 4.3), if antenna 2 is located such that the direction cosines with respect to the three coordinate axes are any combination of $+1/\sqrt{3}$ and $-1/\sqrt{3}$, thereby defining eight directions.

3. Mixed terms: Zero coupling occurs for the sum of $\vec{p}_1$ to $\vec{m}_2$ and $\vec{m}_1$ to $\vec{p}_2$ everywhere.

These three conditions can all be satisfied using the form in condition 2 above. For a given r this gives eight points at the intersections of the $x = \pm z$, $y = \pm x$, and $y = \pm z$ planes (excepting $\vec{r} = \vec{0}$), or for general r, four lines of intersection. For a given r the eight points are the corners of a cube centered at the origin with each face perpendicular to a coordinate axis. The mutual relation of these points is found in the O (octahedral) symmetry group.

Of course, various imperfections in the antennas mean that the coupling is not truly zero. This points out the desirability of closely matching p and m/c and making these at right angles in the antenna design. Likewise accurate relative positioning and orientation of the two antennas is required.

At high frequencies for which the radian wavelength is less than the antenna dimensions, the antennas may not be characterized by only electric and magnetic dipoles. For example, impulse radiating antennas (IRAs) are designed to be highly directive at such frequencies.[10] Fortunately condition 2 above places the antennas such that neither is in the main beam of the other at such high frequencies. Nevertheless, there may be some residual coupling here from the higher-order multipoles. Further studies and perhaps use of absorbers between the two antennas may help. There is also the spacing r between the two antennas that one can choose. Larger r gives smaller coupling (especially for the near-field terms). However, r should be small compared to the distance to the target and large r may present mechanical difficulties.

Note that various other conductors such as coaxial cables connected to the antennas may increase the coupling. Use of chokes (inductance) to suppress currents on such conductors may help. Appropriate positioning of such conductors may also help.

Replication of antenna 2 to occupy two or more of the eight optimal positions (and summing the signals) can add more symmetry to the ensemble and perhaps cancel some of the unwanted coupling. So the present paper is a first step in realizing practical designs.

6. ACKNOWLEDGEMENTS

This work was sponsored in part by the Air Force Office of Scientific Research, and in part by the Air Force Research Laboratory, Directed Energy Directorate.

7. REFERENCES

1. C.E. Baum, Some Characteristics of Electric and Magnetic Dipole Antennas for Radiating Transient Pulses, *Sensor and Simulation Note 125*, Air Force Research Laboratory, Kirtland Air Force Base NM (January 1971).
2. J.D. Quinn and C.E. Baum, Positioning Loops with Parallel Magnetic Dipole Moments to Avoid Mutual Inductance, *Sensor and Simulation Note 303*, Air Force Research Laboratory, Kirtland Air Force Base NM (May 1987).
3. C.E. Baum, General Properties of Antennas, *Sensor and Simulation Note 330*, Air Force Research Laboratory, Kirtland Air Force Base NM (July 1991); *IEEE Trans. EMC*, 18–24 (2002).
4. C.E. Baum, E.G. Farr, and C.A. Frost, Transient gain of antennas related to the traditional continuous-wave (CW) definition of gain, in: *Ultra-Wideband, Short-Pulse Electromagnetics 4*, edited by E. Heyman, B. Mandelbaum, and J. Shiloh (Kluwer Academic/Plenum Publishers, New York, 1999), pp. 109-118; *Sensor and Simulation Note 412*, Air Force Research Laboratory, Kirtland Air Force Base NM (July 1997).
5. C.E. Baum, Symmetry and SAR Antennas, *Sensor and Simulation Note 431*, Air Force Research Laboratory, Kirtland Air Force Base NM (November 1998).
6. C.E. Baum, Unipolarized Currents for Antenna Polarization Control, *Sensor and Simulation Note 437*, Air Force Research Laboratory, Kirtland Air Force Base NM (June 1999).
7. C.E. Baum, Interaction of Electromagnetic Fields with an Object Which Has an Electromagnetic Symmetry Plane, *Interaction Note 63*, Air Force Research Laboratory, Kirtland Air Force Base NM (March 1971).
8. J. Van Bladel, *Electromagnetic Fields* (Taylor & Francis, 1995).
9. C.E. Baum and H.N. Kritikos, Symmetry in Electromagnetics, in: *Electromagnetic Symmetry*, edited by C.E. Baum and H.N. Kritikos (Taylor & Francis, 1995), Ch. 1, pp. 1–90.
10. C.E. Baum, E.G. Farr, and D.V. Giri, Review of Impulse-Radiating Antennas, in: *Review of Radio Science 1996-1999*, edited by W.R. Stone (Oxford University Press, London, 1999), Ch. 16, pp. 403–439.

THE ROLE OF MUTUAL COUPLING IN A 5:1 BANDWIDTH STRIPLINE NOTCH ARRAY

Mark Kragalott, William R. Pickles, and Michael S. Kluskens*

1. INTRODUCTION

The bandwidth broadening effects of time-domain mutual coupling from adjacent elements in a linear array of stripline notch elements was briefly described in previous papers[1-3]. The array was designed at the Naval Research Laboratory (NRL) to demonstrate the feasibility of 5:1 bandwidth wide-scan angle arrays with one-half wavelength spacing at the highest operational frequency. This paper presents an expanded examination of the role of mutual coupling in pulse-excited linear and planar single polarization arrays as well as in a planar dual polarization array. When viewed in the time domain, inspection of mutual coupling leads to a physical understanding of the cause of the ultrawideband (UWB) behavior of the NRL arrays. The finite-difference time-domain (FDTD) method[4,5] is employed to perform the studies, and measurements verify some of the calculations.

Array mutual coupling has been considered by several authors[6-8], with most analyses performed in the frequency domain from experimental or numerical results. Although frequency domain mutual coupling analysis is very useful for a number of performance measures, mutual coupling is typically presented as magnitude and/or phase data with unclear effects on UWB behavior. From a frequency domain perspective, good active array impedance behavior is manifested by the ability of each element to sum the complex-valued mutual coupling components from all active elements to achieve a small return loss relative to the input signal. To achieve good UWB array impedance behavior, return loss suppression must happen for a wide range of frequencies and scan angles. From a time-domain perspective, good UWB active impedance behavior is characterized by small time-domain return loss relative to the input signal. A Fourier transform reveals return loss as a function of frequency.

* The authors are with the Radar Division of the Naval Research Laboratory, Washington, DC 20375. This work was supported by the Office of Naval Research, Washington, DC.

Antenna designers routinely optimize wideband array impedance without a physical understanding of the mechanisms that cause wideband behavior. The culprit is often an infinite array computational electromagnetic code that combines mutual coupling effects so that the impact of a given element on the return loss of other elements is difficult to ascertain. In addition, infinite array analysis can lead to array designs that lose wideband behavior when constructed as smaller arrays (e.g. 10x10 arrays with $\lambda/2$ element spacing at the highest UWB frequency) because edge effects are often not included in the analysis[6]. The NRL array design experience has led us to the conclusion that a combination of finite, semi-infinite, and infinite array analyses is the most effective way to achieve relatively compact UWB array designs.

2. TIME-DOMAIN MUTUAL COUPLING ANALYSIS

The basis for time-domain analysis of mutual coupling of the NRL finite arrays is the superposition of currents from N separate FDTD array calculations, where N is the number of array elements. Figure 1 shows the nth array element is excited with a test pulse voltage source V_n with a series termination impedance Z_0 while all other elements are passively terminated with Z_0. The induced current at each element is received and stored. The FDTD calculation is repeated N times, each time with a different driven element.

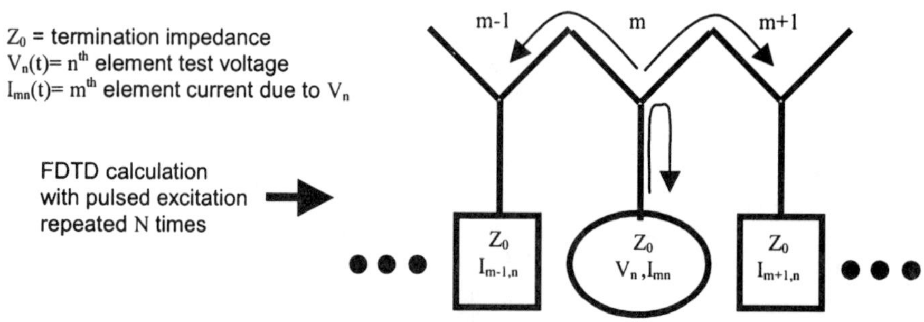

Figure 1. A single element is excited by a loaded FDTD voltage source with all others passively terminated and the time domain current is received at each element.

Once N FDTD calculations have been performed, the active scanned array time-domain current at the mth element can be found from superposition of currents:

$$I_m(t) = \sum_{n=1}^{N} I_{mn}(t + \Delta t_n), \tag{1}$$

where I_{mn} is the current at the mth element due to the pulsed voltage V_n at the nth element, t is time, and Δt_n is the time delay between a reference element input voltage and an input voltage at the nth element. In a uniformly spaced array lattice scanned to a particular angle, adjacent elements differ in time delay by an integer multiple of a fundamental time-delay $\Delta\tau$. The mth element active array impedance at a given frequency ω and spherical coordinate scan angle (θ,ϕ) is determined from the ratio of the Fourier transformed active scan current to the Fourier transformed input voltage pulse:

$$Z_m(\omega,\theta,\phi) = V_m(\omega,\theta,\phi) \,/\, I_m(\omega,\theta,\phi). \tag{2}$$

The active current reflection coefficient is

$$\rho(\omega,\theta,\phi) = -\,[Z_m(\omega,\theta,\phi) - Z_0] \,/\, [Z_m(\omega,\theta,\phi) + Z_0], \tag{3}$$

and the active voltage standing wave ratio is

$$\mathrm{VSWR}(\omega,\theta,\phi) = [1 + |\rho(\omega,\theta,\phi)|] \,/\, [1 - |\rho(\omega,\theta,\phi)|]. \tag{4}$$

A complementary frequency domain procedure relates the total current induced at a given element to the product of a voltage excitation vector and a mutual coupling admittance matrix[1].

3. LINEAR ARRAY COUPLING

The details of the NRL UWB stripline notch printed circuit board element design can be found in recent articles[1-3]. A linear array version is shown in Figure 2(a).

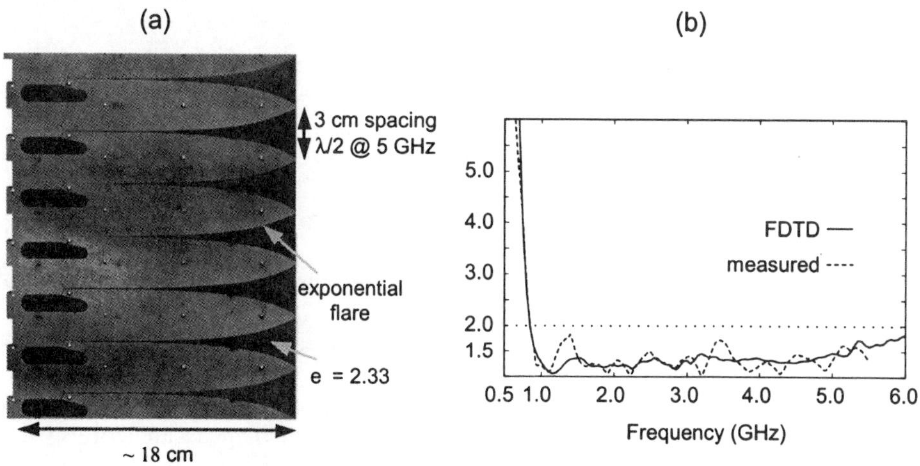

Figure 2. (a) Photograph of a seven element linear array. (b) The measured and calculated broadside scan impedance bandwidth for an eight element linear array is 6:1.

Array elements are spaced at one-half wavelength at the highest frequency of UWB operation (5 GHz) so grating lobes are avoided for all operational frequencies and scan angles. Given this constraint, FDTD analysis was employed to obtain wideband optimization of the broadside scan impedance for a linear array. Optimization of the active impedance (VSWR < 2) both over the operational frequency band and for scan angles out to 45 degrees ensures that over 89 percent of the energy is radiated since less than one percent of the energy is dissipated in element heating. Further, for electrically large arrays a low active VSWR is directly related to embedded element gain so that optimization of low active VSWR over broad scan angles leads to optimized element patterns[1], although the relationship does not strictly hold for small arrays. Several variables were involved in the optimization including flare length, taper function, and YY stripline-to-slotline transition parameters such as its location and open circuit cavity size and shape. Ultimately, as shown in Figure 2(b), a 6:1 bandwidth active impedance match (VSWR < 2) was achieved at broadside scan.

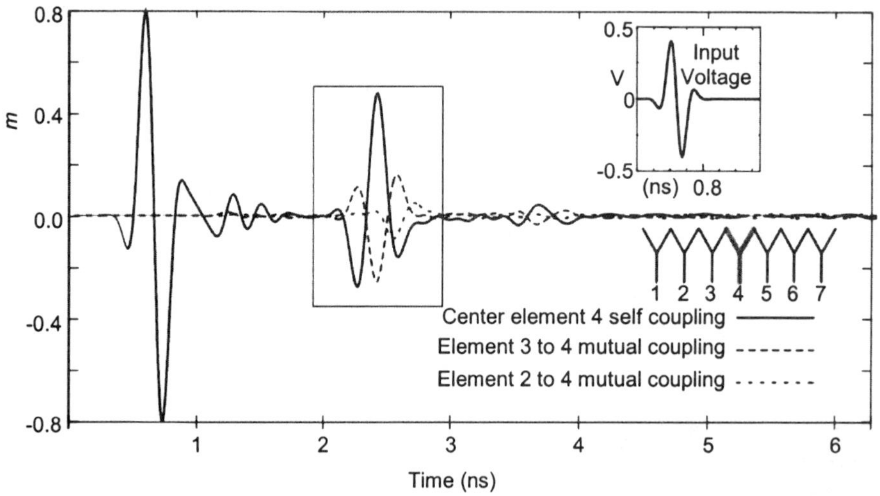

Figure 3. Inspection of the broadside scan current components shows that the sum of mutual coupling from adjacent elements tend to cancel the self-element reflection, which explains ultrawideband impedance behavior.

The physical mechanism for UWB impedance behavior is sought by means of time-domain coupling analysis using a seven element linear array. The total coupling observed at the feed point of the center element 4 is determined by seven separate FDTD calculations in which the receiving element is always element 4 and the lone exciting element is varied. In each case the active element is fed with a modulated Gaussian sine voltage pulse through a series 50 ohm resistance, and all other elements are terminated in 50 ohm loads. The incident pulse is shown in the inset graph of Figure 3. The significant energy content of the pulse is in the 1 to 5 GHz frequency range. The self-coupling current component, which is represented by the solid line in Figure 3, is determined by exciting only element 4 and receiving at element 4. The self coupling is nearly a scaled replica of the input voltage prior to 1 ns as the input voltage pulse emerges from the source location followed by a large reflection that is highlighted by the boxed region between 2 ns and 3 ns. This large reflection is due to the small element aperture size at middle to lower frequencies in the operational band. Thus, element 4 is poorly matched at low frequencies when all other elements are passively terminated in matched loads.

Mutual coupling current components have nearly symmetric contributions for elements on opposite sides of element 4 but equidistant from it. Thus, the coupling at element 3 to 4 is nearly equivalent to the element 5 to 4 coupling, and 2 to 4 coupling is nearly equivalent to 6 to 4 coupling. In addition, active elements 3 (5) and 2 (6) have the most dominant impact on impedance bandwidth. Thus, only the mutual coupling currents from element 3 to 4 and element 2 to 4 are plotted here, but additional current components contribute significantly toward impedance bandwidth. Figure 3 depicts the broadside scan case, which means that all input voltage pulses are fed in time phase. It is apparent from inspection of the boxed highlighted time frame that that the aperture reflection is in time phase opposition to the mutual coupling current from element 3 to element 4. Element 3 to 4 mutual coupling provides only about half of the amplitude that is required to cancel the self-coupling aperture reflection. However, mutual coupling is nearly symmetric about element 4, so the sum of coupling from elements 3 and 5 provides a significant portion of the coupling to cancel the element 4 self-coupling. It is the mutual coupling cancellation of self-coupling aperture reflection that is responsible for the UWB impedance behavior of the NRL arrays.

The cancellation effects of mutual coupling on self-coupling aperture reflections are clearly displayed in Figure 4, which shows the resultant time-domain broadside scan current at element 4 for various numbers of active elements with all non-active elements resistively terminated. Obviously, the reflected voltage present when only element 4 is active, which is indicated by the solid line in the boxed highlighted region in Figure 3, is substantially reduced when elements 3 and 5 are also active, which is indicated in Figure 4 by the dotted curve. When elements 2 and 6 are also activated, the reflections are further reduced (dashed curve). Activation of elements 1 and 7 produces an additional reduction in return loss. The resulting VSWR at element 4 when all seven elements are active is similar to Figure 2(b), which is the VSWR for an eight element linear array. The limitation of mutual coupling on bandwidth broadening was explored[1] for a 13 element linear array at broadside scan. It was found that as the number of active elements is increased from 7 to 13, the low frequency performance extends below 1 GHz and slowly approaches 0.7 GHz, so that the array has a bandwidth of 7:1. This is certainly not a rigid bandwidth limit for all linear arrays, but is merely an observation about the NRL linear arrays.

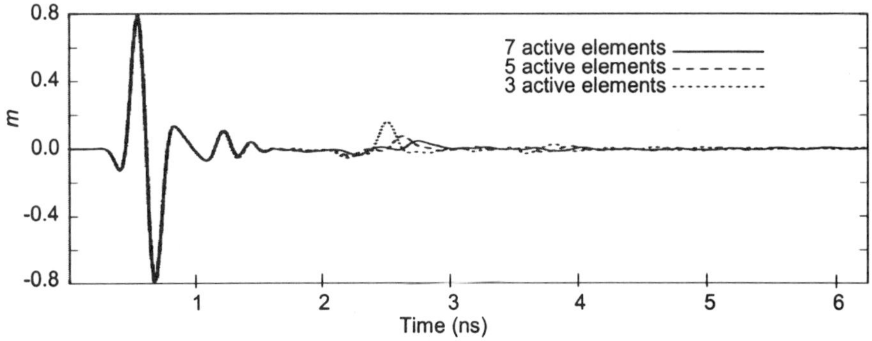

Figure 4. Current at the center element of a 7 element linear array as a function of the number of active elements.

Now that symmetric E-plane mutual coupling has been uncovered as the mechanism for UWB impedance behavior, degradation in the VSWR performance of the linear array both at the edge elements and at wide scan angles can be explained. Edge effects are understood by observing the time-domain coupling at edge element 1 due to self-coupling as well as element 2 to 1 mutual coupling for broadside scan, which is shown in Figure 5(a) for a seven element linear array. Figure 5(a) demonstrates that when the input voltage amplitudes of elements 1 and 2 are identical, element 2 to 1 coupling provides a current that is in time phase opposition to the element 1 self-coupling aperture reflection between 2 and 3 ns. However, the amplitude of the element 2 to 1 coupling is only about half of what is required for significant cancellation of self-coupling. Further, there is no symmetric current coupling component on the opposite side of element 1 providing the additional cancellation required for UWB impedance behavior. Correspondingly, Figure 5(b) shows that the active VSWR of element 1, indicated by the long dashed line, remains well above 2 for most of the operational frequency band when the input voltages of all seven elements have equal amplitudes. The most obvious means of reducing return loss at the edge element is to weight the input voltage amplitudes *only* at edge elements 1 and 7 by a factor of one-half. This will decrease the amplitude of the self-coupling aperture reflection at element 1 so that mutual coupling from unit-weighted element 2 can cancel the majority of it. Figure 5(b) shows that active VSWR at element 1 with edge element input voltages weighted by one half, indicated by the solid line, remains below 2 for nearly all frequencies in the operational 5:1 bandwidth. The application of one half weights to the input voltage amplitudes at the edge elements 1 and 7 does not significantly degrade the impedance at element 2, as is shown by the dotted line in Figure 5(b).

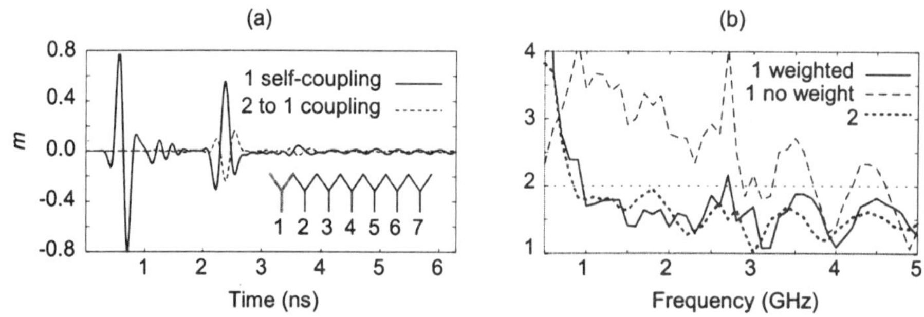

Figure 5. (a) When all element input voltage amplitudes are identical, element 2 to 1 coupling does not have sufficient amplitude to cancel element 1 self-coupling. However, (b) 0.5 weighting of the input voltage amplitudes of edge elements 1 and 7 results in a low active VSWR at elements 1 and 2 over the frequency band.

The self-coupling aperture reflections are not as easily mitigated when the array is scanned to oblique angles in the E-plane of the linear array. When the array is scanned in the time-domain, each element is time advanced or delayed by an integer multiple of a fundamental time delay $\Delta\tau$ relative to its neighboring elements. Increasing the time shift between elements gradually deteriorates the beneficial effects of mutual coupling for reasons that are clear by inspecting Figure 3. Figure 3 shows the self-coupling current of center element 4 of a seven element linear array as well as the mutual coupling current from the two nearest (3 and 5) and next-nearest (2 and 6) elements. As the voltage excitation of the element 3 is shifted by $\Delta\tau$ and the voltage of element 2 is shifted by $2\Delta\tau$, where $\Delta\tau$ is a significant fraction of the input pulse width, their corresponding mutual coupling current components will also shift by $\Delta\tau$ and $2\Delta\tau$, respectively. As the mutual coupling from element 3 to 4 and element 2 to 4 shifts significantly relative to the self-coupling component, then the cancellation of the aperture reflection that occurs between 2 ns and 3 ns will no longer be as effective.

Figure 6 shows the total current (i.e. all coupling components summed) at the center element 5 of a nine element linear array when the array is scanned to broadside, 25 degrees, and 50 degrees. Note that the total reflected current between 2 ns and 3 ns for the nine element array is reduced compared to the seven element array reflected current in Figure 4. In Figure 6, the current return loss between 2 ns and 3 ns increases as the scan angle increases. However, the current reflection between 1.0 ns and 1.5 ns is nearly independent of scan angle because the reflection is primarily due to self-coupling interactions with the YY stripline-to-slotline transition. For linear arrays the scan VSWR remains below 2 for most frequencies and angles less than 45 degrees, although there are some frequencies (e.g. 2.9 GHz and 4.6 GHz) where VSWR rises over 2 for angles greater than 25 degrees[1].

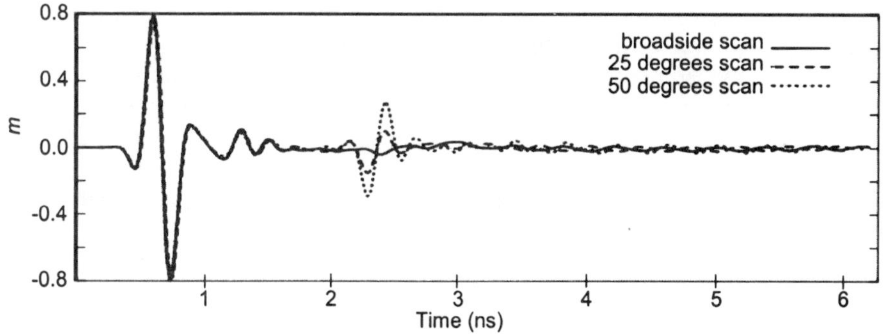

Figure 6. The total current at the center element of a 9 element linear array as a function of scan angle.

4. SINGLE-POL PLANAR ARRAY COUPLING

The NRL single polarization planar arrays are formed by parallel rows of linear elements with each row spaced one-half wavelength at the highest operational frequency (5 GHz) from adjacent rows. This geometrical arrangement alters the element to element coupling characteristic at a given element within a particular row compared to the linear array since there are coupling paths due to multiple bounces from adjacent rows of elements. Owing to the prohibitive computational requirements of even moderate size (7 x 7) arrays, FDTD semi-infinite array H-plane scan simulator analysis[1] was employed. In the H-plane scan simulator all elements within a given row have time synchronized voltage pulses, and although a time delay can exist between rows, we will first consider the broadside scan case where all sources are synchronized.

The coupling from an array of size 5 rows by infinite elements per row is represented in Figure 7, which depicts the time domain current coupling at the center element of row 3. The solid line shows the coupling at the center element of row 3 when only row 3 is active. Compared to the broadside scan current of the center element of the 9 element linear array in Figure 6, the center element current of the 5 row semi-infinite array with only row 3 active, indicated by the solid line in Figure 7, reveals a larger time domain return loss. This is a result of multiple path bounces from adjacent rows of elements. The dashed line in Figure 7 shows the mutual coupling at the center element of row 3 when only row 2 is active, and the dotted line shows the mutual coupling at the center element of row 3 when only row 1 is active. Row 4 coupling to the center element is equivalent to row 2 coupling, and row 5 coupling is equivalent to row 1 coupling. It was previously demonstrated[1] that an even (asymmetric) number of active rows produced a lower return loss than an odd (symmetric) number of active rows. By observing the current coupling in Figure 7, this phenomenon is now readily explained. Since the row 2 coupling time domain return loss between 2 ns and 3 ns is nearly equal in amplitude but opposite in sign to the row 3 return loss, exciting only these two rows produces a very low return loss at the array center element. When row 4 is also active, the return loss increases since the superposition of rows 2 and 3 current coupling components produces a signal too small to cancel row 4 coupling between 2 ns and 3 ns. As row 1 is also activated, row 1 to center element coupling produces a return loss component between 2 ns and 3 ns that is opposite in sign and a factor of approximately 0.65 the amplitude of row 4 coupling. Thus, the total return loss decreases. The activation of row 5 produces only a marginal increase in return loss as shown in the bottom half of Figure 7 for broadside scan, and the VSWR of the fully active array is less than 2 from 1 GHz to 5 GHz. Since the coupled return loss due to any given row is not substantial, the H-plane scan of the array by time delay of one row relative to other rows does not produce a substantial rise in VSWR. The VSWR from 1 GHz to 5 GHz remains below 2 for nearly all H-plane scan angles out to 60 degrees and for most E-plane scan angles out to 45 degrees[1]. The E-plane scan impedance performance in the single-pol planar array is substantially better at oblique angles compared to the linear array.

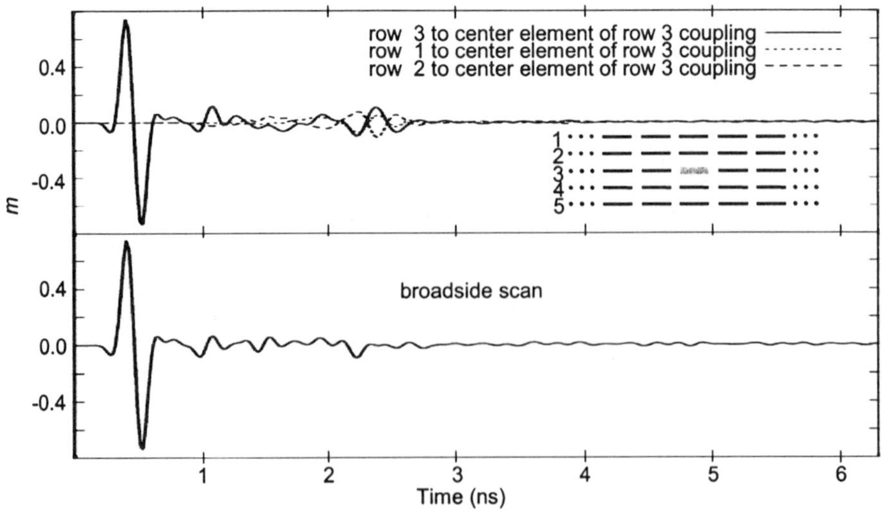

Figure 7. Row to center element coupling for a five row by infinite number of elements per row single-pol planar array, and the total current on broadside scan.

5. DUAL-POL PLANAR ARRAY COUPLING

The NRL dual-pol arrays are formed by orthogonally interlacing two single-pol planar arrays, with the intersection points at the apex of the flares as indicated in Figure 8. Figure 9 shows the calculated self-coupling current at the center element of a 7x3 dual-pol array, as well as the nearest neighbor active element cross-pol mutual coupling. The self-coupling current as a function of time is similar in shape to the self-coupling of the 7 element linear array in Figure 3. However, the presence of parallel and perpendicular elements in the vicinity of the center element reduces the amplitude of the aperture reflection between 2 ns and 3 ns compared to linear array self-coupling. The peak time-domain amplitude of the cross-pol mutual coupling from element 2 to element 1 is a scale factor of approximately 0.6 less than the amplitude of the nearest neighbor co-pol coupling. Hence, nearest neighbor cross-pol elements contribute a significant amount of coupling power. But the dual-pol array has symmetric cross-pol mutual coupling cancellation properties that prevent active VSWR elevation at element 1 for H-plane and E-plane scanning. However, in other scan planes the cross-pol coupling components will not in general cancel.

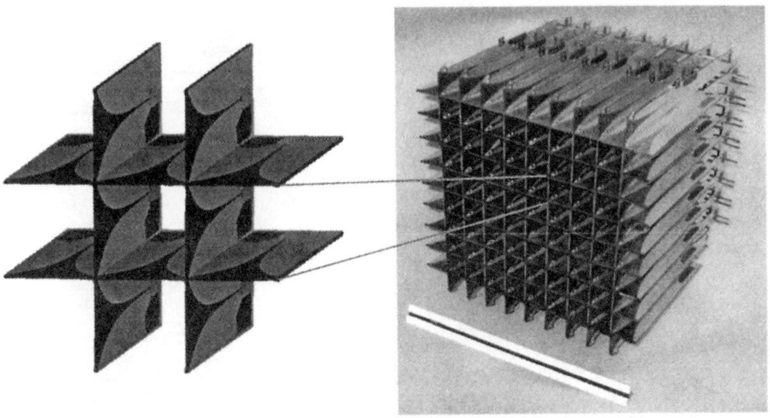

Figure 8. The NRL dual-pol array.

Figure 9 shows that the amplitude and shape of element 2 to 1 cross-pol coupling is nearly equivalent to element 3 to 1 cross-pol coupling, although they differ by a minus sign so when summed on broadside scan they cancel. In addition, Element 5 to 1 coupling is nearly equivalent to element 2 to 1 coupling, and element 4 to 1 coupling is nearly equivalent to element 3 to 1 coupling. Thus, at broadside scan element 4 to 1 coupling cancels element 5 to 1 coupling. When both polarizations are actively scanned in the H-plane of element 1, then elements 2 and 3 are time delayed by the same amount, and their mutual coupling contributions at element 1 cancel. The same cancellation is true of the cross-coupling at element 1 due to elements 4 and 5 for H-plane scan. Thus, interior elements are scarcely affected by the cross-pol coupling when both polarizations are actively scanned in the H-plane of element 1. Further, the amplitude of element 5 to 1 coupling is nearly the negative of element 3 to 1 coupling, and the amplitude of element 4 to 1 coupling is nearly the negative of element 2 to 1 coupling. Thus, if both polarizations are actively scanned in the E-plane of the element 1, then elements 3 and 5 are time delayed by the same amount, and their mutual coupling contributions at element 1 cancel. The same cancellation relationship is true of the coupling at element 1 due to elements 2 and 4 for E-plane scan, and thus interior element active impedance is well behaved on E-plane scan. Although not shown here, calculations of E-plane and H-plane active scan VSWR properties of the semi-infinite dual-pol planar arrays are similar to the scan VSWR of the semi-infinite single-pol planar arrays plotted in Ref. 1.

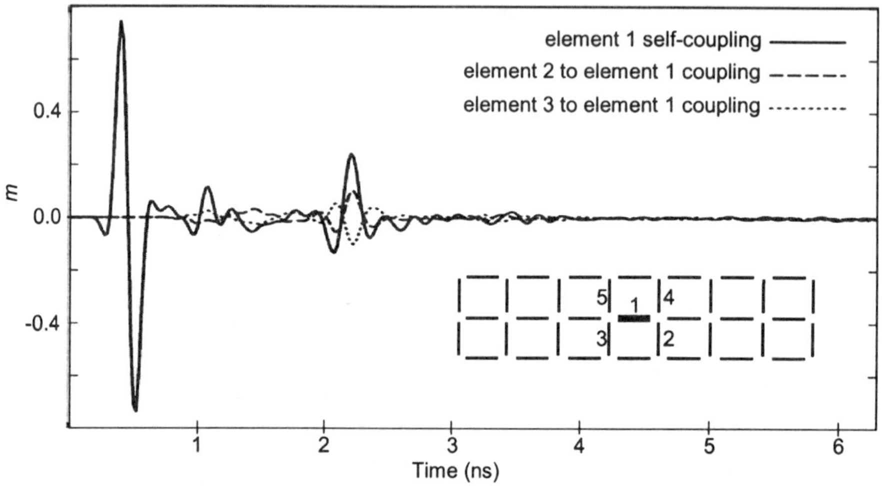

Figure 9. 7x3 dual-pol array center element self-coupling and cross-pol coupling from nearest neighbors.

The situation is altered when both polarizations of the array are scanned in planes different than the E-plane or H-plane. In this case elements 2, 3, 4, and 5 will have varied relative time-delayed signals so that cross-pol coupling cancellation at element 1 occurs only at limited angles and frequencies. In addition, co-pol coupling components will intermingle with cross-pol coupling components to produce a complicated, although predictable, active impedance characteristic. This will be the subject of future publications.

Assembly methods for printed stripline flared-notch dual-pol arrays can have significant impact on the measured VSWR compared to the calculated VSWR. A common way of interleaving dual-pol elements is shown in Figure 10(a), where a forward slot is cut from the apex of a flare of each element and extends parallel along the flare length. The orthogonal elements have a complementary backward slot cut from the feed portion of the element. Although this arrangement is an inexpensive solution for orthogonal intersection of printed dual-pol flared-notch arrays, array performance degradation can occur owing to the destruction of the sensitive mutual coupling relationships that are responsible for UWB performance. As seen in Figure 10(b), the broadside scan VSWR of a 7 element linear array with forward slot discontinuities is elevated for most frequencies in the band of operation compared to the VSWR for a linear array with no slot discontinuity as shown in Figure 1(b). The degradation is a result of the alteration of the amplitude of mutual coupling from adjacent elements which reduces the effectiveness of cancellation of self-element aperture reflection. Normally when the dual-pol array is assembled there will be some restoration of current coupling from adjacent E-plane elements. NRL performed linear array experiments that characterized broadside scan VSWR as a function of the depth of the forward slot[2]. It was found that significant VSWR degradation occurred when the slot depth becomes one-fifth of one wavelength at the highest UWB frequency. NRL constructed a dual-pol array using a modified slot method that attempted to maintain current continuity around the flare apex by the use of conducting tabs. However, it was found that the polarization that contained the forward slot had a degraded VSWR across the frequency band compared to the polarization with the backward slot, although the increase in VSWR was not as significant as that shown in Figure 10(b). Further, the patterns in the dual-pol array were not equivalent from one polarization to the other.

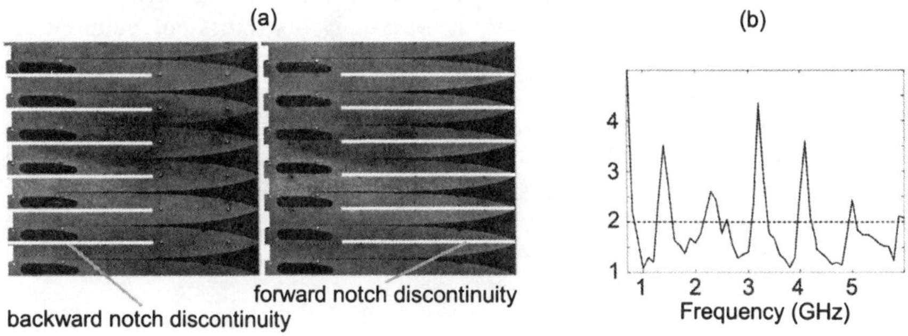

Figure 10. **(a)** The problematical slot assembly method for printed dual-pol arrays and **(b)** the active broadside VSWR for a linear array of 7 elements with the forward notch discontinuity.

As a consequence, NRL has abandoned the slot method for interlacing orthogonal elements in this type of dual-pol array and has adopted the post method. In the post method a grid of vertically parallel metal grooved posts are fastened in a square lattice. Each element is slid between the grooves and snapped into position. The elements are plated along the edges and springs are placed in the grooves to create pressure that promotes continuity. In this case the array has impedance performance consistent with FDTD computations and the patterns from one polarization to the other are nearly equivalent. The performance of this array will be characterized in future publications.

6. CONCLUSIONS

The Radar Division of NRL has designed and constructed linear, planar, and dual-pol printed stripline notch arrays with at least a 5:1 bandwidth. The UWB impedance behavior is understood by analyzing time domain mutual coupling from a combination of finite and semi-infinite FDTD analyses.

Mutual coupling from adjacent elements cancels most of the self-element aperture reflection and extends the broadside scan bandwidth to 7:1 for a linear array of 13 elements. Scan impedance degradation is understood as the reduction in mutual coupling cancellation of the self coupling aperture reflection as time delays are applied to the elements. Linear array edge elements show a rise in active VSWR that is mitigated by the application of one-half weighting of the applied voltage only at the edge elements.

Planar array single polarization mutual coupling was examined from the perspective active row coupling at an element as opposed to element to element coupling. This leads to an understanding of the 5 row array phenomenon that an even (asymmetric) number of active rows creates a lower return loss than an odd (symmetric) number of active rows. The small coupling return loss at the center element of the single-pol planar array due to any given active row explains the very good H-plane and E-plane active scan impedance behavior of the planar array for wide scan angles and for frequencies from 1 GHz to 5 GHz.

The analysis of cross-pol coupling at the center element of a 7x3 element dual-pol array shows that the time-domain amplitude of the coupling is significant in amplitude compared to co-pol coupling. However, nearest neighbor cross-pol coupling nearly cancel each other on broadside scan at the center element owing to a square array lattice. Cross-pol coupling cancellation also occurs for scanning in the E-plane and the H-plane, but not in any other scanning plane.

Construction techniques for dual-pol arrays can be a critical factor in maintaining calculated performance. The forward slot and complementary backward slot method for assembling flared-notch type printed arrays can potentially destroy the sensitive mutual coupling relationships that are responsible for UWB impedance behavior and cause uneven performance from one polarizaton to the other.

7. REFERENCES

1. M. Kragalott, W. R. Pickles, and M. S. Kluskens, Design of a 5:1 Bandwidth Stripline Notch Array from FDTD Analysis, *IEEE Trans. Antennas Propagat.* **48**, 1733-1741 (2000).
2. W. R. Pickles and M. Kragalott, Ultra-broadband Phased Array Antenna, *Proceedings of the 1999 Antenna Applications Symposium*, pp. 310-330 (1999).
3. M. Kragalott, W. R. Pickles, and M. S. Kluskens, A 5:1 Bandwidth Stripline Notch Array Design, *Record of the 45th Annual Tri-Service Radar Symposium* **2**, pp. 165-171 (1999).
4. K. S. Yee, Numerical Solution of Initial Boundary Value Problems Involving Maxwell's Equations in Isotropic Media, *IEEE Trans. Antennas Propagat.* **14**, 302-307 (1966).
5. R. J. Luebbers, *Users Manual for XFDTD*, REMCOM, Inc. (1994)
6. T. H. Chio, D. H. Schaubert, and H. Holter, Experimental Radiation and Mutual Coupling Characteristics of Dual-Polarized Tapered Slot Antenna (TSA) Arrays, *Proceedings of the 1999 Antenna Applications Symposium*, pp. 280-309 (1999).
7. D. H. Schaubert, J. A. Aas, M. E. Cooley, and N. E. Burris, Moment Method Analysis of Infinite Stripline-Fed Tapered Slot Antenna Arrays with a Ground Plane, *IEEE Trans. Antennas Propagat.* **42**, 1161-1166 (1994).
8. D. T. McGrath, Periodic Structures Analysis Using a Hybrid Finite Element Method, *Radio Sci.* **31**, 1173-1179 (1996).

UWB ANTENNA CHARACTERIZATION AND OPTIMIZATION METHODOLOGIES

Glen R. Salo and John S. Gwynne*

1. INTRODUCTION

This paper presents a set of UWB (Ultra-WideBand) antenna performance metrics that have proven to be extremely beneficial at characterizing, and hence, optimizing, both dispersive and non-dispersive UWB antennas for a variety of applications. The UWB antenna performance metrics presented herein are based on a normalization procedure that emphasizes changes in the radiation pattern as a function of frequency, time, angle, and polarization. These performance metrics provide a natural and understandable method to characterize UWB antenna performance and valuable insight into the radiation mechanisms within the antenna.

The UWB antenna performance metrics discussed in this paper are separated into two categories: 1) UWB antenna performance metrics; and 2) UWB/A (Ultra-WideBand/Angle) Synthetic Aperture Radar (SAR) performance metrics. A general description of these performance metrics is presented in Section 2. Examples illustrating the use of these performance metrics are presented in Section 3.

2. UWB ANTENNA PERFORMANCE METRICS

This section provides an overview of the UWB antenna performance metrics used to evaluate, rank, and optimize both dispersive and non-dispersive UWB antennas. First, limitations in analyzing antennas using traditional antenna parameters for UWB applications are identified. Next, the benefits offered by the new performance metrics or descriptors of UWB antenna performance are presented. This section then concludes with a description of the generalized UWB/A and SAR specific performance metrics.

*Glen R. Salo, Mission Research Corporation, St. Louis, Missouri. John S. Gwynne Mission Research Corporation, Dayton, Ohio.

Ultra-Wideband Short-Pulse Electromagnetics 6
Edited by Mokole *et al.*, Kluwer Academic/Plenum Publishers, 2003

2.1. Limitation of Traditional Antenna Descriptors

Traditional antenna parameters are insufficient for specifying the behavior of UWB SAR antennas for three primary reasons: 1) the parameters are defined for a single frequency and do not relate spectral information such as dispersion; 2) the definitions are power based and do not convey phase information; and 3) for UWB SAR, there is no clear mapping between the antenna parameters and the resulting SAR image.

2.2. Improved UWB Antenna Descriptors

The improved UWB antenna descriptors presented here have several benefits. First, they provide a precise and concise means to convey antenna performance in a natural and human readable form.

Second, the new descriptors or performance metrics are more convenient and meaningful for UWB antenna design optimization.

Third, the new descriptor's emphasis on pattern variations complements pre-existing studies on boresight performance and requirements.

Fourth, the new descriptors are consistent with traditional antenna parameters and terminology. As defined in this work, the impedance normalized transmit and receive transfer functions (Section 2.3) are easily reduced to traditional gain and effective aperture parameters. Other concepts such as *beamwidth*, *center of rotation*, and *energy patterns* are closely related to traditional counterparts with simple interpretations.

Finally, included in this set of descriptors are image domain type metrics. Descriptors such as integrated sidelobe energy and peak image response are calculated from synthesized images of ideal point targets incorporating realistic antenna and target motions. The combination of all the descriptors provide more comprehensive insight into the antenna performance as well as diagnostic tools to direct improvements.

The following sub-sections separates the new descriptors into two categories. The first is a generalized class of antenna descriptors useful for dispersive and non-dispersive UWB antennas. The second category of descriptors is directed specifically at UWB SAR applications.

2.3. Generalized UWB and SAR Specific Descriptors

This section presents the generalized and SAR specific UWB antenna performance metrics or descriptors. Please note that in addition to the metrics discussed below, the normalization function must be specified to fully characterize an UWB antenna. The normalization function, generally taken as the boresight response of the dominant polarization, includes both magnitude and phase information unlike the traditional gain parameter. Nevertheless, this normalization function can be related to traditional gain as discussed by Gwynne (1998).

2.3.1. The Generalized UWB Antenna Descriptors

The generalized UWB antenna descriptors provide a convenient and concise set of metrics to compare, rank, and optimize UWB antennas. Key UWB antenna descriptors are described below.

Normalized Spectral Pattern Functions: The normalized spectral pattern function displays the *normalized* radiation response of an antenna as a function of frequency and angle as a two-dimensional color image.

Normalized Transient Pattern Functions: The normalized transient pattern function displays the information contained in the magnitude and phase plots of the normalized spectral pattern function as a function of time and angle in a single color image.

Center of Rotation: The center of rotation concept is extremely important and must be understood and applied correctly in order to use the performance metrics discussed in this paper for identifying the sources of radiation (desirable and well as undesirable) in the antenna. The center of rotation is analogous to the concept of a phase center. The center of rotation is the rotation point which minimizes the time delay change of the impulse waveform peaks over the angular limits of the main beam. It is a single parameter defined for the entire frequency and angle range of interest.

Normalized Energy Pattern Functions: The normalized energy pattern functions provides a means of identifying and quantifying the uniformity or "normalized dispersion" of the antenna. This includes phase and amplitude variation as a function of frequency and polarization.

Energy Magnitude and Polarization Pattern Functions: The energy magnitude and polarization pattern functions provide a means to identify and quantify the polarization uniformity of the antenna.

2.3.2. The SAR Specific UWB Antenna Descriptors

This section presents a set of metrics which relates UWB antenna performance and SAR image formation. These parameters are meant to provide the means to compare and optimize antenna designs for UWB SAR. The parameters are created by simulating a SAR mapping with the antenna under analysis for a specified imaging path. Key performance descriptors are described below.

Image Peak Value: The maximum image domain response of an ideal point target formed by a simulated or actual SAR mapping.

Down and Cross Range Resolution: A measure of the envelop response width in a SAR image of an ideal point target. The width is defined for a given level relative to the peak value. Higher performing antennas have better (i.e., smaller) resolutions or widths.

Total Energy: The total energy is the integrated image domain energy over all space.

Integrated Sidelobe Energy: The integrated sidelobe energy represent the energy in the image domain sidelobes. The integrated sidelobe energy excludes the peak image response region—a region defined by a specified level relative to the peak response.

3. EXAMPLES ILLUSTRATING THE APPLICATION OF THE UWB ANTENNA DESCRIPTORS

The normalized UWB antenna descriptors presented in this paper provide a natural and human readable form to succinctly state and understand many UWB radiation characteristics such as phase dispersion, angular dispersion, and spectral and transient pattern variations. The purpose of this section is to provide examples illustrating how the performance metrics described in Section 2 have been applied to characterize dispersive and non-dispersive UWB antennas, identify deficiencies in the antennas, and optimize

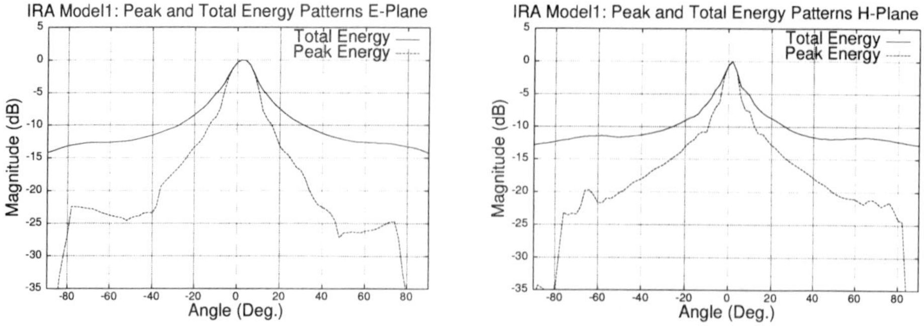

Figure 1. The total and peak energy curves quickly revealed that the E and H planes were not illuminated uniformly.

their performance for specific applications. This section is divided into two parts. Section 3.1 illustrates the types of diagnostics that can be performed. Section 3.2 illustrates an optimization example for UWB SAR.

3.1. Non-Dispersive Antennas

MRC performed comprehensive measurements on four IRAs (Impulse Radiating Antennas) with the goal of providing unprecedented characterization of the antennas' performance and to identify how the antenna could be improved. The techniques described in Section 2.3.1 were then used to characterize the four IRAs and identify the sources of unwanted radiation within the IRA aperture. This process occurred incrementally whereby a single antenna was characterized, the data analyzed, and design changes were recommended before the next antenna was built. Examples illustrating how these methods were used to identify design changes that improve 1) aperture efficiency and 2) polarization purity are presented.

3.1.1. Improved Aperture Efficiency

The first IRA tested was a typical design with four feed arms spaced at 90° intervals. After processing the measured data with the descriptors discussed in this paper, it was observed that the illumination of the aperture was non-uniform. While several descriptors revealed this information, the total and peak energy curves shown in Figure 1 concisely illustrate the different E and H plane energy beamwidths (i.e., the effective beamwidth for all frequencies). These graphs indicate that the aperture in the E-plane was not fully illuminated. By adjusting the feed arms to be closer together in the E-plane, a more uniform aperture illumination was achieved.

The uneven illumination of the aperture was also evidenced in the transient pattern functions, which provide the ability to identify specific sources of radiation within the antenna aperture. A detailed look at the transient pattern function in both the main pulse and subsequent pulse regions was performed. For the main pulse region, seven scattering mechanisms were considered and overlayed onto the transient pattern functions for the E and H planes as shown in Figure 2. The response shown at the angle 0° is the receive boresight response. The E-plane and H-plane patterns are shown to have the direct radiation from the feed (labeled as "Feed") in common but differ in terms of the other dominate features. The results indicate that the illumination of the reflector is not uniformed with

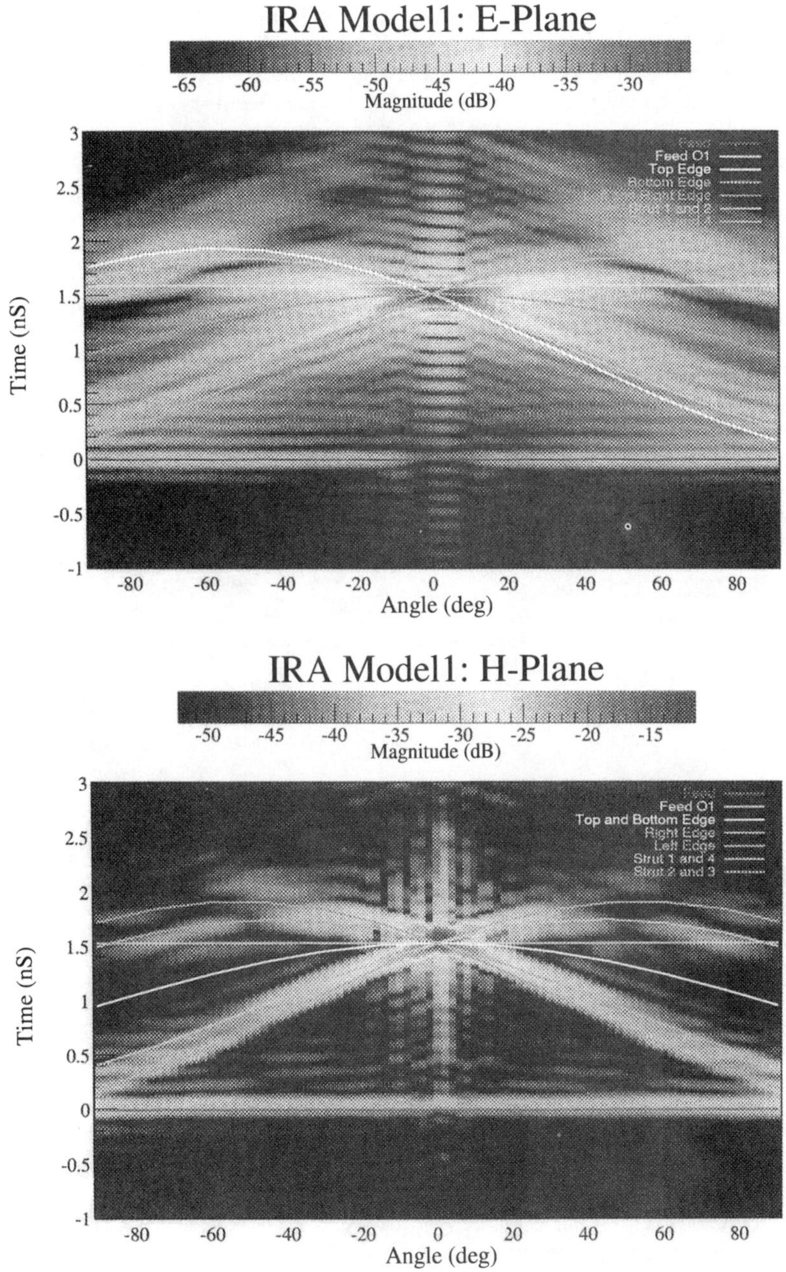

Figure 2. This figure identifies sources of the off-boresight radiation features of IRA antenna in the E and H planes. This figure was generated by overlaying the transient pattern function of the IRA antenna with several analytically derived interaction mechanisms. The interaction mechanisms are discussed in the text. The data shown in this plot have not been normalized by the boresight response, i.e., the 0° response.

respect to the two principal planes. First, the desired radiation (labeled as "Feed O1") has nearly twice the beamwidth in the E-plane which implies the reflector is not as well illuminated in the up/down direction for a vertical polarization orientation. Second, the E-plane shows strong radiation from the strut/reflector interface (labeled as "strut 1 and 4" and "strut 2 and 3") but nearly no scattering from the reflector edges. The reflector edges (labeled as "right edge" and "left edge") are seen in this plane with near grazing incidence near regions ±70°. In contrast, the H-plane shows both strong off-boresight scattering from both the reflector edges and strut/reflector union. In this plane, the edges of the reflector have stronger illumination and are more prominent. These pattern data indicate that for a vertical polarized antenna, the strut feed arrangement does not illuminate the top and bottom portions of the reflector as well as the sides.

3.1.2. Polarization Purity

The magnitude and polarization patterns provided clear evidence that the original IRA did not radiate a polarimetrically pure field throughout it's main beam as illustrated in Figure 3. In subsequent antenna designs, the mechanical tolerances in the manufacturing of the antenna significantly improved the polarization purity as shown in Figure 3. Note how the feed arms are clearly visible in this figure.

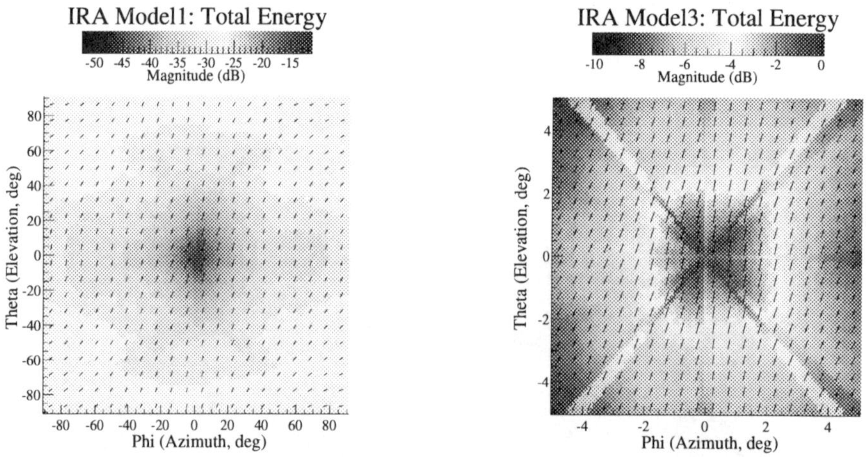

Figure 3. The polarization purity of two different IRA designs (left=poor, right=good) is clearly evident in these figures where a vertical vector signifies the co-polarized energy. Deviations from vertical identify the relative amount of energy in the co-polarized and cross-polarized fields.

3.2. Dispersive Antennas

MRC used the UWB SAR performance metrics discussed in this paper to rank and compare alternative antenna designs for the Army Research Laboratory's BoomSAR (Gwynne and Salo, 2001). Ultimately, the LPDA antenna, shown to radiation uniformly over both angle and frequency, was chosen and optimized using the peak image and integrated sidelobe metrics defined in Section 2.3.2. This optimized LPDA antenna is significantly different in detail to a LPDA antenna optimized for gain as shown in Figure 4. The uniformity of the radiation pattern, however, resulted in excellent imaging capabilities as shown in Figure 5.

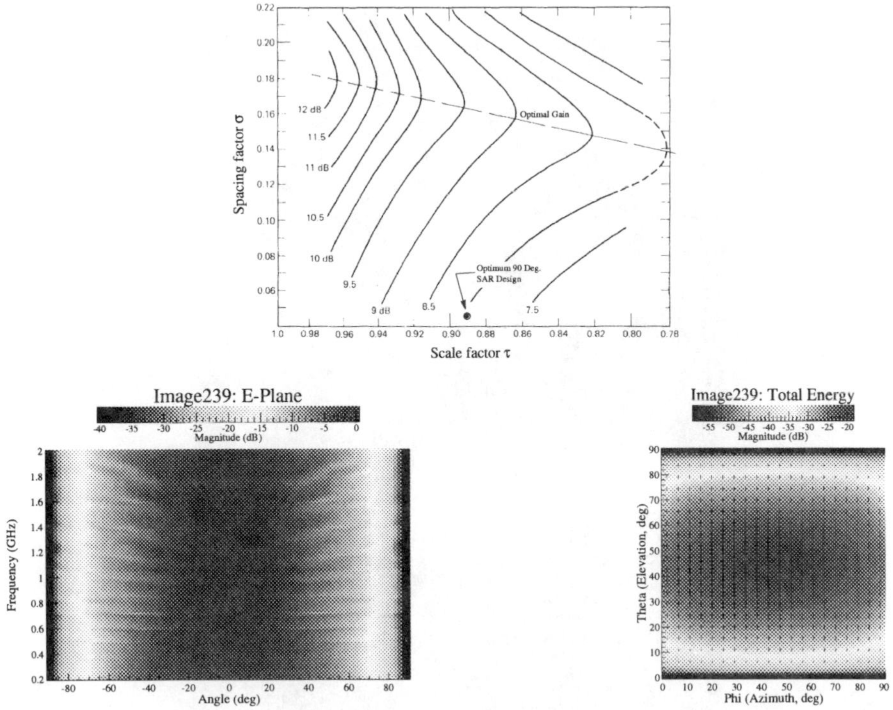

Figure 4. The LPDA optimized for SAR produced a significantly different design than the typical design guidelines used for maximum gain as shown in the top figure (Stutzman (1981) – Reprinted with permission from John Wiley & Sons, Inc. and IEEE), but results in an extremely uniform radiator well suited for SAR imaging (bottom left and right).

4. CONCLUSIONS

This paper presented a set of UWB antenna performance metrics that provide a useful way of characterizing UWB antenna performance. Several examples illustrated the ability of the descriptors to provide a means to identify sources of radiation within the antenna structure and how they manifest themselves in the radiated field. Additionally, an example was presented to illustrate the value of using the descriptors as metrics to optimize antenna performance. In summary, the techniques presented in this paper provide the antenna engineer with powerful tools to rapidly characterize, and hence, optimize, UWB antennas.

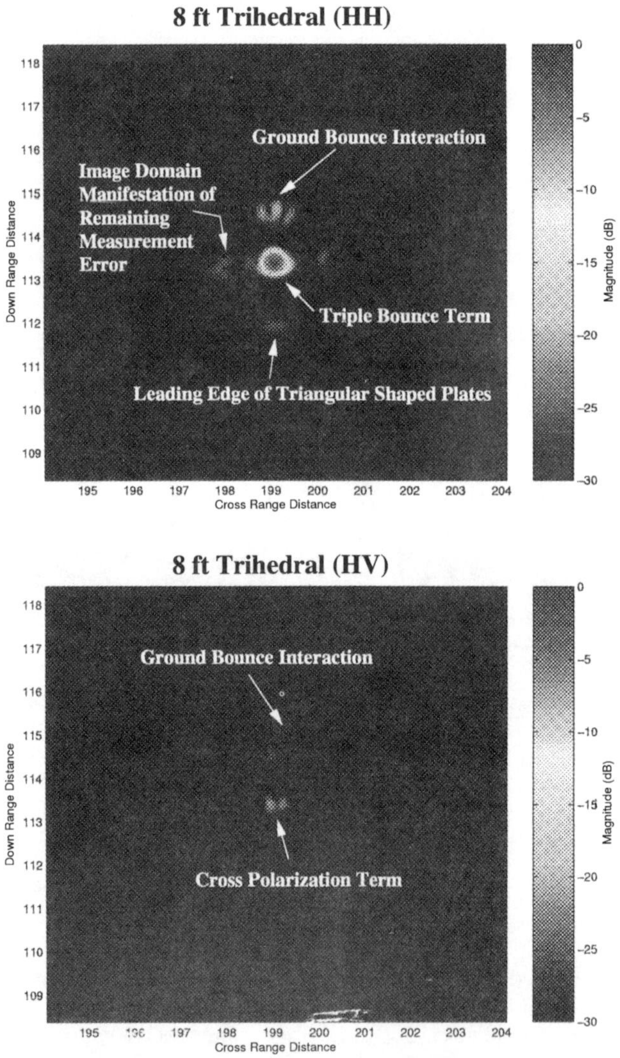

Figure 5. Co-polarized and cross-polarized SAR images formed with ARL's impulse radiating BoomSAR radar and the optimized LPDA antennas of an $8'$ trihedral.

5. REFERENCES

J. S. Gwynne, *The Propagation and Scattering Characteristics of a Forest as Measured by Coherent Ultra-Wideband Foliage Penetration SAR*, PhD thesis, The Ohio State University, Columbus, OH 43212, 1998.

J. S. Gwynne and G. R. Salo, Antenna System for Unmanned Aerial Vehicles, Army Research Laboratory Technical Report No. ARL-CR-392, August 2001.

Warren L. Stutzman and Gary A. Thiele, *Antenna Theory and Design* (John Wiley & Sons, Inc., 1981).

OPTIMIZED ULTRA-WIDEBAND RADIATION OF DIPOLE ANTENNAS WITH TRIANGLE DRIVING PULSES

Anatoliy O. Boryssenko* and Daniel H. Schaubert*

1. INTRODUCTION

This paper addresses optimization of transient radiation from dipole antennas in terms of minimized pulse extent in time or flat power spectral density. The former is a typical requirement for high-range-resolution radars while the latter can meet the Federal Communication Commission (FCC) regulations on spectrum usage. Recent needs and advances in ultra-wideband (UWB) signal processing, target identification, remote and subsurface sensing, reach-frequency tagging, secure communication, and others stimulate a variety of diverse studies in transient and broadband electromagnetics, including pulsed antennas. These generic problems have been widely explored through full-wave numerical analysis, theoretical and asymptotic approaches, nonlinear optimization methods combined with full-wave electromagnetic solvers, the E- and K-pulse signal processing techniques, and other fashions.

Previous work often involves antennas of complicated geometry/structure and/or pulses of complex shapes to drive the antennas' currents. The work reported here is based on three guiding requirements:

1. Geometrically simple antennas like thin-wire linear dipole, flat dipole, bow-tie radiator and conical dipole.
2. Simple triangle-like pulse shapes of voltage source to reduce the complexity of driving electronics.
3. Operation below the first parallel resonance of the antenna so the radiator is not large in size.

The optimal signals for driving of the dipole antennas are realized here in numerical simulation through full-wave method of moment (MoM) numerical analysis, and combined MoM simulation and nonlinear optimization. Results of the electromagnetic analysis are supported with circuitry modeling in both time and frequency domains also.

---

* Antenna Laboratory, University of Massachusetts, Amherst, MA, 01003, USA

2. PROBLEM FORMULATION

It is accepted[1, 2, 3] that conventional wideband antennas like dipoles often do not perform well under wideband excitation because of their dispersive characteristics. In this work, we propose some novel operational modes of those antennas, which enable controlled UWB behavior by using their inherent dispersive properties. The problem can be solved with the help of specially constructed waveforms of driving signals that are matched in some sense with the antennas. The primary intent here is to achieve some specific performance criteria including the radiation of temporally short and/or broad bandwidth pulses. Particular attention is given in this work to transmitting antennas of simple geometries, like center-fed thin-wire and flat dipoles, Figure 1.

It is known[1, 2] that the simplest solution to support short radiated pulses is using an electrically very short antenna whose radiation is proportional to the time derivative of the driving waveform. Resistively loaded antennas[4] can be used to reduce unwanted reflection and associated pulse distortions. However, electrically small antennas and resistive loading are not always applicable due to poor radiation efficiency. Other related studies in the transient antennas involve structures of composite geometry and/or complex waveform for excitation such as log-periodic antennas with chirp pulse[5], V-shape dipoles, TEM-horns, tapered-slot antennas, reflectors, lenses, and others[3]. For the purpose of our project, forward protruding end-fire radiators do not meet the requirement for small volume.

Moreover, the FCC regulates the EM spectrum for UWB devices. Thus, besides flatness, simplicity in geometry, and efficiency in radiation, waveform/spectrum control is required that has to be consistent with realistic solutions for the driving electronics. Various methods based mainly on signal processing like E- and K-pulse techniques are not well suited for direct implementation in antennas' hardware and left beyond the scope of our research. However, we employ widely methods of optimization combined with full-wave numerical analysis[6] for improving the transient characteristics in antennas.

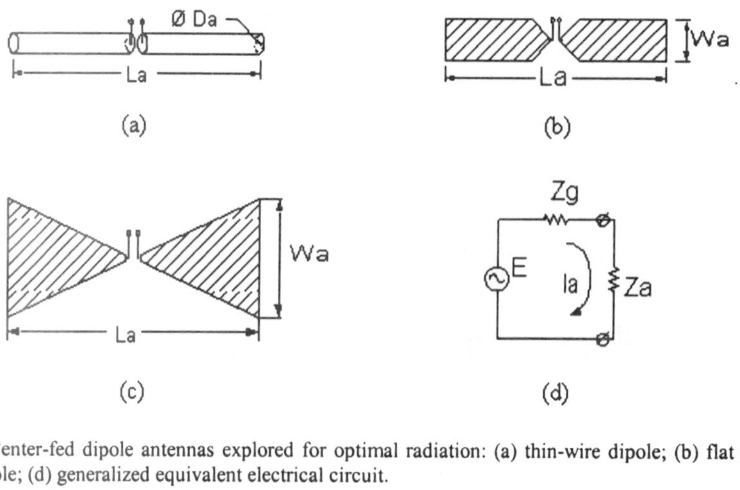

Figure 1. Center-fed dipole antennas explored for optimal radiation: (a) thin-wire dipole; (b) flat dipole; (c) bow-tie dipole; (d) generalized equivalent electrical circuit.

3. ANALYSIS THROUGH TIME-DOMAIN INTEGRAL-EQUATION SOLVER

Full-wave analysis of the antennas in Figure 1 is performed by using a time-domain electric integral equation solver[7] that we developed. It simulates, through marching in time, 3-D electromagnetic problems with thin-wire and surface-patched elements. Usually, when a thin wire dipole is center-fed with an arbitrary exciting signal, the terminal current in the antenna has long oscillations in late time because the antenna has high-Q resonances. Figure 2a shows this result for a sinusoidal burst with Gauss pulse envelope. The spectral maximums for the current correspond to the resonant frequencies covered by the spectrum of the exciting waveform. Surprisingly, a simple asymmetric triangle pulse applied to the same antenna extinguishes the late-time oscillations, Figure 2b, when the appropriate pulse length and ratio of rise-to-fall time is provided. The antenna current resembles a single sinusoidal cycle while the triangular source is applied and a low-amplitude ringing during source-free interval. The power spectrum of the triangular pulse is mostly below the first parallel resonance of the 0.5-meter antenna. The spectral power density of the terminal current is relatively broad and flat and centered near 150 MHz. The source is "matched" to the antenna in the sense that its power spectral density compensates for the low-frequency response of the antenna and excites the first series resonance sufficiently to broaden and flatten the current spectrum.

Figure 3 illustrates computed far-field radiation patterns in time-domain for the same cases as in Figure 2. For the optimal triangle pulse, Figure 2b, the radiated waveform has dominant sinusoidal-like triplet with low oscillations in source-free interval. Optimizations may be performed on the radiated field or the terminal current. Hereafter, we will illustrate the antennas' driving/terminal current.

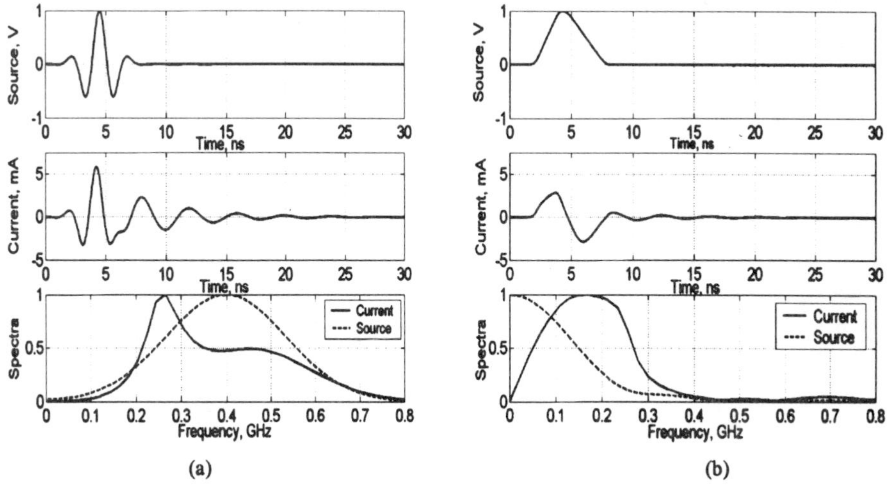

Figure 2. Transient excitation of the center-fed 0.5-meter thin wire dipole in transmit mode by: (a) Gauss pulse: (b) optimized triangle pulse.

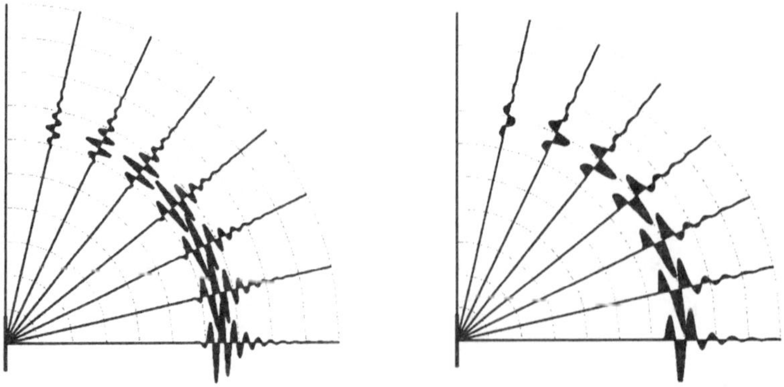

(a) (b)

Figure 3. Far-field radiation pattern for (a) Gauss and (b) triangle pulses.

The shape and spectrum of the optimal triangle pulse are shown in Figure 4. In fact, the triangle shape of the pulse has been discovered by chance during extensive calculations with the MoM code. Similar waveforms can be found for all of linear antennas in Figure 1, but only for the thin wire dipoles the ratio of rise-to-fall time is amazingly close to the golden section number, Figure 4a. However, optimization techniques combined with MoM analysis[6] can be employed effectively to optimize transient behavior of many antennas in the same way. For example, Figure 5 demonstrates this approach applied to the flat dipole. Two principal cases are seen in Figure 5. The first one corresponds to optimization with a goal function to support well-localized spectrum, Figure 5a, while optimization in the second case, Figure 5b, is aimed at obtaining temporally short pulse but by some price of small spreading in the spectrum.

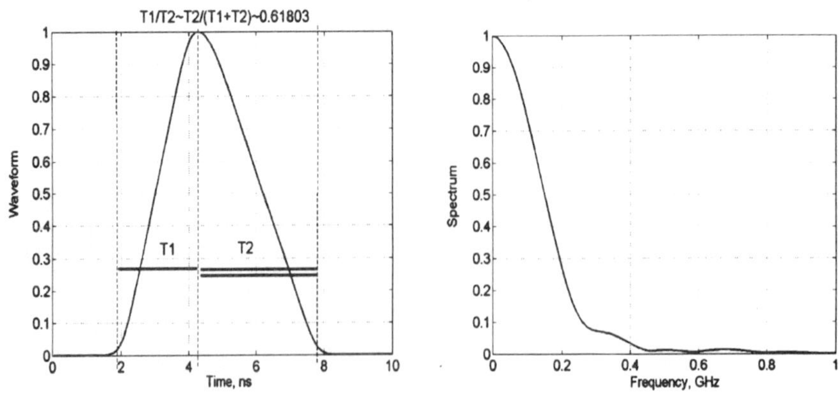

(a) (b)

Figure 4. An optimal triangular pulse for the thin-wire 0.5-m dipole: (a) pulse waveform with rise-to-fall ratio equals about to the golden section number; (c) power spectrum.

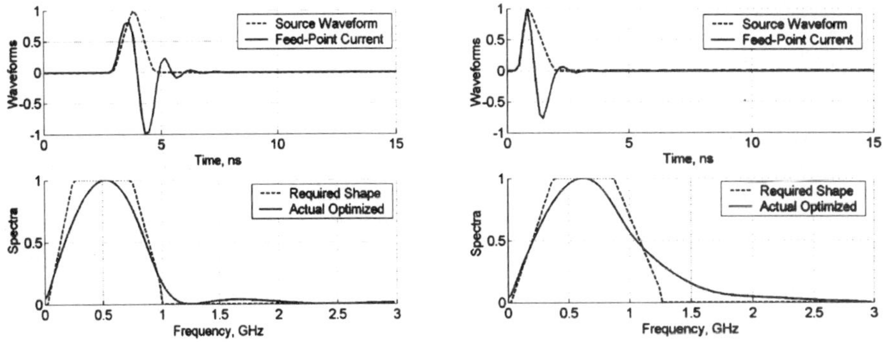

(a) (b)

Figure 5. Some optimization for center-fed bow-tie dipole of 10-cm length and 5-cm width to provide: (a) localized spectrum; (b) shortest pulse duration.

When optimizing for short time duration, greater than 95% of the pulse response energy belongs to sine wave doublet, which coincides with the duration of the triangular voltage pulse applied to the antenna. When optimizing for flat spectrum, a typical 4:1 half-power bandwidth is achieved with nearly all of the power at frequencies below the first parallel resonance of the dipole. The optimization procedure can be repeated for different goal functions tracking good localization for the antenna pulsed response in spectrum or in time or in a mixed mode.

A simple phenomenological interpretation in terms of radiation physics of transient antennas helps to explain the found effect of the triangle pulse. Obviously, the driving waveform delivers the pulsed energy to the antenna. Due to finite lengths of both the pulse and antenna this energy comes into the antenna consecutively through its terminals and propagates as a current/charge wave in it. The portion of the wave belonging to the front of the pulse will be reflected from the open edges of the antenna and returned back to the feed point. Nearly at the time when the reflection happens the second portion of the driving pulse with falling edge comes into the antenna through the terminal and annihilates the back-reflected wave because of phase inversion after reflection from the antenna's open edges. In the end, the current exists in the antenna when the source pulse is present there only, and there is little ringing. The pulse is asymmetric to compensate for dispersion in the antenna.

4. ANALYSIS BASED ON EQUIVALENT ELECTRICAL CIRCUITS

Since we use simple antennas that are electrically small, the radiating currents of the antenna (or the fields if we choose to consider them) are related to the source voltage by a simple equivalent circuit. We can optimize antenna current performance by considering only the impedance of the antenna and the voltage generator. Thus, this problem of antenna optimizing can be reduced to the simpler problem of circuit optimization.

The equivalent circuit of any antenna can be defined through the samples of frequency-domain data on the antenna input impedance by least-square fitting of these actual full-wave data and those from the equivalent circuit. For example, Figure 6 shows the equivalent circuit for bow-tie dipole of the specified geometry and demonstrates excellent agreement between both the models over a very broad bandwidth. This simple equivalent circuit can be used to explore the behavior of the driving electronics over a frequency range broader than the antenna's operating bandwidth.

A simple mathematical model, based on the equivalent circuit in Figure 1d, can be utilized for synthesis of an optimized pulse. A source waveform *E(t)* that provides at the antenna terminals some required spectral properties for the current $I(\omega)$ can be obtained from:

$$E(t) = F_{-}\{I(\omega) \cdot Z(\omega)\} \tag{1}$$

where F_{-} - inverse Fourier transform operator; $I(\omega)$ – required current response, approximated through polynomial or error-function for flat spectrum density and linear phase; $Z(\omega)$ – input impedance of antenna like that presented in Figure 6.

Figure 7 contains typical results for the antenna in Figure 6. Four cases of increasing bandwidth are presented. If the frequency range of the signal is approximately 0.2 – 0.7 of the first parallel resonant frequency of the dipole antenna, Figure 7a, a relatively simple voltage waveform compensates for the variation of the antenna's impedance function, resulting in approximately uniform power spectral density. However, if the signal's frequency range includes the first parallel resonance, Figure 7d, or close to it, Figure 7c, the voltage waveform must be more complicated and contains a large power density near the resonance to offset the reduced response in this frequency range. The intermediate case, Figure 7b, confirms the above observations on the significance of proper choice of bandwidth with respect to the resonant frequency.

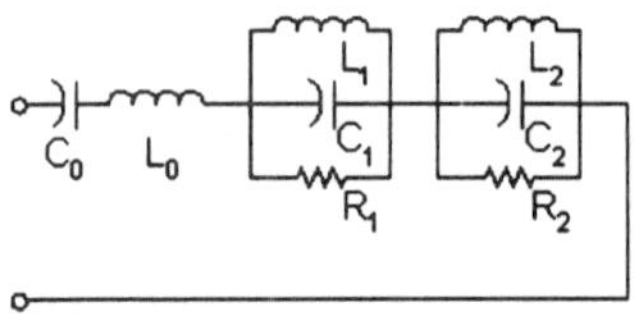

C_0 pF	L0 nH	C1 pF	L1 nH	R1 Oh	C2 pF	L2 nH	R2 Oh
1.5	1.2	0.5	1.2	240	0.3	2.8	118

(a)

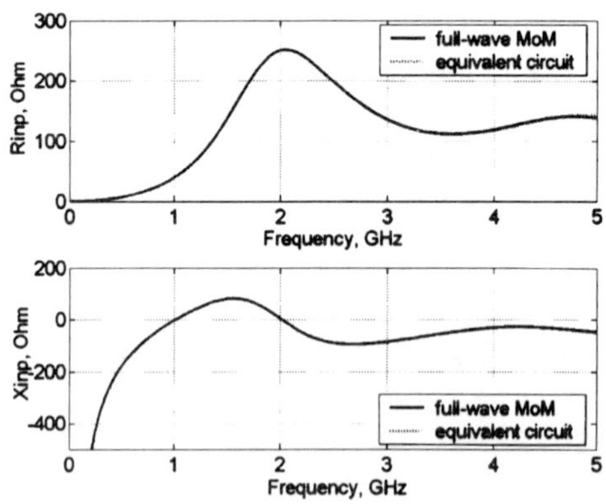

(b)

Figure 6. Bow-tie dipole of 10-cm length and 5-cm width: (a) equivalent circuit; (c) input impedance through the full-wave MoM analysis and the equivalent circuitry.

The antenna current responses in Figure 7 demonstrate several other features worth mentioning:

1. The most useful bandwidth of the antenna ranges up to 0.7-0.8 of its first arallel resonance.
2. Relatively simple voltage waveforms yield moderately well localized current in both frequency and time domains.
3. Bandwidths in excess of 6:1 (-3dB level) can be supported with simple voltage pulses suitable for implementation through driving electronics.
4. Localization of the antenna driving current in time is inversely proportional to the achieved bandwidth.
5. Size scaling would maintain simplicity of voltage waveform while increasing operating frequency.

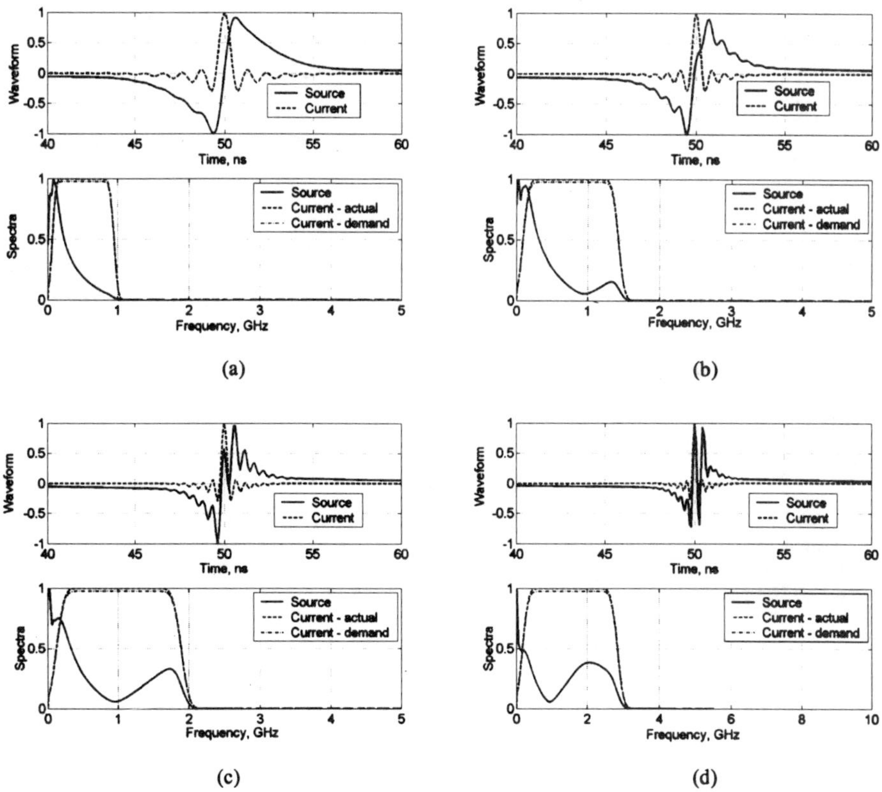

Figure 7. Bow-tie dipole of 10-cm length and 5-cm width with optimizations for bandwidth of about: (a) 0.2-0.8 GHz; (b) 0.2-1.2 GHz; (c) 0.2-1.7 GHz; (d) 0.4-2.5 GHz.

5. CONCLUSIONS AND DIRECTIONS OF FUTURE WORK

Some solutions to support high performance pulse radiation modes in the dipole-like antennas are presented. Two types of waveforms for pulse driving in the antennas are revealed; first, the triangle pulse with DC components in spectrum and odd response function in time and, second, the triangle-like doublet with zero DC and even response function in time. In both cases, simple electronic circuits can generate the voltage driving waveform. In general, the methods presented here yield transient waveforms to meet specific performance criteria for band limited antennas.

The demonstrated analysis and its results are directly applicable to control the shapes of pulses and their spectra for the dipoles, and maybe other antennas, by constructing of the necessary driving signals. For example, the cases shown here exhibit a dominant early-time response (more than 95% of the pulse energy) and very low late-time tails. At the same time, much of energy of antenna response can be effectively localized in a specific frequency band below the first parallel resonance in the antennas. Because antennas of nearly resonance length without resistive loading are employed, radiation efficiency is high and the antennas are of moderate size. Our current and further efforts are aimed at looking for new types of exciting waveforms to be employed in time-hoping UWB wireless systems and realized through CMOS integral circuitry. The actual antenna-driver configuration becomes extremely important to enhance overall performance of the UWB system.

6. ACKNOWLEDGMENTS

This research was supported by the MURI grant DAAD19-01-1-0477, "Short-Range Ultra-Wideband Systems."

7. REFERENCES

1. G. Franceschetti, C.H. Papas, Pulsed Antennas, *IEEE Trans. Ant. Propag.* **22**(5), pp. 651-661 (1974).
2. H. F. Harmuth, *Antennas and Waveguides for Nonsinusoidal Waves* (New York, Academic Press, 1984).
3. C. E. Baum, E. G. Farr, Impulsive Radiating Antennas, in: *Ultra-Wideband Short-Pulse Electromagnetics*, edited by H. Bertoni et al. (Plenum, New York, 1993), pp. 139-147.
4. T. T. Wu, R. W. P. King, The Cylindrical Antenna with Nonreflecting Resistive Loading, *IEEE Trans. Ant. Propagat.* **13**(3), pp. 369-373 (1965).
5. P. S. Excell, A. D. Tinniswood, R.W. Clarke, An Independently Fed Log-Periodic Antenna for Directed Pulsed Radiation, *IEEE Trans. Ant. Propagat.* **41**(4), pp. 344-349 (1999).
6. J. R. Bayard, D. H. Schaubert, Target Identification Using Optimization Technique, *IEEE Trans. Ant. Propagat.* **38**(4), pp. 450-456 (1990).
7. A. O. Boryssenko, D. H. Schaubert, Time-Domain Integral-Equation-Based Solver for Transient and Broadband Problems in Electromagnetics, in: *This publication*.

ANALYTICAL, NUMERICAL, AND EXPERIMENTAL RESULTS FOR WIDE-ANGLE BICONICAL ANTENNAS

E. Mokole, S. Samaddar, G. Sorenson, M. Kragalott, B. Gold, M. Parent*

1. INTRODUCTION

In a previous paper,[1] research on spherically capped, axially symmetric, wide-angle, biconical antennas[2–9] was generalized to bicones with unequal cone angles. In particular, Samaddar and Mokole[1] theoretically analyzed the radiated and received fields of such bicones for ultrawideband (UWB) short pulse excitations, the corresponding radiation patterns versus frequency, and the high-frequency behavior of electrically small and electrically large bicones. In the current discussion, the authors use the theoretical results of the earlier paper to study the input impedance (Z_{in}), the voltage standing wave ratio (VSWR), and the normalized radiated field pattern. The expression for Z_{in} is recast in a more usable form, which is then used to derive an analytical expression for the VSWR. These three quantities are computed numerically by two means: directly from the analytical representations and from Maxwell's equations with a finite-difference time-domain (FDTD) method. The FDTD results compare reasonably well with the analytical representations. Finally, some preliminary measurements of the normalized radiated field patterns for an electrically small ($ka \ll 1$), wide-angle (100°), mono-cone (0.60325-cm radius) with a large, finite, circular ground plane (121.92-cm diameter) are compared to a theoretically derived, analytical representation.[1] Because an infinite ground plane was assumed in the derivations of the analytical expressions of Samaddar and Mokole,[1] the measured patterns for the finite ground plane deviated from the patterns computed with the analytical expressions, as a result of the edge effect of the finite ground plane, especially in the broadside direction.

The geometry of the antenna configuration is shown in Fig. 1. The bicone is assumed to be made of perfectly conducting metal and consists of two cones with coincident vertices and equal lateral sides of length a. The common axis of the two cones is oriented along the z-axis, where the cone angles ψ_1 and ψ_2 are measured from the positive and negative z-axes, respectively, and obey $0 < \psi_i < \pi/2$ and $0 < \psi_1 + \psi_2 < \pi$. In terms of spherical coordinates (r,θ,ϕ), the lateral surfaces of the upper and lower cones correspond

* E. Mokole, G. Sorenson, M. Kragalott, B. Gold, and M. Parent, Radar Division of Naval Research Laboratory, Washington DC 20375. S. Samaddar, SFA, Inc., Largo MD 20774.

to $\theta = \psi_1$ and $\theta = \pi - \psi_2$, respectively, where θ is measured from the positive z-axis. Since the cones are excited symmetrically at the apices, only TEM and TM modes are generated in the antenna region. However, the higher-order TM modes can be neglected for wide-angle bicones ($\psi_1, \psi_2 \geq 40°$).

One of the motivations for studying the wide-angle biconical antenna is that knowledge of its radiation characteristics is helpful in understanding why TEM horns, V-antennas, triangular plates, and bow-tie antennas are ultrawideband.[10,11] Moreover, by a proper choice of ψ_1 and ψ_2, the exit aperture of the antenna can be adjusted so that the radiated energy is focused in a desired direction.

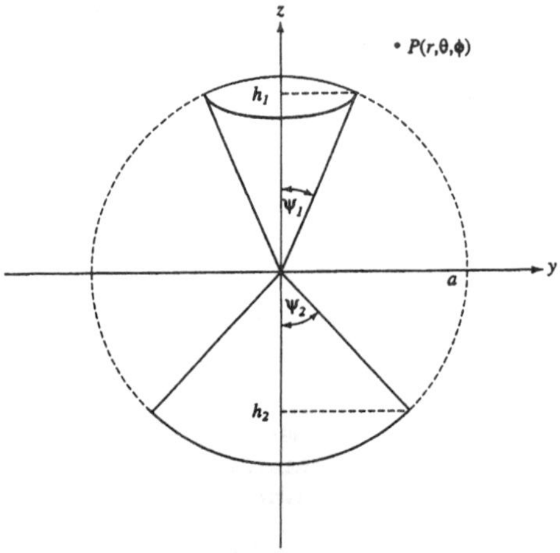

Figure 1. Cross-sectional view of spherically capped, biconical antenna with x-axis pointing out of the page and the observation point P in spherical coordinates (r, θ, ϕ).

2. INPUT IMPEDANCE AND VSWR

By substituting Eq. (34a) into Eq. (29) of Ref. 1, the wide-angle representation for the input impedance is recast as

$$Z_{in}(ka,\psi_1,\psi_2) = R_{in} + iX_{in} = -iZ_C(\psi_1,\psi_2)\frac{\cos(ka) + S(ka,\psi_1,\psi_2)\sin(ka)}{\sin(ka) - S(ka,\psi_1,\psi_2)\cos(ka)}, \quad (1)$$

where Z_0 is the free-space impedance (120π Ohms),

$$S(ka,\psi_1,\psi_2) = \frac{Z_0\Omega(\psi_2)}{4\pi Z_C(\psi_1,\psi_2)}\sum_{n=1}^{\infty}\frac{2n+1}{n(n+1)}G_n^2(\psi_1,\psi_2)\zeta_n(ka), \quad 1.0$$

$$Z_C(\psi_1,\psi_2) = \frac{Z_0}{2\pi}\ln\left(\cot\frac{\psi_1}{2}\right)\ln\left(\cot\frac{\psi_2}{2}\right), \qquad \Omega(\psi_2) = \begin{cases} 1, & \psi_2 \neq \pi/2 \\ 2, & \psi_2 = \pi/2, \end{cases} \tag{3}$$

$$G_n(\psi_1,\psi_2) = P_n(\cos\psi_1) - (-1)^n P_n(\cos\psi_2), \qquad \zeta_n(ka) = \frac{h_n^{(2)}(ka)}{h_{n-1}^{(2)}(ka) - \frac{n}{ka}h_n^{(2)}(ka)}. \tag{4}$$

Note that Z_{in} depends on frequency f ($k = 2\pi f/c$) and the spatial configuration of the bicone (a, ψ_1, ψ_2), where c is the speed of light. As indicated by the series representing S, this dependence is a complicated combination of Legendre polynomials P_n and spherical Hankel functions $h_n^{(2)}$ of the second kind.[12] When ψ_2 is 90°, the equations correspond to the case of an infinite perfectly conducting plane at $z = 0$. In fact, the function $\Omega(\psi_2)$ is a mathematical artifice that was introduced to account for this situation and does not yield the solution of the finite problem for $\psi_2 = 90°$, which is obtained by taking the limit of the solution when $\psi_2 \neq 90°$ as ψ_2 approaches 90°.

The normalized input impedance Z_{in}/Z_c, the normalized resistance R_{in}/Z_c, and the normalized reactance X_{in}/Z_c are plotted versus ka in Fig. 2 for $\psi_1 = \psi_2 = 53.1°$. The cone angle of 53.1° is chosen for comparison with Sandler and King.[9] In these calculations, the series in Eq. (2) is truncated, which raises the issue of how many terms are necessary for convergence. When ka is small, only a few terms are required for the series to converge to the correct answer; however, the number of terms required for convergence grows quickly as ka increases. For example, only 6 terms are needed when $ka = 10$; whereas ten times (60) as many terms are needed when $ka = 100$. The large increase in terms greatly increases the computation time, which appears to grow as the square of the number of terms in the truncated series.

When $f = 0$, no current distribution exists and the constituent cones of the bicone behave as if physically separated. In ordinary circuit analysis, the reactive part of the input impedance is dominated by the capacitive effect at low frequencies. As the frequency increases, the inductive effect increases until it equals the capacitive effect at resonance and finally dominates the reactance above the resonance frequency. For low frequencies, the impedance of the bicone is similarly dominated by the capacitive effect, which is consistent with the analytically derived approximate representation of Z_{in} in Eq. (39b) of Samaddar and Mokole.[1] However, in contrast to the circuit analogy, the effective inductance and the effective capacitance of the bicone at the higher frequencies are much more complicated functions of frequency. Specifically, as the frequency increases, the reactance oscillates about zero with ever decreasing amplitude. For $ka \geq 3$, the reactance is small and corresponds to the resistive part of the impedance decreasing to the antenna's characteristic impedance Z_C in a damped oscillatory manner, which is consistent with theoretical result that Z_{in} approaches Z_C as ka approaches infinity. That Z_{in} is dominated by R_{in} and does not vary much with frequency for $ka \geq 3$ is an indication that the wide-angle bicone is an ultrawideband antenna.

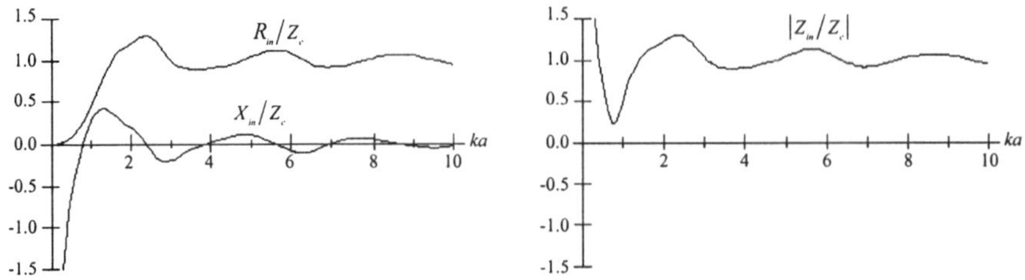

Figure 2. (a) Normalized resistance R_{in}/Z_C and reactance X_{in}/Z_C of a 53.1° bicone versus *ka*. (b) Magnitude of normalized impedance Z_{in}/Z_C of 53.1° bicone versus *ka*.

The usual standard for defining the performance of an antenna is the VSWR as specified by

$$\mathrm{VSWR}(ka,\psi_1,\psi_2)=\frac{1+\left|\dfrac{Z_{in}-Z_C}{Z_{in}+Z_C}\right|}{1-\left|\dfrac{Z_{in}-Z_C}{Z_{in}+Z_C}\right|}=\frac{\left\{\left[\left(1+\dfrac{R_{in}}{Z_C}\right)^2+\left(\dfrac{X_{in}}{Z_C}\right)^2\right]^{\frac{1}{2}}+\left[\left(1-\dfrac{R_{in}}{Z_C}\right)^2+\left(\dfrac{X_{in}}{Z_C}\right)^2\right]^{\frac{1}{2}}\right\}^2}{4\dfrac{R_{in}}{Z_C}}. \qquad (5)$$

The VSWR indicates how much energy is transferred from the transmission line feeding the bicone across its input terminals. In particular, the percentages of transmitted power and reflected power are $(1-|\Gamma|^2)\times100\%$ and $|\Gamma|^2\times100\%$, respectively, where the voltage reflection coefficient Γ is the ratio of $Z_{in}-Z_C$ to $Z_{in}+Z_C$ in Eq. (5). As the VSWR approaches unity, 100% of the power is transferred to the bicone. For example, a VSWR of 1.2 corresponds to a transmitted power of 99.77%.

Substituting $\psi_1=\psi_2=53.1°$ into Eq. (5) yields the VSWR versus *ka* in Fig. 3. As *ka* increases, the VSWR initially decreases to a local minimum value of 1.02 for $ka=4$ and has values between 1.02 and 1.1 for $4\le ka\le 100$. Because the number of terms in the truncated series that is needed for an accurate approximation of Z_{in} increases significantly as *ka* increases, calculating the VSWR with the series respresentation of Z_{in} for $ka>100$ becomes computationally intensive. Since Samaddar and Mokole have proven that $Z_{in}\to Z_C$ as $ka\to\infty$, the VSWR must approach unity from above as $ka\to\infty$. Consequently, the curve in Fig. 3 will asymptotically approach the horizontal axis (unity) if the VSWR is computed with sufficient accuracy. That the VSWR is essentially less than 1.1 for $ka\ge3$ is a demonstration of the ultrawideband performance of wide-angle biconical antennas. Although the plots are not shown, the VSWRs of other equiangular, wide-angle bicones (45°,50°,55°,60°) have good ultrawideband characteristics.

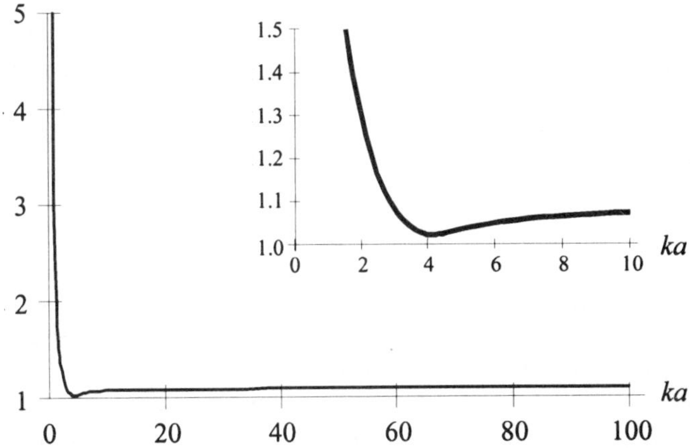

Figure 3. Values of VSWR for $\psi_1 = \psi_2 = 53.1°$. The inset plot provides a magnified view of the VSWR for *ka* between 0 and 10.

3. RADIATION PATTERNS

An interesting aspect of the bicone antenna is the relative field pattern R associated with the E-plane radiation pattern as defined by Papas and King[7] and given by Eq. (6). For each *ka*, the relative pattern is normalized to its value at broadside ($\theta = 90°$). Like the input impedance, the relative radiation pattern is an infinite series involving Legendre polynomials and spherical Hankel functions of the second kind. In addition, the series includes the associated Legendre function $P_n^{\,1}$.

$$R(\theta,ka,\psi_1,\psi_2) = \frac{E_\theta^{rad}(r,\theta,ka,\psi_1,\psi_2)}{E_\theta^{rad}(r,\pi/2,ka,\psi_1,\psi_2)} = \frac{\displaystyle\sum_{n=1}^{\infty} \frac{i^{n-1}(2n+1)}{2n(n+1)} \frac{P_n^1(\cos\theta)\,G_n(\psi_1,\psi_2)}{h_{n-1}^{(2)}(ka) - \frac{n}{ka} h_n^{(2)}(ka)}}{\displaystyle\sum_{n=1}^{\infty} \frac{i^{n-1}(2n+1)}{2n(n+1)} \frac{P_n^1(0)\,G_n(\psi_1,\psi_2)}{h_{n-1}^{(2)}(ka) - \frac{n}{ka} h_n^{(2)}(ka)}}. \tag{6}$$

Explicitly, R depends on the four parameters: θ, ka, ψ_1, ψ_2. As a consequence of the axially symmetric excitation at $z = 0$, R is independent of ϕ. To compare with the numerically based patterns of Sandler and King,[9] the magnitude $|R|$ is plotted in polar format $(|R|,\theta)$ for nine values of *ka*, with $\psi_1 = \psi_2 = 53.1°$ in Fig. 4. Since $\psi_1 = \psi_2$ implies that $G_n(\psi_1,\psi_2) = 0$ for even values of the index n, the terms in both series corresponding to even n are zero. To evaluate $|R|$, each series in Eq. (6) is truncated to the same number of terms M, which varies with *ka*. As *ka* increases, M must increase to achieve a desired accuracy. Since R is independent of ϕ, it has azimuthal symmetry. Moreover, R is symmetric about the xy-plane ($\theta = 90°$). In light of both symmetries, the 3-dimensional pattern can be represented by the 2-dimensional patterns of Fig. 4, where the 3-dimensional patterns are obtained first by rotating the curves in Fig. 4 about the y-axis and then about the z-axis.

For electrically small bicones (ka = 0.0106 and 0.1064), the pattern is similar to that of a short dipole, where the maximum radiation occurs in the broadside direction (θ = 90°). As ka increases through transitional values from 1 to 21.279, the maximum value of $|R|$ usually occurs at broadside, except for values of ka near 3.2, and side lobes appear in increasing numbers. The curves for ka of 1.06, 5.319, and 10.64 in Fig. 4 are identical to those in Fig. (5c) of Sandler and King. As ka increases through the transitional region to the electrically large regime ($ka \gg 1$), the direction of maximum radiation migrates away from broadside and eventually approaches the surface of the bicone. Samaddar and Mokole analytically showed that $R \to 1/\sin\theta$ as $ka \to \infty$. Consequently, in the high-frequency limit, the 2-dimensional pattern is the triangle in the lower-right frame of Fig. 4 (ka = 106.4) that is formed by the vertical line $y = 1$ ($R = 1/\sin\theta$), the line $z = 0$, and the line $z = y\cot(53.1°)$. The pattern for ka = 106.4 is consistent with the derived high-frequency limit ($1/\sin\theta$). For example, most of the energy is radiated for $50° \leq \theta \leq 90°$, and the five local extrema are approximately aligned with the vertical line $y = 1$.

4. FDTD SIMULATION VS ANALYTICAL COMPUTATION

The analytical theory is now compared to results obtained from a purely numerical method for four values of ka in the transitional region. Specifically, the relative patterns and input impedances computed from the analytical expressions in Eqs. (1), (5), and (6) are compared to the same quantities when computed with the Finite-Difference Time-Domain (FDTD) method applied to Maxwell's equations for ka = 1.0640, 5.3198, 7.9800, 10.6395.

For ease of comparison, instead of using the polar format of Fig. 4, the relative patterns in Fig. 5 are plotted versus the spherical observation angle θ, which is varied between 0° and 90° as a result of antenna symmetry. Note that the normalized patterns match to within a few percent for low transitional values of ka. For the larger transitional values of ka, the FDTD patterns are between 5% and 10% larger than those calculated with the analytical method. However, the shapes match quite well, as the FDTD patterns have local extrema that occur at approximately the same angles as the local extrema of the analytically calculated patterns. To achieve good accuracy (pattern convergence), the FDTD solution required 80 cells per wavelength.

For the input impedance and VSWR calculations, two sets of FDTD calculations are obtained: the first set has ka values ranging from 0.34 to 2.78; while the second set of ka ranges from 4.21 to 12.56. The FDTD results of the resistive and reactive components of Z_{in} agree very well with the analytical computations for small values of ka (Fig. 6). Although the values do not match as well once ka gets larger, minima and maxima occur near the same values of ka. The resistance is lower for the FDTD method, and the slightly negative reactance implies a net capacitive effect. The capacitance is a result of the FDTD feed model, which is a voltage source between the centers of four-cell flat-plate regions on the upper and lower cones. Other feed models were attempted that avoided the four-cell flat-plate regions, but they created an inductive effect. A sub-cell FDTD is required for characterizing the feed region accurately without incurring large computational costs, but such a code was unavailable during the writing of this article. Overall, calculations of Z_{in} from the analytical and FDTD methods match very well, although the analytically calculated Z_{in} is likely the more accurate of the two, because the number of terms in the truncated-sum approximation of the infinite-series representation of Z_{in} can be chosen large enough to ensure the accuracy of the approximation without becoming computationally prohibitive. When *ka* is sufficiently large, the expression for Z_{in} asymptotically approaches the analytically determined value of Z_C.

With respect to computation of the relative pattern, the FDTD and analytical models differ markedly in Fig. 6, owing to the FDTD under-estimate of the resistive component of Z_{in} and to the capacitive effect of the FDTD feed model. However, the FDTD results oscillate around the analytical answer, and both methods appear to converge to the similar values as *ka* becomes large.

Generally, the analytical results were considerably more efficient to obtain than the FDTD results. To calculate the input impedance, the VSWR, and the patterns of the bicone for *ka* between 95 and 105, the FDTD method required 20 hours with six 300-MHz MIPS processors on an SGI Origin 2000. By contrast, the computations of the analytical expressions required only 10 hours on a single-processor, 300-MHz, Pentium-II PC over the same range of *ka*.

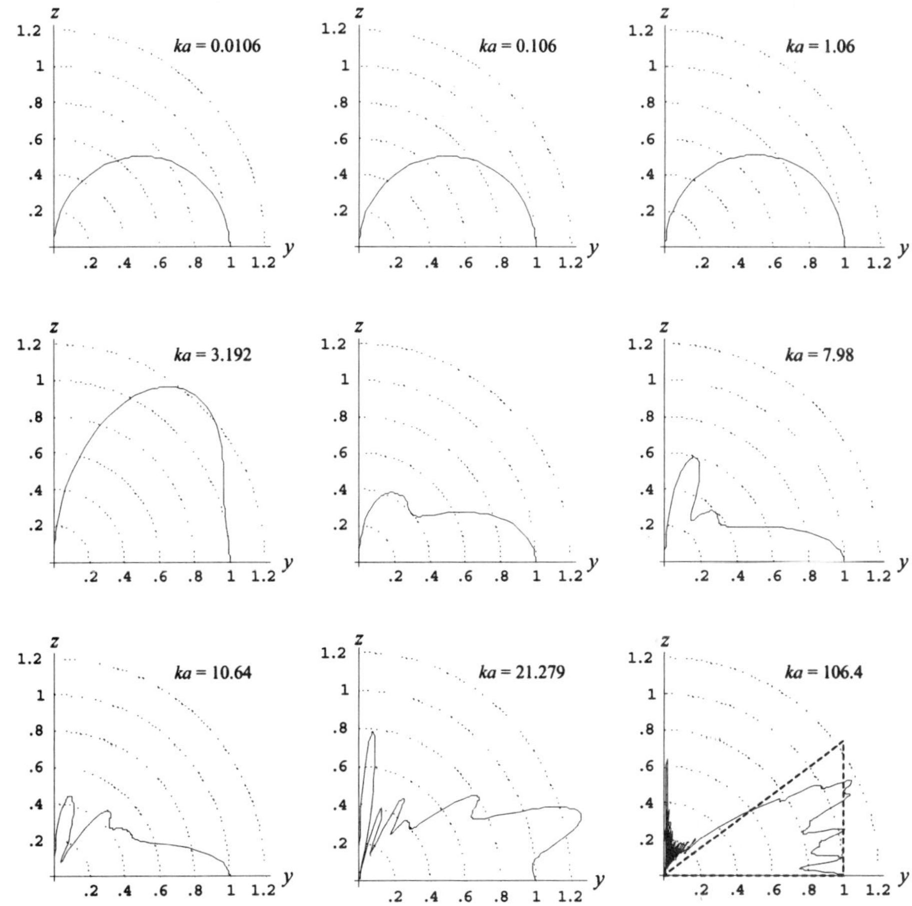

Figure 4. Polar plots of the magnitude of the relative pattern for a 53.1° equiangular biconical antenna for nine values of *ka*: 0.0106, 0.106, 1.064, 3.192, 5.319, 7.98, 10.64, 21.279, 106.4. Vertical and horizontal axes correspond respectively to the z axis and y axis in Fig. 1. The plot for *ka* = 106.4 includes the triangular pattern associated with the high-frequency limit ($ka \to \infty$).

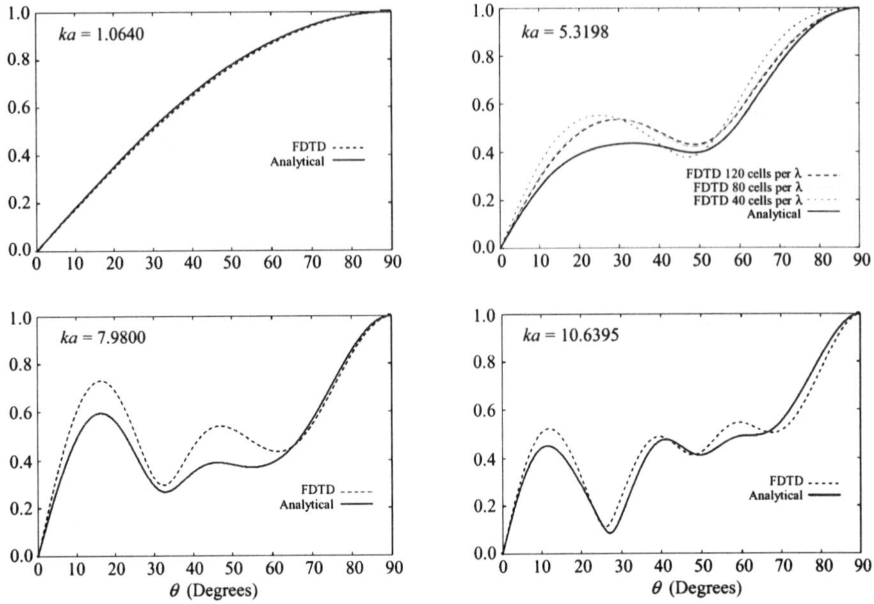

Figure 5. Comparison of analytical (solid curve) and FDTD (dashed curve) calculations of the relative patterns versus θ for a 53.1° bicone at four intermediate values of *ka*: 1.0640, 5.3198, 7.9800, 10.6395.

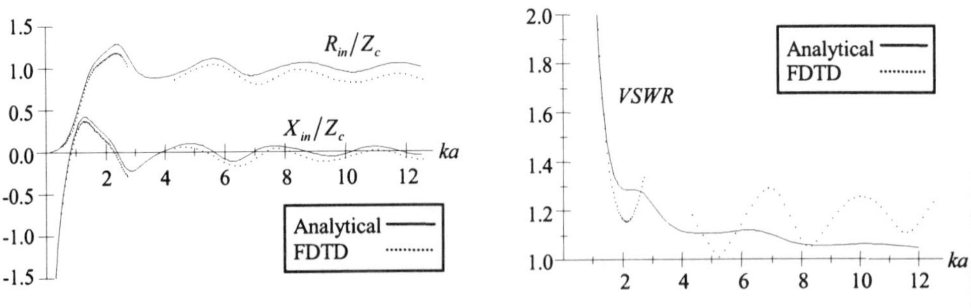

Figure 6. Comparison of analytical (solid curves) and FDTD (dashed curves) calculations of the normalized resistance R_{in}/Z_c, normalized reactance X_{in}/Z_c, and VSWR versus *ka* for a 53.1° bicone.

5. EXPERIMENTAL SETUP AND RESULTS

Some preliminary experiments were performed to compare measurements with the analytical and numerical (FDTD) results. Brass mono-cones were constructed for cone angles ($2\psi_1$) of 60°, 100°, 110°, and 120°. The table provides the lengths of the mono-cones, each approximately 0.6-cm long, as well as the values of *ka* at the frequency limits (2 GHz and 18 GHz) of the experimental data mandated by the measurement system. The difficulty and cost in machining such small mono-cones precluded constructing a 53.1° mono-cone that could be compared directly with the results of Sandler and King. Each mono-cone was mounted on a circular metallic ground plane with a 4-ft diameter and was fed by an SMA jack through a 0.011-inch hole in the apex of the cone. The ground plane was mounted on an elevation-over-azimuth positioner capable of various pattern cuts. Only data for the 100° mono-cone are discussed (Fig. 7).

Table 1. Brass Mono-cones

$2\psi_1$	*a* (cm)	*ka* @ 2 GHz	*ka* @ 18 GHz
60°	0.6045	0.2543	2.289
100°	0.6033	0.2529	2.276
110°	0.6248	0.2619	2.357
120°	0.6426	0.2693	2.424

Figure 7. Brass mono-cone with 100° full cone angle and 0.6033-cm lateral side *a*.

The measurement system, depicted in Fig. 8, consisted of a Scientific-Atlanta 5906M compact range with associated range feeds, an HP8530 microwave receiver system using external mixers, and a Scientific-Atlanta multi-axis positioner. The compact range produces a nearly constant far-field approximation to a plane wave inside an 8-ft cylinder that is centered 8 ft above the floor of the chamber. Measurements were taken from 2 GHz to 18 GHz, with a frequency resolution of 160 MHz (101 data points). As a result of the pattern symmetry of the antennas under test, only elevation patterns were recorded. The antennas, mounted on a 4-ft circular ground plane that was positioned at the center of the quiet zone of the chamber, were moved in elevation by the positioning equipment. For each antenna configuration, patterns were recorded every 1° for vertical polarization and horizontal polarization.

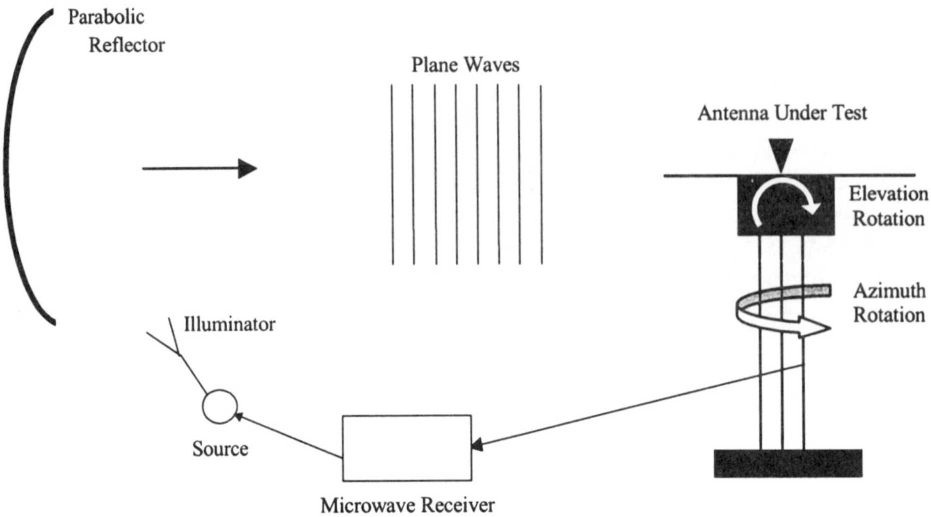

Figure 8. Experimental setup in the Compact Range of the Radar Division of the Naval Research Laboratory.

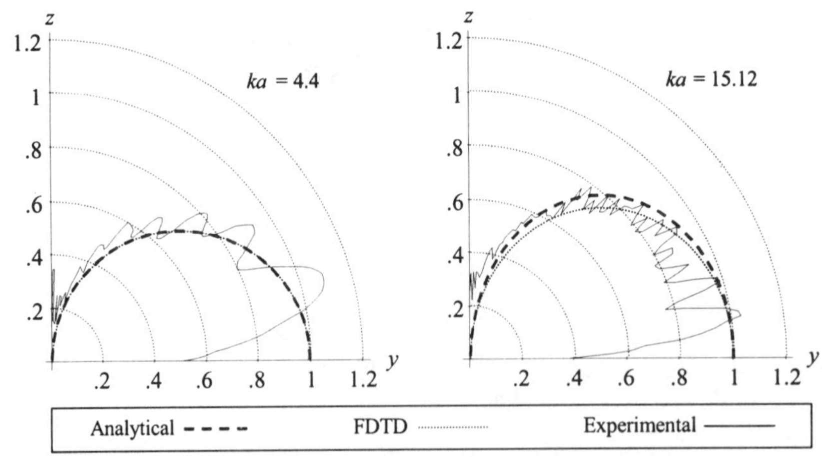

Figure 9. Comparison of analytical calculation, FDTD calculation, and compact-range measurements of the relative patterns versus θ for a 50° mono-cone at two values of *ka*: 4.4 and 15.12. The dashed, dotted, and solid curves respectively correspond to analytical calculations, FDTD calculations, and measurements.

The magnitudes of the relative patterns of the mono-cone with the full cone angle of 100° are displayed (Fig. 9) in polar format for two values of *ka* (4.4 and 15.12). Overall, the measured patterns (solid curves) agree with the theoretical (dashed curves) and FDTD (dotted curves) patterns. However, the measured patterns have two significant differences: (1) they tend to oscillate about the theoretical plots; and (2) they significantly drop off at broadside ($\theta = 90°$).

The discrepancies between the theoretical patterns and the measured pattern are attributable to edge effects induced by the finite ground plane. When the direction of an incident beam is not near broadside, small interference effects between the cone and the ground plane are manifested as the small oscillations about the pattern of the infinite ground plane. The coarse periodicity of this scalloping behavior about the theoretical and numerical patterns arises from the interference pattern induced by the sum of the signals from the direct path to the mono-cone and the diffracted path from the edge of the ground plane. As the observation direction of an incident beam approaches 90°, the effect of the finite ground plane is increasingly significant. This decrease in the relative pattern along the direction of an antenna's finite ground plane is typical of finite antenna structures.

7. SUMMARY

In this article, previous analyses of wide-angle biconical antennas were extended to include the VSWR and the input impedance Z_{in}. The analytical theory shows that Z_{in} is essentially capacitive for $ka < 0.5$ and is essentially resistive for $ka > 2$. Consequently, the bicone is a very poor radiator at low frequencies and a very good radiator at all moderate to high frequencies. The theory also demonstrates that wide-angle bicones have excellent VSWRs (< 1.3) for $ka > 2$. For example, if the length a of the lateral side is 10 cm and if the cone angle is 53.1°, then the VSWR < 1.3 for all frequencies above 955 MHz. This implies that 98.3% of the power is transferred from the feed network to the bicone across its input terminals. The good behaviors of Z_{in} and VSWR across an extremely broad range of frequencies are strong indications that this antenna is ultrawideband. In addition, computations using the analytically derived representation of the relative radiation pattern were compared with calculations from a finite-difference time-domain (FDTD) numerical code with very good agreement.

Finally, measurements of the radiated field for an electrically small ($ka \ll 1$), wide-angle (100°), 0.60325-cm radius, brass, mono-cone above a 121.92-cm diameter circular ground plane compare favorably with the analytical representation of the field and the associated FDTD calculations, except that the data curves mildly oscillate about the theoretical curves and significantly drop off at broadside. On reflection, both differences between measurement and the theoretical predictions are to be expected. The discrepancy at broadside is caused by finite-ground-plane effects like edge diffraction, which are not characterized by an idealized, infinite, perfectly conducting ground plane. Oscillations of the measured data about the theoretically predicted curve are the result of interference arising from the interaction of the mono-cone and the finite ground plane.

Unfortunately, time did not permit a more complete analysis of the measurements. Data from three other bicones (60°,110°,120°) still require investigation and will be the subject of subsequent research. In addition, further measurements are needed to validate the theory for electrically large, wide-angle bicones.

8. ACKNOWLEDGMENTS

The authors thank Albert Moffat for his creativity and diligence in performing the difficult task of designing and building the small mono-cones used in the measurements. This work was supported by the Naval Research Laboratory and the Office of Naval Research in Washington DC.

9. REFERENCES

1. S. N. Samaddar and E. L. Mokole, Biconical antennas with unequal cone angles, *IEEE Trans. Antennas Propagat.* **46**(2), 181–193 (1998).
2. S. A. Schelkunoff, *Electromagnetic Waves* (Van Nostrand, New York, 1943), ch. XI.
3. P. D. P. Smith, The conical dipole of wide-angle, *J. Appl. Phys.* **19**, 11–23 (1948).
4. C. T. Tai, On the theory of biconical antennas, *J. Appl. Phys.* **19**, 1155–1160 (1948).
5. C. T. Tai, A study of the E.M.F. method, *J. Appl. Phys.* **20**, 717–723 (1949).
6. C. T. Tai, Application of variational principle to biconical antennas, *J. Appl. Phys.* **20**, 1076–1084 (1949).
7. C. H. Papas and R. W. P. King, Input impedance of wide-angle conical antennas fed by a coaxial line, *Proc. IRE* **37**, 1269–1271 (1949).
8. C. H. Papas and R. W. P. King, Radiation from wide-angle conical antennas fed by a coaxial line, *Proc. IRE* **3**, 49–51 (1951).
9. S. S. Sandler and R. W. P. King, Compact coaxial antennas for wide-band coverage, *IEEE Trans. Antennas Propagat.* **42**, 436–439 (1994).
10. R. L. Carrel, The characteristic impedance of two infinite cones of arbitrary cross section, *IRE Trans. Antennas Propagat.*, **AP-6**, 197–201 (1958).
11. A. P. Lambert, S. M. Booker, and P. D. Smith, Calculation of the characteristic impedance of TEM horn antennas using the conformal mapping approach, *IEEE Trans. Antennas Propagat.*, **43**(1), pp. 47–53 (1995).
12. H. A. Antosiewicz, in: *Handbook of Mathematical Functions with Formulas, Graphs, and Mathematical* Tables, edited by M. Abramowitz and I. A. Stegun (Dover, New York, 1965), p. 437.

ULTRAWIDEBAND TRANSMITTING ANTENNAS, ARRAYS, AND HIGH-POWER RADIATION SOURCES

Vladimir I. Koshelev, Yury A. Andreev, Yury I. Buyanov,
Vyacheslav V. Plisko, Konstantin N. Sukhushin*

1. INTRODUCTION

To detect and recognize objects moving in the earth atmosphere at a distance up to 100 km, ultrawideband (UWB) 1-ns long radiation pulses with the peak power up to 10 GW are necessary. Creation of radar with these parameters is possible on the basis of multielement steering antenna arrays. For operation in these arrays it is necessary to develop compact high-efficiency antennas capable to radiate pulses with the peak power of ~100 MW. In the previous works[1-3] an approach to creation of this type of antenna was suggested on the basis of the electric monopole and magnetic dipole combination and the sources of UWB radiation with the pulse length of 3-4 ns and power of 0.1-1 GW were created. Necessity to radiate shorter pulses (≤1 ns) resulted in creation of a new antenna[4] on the basis of combination of electric monopole, two magnetic dipoles, and TEM-horn.

Investigation results of the antennas as well as antenna arrays and UWB radiation sources are presented below.

2. TRANSMITTING ANTENNAS

2.1. Design

Fig. 1 presents the geometry of the combined antenna with extended passband. The antenna consists of a case (1), electric monopole (2), two magnetic dipoles (3) and TEM-horn (4). The antenna width and height are h ≅ L, where L is the longitudinal antenna

* Institute of High Current Electronics, RAS, 4 Akademichesky Ave., 634055 Tomsk, Russia.

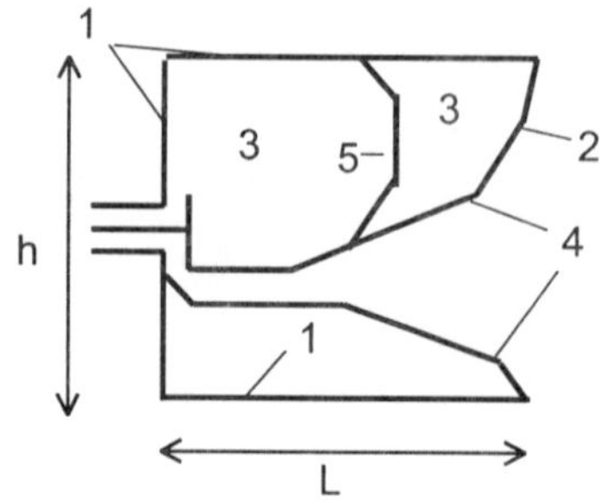

Figure 1. Transmitting antenna geometry.

dimension. The upper part of the TEM-horn is simultaneously the electric monopole and is connected galvanically with the antenna case. The lower part of the TEM-horn is also connected with the antenna case. Perimeter S of a magnetic dipole (neighboring to the antenna input) was changed by the displacement of a plate 5. Maximum dipole perimeter $S=S_0$ takes place at the absence of the plate 5 and, respectively, at the absence of the second magnetic dipole. In the center of the antenna rear wall there is a hole for connection of a 50-Ω feeder.

Two variants of this antenna – A1 and A2 have been developed. The antennas are made on a scale of 1:2. Antenna investigation results are presented below.

2.2. Frequency Domain Measurements

Antenna matching with a 50-Ω feeder was estimated by VSWR. Fig. 2 presents VSWR versus frequency for the antennas A1 and A2 with $S/S_0 \cong 0.6$ and antenna electric monopole of the length L (antenna rear wall is used as a screen). The matching band of the antenna A2 is $f_h/f_l \cong 9$ taken by a maximum VSWR of 3. As it is seen in Fig. 2, the low frequency f_l of the antenna matching band is determined by the low frequency of the

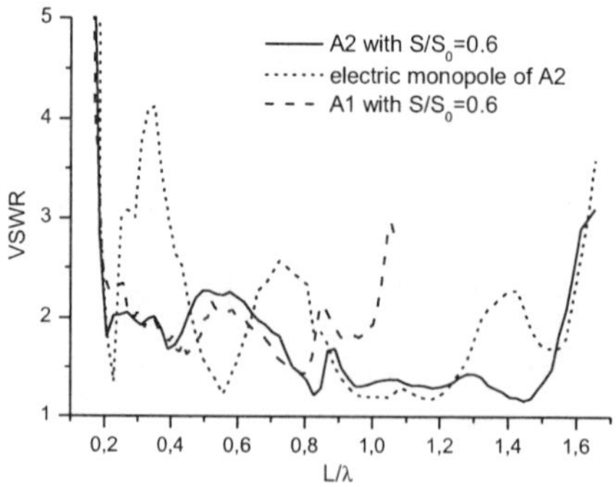

Figure 2. VSWR of combined antennas and electric monopole.

electric monopole matching. The high frequency boundary of the antenna f_h is determined by the antenna input geometry. The matching band for the antenna A1 is $f_h/f_l \cong 5.5$, which is related to the limitation of f_h. Geometry of antenna inputs for the antennas A1 and A2 is not made on the scale 1:2 (as antenna main elements) and differs slightly. This is related to decrease of $L/\lambda_h \cong 1.6$ for A2 to $L/\lambda_h \cong 1.1$ for A1, where λ_h is the wavelength corresponding to f_h. Insignificant f_l shifting to the direction of low frequencies is observed as S/S_0 decreases. So, at $S/S_0 = 1$ $L/\lambda_l \cong 0.2$, and at $S/S_0 = 0.31$-0.4 $L/\lambda_l \cong 0.18$. When the ratio S/S_0 is changed, VSWR in the matching band is changed as well. The best matching of the antenna A2 with a 50-Ω feeder corresponds to $S/S_0 \cong 0.75$ and for A1 at $S/S_0 \cong 0.9$.

Fig. 3 presents the real and image components of complex impedance of the antenna A2 input and electric monopole of the same antenna. As it seen from presented dependences, antenna complex input impedance essentially differs from the electric monopole impedance in the low-frequency band.

When using an antenna as the steering array element the waveform of the radiated pulse should be preserved in the sector of steering angles. Fig. 4 presents the amplitude frequency response and deviation from linear phase frequency response of the antenna A2 ($S/S_0 \cong 0.6$) for the observation angles of 0° and ±30° for the E and H planes. It is seen that already in the direction of φ, δ = 0° (φ is the azimuth angle, δ is the angle of elevation) the antenna passband is limited from above by decay at the frequency $L/\lambda \cong 1.2$ that probably results from mutual influence of magnetic dipoles. The amplitude frequency response of a combined antenna in the main maximum direction depends on the relative dimension of the first magnetic dipole S/S_0 and at $S/S_0 \cong 0.3$ the high frequency boundary of the antenna passband $L/\lambda \cong 1.5$. For the observation angles of ±30° the high frequency boundary decreases to $L/\lambda \cong 0.95$ in the horizontal plane (Fig. 4a). In the vertical plane (Fig. 4b) amplitude frequencies responses for the observation angles of ±30° are not the same, hence, the pattern of the antenna A2 is asymmetrical relative to the horizontal plane. Frequency overlapping for A2 decreases with $f_h/f_l \cong 7$ for φ, δ = 0° to $f_h/f_l \cong 4.5$ for φ = 0°, δ = -30°.

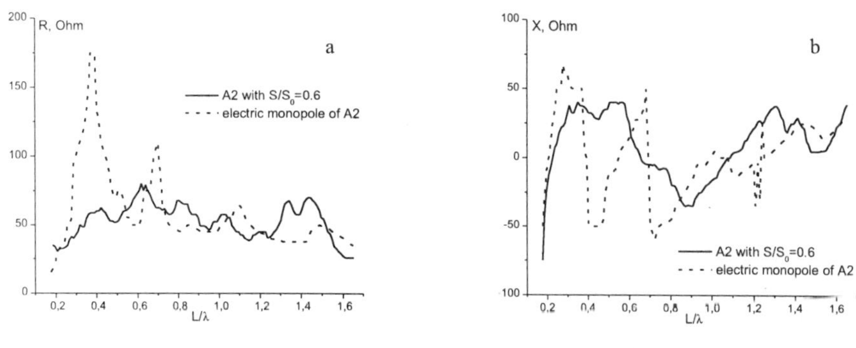

Figure 3. Real and image components of complex input impedance of a combined antenna and electric monopole.

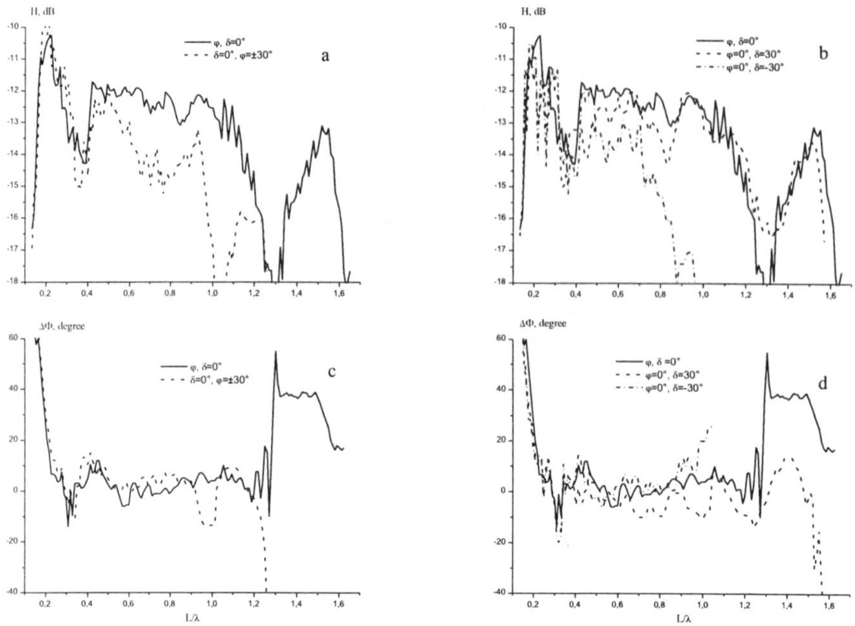

Figure 4. Amplitude frequency response and phase frequency response of the combined antenna A2 for different observation angles.

2.3. Time Domain Measurements

Antennas A1 and A2 were developed to radiate high-voltage (~100 kV) nano- and subnanosecond pulses, but for studying antenna responses in the time domain, low-voltage pulse generators were used. Voltage pulse waveforms at the antennas A1 and A2 inputs and from the receiving antennas outputs are presented in Fig. 5 and Fig. 6 for monopolar and bipolar excitation. The TEM-horn and electric dipole were used as

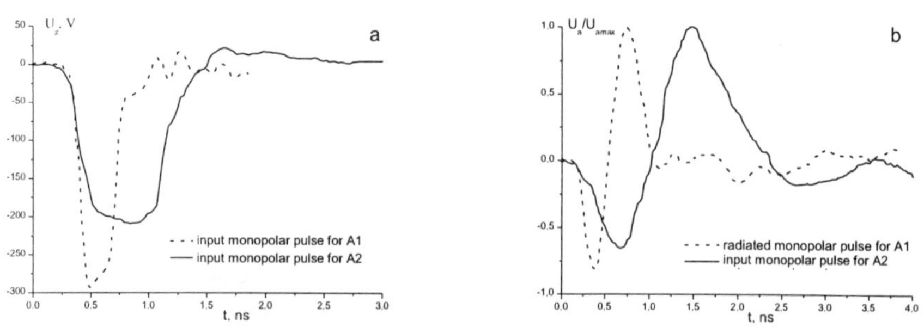

Figure 5. Voltage pulses at the input to the antennas A1, A2 (a); radiated pulses (b) – monopolar excitation.

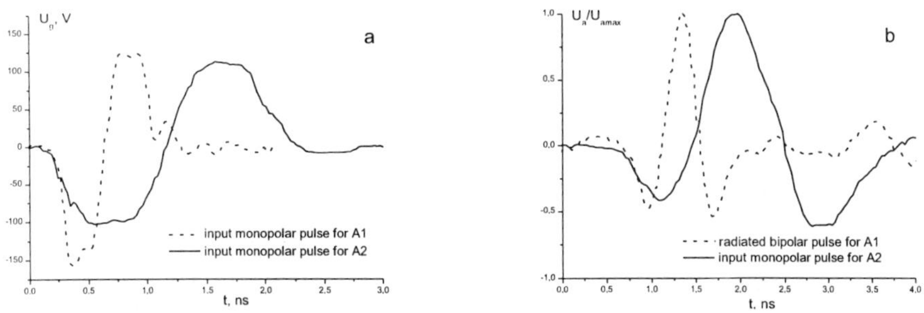

Figure 6. Voltage pulses at the input to the antennas A1, A2 (a); radiated pulses (b) – bipolar excitation.

receiving antennas for the antennas A1 and A2, respectively.

Fig. 7 presents the patterns by peak power of the antennas A1 and A2 ($S/S_0 \cong 0.6$) in the horizontal (Fig. 7a) and vertical (Fig. 7b) planes. Antennas patterns in the vertical plane are really asymmetric relative to the horizontal plane and maximums correspond to $\delta \cong 10°$ for A2 and $\delta \cong 15°$ for A1.

Fig. 8 presents the dependences of energetic efficiency k_w, vertically polarized electric field peak strength, and antenna pattern width by a half-power level for monopolar and bipolar excitation of antennas A1 and A2.

Energetic efficiency was determined as ratio of radiated energy to the energy in the generator pulse $k_w = W_r/W_g$, where radiated energy was determined as $W_r = W_g - W_{ref}$ (W_{ref} is the energy in the pulse reflected from antenna). As it is seen from Fig. 8a, k_w dependence on S/S_0 has a resonance character and reaches maximum at $S/S_0 \cong 0.9$ for A1 and $S/S_0 \cong 0.77$ for A2. Maximum k_w values for antenna excitation by monopolar and bipolar voltage pulses are 0.75 and 0.9, respectively.

Peak strength of the vertically polarized electric field E_p was measured in the direction of φ, $\delta = 0°$ of the antenna pattern. The results were obtained at the antenna excitation by bipolar pulses with the voltage 2U and monopolar pulses with the voltage U. Measurements were carried out at equal distance from the transmitting antenna and

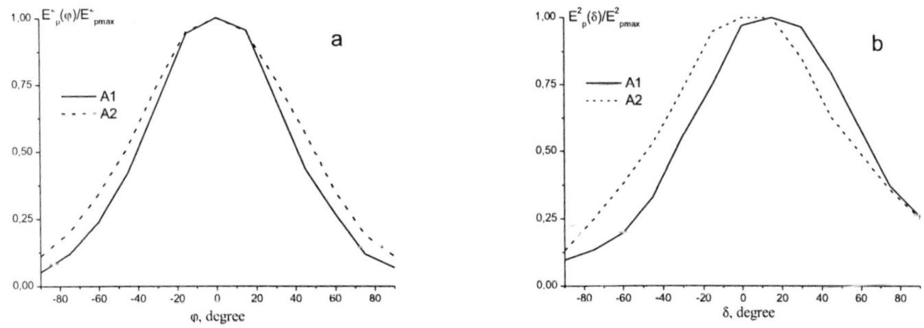

Figure 7. Patterns of antennas A1 and A2 in the horizontal (a) and vertical (b) planes.

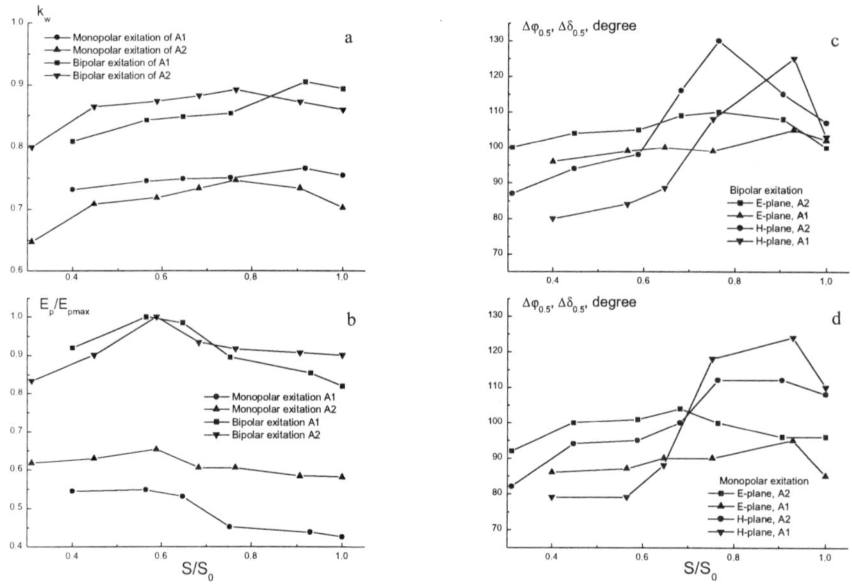

Figure 8. Antennas characteristics versus magnetic dipole perimeter.

results were normalized to maximum values of the electric field strength of radiation for antenna excitation by bipolar pulses. As it is seen from Fig. 8b, E_p dependence on S/S_0 also has a resonance character and reaches maximum at $S/S_0 \cong 0.6$.

The patterns of antennas A1 and A2 are close to cardioid. Radiated electric field polarization is vertical. Fig. 8c and 8d present the curves corresponding to the values of the pattern width of A1 and A2 by the peak power at half-height for antenna excitation by bipolar and monopolar voltage pulses, respectively. As it is seen from Fig. 8c,d, as S/S_0 decreases pattern extension takes place first and then it is narrowing. Maximum width of the antenna patterns takes place at $S/S_0 \cong 0.75$-0.9. Depending on S/S_0, antennas pattern width in the E-plane takes the values of 85°-110°. Depending on S/S_0, antennas pattern width in the H-plane takes the values of 80°-130°.

Antenna efficiency by the peak power k_p equals to P_r/P_g, where P_r is the peak power of vertically polarized radiation and P_g is the peak power in the pulse at the antenna input. For antenna A1 at $S/S_0 \cong 0.6$, that corresponded to the recording maximum electric field strength in the far-field zone $k_p \cong 0.65$ for the antenna excitation by a bipolar pulse.

The wide frequency band of antenna A2 (for $S/S_0 \cong 0.3$, $f_h/f_l \cong 9$) allows radiation of bipolar 1-ns long pulses. Maximum electric field strength increases by ~20% in comparison with the antenna A1 ($S/S_0 \cong 0.6$). At radiation of a bipolar 1-ns long pulse the pattern of the antenna A2 ($S/S_0 \cong 0.3$) is narrowing from 90° to 60° and from 100° to 70° in the horizontal and vertical planes, respectively, in comparison with radiation of a bipolar 2-ns long pulse. Antenna energetic efficiency k_w increased from 0.8 (for the antenna excitation by a 2-ns long bipolar pulse) to 0.87 (for the antenna excitation by a

1-ns long bipolar pulse). However, the passband of A2 is insufficient for radiation of a 1-ns long pulse without distortions. Additional time lobes are present at the signal after the main pulse.

3. ANTENNA ARRAYS

Two-, three-, and four-element antenna arrays designed on the basis of a combined antenna A1 were investigated. Possible variants of a linear array design relative to the plane of vector **E** polarization are shown in Fig. 9: a) horizontal, b) vertical, c) rectangular. Here, d_h and d_v are the distances between the centers of elements in a horizontal and vertical planes, respectively.

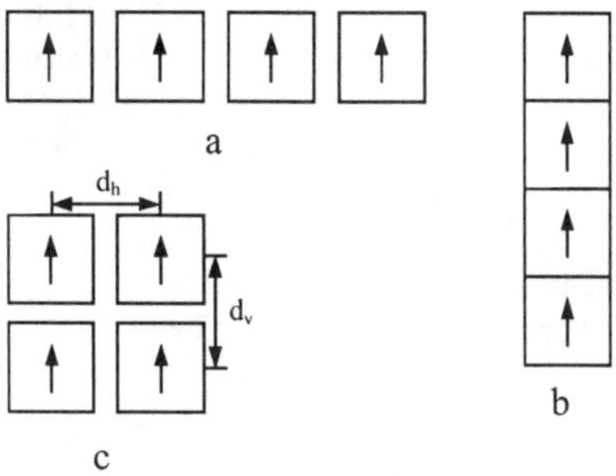

Figure 9. Array design variants.

Investigations of VSWR of the array elements versus the distance between them were carried out. VSWR of elements of a two-element horizontal array insufficiently differs from VSWR of a single element irrespective of the distance between the elements. VSWR of the internal element increases with decrease of the distance between the elements for the three- and four-element horizontal arrays. Fig. 10 presents VSWR of

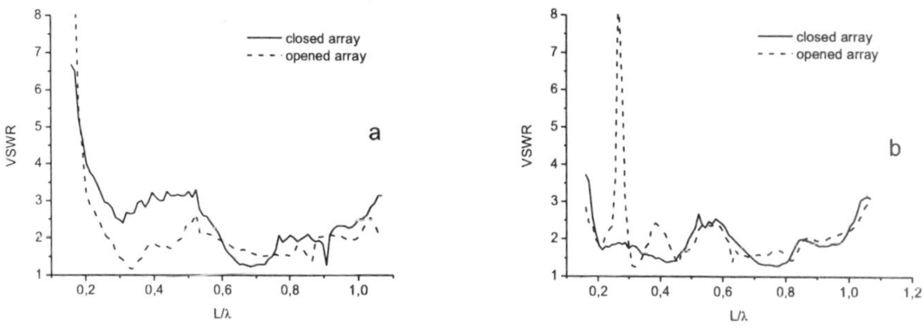

Figure 10. VSWR of internal elements of horizontal (a) and vertical (b) arrays.

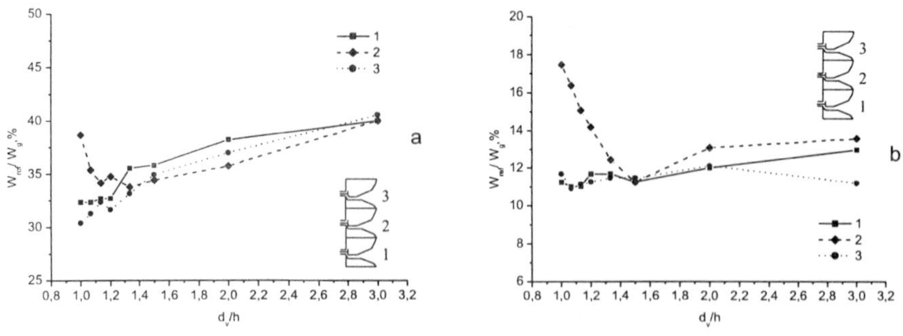

Figure 11. Energy reflected from the array element at excitation by monopolar (a) and bipolar(b) pulses.

antenna elements of horizontal (Fig. 10a) and vertical (Fig. 10b) four-element arrays. A closed array denotes that the array elements are closed between each other galvanically and the distance between the centers of the elements is $d_h=d_v=h$. An opened array in this case has $d_h=d_v=1.2h$. Vertical array has resonance at $L/\lambda=0.25\text{-}0.3$ when the elements are opened.

Investigations of dependence of the energy reflected from the element inputs on the distance between the elements for two-, three-, four-element arrays were made. Reflected energy for a two-element horizontal array weakly depends on the distance between the elements and is equal practically to the reflected energy of a single antenna. Reflected energy for a two-element vertical array decreases from 40% to 32% with decrease of the distance between the elements at excitation by a monopolar pulse and changes slightly at the excitation by a bipolar pulse. Fig. 11 presents dependences of reflected energy on the distance between the elements for a vertical three-element array excited by monopolar (Fig. 11a) and bipolar (Fig. 11b) pulses. Energy reflected from elements for a horizontal three-element array weakly depends on the distance between them. Analogous results were obtained for a four-element linear array.

Fig. 12 presents the patterns for the array of 2x2 for monopolar (Fig. 12a) and bipolar (Fig. 12b) pulses, respectively. Field strength in the main direction in four times

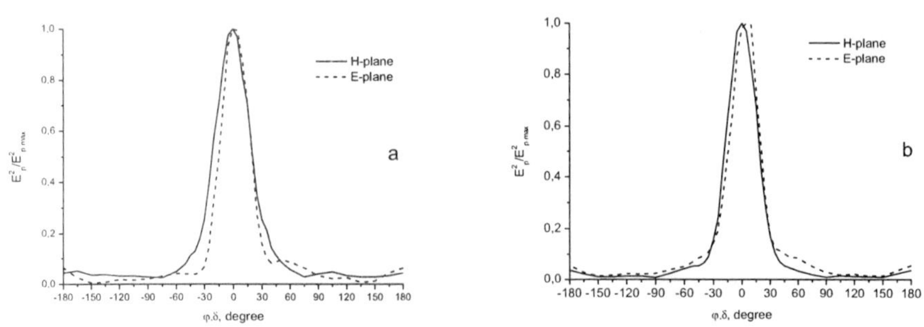

Figure 12. Pattern of the 2x2 array at excitation by monopolar (a) and bipolar (b) pulses.

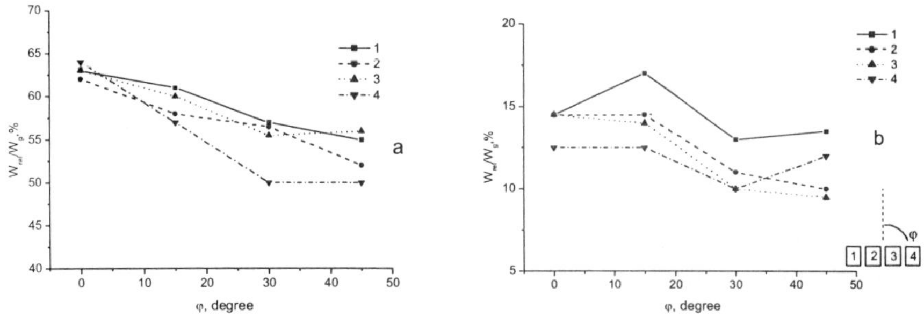

Figure 13. Dependence of reflected energy on the steering angle for a horizontal array excitated by monopolar (a) and bipolar (b) pulses.

higher than field strength from a single element in the array, i.e. fields from the elements are added together. Array efficiency by peak power is κ_{pa}=0,57 that is by 13% less than for a single element.

Fig. 13 presents dependences of reflected energy on the steering angle for monopolar (Fig. 13a) and bipolar (Fig. 13b) pulses for a horizontal four-element array. It is seen that reflected energy on the average decreases as the steering angle increases. Dependence of reflected energy on the angle for a vertical array is more pronounced and depends on the element position in the array (Fig. 14).

Fig. 15 presents the pattern dependences on the steering angle for horizontal (Fig. 15a) and vertical (Fig. 15b) arrays. Pattern maximum decrease with the steering angle increase is related to the influence of the array element pattern.

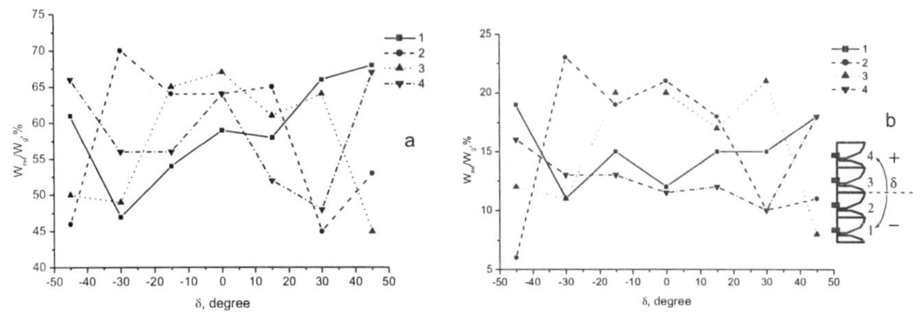

Figure 14. Dependence of reflected energy on the steering angle for a vertical array excited by monopolar (a) and bipolar (b) pulses.

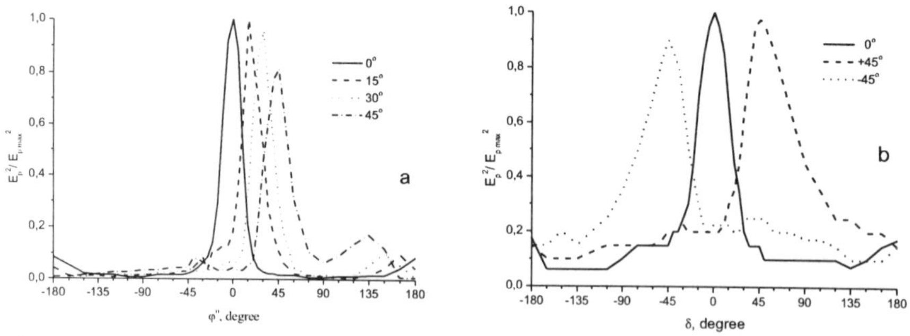

Figure 15. Pattern versus steering angle in horizontal (a) and vertical (b) planes.

4. UWB RADIATION SOURCES

Sources of high-power UWB radiation based on single antennas A1 and A2 excited by high-voltage bipolar pulses with the pulse repetition frequency of 50 Hz were developed. Main efforts were given to development of different variants of bipolar pulse formers with one gas switch. Fig. 16 presents the geometry of the formers on coaxial (Fig. 16a) and radial (Fig. 16b) lines.

The lines are charged from a pulsed source up to the voltage U_0 of the switch (1) self-breakdown. In the transmitting line CL3 with the wave impedance of 50 Ω a bipolar pulse with the voltage of $\pm U_0/2$ is formed. The output voltage pulse is measured by a capacity divider (2). After an insulator (3) a resistive load R_l or a transmitting antenna are installed. Pulse length was changed owing to the lengths of the forming lines and depended on the time of the gas switch commutation (nitrogen up to 40 atm.).

Waveforms of voltage pulses for two formers and electromagnetic pulses radiated by the antenna A2 are presented in Fig. 17. Both formers have shortcomings. In the former based on coaxial lines the amplitudes and lengths of the first and second half-periods differ essentially. In the former based on radial lines after the main bipolar pulse, oscillations due to reflections take place resulting in the radiation spectrum width decrease. For the antenna A2 field strength at a 3-m distance reached 18 kV/m at a 70-kV

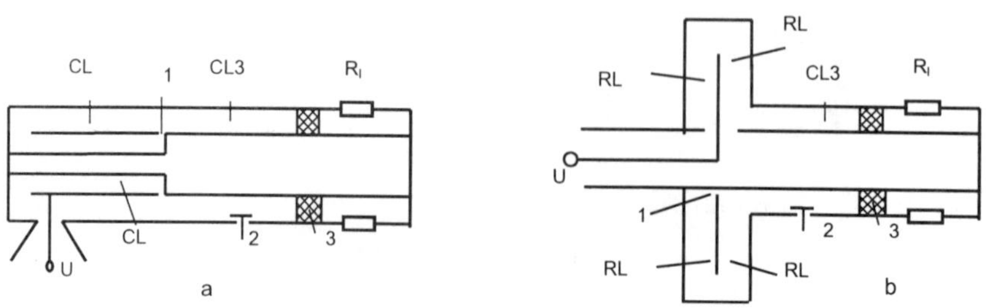

Figure 16. Bipolar pulse former geometries.

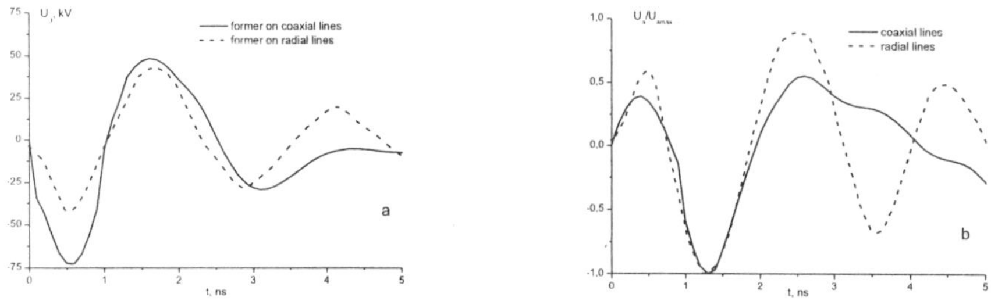

Figure 17. Waveforms of output voltage (a) and radiated (b) pulses.

voltage amplitude. As the pulse length decreases, the difference between half-periods of a bipolar pulse increases for a coaxial former and the pulse amplitude for a radial former sharply decreases.

5. CONCLUSION

A compact combined antennas with a wide passband (ratio of the high and low frequencies up to 9) for radiation of high-power nano- and subnanosecond pulses were developed. The efficiencies by energy up to 90% and by peak power up to 65% were achieved at the antenna excitation by a bipolar pulse.

The arrays on the basis of the created antenna allow realizing wave beam steering in the limits of 30°-45° with no essential changes of features of radiated pulses.

The formers of voltage bipolar pulses based on coaxial and radial lines with one gas switch were investigated and the sources of high-power UWB radiation pulses were created.

6. REFERENCES

1. V.I. Koshelev, Yu.I. Buyanov, B.M. Kovalchuk, Yu.A. Andreev, V.P. Belichenko, A.M. Efremov, V.V. Plisko, K.N. Sukhushin, V.A. Vizir, V.B. Zorin, in: *Proc. SPIE Intense Microwave Pulses V,* edited by H. Brandt, 1997, v. 3158, pp.209-219.
2. Yu.A. Andreev, Yu.I. Buyanov, A.M. Efremov, V.I. Koshelev, B.M. Kovalchuk, K.N. Sukhushin, V.A. Vizir, V.B. Zorin, in: *Proc. 11 IEEE Pulsed Power Conf.*, edited by G. Cooperstein and I. Vitkovitsky, 1997, pp. 730-735.
3. Yu.A. Andreev, Yu.I. Buyanov, A.M. Efremov, V.I. Koshelev, B.M. Kovalchuk, V.V. Plisko, K.N. Sukhushin, V.A. Vizir, V.B. Zorin, in: *Proc. 12 IEEE Pulsed Power Conf.*, edited by C. Stallings and H. Kirbie, 1999, pp. 1337-1340.
4. V.I. Koshelev, Yu.I. Buyanov, Yu.A. Andreev, V.V. Plisko, K.N. Sukhushin, in: *Proc. IEEE Pulsed Power Plasma Science Conf.*, edited by R. Reinovsky and M. Newton, 2001, pp. 1661-1664.

RECENT DEVELOPMENTS IN ULTRA-WIDEBAND SOURCES AND ANTENNAS

William D. Prather, Carl E. Baum, Jane M. Lehr, Robert J. Torres, Tyrone C. Tran, Jeffrey W. Burger, Dean I. Lawry, John A. Gaudet*

1. INTRODUCTION

Ultra-wideband (UWB) sources and antennas are of interest for applications such as the location and identification of hidden objects, that is, land mines and unexploded ordnance (UXO). Much of the current research is being performed at the Directed Energy Directorate of the Air Force Research Laboratory (AFRL) at Kirtland AFB, NM. Although significant progress has been made in UWB technology since 1980s, new challenges continue to arise.[1-6] Recent requirements have included antennas that are lighter weight, have wider bandwidths, and operate at lower frequencies. Noteworthy progress is also evident in the development of fast switching technology. In this paper, we describe recent progress in sources, antennas, and lens technology and new developments in gas and solid-state switching.

2. SOLID STATE SWITCHING

2.1. Delayed Breakdown Device Technology in Silicon.

Ultra-fast solid state switching has evolved through several stages since it began to be seriously investigated in the late 1980s. The developments have evolved from the earliest bulk avalanche devices to the current technology in lateral gallium arsenide (GaAs) switches, a technology that is presently the leader for sheer power. However, these switches still suffer from limited lifetimes, primarily due to problems with heat dissipation.[7-9] For future system development, there is still a definite need for small, inexpensive switches that have long lifetimes. Promising new approaches are based on basic research into the semiconductor switching physics pioneered by I.V. Grekhov and his colleagues at the Ioffe Physical-Technical Institute (PTI) in St. Petersburg in the late

* Air Force Research Laboratory, Directed Energy Directorate, Kirtland AFB, NM 87117-5776.

Ultra-Wideband Short-Pulse Electromagnetics 6
Edited by Mokole *et al.*, Kluwer Academic/Plenum Publishers, 2003

1970s.[10-12] The developments in Dr. Grekhov's group include the delayed breakdown device (DBD), which is a semiconductor device used as an ultra-fast avalanche-breakdown closing switch, and the junction recovery diode (JRD), a similar device used as a nanosecond opening switch. AFRL is continuing its collaboration with Dr. Grekhov and with the University of New Mexico (UNM), both of whom are engaged in studying these devices.[13]

2.1.1. Junction Recovery Diodes

The Junction Recovery Diode (JRD) mode in a high voltage Si diode is achieved by adjusting the plasma wave front during the recovery process to occur precisely at the p-n junction and to use the maximum reverse current allowable. There are two types of diodes that Ioffe scientists have developed based upon the JR mode, the Drift Step Recovery Diode (DSRD) and the Inverse Recovery Diode (IRD). Each is formed by using different concentrations of dopant in the n^+ layer resulting in reverse motion of the plasma wave. With the use of Si-based DSRDs, Ioffe PTI scientists have created pulse generators that are able to form 80 kV, nanosecond pulses at pulse repetition frequencies (PRFs) of 1 kHz. GaAs structures have also been tested to confirm that similar pulse formation is possible in them. It is, but GaAs devices have not been widely optimized or implemented due to the narrow performance advantages, compared to Si, and its much higher cost.

2.1.2. Drift Step Recovery Diodes with Silicon Avalanche Shapers

When a DBD closing switch is combined with a peaking element, it can sharpen the nanosecond pulse generated by a Drift Step Recovery Diode (DSRD) driver to the picosecond range. If the ultra-fast silicon DBD peaking switch is located at the load end of a coaxial transmission line, it is referred to as a silicon avalanche shaper (SAS). The SAS sharpens the transient to a 100 ps rise time at the DSRD pulse amplitude level into a 50 Ω load. The DSRDs have been stacked in such a way as to produce high voltage pulsers producing fast transients by using an SAS shaping head. One such system was a Ioffe PTI PPG-10-5 pulse generator that achieved a pulse amplitude of 10 kV, a rise time of less than 100 ps (with shaping head), a pulse width of 2 ns, and a pulse repetition rate of 5 kHz into a 50 Ω load. Since the stacking limits of the technology have not yet been reached, AFRL is pursuing a design with the Russian group that will produce 100 kV pulses with a rise time of less than 100 ps. Life times greater than 10^{11} shots have been achieved for this system. A miniature 5 kV nanosecond pulser based on a silicon DSRD stack driven by a magnetic switch has also been developed under Air Force sponsorship as have multi-kilovolt pulsers based on silicon DSRD drivers mated with SAS peaking devices.[14-15]

2.1.3. Solid State Shocklines

Since the ultra-fast breakdown characteristics of DBDs are very precise and repeatable, they may be exploited to make miniature pulse-sharpening shockline pulsers having 100 picosecond risetimes at 3.2 kV. The original miniature shockline pulser was precisely triggered at 2 kHz with 16 ps of (rms) jitter.[16] Very recently, this approach has

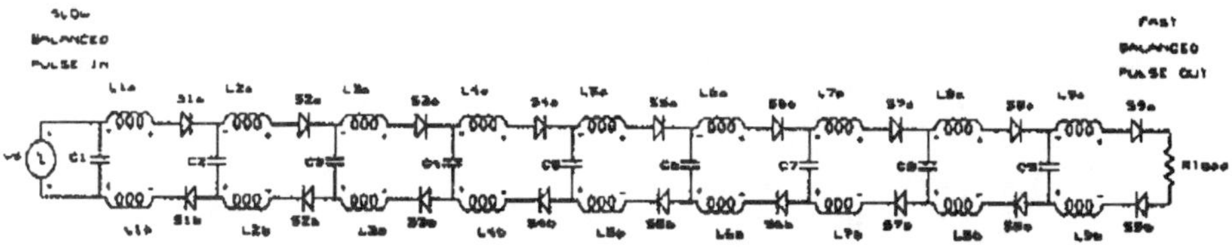

Figure 1. Balanced Shockline Solid-State Pulser

been extended to multiple 6 kV pulsers synchronized to drive a phased array. In this application a command trigger jitter of less than 10 picoseconds was demonstrated.

2.2. Silicon Carbide (SiC) Devices

It has long been desired to find a material that will operate at higher power than is possible with Si or GaAs. Recent research at the Ioffe Institute has focused on the study of SiC as an alternative material. It belongs to a class of semiconductors commonly known as "wide bandgap" materials that are, among other things, less sensitive to increased temperature. SiC devices should be able to operate at 500°C or higher, a range far beyond the range of Si. Also, the thermal conductivity of SiC exceeds that of even copper; so the heat produced by a device is quickly dissipated.

From its material properties, SiC appears to have some definite advantages over Si and GaAs. Since SiC has a higher operating electric field strength than Si, it is expected that when turned on, the resistance of a SiC switch will be about two orders of magnitude lower than Si. This means that a SiC device with the same switching speed as Si should have an operating voltage at least an order of magnitude higher than Si, or stated conversely, a SiC device with a given blocking voltage should switch at least on order of magnitude faster than Si.

The fact that SiC is relatively inert to most chemicals gives SiC devices the potential to operate in very caustic environments. In addition SiC devices are extremely hard to ionizing radiation, so they can be used in close proximity to nuclear reactors or in space environments (i.e., neutron fluence of 10^{14} n/cm^2 and up to several Mrad). Of particular importance for design of electronic devices is SiC's high electric field strength and high saturation drift velocity (2×10^7 cm/s) that allows for smaller, more efficient devices.

The Air Force is currently collaborating with Dr. Grekhov to study the performance of SiC-based JRD devices using 4H-SiC as fast opening switches. Preliminary results indicate that the $p^+p_0n^+$ structure has the best chance of producing picosecond opening times at high voltage.[12] This switch is basically a DSRD using SiC as its switching medium. The Air Force also sponsored studies of SiC closing switches. Multi-kilovolt switching was observed in SiC devices operated in DBD mode.[17] Computer simulations predict rise times of approximately 30 ps at the 12 kV level.

3. GAS SWITCHING

Precision switching is a critical technology for the development of all UWB fast transient systems. In addition to the solid state switching research, AFRL has also invested a considerable amount of effort investigating basic approaches for reducing the triggering jitter of gas switches. Ferroelectric triggering, electrode field enhancement, laser triggering, and overall optimization of system design parameters have all been studied and compared.

3.1. Ferroelectric Triggering

Ferroelectric triggering has been recently demonstrated in the form of a switch known as a Ferratron. It uses a ferroelectric trigger as an electron source which, when combined with high gas flow, provides reliable, low-jitter triggering at high rep rates. Although the design is not yet optimized, the measured jitter of the prototype Ferratron was only 63 ps using a charge voltage of 2.5 kV with a 500 ps risetime. Triggering was done in low-pressure nitrogen gas.[18]

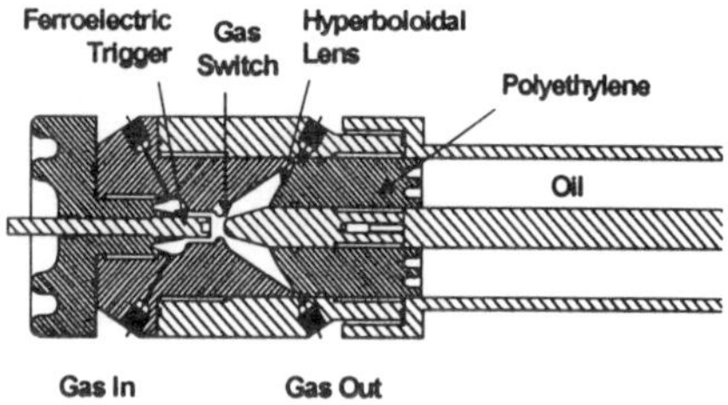

Figure 2. Triggered UWB Switch with a Hyperboloidal Lens

3.2. Enhanced Field Switching

For high current, pulsed power, we have found that a switch with an enhanced electric field, such as a trigatron, gives excellent performance with acceptably repeatable results. As a result, when a very fast, high current switch is required, a trigatron is frequently used instead of a gas thyratron because it can carry more power and is less susceptible to load variations.[19,20]

3.3. Laser Triggering

The most commonly used method of laser triggering introduces the laser beam axially through a hole in one electrode. The laser energy is focused on the opposite electrode, which produces an axial line spark that rapidly propagates to join the electrodes. Some recent investigations have sought to optimize the performance of such switches, but they have only achieved an rms jitter of 130 ps in a 500 kV, 1 cm gap filled with 50 psi of SF_6 gas.[19] More recent research has concentrated, with more success, on enhanced field switching and optimization of parameters.

4.0. DIELECTRICS AND LENSES

4.1. Lens Theory

Many HPM applications require the use of dielectric materials for lenses. Several papers have recently reported the results of theoretical studies of ultra-wideband propagation through dielectric media and techniques useful for the synthesis of radio frequency lenses.[21,22] Work in this area has also continued to increase our understanding of artificial dielectric materials and continually varying dielectric materials for radio frequency (RF) lens applications.

4.2. Artificial Dielectrics

Polyethylene, polypropylene, and polycarbonates have been widely used for RF lenses because they are inexpensive, are easy to machine, and have a dielectric constant close to that of transformer oil ($\varepsilon_r \approx 1.3$). However, these materials are very heavy, and a search has begun for lightweight alternatives. Recent studies focused on a new type of artificial dielectric material made by Cumings. Although it was designed and made for narrow band microwave operation, we are investigating it for use in wideband systems.[23] It is made of aluminum-coated, hollow glass beads that are suspended in a lightweight dielectric medium. The beads can be made in various sizes from a few mm in diameter to less than one mm. Thus, different concentrations of beads can be used to make an artificial dielectric in the classical sense or, alternatively, hollow glass beads without any coating could be used to lower rather than raise the bulk dielectric constant of the material. Research continues in this area. The spacing and geometrical arrangement of the beads or the use of conducting pieces of different shapes may yield improvement in the bandwidth and/or power handling capability of these new materials.

4.3. Continually Varying Dielectrics

Another investigation into the application of dielectrics to UWB systems involves the use of continually varying dielectrics (CVDs) for focusing and guiding UWB signals around bends without the introduction of dispersion. In recent years, many configurations of varying dielectrics have been investigated, and have met with considerable success. The best results to date have been achieved using 5 or 6 parallel channels of different dielectric materials consisting of TiO_2 powder in a polybutyldiene base, that is mixed in increasing concentrations to give dielectric constants (ε_r) varying from 2.5 to 6 across the transmission line as illustrated in Figure 3. Using such dielectrics to control the speed of light in the medium, this technique has demonstrated the ability to guide a 150-ps risetime pulse around a transmission line bend that measures about 3 cm across on a 5 cm radius with virtually no distortion of the pulse. The input and output pulses with and without the CVD lens are shown in Figure 4.[24,25]

Figure 3. Continually Varying Lens in a Square Coaxial Test Fixture

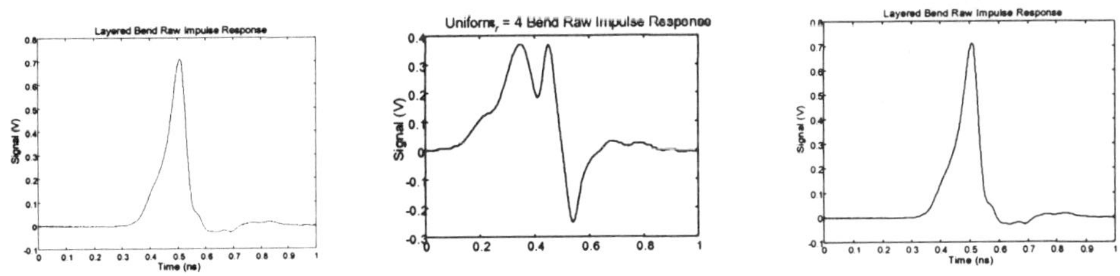

a. Input Waveform

b. Output without Lens

c. Output with Lens

Figure 4. Ultra-Wideband Transient Waveform Transmission Through the CVD Lens.

4.4 Hyperboloidal Lenses

Lenses have also been introduced into coaxial transmission lines to solve the very familiar problem that involves guiding wideband transients into a transmission line when they are emitted radially from an in-line gas switch that is part of the center conductor.[26] Figure 2 above shows a hyperboloidal lens as a part of the Ferratron. We know that the frequency response of a coaxial test apparatus is limited by the dispersion introduced as the wave turns the corner from the switch into the transmission line. Without the use of a lens to reduce the dispersion, this will severely limit our ability to transmit an UWB transient or, for switching research, to observe and record the formation of the gas arc accurately. However, with the introduction of the lens, the dispersion is overcome, and the physics of the arc channel formation may be observed much more clearly.

5. GAS SWITCHED UWB SOURCES

5.1. The IRA II

The Impulse Radiating Antenna (IRA) Ultra-Wideband Source was rebuilt in 1999 and 2000 by adapting the high voltage modulator and TEM arms from the original IRA to a smaller reflector. This version was designated IRA II, and is shown below in Figure 5. The pulsed power source was increased from ± 60 kV to ± 75 kV, and the gas thyratron was replaced with a trigatron. Thus, with the reflector diameter at 2 m and the output voltage at 150 kV, the near electric field is now about 60 kV/m, approximately twice that of the original IRA. The far field, of course, is approximately half that of the original, but for applications such as laboratory tests, which are within the near field zone, this source is very useful. It has the same voltage risetime of 130 ps and 1/e fall time of about 20 ns. The near field, of course, has exactly the same waveform as shown in Figure 6. The radiated far field, being a derivative of the voltage waveform, has a rise time of 85 ps and a pulse width (FWHM) of about 130 ps. The radiated spectrum of the 2-m IRA has been measured to be flat within 3 dB from 70 MHz to 4 GHz.[27-30]

Figure 5. The IRA II

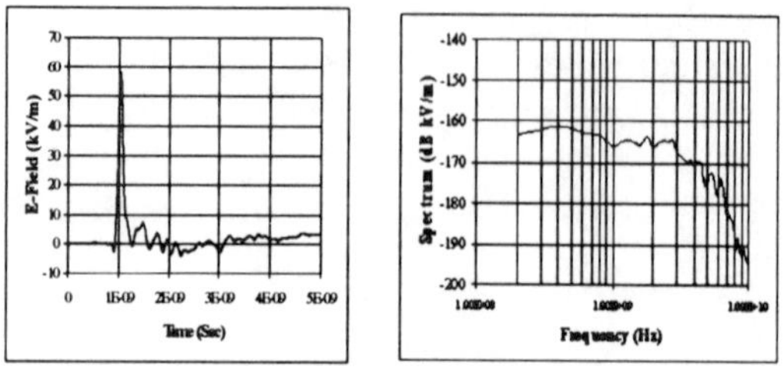

Figure 6. Near Field of the IRA II

5.2. The Half-IRA

Constructing a balanced source such as the IRA II presents some difficulties, particularly for very high voltages where trigger jitter or balun design becomes an issue. In some cases, it is preferable to design a half-IRA over a ground plane so that the voltage signal can be fed directly to the antenna with a coaxial transmission line.[4] One such unit (Figure 7) is part of a system designed and built for mine detection by our colleagues in Munster.[31]

Figure 7. Half-IRA

6. SOLID-STATE UWB ARRAYS

Several types of solid-state switched sources have been built in the last two years using the technology for delayed breakdown devices that is described above. The first one was a solid-state shockline built from standard silicon diodes that operated at 3.2 kV into a single lensed TEM horn.[32-34] More recently, single sources have been assembled into arrays of four horns, in combinations of 1 x 4 and 2 x 2 as seen in Figures 8 and 9.[15]

Figure 8. 1 x 4 Solid-State Array with Aspheric Cylindrical Lenses

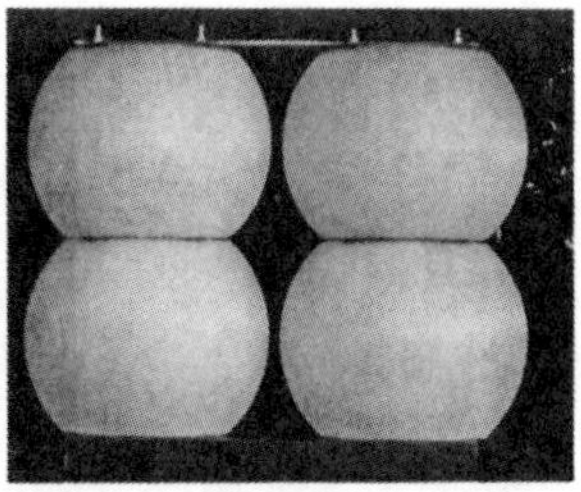

Figure 9. 2 x 2 Solid-State Array with Spherical Lenses

7. UWB ANTENNAS

7.1. Design Optimization

Numerous advances have been made in UWB antenna theory in the last few years, and these have resulted in notable improvements in the performance characteristics of impulse radiating antennas (IRAs) and a better understanding of their radiating characteristics. Parameter studies on the basic IRA shape have allowed us to optimize the placement and shape of the TEM feed arms as well as to examine the effect of mechanical design changes that are sometimes dictated by integration requirements. Also, optimizing the shape and construction of the reflector itself has yielded weight saving design changes in some cases. Since the current on the reflector is not uniformly distributed, parts of the reflector may actually be shaved off without having much effect on the radiating characteristics.[35-37]

7.2 Lightweight UWB Antennas.

UWB antennas that are required to transmit or receive frequencies as low as the VHF and HF bands can be several meters in diameter. If they are made of standard materials, the weight can be prohibitive. Consequently, it is highly desirable to have methods of fabricating large UWB antennas that are very lightweight. Two primary design approaches have been pursued – an umbrella-like construction called a Collapsible IRA (CIRA) as shown in Figure 10, and an inflatable design suitable for space deployment as shown in Figure 11. Both designs are ultra-wideband, collapsible and very lightweight.[38,39]

Figure 10. Collapsible UWB Antenna

Figure 11. Inflatable UWB Space Antenna

7.3. Synthetic Aperture Radar Antennas

The successful demonstration by the Army Research Laboratory of land mine detection using an UWB (transient) synthetic aperture radar (SAR) system high above ground level has raised a considerable amount of interest in UWB technology and particularly in UWB antennas.[5] The SAR community in general has shown interest in increasing the bandwidth of their systems, and in response, AFRL has built and tested several different types of wideband antennas in order to evaluate their suitability for use with SAR systems.[40] The latest in a series of designs, shown in Figure 12, has been measured in the time-domain and frequency-domain facilities in Albuquerque NM and Dayton OH and is scheduled to undergo actual SAR trials in Dayton sometime this year.

Figure 12. Wideband SAR Antenna

8. CONCLUSIONS AND FUTURE WORK

In the last two years, there have been several breakthroughs in fast, high-voltage switching including low jitter, triggered gas switches and SiC-based solid-state switches. This, combined with improvements in UWB lens and antenna technology, provides the opportunity for significant increases in our ability to radiate high-voltage transient (or broadband swept frequency) electromagnetic energy and thereby achieve substantial improvements in UWB sources. As a result of the very low triggering jitter that has been achieved, the ability to array a number of independently controlled elements has been demonstrated, as has electronic beam steering. Further development of this technology is still required to address lifetime and power-handling issues. Future research in this area will focus on continuing to increase our knowledge of the physics of fast solid-state, as well as gas and oil, switching. Increasing voltage and power continue to be primary goals for this technology, as well as reducing trigger jitter and increasing repetition rate. Improvements in antenna efficiency will continue to be explored as will new dielectric materials.

9. ACKNOWLEDGEMENT

This work was sponsored in large part by the Air Force Office of Scientific Research, Washington, DC. In particular, the authors would like to thank Dr. Jack Agee, Dr. Clifford Rhoades, Dr. Arje Nachman, and Dr. Robert Barker for their encouragement and support. Also, thanks to Dr. C.A. Frost and Dr. Everett Farr for many helpful discussions on this subject.

10. REFERENCES[†]

1. W.D. Prather, C.E. Baum, J.M. Lehr, J.P. O'Loughlin, J.S. Tyo, J.S.H. Schoenberg, R.J. Torres, T.C. Tran, D.W. Scholfield, J.A. Gaudet, and J.W. Burger, Ultra-wideband source & antenna research, *IEEE Trans. Plasma Science, 8th Special Issue on High Power Microwave Generation* **28** (5), October 2000.
2. F.J. Agee, J.A. Gaudet, and W.D. Prather, Focus on current topics in Air Force high power microwave (HPM) research, *Proc. International Society for Optical Engineering, SPIE*, Orlando, April 2001.
3. W.D. Prather, C.E. Baum, J.M. Lehr, M.J. Walker, R.J. Torres, T.C. Tran, J.A. Gaudet, and J.W. Burger, Research & development in ultra-wideband sources and antennas, Int'l Conf on EM in Advanced Applications (ICEAA'01), Turin, Italy, September 2001.
4. W.D. Prather, C.E. Baum, J.M. Lehr, R.J. Torres, T.C. Tran, J.A. Gaudet, and J.W. Burger, Recent developments in ultra-wideband sources and antennas, in: *Ultra-Wideband, Short-Pulse Electromagnetics 5*, edited by P.D. Smith and S.R. Cloude (Kluwer Academic/Plenum, New York, 2002), pp. 393–406.
5. L. Carin, N. Geng, M. McClure, J. Sichina, and L. Nguyen, Ultra-wideband synthetic aperture radar for mine field detection, in: *Ultra-Wideband, Short-Pulse Electromagnetics 4*, edited by E. Heyman, B. Mandelbaum, and J. Shiloh (Kluwer Academic/Plenum, New York, 1999), pp. 433–441.
6. C.E. Baum, Singularity Expansion Method, Symmetry and Target Identification, in: *Scattering and Inverse Scattering in Pure and Applied Science*, edited by R. Pike and P. Sabatier (Academic Press, New York, 2002), Chapter 1.6.3.
7. N.E. Islam, E. Shamiloglu, A. Mar, F. Zutavern, G.M. Loubriel, and R.P. Joshi, Characteristics of trap-filled GaAs photoconductive switches used in high gain pulsed power applications, in: *Ultra-Wideband, Short-Pulse Electromagnetics 5*, edited by P.D. Smith and S.R. Cloude (Kluwer Academic/Plenum, New York, 2002), pp. 461–466.
8. J.A. Gaudet, M.C. Skipper, M.D. Abdalla, S.M. Ahern, S.P. Romero, A. Mar, F.J. Zutavern, G.M. Loubriel, M.W. O'Malley, and W.D. Helgeson, Temporal switching jitter in photoconductive switches, *Proc. International Society for Optical Engineering, SPIE* **4031**, April 2000.
9. P. Kayasit, R.P.Joshi, N.E.Islam, E.Schamiloglu, and J. Gaudet, Transient and steady state simulations of internal temperature profiles, *J. Appl. Phys.***89**, 1411–1417 (January, 2001).
10. I.V. Grekhov, Mega and gigawatts-ranges, repetitive mode, semiconductor closing and opening switches, *Proc. 11th IEEE Int. Pulsed Power Conf.*, Baltimore MD, p.010.8 (1997).
11. I.V. Grekhov and G.A. Mesyats, Physical basis for high power semiconductor nanosecond opening switches, *Proc. 12th IEEE Pulsed Power Conference*, Monterey CA (June 1999).
12. I.V. Grekhov, 4H-SiC inverse recovery diodes (IRD) for sub-nanosecond power opening switches, in: *Ultra-Wideband, Short-Pulse Electromagnetics 6*, edited by E.L. Mokole, K. Gerlach, and M. Kragalott (Kluwer Academic/Plenum, New York, 2004).
13. R.J. Focia, E. Shamiloglu, C.B. Fleddermann, F.J. Agee, and J. Gaudet, Silicon diodes in avalanche pulse sharpening for high power microwave applications, *IEEE Trans. Plasma Sci.* **25** (April 1997).
14. C.A. Frost, Private communication (August 2001).
15. C.A. Frost, Multi-kilovolt solid state photoconductive switch studies, *Proc 13th IEEE Pulsed Power Conf*, Las Vegas NV (June 2001).

16. C.A. Frost, T.H. Martin, R.J. Focia, and J.S.H. Schoenberg, Ultra-low jitter, repetitive, solid state, picosecond switching, *Proc. 12th IEEE Pulsed Power Conference*, Monterey CA (June 1999).
17. I.V. Grekhov, Private communication (January 2002).
18. E.G. Farr, C.E. Baum, L.H. Bowen, J.M. Elizondo, D.E. Ellibee, J.M. Lehr, and W.D. Prather, Progress in the development of the ferratron: a novel, repetitively rated, triggered spark gap switch with ultra low jitter, *Proc. International Society for Optical Engineering, SPIE 2000*, Orlando FL (April 2000).
19. D. Giri, V. Carboni, H. Lackner, and J. Lehr, Gas breakdown field and arc risetime dependency on high pressure and short stress times, *Proc. 13th IEEE Pulsed Power Conference*, Las Vegas NV (June 2001).
20. J.M. Lehr, J.W. Burger, M.D. Abdalla, M.C. Skipper, B. Cockreham, and F. Gruner, Development of a hermetically sealed trigatron switch for high repetition rate primary applications, *IEEE Trans. Plasma Science* **28** (October 2000).
21. C.E. Baum and A.P. Stone, Unipolarized generalized inhomogeneous TEM plane waves in differential geometric lens synthesis, *Electromagnetics* **21**, 275–294 (2001).
22. A.P. Stone and C.E. Baum, Generalized TEM, E, and H modes, *Sensor and Simulation Note 453*, Air Force Research Laboratory, Kirtland AFB NM (February 2001).
23. C.A. Frost, Measurement and evaluation of artificial dielectric material, *Measurement Note 57*, Air Force Research Laboratory, Kirtland AFB NM (May 2002).
24. W.S. Bigelow, E.G. Farr, and W.D. Prather, Dispersion reduction in a coaxial transmission line bend by a layered approximation of a graded dielectric lens, in: *Ultra-Wideband, Short-Pulse Electromagnetics 5*, edited by P.D. Smith and S.R. Cloude (Kluwer Academic/Plenum, New York, 2002), pp. 697–706.
25. M.J. Walker and C.E. Baum, Maximizing total and TEM mode power through an azimuthal waveguide bend filled with a graded dielectric medium, *Sensor and Simulation Note 452*, Air Force Research Laboratory, Kirtland AFB NM (January 2001).
26. E.G. Farr, D.E. Ellibee, J.M. Elizondo, C.E. Baum, and J.M. Lehr, A test chamber for a gas switch using a hyperboloidal lens, *Switching Note 30*, Air Force Research Laboratory, Kirtland AFB NM (March 2000).
27. D.V. Giri, B. Cockreham, F.R. Gruner, H. LaValley, T. Smith, D. McLemore, J. Burger, R. Torres, T. Tran, W.D. Prather, and C.E. Baum, Fabrication details of the 6-foot IRA, *Prototype IRA Memo 7*, Air Force Research Laboratory, Kirtland AFB NM (April 2001).
28. M. Abdalla, M. Skipper, D.V. Giri, H. LaValley, T. Smith, D. McLemore, J. Burger, R. Torres, T. Tran, W.D. Prather, and C.E. Baum, Evaluation of the terminating impedance in the conical-line feed of the 6-foot IRA, *Prototype IRA Memo 8*, Air Force Research Laboratory, Kirtland AFB NM (April 2001).
29. W.D. Prather, The 6′ IRA, National Radio Science Meeting, Boulder CO (January 2001).
30. J.S. Tyo, M.J. Baretela, C.J. Buchenauer, L.H. Bowen, and E.G. Farr, Increase in the prompt radiated field from an IRA by aperture design techniques, *Proc. International Society for Optical Engineering, SPIE*, Orlando FL (April 2001).
31. F. Sabath, D. Nitsch, M. Jung, and Th.H.G.G. Weise, Design and setup of a short pulse simulator for susceptibility investigations, *Sensor & Simulation Note 460*, Air Force Research Laboratory, Kirtland AFB NM (October 2001).
32. C.E. Baum, High-power scanning waveguide array, *Sensor & Simulation Note 459*, Air Force Research Laboratory, Kirtland AFB NM (2001).
33. C.A. Frost and M.J. Walker, Ultra-wideband impulse source based on solid-state switched phased array, National Radio Science Meeting, Boulder CO (January 2001).
34. A. Kardo-Sysoev, V. Brylevsky, Y. Lelikov, I. Smirnova, S. Zazulin, and I. Tchashnicov, Ultrawideband solid state pulsed antenna array, in: *Ultra-Wideband, Short-Pulse Electromagnetics 5*, edited by P.D. Smith and S.R. Cloude (Kluwer Academic/Plenum, New York, 2002), pp. 343–349.
35. J.S. Tyo, Optimization of the TEM feed structure for 4-arm reflector impulse radiating antenna, *IEEE Trans. Antennas Propagat.*, (August 2000).
36. C.E. Baum, Modification of TEM-fed reflector for increased efficiency, *Sensor & Simulation Note 458*, Air Force Research Laboratory, Kirtland AFB NM (July 2001).
37. E.G. Farr, L.H. Bowen, G.R. Salo, J.S. Gwynne, C.E. Baum, W.D. Prather, and T.C. Tran, Lightweight ultra-wideband antenna development, in: *Proc. International Society for Optical Engineering: Intense Microwave Pulses VII*, **Vol. 4031**, edited by H.E. Brandt (SPIE, April 2000).
38. L.H. Bowen, E.G. Farr C.E. Baum, W.D. Prather, and T.C. Tran, Experimental results of optimizing the location of feed arms in a collapsible IRA and a solid IRA, *Sensor & Simulation Note 450*, Air Force Research Laboratory, Kirtland AFB NM (November 2000).
39. L.H. Bowen, E.G. Farr, J.P. Paxton, A.J. Witzig, C.E. Baum, D.I. Lawry, and W.D. Prather, Fabrication and testing of a membrane IRA, *Sensor & Simulation Note 464*, Air Force Research Laboratory, Kirtland AFB NM (January 2002).

40. E.G. Farr, L.H. Bowen, G.R. Salo, J.S. Gwynne, C.E. Baum, W.D. Prather, and T.C. Tran, Studies of an impulse radiating antenna and a pulse radiating antenna element for airborne SAR and target identification applications, *Sensor & Simulation Note 442*, Air Force Research Laboratory, Kirtland AFB NM (March 2000).

---

† Copies of Sensor and Simulation Notes are available from the authors or the editor, Dr. C.E. Baum, Air Force Research Laboratory, AFRL/DEHP, 3550 Aberdeen Ave, SE, Kirtland AFB, NM, 87117, USA.

DESIGN AND DEVELOPMENT OF A HIGH VOLTAGE COAXIAL HYDROGEN SWITCH

J.W. Burger, C.E. Baum, W.D. Prather, R. J. Torres, T.C. Tran, M.D. Abdalla, M.C. Skipper, B.C. Cockreham, D.P. McLemore*

1. INTRODUCTION

The high power microwave program at the Air Force Research Lab (AFRL) includes high power source development in narrow band and wideband technologies. The H-2 source is an existing wideband source that was developed at the AFRL. A recent AFRL requirement for a wideband impulse generator for use in materials testing has provided the need to update the old H-2 source. The source is composed of a dual resonant transformer that charges a short length of coaxial transmission line. This is then discharged into a following coaxial transmission line with a self-breaking hydrogen switch. The dual-resonant transformer is driven by a low-inductance, primary capacitor bank that operates through a self-break gas switch. The primary focus of this paper is the upgrade of the coaxial hydrogen output switch. The final design of this switch was developed through the use of extensive electrical and mechanical simulations.

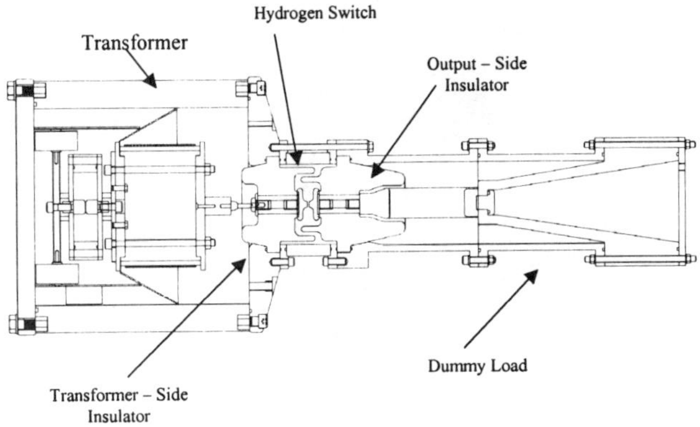

Figure 1. Original H-2 Configuration.

* J.W. Burger, C.E. Baum, W.D. Prather, R.J. Torres, and T.C. Tran: Air Force Research Laboratory, Kirtland AFB NM 87117. M.D. Abdalla and M.C. Skipper: ASR Corporation, Albuquerque NM. B.C. Cockreham: Solid Design, Inc., Albuquerque NM. D.R. McLemore: ITT Industries, AES, Albuquerque NM.

2. TRANSFORMER MODIFICATIONS

The H-2 transformer consists of a single primary winding and a 32-turn secondary with a coupling coefficient of 0.6 to allow dual-resonant operation. A new primary winding was fabricated to alleviate the formation of corona activity in the vicinity of the feed point of the primary winding. During fabrication of the winding, special care was taken to eliminate sharp edges, so the winding band edges were terminated with a ¼″ diameter tube. A new nylon insulator was fabricated to isolate the opposite ends of the primary winding. The length of the Mylar insulation between the primary and secondary windings was increased to prevent any partial discharges. A picture of the revised assembly for the primary and secondary windings is shown in Figure 2.

Table 1. Revised H-2 Transformer Specifications

Open Circuit Resonance	1.13 MHz
Short Circuit Resonance	1.62 MHz
Coupling	
Primary Inductance	74.4 nH
Secondary Inductance	74.3 μH
Primary Turns	1
Secondary Turns	32

The resonant frequency of the secondary winding was measured with and without a shorted primary winding to calculate the coupling coefficient. The primary and secondary inductance of the secondary winding was measured using an LCR meter. The measured parameters are given in Table 1. To estimate the maximum output voltage of the transformer, a Pspice™ model of the transformer was formulated. The expected output charge voltage (hydrogen switch remains open) is shown in Figure 3.

Figure 2. Revised H-2 Transformer Assembly.

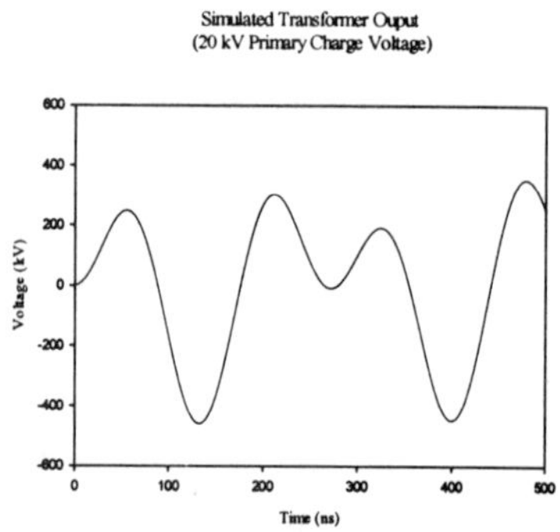

Figure 3. Simulated revised transformer output (H-2 output switch open).

3. MECHANICAL ANALYSIS OF THE COAXIAL PULSE FORMING LINE

The output switch in the H-2 source is a self-breaking hydrogen design and is typically operated at a pressure of 1,000 psig, which had caused mechanical failures in the past (see Figure 4). In the original H-2 design, the dielectric gas containment components were fabricated from Acrylic and the entire switch was mechanically retained with all-thread stock. The components of the outer conductor were fabricated from aluminum; whereas the switch electrodes and the switch-electrode support shafts were made from tungsten and brass, respectively.

A 3-D, finite-element analysis (FEA) package was used to analyze the mechanical stresses in the insulating materials and to estimate the deflections of the components under pressure. Mechanical changes to the original design were made in order to ensure a safety factor of at least 4 for the system, accommodate electric field requirements, and assure sealing at 1,000 psig.

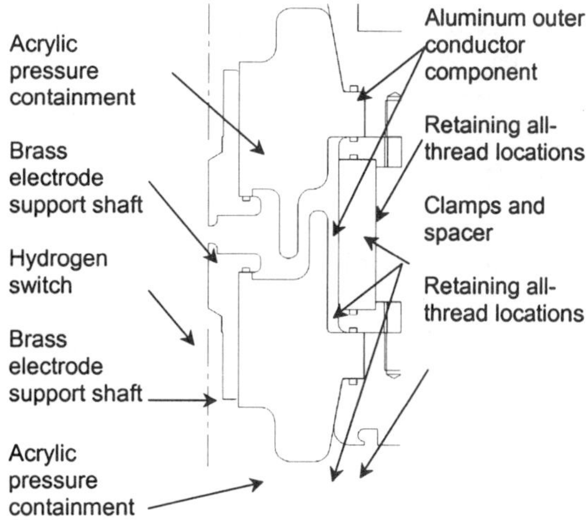

Figure 4. Original H-2 output switch design.

In the original H-2 switch, the pressure containment dielectric components were fabricated from Acrylic, which as a structural material, must be used carefully. Its initial tensile yield strength is approximately 8,000 psi, but when it is left under tension for several hours, the tensile yield strength declines to approximately 4,500 psi. As shown in Figures 9a and 9b, the FEA analysis indicates mechanical stresses of approximately 6,000 psi in the acrylic with a deflection between the electrodes of approximately 0.013". As a result, the safety factor in the Acrylic components is only about 1. The highest mechanical stresses are located in the corners as shown in Figure 9. A visual inspection of a failed H-2 acrylic insulator showed material cracking in this location. In the revised design, the Acrylic is replaced with Ultem 2300, a material that is manufactured by Boedeker. Mechanically, Ultem 2300 has a measured tensile yield strength of approximately 19,000 psi, and it does not change significantly over time. Ultem 2300 was used successfully in the IRA II output switch.[1]

Even though the Ultem 2300 was substantially stronger than Acrylic, additional mechanical modifications were necessary to give the desired safety factor of 4. The geometry associated with the high mechanical stresses in Figure 9 was revised to distribute the mechanical stresses over a smoother surface as shown in Figure 6. The resulting mechanical stresses in the Ultem 2300 are reduced to approximately 5,000 psi, most of which is compressive. The rated compressive strength of Ultem 2300 is approximately 22,000 psi, so the safety factor of 4 was achieved. The deflection between the electrodes with the revised Ultem 2300 design is approximately 9 mils.

Experience with previous high-pressure switches has indicated that it is preferable to place O-rings on surfaces that are not forced apart when high pressure is applied. Several O-rings in the original H-2 design were placed in surfaces that were forced apart as shown in Figure 7. In the original design, as the switch is pressurized, the outer conductor components have a tendency to pull away from each other. If a large enough gap is established, the O-ring will be extruded out from between the components allowing the hydrogen to escape. Such a pressure release could be violent. In the revised design, the O-ring is placed such that as the components move under the force of switch pressure, the O-ring stays in place and continues to seal.

The revised O-ring sealing concept necessitated a complete redesign of the clamp/spacer section of the original design as shown in Figure 8. The revised design is a two-piece screw-together unit that is stronger than the previous design. The highest mechanical stress in the aluminum outer conductor components is approximately 7,000 psi resulting in a safety factor of approximately 10. The mechanical stress and deflection analyses of the revised design are shown in Figures 10a and 10b.

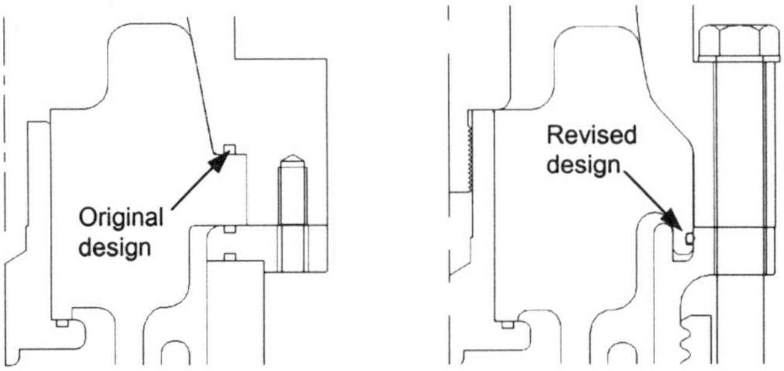

Figure 6. Original and revised mechanical load surfaces in output design.

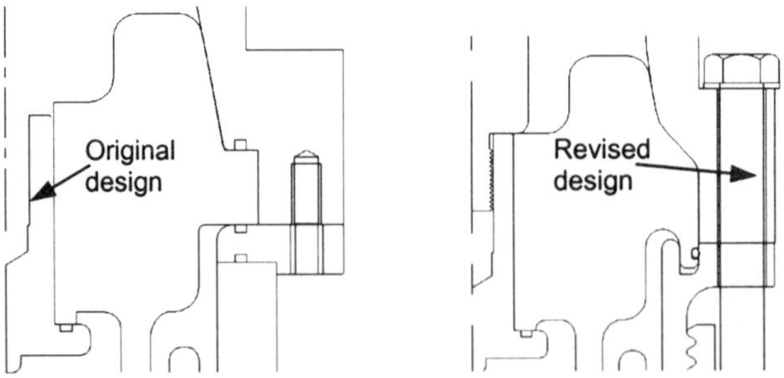

Figure 7. Original and revised O-ring locations.

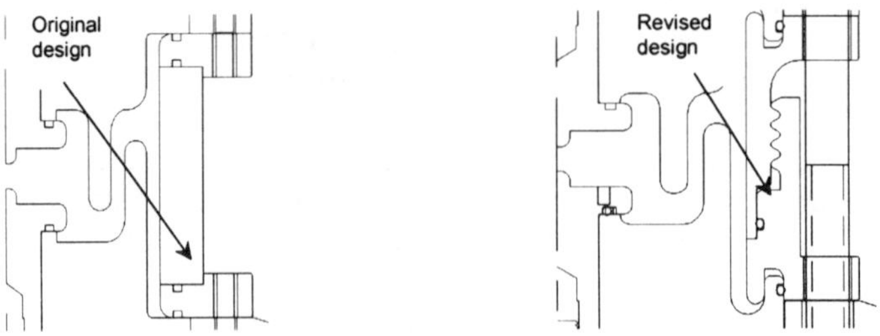

Figure 8. Original and revised clamps and spacer.

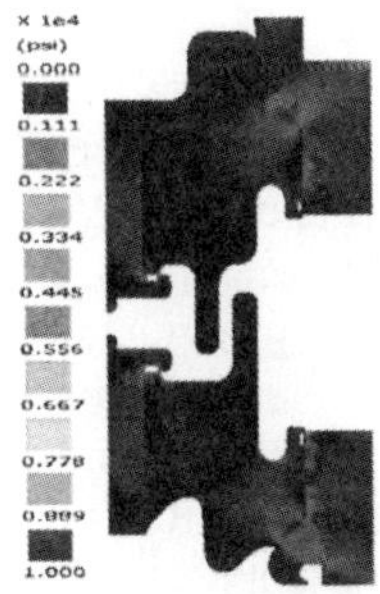

Figure 10a. Mechanical stresses in revised H-2 design.

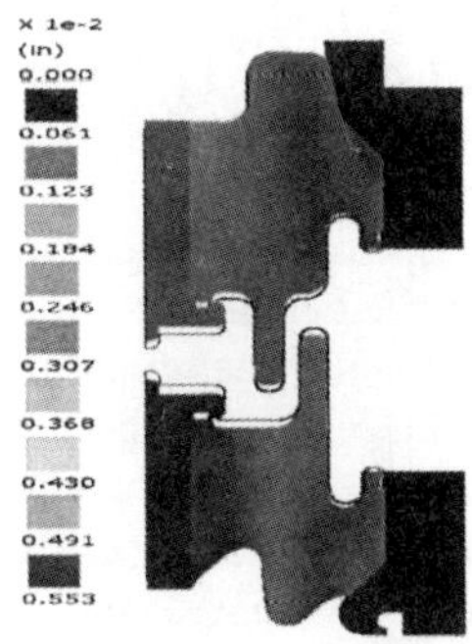

Figure 10b. Mechanical deflection in revised H-2 design.

4. ELECTROSTATIC ANALYSIS OF PULSE FORMING SECTION

The H-2 switch components were electrically analyzed to determine the maximum field levels using Electro™; a two-dimensional boundary element method software package. Based on empirical data obtained during previous experiments, the following criterion was used:

- Field levels at metal/Ultem 2300 interfaces limited to less than 100 kV/cm.
- Field levels at metal/oil interfaces limited to less than 350 kV/cm.
- Tangential field levels along insulator surfaces that bridge a gap between two oppositely charged conductors limited to 160 kV/cm when in oil.
- Tangential field levels along insulator surfaces that bridge a gap between two oppositely charged conductors limited to 130 kV/cm when in hydrogen.

During operation of the H-2 source, the transformer side of the hydrogen switch is charged slowly in approximately 300-400 ns while the output side remains isolated and at ground potential. When the hydrogen switch closes, the output side of the source is charged to a lower peak voltage (approximately half the transformer-side voltage), and the effective stress time is less than 10 ns. Therefore, the primary electrical concern was associated with the switch insulator on the transformer side of the source while the output switch is open.

The unmodified design for the transformer-side insulator used for the electrostatic analysis is shown in Figure 11. The lower end is the source transmission line that is charged by the transformer. A voltage contour plot of the original design is shown in Figure 13, and an E-Field magnitude plot is shown in Figure 27. The high-field areas are clearly evident in Figure 27 with some areas exceeding 250 kV/cm. The four specific areas of concern are the three triple points and the aluminum-Ultem 2300 interface at the center conductor. Shielding techniques were used to move the high field areas into the hydrogen or oil insulated region where higher fields can be sustained without permanent damage to insulator.

The field levels at triple point #1 are actually very low (see Figure 15). However, the field levels in the very near vicinity are in excess of 250 kV/cm (see Figure 27). The field levels at triple point #2 (see Figures 18 and 24) are also in excess of 250 kV/cm. The observed electrical failure of this insulator was initiated and terminated at these two triple points. The aluminum was recessed as is shown in Figure 14 to shield the triple point (#2) and move the high field areas into the oil dielectric. This results in higher fields in the oil region while greatly reducing the field levels at the Ultem/aluminum/oil interface. As is shown in Figure 21 and Figure 22, the field level at triple point #2 is reduced to less than 120 kV/cm. The peak field levels in the oil are increased to almost 180 kV/cm. Note that the o-ring groove in the vicinity of triple point #2 has been eliminated.

Increasing the oil gap between the Ultem 2300 and the metal and changing the angle of incidence for the Ultem 2300 to approximately 45 degrees decreased the field levels at and in the vicinity of triple point #1. The higher dielectric constant of the Ultem 2300 insulator results in the high field region being "pushed" into the oil region. This effect is shown very clearly in the voltage contour plots (see Figure 14).

Using a similar technique, the field levels at triple point #3 were reduced to acceptable levels by moving the triple point "behind" the aluminum. Again, this places the high field area into the higher strength hydrogen insulated area. The O-ring is placed in a very low field area.

As is shown in Figure 15 and Figure 17, the field levels at the aluminum/Ultem interface along the center transmission line conductor are above the acceptable limits for both the original and the revised design. The field levels for the revised design have been reduced slightly by the changes made at the three triple points. The field levels in this area can only be reduced further by changing the ratio of the inner/outer conductor radius, or by increasing the outer conductor significantly while maintaining the same inner conductor radius. A change in either radius will alter the source line impedance and will result in a change in the output pulse characteristics. This would also require a redesign and manufacture of the aluminum components. To reduce the field levels to acceptable levels, the source charge voltage will have to be reduced to 60 % of the expected transformer output of 460 kV (i.e. to 280 kV).

The electric field levels tangential to a dielectric surface bridging two conductors are important to minimize the risk of electrical flashover across the surface. As shown in Figure 23, the tangential field level exceeds 160 kV/cm for the original design. The revised design limits the tangential field level to less than 130 kV/cm (see Figure 25).

The field levels associated with the system after the hydrogen switch closes are much less critical since the stress time is much shorter and the peak voltage is lower. These are shown in Figures 28, 30, and 31, and as can be seen, the levels are well within the guidelines discussed above. By comparison, the Ultem version of the IRA II switch has been tested to more than 1-million shots with no degradation in performance.

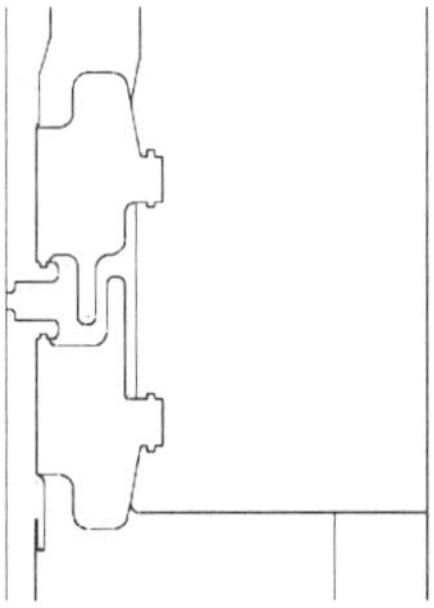

Figure 11. Original H-2 design.

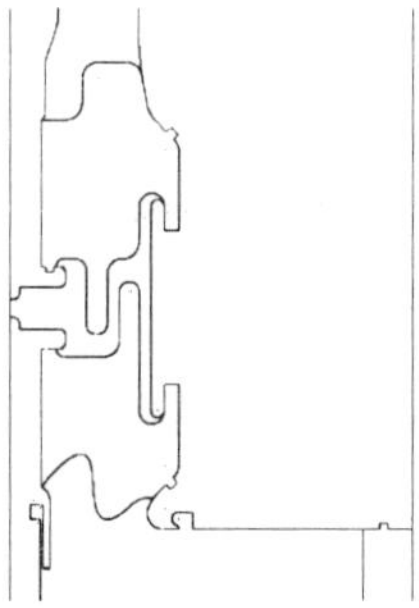

Figure 12. Revised H-2 design.

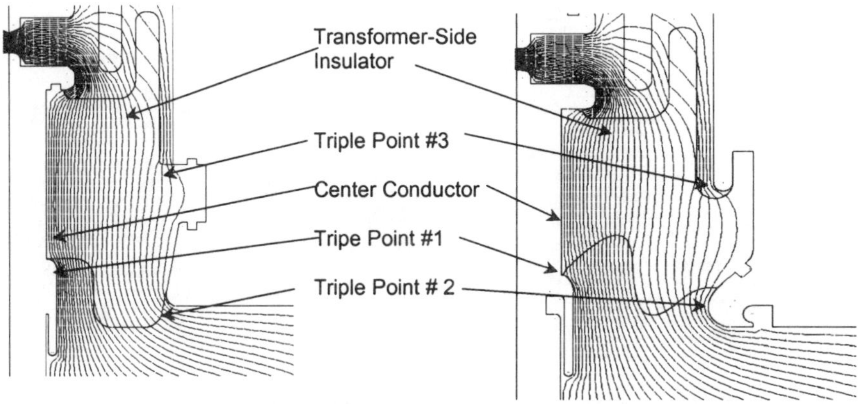

Figure 13. Voltage contour plot for original H-2 design.

Figure 14. Voltage contour plot for the revised H-2 design.

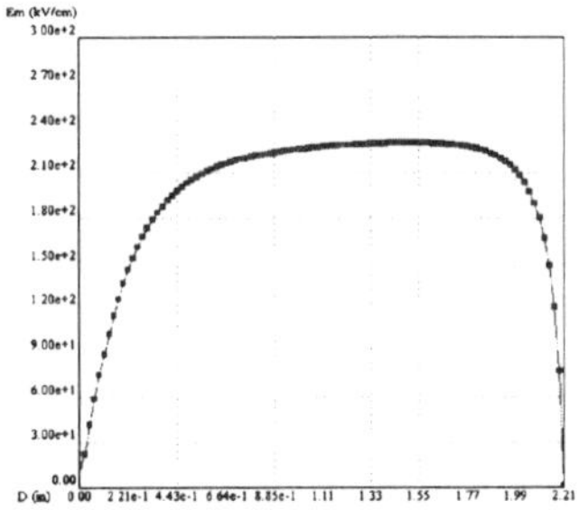

Figure 15. E-field magnitude along conductor /insulator interface for the original H2 design (see Figure 16).

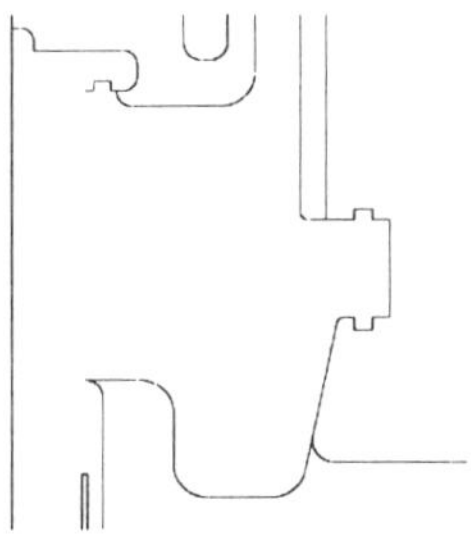

Figure 16. Plot region for Figure 15 (triple point #1, original design).

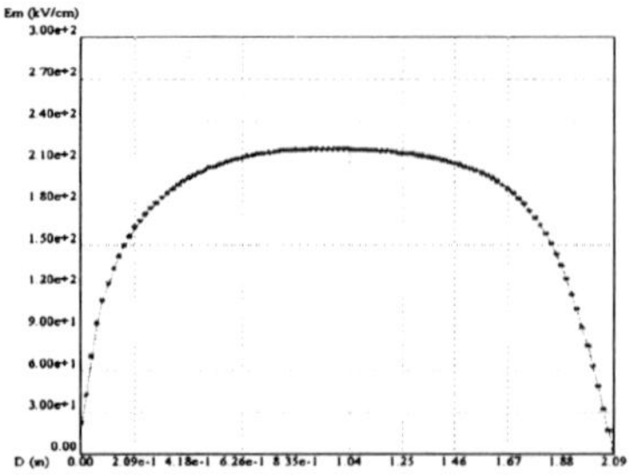

Figure 17. E-field magnitude along conductor/Ultem interface for the revised H-2 design (see Figure 19).

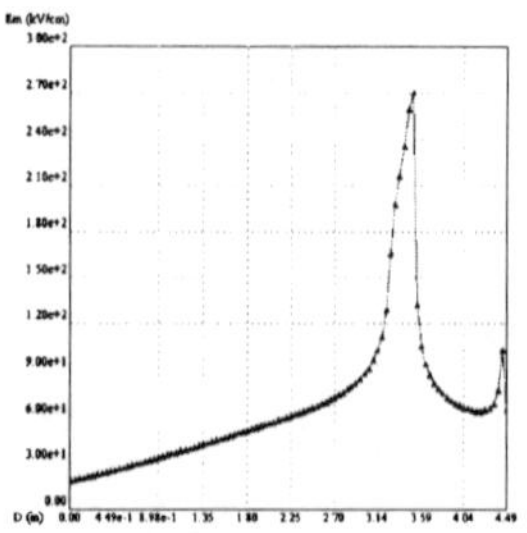

Figure 18. Electric field magnitude at the insulator/aluminum/oil triple point for the original H-2 design (see Figure 20).

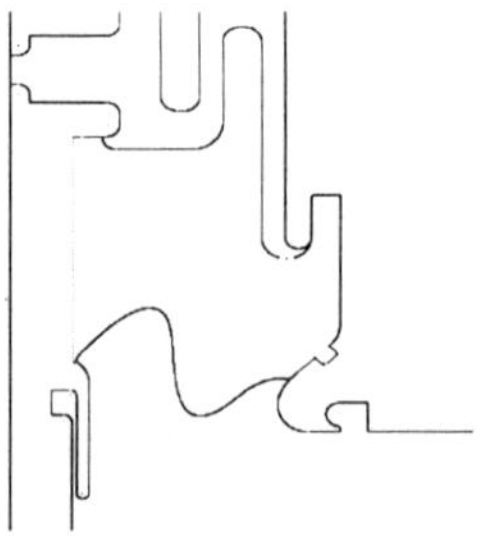

Figure 19. Plot region for Figure 17 (triple point #1, revised design).

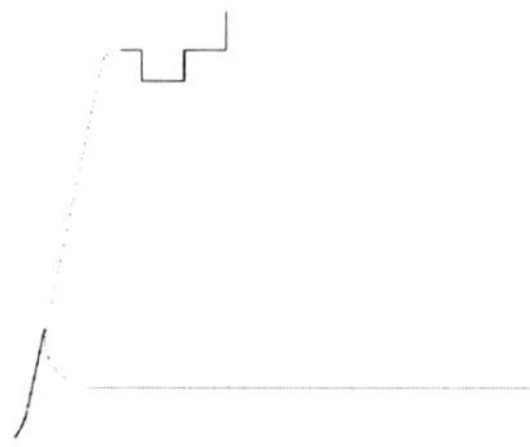

Figure 20. Plot region for Figure 18 (triple point #2, original design).

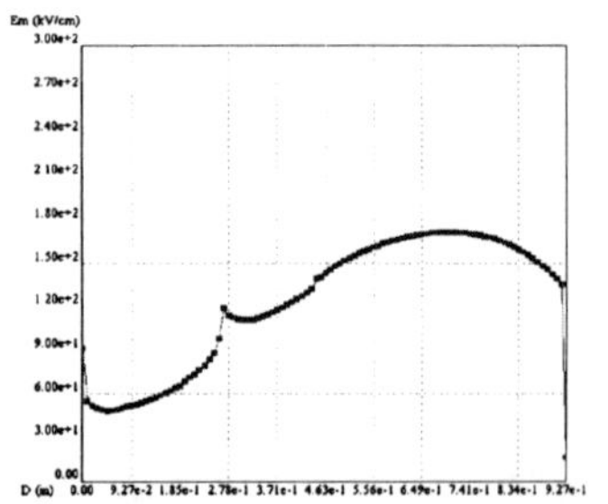

Figure 21. Electric field magnitude at the Ultem/aluminum/oil triple point for the revised H-2 design (see Figure 22).

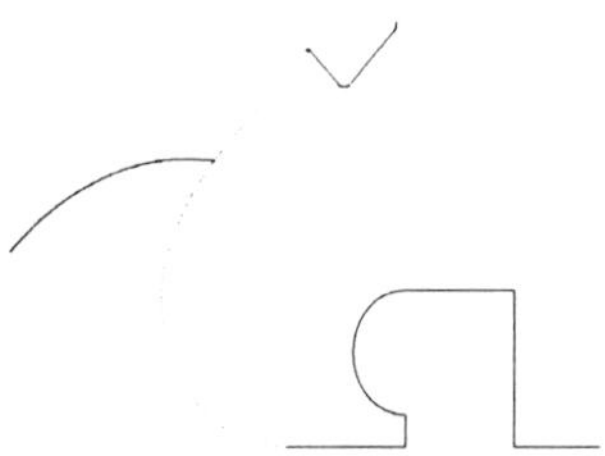

Figure 22. Plot region for Figure 21 (triple point #2, revised design).

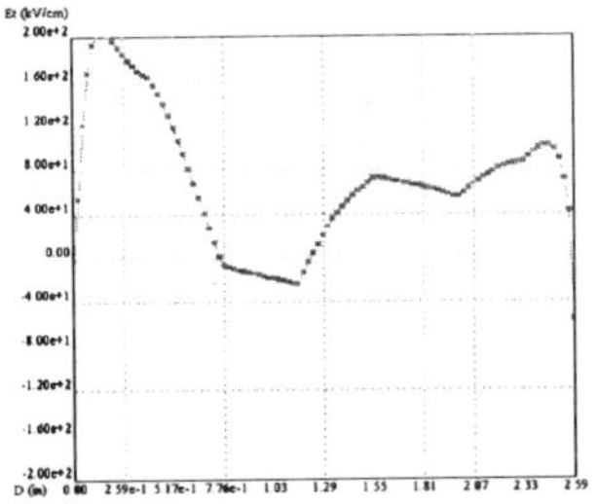

Figure 23. Tangential electric field levels along insulator bridge for the original H-2 design (see Figure 24).

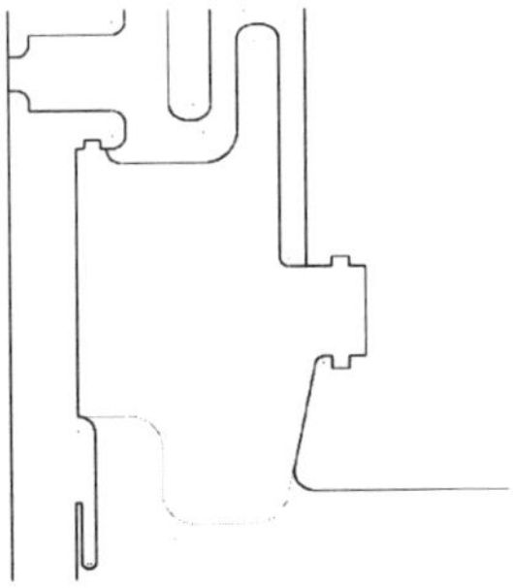

Figure 24. Plot region for Figure 23.

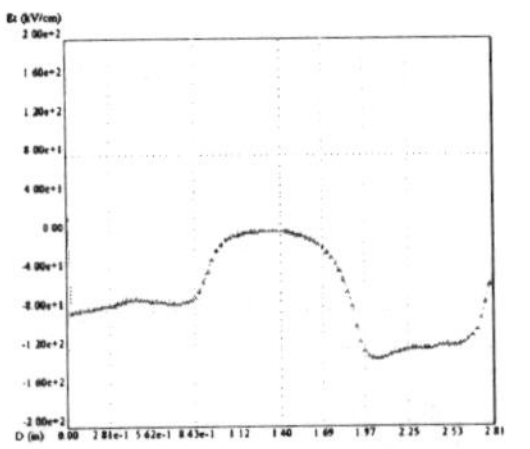

Figure 25. Tangential electric field levels along Ultem bridge for the revised H-2 design (see Figure 26).

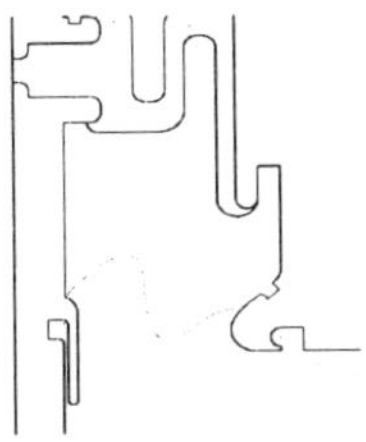

Figure 26. Plot region for Figure 25.

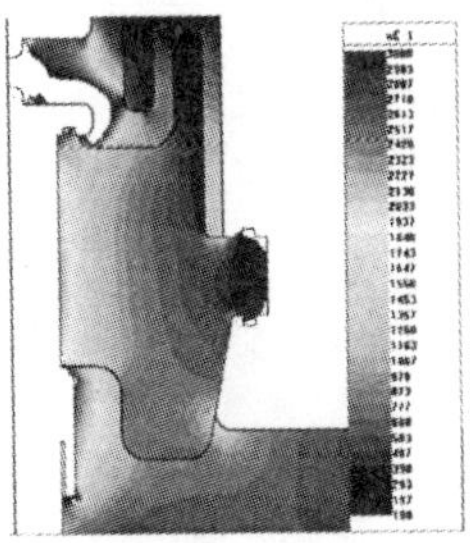

Figure 27. E-Field magnitude plot for original design.

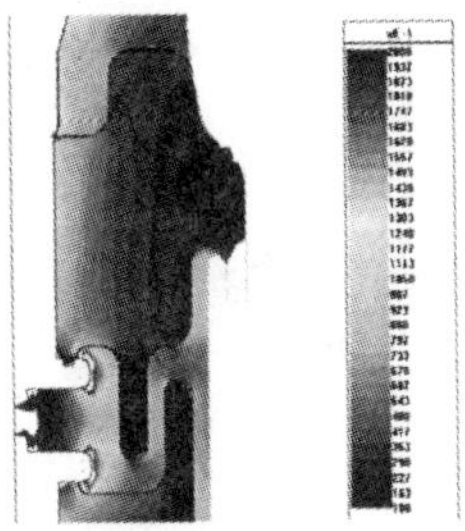

Figure 28. Closed switch E-Field magnitude plot of the output-side insulator for the revised design.

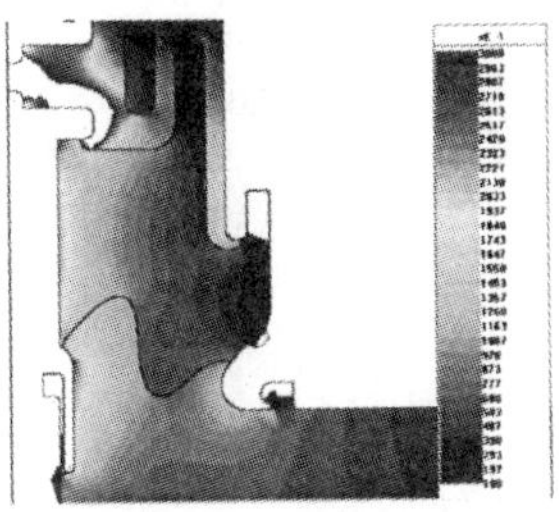

Figure 29. E-Field magnitude plot for revised design.

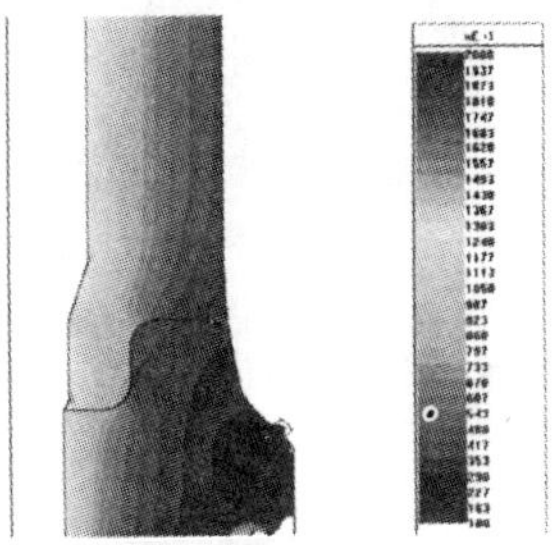

Figure 30. Closed switch E-Field magnitude plot of the output section for the revised design (kV/cm).

5. EXPERIMENTAL DATA

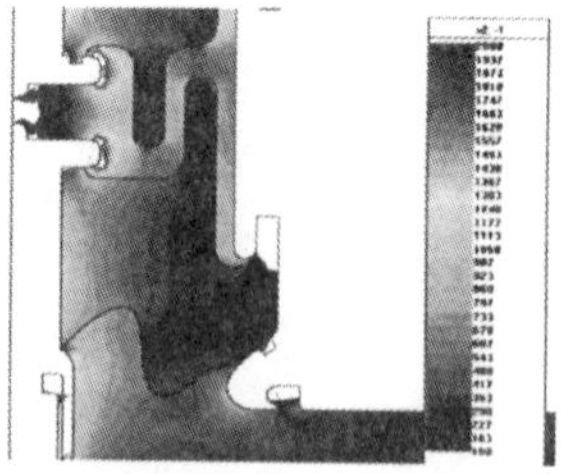

Figure 31. Closed switch E-Field magnitude plot of transformer-side insulator for the revised design (kV/cm).

To insure that the hydrogen switch is closing on the second peak of the transformer output, a Pearson current probe (0.05 V/A) was used to measure the transformer secondary current in the transformer ground strap. The current is then integrated to obtain a representation of the secondary voltage. The magnitude of the secondary voltage cannot be determined using this method since the secondary capacitance is not known, nor can it be measured easily. The primary switch was charged to 12 psi with air and the main switch was set to 450 psi for the experiments. The normalized integrated secondary current (secondary voltage) for a typical shot is shown in Figure 33, and the source output voltage is shown in Figure 34. The hydrogen switch fires at the peak of the second peak of the secondary voltage to maximize the output voltage for a given input voltage. The peak output voltage of approximately 100 kV indicates that the transformer is charging the source capacitance to 200 kV. The maximum repetition rate achieved was 10 Hz for a burst of less than 10 shots, the pulse repetition rate being limited by the low-pressure primary switch.

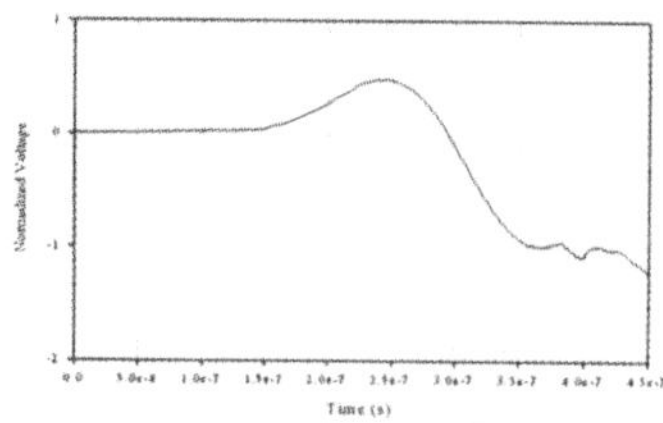

Figure 33. Normalized secondary voltage derived from the secondary current.

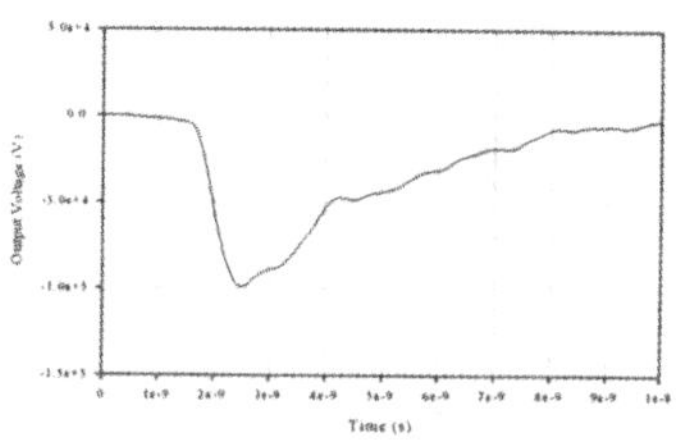

Figure 34. H-2 output sensor voltage.

6. CONCLUSIONS

The H-2 system has been evaluated both mechanically and electrically, and design changes have been made to improve the safety and reliability of the system. The mechanical safety factor was improved from 1 to 4 for long-term operation, and the electrical field levels at the critical triple-point areas have been reduced to acceptable levels.

7. REFERENCES

1. W.D. Prather, C.E. Baum, J.M. Lehr, R.J. Torres, T.C. Tran, J.A. Gaudet, and J.W. Burger, Recent developments in ultra-wideband sources and antennas, in: *Ultra-Wideband, Short-Pulse Electromagnetics 5*, edited by P.D. Smith and S.R. Cloude (Kluwer Academic/Plenum Publishers, New York, 2002), pp. 393–398.

PROGRESS IN DEVELOPMENT AND CHARACTERIZATION OF PHOTOCONDUCTIVELY-SWITCHED STACKED BLUMLEIN DEVICES PRODUCING HIGH POWER NANOSECOND PULSES

Farzin Davanloo, Carl B.Collins, and Forrest J. Agee*

1. INTRODUCTION

To fulfill the demand for pulsed power sources producing several hundred kV pulses at moderately high repetition rates, the University of Texas at Dallas (UTD) first introduced and implemented a new approach to combine the functions of pulse shaping and voltage multiplication using stacked Blumleins. This yielded the development of pulsers which consisted of several triaxial Blumleins stacked in series at one end. The lines were charged in parallel and synchronously commutated with a single switch at the other end. This allowed switching to take place at a low charging voltage relative to the pulser output voltage.[1-4]

The stacked Blumlein pulsers have produced high power waveforms with pulse durations, risetimes and repetition rates in the range of 5-600 ns, 3-50 ns, and 1-300 Hz, respectively using a thyratron or spark gap.[1-4] To generate waveforms with sub-nanosecond risetimes at high repetition rates, photoconductive switches offer significant advantages. Our recent efforts at UTD, for the first time, resulted in implementation and demonstration of several high power stacked Blumlein pulsers commutated by a single photoconductive switch. These devices have a compact geometry and are commutated by a GaAs photoconductive semiconductor switch (PCSS) triggered by a low power laser diode array.[5-7]

Filamentation of the conductivity associated with high gain GaAs switches produces such high current density that the switches are damaged near the metal-semiconductor

* Farzin Davanloo and Carl B. Collins, Center for Quantum Electronics, University of Texas at Dallas, Richardson, Texas, 75080. Forrest J. Agee, Air Force Office of Scientific Research, AFOSR/NE, Arlington, Virginia, 22203.

interface and the lifetime is limited. The semiconductor properties of amorphic diamond can be employed to improve the PCSS longevity by coating the switch cathode or anode areas or both. For example if the switch cathode is coated, the tunneling of electrons from amorphic diamond to GaAs during the off-state stage of PCSS operation provides pre-avalanche sites that diffuse conduction current upon switch activation. This report presents the progress in the development and use of these novel pulsers. A significant improvement in switch lifetime has been demonstrated by testing the diamond-coated switch performance in a stacked Blumlein prototype pulser.

Pulsed power systems to drive UWB sources require the ability to provide fast rising pulses for driving antennas at high voltages.[8] These have been usually limited to the order of 10 kV in the array photoconductive semiconductor switched (PCSS) sources built to date. The product of the field strength times the range is the performance criterion that is used to rate performance of a combination pulser and antenna. The goal is to develop UWB sources in terms that provide sufficient field strength at a given range. The limitation in existing technologies is switch voltage with acceptable switch lifetime. The opposing trend in switch voltage and switch lifetime makes it undesirable simply to operate the switches at higher voltages, as switch lifetime drops dramatically with increased operating voltage.

The work described here is the progress toward developing a technology in pulsed power that can increase the voltage delivered to UWB antennas with reasonable switch lifetimes. Advances in stacked Blumlein technology for voltage multiplication at UTD, together with the results obtained in this study, offer exciting possibilities for a variety of HPM applications especially in array UWB and compact UWB sources. With power levels approaching 150 MW, broad-band HPM sources operating at kilohertz repetition rates can be conceived that match characteristic impedance of the compact stacked Blumlein pulsers to intrinsic impedance of free space through special purpose antenna interfaces.

2. PHOTOCONDUCTIVE SWITCHING OF BLUMLEIN PULSERS

Our recent efforts have resulted in implementation and demonstration of several intense photoconductively switched stacked Blumlein pulsers. Presently, these devices operate with a switch peak power in the range of 50-80 MW and activating laser pulse energies as low as 300 nJ.[6,7] Examinations of output waveforms have indicated pulse durations in the range of 1-5 ns and risetimes as fast as 200 ps. An example of the output voltage generated by a 2-line stacked Blumlein prototype with Blumlein length of 11 cm together with a photograph of the pulser are given in Fig. 1. This pulser was constructed by connecting the corresponding high voltage and ground copper plates for each of the Blumleins to the switch electrode assembly. Blumleins lines with an impedance of about 100 Ω were extended to the proximity of stacking location, 11 cm from the switch electrodes, where the center, high-voltage conductor was terminated. Then the lines angled toward each other along with the dielectric where they were stacked directly on top of one another. The lines were joined in series for about 2.5 cm, and the top and bottom plates were connected to a 200 Ω non-inductive resistive load as shown in photograph of Fig. 1. At the switch assembly, the pulse forming line was screwed to the electrodes in a low profile geometry.

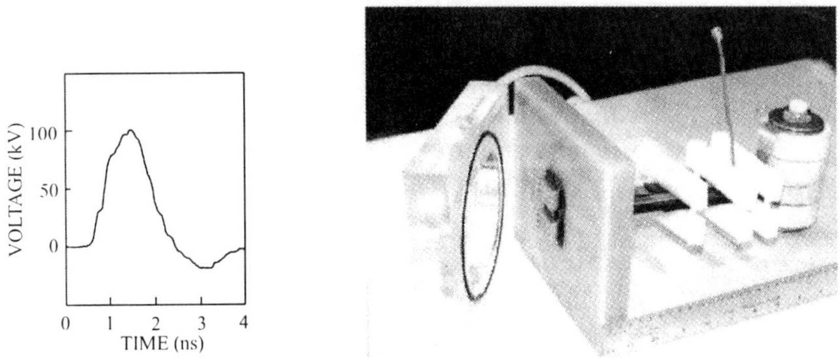

Figure 1. Output voltage waveform generated by the two-line stacked Blumlein prototype pulser shown in this figure. Device was operated with a charging voltage of 50 kV.

A single avalanche GaAs switch in a SF_6 pressurized chamber was used to commutate the pulser as shown in Fig. 1. Pulse durations of voltage waveforms generated were proportional to the Blumlein length corresponding approximately to the two way transit time of Blumleins. Peak voltage value for pulses presented in Fig. 1 was 98 kV and corresponded to a voltage gain of 1.96. Table I present the best simultaneous and individual results obtained, todate, by switching the stacked Blumlein pulsers with a high gain GaAs PCSS.

During the avalanche-mode switching of a pulser, the current is concentrated in filaments that extend from the cathode to the anode across the insulating region of the GaAs switch in a lateral configuration.[9] Carrier recombination results in the emission of characteristic band gap photons in the near infrared region that can be seen by an infrared viewer. We have observed these effects in experiments where laser diodes provided trigger photons for the avalanche commutation of the stacked Blumlein pulsers. As soon as the avalanche is initiated, a single filament can be observed which approximately follows the collimated laser beam that was focused in a line from the cathode to the anode. Multiple branching from the avalanche initiation point has also been observed but with much less probability. Filamentary currents with densities of several MA/cm^2 and diameters of 15-300 μm passing through a narrow channel can cause switch damage, especially at the contact points. A greater number of filaments during each cycle of commutation reduce the stress on the switch, thereby increasing its lifetime.

Our recent efforts have been directed to study and implement the broadening of the current channels in the avalanche photoconductive switch in order to improve lifetime and increase switching peak power. This is being performed by two approaches. The first is control of laser diode beam delivery to the switch, and improvements of switch contacts, pulser configuration and charging mechanism. The second approach is the study and use of amorphic diamond coatings at the switch electrode to enhance operation and lifetime of a PCSS in stacked Blumlein pulse generators.

Table 1. Best simultaneous and individual test results obtained by commutation of 2-line stacked Blumlein pulsers with the high gain GaAs switches.

Parameters	Simultaneous Results		Individual Results
Switch Voltage	60 kV	30 kV	100 kV
Switch Current	1.2 kA	0.6 kA	2 kA
Pulse Risetime	200 ps	200 ps	150 ps
R-M-S Jitter	500 ps	500 ps	200 ps
Optical Trigger Energy	300 nJ	300 nJ	300 nJ
Repetition Rate	10 Hz	20 Hz	200 Hz
Electric Field	60 kV/cm	60 kV/cm	100 kV/cm
Stack Voltage	112 kV	57 kV	175 kV
Switch Lifetime	1×10^4 pulses	1×10^5 pulses	1×10^6 pulses

3. SWITCH TREATMENTS WITH AMORPHIC DIAMOND COATINGS

Basic research at UTD has described[10-17] a conformal coating that has hardness of natural diamond and exceptionally high values of electron emissivity. This material has been termed amorphous ceramic diamond and later shortened to "amorphic diamond" for convenience. Deposited at room temperatures it forms a strong bond to any material onto which it is applied. Such a favorable combination of hardness, chemical bonding and elasticity should translate directly into an increased resistance to abrasive wear of components coated with amorphic diamond. It has been demonstrated that only a 1-3 μm coating of amorphic diamond could protect fragile substrates against erosive environments.[13,16,17]

Analytical techniques have shown amorphic diamond to consist of nodules of tens of nanometers in diameters that are composed of sp3 (diamond) bonded carbon in a matrix of other carbons. The diamond character has been attested by the agreement of structural morphology, density, optical properties, K_α line energies and hardness. Nuclear reaction analysis (NRA) has proven the hydrogen content to be less than 0.5%. The nodules seem to be disordered mixtures of the cubic and the rare hexagonal polytypes of diamond that have no extensive crystalline planes along which to fracture. Since it is condensed from laser plasmas produced under conditions that are also optimal for the growth of interfacial layers, the films of amorphic diamond are strongly bonded to the substrates onto which they are condensed.[17,18]

In the work reported here, nominal amorphic diamond coatings were deposited on one side of semi-insulating LEC grown GaAs substrates with resistivity of about 100 MΩ cm, the type used in our Blumlein pulsers as the photoconductive switch material. Electrode copper foils were attached to either side using conducting silver paint and epoxy as shown schematically in Fig. 2. Electrodes were utilized in opposite configurations and conduction was through the bulk of GaAs with the switch gap settings of 5, and/or 10 mm. It should be noted that the switch configurations with gap settings of 5 and 10 mm are widely used in pulsed power devices utilizing PCSS.[19]

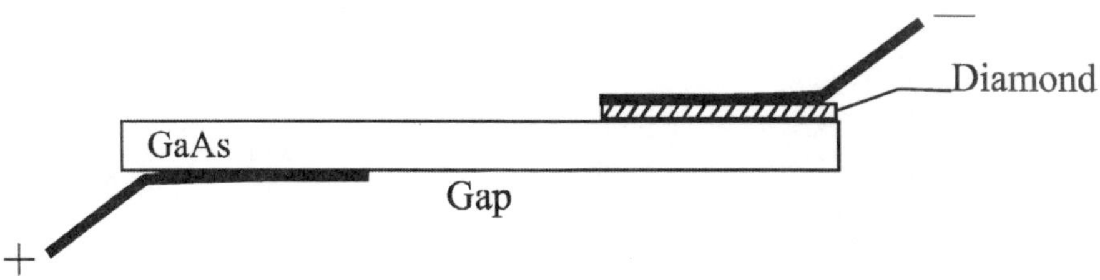

Figure. 2. Schematic presentation of the sample cross section used for the forward I-V measurements.

In all cases studied the electrode copper foils were also attached to similar uncoated GaAs substrates. Electrical properties of these uncoated samples were obtained under similar conditions of measurement to confirm the ohmic characteristics of electrode foil connections. A Keithley 237 high voltage source-measure unit was used to provide constant voltage bias in the range 0-1000 V while monitoring the current through the sample. To avoid heating the sample the measurements were carried out with pulsed voltage sweep having on and off times of 50 and 450 ms, respectively. For measurements of forward diode characteristics, diamond coating was biased negative with respect to the substrate as shown in Fig. 2.

Typical current-voltage characteristics measured for a 1.0 μm nominal amorphic diamond film on the semi-insulating GaAs are shown in Fig. 3 (a). An electrode gap setting of 5 mm was chosen for this measurement. For comparison, the I-V plot for an uncoated similar GaAs substrate is included in this figure. The rectifying behavior for the coated sample seen in Fig. 3 (a) was attributed to the amorphic diamond /GaAs heterojunction because the I-V plot for the uncoated sample showed symmetrical and ohmic character with change in the voltage polarity. The I-V characteristics differ considerably for the diamond-sample, especially for the forward current region where the coating side was biased low. The rapid current increase in forward direction for the coated sample has been attributed to tunneling of electrons from amorphic diamond to the conduction band of GaAs, a process similar to the Fowler-Nordheim tunneling.[19]

In this work, current-voltage measurements were also performed at higher bias voltages with switch samples coated with 1.0 and 2.0 μm amorphic diamond coatings. Results are shown in Fig. 3 (b) for the GaAs switches with a gap setting of 10 mm in the opposed configuration. In appears that the sample with the thicker coating produced enhanced tunneling of electrons from diamond to GaAs conduction band while the thinner coating produced a better rectification under reverse bias voltages. Of importance to this study, is the rapid current increase in forward direction as seen in Fig. 3. The tunneling of electrons from amorphic diamond to the conduction band of the GaAs provides pre-avalanche sites for the operation of diamond coated PCSS and thereby diffuse the conduction current.

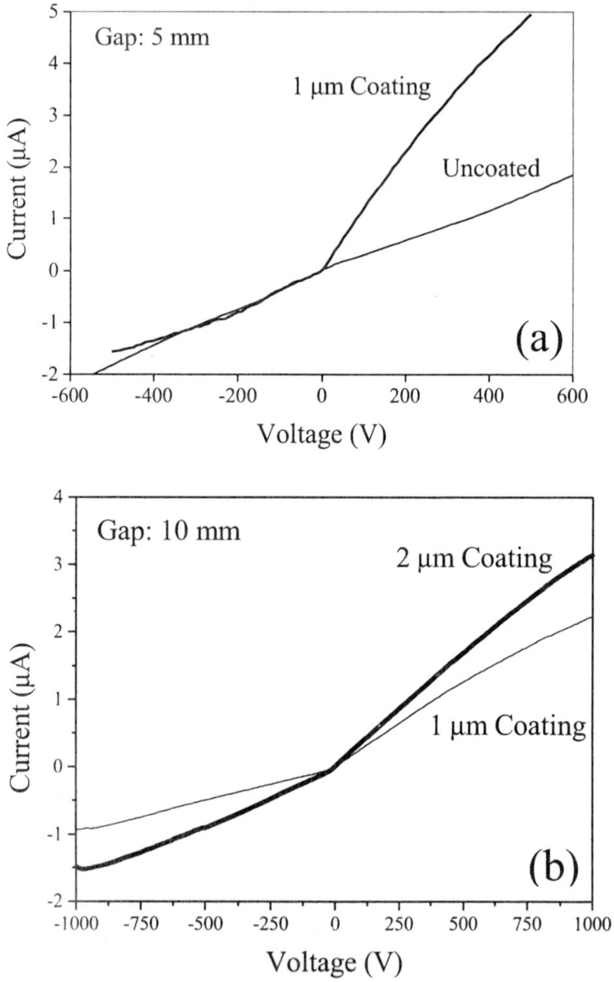

Figure 3. Current-Voltage characteristics measured in the dark. (a) Measurements are for an uncoated and a sample of diamond coated GaAs switch substrates at a gap setting of 5 mm. (b) Measurements are for two samples of diamond coated GaAs switch substrates at a gap setting of 10 mm.

4. SWITCH LIFETIME

In our earlier work, improvements in the PCSS switch operation and lifetime have been examined in a lateral configuration by coating the triggered face of GaAs switch cathodes with strips of highly adhesive films of amorphic diamond. With the application of amorphic diamond, not only the switch lifetime was increased but also the damage at the cathode contact was found to be less than that found for the anode contact.[19,20] This indicated that diamond coating protected and hardened the cathode side.

The switch/electrode configuration used in the PCSS lifetime studies is a part of a low profile switch assembly[6,19] that facilitates the use of a single photoconductive switch in the pulser. The electrode assembly allows for operation of PCSS in either lateral or opposed configurations. Layers of Kapton insulator are placed between the switch and the base electrode to restrict the current path through the top electrode holders in the lateral configuration. In the opposed configuration, switch is placed to contact one of the base electrodes on one side and the other top electrode holder on the other side. Layers of Kapton are used to isolate the switch from the base electrode on one side and the top electrode holder on the other side. The current path is through one of the top electrode holders into the GaAs bulk and through the other base electrode. In Experiments with both switch/electrode configurations, top copper electrode holders are connected to the base electrodes by means of several screws.[6,19]

In the work reported here, the pulse forming lines from a prototype stacked Blumlein pulser were connected to the bottom of base copper electrodes, which were cast in a G-10 plastic plate. Each switch was fabricated from one half of a semi-insulating LEC grown GaAs wafer with a diameter and thickness of 5 cm and 0.5 mm, respectively. It was held in place by means of two copper holders screwed to the electrodes. Commutation of the switch was triggered at 905 nm by focusing the LD-220 laser diode array beam in two straight lines across the switch gap from cathode to anode.

It is anticipated that, by depositing films of amorphic diamond near the PCSS cathode electrode contact, the number of carriers and avalanche sites increase thereby improving the switch lifetime and performance. As described earlier, there is a rapid current increase in forward direction of an amorphic diamond/ GaAs heterojunction where the diamond coating is biased negative with respect to the GaAs substrate. In this work, we studied the lifetime of a GaAs switch with a 1-cm strip of 0.5 μm amorphic diamond coating deposited on the switch at the cathode location. The switch was installed in an opposed configuration as shown in Fig. 2 and a switch gap of 5 mm was chosen for this study. The switch was tested at 18 MW (0.6 kA current and 30 kV voltage) for 10^5 shots after which it failed. Under similar conditions of experiments and switch configuration, we performed the switch longevity experiment with a bare GaAs switch. The uncoated switch lifetime was found to be 1×10^4 commutation cycles.

The schematic drawing of the switch damage is presented in Fig. 4. The switch gap is shown by the two horizontal lines of damage located at the front and the back of the switch corresponding to the limit of the copper plate electrodes. For comparison the negative photographs of the switch electrode areas after 1×10^4 shots are also shown in Fig. 4. It appears that the damage at the electrode contact areas eventually caused thermal runaway at the switch gap surface and produced a crack that shorted the gap.

With the application of amorphic diamond, not only the switch lifetime was increased by an order of magnitude but also the damage at the cathode contact was found to be less than that found for the anode contact as seen in Fig. 4. This indicated that diamond coating protected and hardened the cathode side. In addition, in experiments where the switch had no diamond coating, the majority of the time, a single current filament commutated the switches. The filament was initiated near the cathode and followed, approximately, one of the laser beams to the anode. Multiple branching was rarely seen.[5] In the case of switch with the diamond coating, the multiple branching was observed more often indicating an increase of pre-avalanche sites.[19,20] This produced conditions that even the anode area of the diamond treated switch damaged less than untreated switch as shown in Fig. 4. The amorphic diamond / PCSS junction properties discussed earlier in this paper may have aided longevity of the PCSS in these studies.

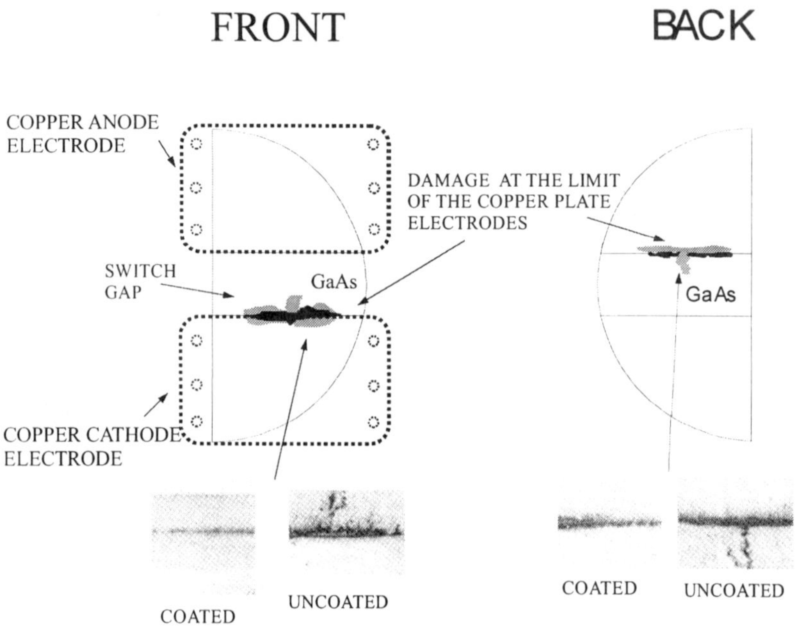

Figure 4. Schematic drawing of the appearance of damage on both sides of switch used in these studies. For comparison the negative photographs of the switch electrode areas for diamond-coated and uncoated switches are shown after 1×10^4 commutation cycles.

5. SUMMARY AND CONCLUSION

The semiconductor properties of amorphic diamond[19,21] can be employed to improve the PCSS longevity by coating the switch cathode or anode areas or both. If the switch cathode is coated, the tunneling of electrons from amorphic diamond to GaAs during the off-state stage of PCSS operation provides pre-avalanche sites that may diffuse conduction current upon switch activation. On the other hand, diamond coating of the switch anode area may result in increased hold-off characteristics and longer switch lifetimes by blocking the leakage current. In this case the rectifying behavior of the amorphic diamond/ GaAs heterojunction operating under reverse bias condition inhibits the conduction until very high fields are reached. However, the critical issue to resolve is the switch design options that make optimal use of amorphic diamond coatings for long life operations of avalanche PCSS in stacked Blumlein pulsers. Design options include: switch configuration, switch gap setting, diamond film thickness, exact locations of diamond coatings and film qualities.

In this work, we studied the use of amorphic diamond film as photoconductive switch coating to improve the longevity and operation of PCSS devices in stacked Blumlein pulsers. Switch longevity was examined by fabrication and testing the PCSS treated with amorphic diamond under different switch design options such as switch configuration, switch gap setting, and diamond film thickness. A switch lifetime of 1 x 10^5 cycles of commutation was realized by testing the diamond-coated switch performance in a prototype pulser. It was shown that not only the diamond coating protected and hardened the cathode side but it also helped branching the filaments conducting high currents. Further experiments are underway to perform switch lifetime test with additional switch design options, including diamond coating of both anode and cathode areas. In addition we plan to adjust the conductivity and qualities of amorphic diamond coatings applied to PCSS electrodes in order to further improve the switch performance.

In addition to UTD effort using amorphic diamond, there are intensive research programs in direction of improving PCSS lifetime in pulsed power systems. The longevity of the PCSS in pulsed power applications has been improved by doping the semi-insulating GaAs underneath the metal contacts.[22] Other efforts have focused on an opposed-contact PCSS with an n^+ region next to the cathode electrode. Increased hold-off characteristic of this class of devices has been analyzed by simulation and it is shown that the introduction of the n^+ region inhibits the flow of electrons at the interface until very high fields are reached.[23] Photoconductive switch enhancements realized in this work are well complementary and strengthen these efforts by utilizing unique semiconductor properties of amorphic diamond.

6. ACKNOWLEDGMENTS

This work was supported by the Air Force Office of Scientific Research (AFOSR) under Grant F49620-00-1-0296. The authors wish to acknowledge the valuable contributions of T.S. Bowen, J.J. Coogan, J.D. Bhawalkar, D.L. Borovina, E.M. Juengerman, D.R. Jander, T.J. Lee, H. Park, J.H. You, W.R. Osborn, R.K. Krause and M.C. Iosif in the development and characterization of stacked Blumlein pulsers and amorphic diamond coatings.

7. REFERENCES

1. F. Davanloo, J. D. Bhawalkar, C. B. Collins, F. J. Agee, and L. E. Kingsley, High power, repetitive stacked Blumlein pulse generators commuted by a single switching element, in: *Conference Record of the 1992 Twentieth Power Modulator Symposium* (IEEE, Piscataway, 1992), pp. 364–367.
2. J. D. Bhawalkar, F. Davanloo, C. B. Collins, F. J. Agee, and L. E. Kingsley, High power repetitive stacked Blumlein pulse generators producing waveforms with pulse durations exceeding 500 nsec, in: *Proceedings of the 9th International Pulsed Power Conference* (IEEE, Piscataway, 1993), pp. 857–860.
3. J. D. Bhawalkar, D. L. Borovina, F. Davanloo, C. B. Collins, F. J. Agee, and L. E. Kingsley, High power repetitive stacked Blumlein pulse generators, in: *Proceedings of the International Conference on Lasers '93*, edited by V. J. Corcoran and T. A. Goldman (STS Press, Mclean, 1994) pp. 712–717.

4. F. Davanloo, D. L. Borovina, J. D. Bhawalkar, C. B. Collins, F. J. Agee, and L. E. Kingsley, High power repetitive waveforms generated by compact stacked Blumlein pulsers, in: *Conference Record of the 1994 Twenty-First Power Modulator Symposium* (IEEE, Piscataway, 1994), pp. 201–205.
5 F. Davanloo, C. B. Collins, and F. J. Agee, High power, repetitive-stacked Blumlein pulsers commutated by a single switching element, *IEEE Trans. Plasma Sciences*, **26(5)**, 1463–1475 (1998).
6. F. Davanloo, D. L. Borovina, J. L. Korioth, R. K. Krause, C. B. Collins, F. J. Agee, J. P. Hull, J. S. H. Schoenberg and L. E. Kingsley, High power, sub-nanosecond rising waveforms created by the stacked Blumlein pulsers, in: *Ultra-Wide-Band, Short-Pulse Electromagnetics 3*, edited by C. E. Baum, L. Carin, and A. P. Stone (Plenum Press, New York, 1997), pp. 31–38.
7. F. Davanloo, R. Dussart, M. C. Iosif, C. B. Collins and F. J. Agee, Photoconductive switch enhancements and lifetime studies for use in stacked Blumlein pulsers, in: *Proceedings of the 12th International Pulsed Power Conference*, edited by C. Stallings and H. Kirbie (IEEE, Piscataway, 1999), pp. 320–323.
8. F. J. Agee, Current issues in high power microwaves, in: *Conference Record of the 1996 Twenty-Second Power Modulator Symposium* (IEEE, Piscataway, 1996), pp. 1–4.
9. F. J. Zutavern, G. M. Loubriel, W. D. Helgeson, M. W. O'Malley, R. R. Gallegos, A. G. Baca, T. A. Plut, and H. P. Hjalmarson, Fiber-optic control of current filaments in high gain photoconductive semiconductor switches, in: *Conference Record of the 1994 Twenty-First Power Modulator Symposium* (IEEE, Piscataway, 1994), pp. 116–119.
10. C. B. Collins, F. Davanloo, D. R. Jander, T. J. Lee, H. Park, and J. H.You, Microstructure of amorphic diamond films, *J. Appl. Phys.* **69**(11), 7862–7870 (1991).
11. C. B. Collins, F. Davanloo, T. J. Lee, J. H. You, and H. Park, Amorphic diamond films produced by laser ablation, *Mat. Res. Soc. Symp. Proc.* **285**, 547 (1993).
12. C. B. Collins, F. Davanloo, T. J. Lee, H. Park, and J. H.You, Noncrystalline films with the chemistry, bonding, and properties of diamond, *J. Vac. Sci. & Technol.* **B 11**(5), 1936–1941 (1993).
13. F. Davanloo, T. J. Lee, D. R. Jander, H. Park, J. H. You, and C. B. Collins, Adhesion and mechanical properties of amorphic diamond films prepared by a laser plasma discharge source, *J. Appl. Phys.* **71**(3), 1446–1453 (1992).
14. C. B. Collins, F. Davanloo, T. J. Lee, D. R Jander, J. H. You, H. Park, and J. C. Pivin, The bonding of protective films of amorphic diamond to titanium, *J. Appl. Phys.* **71**(7), 3260–3265 (1992).
15. C. B. Collins, F. Davanloo, J. H. You, and H. Park, Nanophase diamond films produced by laser deposition, in: *Proc. SPIE: Laser Applications*, edited by Arthur A. Mak, **2097**, 129 (1994).
16. F. Davanloo, T. J. Lee, J. H. You, H. Park, and C. B. Collins, Progress in the characterization of nanophase diamond films uniquely produced by a laser plasma discharge system, *Surf. Coat. Technol.* ***62***, 564–569 (1993).
17. F. Davanloo, H. Park and C. B. Collins, Protective coatings of nanophase diamond deposited directly on stainless steel substrates, *J. Mat Res.* **11**(8), 2042–2050 (1996).
18. F. Davanloo, C. B. Collins and K.J. Koivusaari, in: *Adhesion Measurements of Films & Coatings, Vol. 2*, edited by K.L. Mittal (VSP, Netherlands, 2001), pp. 141–157.
19. F. Davanloo, R. Dussart, K.J. Koivusaari, C. B. Collins, and F. J. Agee, Photoconductive switch enhancements and lifetime studies for use in stacked Blumlein, *IEEE Trans. Plasma Sciences*, **28**(5), 1500–1506 (2000).
20. F. Davanloo, H. Park, C. B. Collins and F. J. Agee, Photoconductive switch enhancements for use in Blumlein pulse generators, in: *AIP Conference Proceedings*, edited by J. L. Duggan and I. L. Morgan, **475** (AIP, New York, 1999), pp. 918–921.
21. F. Davanloo, M. C. Iosif, T. Camase, C. B. Collins and F. J. Agee, High power photoconductive semiconductor switches treated with amorphic diamond coatings, in: *AIP Conference Proceedings,* edited by J. L. Duggan and I. L. Morgan, **576** (AIP, New York, 2001), pp. 1047–1050.
22. A. Mar, G. M. Loubriel, F. J. Zutavern, M. W. O'Malley, W. D. Helgeson, D. J. Brown, H. P. Hjalmarson, A. G. Baca, R. L. Thornton, and R. D. Donaldson, Longevity improvement of optically activated, high gain GaAs photoconductive semiconductor switches, in: *Conference Record of the 2000 Twenty-Fourth International Modulator Symposium* (IEEE, Piscataway, 2000), pp. 69–72.
23. N. Islam, E. Schamiloglu, C. B. Fleddermann, J. S. H. Schoenberg, and R. P. Joshi, Improved hold-off characteristics of GaAs photoconductive switches used in high power applications, in: *Proceedings of the 12th International Pulsed Power Conference*, edited by C. Stallings and H. Kirbie (IEEE, Piscataway, 1999), pp. 316–319.

THE EFFECTS OF PASSIVATION LAYERS ON THE RESPONSE OF PHOTOCONDUCTIVE SWITCHES FOR ULTRA-WIDEBAND HIGH-POWER MICROWAVE APPLICATIONS

N.E. Islam, G. Tzeremes, E. Schamiloglu, and J. Gaudet*

1. INTRODUCTION

High resistivity semiconductor materials have been used as a closing switch in a configuration for generating ultra-wideband high-power microwave radiation (Fig.1).[1] Compared to a conventional switch such as a spark-gap, the semiconductor switch has low jitter and a faster risetime and is comparatively compact in size. It also has low inductance and in the non-linear mode of operation it requires lower laser energy for switching.[2] Because of its superior electrical characteristics GaAs is preferred over Si in applications where fast, repetitive switching action is required. In most cases, the material of choice is the semi-insulating GaAs with resistivity exceeding 10^7 Ω-cm, usually grown through the compensation process.[3] The compensated materials have been known to have characteristics similar to doped semiconductors where the drift-diffusion process is the dominant mode of carrier transport.[4,5] The response of the device, however, is also affected by the compensation technique.[6]

Premature breakdown or 'surface flashover' is one of the mechanisms that limit the switch applications to voltages below the material breakdown voltage.[7] This mechanism has been studied and a new design improvement has been suggested.[8] Since the switches operate at fields exceeding 80 kV/cm, the fabrication process itself is an important area that could also influence the performance of the switches at high bias. Minor fabrication flaws can alter the delicate balance in the compensated material that can lead to failure.

* N.E. Islam, G. Tzeremes, and E. Schamiloglu, Department of Electrical and Computer Engineering, University of New Mexico, Albuquerque, NM 87131. John Gaudet, Air Force Research Laboratory, Directed Energy Directorate, Kirtland AFB, NM 87117-5776.

Hence, it is important to understand the fabrication processes which govern the behavior of the PCSS during high-power operations.

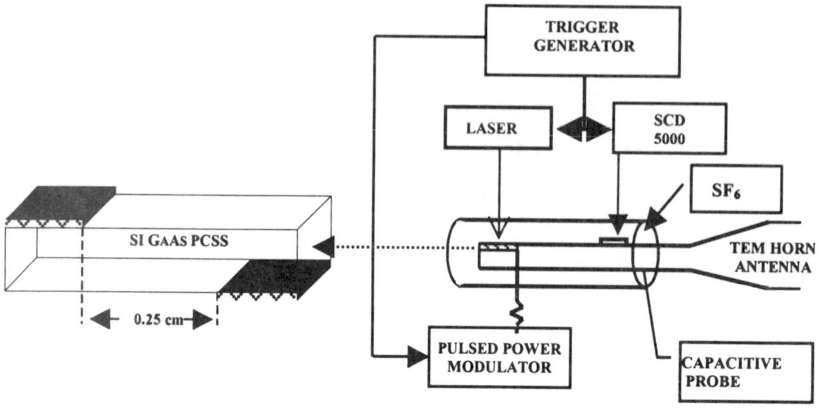

Figure 1. Setup for ultra wideband high power microwave generation showing the opposed contact SI-GaAs switch.

In this paper we highlight some of the issues related to the fabrication process that might affect the device performance at very high bias. We also present simulation results using state-of-the-art TCAD tool for semiconductor device and material analysis. It is expected that a better understanding of the fabrication process would help in identifying critical parameters in the process flow and result in the fabrication of more reliable switches in the future. Even though we have analyzed switches used in the experiment described in Ref. 1, we expect that the results and conclusion would be applicable to any other application where a high field is one of the operational requirements.

2. FABRICATION AND TESTING

The devices were fabricated at the Center for High Technology (CHTM) facility at the University of New Mexico. Two types of photoconductive switches called the baseline and the prototype were fabricated for analysis. The baseline switches are the ones currently used in experiments at AFRL (Figure 1) while the prototype is a new design for better hold-off characteristics that has been proposed through simulation studies.[9] Liquid encapsulated Czochralski (LEC) grown, semi-insulating GaAs material was used to fabricate the 2.0 cm by 1.3 cm by 0.063-cm thick PCSS. The surface area of the device not covered by a contact was protected by a 1000 to 1500A^0 layer of silicon nitride (Si_3N_4). The nitride acts as both a passivation and encapsulating layer. Passivation prevents surface charge formation and the encapsulating layers stops any possible outgassing of arsenic during processing. Encapsulation also prevents the possible removal of EL2 centers near the surface during high-temperature hydrogen anneals.

The prototype model has an n^+ doped layer of 15 to 20 μm below the surface, underneath the cathode. Since diffusion and ion implantation of a 15 to 20 μm n^+ region was not feasible, it was grown through MBE as a doped layer above the sample surface, followed by patterning and etching to remove the unwanted n^+ material. The I-V characteristics of both the prototype and the baseline PCSS were tested at AFRL. Low-voltage tests showed that the prototype device has improved hold-off characteristics over the baseline switch (Figure 2), as predicted by simulations. High-voltage testing (> 1 kV) showed that the baseline device was also able to hold off a voltage of 16 kV.

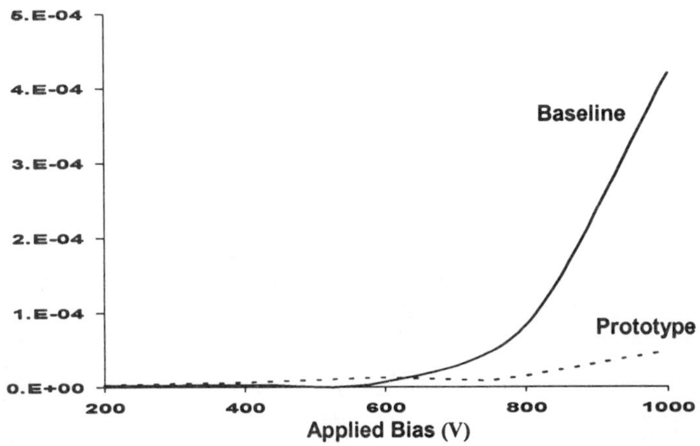

Figure 2. The I-V characteristics of the baseline and the prototype PCSS showing improved performance of the prototype device.

This switch achieved lock-on mode and had a lifetime of 500 shots. The prototype device which had showed better characteristics than the baseline switches, however, was only able to hold off a voltage of 7.5 kV. This was not a high enough voltage to achieve lock-on mode. Another batch of baseline and prototype devices made from vertical gradient freeze (VGF) materials did show similar results. It was postulated earlier that the emission of electrons into the bulk and its subsequent capture results in inhomogeneous material characteristics that interferes with the filamentation process. Breakdown occurs as a result.[7] Mechanism that inhibit the injection of electrons into the bulk would thus contribute to better lifetime for the switch, The n^+-bulk interface achieves this through the creation of a depletion region where the E-field is directed against the bulk electric field. Thus the image force lowering effect of the potential energy for charge carrier emission at the cathode is reduced and carrier injection takes effect at higher bias. Measurements at low voltages support this postulation. Failure of the devices at very high bias could be attributed to the fabrication process itself, as explained below.

3. FABRICATION PROCESS AND ISSUES

Study has shown that active defects introduced during e-beam evaporation are caused by stray electrons originating at the filament during metal deposition. The concentration of these defects increases with increasing exposure to stray electrons and the degrading effect increases with the free carrier density of the epilayer. The formation of electrically active defects is linked to the dopant atoms in the GaAs crystal.[10] The establishment of charge imbalances at the SI-n^+ junction due to the defect formation thus counters the local electric field in this region. It effectively causes the depletion layer at the surface of the lightly doped side to decrease while the depletion layer on the heavily doped side expands. If the diffused side of the junction is heavily doped, the depletion layer shrinkage on the slightly doped side will have the dominant influence. Since the bias across the junction is being supported across a narrower depletion layer at the surface than in the bulk, the electric fields at the surface can be expected to be higher than the field in the bulk material. Electron injection, formation of trap filled region and subsequent breakdown occurs as a result.

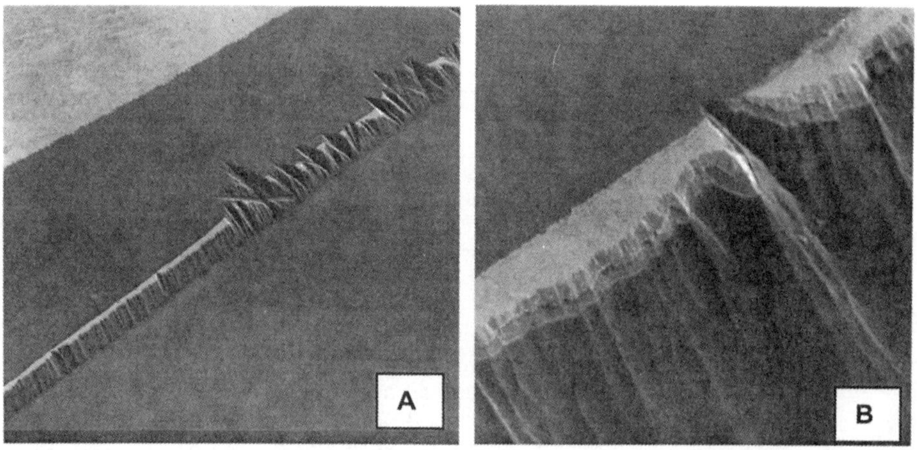

Figure 3. Effects of wet etch [A] and metal deposition [B] on the n^+ region of the photoconductive switch.

In addition, any change in the uniformity of the step, such as undercutting (Fig. 3A) or strip of metal deposition (Fig. 3B), will have an adverse effect on the performance of the device, specifically at very high voltages. Studies have also shown that the different etch rate between Ga (1 1 1) and the other planes with an H_3PO_4 solution also leads to a severely undulated GaAs surface.[11] The H_3PO_4 treated GaAs surface show elongated hollows and hills in one direction. The hill region is mainly composed of Ga (1 1 1) facets arrayed in the [-1 1 0] direction and contained a little more Ga than the valley. Such surface irregularities and deep furrows could result in electric field crowding, and surface breakdown will likely precede bulk breakdown in the n^+ step area. This rougher surface could also adversely affect the contact quality at the anode where etching was done. Studies have also shown that one-sided, abrupt (1 1 1) GaAs junctions have lower breakdown voltages than (1 0 0) GaAs junctions. Since the region underneath the contact at the anode is highly doped with Ge, the (1 1 1) facets resulting from the etch may have an additional detrimental effect on contact quality.[12]

4. PROCESS AND DEVICE SIMULATION RESULTS

Insulating thin films, like the nitrides in our case, have an important role in semiconductor device fabrication and operations. In an ideal situation, device characteristics are based on very little or no conductance in these layers. However at high electric fields and temperatures, carrier conduction can occur which may lead to device performance degradation, instability, hysteresis effects, and breakdown. The transport process in these layers may occur through mobile ions (for example, sodium) or through the carrier tunneling processes at very high fields. The conduction is generally high in Si_3N_4 as compared to SiO_2. Thus, conductance in insulating films in photoconductive switch is an undesirable phenomenon. The presence of traces of nitrides under the contact may have a negative effect on the switch performance as shown by our simulations below.

For simulation, the PCSS was fabrication through the process simulator ATHENA of the SILVACO simulation suite. The device was simulated through the BLAZE module, which caters to III-V materials. Both the modules can be activated through the ATLAS interface. In order to reduce simulation time and to avoid the complexity of an opposed-contact deposition scheme, a scaled down, same side contact PCSS was fabricated for analysis. Also, the material that is in intimate contact with the device (Pd) was used as the contact for simulations, even though the actual PCSS has a refractory material contact having four layers of materials deposited on top of each other in semi-circular form (Rogowaski profile). We did not anticipate any drastic change in the results as a consequence of choosing this option, since the conduction properties are essentially dominated by characteristics of the material that is in intimate contact with the semiconductor.

The simulation processing steps were simple. A GaAs substrate was doped with Carbon and a 0.1-μm thick nitride layer was deposited. The nitride was then etched so that the contact could be deposited at one region while the other contact had a thin layer of nitride under it. For material properties manufacturer-supplied data and published reports were the sources for the input. A simple 7.5 Ω resistor (for the 1/10 scaled down device size) was attached in series with the photoconductive switch. The SI-GaAs material was assumed to have an n-type donor (EL2) density of $3.1\text{x}10^{15}$ cm^3, compensated by a shallow acceptor concentration of $3\text{x}10^{16}/cm^3$. The trap level was taken to be 0.6 eV, with electron and hole capture cross sections of $1\text{x}10^{15}$ cm^2 and $1\text{x}10^{17}$ cm^2, respectively.

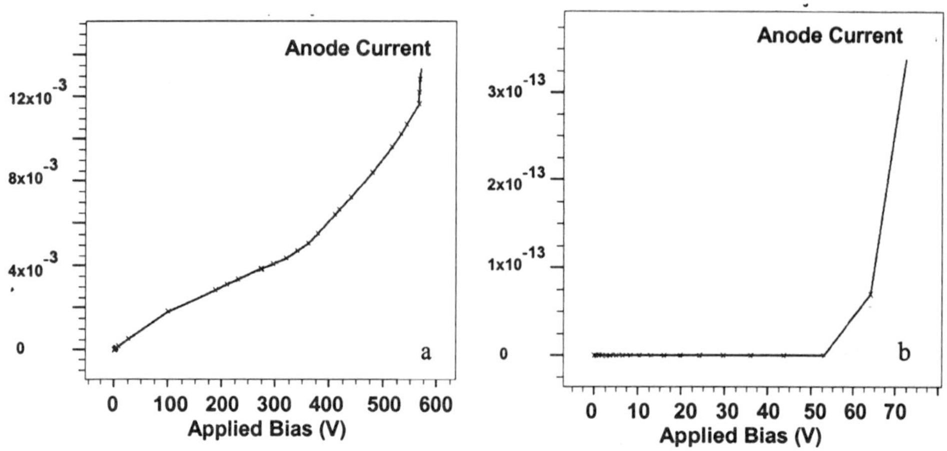

Figure 4. Simulated I-V characteristics of the switch: (a) without the nitride layer; (b) with 0.01 μm nitride.

The I-V characteristics of the devices with and without the nitride layers under the contact were plotted, and the electron density level was monitored for charge crowding effects. Figures 4a and 4b show the I-V characteristics of the device without the nitride layer below the contact and with 0.01 μm of nitride layer below the cathode, respectively.

As seen in Figure 5, the PCSS with the nitride layer breaks down at a very low voltage as compared to the one without the layer. The breakdown may be attributed to the crowding of carriers (electrons) near the contact and the formation of trap-filled regions near it. Trap-filled regions change the basic characteristics of the PCSS material and interfere with its conduction process. For the PCSS with nitride between the semiconductor and the contact, charge crowding occurs at much lower bias than with the switch with no residual nitride under the contact. Figures 5a and 5b show the electron profile just prior to the breakdown voltages for the switches with nitride and without the nitride layer, respectively.

5. DISCUSSION AND CONCLUSION

Fabrication and process simulation analysis of photoconductive semiconductor switches used in ultra-wideband high-power microwave applications suggests that device response, specifically at high bias, depends to a large extent on the fabrication process itself. The wet etching process, the silicon nitride passivation and encapsulation layer deposition scheme, the contact deposition scheme, and the stray electrons injected during contact deposition are some of the processing steps that affect the device surface and regions near the contacts. As a result, the material composition and texture and its properties are affected. At very high bias these defects contribute to the device failures.

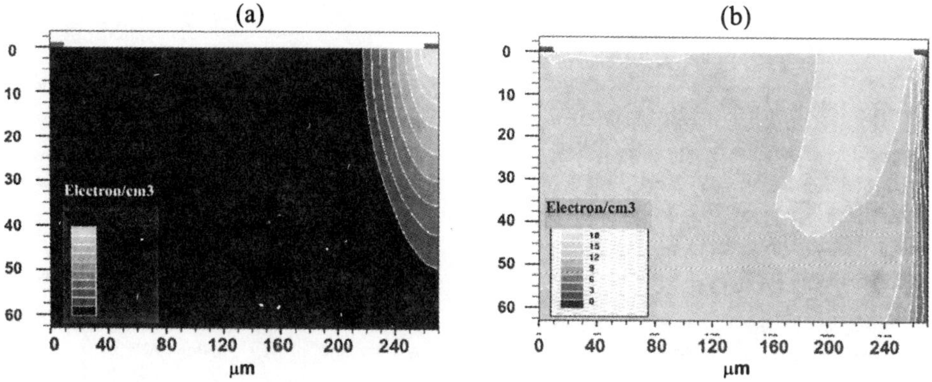

Figure 5. Current (electron) crowding at the electrode prior to breakdown for PCSS: (a) with the nitride layer; and (b) without the nitride layer.

Switches made with a highly doped n^+ layer next to the cathode contact show better breakdown characteristics than without the layer as postulated through simulations. The fabrication of these switches, however, introduces new parameters that make it more vulnerable to the high field effects than the switches without this layer. Finally, photoconductive semiconductor switches for high power microwave applications require stringent process controls than those required for low power.

ACKNOWLEDGMENTS

We would like to acknowledge the help extended by the researchers of the Air Force Research Laboratory (AFRL) at Kirtland Air Force Base NM for testing the switches and for their useful comments. We also acknowledge the support by the Air Force Office of Scientific Research (AFOSR) for this work.

REFERENCES

1. J.S.H. Schoenberg, J.W. Berger, J.S. Tyo, M.D. Abdalla, M.C. Skipper, and W.R. Buchwald, Ultra-wideband source using gallium arsenide photoconductive semiconductor switches, *IEEE Trans. Plasma Sci.* **25**, 327–339 (1997).
2. Buchwald, A. Balekdjian, J. Conrad, J.W. Burger, J.S.H. Schoenberg, J.S. Tyo, M.D. Abdalla, S.M. Aahern, and M.C. Skipper, Fabrication and design issues of bulk photoconductive switches used for ultra-wideband microwave generation, *11th IEEE International Pulsed Power Conference* **2**, Baltimore MD, 970–974 (1997).
3. E. Schamiloglu, N.E. Islam, C.B. Fledderman, B. Shipley, R.P. Joshi, and Z. Zheng, Simulation, modeling, and experimental studies of high-gain gallium arsenide photoconductive switches for ultra-wideband applications, in: *Ultra-Wideband, Short-Pulse Electromagnetics 4*, edited by E. Heyman, B. Mandelbaum, and J. Shiloh (Kluwer Academic/Plenum Publishers, New York, 1999), pp. 221–228.
4. B.T. Cavicchi and N.M. Haegel, Experimental evidence for relaxation phenomena in high purity silicon, *Phys. Rev. Lett.* **63**, 195–197 (1989).
5. N.M. Hagel, Relaxation semiconductors: in theory and in practice, *Appl. Phys. A* **53**, 1–7 (1991).
6. R.B. Darling, Electrostatic and current transport properties of n+/semi-insulating GaAs junctions *J. Appl. Phys.* **74**(7), 4571–4589 (1993).

7. N.E. Islam, E. Schamiloglu, and C.B. Fledderman, Characterization of a semi-insulating GaAs photoconductive semiconductor switch for ultra wide band high power microwave applications, *Appl. Phys. Lett.* **73**, 1988–1991 (October 1998).
8. N.E. Islam, E. Schamiloglu, C.B. Fledderman, J.S.H. Schoenberg, and R.P. Joshi, Analysis of high voltage operation of gallium arsenide photoconductive switches used in high power applications, *J. Appl. Phys.* **86**, 1754–1758 (August 1999).
9. N.E. Islam, E. Schamiloglu, J. Schoenberg, and R.P. Joshi, Compensation mechanism and the response of high resistivity GaAs photoconductive switches during high-power applications, *IEEE Trans. Plasma Sci.* **28**(5), 1512–1519 (2000).
10. G. Myburg and F.D. Auret, *J. Appl. Phys* **71**(12), 6127 (1992).
11. M.G. Kang, S.H. Sa, H.H. Park, K.S. Su, and K.H. Oh, Influence of the electron beam evaporation rate of PT and the semiconductor carrier density on the characteristics of PT/Normal GaAs Schottky contacts, *Thin Solid Films* **634**, 308–309 (1997).
12. B. Shipley, *The Development and Characterization of a Unique High Voltage GaAs Photoconductive Switch*, MS Thesis, Department of Electrical Engineering, University of New Mexico (2000).

NANOSECOND PULSED BREAKDOWN FOR POINT-PLANE GEOMETRIES AT MODERATE VOLTAGE

H. Krompholz, L. Hatfield, M. Haustein, J. Spears, M. Kristiansen*

1. INTRODUCTION

Gaseous breakdown in the sub-nanosecond regime is of interest for fast pulsed-power switching, short-pulse electromagnetics, ultra-wideband radar, and for plasma limiters[1] to protect electronic devices from high-power microwaves. Previous work on sub-nanosecond breakdown [2,3,4,5] concentrated on high-pressure gases or liquids, with applied voltages in excess of 100 kV for quasi-homogeneous electric fields. For practical applications and high-pressure gases as switching media, the required high voltages, along with the high-pressure setup, are technologically demanding. In this paper, we investigate breakdown in point-plane geometry, where the field enhancement is sufficient to provide breakdown for applied voltages of around 10 kV for gap distances between 0.5 to 4 mm. The setup uses transmission lines and fast sensors, which allow a simultaneous measurement of voltage and current across the gap and characterizations of current amplification time constants as a function of the applied instantaneous voltage.

Pulsed breakdown with voltage risetimes of about 1 ns is investigated for point-plane geometries at moderate (< 15 kV) breakdown voltages. The setup consists of a pulser (Russian SOS PNG-10 pulser) with a risetime of 1 ns, pulse width 1.5 ns, a transmission line connected to an axial gap with variable gas pressure, and a load line. This impedance-matched setup allows electrical measurements with resolutions of 300 ps. Voltage measurements outside the gap with capacitive voltage dividers yield direct information about voltage across and current through the gap, with the same temporal resolution. Breakdown current and voltage waveforms have been measured for a tip radius of < 0.5 μm, a gap width of 0.1 to 4 mm, as a function of pressure between 10^{-6} and 600 torr, in argon. The minimum pressure for breakdown is around 5 torr. Breakdown occurs always during the rising edge of the pulse, with delays of 1 ns at 5

* Center for Pulsed Power and Power Electronics (P^3E), Departments of Electrical & Computer Engineering and Physics, Texas Tech University, Lubbock, TX 79409-3102.

torr, and 0.6 ns at 600 torr, with a slight decrease of the breakdown delay with decreasing gap distance. Typical current amplification times $I/(dI/dt)$ are on the order of 10^{-10} s. Statistical time delays are not observed in these experiments. The observations point to explosive field emission as the discharge starting mechanism and gaseous carrier multiplication as the main amplification mechanism.

2. EXPERIMENTAL SETUP

As noted, a Russian PNG-10 solid-state pulser with a voltage amplitude of 7.5 kV has been used for all experiments. The pulser is connected via a high-voltage transmission line to a test gap in vacuum (Fig. 1). The other side of the test gap is terminated by another transmission line. A mesh outer conductor surrounds the test gap, and the overall system has an impedance of 50 Ω. One of the hemispherical electrode tips holds a tungsten needle in the center, and the needle position is controlled by a micrometer, with the whole electrode assembly outside the vacuum system. The gap distance is adjusted using a microscope, and the needle position is fixed by a lateral set screw and an axial spring. This method allows us to set the gap distance reproducibly to values between 0.1 and 4 mm. The test gap consists of the needle tip with a radius of curvature of 0.5 μm as the cathode and a brass electrode with a radius of curvature of 5 mm as the anode. The maximum electric field is estimated to about 500 MV/m at the needle tip. At pulse amplitudes of 7.5 kV, the needle tip is destroyed after one discharge, that is, the tip melts and resolidifies with a radius of curvature of several micrometers.

Diagnostics include several transmission line current sensors[6] and capacitive voltage dividers integrated into the transmission line on both the input and output sides. For a quantitative analysis, only one voltage divider on the input side, at a distance of 70 cm from the test gap, is used. Measured risetimes of the electric probes are less than 300 ps.

This voltage divider measures the input pulse and the pulse reflected at the test gap, which allows us to find the voltage $V_g(t)$ and the current $I_g(t)$ at the gap (Fig. 1). From the

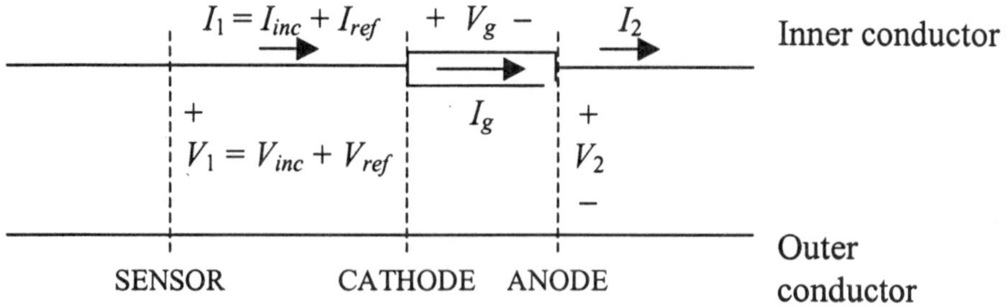

Figure 1. Voltages and currents across the gap and transmission line.

circuit in Fig. 1, it follows that

$$V_g(t) = 2V_{ref}(t+\Delta), \quad I_g(t) = \frac{V_{inc}(t-\Delta) - V_{ref}(t+\Delta)}{Z},$$

where V_{inc} and V_{ref} are the measured incident and reflected waveforms, Δ is the one-way transit time from sensor to gap, and Z is the line impedance of 50 Ω. The accuracy of these relationships is limited by minor reflections at the feedthrough and by digitizer noise. Furthermore, a comparison of the waveforms without breakdown at low pressure (< 0.5 torr) with the waveforms with breakdown allows a separation of the displacement current CdV_g/dt and the conduction current through the gap (C is the gap capacitance).

3. RESULTS

Figure 2 shows an example of a superposition of the measured voltage waveforms for different pressures. The highest reflected pulse represents a case without breakdown, and the difference between this pulse and the individual pressure-dependent pulses represents the conduction current as a function of time. Figure 3 gives an overview on breakdown

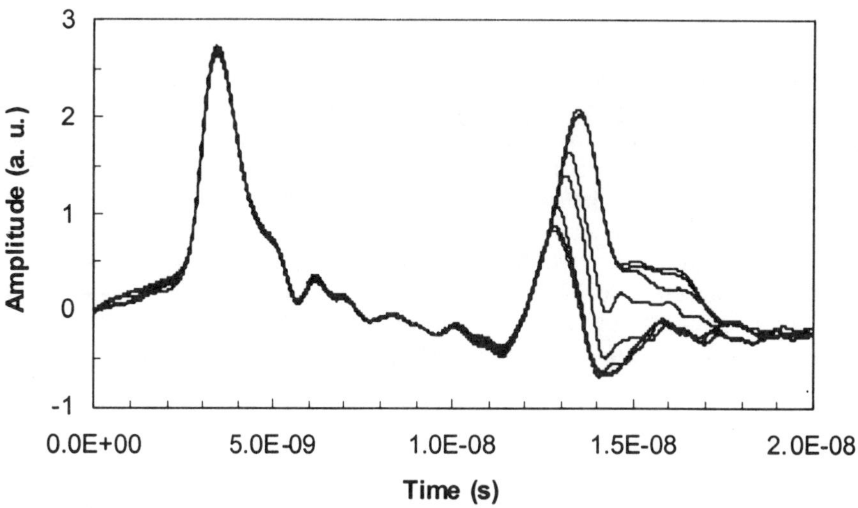

Figure 2. Measured voltage waveforms. The first pulse (left, with a peak at about 3 ns) represents the incident pulse (V_{inc}), the group of pulses on the right (between approximately 12 and 15 ns) shows the reflected pulses (V_{ref}) for different pressures in argon (highest pulse 4.6×10^{-6}, 0.5 and 5 torr, with descending pulse height 25, 50, and 150 torr, lowest 300 and 600 torr). The maximum relative voltage amplitude of 2.7 in this figure corresponds to a pulse height of 7.5 kV of the incident pulse.

voltages and delay times, defined as the time from the minimum of the overall waveform at about 12 ns to the separation of the low-pressure reflected pulse from the individual pulses.

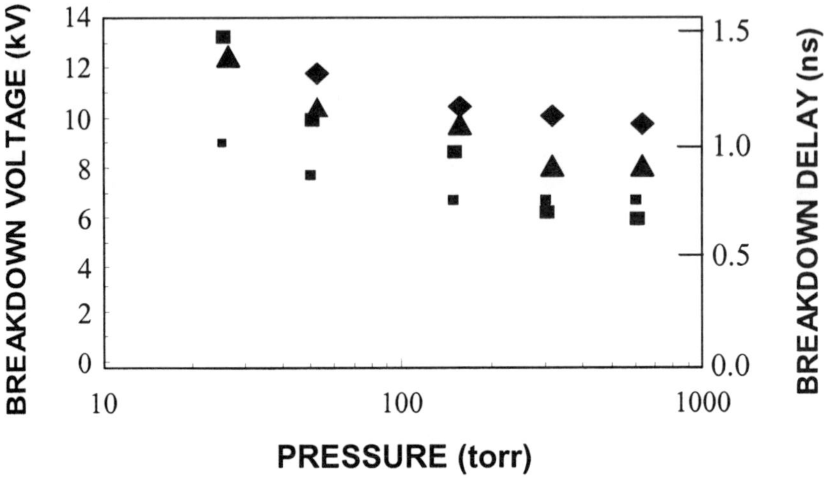

Figure 3. Breakdown voltages and delay times as functions of pressure in argon for different gap distances (small squares: 0.1 mm; large squares: 0.5 mm; triangles: 1 mm; diamonds: 4 mm).

Figure 3 represents overall delay times, from application of the voltage pulse to the start of a measurable current. Formative delay times are usually defined as current risetimes for cases in which the applied voltage remains constant, which is not the case for these fast pulsed discharges. Here, the current attains its maximum (for pressures larger than 150 torr, with the value of the maximum equal to the impedance-limited current) at a time when the applied gap voltage V_g is nearly zero (Fig. 4). Nevertheless, current risetimes (from the onset of a measurable current to its maximum value) are less than 500 ps for low pressures (25 and 50 torr), and reach about 1 ns for 600 torr. For this case, with a rapid change in voltage, a characterization of the discharge development in terms of the current amplification parameter β is more suitable. This parameter is defined as

$$\beta = \alpha v_D = \frac{1}{I_g(t)} \frac{dI_g(t)}{dt},$$

where α is the number of electron-ion pairs created per unit length, and v_D is the drift velocity. Due to a high level of digitizer noise, no systematic evaluation of this parameter has been done, but typical values on the order of $10^{10} s^{-1}$ are estimated from the current curves (cf. Fig. 4), which means e-folding time constants on the order of 100 ps for short gap distances (0.1 mm) and high pressure (600 torr).

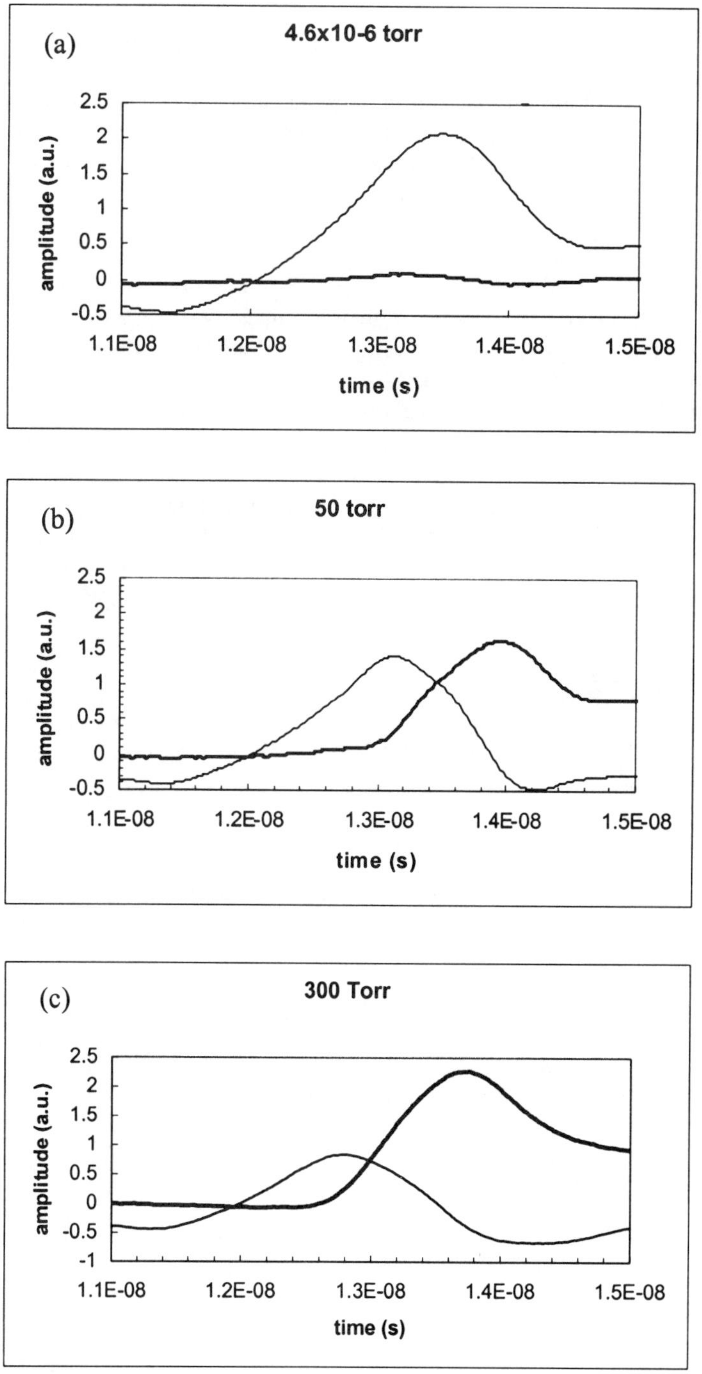

Figure 4. Examples for the voltage V_g across the gap (light line, one unit corresponds to 5.55 kV) and current I_g through the gap (heavy line, one unit corresponds to 55.5 A) from Fig. 2 for the pressures: (a) 4.6×10^{-6} torr (no breakdown); (b) 50 torr; and (c) 300 torr.

4. CONCLUSION

Breakdown times in the sub-nanosecond regime can be achieved with applied voltage amplitudes of less than 10 kV in point-plane geometries. With radii of curvature of the anode on the order of 0.5 μm, breakdown occurs during the rising part of the applied voltage pulse within a fraction of a nanosecond for pressures exceeding a threshold of several torr.

The breakdown mechanism is far from being clear. An extremely inhomogeneous field and a fast time variation of the field, probably with a nonstationary electron energy distribution, would make any modeling rather complex. We assume that field emission[7] starts the discharge, but gaseous amplification and associated feedback mechanisms play a major role for the development, as seen from the pressure dependence.

For practical applications as plasma limiters, the tip plane geometry with radii of curvature below 1 μm seems to be a promising scheme. It is expected that a low-impedance discharge can be ignited by incoming microwaves by the first cycle of GHz radiation (for example), which would prevent further microwave propagation in a waveguide. For lower power densities than the ones which correspond to the pulse amplitudes reported here, the scheme is expected to work without destruction of the needle tip in one discharge.[8]

5. ACKNOWLEDGEMENT

This work was supported by the U.S. Army Space and Missile Defense Command.

6. REFERENCES

1. A. Kikel, L. Altgilbers, I. Merritt, M. Brown, L. Ray, and T.X. Zhang, Plasma limiters, *Proc. 29th Plasmadynamics and Lasers Conference*, American Institute of Aeronautics and Astronautics, Albuquerque NM, June 15-18, 1998, paper AIAA-98-2564.
2. T.H. Martin, An empirical formula for gas switch breakdown delay, *Proc. 7th IEEE International Pulsed Power Conference*, Monterey CA, pp. 73–79 (1989).
3. P. Felsenthal and J.M. Proud, Nanosecond pulse breakdown in gases, *Phys. Rev. A* **139**, 1796–1804 (1965).
4. E. Kuhnhardt, Nanosecond pulsed breakdown of gases, in: *Electrical Breakdown and Discharges in Gases* (Plenum Press, New York, NATO ASI Series, Series B: Physics, Vol. 89a, 1983), pp. 241–263.
5. J. Mankowski, J. Dickens, and M. Kristiansen, High voltage subnanosecond breakdown, *IEEE Trans. Plasma Sci.* **26**, 874–881 (1998).
6. H. Krompholz, K. Schoenbach, and G. Schaefer, Transmission line current sensor, *Proceedings of the IEEE Instrumentation and Measurement Technology Conference*, Tampa FL, pp. 224–227 (1985).
7. G.A. Mesyats, *Explosive Electron Emission* (URO Press, Ekaterinburg, 1998), p. 24.
8. H. Krompholz, L.L. Hatfield, M. Kristiansen, D. Hemmert, B. Short, J. Mankowski, M. Brown, and L. Altgilbers, Gas breakdown in the sub-nanosecond regime with voltages below 15 kV, *Proc. 28th IEEE Pulsed Power Conference*, Las Vegas NV, pp. 487–490 (June 2001).

DESIGN AND TESTING OF A COMPACT SUB-GW, SUBNANOSECOND PULSER

Yefim Yankelevich, Alex Pokryvailo, and Moshe Shapira[*]

1. INTRODUCTION

Subnanosecond high-power pulsers find applications in diversified fields, such as generation of UWB signals and X-ray pulsed radiation, gas-discharge physics, study of rapid processes, etc. Probably the most advanced reliable compact pulsers were designed by the Ekaterinburg group.[0-0] Sub-GW levels were achieved in a modest-size package of several tens of kg. The approach of Refs. 1-3 was to use fast spark gaps (SG) for pulse sharpening. The underlying philosophy is to overvolt both the SGs and the insulation of the energy storage components. At a higher overvoltage, the first break down faster, while the second withstands higher stresses without breakdown. Both these effects are well known in high voltage engineering and discharge physics.[0,0] Rukin et al.[0] describe all solid-state subnanosecond switching at a power level comparable to that of SG switching, with all the benefits of high repetition rate, longevity and stability. The shortcomings of this technique are low efficiency and relatively large transition times.

In Refs. 1-3, the salient features of the electrical and mechanical design of hardware were described, and extensive data from laboratory testing were reported. By utilizing their approach, we developed and tested a subnanosecond pulser and added detailed considerations for aspects of the electrical design, diagnostics, and testing techniques.

2. PULSE COMPRESSOR DESIGN

2.1. Circuit Simulation

A simplified electrical circuit diagram of the pulser used for simulations is shown in Figure 1. It comprises an equivalent circuit of an SOS pulser, model SM-4N, (for a similar device see Ref. 5) acting as a fast charger, a pulse compressor, and a load. The SOS-pulser provides an output voltage of 100 kV at a pulsewidth of 22 ns with a risetime of approximately 15 ns for a 200-Ω load and allows a pulse repetition rate of 50

* Propulsion Physics Laboratory, Soreq NRC, Yavne 81800, Israel.

Hz continuously. The pulse compressor comprises a short transmission pulse-forming line (TPFL), a main closing switch (CS) U3 and an auxiliary CS U2 serving as a crowbar for adjusting the pulsewidth of the output pulse. The load is a matched long transmission line (TL) terminated by a resistor for energy dissipation. A blocking inductor Lblock serves a dual purpose: it decouples the TPFL from parasitic capacitances of the SOS-pulser and boosts the TPFL voltage owing to a resonant character of the charge. This circuit is similar to that of Mesyats et al.,[0] except for the charger: a typical Radan-based subnanosecond pulser is essentially a capacitive source, while an SOS-pulser is an inductive source. This results in a different matching between the charger and the compression stage. In Ref. 3, the charger is an SOS source, but the resonant charge is not used.

Thus, the first objective was to find a value of Lblock that sufficed for an efficient charging but was small enough to be realized as an inductor with lumped parameters. This was done numerically using a PSpice platform.

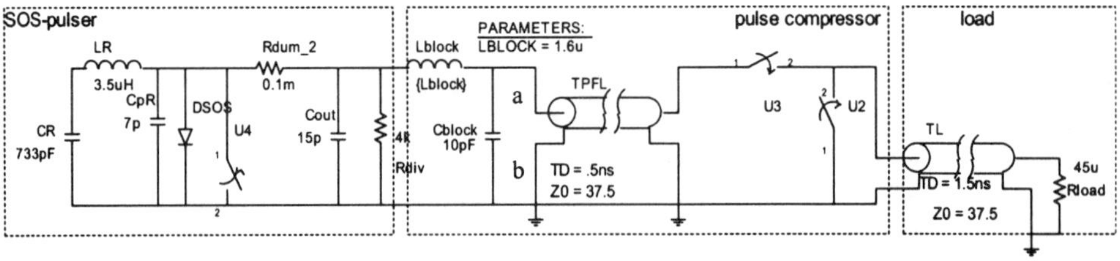

Figure 1. Simplified electrical circuit diagram of subnanosecond pulser. TD is the *delay* time. Load voltage pulsewidth is 2TD.

The charge voltage of a 1-ns TPFL for several values of Lblock is shown in Figure 2. One may note that Lblock = 1.6 μH provides a substantially higher voltage than smaller values of 0.5 μH and 1 μH, while raising it to 3 μH yields no further voltage increase. Therefore, the value of 1.6 μH was adopted for the further design of a *one-stage* pulse compressor.

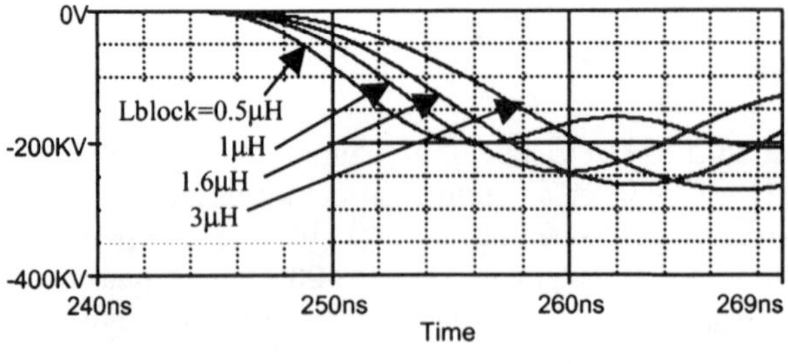

Figure 2. TPFL charge voltage. CSs not activated.

Figure 3 shows the simulation traces of the SOS-generator output voltage V(Cout), the voltage V(TPFL) at the input of TPFL, and the load voltage V(TL).

The TPFL charge lasts approximately 15 ns to a voltage that is 60% greater than the charger voltage, after which U3 closes, and the TPFL generates across the matched load a square 1-ns pulse that can be sliced into a shorter pulse by the CS U2.

It is possible to add a compression stage to boost further the output voltage. The procedure is similar to that described above. A part of the simulation circuit (compare to Figure 1) is shown in Figure 4. Again, a TPFL (a 2-ns line, to account for a practical implementation) is resonantly charged through Lblock. When the TPFL voltage is close to maximum, CS U3 closes and resonantly charges the second line (TPFL2) to an even higher voltage (Figure **5**). A 0.12-μH inductance is sufficient, and this value was used in the *second* compression stage.

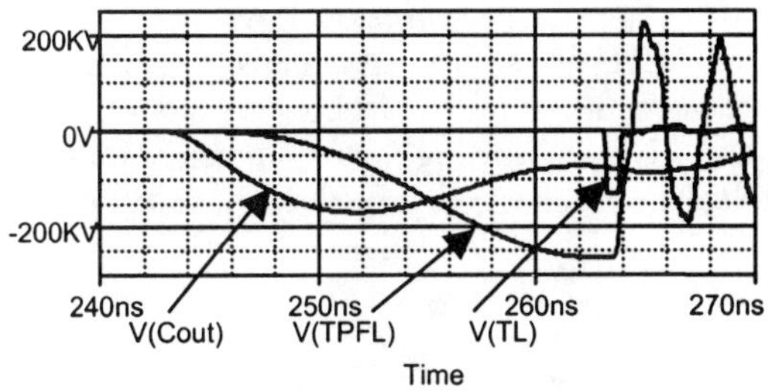

Figure 3. PSpice simulation of one-stage compression.

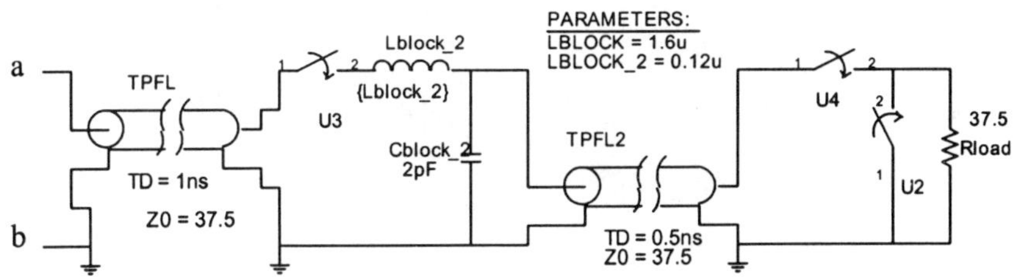

Figure 4. Part of two-stage compression circuit. Points of connection to the rest of simulation circuit are designated as a, b (see Figure 1). Rload is matched to PFLs.

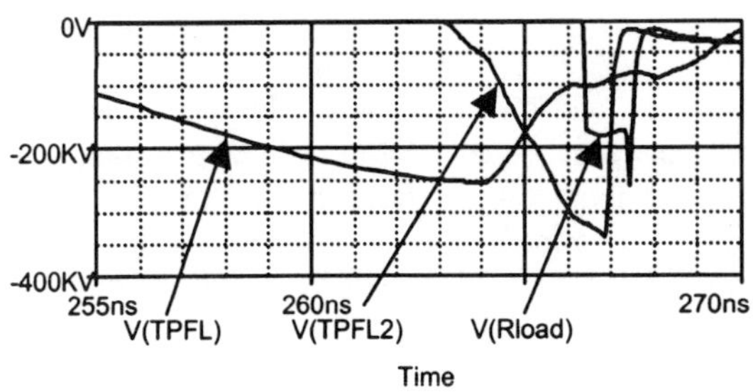

Figure 5. PSpice simulation of two-stage compression.

2.2. Switching System

Figure 6 shows schematically the arrangement of a three-electrode spark-gap switching system. The breakdown voltages of SG1 and SG2 are regulated by varying the gaps and gas pressure. The main gap SG1 breaks down first, and the highly-overvolted crowbar SG2 sparks after a certain delay. The time lag to its breakdown sets the pulsewidth. At negative polarity, the HV electrode of SG1 serves as a cathode, while its low-voltage counterpart acts as an anode for *SG1* and as a cathode for *SG2*. Note that the physical and electrical lengths of the switching system are comparable to those of PFLs. This results in a pulse widening.

The SGs operate in nitrogen at a pressure of up to 60 bar. Mesyats et al.[0] investigated the operation of spark gaps having electrodes made from different materials. The electrodes made of copper-tungsten pseudoalloys and of pure copper exhibited the least erosion compared to stainless steel electrodes. Other researchers[0, 0] studied the erosion of electrodes in vacuum and found that copper electrodes had the smallest erosion rate, except for gold and silver. Following the recommendations of Refs. 0, 0, and 0, we also used a copper-tungsten pseudoalloy, ELRO H 30 (30% copper-70% tungsten) and copper.

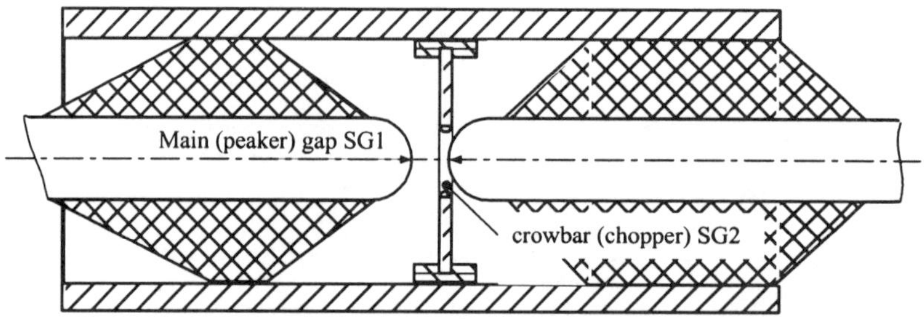

Figure 6. Arrangement of three electrode spark gap switching system.

The hemispherically tipped electrodes of the peaker gap had annuluses as the working surfaces. Electrodes with this shape stabilize the operation owing to a diffuse mode of the discharge, and thus gas circulation through the gap is not required.

2.3. Load

The simplest load for testing would be a matched resistor. Although many non-inductive HV resistors are available, none of them are small enough to function as a component with lumped parameters, and for that reason alone good matching is practically unattainable. Therefore, similar to Shpak et al.,[0] the equivalent load for testing was a long-matched oil-filled TL, terminated by a bulk resistor that served for energy dissipation only.

3. DIAGNOSTICS

It is a common practice to measure high pulsed voltages by capacitive dividers (CD). We have employed workshop-made strip capacitive dividers, because they are readily incorporated into coaxial structures. The high-voltage arm C1 (Figure 7) is formed by the capacitance between the HV and low-voltage electrodes. Kapton or Mylar film insulates a low-voltage copper-foil electrode from the grounded electrode to form the low-voltage arm C_2, which is between two-three orders of magnitude greater than its HV counterpart. To ensure faithful measurements in the subnanosecond range, the low-voltage arm of a CD must conform, at the least, to two requirements: 1) the electrical length of the strip line must be smaller than the pulse risetime; 2) C_2 must be sufficiently large to obtain an acceptable division ratio and to satisfy the condition, $\tau_{in} = Z_{scope}C_2 >> \tau_p$, to keep the pulse top flat. Here τ_{in} is the time constant of the divider-scope input circuit, Z_{scope} is the scope impedance, and τ_p is the pulsewidth of the measured signal. For recording undistorted pulses having a full-width half maximum (FWHM) of about 1 ns and a risetime as small as 100 ps, the characteristic dimension of the strip line has to be approximately 2 cm.

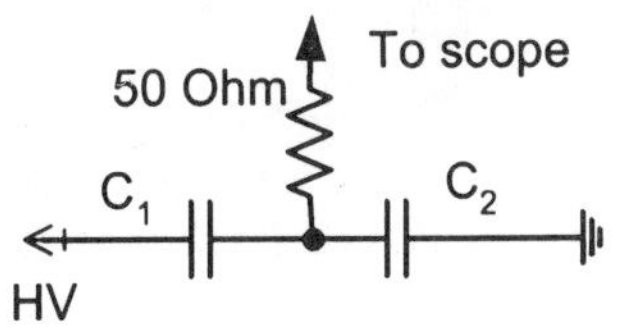

Figure 7. Electrical circuit of capacitive divider.

It is difficult to account for the glue thickness and the variation of the dielectric constant of C_2. Therefore, each CD has to be calibrated, especially in the subnanosecond range. The calibration procedure, which is far from straightforward, is briefly described below.

For current measurement, both with the equivalent load and with an air load, we used workshop-made Rogowski coils set in special flanges. Recommendations on design of Rogowski coils[0] are not altogether applicable to nanosecond and subnanosecond ranges. The main limitation, again, is the electrical length of the wire that must be shorter than the fastest portion of a transient signal. A classical Rogowski toroidal coil surrounds the current-carrying conductor. Obviously, a coil having a diameter of 5 cm (ID of the PFLs) will be too long to record a subnanosecond pulse. Therefore, we employed a one-turn coil that, in fact, was a "B-dot"-probe. The integration was performed numerically.

For calibration of CDs, Rogowski coils, measurement cables, and fast all-solid state pulsers were employed. For calibration of probes and cables in *nanosecond* ranges, we used an AVTECH pulser having a risetime of 250 ps and up to a 200-ns flat top pulse. An HV PPG-10 pulser, manufactured by PSG Ltd. (Russia), provided a 1-kV output at a 50-Ω load with a risetime of 80 ps that permitted calibration of the probes and cables in the *subnanosecond* range. A Tektronix TDS-820 scope coupled to signal sources via broadband HV attenuators was used for fast measurements. As an example, Figure 8 shows the smearing that occurs in the PPG output pulse from cables of various lengths. The scope reading should be corrected using a proper coefficient taken from Table 1. Note that this table, rigorously speaking, is applicable only to signals having a shape identical to the calibration pulse. At longer times, the corresponding attenuation factors are smaller.

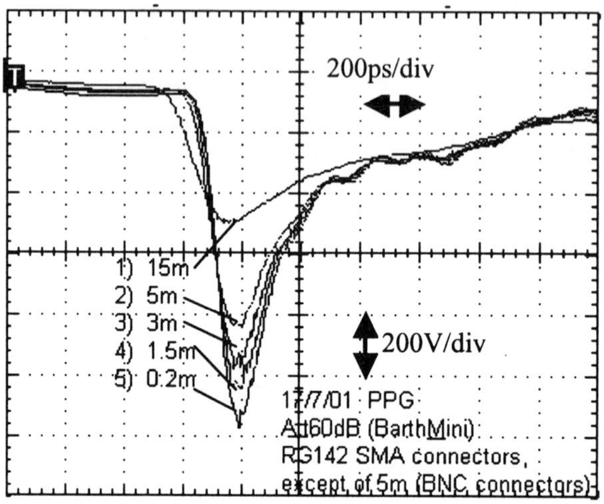

Figure 8. PPG output pulse attenuated by 60 dB attenuator and further smeared by RG-142 cables of various lengths.

Table 1. Attenuation coefficient of RG-142 cable.

Cable length [m]	0.2	1.5	3	5	15
Attenuation times [dB]	1	1.1	1.27	1.43	2.44

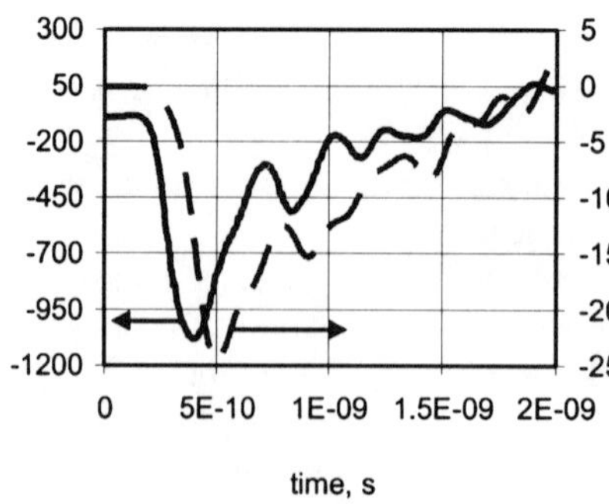

Figure 9. Waveforms from CD5 and RN2. Calibration source PPG-10.

Figure 9 shows the voltage and the current (monitored by the B-dot probe) at the input of the TL fed by the PPG-10 via a matched conical adapter. These wave shapes are similar, which can be considered a good result for these kinds of measurements.

4. EXPERIMENTS WITH EQUIVALENT LOAD

We started the experiments with one-stage pulse compression. Figure 10 shows the voltage traces at the SOS-pulser output and at the input of the 1-ns PFL. In accordance with the PSpice simulation (Figure **3**), the PFL voltage is higher than at the SOS-pulser output owing to the resonant character of the charge, although the simulation predicts a more modest boost. The PFL voltage is chopped sharply by the SG1; the ringing is caused by the reflections from the short-circuited end of the PFL after the SG2 has closed. Note that SG1 breaks down at a voltage of 240 kV, while its static breakdown voltage should be about 60 kV (see, for example, Meek et al.[0]). Such an overvoltage results in a short formative time lag.

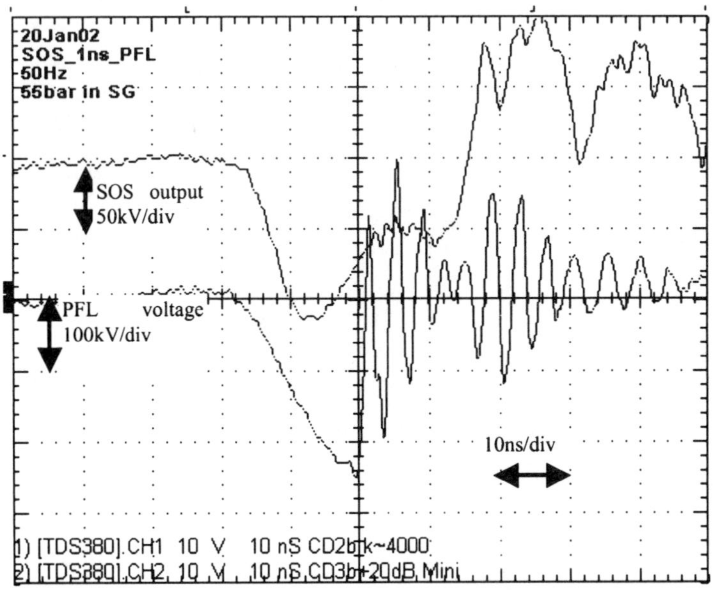

Figure 10. Voltage traces at the SOS-pulser output (CD2) and at the 1ns PFL input (CD3c).

The load voltage for different electrode gaps of the chopper SG2 (Figure 6) is shown in Figure 11. By varying the SG2 gap, the pulsewidth can be regulated from 1.2 ns to less than 200 ps. The leading edge is recorded very clearly (somewhat sloped by a 4.5-m long cable), while the trailing edge has a considerable jitter, which results in a smeared pattern. At a voltage of 125 kV (the widest pulse in the graph), the peak power is

0.42 GW, while the pulse energy is 0.4 J. The efficiency of the compression is approximately 40%, considering the amount of energy at the output of the SOS-pulser[0] of 1 J; the power compression factor is around ten.

A life test was performed with the one-stage compression system, both with ELRO and copper electrodes. The grounded electrode of the chopper was manufactured from ELRO in both cases. After the assembly, the SGs were conditioned by 25,000 pulses, and then the pulse shapes were recorded. The initial peak voltage was 116 kV at a 460-ps FWHM for ELRO and 112 kV at a 520-ps FWHM for copper electrodes. With the ELRO electrodes, the pulser worked quite stably after 1.2 million pulses at 50 Hz (the load and SGs were cooled by an external fan), but the output pulse was considerably modified. Due to erosion, the gaps of both SG1 and SG2 increased, and the voltage reached 143 kV at a 920-ps FWHM (Figure 12). Similar results were obtained with copper electrodes, but the parameters of the output voltage changed more slowly owing to a weaker erosion: after 2.5 millions shots, the peak voltage was 134 kV at a 750-ps FWHM.

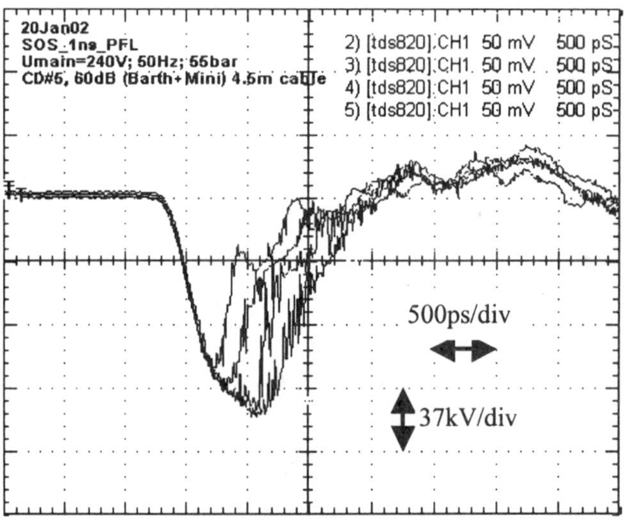

Figure 11. Load voltage for different positions of chopper SG2 (one-stage compression).

The erosion rates for copper electrodes are shown in Table 2. Compared to the data of Ref. 2, they are three times higher. We ascribe the discrepancy to the difference in the electrodes' working area: the SG electrodes were larger,[0] and thus the discharge could be more diffuse, resulting in weaker erosion.

Table 2. Erosion rates for copper electrodes.

Electrode	Erosion rate [mg/C]
SG1 cathode	3.7
SG1 anode	3.7

Grounded electrode of SG2 (ELRO)	4.1

A further pulse compression was realized by adding a second stage. A 2-ns PFL served as the energy store of the first stage, while the same 1-ns PFL was used in the second stage. The gas pressures in the spark gaps of the first and the second stages were 42 bar and 56 bar, respectively. Such an arrangement is far from optimum; the experiment's goal was to show the feasibility of further subnanosecond compression.

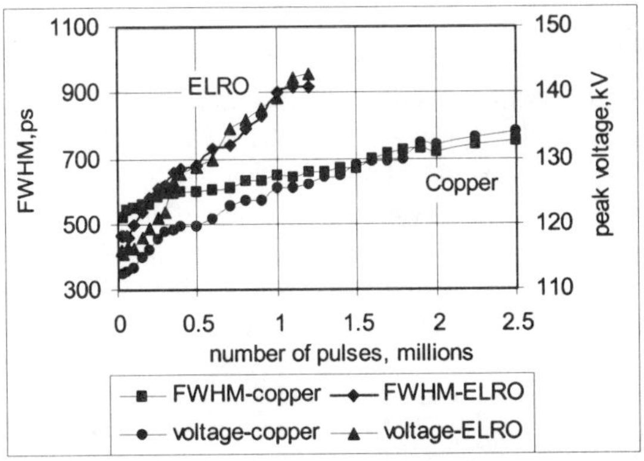

Figure 12. Change of peak load voltage and FWHM for ELRO and copper electrodes (one-stage compression).

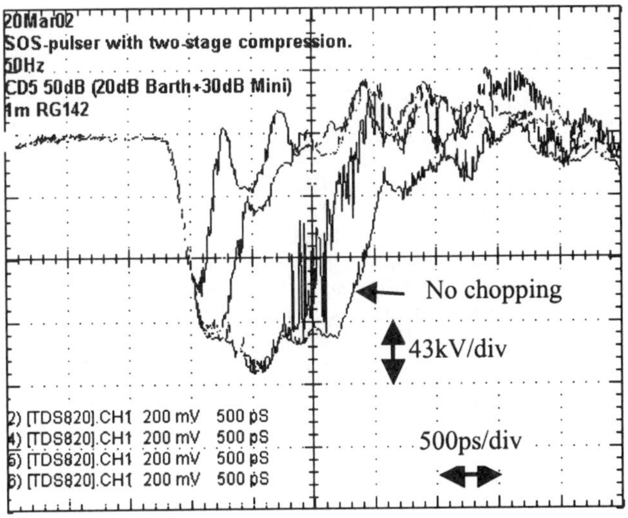

Figure 13. Load voltage for different positions of chopper SG2 (two-stage compression).

Without chopping, the pulse width was 1.3 ns, longer than should be provided by a 1-ns PFL. This widening is caused by the discharge of the energy stored in the second SG that acts as a slightly mismatched line. Both the leading and the trailing edges are stable. The voltage reaches a peak of 160 kV (peak power of 0.68 GW); the energy content in the pulse is 0.6 J, which corresponds to a conversion efficiency of approximately 60%. The rise time is less than 180 ps, and the fall time is around 100 ps. With the chopper slicing a portion of the pulse, the trailing edge jitters; however, the jitter is less than that in a similar experiment with one-stage compression. The pulser operation, stable up to a frequency of 300 Hz in bursts of 30 s, was limited by the SOS-charger performance.

5. CONCLUSIONS

A compact subnanosecond pulser, comprising an SOS-generator acting as a charger and one or two pulse-compression stages based on high-pressure SGs, has been developed and tested. Voltage and current measurement means, including capacitive dividers and Rogowski coils, were developed and time-domain calibrated. Weighing less than 50 kg, the pulser provides up to 160-kV subnanosecond pulses at a 37.5 Ω load at a repetition rate of up to 300 Hz. The pulsewidth can be regulated between 200 ps and 2 ns. The rise time is less than 180 ps, and the fall time is approximately 100 ps for a two-stage compression. An energy efficiency of up to 60%, counting from the input to the output of the compressor, was obtained. It was confirmed that copper was a material of choice for subnanosecond SGs. With copper electrodes, the pulser yielded 2.5 million pulses, at which point the FWHM and the peak voltage increased by 30% and 16%, respectively.

6. ACKNOWLEDGMENTS

The authors thank Mr. E. Shviro for his invaluable help in assembly works and conducting the experiments.

7. REFERENCES

1. V. Shpak, M. Yalandin, S. Shunailov, and M. Ulmaskulov, Generation of high-power UWB signals in subnanosecond range, *Izvestia Vyshikh Uchebnykh Zavedenii, Fizika* **12**, 119–127 (1996).
2. G. Mesyats, V. Shpak, M. Yalandin, and S. Shunailov, Compact high-current accelerators based on the Radan SEF-303 pulsed power source, *Proc.of Ninth IEEE Int. Pulsed Power Conf.*, 835–838 (1993).
3. M.I. Yalandin, S.K. Lyubutin, M.R. Oulmascoulov, S.N. Rukin, V.G. Shpak, S.A. Shunailov, and B.G. Slovikovsky, Subnanosecond modulator possessing a 700 MW peak power and average power of 1.5 kW at repetition frequency of 3.5 kHz, *Proc. PPPS-2001*, 1630–1633 (2001).
4. S.N. Rukin, E.A. Alichkin, S.K. Lyubutin, G.A. Mesyats, and B.G. Slovikovsky, Ultra-high-power repetitive solid state DBD-based switching, *Proc. PPPS-2001*, 329–332 (2001).
5. S.N. Rukin, S.N. Lyubutin, V.V. Kostirev, and V.A. Telnov, Repetitive 200kV nanosecond all-solid state pulser with a semiconductor opening switch, *Proc. of Tenth IEEE Int. Pulsed Power Conf.*, 1211–1214 (1995).
6. J.M. Meek and J.D. Craggs, *Electrical Breakdown of Gases* (Wiley, New York, 1978).
7. E. Kuffel and W.S. Zaengl, *High Voltage Engineering* (Pergamon Press, Oxford, 1984).
8. G.A. Mesyats and D.I. Proskurovski, *Pulsed Electrical Discharge in Vacuum* (Springer-Verlag, Berlin, 1989).
9. E.A. Litvinov, D.I. Proskurovsky, and Y.B. Yankelevich, Study of the processes at point cathodes responsible for the cathode lifetime and the stability of its operation in the explosive emission mode, in: *Development and Application of Intense Electron Beam Sources*, edited by G.A. Mesyats (Nauka, Moscow, 1976), pp. 96–105.
10. A.J. Schwab, *Hochschpannunsmestechnik, 2. Aufl.*, (Springer, Berlin, 1981).

11. A. Pokryvailo, Y. Yankelevich, A. Bornstein, R. Shpitalnik, and E. Shviro, Investigation of electrical parameters and chemical activity of high-power short pulsed corona discharge, *Proc. 12th IEEE Int. Conf. on Pulsed Power,* 1349–1352 (1999).

RESONANCE PULSE COMPRESSION AS AN EFFECTIVE METHOD FOR GENERATION OF UWB HIGH REPETITIVE MICROWAVE PULSES

P.Yu. Chumerin, A.N. Didenko, S.A. Novikov, and Yu.G. Yushkov*

1. INTRODUCTION

A simplified version of the compressor operation is the storing of electromagnetic energy in a resonant cavity and the following rapid extraction of this energy to some kind of load. The attractive feature of resonant microwave compressors (RMC) is that the traveling-wave power in a common, single-mode, copper cavity with a double traveling time of 10^{-9} – 10^{-8} s per unit length of the cavity can exceed the power of an exciting generator by 10 – 20 dB and a multimode cavity by 20 – 30 dB or more.[1] In superconducting cavities, the increase may achieve 50 – 60 dB and 60 – 70 dB, respectively. With the best existing technology, electrical field strengths in microwave elements or systems up to several hundreds of kV/cm can be achieved, and the corresponding traveling-wave power could be as much as 100 MW in single-mode copper cavities and 1 GW in multimode cavities. These levels of power will be followed by an output pulse power if one makes available an extraction time close to the double traveling time per unit length of a cavity. An increase of an extraction time will lead to a decrease of output pulse power. Therefore, short extraction times are the main objectives for elaboration of microwave compressors.

2. STORAGE CAVITIES

Schematically, a multimode cavity is depicted in Fig. 1. The normalized wave amplitudes (denoted by a_i, b_i, and d_i) are related by the following transition-matrix equations at the locations of discontinuity:

* P.Yu. Chumerin, S.A. Novikov, and Yu.G. Yushkov, Nuclear Physics Institute at Tomsk Polytechnical University, Tomsk, Russia, 634050. A.N. Didenko, Russian Academy of Science, Moscow, Russia.

$$\begin{vmatrix} b_1 \\ b_2 \end{vmatrix} = \begin{vmatrix} \sqrt{1-k^2} & jk \\ jk & \sqrt{1-k^2} \end{vmatrix} \begin{vmatrix} a_1 \\ a_2 \end{vmatrix}, \quad \begin{vmatrix} b_3 \\ b_4 \end{vmatrix} = \begin{vmatrix} \sqrt{1-h^2} & jh \\ jh & \sqrt{1-h^2} \end{vmatrix} \begin{vmatrix} a_3 \\ a_4 \end{vmatrix}, \quad \begin{vmatrix} d_1 \\ d_2 \\ d_3 \end{vmatrix} = \begin{vmatrix} -1/2 & 1/2 & 1/\sqrt{2} \\ 1/2 & -1/2 & 1/\sqrt{2} \\ 1/\sqrt{2} & 1/\sqrt{2} & 0 \end{vmatrix} \begin{vmatrix} c_1 \\ c_2 \\ c_3 \end{vmatrix},$$

where k and h are parameters of the cavity coupling to the feed line and the interference switch, respectively.

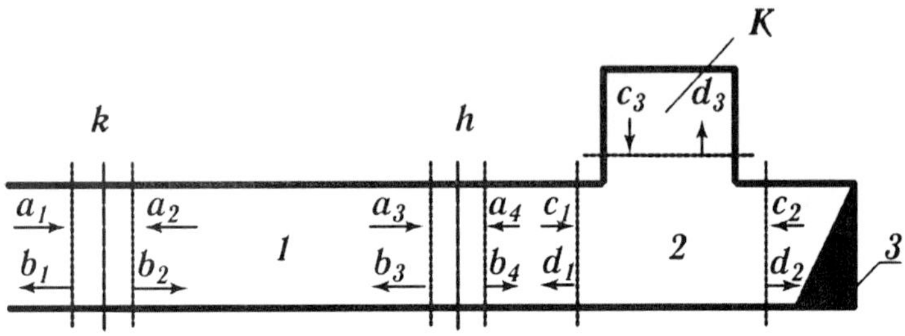

Figure 1. Cavity schematic: Section *1* is the storage section; Section *2* is the waveguide output; Section *3* is the load; Section *K* is the area of a switch position.

In regular sections the relationship is defined by corresponding time constants of wave attenuation and phase shift. Then with due account of the wave traveling time along the regular sections, a wave amplitude value at some moment could be expressed relative to the same wave amplitude at that moment with a doubling of the traveling time along the cavity. This leads to approximate differential equations for variables $b_2(t)$ and $d_2(t)$, which define the transient in a cavity and the envelope of the output pulse, respectively, where t is the current time. The field energy in the cavity is determined by

$$W_c = |b_2(t_g)|^2 \, T_1/2,$$

where t_g is the pulsewidth of the input and T_1 is double traveling time along the main section. The energy that is transferred from the cavity to the load is

$$W = \frac{1}{2}\int_0^\infty |d_2(t)|^2 \, dt.$$

The energy storing efficiency η_c is $W_c/(P_g \cdot t_g)$, the total microwave efficiency η is $W/P_g \cdot t_g$, and the power gain is $M^2 = |d_{2max}(t)|^2 \div (a_1)^2$. Calculated maximum values of M^2 and η are plotted as functions of t_g in Figs. 2 and 3, where Q_0 is the internal Q-factor and τ_1 is the output pulsewidth of the compressor.

For a single-mode cavity, the cross sections of the output element waveguide and the storage resonant volume are equal, and the extraction time equals the double traveling time of the wave along the resonant volume. The amplification coefficient M^2 and the efficiency η are plotted versus the relative input pulsewidth in Figs 2a and 2b, respectively.

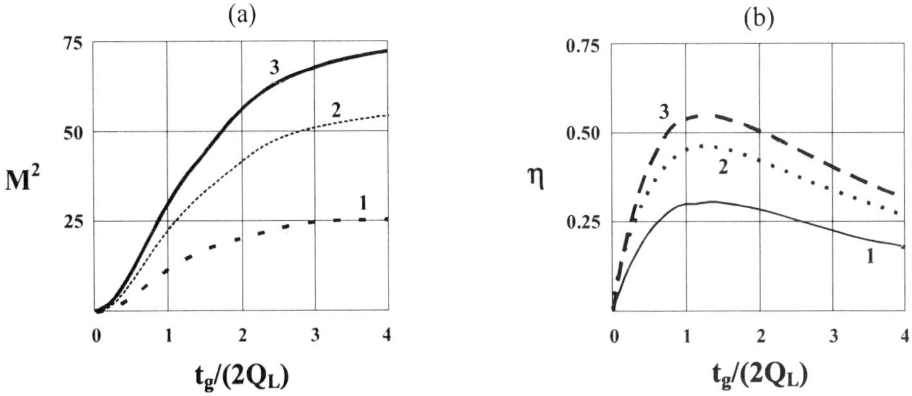

Figure 2. (a) Amplification M^2 as a function of the relative input pulsewidth. (b) Efficiency η as a function of the relative input pulsewidth. Plots are for three ratios Q_0/Q_L of the intrinsic Q-value to the loaded Q-value: designated by 1, 2, and 3 for $Q_0/Q_L = 2$, 4, and 10, respectively.

3. SINGLE-MODE CAVITY RMC

Four typical schematic designs of a single-mode cavity RMC are shown in Fig. 3. Each design has an output T element with one short-circuited arm and a switch that is located one-fourth of the waveguide's wavelength from the short-circuit plate. The switch changes the compressor between storing and extraction modes. The RMC for this design operates for wavelengths between 3 and 100 cm. The storage cavities are manufactured from standard waveguides for the 3-cm and 10-cm wavelength bands and from coaxial lines with coaxial interference switches for the 30-cm and 100-cm bands.

Decoupling the cavity and the load during the energy accumulation process is achieved by choosing a defined length of the short-circuited arm of the tee to account for situations when the phases of the waves propagating to the load from the cavity volume and the tee arm are reversed. When the switch is turned on, the electrical length of the arm is changed by one quarter of a wavelength, which results in in-phase waves and the extraction of energy. The switches are usually gaseous waveguide types, where electrical isolation is maintained by gas (nitrogen, argon, SF_6, air) with an excess pressure of several atm. Table 1 presents typical parameters of RMCs for the geometry of Fig. 3: f is frequency; P_m is the input power; P_{out} is the output pulse power; T_{out} is the output pulsewidth; M^2 is the power gain; Q_0 is the intrinsic quality factor; η is the efficiency: and L is the length of the cavity.

Table 1.

f [GHz]	P_m [MW]	P_{out} [MW]	T_{out} [ns]	M^2 [dB]	$Q_0 \times 10^4$	η [%]	L [m]
1.0	0.30	20	10	18.5	2.0	21	1.50
2.8	1.30	190	4.2	21.6	1.2	22	0.90
2.8	1.00	22	20	13.4	1.4	18	2.50
9.4	0.05	2.2	3.2	17.0	1.4	18	0.35

As Table 1 indicates, for output pulsewidths between 3 ns and 20 ns, P_{out} can reach several megawatts at X-band, tens of megawatts at L-band, and hundreds of megawatts at S-bands. The transient attenuation of the cavity's output element during the excitation process is –45 dB to –45 dB. The relatively low η of this type of RMC can be attributed to the energy reflection during the transient excitation process and to losses in the cavity walls and the switch. Typical traces of microwave pulses obtained with this method are shown in Fig. 4 for two output powers (25 MW and 50 MW) at 880 MHz.

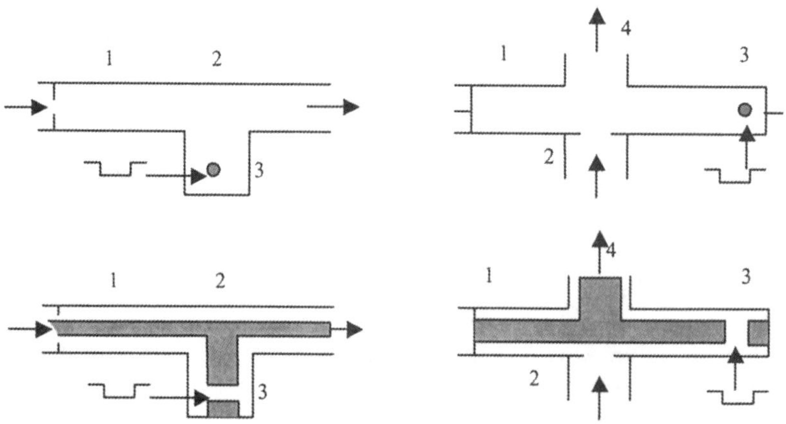

Figure 3. Single-mode cavity: 1 – cavity; 2 – T-junction, 3 – switch, 4 – output arm.

(a) P_{out} = 50 MW (b) P_{out} = 25 MW

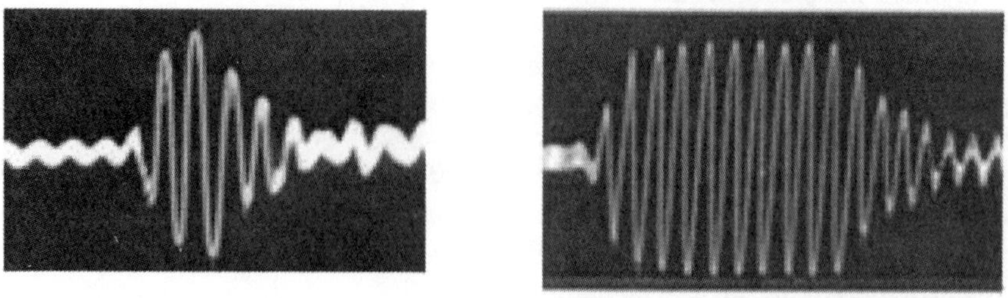

Figure 4. Traces of an RMC microwave pulse at f = 880 MHz for two output power levels.

4. TWO-STAGE SERIES COMPRESSION

A scheme (Fig. 5) that produces gigawatt, nanosecond, microwave pulses with an industry-issued magnetron driver was developed on the basis of series compression.[2] The magnetron driver is a common transmitter for S-band radar and has a 2.5-MW peak power, a 3-μs pulsewidth at the –3 dB level, a tunable working frequency from 2.79 GHz to 2.81 GHz, and a 400-Hz repetition rate. The compressor was a resonant load for the magnetron tube, and their matching was achieved by the ferrite isolator and by fixing a special length of the waveguide line.

When the compressor output peak power exceeds 1 GW, the corresponding electrical field strength is between 300 and 1000 kV/cm in standard waveguides. Problems associated with the electric field strength are solved using some technical procedures for interior surface treatment and purification of gases or using special design and circuitry of compression systems. In this case, the solution of the construction elements insulation problem was attributable to the known fact that the electrical strength increases as the time an element or gaseous volume is exposed to the electrical field decreases. Specifically, this is implemented in the second stage of a two-stage series compression circuit.

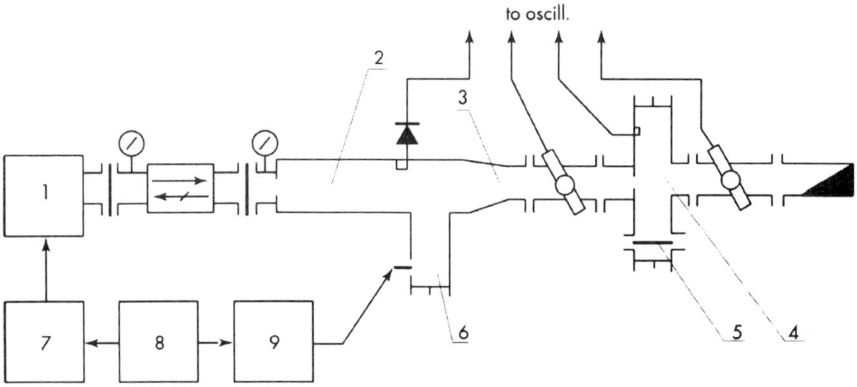

Figure 5. Two-stage series compression scheme: 1 – magnetron; 2 – first-stage cavity; 3 – circular-rectangular waveguide transition; 4 – second-stage cavity; 5 – quartz tube of switch; 6 – short-circuited arm of first-stage interference switch; 7 – high-voltage magnetron modulator; 8 – controlling generator; 9 – high-voltage triggering generator.

Experience has shown that field strength values from 100 to 150 kV/cm can be obtained without difficulty. Cheap insulation with gaseous air or nitrogen may be used. The dimensions of the first-stage cavity were chosen to meet the electrical-field-strength requirement, $E \leq 150$ kV/cm, and to provide the loaded Q_L value satisfying the equation, $\tau = 2\, Q_L / \omega \geq t_g/(3\ldots5)$. The cavity was made from a circular copper waveguide with a 90-mm diameter. The measured internal Q-factor was $2 \cdot 10^4$. The calculated first-stage cavity length yielded an 11-ns output pulsewidth with a power gain of 82 dB. The interference switch of the first stage contained the short-circuited section of rectangular waveguide of 34×72 mm^2 cross-sectional area, which is set at right angles to the circular waveguide axis. The microwave gaseous switch was triggered by the additional spark switch, lighting the waveguide gap through the hole in the short-circuited section broad wall. The switch operation with a 400 Hz repetition rate was supported by continuous gas circulation in the region of the waveguide and triggering gaps. Triggering pulses had voltages up to 10 kV. The second-stage cavity was a double-waveguide tee with short-circuited symmetrical side arms. The waveguide cross section was 34×72 mm^2. The internal Q-factor, the power gain, and the estimated pulsewidth respectively were $9 \cdot 10^3$, 10, and 0.8 ns. The switch included a radio-transparent quartz tube (filled with argon at the pressure of 7 bar), was set at right angles to the broad wall planes at the distance $\lambda_w/4$ from the short-circuit plate, and operated with self-discharges.

The first-stage cavity, the second-stage cavity, the waveguide elements, and the matched load were filled with nitrogen at very high pressure. Water cooling was arranged around the ferrite isolator and the load. The pulsed output power was measured by a calorimetric meter that indicated the average power. The directional-coupler calibration and the transmission-factor measurement were made when the couplers were connected to the output magnetron line for the traveling-wave regime.

The envelopes of first-stage and second-stage output pulses are shown in Fig.6a and Fig.6b, respectively. The first-stage pulse had a 150-MW peak power and an 11-ns pulsewidth. In contrast, for the second stage, the peak power increased to 1 GW, and the pulsewidth decreased to roughly 1 ns.

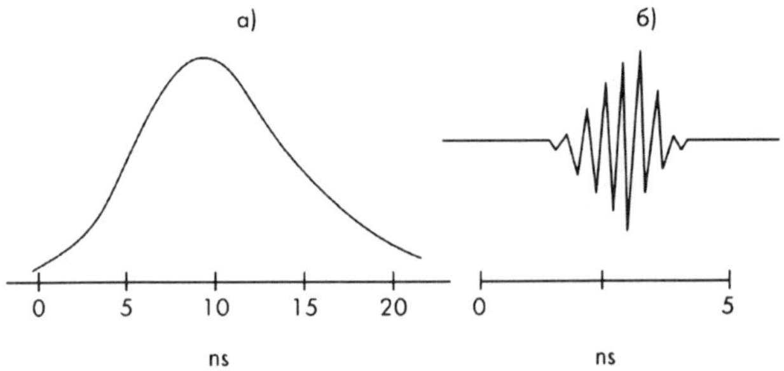

Figure 6. Output pulse envelopes of two-stage series compression: (a) first stage; (b) second stage.

5. COMPRESSION OF RELATIVISTIC MAGNETRON RADIATION

Relativistic magnetrons have relatively short pulsewidths, usually about 100 ns, which do not permit high-power compressor gains. Typical compressor gains for relativistic magnetrons range between 5 dB and 20 dB. On the other hand, an over-coupled storage cavity excited by such pulses has higher efficiencies (up to 80%) than a critically coupled cavity can provide (≈40%) when excited by microsecond pulses.[3,4] Furthermore, a relativistic magnetron has low frequency stability, caused by an explosive-emission cathode operation with high current values of several kiloamperes. If the high Q-factor storage cavity of a compressor was connected to a relativistic magnetron, it would lead to frequency control due to frequency pulling.

So the joint operation of a compressor and a relativistic magnetron is not obvious, as real microwave pulses have amplitude modulation and frequency instability, which means that the frequency can go beyond the working band of a compressor cavity. The circuits where a magnetron is coupled with a resonant load require ferrite isolators or circulators for steady operation, but they have not been developed to date for relativistic-tube power levels. Consequently, another objective is to demonstrate the possibility of joint operation between a microwave compressor and a relativistic magnetron as a pumping source.

The setup schematic is shown in Fig. 7. The S-band relativistic magnetron had a 120-MW radiation power, a 120-ns pulsewidth, and a 10-Hz repetition rate. The magnetron was fed by an LIA-type high-voltage source that supplied 400 kV at 4 kA.

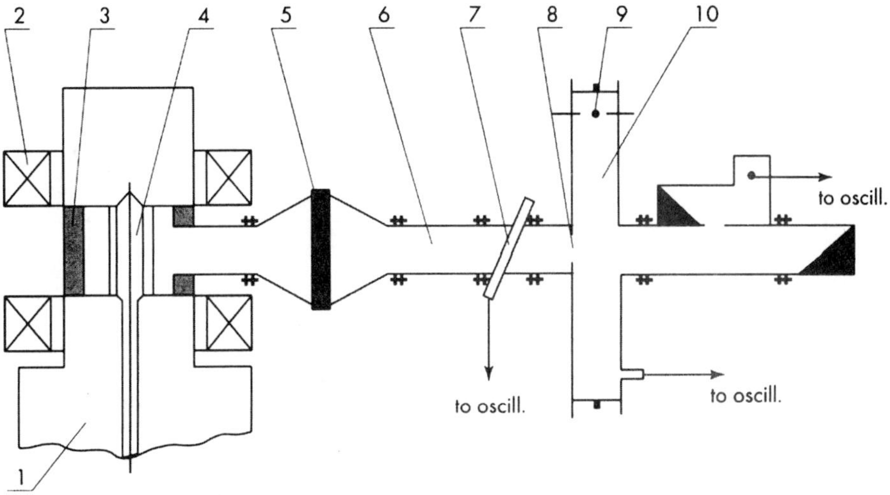

Figure 7. Compression circuit including the relativistic magnetron as the primary microwave source: 1 – coax high-voltage line; 2 – magnet; 3 – anode; 4 – cathode; 5 – dielectric window; 6 – waveguide line; 7 – directional coupler; 8 – input iris; 9 – switch; 10 – storage cavity.

The storage cavity was designed as a double waveguide tee (magic tee) with symmetrical short-circuited side arms and was manufactured from waveguides with a 72×34 mm^2 cross section. The cavity was excited through the coupling hole in the broad waveguide wall that is in E-plane. The energy was extracted through the output in the H-plane as soon as the gaseous discharge commutator was switched. The commutator was installed at the quarter-waveguide-wavelength distance from the short-circuit end plate. The minimum pulsewidth value was determined by the double travelling time of a wave propagating with the group velocity along one of the side arms. The cavity was over coupled and its loaded Q-factor was about 750; consequently, the excitation period decreased to approximately 120 ns. The interior cavity volume was filled with a gas at 6-bar pressure to ensure the electrical strength. The dielectric window, made of lucid, had a 250-mm diameter and separated the pressure and the magnetron-evacuated units.

The operation that matched the magnetron and compressor was adjusted experimentally by varying the length of the waveguide line between the magnetron and the excitation hole. The ferrite isolator was tested as a discoupling element as well. The switching of the gaseous commutator was spontaneous when the microwave field reached breakdown level. Output microwave pulses were transmitted to the matched load. The measuring circuit contained the directional coupler, the attenuator, and the wideband oscilloscope.

A typical output-pulse envelope when the relativistic magnetron is operated with a matched load is shown in Fig. 8a, where the pulse had amplitude modulation and the degree of modulation changed from pulse to pulse. Figure 8b shows the effect on the pulse envelope by varying the field strength in the cavity without switching. The compressor output pulse had a 1.1-GW peak power and a 5-ns pulsewidth at the –3 dB level. One may achieve a further increase of the parameters by decreasing output-pulse amplitude modulation of the magnetron.

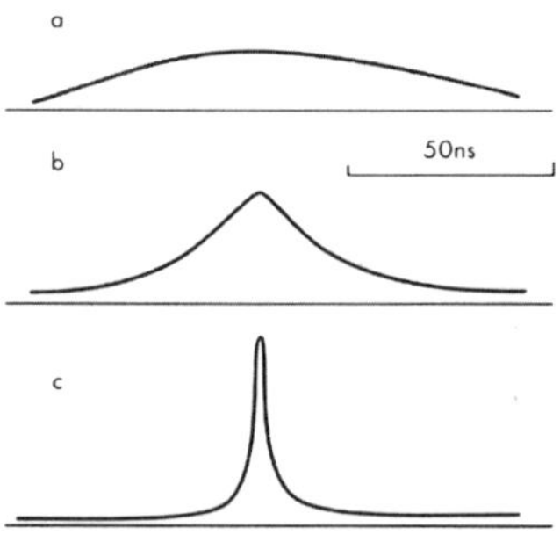

Figure 8. Microwave pulse envelopes of relativistic magnetron compression setup: (a)

6. CONCLUSION

Sources that use the method of resonant pulse compression can produce wideband and UWB pulses at microwave frequencies with power up to several GW and with about a 100- Hz repetition frequency. The repetition frequency will be limited by the capability of the primary exciting source.

7. REFERENCES

1. A.N. Didenko and Yu.G. Yushkov, *Higher Microwave Pulses of Nanosecond Width* (Energoatomizdat, Moscow, 1984).
2. A.N. Didenko, S.A. Novikov, S.V. Razin, and Yu.G. Yushkov, Pulse shaping of high power UWB RF signals by successive time compression, *Sov. Phys. Dokl.* **36**(11), 792–793 (1991).
3. A.N. Didenko, A.G. Zherlitsin, A.V. Melnikov, and Yu.G.Yushkov, Forming high-power electromagnetic nanosecond radiation in a triode with virtual cathode, *Dokl. Akad. Nauk.* **309**(5), 1117–1119 (1989).
4. A.N. Didenko, I.I.Vintizenko, A.I. Ryabchikov, and Yu.G. Yushkov, Resonant compression of microwave pulses at relativistic magnetron output, *Dokl. Akad. Nauk.* **336**(5), 619–621 (1999).

EXCITATION OF SUB-STRUCTURE RESONANCES USING A T-PULSE

Zhong Ji, Tapan K.Sarkar, Baek Ho Jung, and Carl E. Baum*

1. INTRODUCTION

If the signal is constrained to last only for a single baud, the optimum signal is the well-known "Prolate Spheroidal Wave Function"[1]. However, for signals of more than one baud time duration, it is necessary to remove intersymbol interference. There are two common constraints that are used to remove intersymbol interference. The first constraint is that the desired signal has zero values at all sampling instances but one. This is often referred to as the zero crossing property. The other constraint is that the desired time limited signal is orthogonal to its shifted versions i.e., its autocorrelation function is a Nyquist signal. It is called orthogonal Nyquist signals. We call it a T-pulse. The term T-pulse implies that in time domain it is of finite duration. A time limited signal cannot simultaneously be band limited. However, in addition to the pulse being time-limited we focus 99.97 percent of its energy in a very narrow band. Therefore, a T-pulse is also practically band limited. By optimizing a cost function, the discrete sequence of orthogonal T-pulse can be generated. In this paper, a modulated non-zero-mean T-pulse is used as incident wave for a combined field integral equation (CFIE) method[2,3].

In this paper, we apply the concept of a T-pulse as an incident wave to excite substructure resonances. The goal here is to excite a resonance effect, which may be small in amplitude as compared to the dominant resonance and thereby mark the small characteristic effects.

When applying time-domain CFIE model, almost all researchers use a Gaussian pulse as incident wave. It is effectively time and band limited and is well suited to numerical computation. It has a rapid decay to a negligible value in both time and frequency domains. In this paper, a narrow bandwidth modulated T-pulse is instead used as the incident wave. Numerical examples are presented. The result shows that a small

---

* Zhong Ji and Tapan K.Sarkar, Syracuse University, Syracuse, New York, 13244. Baek Ho Jung, Hoseo University, Asan, South Korea ,336-795. Carl E. Baum, Air Force Research Laboratory, Kirtland Air Force Base.

Ultra-Wideband Short-Pulse Electromagnetics 6
Edited by Mokole *et al.*, Kluwer Academic/Plenum Publishers, 2003

change of object can be recognized by using a T-pulse as an incident wave. Hence, a T-pulse only excites a narrow spectrum which may be the finger print of a target without exciting the dominant resonances which may mark the smaller effects.

In this paper, how to get T-pulse is presented in Section 2. Section 3 describes CFIE formulations. In Section 4, the numerical results are presented. Conclusions are concluded in Section 5.

2. HOW TO GENETRATE A T-PULSE

Let us assume a discrete signal sequence $f(n)$, which is defined for $n = 1,2,\Lambda\ N_n$, and is identically zero outside these N_n values. Let us assume there are N_s samples in one baud time. Then the total number of baud times N_c is

$$N_c = N_n / N_s \tag{1}$$

The discrete Fourier transform (DFT) of the signal $f(n)$ is defined by[4]

$$F(l) = \frac{1}{\sqrt{N_l}} \sum_{n=1}^{N_n} f(n) \exp[\frac{-j2\pi(n-1)(l-1)}{N_l}] \qquad l = 1,2,\Lambda\ N_l \tag{2}$$

and its inverse transform is

$$f(n) = \frac{1}{\sqrt{N_l}} \sum_{l=1}^{N_l} F(l) \exp[\frac{+j2\pi(l-1)(n-1)}{N_l}] \qquad n = 1,2,\Lambda\ N_n \tag{3}$$

In the frequency domain, N_l is the total number of samples of the DFT sequence $F(l)$. If we assume there are N_r samples per baud rate (the inverse of baud time), then

$$N_l = N_r N_s \tag{4}$$

Note that $N_r \geq N_c$, and increasing N_r increases resolution in frequency domain.

Our objective is to build a cost function, which can satisfy the orthogonal constraint. Firstly, the signal must concentrate energy in a given bandwidth. That is, maximize the energy in the set $\Phi = \{-N_b \leq l \leq N_b\}$, which corresponds to a frequency band of N_b / N_r baud rate, or equivalently, minimize the energy out of the above set. The energy out of the band can be given by

$$g_1 = \sum_{l \notin \Phi} F(l)^2 = \sum_{n=1}^{N_n} f(n)^2 - \sum_{l=-N_b}^{N_b} F(l)^2$$

$$= \sum_{n=1}^{N_n} f(n)^2 - \frac{1}{N_l} \sum_{n=1}^{N_n} \sum_{m=1}^{N_n} f(n) f(m) \frac{\sin[\frac{2\pi(n-m)}{N_l}(N_b + \frac{1}{2})]}{\sin[\frac{\pi(n-m)}{N_l}]} \tag{5}$$

Secondly, the desired signal must be orthogonal to its shifted version in order to minimize the intersymbol interference. This implies that

$$\sum_{n=1}^{N_n-(p-1)N_s} f(n) f[n+(p-1)N_s] - \delta(p) = 0 \qquad p = 1,2,\Lambda\ N_c \tag{6}$$

where $\delta(p)$ is the delta function. Note that when $p = 1$, it is the square of the function itself, and no constraint needs to be put on that. In order to include this constraint into the cost function, the second term can be defined as

$$g_2 = \sum_{p=1}^{N_c} w_p \left[\sum_{n=1}^{N_n-(p-1)N_s} f(n) f[n+(p-1)N_s - \delta(p) \right]^2 \tag{7}$$

where w_p ($p = 1,2,\Lambda\ N_c$) are the weights to the errors.

Thirdly, sometimes zero mean signals are required as an antenna does not radiate any dc components. That is,

$$\frac{1}{N_n} \sum_{n=1}^{N_n} f(n) = 0 \tag{8}$$

Therefore, the third term of the cost function can be given by

$$g_3 = w_n \left[\frac{1}{N_n} \sum_{n=1}^{N_n} f(n) \right]^2 \tag{9}$$

where w_n is the weight, too.

Then the cost function can be written as the sum of the three terms

$$g = g_1 + g_2 + g_3 \tag{10}$$

When non-zero-mean is required, the third term g_3 is not necessary. Minimizing the cost function will obtain the optimum signal. Some other constraints can be added to the cost function to get special T-pulse. For example, we can force the signal be zero at the beginning and at the end. Figure 1 and Figure 2 show the zero-mean and non-zero-mean T-pulse, respectively. Figure 3 and Figure 4 show their spectrum, respectively.

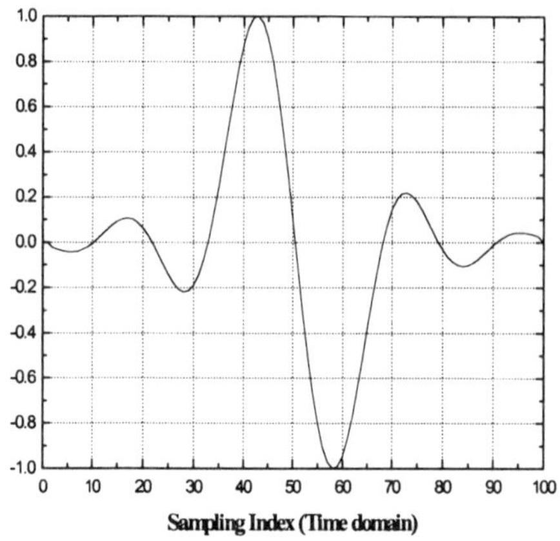

Figure 1. Zero-mean T-pulse

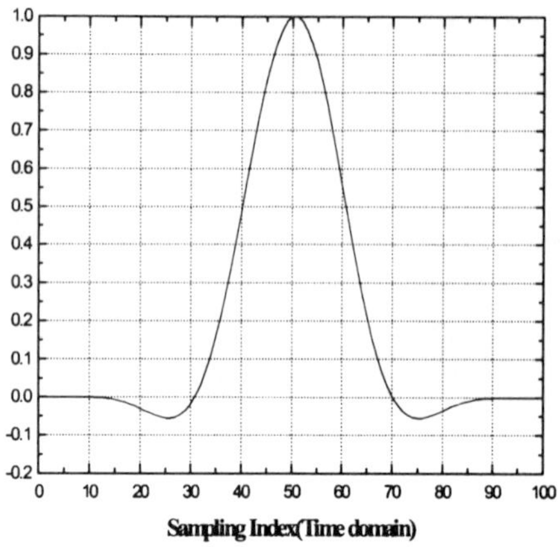

Figure 2. Non-zero-mean T-pulse

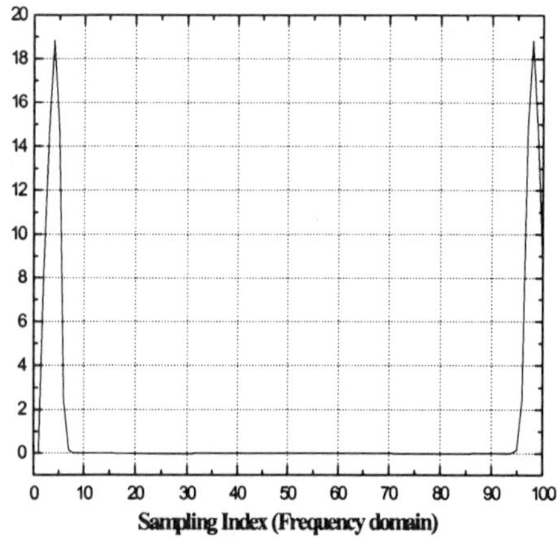

Figure 3. Spectrum of Zero-mean T-pulse

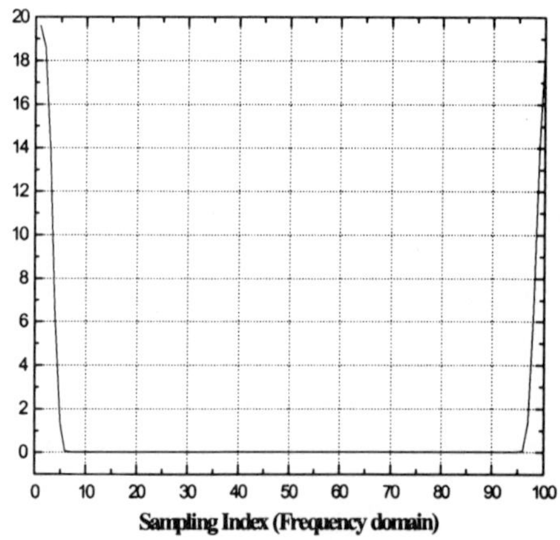

Figure 4. Spectrum of Non-zero-mean T-pulse

3. CFIE FORMULATIONS

3.1. Electric Field Integral Equations (EFIE)

Let S denote the surface of a perfectly conducting body, which may be closed or open, illuminated by a transient electromagnetic wave. This incident wave induces a surface current $\underline{J}$ on S. We have

$$\underline{E}^s(\underline{J}) = -\frac{\partial \underline{A}}{\partial t} - \nabla\phi \tag{11}$$

where $\underline{A}$ and ϕ are the magnetic vector potential and the electric scalar potential given by

$$\underline{A}(\underline{r},t)=\frac{\mu}{4\pi}\int_S\frac{\underline{J}(\underline{r}',\tau)}{R}dS' \tag{12}$$

$$\phi(\underline{r},t)=-\frac{1}{4\pi\varepsilon}\int_S\int_0^\tau\frac{\nabla'\cdot\underline{J}(\underline{r}',t')}{R}dt'dS' \tag{13}$$

Where $\tau=t-R/c$ is a retarded time, $R=|\underline{r}-\underline{r}'|$ represents the distance between the observation point $\underline{r}$ and the source point $\underline{r}'$, c is the velocity of wave propagation in the surrounding medium, and μ and ε are permeability and permittivity of that space, respectively.

Since the total tangential electric field is zero on the conducting surface, we have

$$[\underline{E}^i+\underline{E}^s(\underline{J})]_{\tan}=0,\ \underline{r}\in S. \tag{14}$$

Combining Eq.(11) and Eq.(14) gives

$$\left[\frac{\partial\underline{A}}{\partial t}+\nabla\phi\right]_{\tan}=\left[\underline{E}^i\right]_{\tan},\ \underline{r}\in S \tag{15}$$

where $\underline{E}^i$ is the incident electric field on the scatter and the subscript 'tan' denotes the tangential component. Equation (15) with Eq.(12) and Eq.(13) constitutes a time-domain EFIE from which the unknown current $\underline{J}(\underline{r},t)$ may be determined.

For the numerical solution of the time-domain integral equation we divide the time axis into equal intervals of segment Δt and define $t_i=i\Delta t$. The time derivative in Eq.(15) can be approximated by the forward or backward finite difference for the explicit and implicit solutions of a EFIE[5,6]. We now present a formulation to improve the accuracy and stability of the implicit solution. We write Eq.(15) at time $t=t_{i-1/2}$

$$\left[\frac{\underline{A}(\underline{r},t_i)-\underline{A}(\underline{r},t_{i-1})}{\Delta t}+\frac{\nabla\phi(\underline{r},t_i)+\nabla\phi(\underline{r},t_{i-1})}{2}\right]_{\tan}=\left[\underline{E}^i(\underline{r},t_{i-1/2})\right]_{\tan} \tag{16}$$

where we use the central finite difference for the time derivative associated with the vector potential term and the time averaging for the scalar potential term.

3.2. Magnetic Field Integral Equation (MFIE)

Similar as EFIE, from the boundary conditions, we can obtain MFIE

$$\frac{\underline{J}(\underline{r},t)}{2}-\hat{n}\times\frac{1}{4\pi}\int_{S_0}\nabla\times\frac{\underline{J}(\underline{r}',\tau)}{R}dS'=\hat{n}\times\underline{H}^i(\underline{r},t) \tag{17}$$

In Eq.(17), the curl operator is taken inside the integral, which is allowed as long as $R \neq 0$.

3.3. Combined Field Integral Equation (CFIE)

The integral equation is obtained by imposing the tangential boundary condition to either the electric field or the magnetic field, resulting Eq.(15) or Eq.(17), respectively. Both these equations, the EFIE and MFIE, fail to give a unique solution when the frequency spectrum of the incident wave includes internal resonant frequencies of the object. To overcome this problem, CFIE can be used. The time-domain CFIE is obtained by means of a linear combination of the EFIE with the MFIE through

$$(1-\kappa)[-\underline{E}^s(\underline{J})]_{\tan} + \kappa\eta[\underline{J} - \hat{n}\times\underline{H}^s(\underline{J})] = (1-\kappa)[\underline{E}^i]_{\tan} + \kappa\eta[\hat{n}\times\underline{H}^i] \tag{18}$$

Where κ is the parameter of linear combination, which is between 0 (EFIE) and 1 (MFIE), and η is the wave impedance of the space.

Using planar triangular patches to approximate the structure. The triangular patches have the ability to conform to any geometrical surface of the boundary. As in Ref. 7 and applying Galerkin's method to Eq.(15) and Eq.(17) choosing testing functions are same as the expansion functions. Details can be obtained from Ref. 8. We can obtain a matrix equation for the time-domain CFIE

$$[\alpha_{mn}][I_n(t_i)] = [\beta_m(t_i)] \tag{19}$$

Where

$$\alpha_{mn} = (1-\kappa)\alpha^E_{mn} + \kappa\eta\alpha^M_{mn} \tag{20}$$

$$\beta_m(t_i) = (1-\kappa)\beta^E_m(t_i) + \kappa\eta\beta^M_m(t_i) \tag{21}$$

Also the matrix $[\alpha_{mn}]$ is not function of time, and hence the inverse of the matrix need to be computed only once at the beginning of the computation step.

We now use the T-pulse as an excitation for the solution of the electric currents for the CFIE.

4. NUMERICAL RESULTS

First, we use the modulated T-pulse as an incident wave to analyze the scattering from some strips and poles and then use the matrix pencil method[9] to determine the resonant frequencies of the later time current at an edge. Here, the unit of time is light meters (lm). It means the time for light to travel a 1-meter distance. Table 1 shows the results. Figure 5 and Figure 6 show the structures with one ($Plate$: $0.64m\times0.64m$, $Pole$: $0.375m\times0.01m\times0.01m$) and two poles ($Plate$: $1.28m\times0.64m$, $Pole1$: $0.375m\times0.005m\times0.005m$, $Pole\ 2$: $0.5m\times0.005m\times0.005m$) mounted on a plate, respectively. Table 2 and Table 3 describe the results for these two structures, respectively. All the current results agree well with the inverse discrete Fourier transform (IDFT) results. The results show that the resonant frequency of the poles can be recognized.

5. CONCLUSION

In this paper, a time limited pulse with a practically band limited spectrum can be used as an incident wave. The resonant frequencies of characteristics features of the object can be excited using this method. The sub-structure resonances can be excited in this fashion.

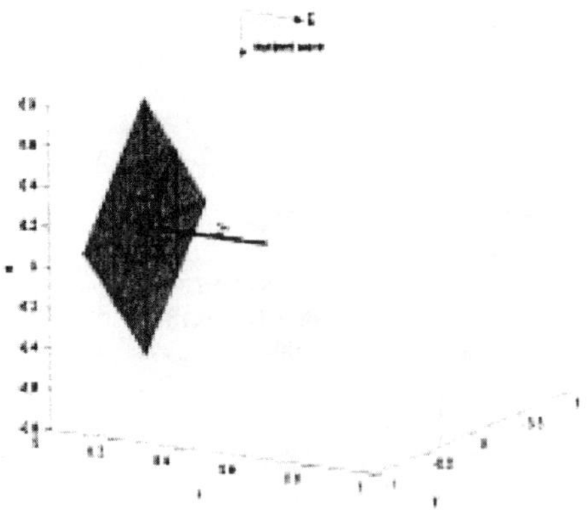

Figure 5. Structure 1

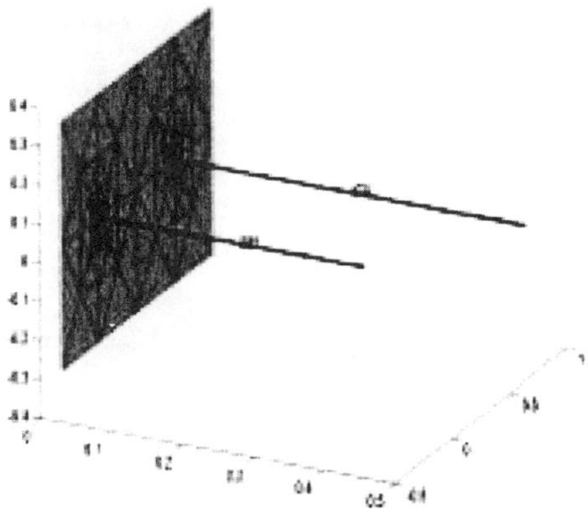

Figure 6. Structure 2

Table 1. Results for different size of strips and poles

Type	Size	Resonance (MHz)
Strip (open)	$1.5m \times 0.02m$	93.52
	$0.75m \times 0.02m$	185.24
	$0.5m \times 0.02m$	275.00
	$0.375m \times 0.02m$	363.32
	$0.3m \times 0.02m$	450.35
	$0.375m \times 0.04m$	352.31
	$0.375m \times 0.01m$	370.49
	$0.375m \times 0.002m$	379.25
Pole (closed)	$1.5m \times 0.02m \times 0.02m$	90.08
	$0.75m \times 0.01m \times 0.01m$	180.40
	$0.5m \times 0.01m \times 0.01m$	266.36
	$0.375m \times 0.005m \times 0.005m$	360.81

Table 2. Results of Structure 1

Start Time(lm)	Sample points	Resonance (MHz)	Corresponding Amplitude $\times 10^3$
8	45	**275.00**, 766.01	**17.6699**, 0.0014
10	45	**275.00**	**5.3554**

Table 3. Results of Structure 2 (45 sample points)

Edge	Start time (lm)	Resonance (MHz)	Corresponding Amplitude $\times 10^3$
355	10	**144.69** 628.22 **190.23** 433.81	**0.6661** 0.0089 **15.1441** 0.0866
	12	546.09 **146.61** **190.21** 431.68	0.0070 **0.4837** **9.1540** 0.0370
476	10	621.13 **146.10** **189.89** 433.26	0.0046 **9.4869** **2.1559** 1.5462
	12	548.91 **146.31** **190.52** 433.33	0.0068 **6.4305** **1.2481** 0.7566
15 (on the plate)	10	**145.00** **190.24** 433.30	**0.4766** **10.3963** 0.0447
	12	**146.67** **190.18** 432.42	**0.3644** **6.2833** 0.0208
183 (on the plate)	10	608.49 **146.01** **189.36** 433.30	0.0060 **6.3871** **1.5591** 1.0503
	12	539.43 **146.38** **190.20** 433.79	0.0101 **4.3551** **0.8925** 0.5212

6. REFERENCES

1. D.Slepian and H. Pollak, Prolate Spheroidal Wave Functions Fourier Analysis and Uncertainty I, *Bell Syst. Tech. J.*, **40**(1), 43-63(1961).
2. T.K.Sarkar, W. Lee, and S.M.Rao, Analysis of transient scattering from composite arbitrary shaped complex structures, *IEEE Trans. Antennas Propagat.*, **48**(10),1625-1634(2000).
3. D. A. Vechinski and S. M. Rao, Transient scattering by dielectric cylinders: E-field, H-field, and combined field solutions, *Radio Sci.*, **27**(5), 611-622(1992).
4. Y. Hua and T. K. Sarkar, Design of optimum discrete finite duration orthogonal Nyquist signals, *IEEE Transactions on Acoustics, Speech and Signal Processing*, **36**(4), 606-608(1988).
5. S. M. Rao and T. K. Sarkar, An alternative version of the time-domain electric field integral equation for arbitrarily shaped conductors, *IEEE Trans. Antennas Propagat.*, **41**(6), 831-834(1993).
6. S. M. Rao and T. K. Sarkar, An efficient method to evaluate the time-domain scattering from arbitrarily shaped conducting bodies, *Microwave Opt. Technol. Lett.*, **17**(5), 321-325(1998).
7. S. M. Rao, D. R. Wilton, and A. W. Glisson, Electromagnetic scattering by surfaces of arbitrary shape, *IEEE Trans. Antennas Propagat.*, **30**(5), 409-418(1982).
8. S. M. Rao, *Time Domain Electromagnetics* (Academic Press, 1999).
9. T.K. Sarkar, Sheeyun Park, Jinhwan Koh and S.M. Rao, Application of the matrix pencil method for estimating the SEM (singularity expansion method) poles of source-free transient responses from multiple look directions, *IEEE Transactions on Antennas and Propagation*, **48**(4), 612 –618(2000).

EXPLOITING EARLY TIME RESPONSE USING THE HALF FOURIER TRANSFORM (HFT)

Seongman Jang, Tapan K. Sarkar, Kyungjung Kim, and Carl E. Baum *

1. INTRODUCTION

The singularity expansion method (SEM) proposed by Baum[1] has been applied to quantify an electromagnetic response in an expansion of complex resonances of the system.[2] It has been shown that the dominant complex natural resonances of a system are a minimal set of parameters that define the overall physical properties of the system.[3] So, a transient scattering response is analyzed in terms of the damped oscillations corresponding to the complex resonant frequency of the scatterer or target. In general, the signal model of the observed late time of an electromagnetic-energy-scattered response from an object can be written as

$$y(t) = x(t) + n(t) \approx \sum_{m=1}^{M} R_m \exp(s_m t) + n(t); \quad 0 \le t \le T, \tag{1}$$

where $y(t)$ is the observed time domain response, $n(t)$ is the noise in the data, $x(t)$ is the signal, R_m is the mth residue or complex amplitude, $s_m = -\alpha_m + j\omega_m$, α_m is the mth damping factor, and ω_m is the mth angular frequencies ($\omega_m = 2\pi f_m$).

After sampling, the time variable t is replaced by kT_s, where T_s is the sampling period. The sequence can be rewritten as

$$y(kT_s) = x(kT_s) + n(kT_s) \approx \sum_{m=1}^{M} R_m z_m^k + n(kT_s) \quad \text{for } k = 0,\ldots,N\text{-}1, \tag{2}$$

* Seongman Jang, Tapan K. Sarkar, and Kyungjung Kim , Department of Electrical Engineering and Computer Science, Syracuse University, 121 Link Hall, Syracuse, NY 13244. Carl E. Baum, Air Force Research Laboratory, Kirtland Air Force Base, New Mexico.

$$z_m = e^{s_i T_s} = \exp(-\alpha_m + j\omega_m)T_s \text{ for } m = 1,2,\ldots,M. \tag{3}$$

Since the resonance describe global wave fields that encompass the scattering object as a whole, the SEM series representation encounters convergence difficulties at early times when portions of the objects are not yet excited. Because of this difficulty, most previous techniques used just late time signals only. The most well-known pole retrieval methods are Prony's method[4] and Matrix Pencil Method.[5] The Matrix Pencil Method is more robust to noise and does not require an *a priori* knowledge of the model order. But the SEM representation does not account for the impulsive portion of the early time behavior of the system response.

In this paper, the transient electromagnetic response is considered in the time domain and in the fractional Fourier Transform (FrFT) domain using the entire data set. The FrFT is a generalized Fourier transform. When using the FrFT, it is possible to discriminate an impulse or a Gaussian pulse from the other components of a signal. Because of this capability, impulse-like early-time components can be separated from the damped exponentials. To describe the early-time response, a Gaussian pulse is selected. A Gaussian pulse is an entire function[6] and is quite adequate to describe the pulse-like components in early time. Complex exponentials are used to describe the late-time signals. The concept of a 'Turn-on time'[7] is utilized to consider a time when the fully excited resonance can be used formally. In this study a Quasi-Newton approach, augmented by a model trust strategy, is used to optimize the parameters. Initial guesses for the damped exponentials were determined by the Matrix Pencil Method (MPM).

2. FRACTIONAL FOURIER TRANSFORM (FrFT)

In 1980, Victor Namias defined a fractional-order Fourier transform using Hermite polynomials, called the Fractional Fourier transform (FrFT),[8] as

$$F_\alpha\left\{e^{-x^2/2}H_n(x)\right\} = e^{-jn\alpha}e^{-x^2/2}H_n(x)\,, \tag{4}$$

where F_α is a Fourier transform operator, α is a continuous transformation parameter, and $H_n(x)$ is the Hermite polynomial of order n. It can be shown that the Hermite polynomials are eigenfunctions of the Fourier transform operator F_α and its eigenvalues are $e^{-jn\alpha}$. A short time later A. C. McBride and F. H. Kerr made Namias' definition more clear and rigorous,[9] and Luis B. Almeida used the engineering convention of Fourier transform to define FrFT and represented it as a rotation in the time-frequency plane.[10] Almeida's convention for FrFT and FT are used in this paper, so that a Fourier transform pair of $x(t)$ is defined as

$$F_{\pi/2}\{x(t)\}(\omega) = X(\omega) = \frac{1}{\sqrt{2\pi}}\int_{-\infty}^{\infty} x(t)e^{-j\omega t}\,dt\,. \tag{5}$$

The Fourier Transform of a signal can be interpreted as a $\pi/2$ angular rotation of the signal in the time-frequency plane. The FrFT is then developed and interpreted as the rotation of a signal to any angle α with the time axis in the time-frequency plane

(Fig.1).[10] It is known that the operators $F_{\pi/2}$ and $F_{-\pi/2}$ are complex conjugates and satisfy the relations $F_{\pi/2}F_{-\pi/2} = F_{-\pi/2}F_{\pi/2} = 1$. Also note that the additivity property of the fractional Fourier transform operator holds, that is,

$$F_\alpha + F_\beta = F_{\alpha+\beta}\,. \tag{6}$$

The FrFT is a linear transform defined by means of the transformation kernel as[10]

$$K_\alpha(t,u) = \begin{cases} \sqrt{\dfrac{1-j\cot\alpha}{2\pi}}\,e^{j\frac{t^2+u^2}{2}\cot\alpha - jut\csc\alpha}, & \text{if } \alpha \text{ is not a multiple of } \pi, \\ \delta(t-u), & \text{if } \alpha \text{ is not a multiple of } 2\pi, \\ \delta(t+u), & \text{if } \alpha+\pi \text{ is not a multiple of } 2\pi, \end{cases} \tag{7}$$

where $j=\sqrt{-1}$ and α is the rotation angle of a transformed signal. The forward and backward FrFTs are defined as

$$(F_\alpha x)(u) = X_\alpha(u) = \int_{-\infty}^{\infty} x(t)K_\alpha(t,u)dt\,, \tag{8}$$

$$(F_{-\alpha} X)(t) = x(t) = \int_{-\infty}^{\infty} X_\alpha(u)K_{-\alpha}(t,u)du. \tag{9}$$

So, if α is not a multiple of π, the forward FrFT is defined as

$$X_\alpha(u) = \sqrt{\frac{1-j\cot\alpha}{2\pi}}\,e^{j\frac{u^2}{2}\cot\alpha}\int_{-\infty}^{\infty} x(t)e^{j\frac{t^2}{2}\cot\alpha}\,e^{jut\csc\alpha}dt\,. \tag{10}$$

Thus the computation of the FrFT can be understood from the following steps: 1) multiply the signal by a chirp; 2) take the Fourier transform with its argument scaled by $\csc\alpha$; 3) then another product is taken with a chirp; 4) this product is scaled by a complex amplitude factor. From Eq. (9), the FrFT consists of expressing $x(t)$ using a basis formed by the set of functions $K_{-\alpha}(u,t)$, with u acting as a parameter for spanning the set of basis functions. Not only are the basis functions chirps, that is, complex exponentials with linear frequency modulation, they are also orthonormal. A number of important properties of the FrFT and their proofs can be found in Ref. 10. Those properties are extensions of the corresponding properties of the Fourier transform. Table 1 gives the half Fourier transform (HFT) of a number of special functions. The Half Fourier transform is a special case of the FrFT when $\alpha = \pi/4$.

Table 1. HFT of some special functions

	Signal	HFT ($\alpha = \pi/4$)
11	$\delta(t-\tau)$	$\sqrt{\frac{1-j}{2\pi}} e^{-\frac{j}{2}\tau^2} e^{\frac{j}{2}(u-\sqrt{2}\tau)^2}$
12	e^{jvt}	$\sqrt{1+j}\, e^{\frac{j}{2}v^2} e^{-\frac{j}{2}(u-\sqrt{2}v)^2}$
13	$e^{-c\frac{t^2}{2}}$	$\sqrt{\frac{1-j}{c-j} \exp\left\{\frac{ju^2}{2}\frac{c^2-1}{c^2+1}\right\} \exp\left\{-\frac{u^2}{2}\frac{2c}{c^2+1}\right\}}$
14	$e^{-c\frac{(t-\tau)^2}{2}}$	$\sqrt{\frac{1-j}{c-j}} \exp\left\{-\frac{c}{c^2+1}\left(u-\frac{\tau}{\sqrt{2}}\right)^2\right\} \exp\left\{\frac{j}{2}\left[\frac{c^2-1}{c^2+1}\left(u-\frac{\tau}{\sqrt{2}}\right)^2 + \frac{\tau^2}{2} - \sqrt{2}u\tau\right]\right\}$

Note: n = a nonnegative integer; u is the HFT domain variable; c is a constant; and $H_n(u)$ is the nth order Hermite polynomial.

In this research, the HFT is used for parameter optimization. Since many impulse-like components are present in the early time of a scattered transient response, it is difficult to separate each impulse-like component in the time domain. But it is possible to tell when those pulses occur in the HFT domain because the HFTs of an impulse and a Gaussian pulse, Eq. (11) and Eq. (14), have much more information about the times when they first appear. Components such as an impulse or a Gaussian pulse can be used to characterize the early-time representations and the complex exponentials to approximate late-time responses. Their functional representations are almost the same in the HFT domain. It is clear from Eqs. (11) and (12) that the HFT of the shifted impulse and the damped exponential have similar functional representations, except for the coefficient and the sign of the power of the exponent. So it is possible to separate the impulse-like component in the half Fourier transform domain quite easily.

Usually signals that we encounter in real life are causal, that is, $x(t) = 0$ if $t \leq 0$. Furthermore, the initial time at which the resonant component starts is different from component to component. So it is possible to assume that each resonant component has a different 'turn-on time'. In this study, the phrase 'turn-on time' is the time when each resonant component can be used formally to represent the temporal data. To apply FrFT to those causal signals, we present some formulas in Table 2. A detailed derivation of the FrFT of function (16) in Table 2 is shown next.

Table 2. HFT of some causal functions for t > 0

	Signal	HFT, $\alpha=\pi/4$
15	$e^{jvt}u(t)$	$\frac{\sqrt{1+j}}{2} e^{\frac{j}{2}v^2} \exp\left\{-\frac{j}{2}\left(u-\sqrt{2}v\right)^2\right\}\left[1-\Phi\left(\gamma\sqrt{\beta}\right)\right], \quad \gamma = j\left(\sqrt{2}u - v\right)$
16	$e^{jvt}u(t-\tau)$	$\frac{\sqrt{1+j}}{2} e^{\frac{j}{2}v^2} \exp\left\{-\frac{j}{2}\left(u-\sqrt{2}v\right)^2\right\}\left[1-\Phi\left(\gamma'\sqrt{\beta}\right)\right], \quad \gamma' = j\left(\sqrt{2}u - v - \tau\right)$

v is complex, $u(t)$ is unit step function, u is FrFT variable, $\beta = j/2$, a > 0, and $\Phi(z) = \int_0^z e^{-\lambda^2/2} d\lambda / \sqrt{2\pi}$.

For the signal $x(t) = e^{jvt}u(t-\tau)$, we have $u(t-\tau)$ is a unit step function and τ is a 'Turn-on time'. By using the definition of the FrFT in Eq. (15), we obtain

$$X_\alpha(u) = \sqrt{\frac{1-j\cot\alpha}{2\pi}}e^{\frac{j}{2}u^2\cot\alpha}\int_{-\infty}^{\infty} x(t)\exp\left\{\frac{j}{2}t^2\cot\alpha - jut\csc\alpha\right\}dt$$
$$= \sqrt{\frac{1-j\cot\alpha}{2\pi}}e^{\frac{j}{2}u^2\cot\alpha}\int_{\tau}^{\infty}\exp\left\{-\frac{t^2}{4\beta} - \gamma t\right\}dt \tag{17}$$

where

$$\beta = (j/2)\tan\alpha \qquad \text{and} \qquad \gamma = j(u\csc\alpha - v). \tag{18}$$

Now make the change of variable $y = t - \tau$ to yield

$$X_\alpha(u) = \sqrt{\frac{1-j\cot\alpha}{2\pi}}e^{\frac{j}{2}u^2\cot\alpha}\exp\left\{-\frac{\tau^2}{4\beta} - \gamma\tau\right\}\int_0^{\infty}\exp\left\{-\frac{y^2}{4\beta} - \left(\gamma + \frac{\tau}{2\beta}\right)y\right\}dy$$
$$= \sqrt{\frac{1-j\cot\alpha}{2\pi}}e^{\frac{j}{2}u^2\cot\alpha}\exp\left\{-\frac{\tau^2}{4\beta} - \gamma\tau\right\}\int_0^{\infty}\exp\left\{-\frac{y^2}{4\beta} - \gamma' y\right\}dy, \tag{19}$$

where

$$\gamma' = j(u\csc\alpha - v - \tau\cot\alpha). \tag{20}$$

From the integral in Eq. 3.322.2 of Ref. 11, we get

$$X_\alpha(u) = \frac{\sqrt{1+j\tan\alpha}}{2}e^{\frac{j}{2}u^2\cot\alpha}\exp\left\{-\frac{\tau^2}{4\beta} - \gamma\tau\right\}\exp\left\{\frac{j}{2}\tan\alpha\left(\gamma + \frac{\tau}{2\beta}\right)^2\right\}\left[1 - \Phi\left(\gamma'\sqrt{\beta}\right)\right], \tag{21}$$

where $\Phi(z)$ is the complex error function. After substituting β and γ from Eq. (18) into Eq. (21) and performing some manipulations, one obtains

$$X_\alpha(u) = \frac{\sqrt{1+j\tan\alpha}}{2}\exp\left\{-\frac{j}{2}\left[\left(u^2 + v^2\right)\tan\alpha - 2uv\sec\alpha\right]\right\}\left[1 - \Phi\left(\gamma'\sqrt{\beta}\right)\right], \tag{22}$$

where $v = \omega + ja$, $\gamma' = j(u\csc\alpha - v - \tau\cot\alpha)$, $\beta = (j/2)\tan\alpha$, and a>0. (23)

3. CALCULATION OF THE FrFT OF AN ARBITRARY INPUT SIGNAL

Namias[8] defined the generalized Fourier Transform operator F_α by

$$F_\alpha\left\{e^{-x^2/2}H_n(x)\right\}=e^{-jn\alpha}e^{-x^2/2}H_n(x), \tag{24}$$

where α is variable parameter, $-\pi\le\alpha\le\pi$, with negative values of α corresponding to the inverse transform and $H_n(t)$ is the nth order Hermite polynomial. Therefore, the usual Fourier Transform operator can be written as $F_{\pi/2}$ to denote the eigenvalue $e^{-jn\pi/2}$ associated with the orthogonal Hermite functions $e^{-x^2/2}H_n(x)$. The Associated Hermite (AH) polynomials $h_n(t,\lambda)$ are defined in terms of the Hermite polynomials $H_n(t,\lambda)$.[12]

$$h_n(t,\lambda)=\frac{1}{\sqrt{2^n n!\sqrt{\pi}\,\lambda}}H_n\left(\frac{t}{\lambda}\right)\exp\left(-\frac{t^2}{2\lambda^2}\right),\qquad n\ge 0, \tag{25}$$

where λ is a scaling factor. For $n \ge 2$, the Hermite polynomials are generated recursively through

$$h_n(t)=\frac{1}{\sqrt{n}}\left[\sqrt{2}th_{n-1}(t)-\sqrt{n-1}h_{n-2}(t)\right]. \tag{26}$$

The AH polynomials are orthogonal to each other and form a complete basis on the interval $[-\infty,\infty]$. If $x(t)$ is a piecewise smooth function in every finite interval $[-p,p]$ and $\int_{-\infty}^{\infty}e^{-t^2}x^2(t)dt < \infty$, then $x(t)$ can be expanded using the AH series as

$$x(t)=\sum_{n=0}^{\infty}a_n h_n(t), \tag{27}$$

$$a_n=\int_{-\infty}^{\infty}x(t)h_n(t)dt, \tag{28}$$

for $-\infty < t < \infty$. Applying Eq. (24) to Eq. (27) with Eq. (28) yields

$$F_\alpha x(t)(u)=\sum_{0}^{\infty}a_n e^{-jn\alpha}h_n(u). \tag{29}$$

The HFT can be obtained by substituting $\pi/4$ for α.

4. OPTIMIZATION

The back-scattered field of an object or target can be described by summations of complex exponentials and entire functions. To represent the pulse-like early-time components of such fields, an entire function is needed. A function is entire if it is analytic over the complex plane; that is, it has no singularities. In this study, a Gaussian pulse, which is entire function and time limited,[6] is used for the early-time description. So, it is appropriate to represent superimposed impulse-like signals at early times of a scattered response. Complex natural frequencies occur in complex conjugate pairs and lie in the left half of the complex plane with nonzero real part. To represent real signals, we treat two conjugate poles together. So, the scattered field can be represented by

$$x(t)=\frac{1}{2}\sum_{m=1}^{M}c_m e^{-\alpha_m t}\left\{e^{j(\omega_m t+\phi_m)}+e^{-j(\omega_m t+\phi_m)}\right\}u(t-\tau_m)+\sum_{n=1}^{N}A_n\exp\left\{-C_n\frac{(t-B_n)^2}{2}\right\}, \quad (30)$$

where $t\geq 0$, $\tau\geq 0$, $\alpha_m>0$. The parameters c_m, ϕ_m, α_m, and ω_m of the first summand are the amplitude, phase, damping factor, and frequency, respectively. The constants A_n, B_n, and C_n are the amplitude, time shift, and a coefficient representing the temporal duration, respectively, of the nth Gaussian pulse. The numbers M and N respectively are the number of sinusoidal signals and the number of Gaussian pulses. To apply a parameter identification algorithm, the parameter set is defined by

$$\underline{p}=[\omega_1\ \phi_1\ c_1\ \alpha_1\ \tau_1\cdots\omega_M\ \phi_M\ c_M\ \alpha_M\ \tau_M\ ;B_1\ A_1\ C_1\cdots B_N\ A_N\ C_N], \quad (31)$$

and the residual vector is defined by

$$r=\frac{1}{2}\left\|\bar{G}(u)=\bar{G}_M(u,p)\right\|_2^2, \quad (32)$$

where $\bar{G}(u)$ is the HFT of the measured signal, $\bar{G}_M(u,p)$ is the HFT of the damped sinusoidal model including the pulses, and $\|\cdot\|_2^2$ is the square of the 2 norm. From Eq. (14) of Table 1 and Eq. (16) of Table 2, $\bar{G}_M(u,p)$ is constructed as follows:

$$\begin{aligned}\bar{G}_M(u,p)=&\frac{1}{2}\sum_{m=1}^{M}c_m\frac{\sqrt{1+j}}{2}\left\{e^{j\phi_m}e^{\frac{j}{2}v_{m1}^2}\exp\left\{-\frac{j}{2}\left(u-\sqrt{2}v_{m1}\right)^2\right\}\left[1-\Phi\left(\gamma_{m1}\sqrt{\beta}\right)\right]\right.\\&\left.+e^{-j\phi_m}e^{\frac{j}{2}v_{m2}^2}\exp\left\{-\frac{j}{2}\left(u-\sqrt{2}v_{m2}\right)^2\right\}\left[1-\Phi\left(\gamma_{m2}\sqrt{\beta}\right)\right]\right\}\\&+\sum_{n=1}^{N}A_n\sqrt{\frac{1-j}{C_n-j}}\exp\left\{-\frac{C_n}{C_n^2+1}\left(u-\frac{B}{\sqrt{2}}\right)^2\right\}\exp\left\{\frac{j}{2}\left[\frac{C_n^2-1}{C_n^2+1}\left(u-\frac{B}{\sqrt{2}}\right)^2+\frac{B^2}{2}-\sqrt{2}uB\right]\right\},\end{aligned} \quad (33)$$

where $v_{m1}=b+ja$, $v_{m2}=-b+ja$, $\gamma_{m1}=j\left(\sqrt{2}u-v_{m1}-\tau_m\right)$, and $\gamma_{m2}=j\left(\sqrt{2}u-v_{m2}-\tau_m\right)$.

To increase the speed of convergence of the optimization process and to avoid the convergence to undesirable local minima, initial guesses of the parameters are important. In this study, the TLS-MPM is used for extracting the parameters of complex exponentials as initial guesses. The TLS-MPM approach is the most efficient and robust to fit noisy data with a sum of complex exponentials.[5]

5. EXAMPLE

In this section, one example is presented to validate the above technique: a conducting sphere of radius 1 cm. The back-scattered field has been computed using WIPL-D[13] in the frequency domain. The incident field is coming from the top and is x-polarized. A Gaussian window is applied to limit the bandwidth. The inverse transformed time-domain data is shown in Fig. 2 as a solid line. This data exhibits two kinds of waves: directly reflected pulses and creeping waves. The time difference in the location between those two pulses is about 0.176 ns, which is good agreement with the calculated value of 0.1714 ns $\{=(2+\pi)\, r/(3\times10^8)\}$. The order of the Hermite polynomial is also determined using the time-bandwidth product rule of $2BT+1$, where B is the bandwidth in the frequency domain and T is the time duration of the signal. Therefore, to generate the same waveform of a signal of duration T and practically band limited by B with a set of orthonormal basis functions in the time domain, at least $(2BT+1)$ basis functions are necessary from a mathematical point of view. In this example, the frequency band B is 100 GHz and the time duration T is 0.65 ns. So in this case $N = 2\times100\times0.65+1 = 131$. But $N = 301$ has been chosen to guarantee convergence of the solution.

By using the Matrix Pencil Method, 21 pole-pairs are selected as an initial guess of the complex exponentials. The solid line in Fig. 4 shows the half Fourier transform of the back-scattered field, which is calculated using the expansion coefficient of the Associated Hermite function. Parameter optimization is performed simultaneously in the HFT and time domains to select the optimum set of parameters of the Gaussian-like impulses and the decaying exponentials. In this example, 3 Gaussian pulses have been obtained which are used to represent an early-time signal. Those pulses are indicated by the asterisks (*) in Fig. 2. The asterisks represent the amplitude and position of the pulses. It appears that there is a pulse doublet. This is obtained from a derivative of the incident pulse. The last one is the creeping wave. Another creeping wave can be found around 0.4 ns, but its amplitude is very small.

Reconstructed time domain and HFT domain signals are shown in Fig. 2 and in Fig. 3 as shown by dots. The root mean square error for the time-domain data is 0.0036 and that for the HFT domain data is 0.0026. Figure 4 shows the estimated poles and compares them with the analytic values. Analytical data for the poles were obtained from Ref. 14. In this figure, a is the radius of the sphere and c is speed of light. They show good agreement even for the damping constants.

6. CONCLUSIONS

A new technique is presented for extracting the parameters of the damped sinusoids and the early-time impulse-like responses using the entire transient response data set, which includes both the early-time and late-time responses. Early-time responses have pulse-like components that make it hard to analyze the complete signal. In this study, Gaussian pulses have been used to describe the early-time portion of the dat, and a fractional Fourier transform is used to separate pulse-like components from damped exponentials. An initial guess for the complex exponentials using the Matrix Pencil

Method gives a good starting point for the optimization process. We show that the Half Fourier Transform is very effective when the transient response from a target has pulse-like components. The results for the conducting sphere show that the proposed technique performs well in extracting the aspect-independent complex-resonance information of the object. Analytical results from the literature[14] show that there are multiple layers of poles, but only the poles from the first layer are observed in this study.

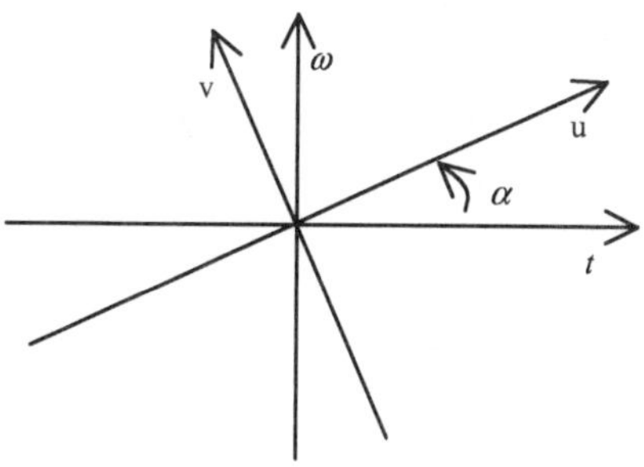

Figure1. Time-frequency plane and a set of coordinates (u,v) rotated by an angle α relative to the original coordinates (t,ω).

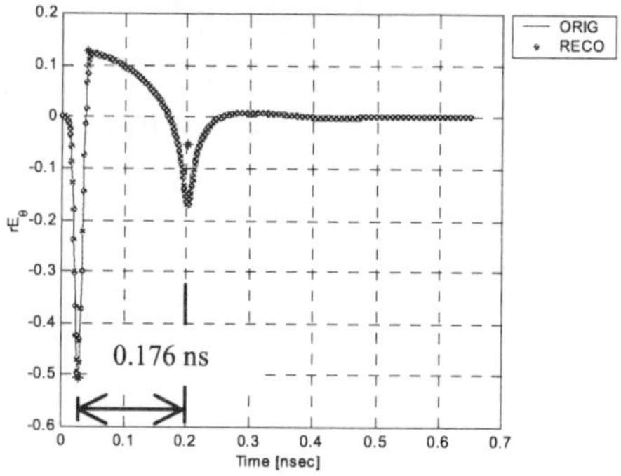

Figure 2. Back-scattered field of a conducting sphere in the time domain.

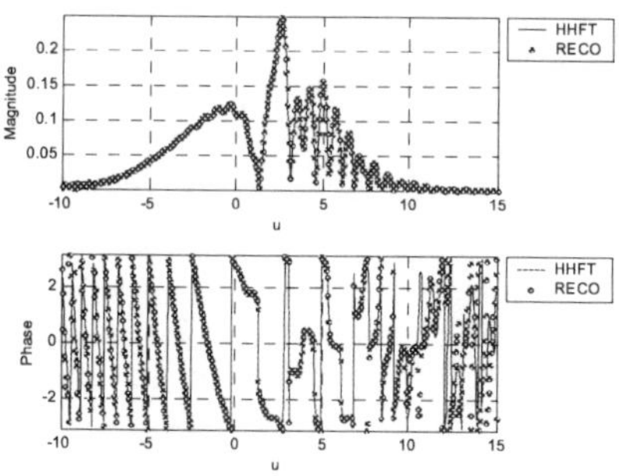

Figure 3. Half Fourier Transform (HFT) domain signals. Dot is the reconstructed signal using optimized parameters.

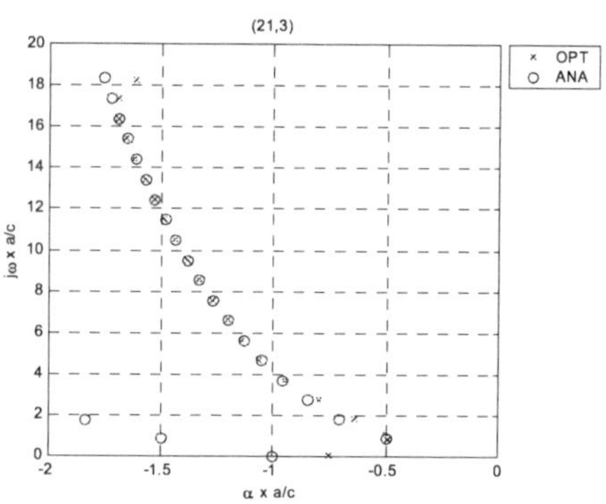

Figure 4. Pole plot of the Conducting sphere.

REFERENCES

1. C.E. Baum, On the singularity expansion method for the solution of electromagnetic interaction problems, *EMP Interaction Notes 88*, Air Force Weapons Laboratory, Kirtland Air Force Base NM (1971).
2. C.E. Baum, Emerging technology for transient and broad-band analysis and synthesis of antenna and scatterers, *Proc. IEEE* **64**(11), 1598–1616 (1976).
3. D.L. Moffatt and R.K. Mains, Detection and discrimination of radar targets, *IEEE Trans. Antennas Propagat.* **AP-23**(3), 358–367 (1975).
4. M.L. VanBlaricum and R. Mittra, A technique for extracting the poles and residues of a system directly from its transient response, *IEEE Trans. Antennas Propagat.* **AP-23**(6), 777–781 (1975).
5. Y. Hua and T.K. Sarkar, Matrix pencil method for estimating parameters of exponentially damped/undamped sinusoids in noise, *IEEE Trans on Acoustics, Speech, Signal Proc.* **38**(5), 814–824 (1990).
6. C.E. Baum, Representation of surface current and far scattering in EEM and SEM with entire functions, *EMP Interaction Note 486*, Air Force Weapons Laboratory, Kirtland Air Force Base NM (1971).
7. E. Heyman and L.B. Felsen, Wavefront interpretation of SEM resonances, turn-on time, and entire functions, in: *Hybrid Formulation of Wave Propagation and Scattering*, edited by L.B. Felsen (The Hague, Nijhoff, Netherlands, 1984), pp. 253–267.
8. V. Namias, The fractional order Fourier transform and its application to quantum mechanics, *J .Inst. Math. Applic.* **25**, 241–265 (1980).
9. A.C. McBride and F.H. Kerr, On Namias's fractional Fourier transforms, *IMA J. Appl. Math*, **39**, 159–175 (1987).
10. L.B. Almeida, The fractional Fourier transform and time-frequency representations, *IEEE Trans. On Signal Proc.* **42**(11), 3084–3091 (1984).
11. I.S. Gradshteyn and I.M. Ryzhik, *Table of Integrals, Series, and Products* (Academic Press, New York, 1980), p. 307.
12. L.R. Conte, R. Merletti, and G.V. Sandri, Hermite expansions of compact support waveforms: applications to myoelectric signals, *IEEE Trans. Biomedical Eng.* **41**(12) (1994).
13. B.M. Kolundzija, J.S. Ognjanovic, and T.K. Sarkar, *WIPL-D for Windows Manual* (Artech House, Norwood, 2000).
14. K.M. Chen and D. Westmoreland, Impulse response of a conducting sphere based on singularity expansion method, *Proc. IEEE* **69**(6), 747–750 (1981).

DETECTION OF A BURIED OBJECT WITH PULSE-COMPENSATED WIRE ANTENNAS

Stefan H.J.A. Vossen, Anton G. Tijhuis, Eugène S.A.M. Lepelaars and A. Peter M. Zwamborn*

ABSTRACT

For the detection of a buried object we consider two straight thin-wire antennas above an interface between two homogeneous dielectric half spaces. One antenna is a transmitting wire and the other is a receiving wire. Our aim is to use this simple antenna set up for the detection of buried objects without applying pre and post processing of the received signal. The buried wire can be detected directly from the shape of the current at the receiving antenna. To minimize the effects of repeated reflections at the end faces of the wire antennas, pulse compensation is introduced. Numerical results show that with the aid of pulse compensation a buried wire can easily be detected.

1. INTRODUCTION

We consider the feasibility of detecting a buried object with the aid of two straight thin-wire antennas above an interface between two homogeneous dielectric half spaces. One antenna is transmitting, and the other one is receiving. Our aim is to use this simple detection set up without applying pre and post processing to the received signal. As an initial test object we consider a buried straight thin-wire antenna, as shown in Figure 1. In the examples, all wires have a length $L = 1$ m and radius $a = 2$ mm. The distance between the wires of the detection setup is $d = 1$ m and they are located at a height $z_1 = -0.1$ m. The buried wire is located at a depth z_2 with a lateral distance $d_2 = 0.5$ m with respect to wire 1. The upper half space is vacuum. The lower half space has a permittivity ε_2, permeability $\mu_2 = \mu_0$ and a conductivity σ_2.

*Stefan H.J.A. Vossen, Eugène S.A.M. Lepelaars and A. Peter M. Zwamborn, TNO Physics and Electronics Laboratory, P.O. Box 96864, 2509 JG 's-Gravenhage, The Netherlands. Anton G. Tijhuis, Eindhoven University of Technology, Electrical Engineering, P.O. Box 513, 5600 MB Eindhoven, The Netherlands.

Ultra-Wideband Short-Pulse Electromagnetics 6
Edited by Mokole *et al.*, Kluwer Academic/Plenum Publishers, 2003

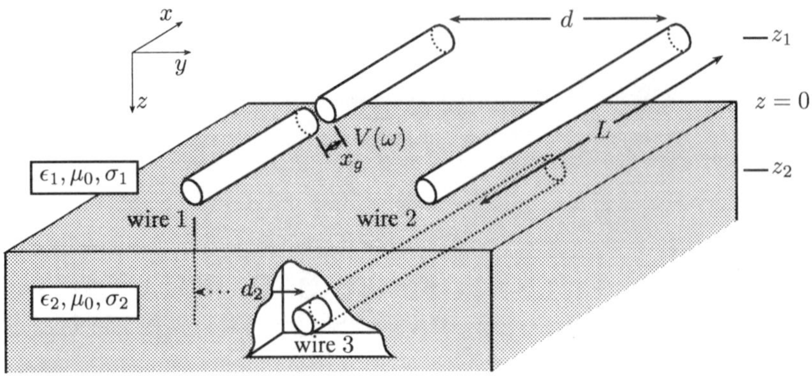

Figure 1: Two wires above an interface between two half spaccs and one buried wire.

2. THREE COUPLED WIRES

The problem of three coupled wires in two homogeneous half spaces is solved numerically with the aid of coupled Hallén equations and a dedicated spectral approach for obtaining the Green's functions.

The theory involved in the modeling of the mutual coupling between the transmitting and receiving antennas as well as the reflected and transmitted fields at the interface has already been described in previous work [1,2]. The transmitting wire is excited at its center by a transient Gaussian voltage $\mathcal{V}(t) = \mathcal{V}_0 \exp[-((t - t_1)/\tau)^2]$, where we have chosen $\mathcal{V}_0$=1 V, t_1=2 ns, τ=0.5 ns. This pulse generates traveling current waves along both wires. For conducting wires, it was observed that repeated reflections of these current waves at the end faces of both wire antennas result in a strongly oscillating behavior at late times. The presence of a buried wire in the lower half space results in a small amplitude change of the current along the receiving wire. Therefore, a direct observation of the effect of a buried wire from the shape of that current is very difficult.

3. PULSE COMPENSATION

To suppress the end reflections, the transmitting and receiving wires can be coated with a so-called Wu-King resistive profile [1,2]. Repeating the numerical experiments for the coated wires showed that the presence of a buried wire in the lower half space is directly visible in the shape of the current along the receiving wire. A major disadvantage of a resistance profile is the high power consumption.

As an alternative method for suppressing the repeated reflections, pulse compensation is introduced. For the entire configuration of the three coupled wires of Figure 1, pulse compensation is carried out in three stages.

3.1. Compensation of the current along the transmitting wire

First we consider only the transmitting antenna in free space. After the initial voltage pulse, a second voltage pulse is impressed with a carefully chosen amplitude A_1 and time delay T_1. Combining the currents generated by the original voltage pulse and the compensation pulse results in a single current pulse along the transmitting wire:

$$\mathcal{I}_{\text{tot}}(x,t) = \mathcal{I}(x,t) + A_1\mathcal{I}(x,t-T_1), \tag{1}$$

where $\mathcal{I}(x,t)$ is the current along the wire pertaining to a single voltage pulse $\mathcal{V}(t)$. The parameters A_1 and T_1 are determined by minimizing the integral

$$\int_{t_{\text{min}}}^{t_{\text{max}}} \mathcal{I}^2_{\text{tot}}(x_g,t)dt, \tag{2}$$

where $x_g = L/2$ is the center of the wire and $[t_{\text{min}}, t_{\text{max}}]$ is the time interval where the compensation should take place. It turns out that the parameters A_1 and T_1 are hardly influenced by the end point t_{max}, but the choice of t_{min} is very critical. Before the actual compensation can start, the complete current pulse should travel along the wire at least once, see Figure 2. For a wire with length $L = 1$ m and radius a=2 mm,

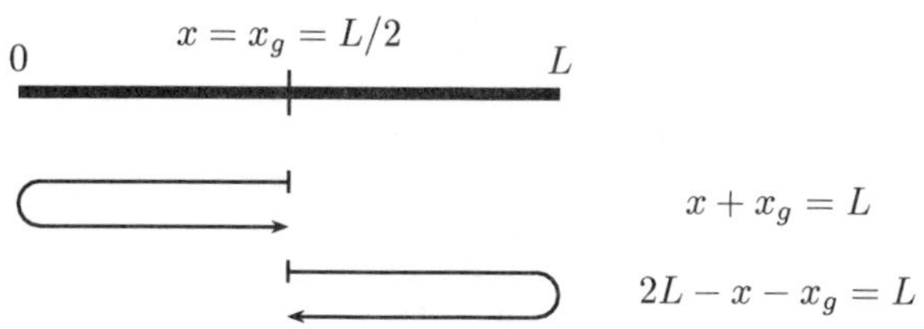

Figure 2: The current waves arriving at $x = x_g = L/2$ after one reflection.

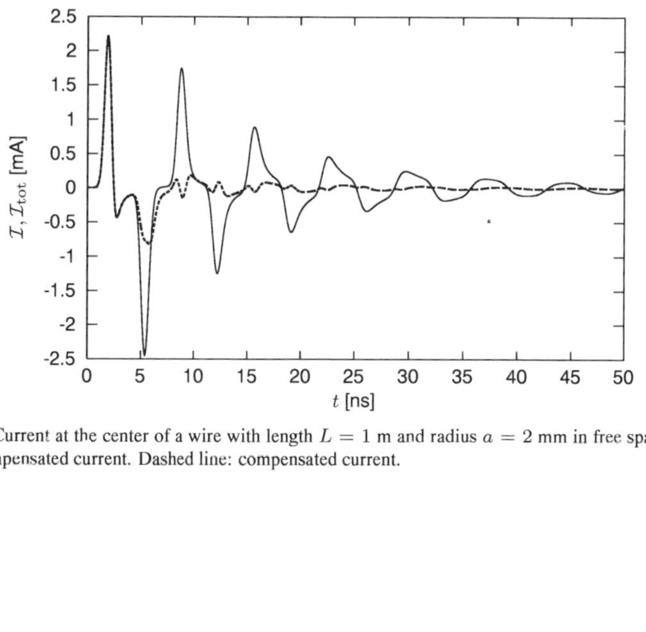

Figure 3: Current at the center of a wire with length $L = 1$ m and radius $a = 2$ mm in free space. Solid line: uncompensated current. Dashed line: compensated current.

excited by the Gaussian voltage pulse as mentioned before, the minimization with $t_{\min}$=7 ns and $t_{\max}$=80 ns results in the parameters A_1=0.7590 and T_1=3.4637 ns.

In Figure 3 the effect of pulse compensation for a single wire with these parameters is demonstrated. Since all calculations are carried out in the frequency domain, it is possible to choose a complex-valued parameter A_1. The imaginary part of A_1 then can be interpreted with the aid of the theory of dual analytic signals [3] and provides an extra minimization parameter. In the time domain the total current results in

$$\mathcal{I}_{\rm tot}(x,t) = \mathcal{I}(x,t) + ({\rm Re}A_1)\mathcal{I}(x,t-T_1) + ({\rm Im}A_1)\mathcal{I}_H(x,t-T_1), \qquad (3)$$

where $\mathcal{I}_H$ denotes the Hilbert transform of $\mathcal{I}$. After minimizing (2) with the same time interval $[t_{\min}, t_{\max}]$ as mentioned earlier, the compensation parameters are found as ${\rm Re}A_1 = 0.7538$, ${\rm Im}A_1 = 0.1315$ and $T_1 = 3.379$ ns. The effect of the extra minimization parameter is shown in Figure 4.

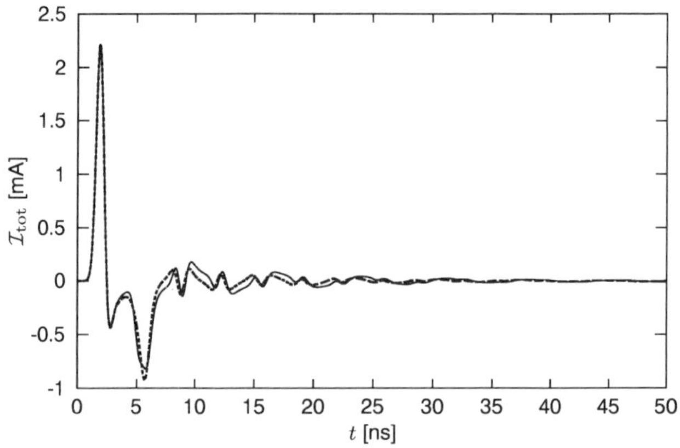

Figure 4: Current at the center of a wire with length $L = 1$ m and radius $a = 2$ mm in free space. Solid line: Compensated current with real-valued A_1. Dashed line: Compensated current with complex-valued A_1.

It is observed that the complex-valued A_1 minimizes the integral (2) slightly better.

3.2. Compensation of the current along the receiving wire

With the above pulse compensation, reflections are suppressed and the electromagnetic field of the transmitting wire therefore almost consists of a single pulse. To demonstrate this, the transmitted electric field from wire 1, $\mathcal{E}_1^{\rm t}$ is calculated at the position of wire 2 for $x = 0$, $x = L/4$ and $x = x_g = L/2$, respectively. The results are depicted in Figure 5. It is observed that the transmitted electric field consists mainly of a single current pulse. Therefore, the current along wire 2 is attenuated considerably due to the compensation of wire 1.

Now we return to the three-wire configuration of Figure 1. The transmitted pulsed field will result in repeated reflections along the receiving wire and an unwanted reflection at the interface. These effects can be compensated in a similar fashion as for

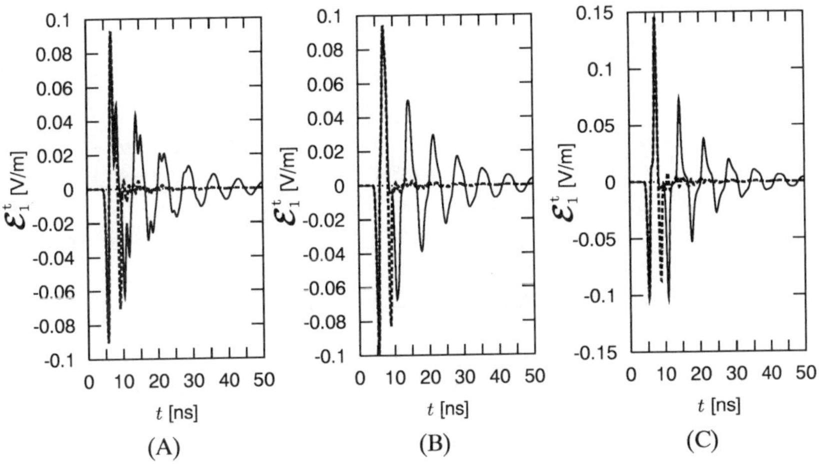

Figure 5: The transmitted electric field of wire 1 with length $L = 1$ m and radius $a = 2$ mm in free space at $z = z_1$, $y = d$ and $x = 0$ (A), $x = L/4$ (B) and $x = L/2$ (C). Solid line: Uncompensated current. Dashed line: Compensated current.

a single thin wire. Each effect is compensated by one specific set of compensation parameters.

Let us focus on the repeated reflections along the receiving wire. To this end, we only consider the parallel transmitting and receiving wires in free space. The incident field at each point x of this wire may be regarded as a small voltage pulse. After two reflections, see Figure 6 and a total travel distance of $2L$, the pulse reaches the same point x again. And this holds for each point x. The compensation parameters for wire

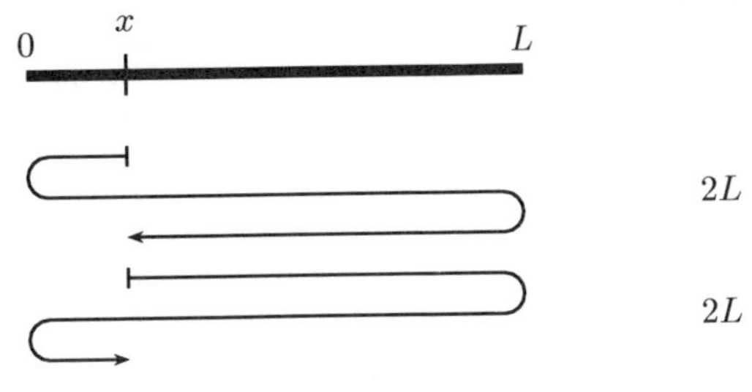

Figure 6: The current waves arriving at a point x after two reflections.

2 in free space are based on two reflections and a $2L$ travel distance. The total current along the receiving wire can thus be written as

$$\mathcal{I}_{2,\mathrm{tot}}(x,t) = \mathcal{I}_2'(x,t) + (\mathrm{Re}A_2)\mathcal{I}_2'(x,t-T_2) + (\mathrm{Im}A_2)\mathcal{I}_{2,H}'(x,t-T_2), \qquad (4)$$

where $\mathcal{I}_2'(x,t)$ is the current along the receiving wire when the current along the receiving wire is pulse compensated. Since both the transmitting and receiving wires are assumed to be identical, the following relations between the compensation parameters is almost valid:

$$A_2 = -A_1^2, \qquad T_2 = 2T_1. \tag{5}$$

These relations are not exact because coupling effects between the wires are not yet taken into account, but they serve as a good initial estimate for the minimization procedure of the compensation parameters of the receiving wire. Figure 7 shows the current along the receiving wire. Pulse compensation of wire 1 attenuates the current

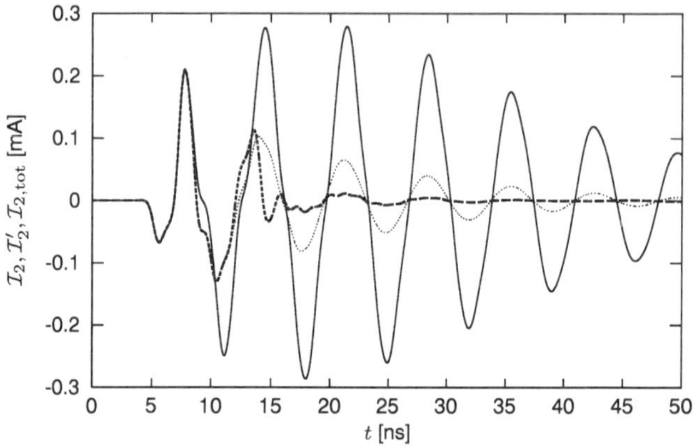

Figure 7: Current at the center of the receiving wire. The solid line represents the current without any compensation. The dotted line shows the current after pulse compensation of the transmitting wire only. The dashed line shows the current after compensation of both wires.

along wire 2 considerably. Again, pulse compensation of wire 2 suppresses repeated reflections of the current along wire 2 quite well.

3.3. Compensation of interface reflections

As mentioned above, the reflection at the interface causes an unwanted effect on the current along wire 2. In a similar manner as described before, the effect of this reflection on the current along the receiving wire can be suppressed.

To this end, an extra set of compensation parameters A_3, T_3 is introduced. The values of these parameters depend on the height z_1 of the detection setup, the distance d between the transmitting and receiving wire and on the material parameters ε_2 and σ_2 of the lower half space. An additional minimization procedure makes it is possible to find the necessary parameters to compensate the effects of interface reflections on the current along the receiving wire.

An example of the additional compensation for interface reflections is given in Figure 8. The lower half space is lossless with a relative permittivity $\varepsilon_{2r} = 3$. It is observed that the interface indeed has a strong effect on the current along wire 2, which was compensated with the free space parameters. Once more, the additional third compensation decreases the effects of the interface on the current along wire 2.

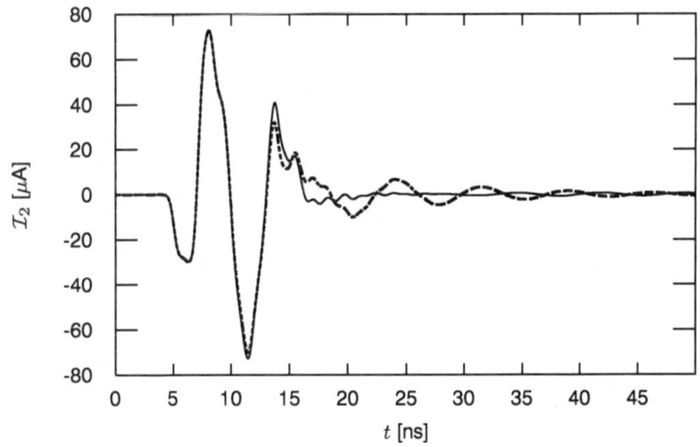

Figure 8: Current at the center of the receiving wire. The solid line represents the current without the compensation of interface reflections. The dashed line shows the current after this compensation.

4. RESULTS

In previous sections, we observed that the current along the wires of the detection set up can be compensated quite well. The compensation of the receiving wire yields two sets of parameters. One set to compensate for the free space contribution of wire 1 and one set for the compensation of interface reflections.

Based on results for wires with a Wu-King resistive profile [2], it is now possible to observe the effects of a buried wire directly from the current along the receiving wire. As an example, Figure 9 shows the current along the receiving wire antenna with and without a buried wire. The buried wire is located at a depth of $z_2 = 1$ m. The upper half space is vacuum, the lower half space has a relative permittivity ε_{2r}=9. The lower half space is assumed to be lossless. It should be pointed out that the wire can be detected so well because it is buried relatively deep. Therefore the total travel time of the transmitted electromagnetic field to and from the buried wire is large enough for the pulse compensation to be effective. This remarkable fact makes pulse compensation an excellent way to detect buried objects, especially when they are relatively deep.

As a final example, both a B-scan and a synthetic seismogram of a lossless lower half space with permittivity $\epsilon_{2r} = 9$ and a buried wire are visualized in Figure 10. This figure shows the current at the center of the receiving wire as a function of time and lateral distance d_2 (see Figure 1). The buried wire is clearly visible from the non-uniformity in the d_2-direction.

5. CONCLUSION

We obtained an efficient and simple way to suppress unwanted current reflections at the end of wire antennas. Next to the repeated reflections, effects of an interface on the current along wire antennas can be compensated as well.

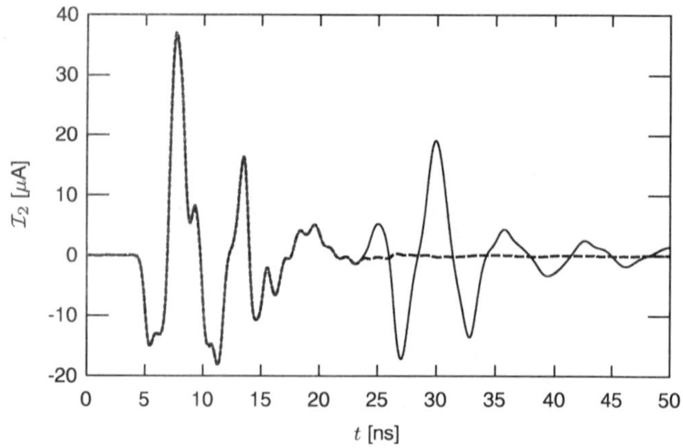

Figure 9: The induced current at the center of the receiving wire with (solid) and without (dashed) a buried wire. All wires have a length $L = 1$ m and radius $a = 0.002$ m. The distance between the transmitting and receiving wires is $d = 1$ m and the height of the detection set up is $z_1 = -0.1$ m. The buried wire is located at a depth of z_2=1 m.

Therefore, it is possible to detect a buried object directly from the received current signal without applying pre and post processing. Unlike a Wu-King resistance profile, pulse compensation does not have the disadvantage of a high power consumption. In addition, the necessary processing can be carried out either in software or in hardware, which will result in a low-cost detection set up.

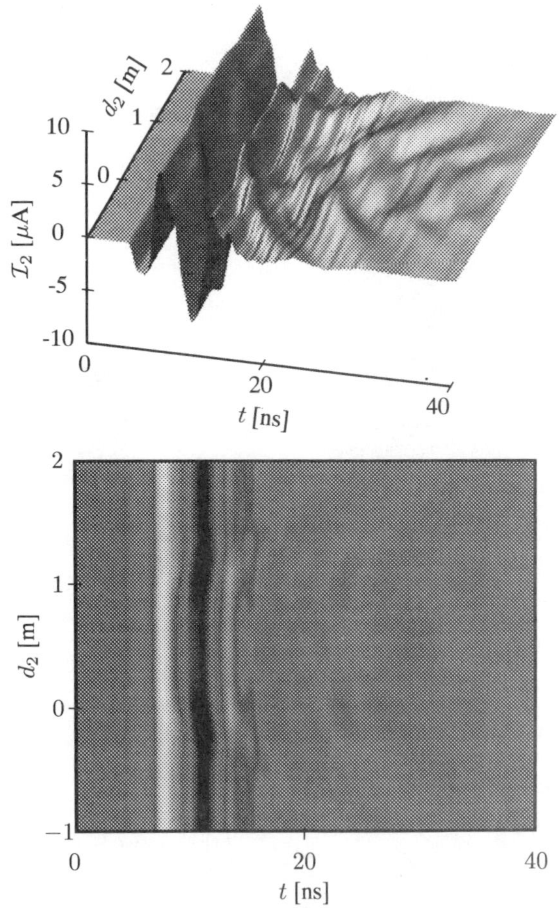

Figure 10: The induced current at the center of the receiving wire. All wires have a length $L = 1$ m and radius $a = 0.002$ m. The distance between the transmitting and receiving wires is $d = 1$ m and the height of the detection set up is $z_1 = -0.1$ m. The buried wire is located at a depth of z_2=1 m. The distance between the buried wire and wire 1 is taken as $-1 < d_2 < 2$.

6. REFERENCES

1. A. Rubio Bretones, A.G. Tijhuis, R. Gómez Martín and S.H.J.A. Vossen, "Transient excitation of two loaded wires above a half-space with a cover layer", to be published in *Journal of Electromagnetic Waves and Applications*.
2. S.H.J.A. Vossen, A.G. Tijhuis, E.S.A.M. Lepelaars and A.P.M. Zwamborn, "Detection of a buried wire with two resistively loaded wire antennas", accepted for publication in *Radio Science*.
3. M.Born and E.Wolf, *Principles of Optics* (Pergamon Press, Oxford, 1975).

TARGET RECOGNITION VIA LATE TIME SCATTERING

K.J. Pascoe, W.D. Wood, P.S. Maybeck, and A.W. Wood*

1. INTRODUCTION

Since the development of the Singularity Expansion Method (SEM), several authors have attempted to identify targets via their SEM parameters.[1] There have been several approaches using resonance annihilation filters, known as the E-pulse, K-pulse, S-pulse, etc.[2-5] There have also been some attempts using signal-processing techniques such as ARX, IV, ARMAX,[6,7] and GLRT.[8]

In this paper we describe a new architecture for target recognition that has the potential for near-real-time application. We separate the target recognition problem into two stages, library construction and library reading. Several techniques are of potential value in the library construction phase. For the library reading phase, the authors introduce the use of a bank of Kalman filters, each of which corresponds to the SEM response from a target type.

In the next section we provide background information on SEM. Following that, we briefly discuss possibilities for building the library. In Section 4 we develop a technique to read from this library using Kalman filters. We present results in Section 6.

2. SINGULARITY EXPANSION METHOD

According to the Singularity Expansion Method developed by Baum,[9,10] late-time scattering from short, UWB pulses can be represented by an infinite series of poles in the complex frequency s-plane. In the s-plane, or Laplace domain, poles will appear either on the real axis or as conjugate pairs in the left half of the s-plane. If the target is a finite, perfectly conducting scatterer, the poles will all be simple, first-order poles. These poles have the form $c / (s - s_n)$, where c and s_n are complex constants and s is a complex

* Air Force Institute of Technology, Wright-Patterson AFB OH, 45433, USA.

frequency variable. The pole location or natural frequency is $s_n = \Omega + j\omega$ and the magnitude (residue) associated with it is $c = a + jb$. For each pole the complex conjugate pole $c^* / (s - s_n^*) = (a - jb) / (s - \Omega + j\omega)$ is also included in the series. In late time, the pole series converges, allowing its approximation via a truncated series.

Transforming this pole-series model into the time domain, we find that it corresponds to a series of damped sinusoids with frequencies w and damping rates Ω determined by the locations of the poles in the s-plane. We now have a late time return composed of several terms of the form

$$x(t) = 2 \exp(\Omega t) [a \cos(\omega t) - b \sin(\omega t)], \quad t > 0. \tag{1}$$

A key observation of Baum is that the complex frequencies do not depend on the aspect from which the target is illuminated or observed. This is the feature that has drawn much attention. However, the weight/phase coefficients (a and b) do depend on the aspect in general.

Scattering calculations using SEM involve Green's functions, as do other analytic representations of scattering. Thus, although an SEM response function exists, it may be impractical or even impossible to find analytically. So, to find the poles, we must first find the scattering with the standard tools of electromagnetics: asymptotic techniques, first-principle techniques, and scattering measurements. We reject asymptotic techniques because we must consider frequencies low enough that asymptotic assumptions are invalid. Either computational techniques or measurements are suitable if the resulting data is appropriately transformed and windowed to focus on the late time response. We consider measurements.

3. TARGET RECOGNITION ARCHITECTURE

Given an unknown target measured via an ultra wideband, short pulse radar, what is the best approach to identify it? Note that the target has already been detected. For accuracy and confidence, we need a technique which can find the best possible estimate of both damping rates Ω and oscillating frequencies w of the signal embedded in the noisy measured data. We need a technique that is fast and computationally efficient, because we require a near-real-time answer. The two requirements of optimality and speed are in conflict, which may lead to an approach that is neither fast enough for near-real-time use nor effective enough to identify targets with confidence.

One solution is to decouple the requirements. If we are able to prepare for a target recognition session in advance, we can construct a library of target signatures based on the best techniques. The data we use can be measured in the best available range or computed on the best available computer network with the most appropriate scattering code. When attempting to recognize a target, we are now armed with a library of likely signatures. We can measure our unknown target with our near-impulse radar and quickly determine which signature in our library matches it best, or learn that it matches none of them well. In this way, we are limited to positively identifying only those targets within our library, but the memory and computational requirements for even a large library are small given the form of the data required.

4. BUILDING THE LIBRARY

This paper will not go into great detail regarding the various methods used to build the library. It is sufficient for our purposes to point out a few things. To start building the library, we determine late-time scattering on a target via computational means or measurement. Given noisy late time data on an object, we may attempt to reduce the noise in it via a reduced-rank Hankel approximation.[11]

We estimate the complex frequencies $\Omega + j\omega$ via a linear predictor such as the Matrix Pencil method[12] or a similar method such as ESPRIT.[13] We may use this information directly or use it as input into an iterative maximum likelihood method.[14,15] Gcnctic algorithms (GA) are another possible method. To our knowledge, neither ESPRIT nor GA has been applied to this application in the literature to date.

The result of this is either a set of poles which best fit each measurement, or a single set which is the best fit over all measurements of the target.[16] We now establish an algorithm to distinguish targets based on our library.

5. READING FROM THE LIBRARY

We return to the Laplace-domain form of the theoretical late time signal in Eq. (1). If we take each pole pair in the series by itself, we find that the pair can be manipulated into the form of the transfer function below:

$$T(s) = (2as - 2a\Omega - 2b\omega) / (s^2 - 2\Omega s + \Omega^2 + \omega^2). \tag{2}$$

From here we can develop a state space representation of the pole pair. We use the denominator of the right side of Eq. (2) to form a square (2x2) matrix F, known as the system dynamics matrix. If our signal is formed from a single pole pair, F has a top row of [0, 1] and a bottom row of $[(-\Omega^2 - \omega^2), 2\Omega]$. The numerator is used to form the measurement matrix H, which is $[-2(a\Omega + b\omega), 2a]$. There is a control input matrix B with zero in the top row and one in the bottom. For N pole pairs, we arrange N F(2x2) matrices along the main diagonal of an F(2Nx2N) matrix. Other entries are zero. We then concatenate H(1x2) matrices into an H(1x2N) matrix and B(2x1) into a B(2Nx1) matrix, as shown below.

$$\mathbf{F} = \begin{bmatrix} F_1 & 0 & 0 \\ 0 & F_2 & 0 \\ 0 & 0 & F_3 \end{bmatrix} \qquad \mathbf{B} = \begin{bmatrix} B_1 \\ B_2 \\ B_3 \end{bmatrix}$$

$$\mathbf{H} = [\,H_1 \;\; H_2 \;\; H_3\,] \tag{3}$$

We now have a linear state space model. Given a noisy measurement of a signal described by a linear state space model, the optimal estimator for the true signal is a Kalman filter.[17] This filter is designed for online, real-time use in control systems. We won't approach real-time application in this algorithm, but will enjoy the ease of application. In this case the model is time invariant, making computation easier. The continuous-time filter is based on the system dynamics equation

$$dx(t) / dt = F\, x(t) + B\, u(t) + G\, q(t) \tag{4}$$

and the measurement equation

$$y(t) = H\, x(t) + r(t), \tag{5}$$

where x is the state vector, u the control input vector, q system noise with strength matrix Q, and r measurement noise with covariance R. Noises are considered additive, white, and Gaussian.

In our case the data is sampled. So the F and B matrices developed above must be transformed into their sampled-data counterparts, Φ and B_D. The measurement equation is basically unchanged, but the sampled data dynamics equation for time t_n becomes

$$x(t_n) = \Phi\, x(t_{n-1}) + B_D\, u(t_{n-1}) + G_D\, q_D(t_{n-1}). \tag{6}$$

When F is linear and time invariant, the state transition matrix Φ is defined as the Inverse Laplace Transform of the resolvant,

$$\Phi(t_n - t_{n-1}) = \mathcal{L}^{-1}\, [\Phi(s)] = \mathcal{L}^{-1}\, [\, (s\, I - F)^{-1}], \tag{7}$$

evaluated at $t_n - t_{n-1}$. The Φ matrix is used to define the discrete-time matrices B_D, G_D and noise covariance Q_D. They are

$$B_D\, u(t_{n-1}) = \int \Phi(t_n - \tau)\, B\, u(t)\, d\tau, \tag{8}$$

$$G_D\, Q_D(t_{n-1})\, G_D = \int \Phi(t_n - \tau)\, G\, Q(\tau)\, G^T\, \Phi^T(t_n - \tau\ )\, d\tau. \tag{9}$$

The integrals in Eqs. (8) and (9) are taken over one sample period, from t_{n-1} to t_n. While Φ, B_D, and G_D are time varying in general, in this case they are constant.

The Kalman filter accepts a measurement vector (a scalar in this case) in every time step and compares the measurement to a prediction of that measurement based on a model of the system's behavior. It generates its own estimates of covariance of both the system state vector and the measurement. Using these estimates, it forms an optimally weighted average of the measurement and the model. We initialize the state vector x(0) and state covariance matrix P(0). After that and after each update, we propagate them forward in time to the end of the sample period, t_n^-. The propagation equations for the state vector and its covariance are

$$x(t_{n-1}^-) = \Phi\, x(t_{n-1}^+) + B_D\, u(t_{n-1}), \tag{10}$$

$$P(t_n^-) = \Phi\, P(t_{n-1}^+)\, \Phi^T + G_D\, Q(t_{n-1})\, G_D^T. \tag{11}$$

The measurement y and control u come in at time t_n. We find the post-update (t_n^+) values via the update equations

$$A(t_n) = H\,P(t_n^-)\,H^T + R(t_n), \quad (12)$$

$$K(t_n) = P(t_n^-)\,H^T\,A(t_n)^{-1}, \quad (13)$$

$$x(t_n^+) = x(t_n^-) + K(t_n)\,[\,y(t_n) - H\,x(t_n^-)\,], \quad (14)$$

$$P(t_n^+) = P(t_n^-) - K(t_n)\,H\,P(t_n^-). \quad (15)$$

The propagation and update equations are iterated in each sample period. We let the incident wave be the control input u, and the scattered field seen by the radar be the measurement y. If a filter has the correct state transition Φ, based on the SEM poles discussed in Section 2, its model $[H\,x(t_n^-)]$ will be able to follow the measurement y(n) closely, resulting in a small residual $[y(t_n) - H\,x(t_n^-)]$. We can implement a bank of Kalman filters operating in parallel. Each has a state transition matrix based on the SEM poles of a target type of interest. Each sample period, in addition to propagating and updating each filter, we evaluate each residual. The residuals are scaled by $A(t_n)$ and used to find the probability that each filter is the correct one. First, we find the conditional probability density function for each filter via

$$f(t_n) = [(2\pi)^{m/2}\,|A(t_n)|^{1/2}]^{-1}\exp\{-\,[y(t_n) - H\,x(t_n^-)]^T\,A(t_n)^{-1}\,[y(t_n) - H\,x(t_n^-)]\,/2\}. \quad (16)$$

Each filter's f is the probability density of the current measurement, on the condition that the previous measurements happened as they did, and that the filter's model is correct. The m superscript in Eq. (16) refers to the size of the measurement vector. Our measurement is scalar, so m=1. We use f to find the hypothesis conditional probability $p(t_n)$ of each filter. Given the measurements actually received, $p_i(t_n)$ is the probability that filter i correctly describes the target. Each p_i is initialized such that all filters are considered equally likely to be correct, i.e. $p_i(0) = 1/J$ for a system with J filters. After each filter update, we calculate a new hypothesis conditional probability for filter i via

$$p_i(t_n) = f_i(t_n)\,p_i(t_{n-1})\,/\,[\Sigma_k\,f_k(t_n)\,p_k(t_{n-1})]. \quad (17)$$

So far, what we have described applies to either the standard form or the Maximum a Posteriori (MAP) form of the Multiple Model Adaptive Estimation algorithm (MMAE).[18] The standard MMAE is used in aircraft flight control systems that conduct failure detection and estimation, among other applications.[19] We apply the MAP form.

For best results, each filter's noise covariances Q_d and R should be set carefully. This is commonly called filter tuning. We tune each filter so that $H\,Q_d\,H^T \ll R$, to force it to follow its model more closely than the incoming measurements. That helps the overall system distinguish correct from incorrect models. Also note that H is based on the unknown coefficients a, b. We set a = b = 1 for each pole pair.

Our algorithm arrives at a decision very early, shutting out information from the incorrect filters. To force it to use all available data, we give each filter a minimum HCP of L. We must then reduce each filter's maximum HCP from 1 to 1 – (J – 1)L. For our four-filter system we set L = 0.01 and set each filter's maximum HCP at 0.97. To make the final decision, we form a weighted time-average of the HCP for each filter. The filter with the highest average HCP corresponds to the target. If all filters have nearly equal average HCP, the algorithm is unable to identify the target.

6. RESULTS AND CONCLUSION

Figures 1 and 2 show the output of the component filters of a four-filter bank. Figure 1 shows the filter with the correct model. The filter follows the measurement fairly well. The weighted average HPC for this filter was 0.6330. We added white, Gaussian noise to the radar signal to test the noise rejection properties of our technique. The signal to noise ratio was 8.25 dB.

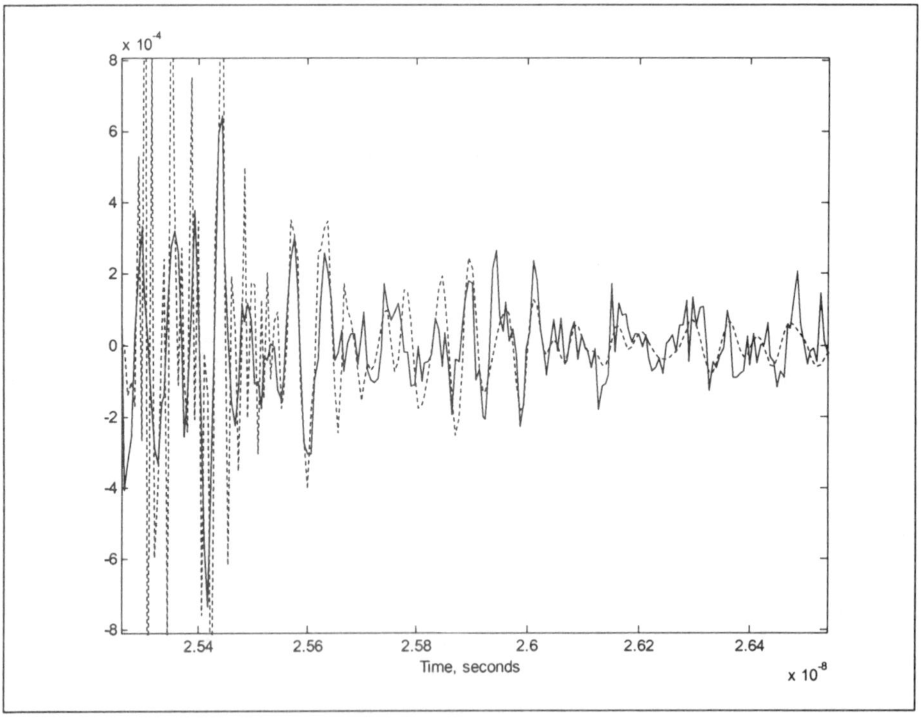

Figure 1. Output of filter with correct model: y solid; Hx dashed.

The solid line is measured late-time radar data. The dashed line is the filter model Hx. The areas where the solid and dashed lines diverge are evidence that the model could be improved, possibly by adding pole pairs or refining the pole locations. Another area of improvement is the coefficients. We do not know the correct values of the coefficients a, b associated with each pole pair, but we do know they are constant over time. Better estimates of each a, b would lead to a better H matrix and a model that better fits measurements.

Figure 2 shows a filter with the incorrect model. It is unable to follow the radar signal. This filter's residuals remain high, driving down the hypothesis conditional probability and preventing our algorithm from choosing it. This filter's average HPC was 0.1633. It was the second best performing of the four filters.

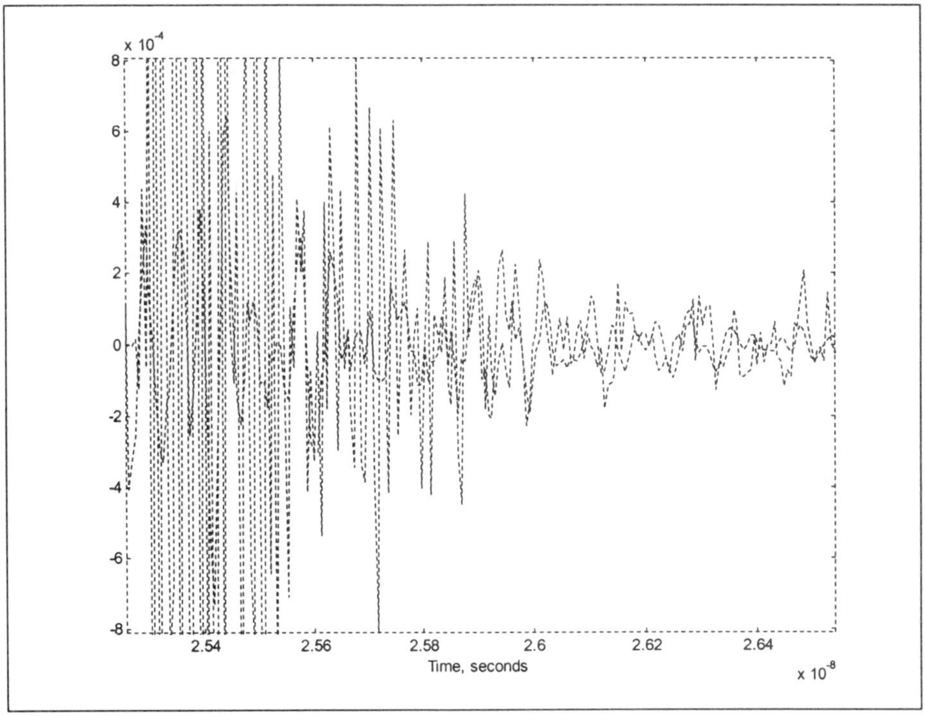

Figure 2. Output of filter with incorrect model: y solid; Hx dashed.

The target recognition algorithm described here is effective and noise resistant. Because it relies on Kalman filters, it is flexible and supported by a rich body of literature on state-space control and estimation of stochastic systems. Our main innovation is applying MMAE to target recognition via SEM parameters. A minor innovation is using the weighted average of the entire time window. We note that this paper represents a rare union of electromagnetics with control theory and hope that it sparks interest among practitioners in both fields.

The views expressed in this article are those of the author and do not reflect the official policy or position of the United States Air Force, the Department of Defense, or the United States Government.

7. REFERENCES

1. C. E. Baum, E. J. Rothwell, K. M. Chen, and D. P. Nyquist, The singularity expansion method and its application to target identification, *Proc. IEEE* **79**(10), 1481–1492 (1991).
2. E. M. Kennaugh, The K-pulse concept, *IEEE Trans Antennas Propagat.* **35**(3), 327–331 (1981).
3. P. Ilavarasan, J. E. Ross, E. J. Rothwell, K.-M. Chen, and D. P Nyquist, Performance of an automated radar target discrimination scheme using E pulses and S pulses, *IEEE Trans Antennas Propagat.* **41**(5), 582–588 (1993).
4. D. P. Ruiz, A. Gallego, and M. C. Carrion, Scattered signal enhancement algorithm applied to radar target discrimination schemes, *IEICE Trans. Communications* **E82-B**(11), 1858–1866 (1999).
5. C. E. Baum, Direct Construction of a Xi-Pulse from Natural Frequencies and Evaluation of the Late-Time Residuals, *Interaction Note 519*, Air Force Research Laboratory, Kirtland AFB NM (1996).
6. J. W. Brooks and M. W. Maier, Object classification by system identification and feature extraction methods applied to estimation of SEM parameters, in: *Proceedings of the IEEE National Radar Conference*, (IEEE Press, Piscataway, NJ, 1994), pp. 200–205.
7. J. W. Brooks, M. W. Maier, and S. R. Vechinski, Applying system identification and neural networks to the efficient discrimination of unexploded ordnance, in: *Proceedings of the IEEE Aerospace Conference*, (IEEE Press, Piscataway, NJ, 1997), pp. 449–467.
8. J. E. Mooney, Z. Ding, and L. Riggs, Robust target identification using a generalized likelihood ratio test, in: *Ultra-Wideband, Short-Pulse Electromagnetics 3*, edited by ????? (Plenum Press, New York, 1997), pp. 343–350.
9. C. E. Baum, On the Singularity Expansion Method for the Solution of Electromagnetic Interaction Problems, *Interaction Note 88*, Air Force Weapons Laboratory, Kirtland AFB NM (1971).
10. C. E. Baum, Title???. in: *Transient Electromagnetic Fields*, edited by L. B. Felsen (Springer-Verlag, New York, 1976), pp. 129–177.
11. B. Lu, D. Wei, B. L. Evans, and A. C. Bovik, Improved matrix pencil methods, in: *Conference Record of the Thirty-Second Asilomar Conference on Signals, Systems and Computers* (IEEE Press, Piscataway, NJ, 1997), pp. 1433–1437.
12. T. K. Sarkar and O. Pereira, Using the matrix pencil method to estimate the parameters of a sum of complex exponentials, *IEEE Antennas Propagat. Mag.* **37**(1), 48–55 (1995).
13. R. Roy and T. Kailath, ESPRIT - estimation of signal parameters via rotational invariance techniques, *IEEE Trans. Acoustics, Speech, Sig. Proc.* **37**(7), 984–995 (1989).
14. Y. Bresler and A. Macovski, Exact maximum likelihood parameter estimation of superimposed exponential signals in noise, *IEEE Trans. Acoustics, Speech, Sig. Proc.* **34**(5), 1081–1089 (1986).
15. I. Ziskind and M. Wax, Maximum likelihood localization of multiple sources by alternating projection, *IEEE Trans. Acoustics, Speech, Sig. Proc.* **36**(10), 1553–1560 (1988).
16. T. K. Sarkar, S. Park, J. Koh, and S. M. Rao, Application of the matrix pencil method for estimating the SEM (singularity expansion method) poles of source-free transient responses from multiple look directions, *IEEE Trans Antennas Propagat.* **48**(4), 612–618 (2000).
17. P. S. Maybeck, *Stochastic Models, Estimation, and Control, Volume I* (Navtech, Arlington VA, 1994).
18. P. S. Maybeck, *Stochastic Models, Estimation, and Control, Volume II* (Navtech, Arlington VA, 1994).
19. P. K. Eide and P. S. Maybeck, An MMAE failure detection system for the F-16, *IEEE Trans. Aero. & Elect. Sys.* **32**(3), 1125–1136 (1996).

DETECTION OF WIRES USING UWB RADAR

G.N. Crisp, C.R. Thornhill, and G. Gooding-Williams*

1. INTRODUCTION

UWB radar has been proposed as one of the primary sensors for stand-off detection of landmines both for civil (humanitarian) and military application. It has the advantage that it presently provides the only direct means of remote detection of buried targets, clearly relevant to the mine detection problem. Pressure activated mines do not provide the only threat relevant to such systems and the capacity to detect off route mines is an important factor for all proposed mine detection systems. Off route mines may be activated by a variety of mechanisms and one commonly employed method is the use of trip wires and break wires. These are generally long fine wires deployed across the path of the intended target and are remarkably difficult to detect visually, even under good lighting conditions. It is therefore useful in the context of mine detection systems to be able to detect such wires. This paper addresses the use of Ultra-Wideband Radar (UWBR) for this purpose.

It should be noted that in contrast to other previous work concerned with the UWBR response to chaff,[10,11] the present paper is concerned with the response of single long wires rather than the response due to an ensemble of relatively short wires.

2. RADAR CHARACTERISTICS OF WIRES

Just as mines themselves encompass a wide diversity of types varying in size, shape and material of construction, so trip and break wires also vary in characteristics. They may be made from dielectric materials as well as from metal, and may be deployed at various heights above the ground. However trip wires are generally many metres in length, sub-millimetre in diameter, and are commonly made from steel. In the case of wires use to trigger anti-tank mines, which provide the most serious threat to dedicated mine detection vehicles, the wire is usually deployed taught approximately parallel with

* QinetiQ, Malvern Technology Park, St. Andrews Road, Malvern, WORCS, WR14 3PS, UK, gncrisp@qinetiq.com.

the ground surface and directly across the path of the intended victim. They are generally located some height above the ground surface in order to avoid accidental initiation of the mine by small mammals. Thus a perfectly conducting long circular cylinder provides an idealised first approximation to practical metal tripwires. In this section of the paper we consider the radar characteristics of such idealised "wires".

3. ANALYTICAL MODELS

The two-dimensional treatment of electromagnetic scattering by a perfectly conducting body of circular cross section is exactly soluble and was first treated by Lord Rayleigh,[1,2] and was later generalised.[3-5] The results for the two-dimensional backscatter problem can be expressed in terms of a scattering width (analogous to radar cross section). In the limit of wave length << wire diameter, the results are:

(a) for incident electric field (E) parallel to the cylinder's axis,

$$\sigma = \frac{\pi^2 a}{ka\left[\ln^2(0.8905ka) + \frac{\pi^2}{4}\right]};$$

(b) forr incident electric field perpendicular to the cylinder's axis,

$$\sigma = \pi^2 a\left[\frac{9}{4}(ka)^3\right].$$

The corresponding results for non-magnetic homogeneous dielectric cylinders[1,2] are:

(c) for E parallel to the cylinder axis,

$$\sigma = \frac{a\pi^2}{4}(ka)^3\left[\frac{\varepsilon_1}{\varepsilon_2} - 1\right]^2;$$

(d) for E perpendicular to the cylinder axis;

$$\sigma = \frac{a\pi^2}{4}(ka)^3 2\left[\frac{\varepsilon_1 - \varepsilon_2}{\varepsilon_1 + \varepsilon_2}\right]^2.$$

For these result, $k = 2\pi/\lambda$, λ is the wavelength, a is the cylinder radius, σ is the scattering width and ε_1 and ε_2 are the dielectric constants of the surrounding medium and of the dielectric cylinder.

It is instructive to consider numerical values pertinent to the detection of fine wires, for which we take as an example a wire diameter of 0.5 mm and length 10 m, interrogated at a frequency of 1 GHz. The corresponding scattering widths are summarised in Table 1 below. From the data shown it is clear that the scattering cross section of metal wires observed with the electric field parallel to the wire axis greatly exceeds the cross section for perpendicular polarisation. Dielectric wires of the small diameter considered have very small cross sections irrespective of polarisation.

Table 1. Scattering widths of metal and plastic cylinders at 1 GHz.

Cylinder	Polarisation	Diameter (m)	Relative Dielectric Constant	Scattering Width (m)
Metal	E Parallel to axis	0.0005	Not Applicable	0.015
Metal	E Perpendicular to axis	0.0005	Not Applicable	8×10^{-10}
Glass	E Parallel to axis	0.0005	7	3×10^{-9}
Glass	E Perpendicular to axis	0.0005	7	9×10^{-11}

The three dimensional scattering by finite length wires with the electric field was treated by Van Vleck et al.[6] whose formulation is valid for wires with lengths exceeding half a free space wavelength. Their result is a complicated expression, which, normalised to the square of the wire length, is plotted in Figure 1. In the limit of long wires at normal incidence their result reduces, for a wire of length l, to:

$$\sigma = \frac{4\pi l^2}{\ln^2(0.8905ka) + (\pi/4)^2}.$$

Figure 1 shows that the long wire limit is approached rapidly and is a good approximation for wires exceeding $5\lambda/2$ in length. It should be noted that surface laid landmines have radar cross-sections (RCS), which are typically of the order –25 dBm^2. Evidently at frequencies typical for UWBR radars even relatively short lengths of wire, viewed close to normal incidence, are expected to exhibit large signatures in comparison to AT mines.

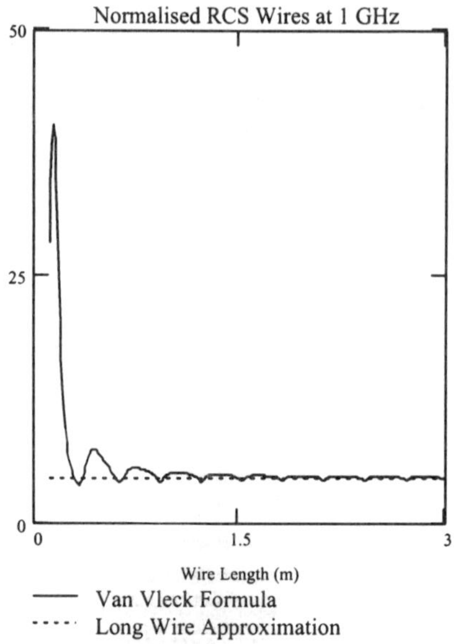

Figure 1. Normalised RCS of finite thin conducting wire of 0.5 mm diameter.

4. FINITE DIFFERENCE TIME DOMAIN (FDTD) SIMULATIONS

FDTD simulations have been conducted to predict the impulse response of horizontal metal wires. The incident field comprised a unit magnitude horizontally polarised Gaussian pulse of 150 ps full height half-width duration. The simulations employed first order one-way wave equation boundary conditions at the boundary of the computational volume, and a sub cell wire segment to provide accurate representation of the wire whilst maintaining a manageable grid spacing. Figure 2 below shows the computed scattered far field response of a 1.6 m length of wire viewed at normal incidence.

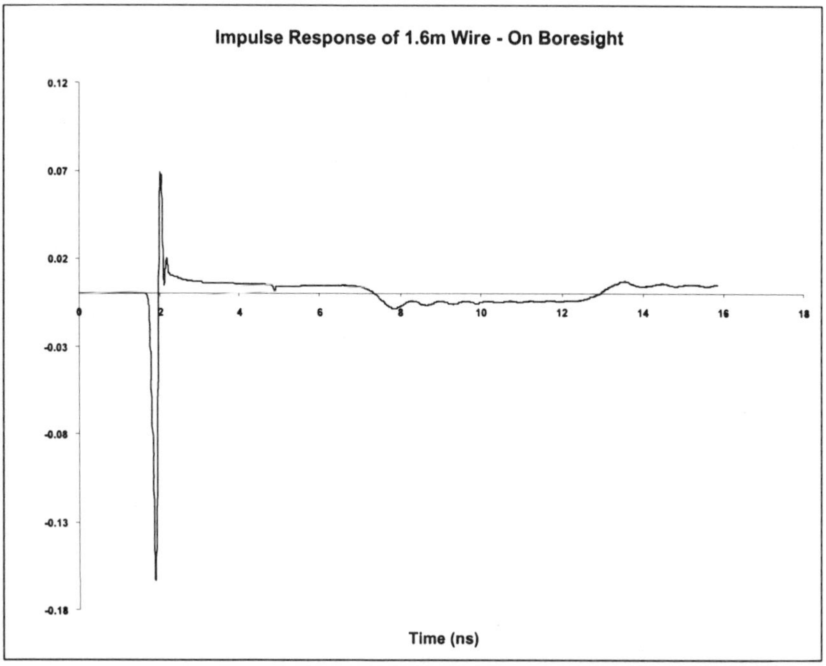

Figure 2. FDTD far field simulation for 1.6 m long wire viewed at normal incidence. The computed field amplitude is shown normalised to 1 m range.

The computed peak response at normal incidence corresponds to a scattering parameter of ~0.15 m and the peak response is equivalent to a radar cross section of –5.5 dBm^2. This is in reasonable agreement with the analytical model due to Van Vleck discussed above as can be seen from Figure 3 which shows a direct comparison of the two calculations made using a variety of wire lengths. It should be noted that the response shown in figure 2 is inverted with respect to other results presented below in Figure 4.

A series of simulations were conducted to predict the variation of response with incidence angle. The results are shown in Figure 4. It can be seen that while the peak response reduces rapidly with increasing incidence angle, the response also broadens so that the integrated energy in the first peak reduces relatively slowly with incidence angle. Similar simulations carried out at 3° incidence angle show that the peak far field response remained almost unaltered when the wire length was varied from 0.3 m to 3.6 m. This insensitivity to length is attributed to a balance between the increase in cross section with increasing length, and narrowing of the beam width of the reflected field as the wire length is increased.

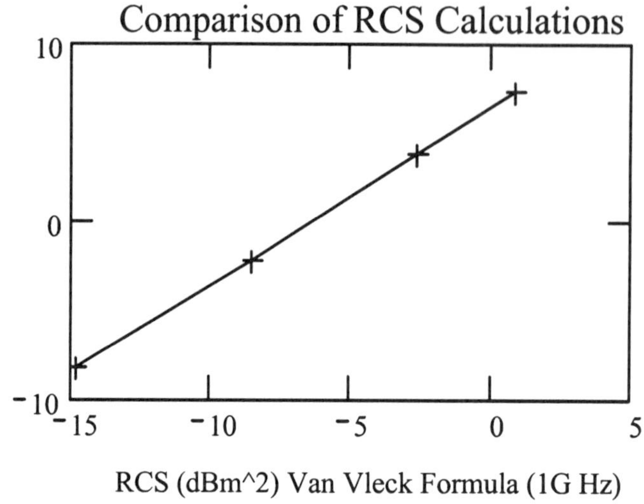

Figure 3. Comparison of FDTD simulation results with analytical model.

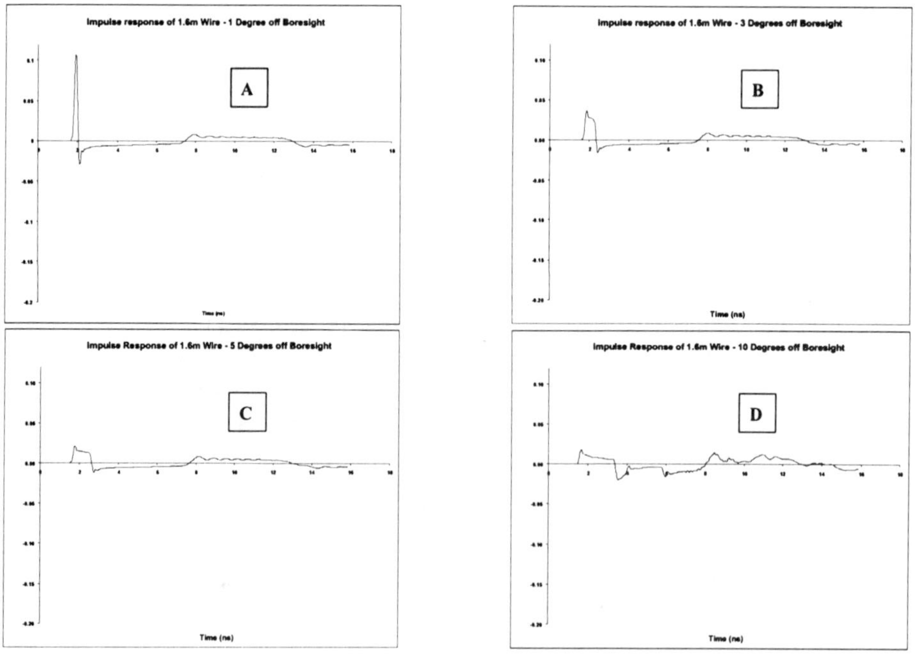

Figure 4. FDTD simulations at various incidence angles: A. 1°; B. 3°; C. 5°; D. 10°.

5. EXPERIMENTAL OBSERVATIONS

Experimental observations were made using a four element sparse array impulse UWBR arrangement described elsewhere.[7] The radar is mounted on a mobile platform and performs measurements in a forward looking direction as the platform approaches the target. Two antennas are mounted at each end of a short baseline B, orthogonal to the direction of platform motion, as shown in Figure 5.

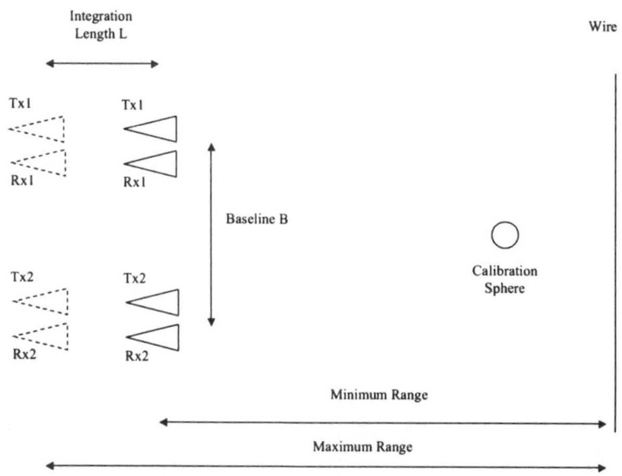

Figure 5. Experimental arrangement.

As the radar approaches the target, impulse sources connected to antennas Tx1 and Tx2 are alternately fired and data recorded using receivers attached to antennas Rx1 and Rx2. A motion sensor attached to the platform allows the radar position to be measured. Each pair of transmissions can be used to form a two dimensional image of the scene in front of the radar using back projection algorithms similar to those described by McKorkle and others.[8,9] In the system described here a series of such images is superimposed to generate an integrated image with improved signal to interference ratio and reduced clutter variance. For the experiments described below images are integrated over 2 m of platform travel, and were obtained using data measured at a fixed range in front of the radar. Figure 6 below shows the resulting image obtained when a wire and a calibration target were deployed in the test area. The wire concerned was of approximately 0.5 mm diameter and 12 m length. It was deployed approximately normal to the radar look direction at about 0.4 m above the ground surface.

The response of the wire shown in Figure 6 is of significantly larger amplitude than the response of the calibration sphere. This can be seen more clearly in Figure 7 which shows a section through the radar image at 0-m cross track. The peak response of the wire exceeds that of the calibration sphere by a factor of 3.5 indicating a wire RCS of the order –10 dbm^2. Assuming that the wire RCS varies as the square of its length, as predicted by the Van Vleck model, the normal incidence FDTD calculation presented above corresponds, for a 12 m long wire, to a far field equivalent RCS of +12 dBm^2. However the experimental conditions do not correspond to far field observation of the wire and it should therefore be expected that only a small portion of the wire would be effective in scattering the incident pulse back in the direction of observation. Careful analysis of the data shown in Figures 6 and 7 shows that the cross track extent of the wire response is ~2.5 m (half height full width) and that the duration of the initial response is ~400 ps. An estimate of the wire RCS, taking the viewing geometry into account, is only –7 dBm^2. In view of the likely experimental errors, this agreement with the observed value of –10 dBm^2 is remarkably good.

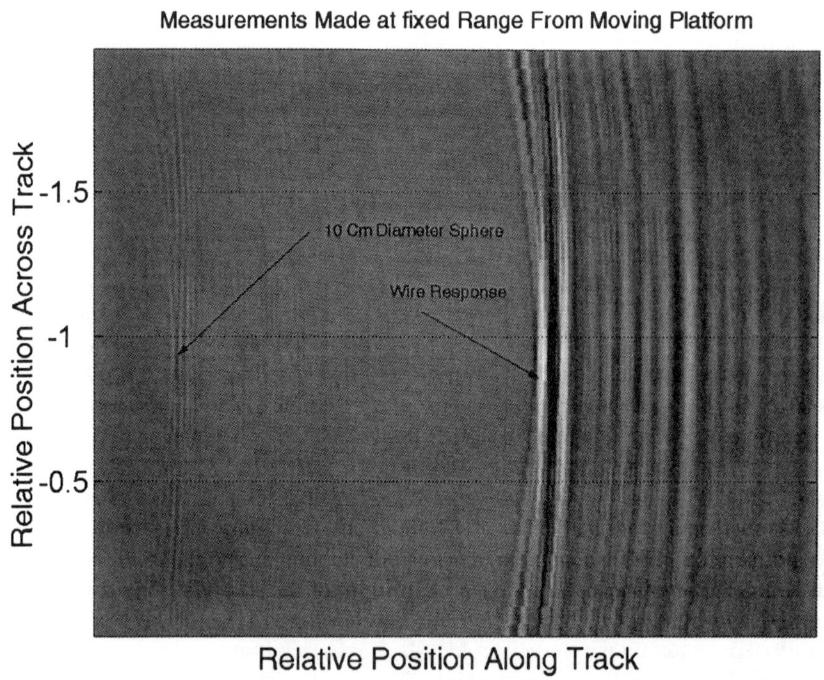

Figure 6. Radar image of calibration target and wire.

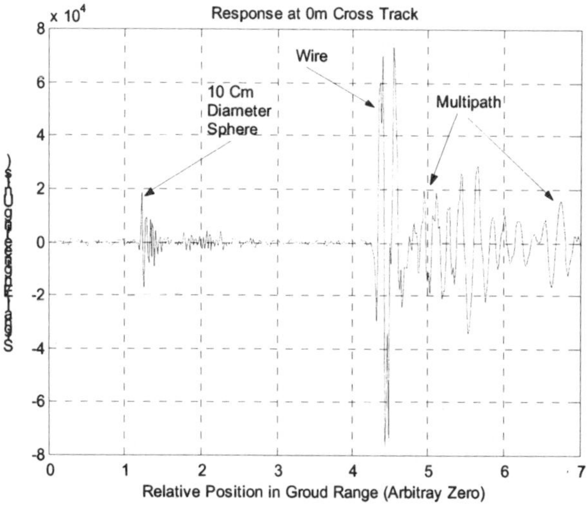

Figure 7. Signal versus ground range at 0-m cross track.

6. CONCLUSIONS

Theoretical analysis of the scattering by thin wires indicates that with the electric field polarised parallel to the wire axis significant reflection coefficients are expected for metal wires. In the orthogonal polarisation the predicted cross sections are reduced by a factor $\sim 10^7$. Fine dielectric wires are also predicted to have very small cross sections irrespective of polarisation. Thus while the detection of thin metal wires in the field using UWB radar is practicable, fine dielectric wires are expected to be undetectable.

Practical measurements have been conducted using an imaging, horizontally polarised, UWBR system. Fine wires of 0.5 mm diameter were detected at stand-off range. Analysis of the resulting signature shows that near field effects are important, and once these are taken into account, the magnitude of the observed response is comparable with that expected theoretically. A strong specular reflection is observed from only part of the wire, as is expected on the basis of physical optics.

The experiment described here was conducted using a wire deployed 0.4 m above the ground surface. Multipath reflections were observed, as would be expected when making shallow depression angle measurements in HH polarisation. It is therefore expected that for wires deployed very close to the ground surface target/ground interactions may become important. More detailed study of wires deployed close to the ground surface is therefore required.

The observed response normal to the wire was large compared to that expected from surface laid landmines. It is therefore expected that appropriately polarised UWBR systems designed for stand-off mine detection will have adequate sensitivity to detect metal trip wires, provided part of the wire concerned lies parallel to the radiated wave front. FDTD modelling shows that as the incidence angle at the wire increases the signal falls. For a 1.6-m length of wire at 10 degrees incidence angle the signal amplitude was reduced by approximately one order of magnitude compared with the signal at normal incidence. In forward looking vehicle detection scenarios, where wide beamed antennas are used, it is likely that some part of any tripwire deployed across the vehicles route will be illuminated close to normal incidence. We therefore conclude that UWBR has a useful potential against the threat of metal trip wire activated landmines.

REFERENCES

1. Lord Rayleigh, On the electromagnetic theory of light, *Phil. Mag.* **XII**, 81 (1881). Also in *Scientific Papers Vol. I* (Dover Publications, New York, 1964), p. 518.
2. Lord Rayleigh, The dispersal of light by a dielectric cylinder, *Phil. Mag.* **XXXVI**, 365 (1918). Also in *Scientific Papers Vol. VI* (Dover Publications, New York, 1964), p. 554.
3. R.W.P. King and T.T. Wu, *The Scattering and Diffraction of Waves* (Harvard University Press, Boston, 1959), p. 22.
4. R.F.Harrington, *Time Harmonic Electromagnetic Fields* (McGraw-Hill, New York, 1961), p. 232.
5. C.C.H. Tang, Backscattering from dielectric-coated infinite cylindrical obstacles, *J. Appl. Phys.* **2**, 628 (1957).
6. J.H.Van Vleck, F. Bloch, and M. Hamermesh, Theory of radar reflection from wires or thin metallic strips, *J. Appl. Phys.* **18**, 274 (1947).
7. G.N. Crisp, C.R. Thornhill, R. Rowley, and J. Ratcliffe, *Ultra-wideband sparse array imaging radar, in Ultra-Wideband, Short-Pulse Electromagnetics 5*, edited by P.D. Smith and S.R. Cloude (Kluwer Academic/Plenum Publishers, New York, 2002), pp. 399–406.
8. J.W. McKorkle, Focusing of synthetic aperture ultra-wideband Data, *Proc. IEE International Conference on Systems and Engineering* (August 1991).
9. R. Benjamin, I.J. Craddock, G. Hilton, S. Litobarski, E. McCutcheon, R. Nilavalan, and G.N. Crisp, Microwave detection of buried mines using non-contact synthetic near field focussing, *IEE Proceedings on Radar Sonar and Navigation* **148**(4), 233 (August 2001).
10. Y. Guo and H. Uberall, Short pulse scattering from wires and chaff clouds, Ultra-Wideband, Short-Pulse Electromagnetics 2, edited by L. Carin and L.B. Felsen (Plenum Press, New York, 1995), pp. 439–445.
11. H. Uberall and Y. Guo, Scattering of short radar pulses from multiple wires and from a chaff cloud, Ultra-Wideband, Short-Pulse Electromagnetics 3, edited by C.E. Baum, L. Carin, and A.P. Stone (Plenum Press, New York, 1997), pp. 271–278.

COMPARISON OF TIME, FREQUENCY AND TIME-FREQUENCY ANALYSIS OF BACK-SCATTERED ECHOES FROM RADAR TARGETS BURIED IN REALISTIC HALF-SPACES

G. C. Gaunaurd, H. C. Strifors, and A. Sullivan*

1. INTRODUCTION

In recent years, electromagnetic characterization of subsurface targets and its application to detection and identification using ultra-wideband (UWB) ground penetrating radar (GPR) have attracted great interest from the research community.[1–6] The targets are stationary and located in clutter environments with large false-alarm rates as a universal consequence. This, in turn, makes the target identification capability of GPR systems indispensable.

In the present work, we focus the interest on the electromagnetic interaction between targets and the soil they are buried in. More specifically, we investigate the backscattered echoes from a simple-shaped conducting or dielectric target contained in the lower half-space when it is illuminated by a GPR located in the upper half-space. We examine the influence on the returned echoes by the moisture content in the soil and burial depth of each target. The frequency content of the projected pulses is designed to match the size and kind of targets and environments. To reach deep penetration into the ground, low frequencies must be used, particularly in moist soils. For detailed information about the interrogated target, high-frequency pulses are desired.

The permittivity of soils varies with frequency and the moisture content in the soil. Data of the real and imaginary parts of these permittivities were previously obtained at the U. S. Army Yuma Proving Grounds, and are used as inputs in our study. The echoes returned by buried dielectric or conducting cylinders are simulated here by a Method-of-Moments[7] (MoM) program that was tailor-made in past years for bodies of revolution[8] (BOR) contained in the dielectric half-spaces. The returned echoes from the targets are

* G. C. Gaunaurd and A. Sullivan, Army Research Laboratory, 2800 Powder Mill Road, Adelphi, MD 20783-1197. H. C. Strifors, Swedish Defense Research Agency (FOI), P. O. Box 1165, SE-58111, Linköping, Sweden.

then analyzed in three domains. In the frequency domain we generate the respective radar cross-sections (RCS) of the targets buried in the various half-spaces. In the time domain we study the shape and time delays of the returned pulses. Finally, in the time-frequency domain, we generate time-frequency distributions (TFD) of the general Cohen class.[9] We have previously found that for sufficiently broadband signals returned from targets buried at different depths a pseudo-Wigner distribution (PWD) is capable of generating signature features that could be used for target recognition purposes.[3,4] Here, we use a PWD and, the even more robust, Baraniuk-Jones optimal kernel distribution[10] (BJOKD).

2. THEORETICAL MODELS

A BOR-MoM analysis of the scattering problem is applied in the frequency domain. This analysis uses a mixed-potential integral-equation[11] in terms of the space-domain dyadic Green's function for a general layered medium. Each one of the frequency dependent components of the Green's function involves Sommerfeld integrals, which are solved using the method of complex images.[12] The resulting code is then used to predict echoes from the finite-height cylinders used here to simulate mines (Fig. 1). The time-domain response is then calculated using an inverse fast Fourier transform (FFT).

Bilinear TFDs are a primary signal analysis tool in the time-frequency domain. Any member of the general class (or Cohen class) of bilinear distributions can be interpreted as the two-dimensional Fourier transform of a weighted version of the (symmetric) ambiguity function (AF) of the signal:

$$TFD_f(t,\omega;\Phi) = \frac{1}{2\pi} \int_{-\infty}^{+\infty}\int_{-\infty}^{+\infty} A(\theta,\tau)\Phi(\theta,\tau)\, e^{-i\theta t - i\tau\omega}\, d\theta\, d\tau, \tag{1}$$

where $A(\theta,\tau)$ is the AF of the signal $f(t)$ given by

$$A(\theta,\tau) = \int_{-\infty}^{+\infty} f(t+\frac{\tau}{2})\, f^*(t-\frac{\tau}{2}) e^{i\theta t}\, dt, \tag{2}$$

where an asterisk denotes complex conjugation.

The kernel function $\Phi(\theta,\tau)$ determines the specific properties of the distribution. Particular choices of the kernel function define the various TFDs that have been introduced. The kernel of the Wigner distribution (WD) is constant: $\Phi_{\text{WD}}(\theta,\tau) = 1$. Incorporating a smooth window function in the WD gives a PWD in the form:

$$PWD_f(t,\omega) = \int_{-\infty}^{+\infty} f(t+\frac{\tau}{2})\, f^*(t-\frac{\tau}{2})\, w(\frac{\tau}{2})\, w^*(-\frac{\tau}{2}) e^{-i\omega\tau}\, d\tau, \tag{3}$$

where we select a Gaussian window of the form: $w(t) = \exp(-\alpha t^2)$. The parameter α controls the width of the window and, accordingly, the degree of cross-terms suppression, or smoothing.

A "radially Gaussian kernel" is defined by

$$\Phi(\theta,\tau) = e^{-(\theta^2+\tau^2)/2\sigma^2(\psi)}, \tag{4}$$

where the "spread function" $\sigma(\psi)$ controls the spread of the Gaussian at the radial angle $\psi = \arctan(\tau/\theta)$. Baraniuk and Jones[10] then define the optimal kernel as the radially Gaussian function whose spread function $\sigma_{opt}(\psi)$ solves the optimization problem:

$$\max_{\psi} \int_0^{2\pi}\int_0^{\infty} |A(r,\psi)\Phi(r,\psi)|^2 \, r\,dr\,d\psi, \quad r = \sqrt{\theta^2+\tau^2}, \tag{5}$$

subject to the volume constraint,

$$\frac{1}{2\pi}\int_0^{\pi} \sigma^2(\psi)\,d\psi \le \alpha. \tag{6}$$

This is a restriction on how concentrated, or peaked, around the origin the kernel should be. The larger the value of the parameter α, the less the cross-terms suppression will be. We note that when the kernel function is signal dependent, the TFD is no longer bilinear.

3. THEORETICAL DATA

The targets used in the analysis are either a perfectly conducting (metal) cylinder or a dielectric (plastic) cylinder, both of diameter 11 cm and height 3 cm. The dielectric target has the relative dielectric permittivity of ε_r = 2.7, which is a typical value for a plastic mine. Each target is buried in Yuma soil to the depths 2, 4, or 8 cm as shown schematically in Fig. 1, left plot. The electric properties of the soil are shown in Fig. 1, right plot, for different moisture content by percentage weight.[6] In the analysis, moisture contents of 0, 10, or 15% are used.

For each target, each burial depth, and each moisture content the returned echo from the target is computed using the BOR-MoM when the target is illuminated from above with a continuous waveform of 255 frequencies in the band 25–6,375 MHz (in increments of df = 25 MHz). This very broad bandwidth covers those used by most GPR systems in practical use. Using zero-padding and folding of the complex conjugate sequence about the Nyquist frequency, $f_N = 25{,}600\,\text{MHz}$, we get a frequency vector of length 2048 elements in the band 0–51,175 MHz. This conjugate symmetric frequency series is then used to synthesize the (real-valued) target response in the time domain using an inverse FFT.[13] The duration of each synthesized time-series is then: $T = 2048/(2f_N) \equiv 1/df = 40\,\text{ns}$.

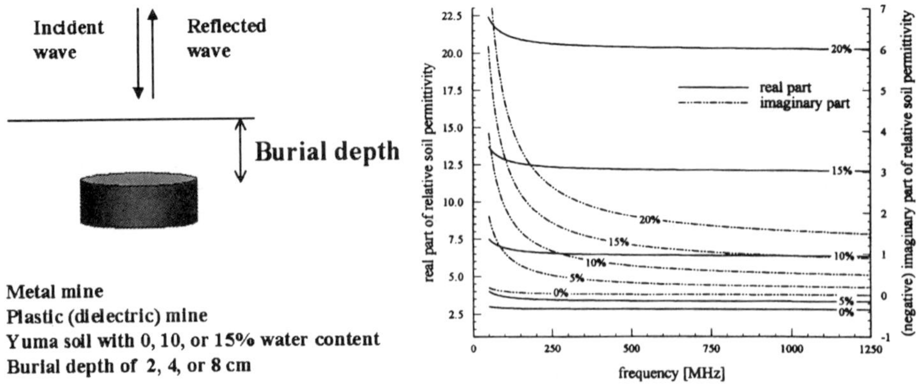

Figure 1. Schematic diagram of a mine-like cylindrical target buried in a homogeneous soil (left), and (right) the measured real and imaginary parts of the frequency dependent dielectric permittivity of Yuma soil for different water contents (percentage by weight).

4. FREQUENCY SIGNATURES OF THE TARGETS

We examine the radar cross-section (RCS) of each target when they are buried at three different depths in Yuma soil of three different moisture contents. Figure 2 displays (the modulus of) the RCS in a linear unit (cm^2) for the metal mine, left plots, and plastic mine, right plots, when the targets are buried at the depth of either 2, 4, or 8 cm, top, middle, or bottom plots when the moisture content is either 0, 10, or 15% by weight. Measuring the targets' backscattering responses in a practical situation using a GPR, two main difficulties would be encountered. First, the ground surface reflection (not present in our theoretical results) must be taken into consideration. Also, the clutter from the inhomogeneity of the soil and from adjacent objects would blur the target response. These difficulties would make detection and identification of the targets more difficult. In dry soil, the plastic mine (RCS < 1.5 cm^2) could be virtually impossible to detect. (The solid line in the right plots almost coincides with the frequency axis.) We notice in Fig. 2, right top plot, that, over the entire frequency band, the RCS of the plastic mine is larger for the larger moisture content, which means that the effect of the larger contrast dominates over the effect of electric energy loss in the moist soil at this small burial depth of 2 cm. Comparing Fig. 2 (top, middle, and bottom plots) of both columns shows that the RCS of both targets decrease with increasing depth in moist soil. However, for the plastic mine, the combined effects of improved contrast and larger electric energy loss are balanced when the moisture content is increased from 10% to 15% at the burial depths of 4 and 8 cm as seen in Figs. 2, middle and bottom right plots. Finally, on comparing the left plots of Figs. 2, we notice the prominent effects of both the burial depth and the soil's moisture content on the occurrence of the peaks and dips of the RCS as a function of frequency. These effects are present, although in a lesser degree, also in the RCS of the plastic mine.

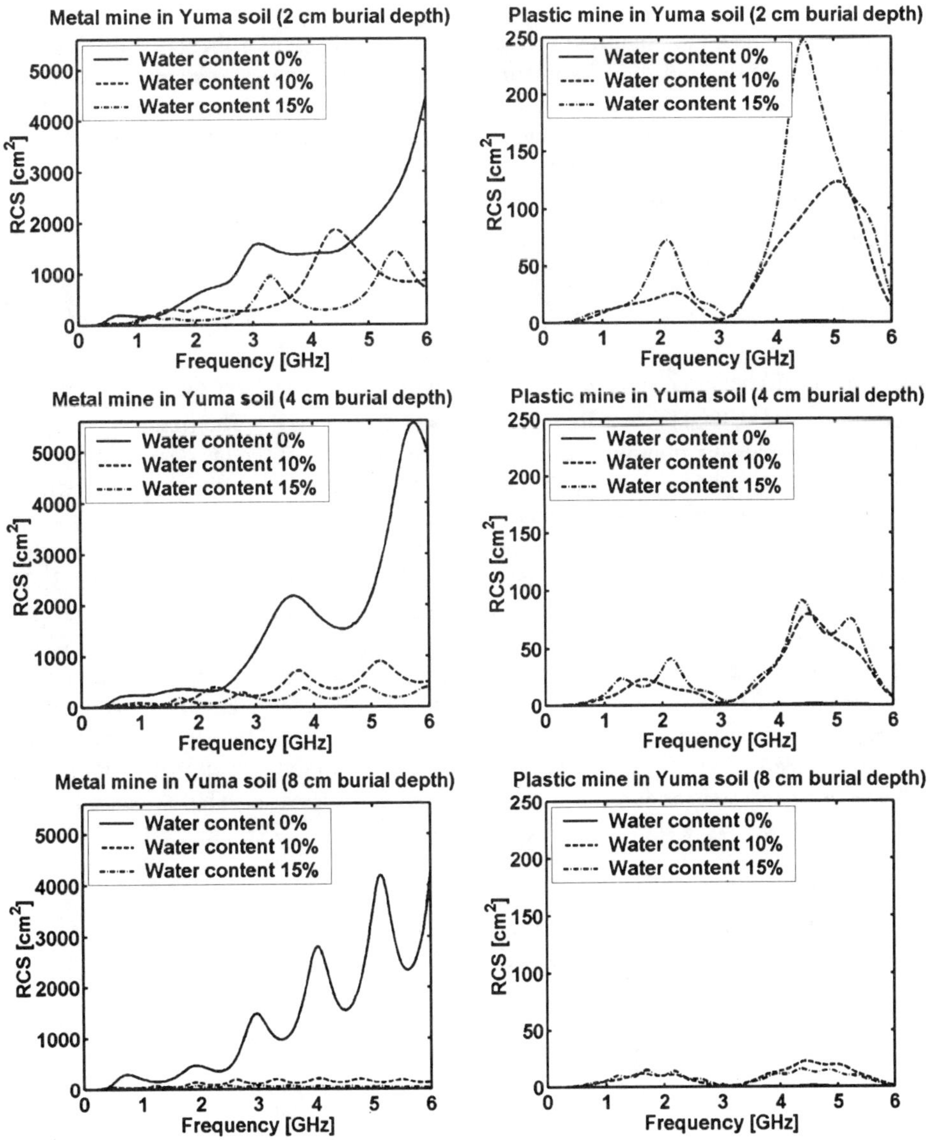

Figure 2. Left (right) plots display the RCS (cm^2) of a metal (dielectric) cylinder of diameter 11 cm and height 3 cm, when it is buried at a depth of 2 cm (top plots), 4 cm (middle plots), or 8 cm (bottom plots) in Yuma soil.

5. TIME SIGNATURES OF THE TARGETS

To obtain broadband returned echoes in the time domain, a theoretically designed, fictitious but realistic, pulse is illuminating the targets. The pulse is generated by filtering an ideal impulse (i.e., a delta function pulse) through a Butterworth bandpass filter.[13] Bandpass filters of this type have the property that their frequency response is maximally flat in the passband and monotonic in the passband, transition band and stopband.

The transfer function of a digital filter has the general form

$$H(z)=\frac{Y(z)}{X(z)},\quad Y(z)=\sum_{m=0}^{M} b_m z^{-m},\quad X(z)=1-\sum_{m=0}^{M} a_m z^{-m}, \tag{7}$$

where the coefficients a_m and b_m should be obtained by the filter design. We select a filter of order $M = 8$, and cutoff frequencies 200 and 4,000 MHz. The frequency response of the filter that results from this design is also the spectrum of the incident pulse generated by filtering an ideal impulse using this filter. The resulting waveform and spectrum of the designed incident pulse are displayed in Fig. 3.

The top left (right) plot of Fig. 4 shows the returned waveform for a 2 cm deep metal (dielectric) mine in Yuma soil, illuminated from above with the waveform of Fig. 3. The moisture content is 0, 10, or 15% by weight. The scattered field from the metal target is characterized by a strong specular return from the target's flat top surface. This primary portion of the return has about the same duration as the strong portion of the incident pulse (or about 0.5 ns). The specular return is closely followed in time by a much weaker, secondary return that is mainly caused by reverberation between the air-soil interface and the top surface of the target. We also note that the returned waveforms become more delayed in time when the moisture content in the soil increases. This is caused by the successively lower wave speed in the soil.

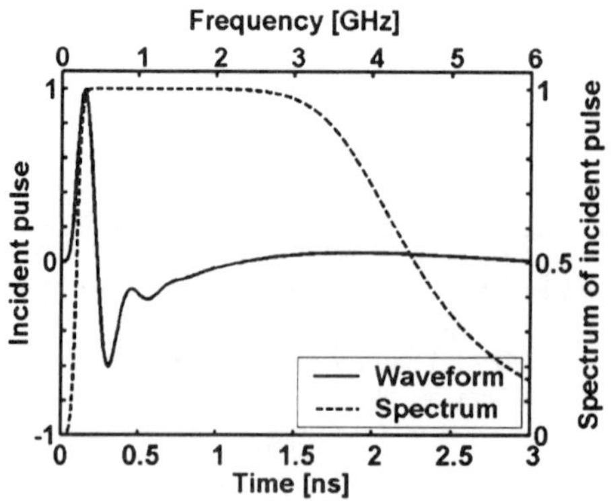

Figure 3. The waveform and spectrum of the designed incident pulse.

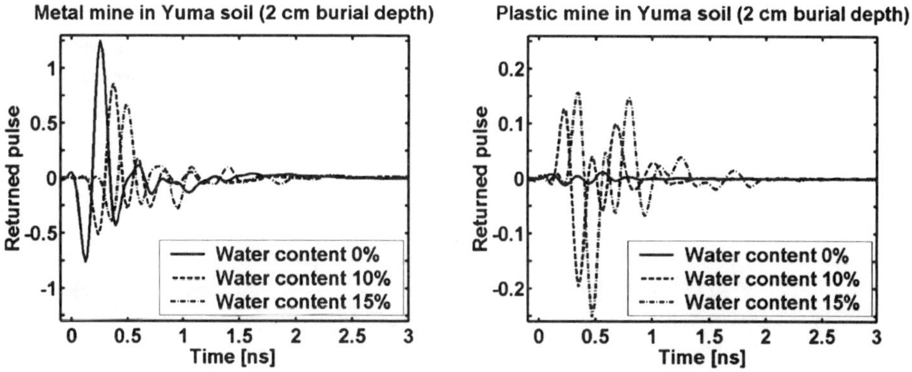

Figure 4. Left (right) plots display the pulse backscattered from a metal (dielectric) cylinder of diameter 11 cm and height 3 cm, when it is buried at a depth of 2 cm in Yuma soil.

An initial specular return, however much weaker than for the metal target, is also characteristic of the dielectric target (Fig. 4, right plot). A large portion of the incident pulse penetrates this target to be, to some degree, returned from the flat bottom surface of the target. In a similar way, the two flat horizontal boundary surfaces of the dielectric target contribute to the reverberation-generated returns. The combined effect is a more prolonged returned pulse from the dielectric target, where the initial and later portions are less distinguishable. As can be seen in Fig. 4, top right plot, the various portions of the returned signal become particularly blurred when the dielectric target is buried in dry soil.

The effect of electric energy dissipation due to the conductivity of the soil can be seen as a successively decreasing strength of the returns from the metal mine case in Fig. 4, left plot. In the case of the plastic mine, Fig. 4, right plot, the foremost effect of increased water content in the soil is a larger contrast, which gives the opposite trend from that of the metal mine. The peak return amplitude is also seen to increase as the moisture content increases.

6. TIME-FREQUENCY SIGNATURES OF THE TARGETS

In the time-frequency domain we compute the BJOKD and the PWD using a suitable portion (256 elements) of the synthesized, backscattered waveform used in the previous section and a value of the parameter α (in both cases) that permits some amount of cross-terms to develop. For the computation of the BJOKD we use software developed by the Digital Signal Processing group at Rice University. The results are displayed as image plots of the modulus of the TFD using progressively darker shades of gray (pseudo-colors). Each TFD is normalized by setting its maximum value to unity, which corresponds to black in the images below.

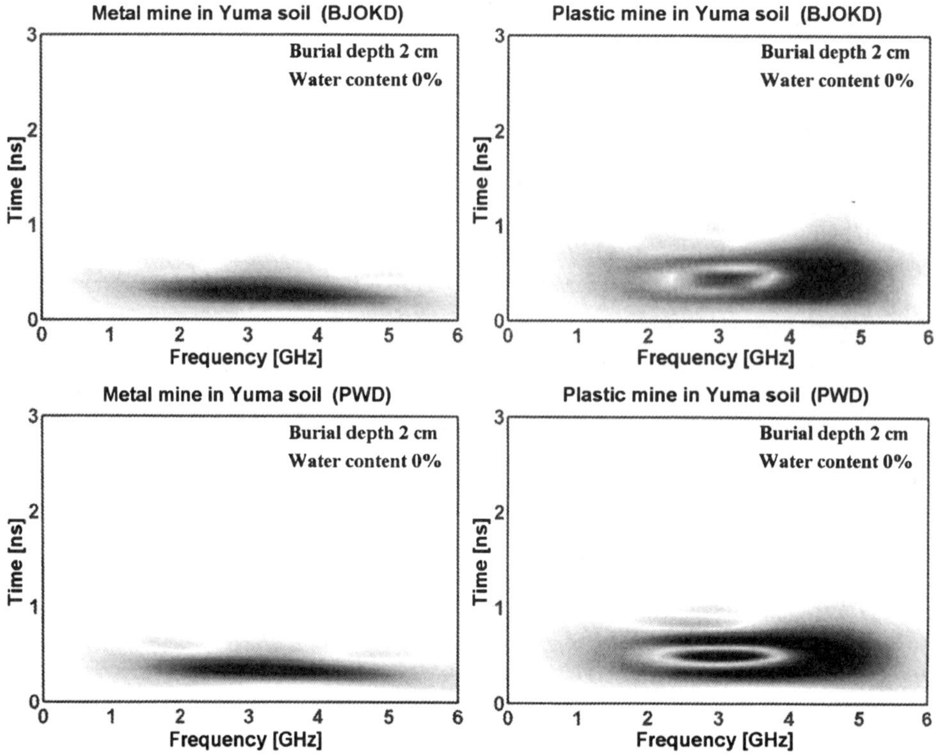

Figure 5. Left (right) plots display the TFDs of a pulse backscattered from a metal (dielectric) cylinder of diameter 11 cm and height 3 cm, when it is buried at a depth of 2 cm in dry Yuma soil.

Figure 5, left (or right) plots display the TFDs of the metal (or dielectric) target when it is buried at a depth of 2 cm in dry Yuma soil. The top plots show the BJOKD of the returned echo and the bottom plots the PWD. The effect of cross-terms, the ellipsoid-like features centered about 3 GHz, are clearly noticeable in the TFDs for the dielectric mine (Fig. 5, right plots) and, much less so, in the TFDs for the conducting mine. The occurrence of cross-terms is related to the locality of the particular time-frequency distribution, and the cross-terms convey information about the processed function.[12] By generating features that occur in between two self-terms in both time and frequency the cross-terms contribute signature information that can be used for target identification purposes.

Figure 6 displays the TFDs of the two targets when they are buried at a depth of 2 cm in Yuma soil of moisture content 15%. As in Fig.5 the top plots show the BJOKD of the returned echo and the bottom plots the PWD. At the present moisture content, the features generated by the cross-terms are clearly noticeable for both targets. Comparing the plots of Figs. 5 and 6, we also notice that the TFDs, as expected, occur at later times when the moisture of the soil content increases.

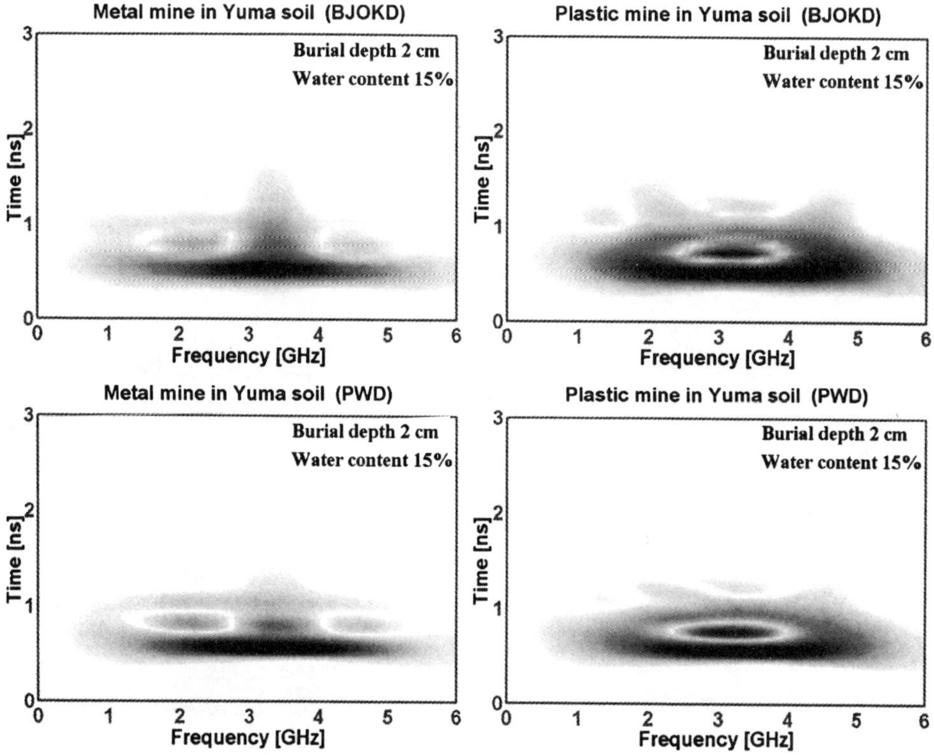

Figure 6. Left (right) plots display the TFDs of a pulse backscattered from a metal (dielectric) cylinder of diameter 11 cm and height 3 cm, when it is buried at a depth of 2 cm in Yuma soil with moisture content of 15%.

By comparing the top plots of either Fig. 5 or 6 with the bottom plots we see that the two TFDs generate similar features. Which one of the TFDs, if any, would be preferable for target identification depends on its robustness when it is applied to targets in a realistic clutter environment. The results obtained in this study does not permit any definite conclusion in this regard.

We then compare the results obtained when the targets are buried at the three depths of 2, 4, or 8 cm in Yuma soil with a moisture content of 15%. The target signatures generated by the BJOKD are displayed in Fig. 7. Consistent with the results obtained in the time-domain, the image of each TFD starts to develop at a successively later time when the moisture content in the soil increases. We also notice that the general shape of the image is preserved for the two targets. However, there is an evident trend in the images to shift their main emphasis toward lower frequencies when the moisture content increases, which is a consequence of the lower electric loss at lower frequencies.

More specifically for the TFD images of the metal mine, the barely perceptible hint of cross-term features at about 2 GHz and 4.5 GHz in Fig. 5, left plots, is much more pronounced in Fig. 7, top left plot, but not at the larger burial depths of Fig. 7, middle and bottom left plots. These effects are related to the levels of the peaks of the corresponding frequency responses (Fig. 2, left plots) weighted by the spectrum of the incident pulse (Fig. 3).

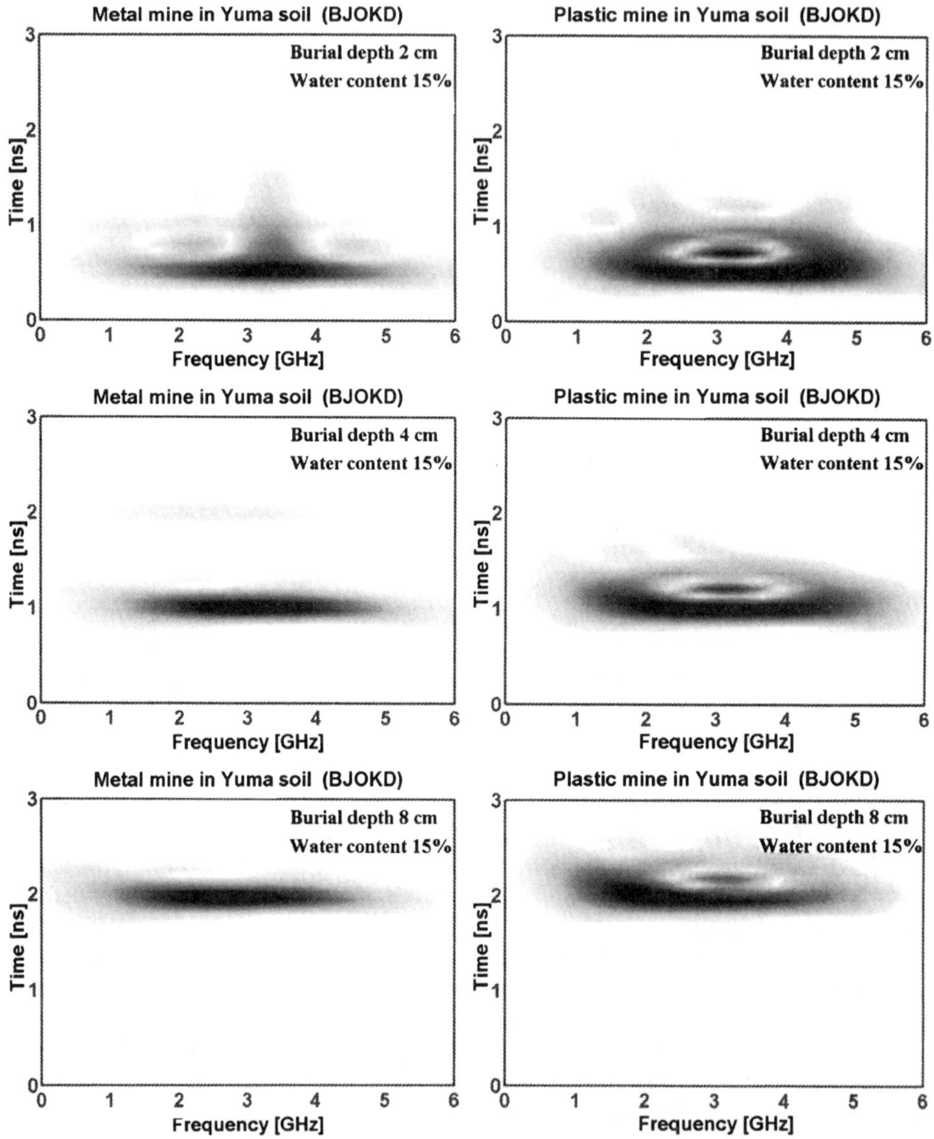

Figure 7. Left (right) plots display the TFD of a pulse backscattered from a metal (dielectric) cylinder of diameter 11 cm and height 3 cm, when it is buried at a depth of 2 cm (top plots), 4 cm (middle plots), or 8 cm (bottom plots) in Yuma soil with moisture content of 15%.

7. SUMMARY AND CONCLUSIONS

We have studied the backscattered echoes returned by perfectly conducting and dielectric cylindrical mine-like objects buried at three representative depths in a soil that has three different moisture contents. The moisture is uniformly distributed in both the horizontal and vertical directions. The backscattered signals were simulated using a Method-of-Moments code. The incident pulses were Dirac delta pulses passed through an eighth-order Butterworth bandpass filter as illustrated in Fig. 3. The backscattered responses were studied in three domains: the frequency domain, time domain, and time-frequency domain.

In the frequency domain, we have seen that as the burial depth increases, the magnitude of the RCS of the metal target is not appreciably affected in very dry soils but decreases in moist soils. However, the number of peaks and dips and their occurrences depend on the burial depth. The RCS of the plastic target is negligible for very dry soils and, as the metal target, becomes independent of burial depth. Like the metal target, the RCS of the plastic target decreases with burial depth in moist soils.

In the time domain, for metal mines there is a strong specular return, which has almost the same duration as the incident, interrogating pulse. A second, weaker return is caused by multiple reflections between the mine's top surface and the ground-air interface. As the moisture content in the soil increases, the strength of the returns is seen to decrease. For the plastic mine, the initial specular return is much weaker than for the metal mine. In dry soil, due to the target's low dielectric contrast the specular return is extremely weak. As the moisture content increases, the contrast increases and so does the peak strength of the return amplitude. For both mines analyzed, the returned pulses appear more delayed in time as the moisture content increases. For the moist soils, the returned pulses are weaker as the burial depth increases.

The time-frequency domain shows results that combine much of the above observations. For plastic mines, the TFD images appear more elongated along the time-direction than for metal ones. The images begin to develop at successively later times as the moisture content increases, and also as the burial depth increases, for reasons given earlier. The images also spread out to lower frequencies as the moisture content or burial depth in moist soils, or both, increases. In spite of this frequency shift, the overall characteristics of the TFD images of any target seem to be much more stable than the signature features in the time domain or in the frequency domain. Both the BJOKD and the PWD generate similar features and which one to use in practice will depend on its robustness against targets in clutter. These observations will become useful for the later applications to target identification.

8. ACKNOWLEDGMENTS

The authors thank the Digital Signal Processing Group at Rice University for permission to use the MATLAB software they have developed for computing the signal adaptive time-frequency distribution by Baraniuk and Jones. The authors gratefully acknowledge the partial support of their respective Institutions. This work was partially supported by the Office of Naval Research and the Swedish Rescue Service Agency.

9. REFERENCES

1. R. W. P. King and C. W. Harrington, The transmission of electromagnetic waves and pulses into the earth, *J. Appl. Phys.* **39**, 4444–4452, 1968.
2. D. L. Moffat and R. J. Puskar, A subsurface electromagnetic pulse radar, *Geophys.* **41**, 506–518, 1976.
3. H. C. Strifors, B. Brusmark, S. Abrahamson, and G. C. Gaunaurd, Signature features in time-frequency of simple targets extracted by ground penetrating radar, in: *Automatic Object Recognition V*, (*Proc. International Society for Optical Engineering, SPIE 2485*) edited by F.A. Sadjadi, Orlando, April 1995, pp.175–186.
4. S. Abrahamson, A. Ericsson, A. Gustafsson, H. C. Strifors, and G. C. Gaunaurd, Comparison of fuzzy-cluster based time-frequency signatures for automatic classification of underground mine-like targets, in: *Automatic Target Recognition IX*, (*Proc. International Society for Optical Engineering, SPIE 3718*) edited by F.A. Sadjadi, Orlando, April 1999, pp. 128–139.
5. D. J. Daniels, *Surface-Penetrating Radar* (IEE, London, UK, 1996).
6. S. Vitebskiy, L. Carin, M. A. Ressler, and F. H. Le, Ultra-wideband, short-pulse ground-penetrating radar: Simulation and measurement, *IEEE Trans. Geosci.Remote Sensing* **35**, 762–772, 1997.
7. R. F. Harrington, *Field Computation by Moment Methods* (MacMillan, New York, NY, 1968).
8. N. Geng and L. Carin, Wide-band electromagnetic scattering from a dielectric BOR buried in a layered lossy dispersive medium, *IEEE Trans. Antennas Propagat.* **47** (4), 610–619, 1999.
9. L. Cohen, *Time-Frequency Analysis* (Prentice Hall, Englewood Cliffs, NJ, 1995).
10. R. G. Baraniuk and D. L. Jones, Signal-dependent time-frequency analysis using a radially Gaussian kernel, *Signal Processing* **32**, 263–284, 1993.
11. K. A. Michalski and D. Zheng, Electromagnetic scattering and radiation by surfaces of arbitrary shape in layered media, Part I: Theory, *IEEE Trans. Antennas Propagat.* **38**, 335–344, 1990.
12. J. J. Yang, Y. L. Chow, and D. G. Fang, Discrete complex images of a three-dimensional dipole above and within a lossy ground, *Proc. Inst. Elect. Eng.* **H 138**, 319–326, 1991.
13. A. V. Oppenheim and R. W. Schafer, *Digital Signal Processing* (Prentice-Hall, Englewood Cliffs, NJ, 1975).

DETECTION OF TARGETS BEHIND WALLS USING ULTRA-WIDEBAND SHORT PULSE

Walid A. Chamma and Satish Kashyap*

1. INTRODUCTION

Impulse radar can be used to detect the presence and movement of targets behind walls. To be an effective detection system, the radar should have the transmitted signal at a frequency low enough to be able to penetrate walls and have a very wide bandwidth so that targets behind walls are clearly identified. Bandwidths need to be several gigahertz to achieve high resolution of the order of a fraction of a meter. An ultra-wide band (UWB) radar system satisfies these low frequency and large bandwidth requirements. UWB radars are defined as those for which the relative bandwidth is equal or greater than 25%. The UWB transmitted pulse usually consists of a very short pulse train of just a few cycles. The EM scattered fields from a target, illuminated by the UWB radar, are received at several locations and then processed to construct their corresponding images. UWB radar systems have been used for a wide variety of civilian and military applications. M. Skolnik et al.[1] outlined considerations that characterize the design of UWB radar for the detection of low-altitude missiles over the sea. He discussed the factors that enter into the choice of frequency, the selection of the type of the transmitter, antenna, and the receiver as well as signal processing issues. UWB for minefield detection (ground penetrating radars) was also investigated by L. Carin et al.[2], where a full-wave model for EM scattering from buried targets is developed. Such systems were also studied thoroughly by E. K. Walton and his group at the Ohio State University Electro Science Laboratory[3].

In another unique application, a UWB radar system is used to detect and predict images of moving targets behind walls or non-metallic visually opaque boundaries. Nag et al.[4,5] introduced a hand-held wall penetrating radar unit that transmitted time-

*Authors are with the Radar Electronic Warfare Section, Defence R&D Canada – Ottawa, Department of National Defence, 3701 Carling Avenue, Ottawa, Ontario, CANADA K1A 0Z4.

modulated UWB signals through barriers. Such a system is very useful in forced entry and hostage situations, as well as emergencies where victims under debris can be detected.

This work describes the use of a full-wave time-domain numerical procedure, based on the finite-difference time-domain[6] (FDTD) method, to investigate the capabilities and limitations in the use of a UWB radar system to detect moving targets behind walls. A room with a concrete floor and wooden walls is modelled inside the FDTD lattice. The modelling assumes a transmitting antenna located outside the room at a 1.0m height and transmits a UWB pulse with a centre frequency of 2GHz. A conducting cube target is placed at a 1.0m height inside the room. The scattered field due to the box is calculated using FDTD and is recorded at several receiver points outside the room. The range location of the target can be determined by measuring the propagation delay time from the transmitting antenna to the target and back to the receiver point after accounting for the presence of the wall. Using the received field data at several observation points, an image of the target can be constructed. The above investigation is then repeated with the single target placed at different locations inside the room. Using the field information at various points outside the room, the movement of an object inside the room is traced and displayed.

This paper describes the effect of various parameters on the performance of the detection system. These include the number of receiver and transmitter points and their separation distances. Additional simulation is then conducted for two conducting cubes inside the room to investigate the capability to locate and distinguish their corresponding images using the received scattered field information.

2. FDTD MODELLING

The finite-difference time-domain (FDTD) numerical method is a direct solution of the time-dependent Maxwell's curl equations in the time-domain using the finite-difference technique. It is analogous to the existing finite-difference solution of fluid flow problems encountered in computational aerodynamics, where a numerical model is based on a direct solution of the corresponding partial differential equation. In FDTD the propagation of an EM wave into a volume of space containing a dielectric or a conducting structure (or a combination of both) is modelled. By time stepping, the incident wave is tracked as it first propagates to the structure and then interacts with the latter through current excitation, scattering, multiple scattering, penetration and diffraction.

A 2.36m x 3.59m x 2.62m room with a concrete floor and wooden walls is modelled inside the FDTD lattice and is shown in Fig. 1. The wooden walls and the concrete floor are represented inside the FDTD lattice by their physical parameters, namely, a relative permittivity $\varepsilon_r = 2.75$ and conductivity $\sigma = 3.0\text{x}10^{-3}$ S/m for the wooden wall and $\varepsilon_r = 7.0$ and conductivity $\sigma = 5.0\text{x}10^{-2}$ S/m for the concrete floor. These values are chosen for an operating frequency $f = 2$GHz. In this simulation the antenna source is an x-directed 9cm centre-fed dipole located outside the room at an elevation $z = 1.0$m as illustrated in Fig. 1. Initially, the target inside the room consists of a 10cm conducting cubic box located inside the room at x_b = -30cm, y_b= 94cm, and z_b = 100cm, where (x_b, y_b, z_b) are the coordinates of the conducting box centre. Fig. 1 also shows the FDTD spatial and temporal resolution parameters used in the modelling and simulation of the target EM illumination.

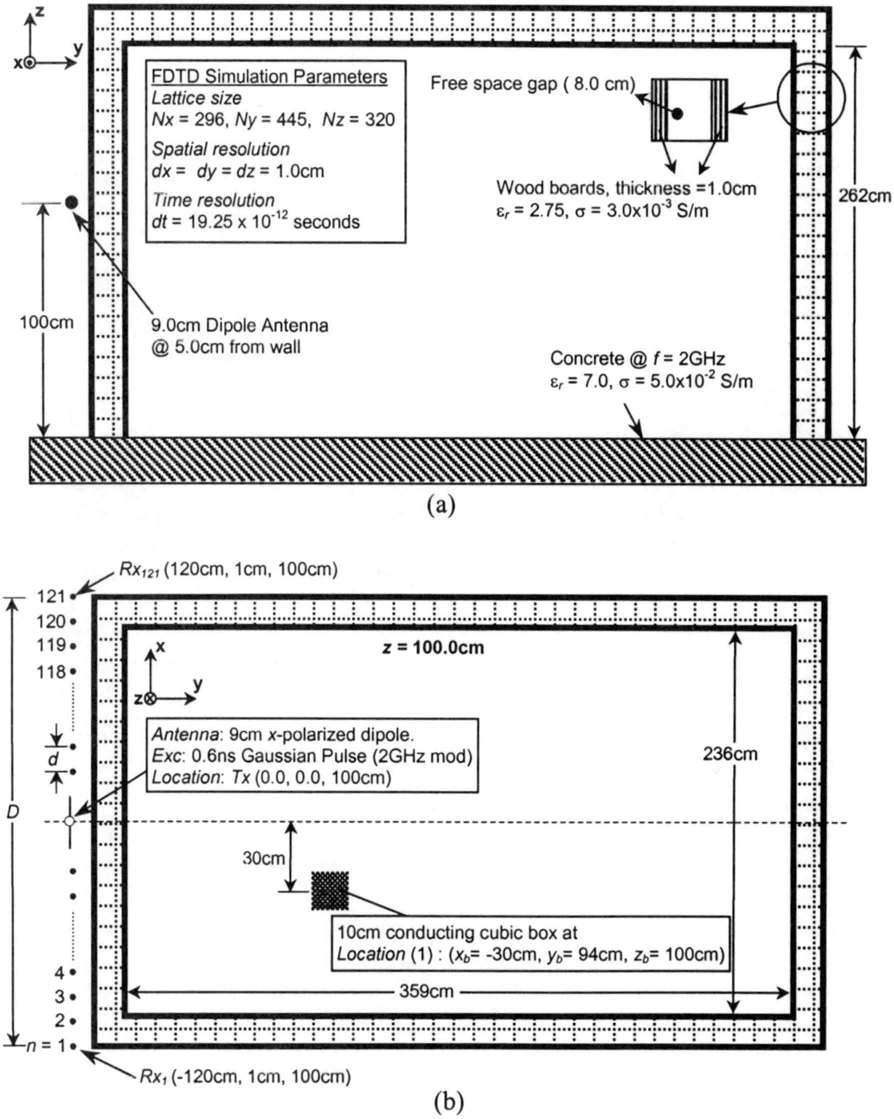

Figure 1. Room geometry and the UWB radar setup

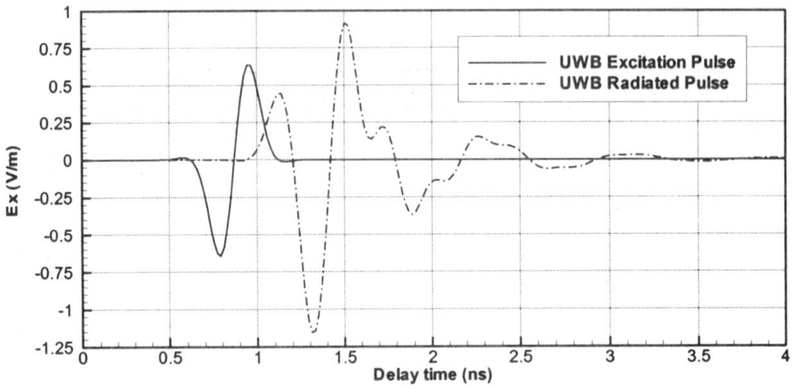

Figure 2. Antenna excitation and radiated UWB pulse

3. UWB EXCITATION PULSE

To model the EM illumination of the conducting box target by a UWB short pulse, the dipole antenna is fed by a 0.6ns gaussian pulse modulated by a 2GHz sine wave, and is shown in Fig. 2. Since the dipole antenna used is not designed to have a very wide bandwidth, the resulting radiation pulse will be distorted (convolved) with the antenna's own impulse response. Fig. 2 also shows the near-field radiated UWB pulse in free space from the dipole antenna. Fig. 3 shows the frequency response of the UWB excitation pulse and the radiated pulse, respectively. For the case of a plane wave with the UWB excitation pulse of Fig. 2 as temporal distribution, the pulse will propagate in free space without distortion or attenuation. Hence, the frequency-domain transfer function, $H_t(\omega)$, of the dipole antenna used in our simulation can be obtained through:

$$H_t(\omega) = H_{rad}(\omega) / H_{UWB}(\omega) \tag{1}$$

Here $H_{rad}(\omega)$ is the frequency response of the radiated field from the antenna at a

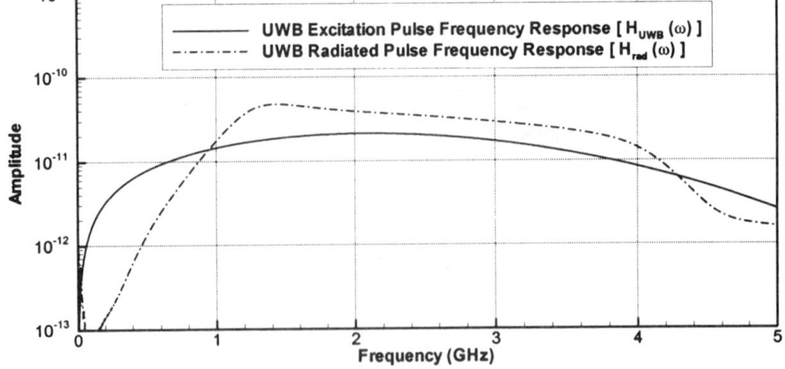

Figure 3. Antenna excitation and radiated UWB pulse frequency response.

specified observation location in the near-field, and $H_{UWB}(\omega)$ is the frequency response of the radiated field due to the plane wave at the same observation location. Fig. 4 shows the frequency-domain transfer function of the antenna $H_t(\omega)$. The knowledge of the impulse response of the antenna is necessary information that will be used to obtain the impulse response the target detected inside the modelled room.

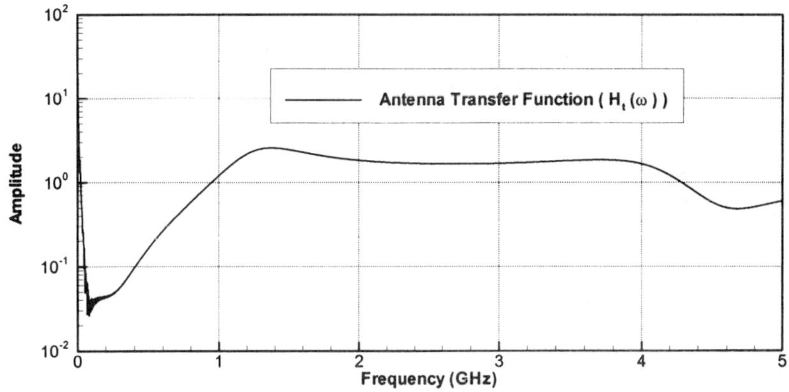

Figure 4. Frequency-domain transfer function of the excitation antenna (9.0cm dipole)

4. SIMULATION RESULTS & DISCUSSION

Our initial investigation considers a 10cm conducting cubic box inside the room shown in Fig. 1. A setup of one x-directed transmitter antenna, represented by the 9cm dipole centred at (0.0,0.0,100cm), and 121 receivers are considered. The receiver points in the FDTD simulation are observation points where the EM fields are recorded and saved for post-processing. The receiver points Rx_1 to Rx_{121} are distributed along the side of the room in the x-direction with coordinates x_1 = -120cm, x_2 = -118cm, . . . , x_{121} = +120cm (a resolution d = 2.0cm between each of the receiver points, with y = 1.0cm and z = 100cm), and an aperture length D = 240cm. The room wall begins at y = 5.0cm, so the transmitter antenna at Tx = (0.0, 0.0, 100cm) is 5.0cm from the wall, and the receivers points are 4.0cm from the, wall as illustrated in Fig. 1.

Fig. 5 shows the isolated recorded response at Rx_{41} (x_{41} = -40cm, y_{41} =1.0cm, z_{41} = 100cm) of the conducting box at locations (1) and (2) inside the room at (x_b = -30cm, y_b = 94cm, z_b = 100cm) and (x_b = -30cm, y_b = 294cm, z_b = 100cm) respectively. The isolated recorded response of the conducting box refers to the difference between the responses recorded at Rx_{41} with and without the presence of the conducting box. The responses shown in Fig. 5 are those without the de-convolution of the antenna impulse response. Fig. 6 shows the responses with the de-convolution of the antenna impulse response implemented, resulting in the impulse response of the conducting box at the locations considered. A spatial shift is observed in the impulse response of Fig. 6 as compared with that of Fig. 5. This will result in the image of the conducting box at the correct location inside the room image plane. The scale difference in the Ex field magnitude of Fig. 5 and 6, is due to the de-convolution process and will not effect the image generation procedure.

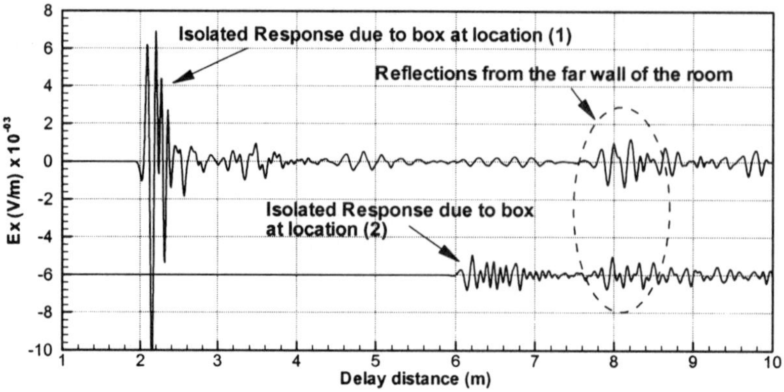

Figure 5. Isolated response of the 10cm conducting box at locations (1) and (2) recorded at receiver Rx_{41} obtained without de-convolution of the antenna impulse response.

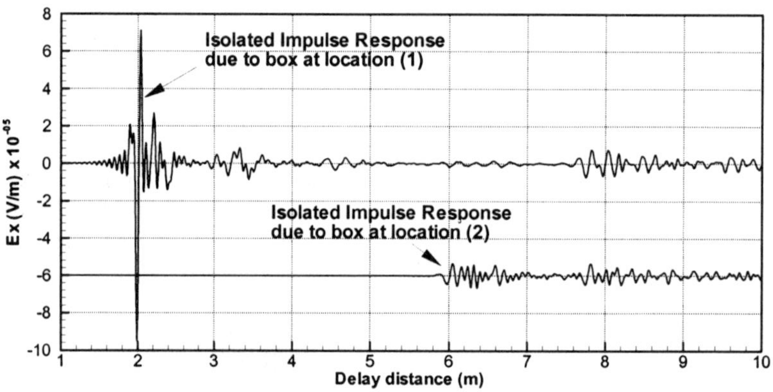

Figure 6. Processed isolated response of the 10cm conducting box at locations (1) and (2) recorded at receiver Rx_{41} obtained with de-convolution of the antenna impulse response.

The transmitter and the receiver points of the above setup constitute a bistatic radar system. The processing of the received signals to produce the target image is done by using the time-domain back-projection technique[7]. The time-domain response data recorded at each receiver point are generated directly by the FDTD method with no need for any Fourier Transformation. The back-projected signal at pixel (x_i, y_j) in the room image plane due to the recorded field E at receivers Rx_n $\{n = 1 \text{ to } 121\}$ at location (x_{Rxn}, y_{Rxn}) and a single transmitter Tx at (x_{Tx}, y_{Tx}) is given by:

$$I(x_i, y_j) = \sum_n E(t_{ij}(n), n) \tag{2}$$

where

$$t_{ij}(n) = \left(Tx_{ij} + Rx_{ij}(n)\right)/v \tag{3}$$

v is the speed of light in the propagation media and $t_{ij}(n)$ is the total delay recorded for the transmitted signal to travel from the transmitter location to pixel (x_i, y_j), Tx_{ij}, plus the delay from the pixel (x_i, y_j) to receiver n, $Rx_{ij}(n)$. Tx_{ij} and $Rx_{ij}(n)$ are obtained from the following expressions:

$$Tx_{ij} = \sqrt{\left(x_i - x_{Tx}\right)^2 + \left(y_j - y_{Tx}\right)^2} \tag{4}$$

$$Rx_{ij}(n) = \sqrt{\left(x_i - x_{Rxn}\right)^2 + \left(y_j - y_{Rxn}\right)^2} \tag{5}$$

Hence, the target function at a given pixel (x_i, y_j) in the image plane is formed from the data in the received responses at specific delay (*time bins*) corresponding to the distance of this pixel point to each of the receiver points. Interpolation is used at instances where the delay time calculated does not correspond to a data point in the recorded responses. Using Hilbert's Transformation[7], G, the power envelope image can be obtained by substituting $E(t,n)$ with $\hat{E}(t,n)$ where

$$\hat{E}(t,n) = E(t,n) + jG(E(t,n)) \tag{6}$$

A power envelope image can be found by taking the amplitude of the complex image $\left|\hat{E}(t,n)\right|$ in Eq. (6).

4.1 Bistatic UWB Radar System

Using the above procedure the image of the conducting box inside the room of Fig. 1 is generated. Fig. 7 shows the generated images of the conducting box inside the room at locations 1 – 5. The coordinates of these locations are also shown in Fig. 7. The box image for each location is displayed on a single image plane for comparison purposes. In addition, for clarity, the sidelobes of the generated images are suppressed by setting the threshold level in the plotting program to display only the peak magnitudes. These images were generated using only 25 receiver points. These receivers are separated by distance d = 10cm and have an array aperture length of D = 240cm. For location (1), the box image is displayed more clearly due to stronger return signal to the receivers on the side of the room, as compared to that of the other locations (2 – 5). As the conducting box moves further inside the room and further away from the UWB radar setup, the generated image is more out-of-focus than that of location (1). This is attributed to the decreasing magnitude of the reflected signal, the dispersion of the propagating pulse, and the magnitude of the reflected signal from the box at comparable levels to reflections from the room structure. Figs. 5 and 6 demonstrate such reasoning.

In the present radar setup, the parameters that influence the resolution at which targets are detected are similar to those that influence a Synthetic Aperture Radar (SAR) system. In SAR, the range resolution ΔR (y-axis in the room plane) is given by:

$$\Delta R = c/(2B) \tag{7}$$

where c is the speed of light in free space and B is the bandwidth of the transmitted signal. The azimuthal resolution (x-axis in the room plane), $\Delta\theta$ for an array of transceivers is given by

$$\Delta\theta = \lambda_o / (2D) \tag{8}$$

where λ_o is the wavelength of the centre frequency and D is the array aperture length. Although our current radar setup is bistatic, the expressions in Eqs. (7) and (8) still apply.

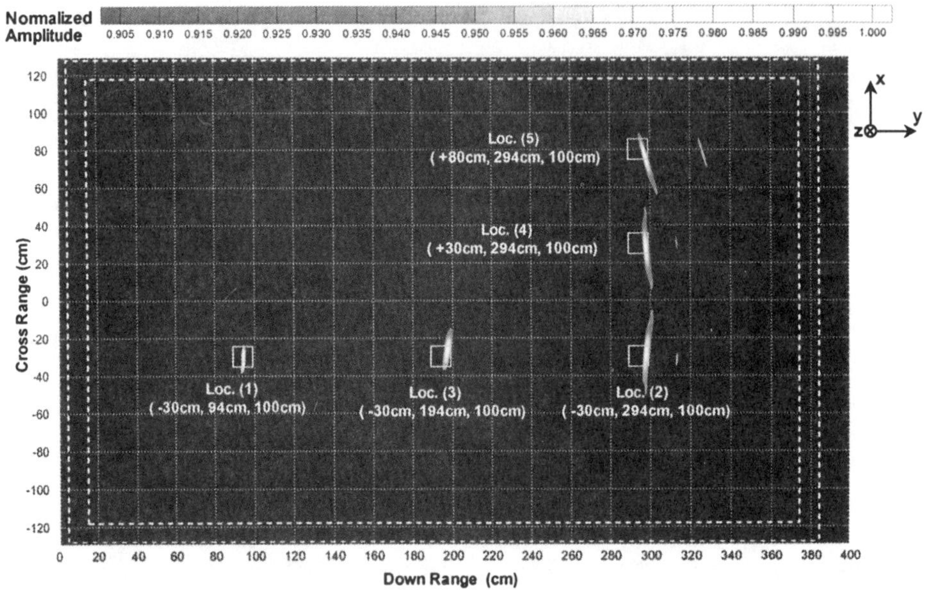

Figure 7. Generated image of the 10cm conducting box at locations (1 – 5) inside the room detected by a bistatic UWB radar setup ($n = 25$, d = 10cm, D = 240cm).

The following section investigates the effect in changing the resolution distance d between the receiver points, and the array aperture length D on the quality of the detected image of the conducting box at location (1) inside the room. In Fig. 8 the image of the conducting box is generated for different values of d but for a fixed value of D = 240cm. The number of receivers n used for each value of d range from 121 to 16 receivers, and are noted in Fig. 8. With aperture length D = 240cm, the resolution of the generated images, representing the conducting box, are the same for all the cases studied. However, for larger resolution distance d corresponding to lower number of receivers, higher sidelobes are generated. Fig. 9, on the other hand, shows the effect of changing the array aperture length D for a fixed value of resolution distance d at 4.0cm. Here, as expected and illustrated in Eq. (8), the image representing the conducting box becomes more unfocused for smaller values of D. In addition, ghost images of the box at lower intensity start to show alongside the main image with decreasing value of D. In both Figs. 8 and 9, the transmitter is located at the centre of the room at (0.0,0.0,100cm).

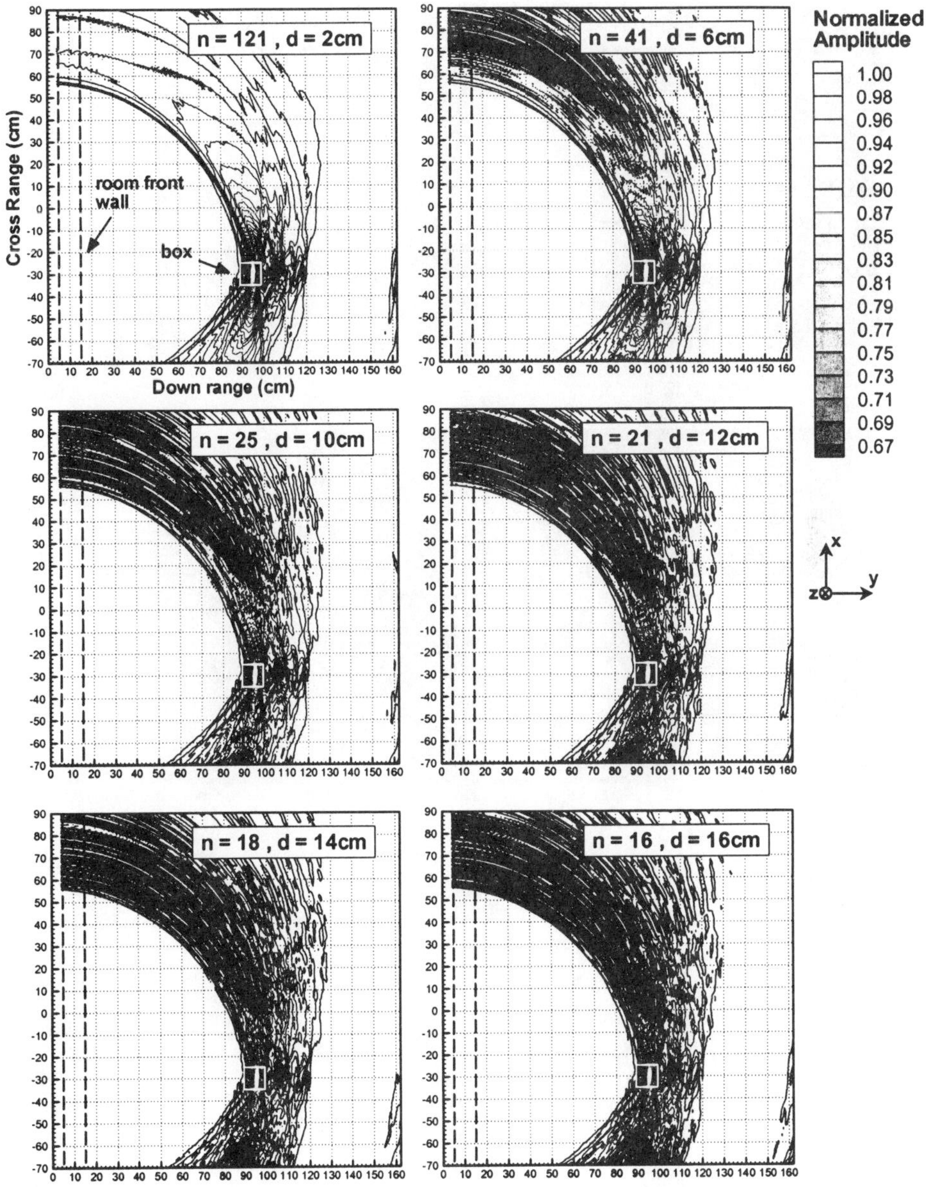

Figure 8. Generated image of a 10cm box at location (1) inside the room detected by a bistatic UWB radar setup for different values of n and d. (D = 240cm).

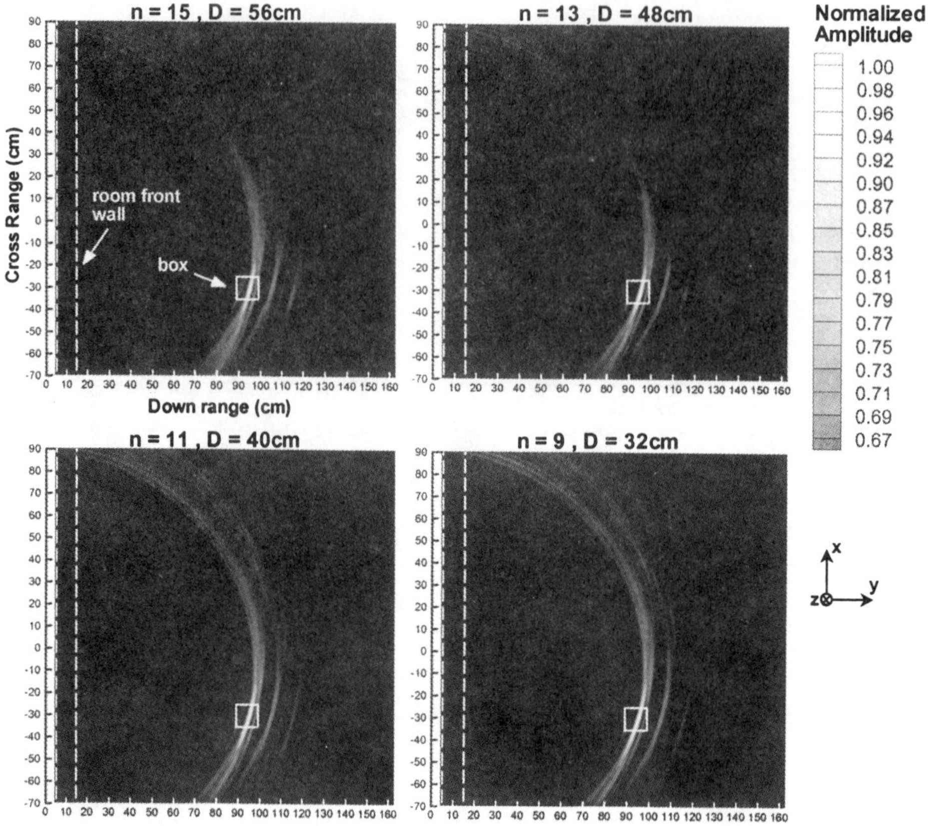

Figure 9. Generated image of a 10cm box at location (1) inside the room detected by a bistatic UWB radar setup for different values of *n* and *D*. (d = 4cm).

4.2 Monostatic UWB Radar System

Next, we consider a monostatic UWB radar setup consisting of a maximum of *nine* transmitters/receivers (*Tx/Rx*). In this case Eqs. (3) and (4) become

$$t_{ij}(n) = \left(Tx_{ij}(n) + Rx_{ij}(n)\right)/v \tag{10}$$

and

$$Tx_{ij}(n) = \sqrt{\left(x_i - x_{Txn}\right)^2 + \left(y_j - y_{Txn}\right)^2} \tag{11}$$

These are used in calculating the image intensity at pixel (x_i, y_j) in the expression of Eq. (2).

This setup has a resolution distance d = 10cm and an array aperture length D = 80cm centred at the room wall as before and at a height z = 100cm. The transmitters consist of *nine* x-directed 9cm dipole antennas excited by the UWB pulse, while the receivers consist of response recording observation points. Fig. 10 shows the generated image of the same 10cm conducting box at location (1) for different values of D and the number of Tx and Rx. Here the generated image is displayed at the correct location even with small number of transmitters and receivers. For the case of D = 20cm, and Tx = Rx = 3 elements, ghost images alongside the main image start to show as predicted by Eq. (8). As expected, the sidelobe levels are relatively high when compared to thosc of Fig. 8 with d < 10cm. By setting the intensity display threshold level in the plotting software, these sidelobes can be filtered out and a clear image representing the box inside the room is obtained.

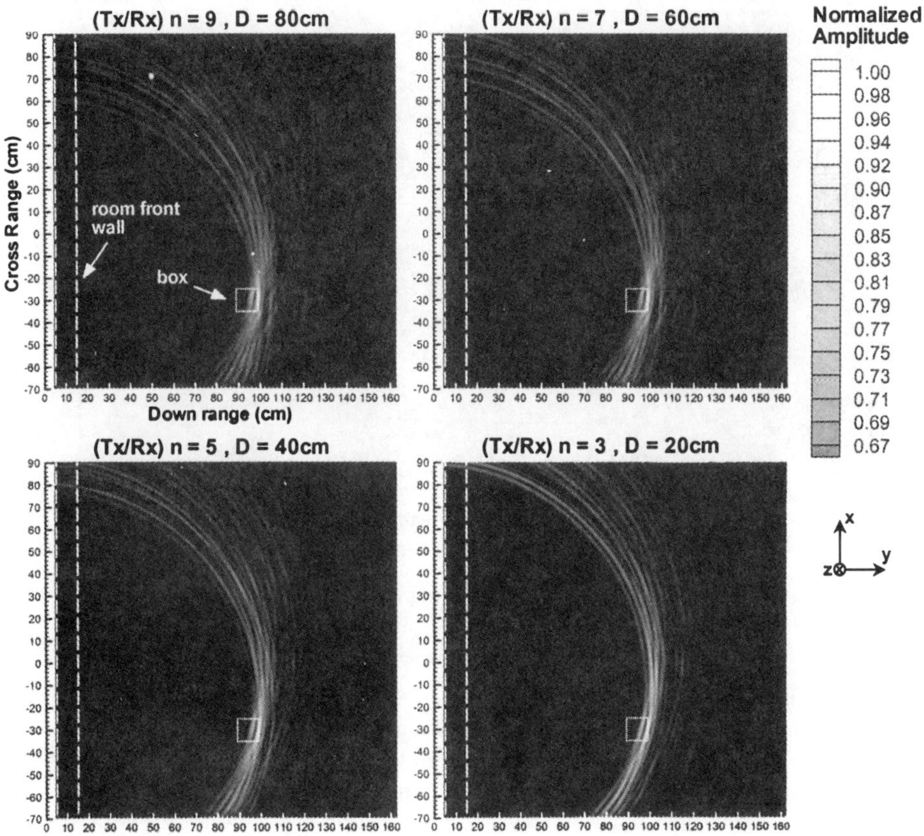

Figure 10. Generated image of a 10cm box at location (1) inside the room detected by a monostatic UWB radar setup for different values of n and D. (d = 10cm).

4.3 Multiple Target Detection

The above discussion involves only a single 10cm conducting cubic box, in this section a 30cm conducting cubic box is added inside the room at coordinates (65cm, 297cm, 100cm) at the back of the room. Using the monostatic UWB radar setup ($Tx = Rx = 9$, d = 10cm, D = 80cm) and no Hilbert transformation, the generated image is shown in Fig. 11. In this Figure, the 10cm box can be identified very clearly, however, the 30cm box image is generated with much lower intensity. This is attributed to the magnitude of the radar signal reaching the 30cm box being less than that reaching the 10cm box at location (1). The scattered signal from the 30cm box is further attenuated as it travels back to the receivers.

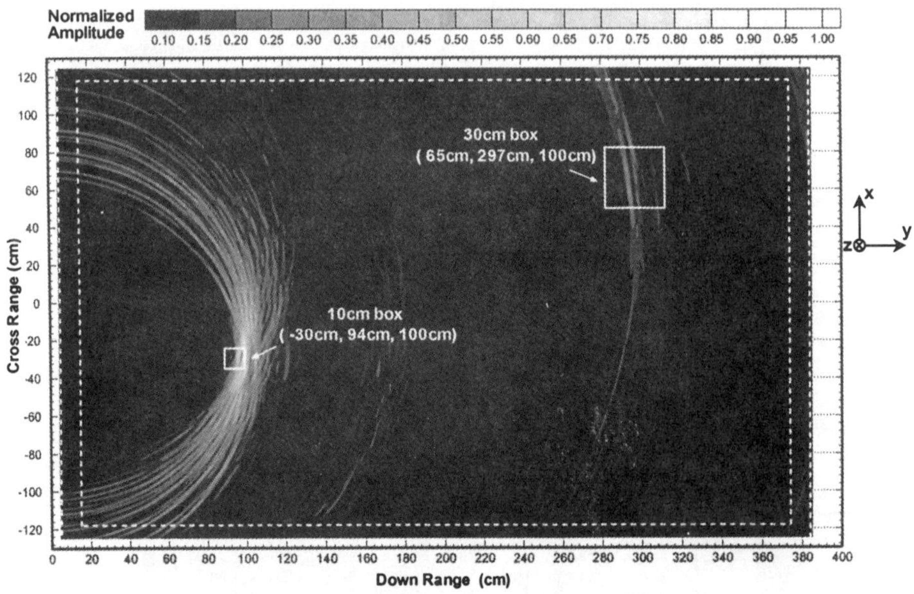

Figure 11. Generated images of a 10cm box and a 30cm box inside the room detected by a monostatic UWB radar setup (n =9, d = 10cm, D = 80cm).

Fig. 12 shows the isolated signatures received at the centre element of the UWB radar with and without the 30cm box. This Figure shows clearly a lower magnitude response resulting from the 30cm box than from the 10cm box. By gating the received responses in the image processing part to neglect responses from targets at distances that are relatively closer to the radar system, and normalizing to a new image peak value, a clearer image of the targets at the far side of the room can be obtained. By superimposing the image obtained from the gated response with that of Fig. 11, a clearer display of the targets inside the room is obtained. Fig. 13 shows the resulting display.

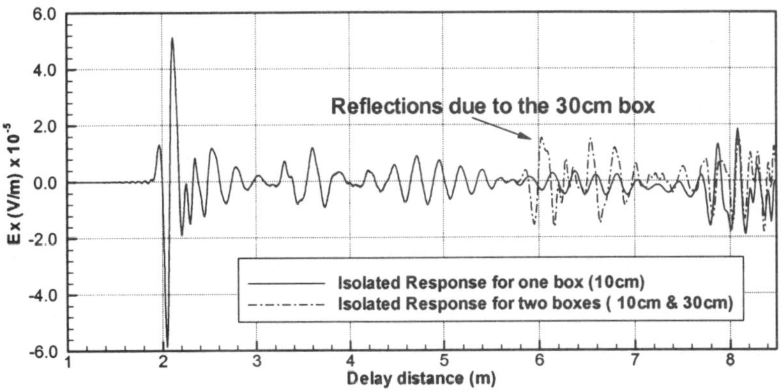

Figure 12. Isolated impulse response recorded at *Rx* (0.0, 1.0cm, 100cm) / *Tx* (0.0, 0.0,100cm) of a monostatic UWB radar setup for the cases of one and two boxes inside the room.

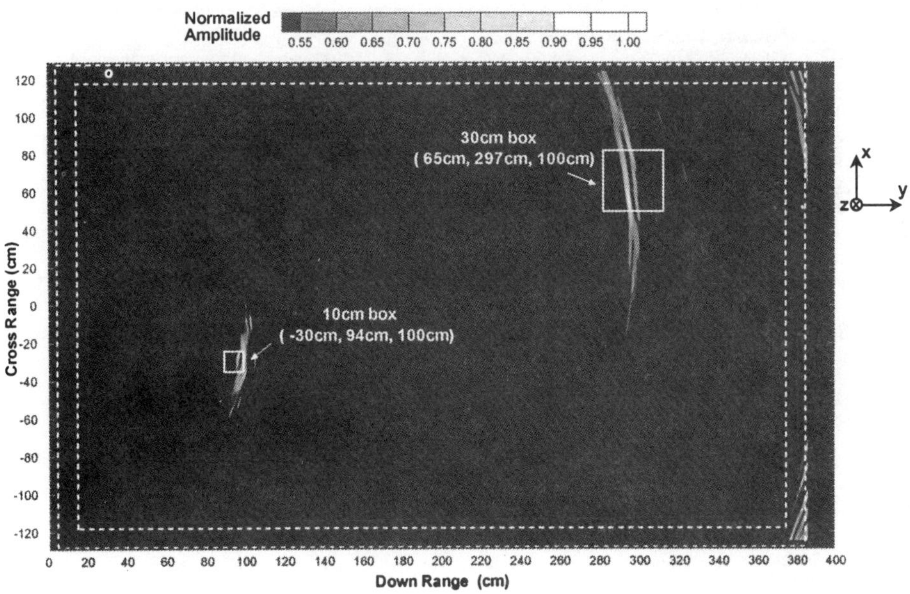

Figure 13. Generated images of a 10cm box and a 30cm box inside the room obtained by the original and the gated impulse responses.

5. CONCLUSIONS

In this work, a bistatic and a monostatic UWB radar setups used for the detection of targets behind walls are investigated via numerical simulation using the finite-difference time-domain (FDTD) method. For the case of a 2.36m x 3.59m x 2.62m room with wooden walls and a concrete floor, the radar setup consisted of one transmitter and 121 receivers with an array aperture length D = 240cm, and resolution d = 2cm. This bistatic radar is used to detect the location of a 10cm conducting cubic box inside the room. The effect of changing the number of receivers, array aperture length, and receivers resolution on the image quality of the detected structure is investigated. It was observed that larger array aperture length, D, produced a better-focused image. However, larger values of resolution distance, d, produced higher sidelobes.

For the monostatic radar setup case, a clear image is generated for a small number of Tx and Rx and small array aperture length D. Also, a relatively clear image is generated for the case of Tx and Rx as low as *three* elements, D = 20cm, and d = 10cm. For a room with two targets, it was observed that lower image intensity is generated for the target that is further away from the radar setup. However, gating the received responses to ignore early time data and overlapping of the generated images (generated using original and gated responses) produce better images for the detected targets.

6. REFERENCES

1. M. Skolnik, G. Andrews, and J.P. Hansen, Ultra-wideband microwave-radar conceptual design, *IEEE AES Systems Magazine*, 25–29 (October 1995).
2. L. Carin, N. Geng, M. McClure, J. Sichina, and L. Nguyen, Ultra-wideband synthetic aperture radar for mine-field detection, *IEEE Antennas Propagat. Mag.* **41**(1), 18–33 (February 1999).
3. E.K. Walton and S. Gunawan, Comparative analysis of UWB underground data collected using step-frequency short pulse and noise wave form, in: *Ultra-Wideband, Short-Pulse Electromagnetics 3*, edited by C.E. Baum, L. Carin, and A.P. Stone (Plenum Press, New York, 1997), pp. 511–516.
4. T. Payment, A low power, ultra-wideband radar testbed, in: *Ultra-Wideband, Short-Pulse Electromagnetics 5*, edited by P.D. Smith and S.R. Cloude (Kluwer Academic/Plenum Publishers, New York, 2002), pp. 235–245.
5. S. Nag, H. Fluhler, and M. Barnes, Preliminary interferometric images of moving targets obtained using a time-modulated ultra-wideband through-wall penetration radar, *Proc. of IEEE Radar Conf.*, 64–69, Atlanta GA (May 2001).
6. A. Taflove and S.C. Hagness, *Computational Electrodynamics: The Finite-Difference Time-Domain* (Artech House, Boston, 2000).
7. D. Mensa, *High Resolution Radar cross-section Imaging* (Artech House, Boston, 1991).

INTEGRATED DIGITAL UWB-RADAR

Jürgen Sachs, Peter Peyerl, Peter Rauschenbach, Frantisek Tkac, Martin Kmec, Stephen Crabbe*

1. INTRODUCTION

Ultra wideband (UWB) radar and impedance spectroscopy are of great interest for a vast number of applications such as surface penetrating radar, surveillance and emergency radar, medical instrumentation, non-destructive testing in civil engineering and the food industry, industrial sensors and microwave imaging and many others. The fractional bandwidth of the sounding waves for such types of applications should be as close as possible to 200 % resulting in a high spatial resolution and good penetration in materials. An UWB radar is able to detect hidden objects and a high bandwidth not only results in good spatial resolution but also in improved capabilities for object recognition. Concerning the impedance spectroscopy, a large bandwidth covers different relaxation phenomenon of matter so that more information is available for the substance characterisation.

This article will be concentrated on UWB radar rather than on problems of impedance spectroscopy even if both measurement tasks are closely related. The UWB radar measures the time variant impulse response function (IRF) of the scenario of investigation which is determined by the arrangement and movement of scatterer and the illumination of the scene by the antennas. The classical methods of determining a systems impulse response function is based on pulse excitation methods as the function's name implies. Other methods deal with the determination of the frequency response function (FRF) – a Fourier transformed version of the IRF - over a very large bandwidth. These methods typically use sine-waves which are stepped or swept through the band of interest or a Multi-Sine concept is applied. However all these methods are burdened by critical disadvantages concerning their technical parameters or/and the price (or/and size) of the equipment. In the opinion of the authors, this is the main explanation for the gap between

* Jürgen Sachs, Peter Rauschenbach, Frantisek Tkac, Martin Kmec, Technische Universität Ilmenau, 98684 Ilmenau, Germany; Peter Peyerl, Stephen Crabbe, MEODAT GmbH, 98693 Ilmenau, Germany.

the current status in the practical use of UWB sensor technique in industry and its actual physical potential.

The article will present a UWB-measurement concept which will reduce the above mentioned gap. The principle was tested for the first time in an anti-personnel landmine (APL) detection radar. The radar profits from the use of UWB pseudo random binary sequences which may be generated up to a very large bandwidth and which have high energy even at low signal levels. This favours the noise suppression as well as the system integration on low cost RF-circuits using SiGe-technology. Furthermore, nearly the complete system conception is based on digital circuit schemes promoting high flexibility and cost efficient implementation for applications. In what follows, the basic principle will be introduced and the main design rules will be described. In contrast to simple pulse systems, the presented principle is robust against drift appearances as sine-wave systems. Thus, the reduction of systematic errors by calibration procedures is an interesting feature to improve the performance. Two basic methods will be mentioned. Some results from the first practical implementation will close the article.

2. THE BASIC PRINCIPLE

The key for an efficient system configuration as well as its technical implementation is the appropriate choice of the test signal. Referring to UWB measurements, the theory of linear systems does not suppose a specific shape of stimulus rather it only requires that the signal has a very short auto correlation function (ACF) i.e. a large bandwidth. On this basis, one is not exclusively tied to pulses or sine-waves which increases the degree of freedom in the system conception. Now, aspects of the technical implementation may be moved into the foreground of interest.

Avoid high signal peaks within the analogue part of the electronics: Low level signals do not burden the electronics extremely which provides for stable and fast (high bandwidth) operation. Furthermore, low level signals may be constructed in integrated RF-circuits promoting a further improvement in bandwidth. The signal must however also contain as much energy as possible in order to provide good suppression of perturbations. Thus, the instant signal power must be distributed equally over time i.e. the crest factor should be close to one. However, the reaction of a device under test on such a signal does not correspond to the IRF. Therefore, the reaction of the device must be compressed in the receiver in order to provide the IRF.

Use under-sampling: By under-sampling the speed of data gathering can be reduced to a value that is determined by the time variance of the test scenario (for example the maximum speed of a target) rather than the bandwidth of the sounding waves. This greatly simplifies the electronics and reduces costs. Under-sampling may be applied in connection with any periodic stimulus signals. However, attention must be paid to provide an efficient and stable control of the sampling instant.

Use digital impulse compression: Impulse compression may be performed by matched filtering or correlation. Most principles use analogue techniques which is the usual method to improve the signal to noise ratio. This method is however only rarely applied in UWB systems since it is very difficult to build analogue systems with a large fractional bandwidth and a high compression factor. A high compression factor requires the storing of a huge amount of energy distributed over a wide spectrum within the impulse compression systems. Herewith, analogue solutions quickly meet with their technical limits. Moreover, the output signal of an analogue impulsion compression is again a high crest factor signal which should be avoided in analogue circuits (see above). The digital impulse compression is not sensible to these effects. The "energy storage" is a question of memory size and a high crest factor signal supposes an appropriate word length respectively number format. Both points, may be respected by a corresponding design. They are underlying less critical technical constraints as for analogue system solutions.

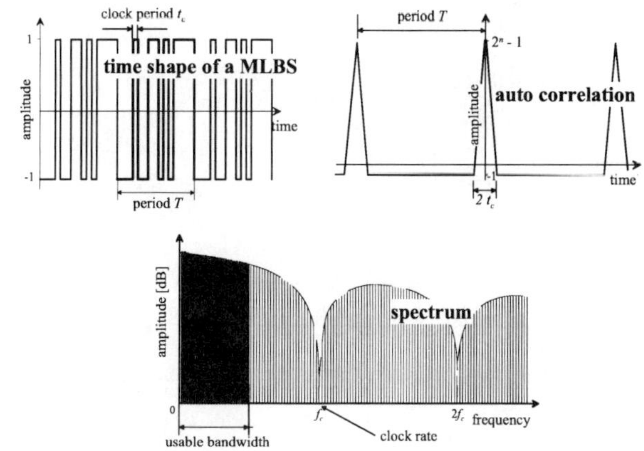

Figure 1: Time shape, auto correlation function and spectrum of a MLBS. For a better graphical representation the shown spectrum applies to a MLBS of higher order than used in the upper diagrams.

A good choice for a test signal under these circumstances is the Maximum Length Binary Sequence (MLBS or M-sequence). It is a periodic wideband signal having a short ACF which is free of side lobes (in the ideal case). Figure 1, shows an example of its idealised time shape, spectrum and ACF. The MLBS stimulus may be generated easily by a digital shift register with appropriate feedback loops even at clock rates in the GHz domain. It is a periodic signal permitting under-sampling and synchronous averaging for noise suppression. Interestingly, there is a very simple and stable way to control the sampling instants for the data gathering. If one is satisfied with an equivalent sampling rate of f_c (i.e. one sample per chip), a simple binary divider is able to provide a stable sampling clock. This sampling principle exploits the fact that the number of chips in a MLBS is always one less than a power of two. Due to the Nyquist sampling theorem the usable bandwidth is limited to half the clock rate (see figure 1) with this kind of sampling. However the lower part of the spectrum contains the majority (more than 70 %) of signal energy. Thus, an increased equivalent sampling rate will only scarcely improve the usable frequency band because the immunity against noise will rapidly decrease by going beyond the $f_c/2$ value. Finally, there are simple and fast algorithms to compress the MLBS response of a device under test (DUT) in order to receive its IRF.

The basic structure of the MLBS front-end which is aimed to measure the IRF of a DUT can be derived in a straight forward way as mentioned above, figure 2 represents the result. A RF single-tone clock generator pushes in parallel the shift register which generates the stimulus signal and the binary clock divider, which controls the data gathering. This guarantees a very stable synchronism between signal generation and signal capturing. The input signals are sampled by a T&H or S&H circuit before they are transformed into the digital domain. The signals are then synchronously averaged to increase the dynamic range and to reduce the data rate. Finally, the impulse compression is performed either by a matched (FIR) filter or by the Fast Hadamard Transform (FHT). The FHT algorithm is organised in a very close manner as the FFT algorithm but it only requires additions thus it can be implemented in a very fast way. It should also be noted that no high crest factor signals are loading the analogue circuits. High signal peaks only appear after the FHT respectively the FIR-filter.

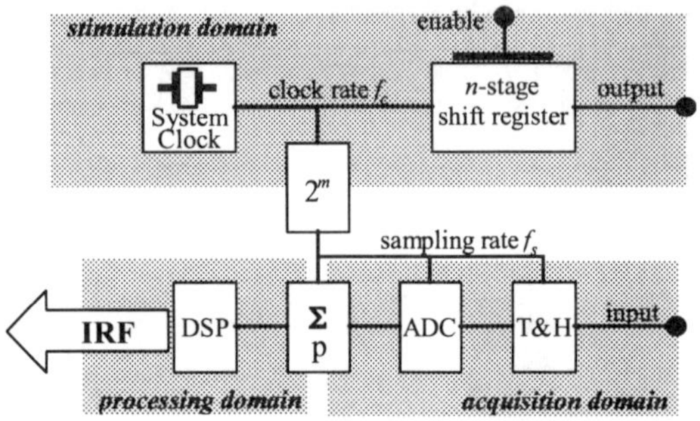

Figure 2: Basic block schema of the UWB front-end.

Considering figure 2 it does not provide any difficulties to expand the principle to a multi-channel measurement system as required in UWB arrays. As a matter of principle any number of measurement channels may be synchronously driven by the binary divider. The same is for the shift registers so that several stimulation channels are available. However, usually only one signal generator should be active at a certain time. This is controlled by enabling/disabling the individual shift registers[1]. There is no need for RF-switches.

3. THE KEY FIGURES

Most of the technical parameters of the MLBS front-end are determined respectively influenced by easily controllable values. This opens up the opportunity to adapt the system performance by simple means to the requirements of different applications. The concerning quantities are the number n of shift register stages, the number m of binary divider stages, the number of synchronous averages p and the RF-clock rate f_c. In what follows, a short overview about the most important technical parameters and their relations to the design parameters shall be given.

Bandwidth, range resolution: The usable bandwidth B and by that the range resolution δ_r in case of radar applications are fixed by the Nyquist sampling theorem.

Supposing an equivalent sampling rate of f_c as in our case they result in (c – propagation speed of the waves):

$$B = \frac{1}{2} f_c \qquad\qquad \delta_r = \frac{c}{f_c}$$

It should be noted that for most applications which require a certain digital post-processing one has to see that the spectral components of the received signal are really limited to the range between DC and $f_c/2$. If the DUT by itself does not limit the upper frequency parts, an anti-aliasing filter has to force it.

Time window, ambiguity range: The length of the IRF T i.e. their number of points depends upon the period length of the MLBS and this is given by the shift register length. Since the measurement results are periodic in T, the radar ambiguity range R corresponds to that period.

$$T = \frac{2^n - 1}{f_c} \qquad\qquad R \approx \frac{2^{n-1} c}{f_c}$$

Measurement time/rate, maximum target velocity: The measurement time T_{obs} is the time over which the object is "observed" to get one complete IRF. It should be noted that the measurement principle supposes time invariance of the DUT during the data gathering . From this it follows:

$$T_{obs} \approx \frac{p\,2^{n+m}}{f_c} \qquad\qquad 2\,T_{obs}\,v_{\max} \le \delta_r$$

If there is no lag in data processing or data transport, the measurement rate r_m is the inverse of T_{obs}.

Signal to Noise Ratio (SNR): The SNR-value is given by the quotient between the maximum possible value of the IRF referred to the RMS noise level. Supposing the dynamic of the analogue input part (amplifier, T&H, ADC) is expressed by ENOB (effective number of bits), the SNR-value L_n [dB] can be approximated by:

$$L_n[dB] \approx 6\,ENOB + 3n + 10\log_{10} p$$

Peak to side lobe ratio: The ideal MLBS has an infinity peak to side lobe ratio but by the required band limitation because of the limited equivalent sampling rate side lobes appear in the system IRF. Then, the actual size of the side lobes mainly depends upon the anti-aliasing filter which the system designer can largely influence. By increasing the equivalent sampling rate, the need of a band limitation will disappear. Thus the resulting side lobes will increasingly vanish.

Peak to spurious ratio: The peak to spurious ratio expresses the influence of non-linear effects onto the IRF. In the case of the MLBS front-end such effects may be observed mainly in the shift register which generates the stimulus signal. Thus, the actual created test signal differs from the supposed one. This is expressed in small peaks which are distributed in a defined manner over the basis of the system IRF[2]. By an appropriate signal correction these kind of perturbations may be mostly removed (see below).

Jitter, Drift: Most of impulse systems applying under-sampling are sensible to drift and jitter. This is not the case here. The complete control of the measurement process is done by steep signal flanks which suppress the effect of additive noise and temperature dependent offset voltages on the trigger events. Thus, the system is as stable as a network analyser. An influence of the shift register, binary divider and the T&H onto the jitter behaviour could not be observed i.e. the measurements indicated that their effect must be less than 140 fs of RMS jitter. The time axis representation of the measurements is absolutely linear because of the digitally controlled sampling. The only source of jitter and drift which could be observed is provoked by the single tone RF-source but a variety of commercial solutions are available for this. In the end, the application dictates the requirements on the RF clock generator and finally its price. The higher the ambiguity range is the lower the phase noise of the RF source must be. Its medium term frequency drift limits the maximum averaging time and thus the maximum dynamic range and its long term drift reduces the efficiency to remove systematic errors by calibration.

4. CORRECTION OF SYSTEMETIC ERRORS

The ability to design a stable system opens up interesting opportunities to correct systematic errors by software. Since decades, this is for example a usual method to improve the system performance of network analysers. In our case, more or less extensive methods can be applied which again are dependent on the requirements posed by a specific application. By concentrating on the systematic errors of the MLBS front-end and omitting the influence of other components (antennas, cables, ...), the difference between the real form of the MLBS and the ideal one cause the most serious deviations. However they are of deterministic nature so they may be eliminated to certain extent. Two different methods of error correction will be shortly mentioned here which consider the same problem from different sides. A third one refers to Volterra Filtering which will be omitted here (see paper [3] for more information).

Source Calibration: The measurement method as represented in figure 2 is based on an impulse compression referring to an ideal reference signal rather than the actual stimulus. Thus it is not surprising that spurious signals appear after the compression but they may be removed if the correct stimulus signal is used for impulse compression. For that purpose calibration measurements must be undertaken by using DUTs with known behaviour. This method is well known from the network analyser technique and must not be explained in detail here. Figure 3 demonstrates the results of a simple response and isolation calibration.

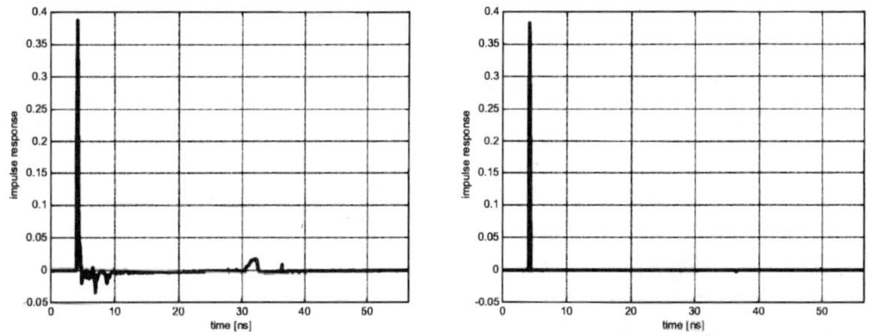

Figure 3: The system IRF before and after calibration. The FWHM value of the main peak is 170 ps.

Reference Channel: As with a network analyser, the actual stimulus signal will be captured by a second measurement channel. This signal will act as reference for the impulse compression. Such an arrangement provides the best results because it must no longer (as in the case above) be supposed that MLBS generator and sampling unit are absolutely stable over time. It is rather enough to require that only the difference between both measurement channels must not vary. This is technically quite less critical than to fix their absolute parameters particularly if one considers that both measurement channels are constructed in the same way. Nevertheless, a usual (network analyser) calibration technique will further improve the results.

5. EXPERIMENTAL RESULTS

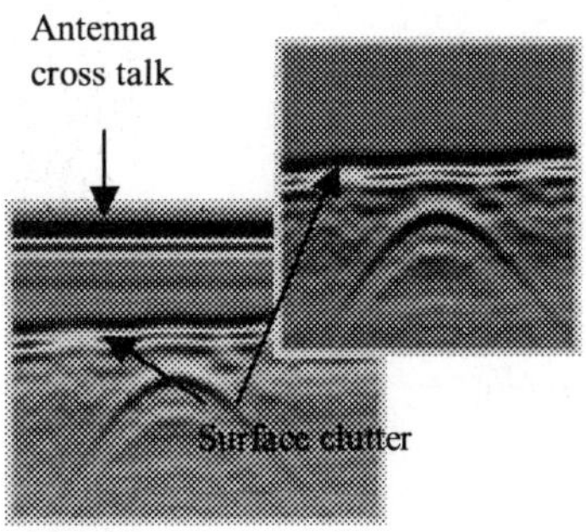

Figure 4: B-scan of a Tx-Rx-antenna pair indicating a buried object close to the surface. The left image represents the original data. On the right, the antenna cross-talk is removed. The perfect removal of antenna cross talk underlines the stability of the electronics.

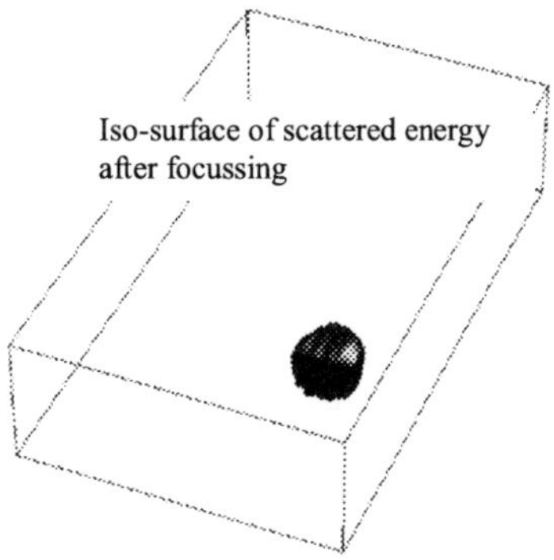

Figure 5: Results of data processing showing a small buried object (PPM2-mine). The data processing was performed by Vrije Universiteit Brussel.

A first integrated version of the UWB front-end was built for a GPR array with 6 transmitter and 6 receiver antennas for mine detection purposes (EU-Project: ESPRIT 29902 DEMINE[4]). The shift register, the binary divider and the T&H-circuit were integrated in 0.8 μm SiGe-HBT-technology. The complete analogue part of the device was designed by differential circuits. The wiring of the RF components was performed by four layer low temperature co-fired ceramics (LTCC). A Video-ADC captures the sampled data which were either stored directly on a PCI-board or they passed through a fast averaging circuit (FPGA) and the impulse compression (FHT-algorithm running on a DSP) to the main processor. The shift register length was $n = 9$ resulting in a 511 points IRF. The RF-clock rate was $f_c = 9$ GHz i.e. the useful bandwidth was approximately 4.5 GHz. The maximum measurement rate was approximately 30 000 IRF/s if only data storage and off-line processing was applied.

The figures 4 and 5 summarise some measurement examples coming from a measurement campaign at a mine test field in Angola. Figure 4 represents a 2.5 m long B-scan measured along a demining lane. The GPR array gathered 21 of such B-scans at the same time each measured between different array elements. From this data the scattering energy of buried objects were calculated (see figure 5).

6. CONCLUSIONS

The function principle of a UWB front-end were explained and some experimental results were shown. The method works with low voltage MLBS signals and it is widely based on digital circuit conceptions and software routines. This promotes flexible system design, stable operation and forms the basis for a largely integrated system. Thus, the prerequisite is met for large scale applications of UWB-sensors for radar purposes as well as for impedance spectroscopy.

7. REFERENCES

1. J. Sachs, P. Peyerl, M. Roßberg: A New UWB-Principle for Sensor-Array Application. Proc. of IMTC/99, vol. 3, p. 1390-5
2. H. Alrutz: Über die Anwendung von Pseudorauschfolgen zur Messung an linearen Übertragungssystemen. Dissertation 1983, Georg-August-Universität zu Göttingen, Germany
3. E.Roy, R.W. Stewart, T.S. Durrani: Theory and Application of adaptive second order IIR Volterra filters. Proc. Int. Conf. On Acoustics , Speech and Signal Processing, pp. 1597 –1600, Atlanta 1996
4. ESPRIT Project: 29902 DEMINE (Improved Cost-Efficient Surface Penetrating Radar Detector with System on Chip Solution for Humanitarian Demining).

VECTOR RECEIVING ANTENNA FOR ULTRAWIDEBAND RADAR POLARIMETRY

V.I. Koshelev, E.V. Balzovsky, and Y.I. Buyanov*

1. INTRODUCTION

Electromagnetic field polarization structure (PS) knowledge is of special interest for radar since it contains information concerning an object. Conventional radar uses for this purpose antenna allow recording coordinate vector components of electric and magnetic field. This type of antenna has got the name of vector receiving antenna (VRA).[1] In ultrawideband radar, VRA should simultaneously and independently record all three components of electric field with minimum distortions. This is possible if the VRA passband exceeds the frequency band occupied by the recorded signal spectrum and antenna inserts minimum distortions into the measured field structure. To record the electric field orthogonal components, a VRA consisting of three mutually perpendicular dipoles with superposed phase centers can be used.

2. DIPOLE RECEIVING ANTENNA

Choice of a dipole as the VRA element is based on possibility to join the phase centers of three mutually orthogonal dipoles, permanence of a phase center, and possibility to smooth the amplitude- and phase-frequency responses in the frequency band occupied by a pulsed signal. When determining the dipole parameters and optimum load values, a traveling wave superposition method[2] was used allowing calculation of the current distribution in the receiving dipole and voltage at its load in the receiving regime. Calculations were carried out in a frequency domain.

According to the electromagnetic induction law, at the elementary part of the conductor dz the electric field component E_z tangential to the conductor induces electromotive force $d\varepsilon_z = E_z dz$ which generates in the conductor two current waves – one traveling from the considered part to the left and another traveling to the right. Current

* Institute of High Current Electronics, RAS, 4 Akademichesky Avenue, 634055 Tomsk, Russia.

direction of both waves coincides with direction of the projection E_z, and dependence on time coincides with $E_z(t)$. Let electromagnetic wave (Figure 1) excites the L-long wire (model of the dipole arm) with the wave impedance ρ and propagation constant γ under the angle θ. Current waves generated by elementary electromotive force $d\varepsilon_z$ applied to the wire part dz resulting from multiple reflections from the wire ends loaded to the impedances Z_1 and Z_2 create currents dI_1 (for $\zeta < z$) and dI_2 (for $\zeta > z$) at the point ζ:

$$dI_1 = \frac{E_z}{2\rho} e^{i\kappa z\cos\theta} \frac{e^{-i\gamma(z-\zeta)} + q_1 e^{-i\gamma(z+\zeta)} + q_2 e^{-i2\gamma L}\left(e^{i\gamma(z+\zeta)} + q_1 e^{i\gamma(z-\zeta)}\right)}{1 - q_1 q_2 e^{-i2\gamma L}} dz,$$

$$dI_2 = \frac{E_z}{2\rho} e^{i\kappa z\cos\theta} \frac{e^{i\gamma(z-\zeta)} + q_1 e^{-i\gamma(z+\zeta)} + q_2 e^{-i2\gamma L}\left(e^{i\gamma(z+\zeta)} + q_1 e^{-i\gamma(z-\zeta)}\right)}{1 - q_1 q_2 e^{-i2\gamma L}} dz,,$$

$$I(\zeta) = \int_0^\zeta dI_1 + \int_\zeta^L dI_2.$$

For a linear wire with the TEM current wave $q_{1,2} = (\rho - Z_{1,2})/(Z_{1,2})$. If the end load is absent, then $Z_2 = \infty$, and $E_z = E_0 \sin\theta \, exp(ikz \cos\theta)$, the current in the load $Z_1(\omega)$ is described by the expression:

$$I(0,\omega) = \frac{E_0(\omega)\gamma}{Z_1(\omega) - i\rho \, ctg\gamma L} \frac{e^{ikL\cos\theta} - \cos\gamma L - i\frac{k}{\gamma}\cos\theta\sin\gamma L}{(\gamma^2 - k^2\cos^2\theta)\sin\gamma L}\sin\theta,$$

where E_0 is the amplitude of the incident wave electric field, $k = \omega/c$, c is the velocity of light, $\gamma = (k + \alpha^2 L/8) - i\alpha$, $\alpha = R'_\Sigma(\omega)/\rho$, and $R'_\Sigma(\omega)$ is the resistance of the backscatter radiation losses per unit of length.

The obtained result is applied to the case of a symmetric dipole. The tangential field component for the second arm is equal to $E_z = E_0 \sin\theta \, exp(-ikz'\cos\theta)$. If the dipole arms are identical and $Z'_1(\omega) = Z_1(\omega)$, then current in the load impedance $Z_L = 2Z_1 = const$ between the dipole clamps is determined by the expression:

$$I_L(\omega) = \frac{E_0(\omega)\gamma}{Z_L + Z_D(\omega)} \frac{\cos(kL\cos\theta) - \cos\gamma L}{(\gamma^2 - k^2\cos^2\theta)\sin\gamma L}\sin\theta,$$

where $Z_D(\omega) = -i\rho\,\mathrm{ctg}\gamma L$ is the dipole impedance attributed to its clamps. In this case the dipole load voltage is equal to

$$U_L(\omega) = I_L(\omega)\, Z_L = E_0(\omega) \frac{1}{1 + \frac{Z_D(\omega)}{Z_L}} \frac{\gamma(\cos(kL\cos\theta) - \cos\gamma L)}{(\gamma^2 - k^2\cos^2\theta)\sin\gamma L}\sin\theta.$$

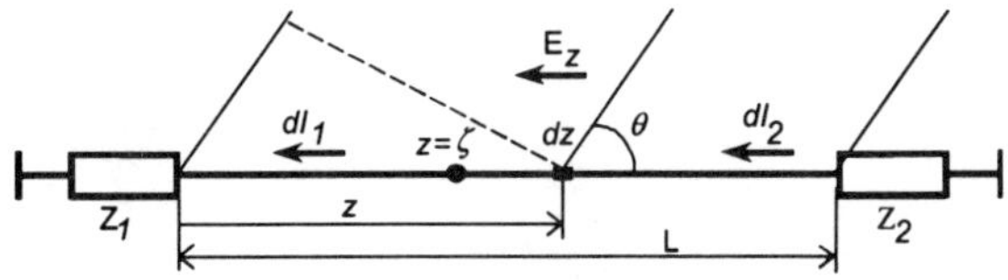

Figure 1. Problem geometry.

For decreasing the output signal $U_L(\omega)$ distortions, it is necessary to decrease the influence of the dipole impedance $Z_D(\omega)$ and attenuation constant $\alpha(\omega)$. For this, it is necessary to increase Z_L, to decrease ρ, and to increase the dipole material resistance per unit of length R'. Fig. 2 presents the calculated signals at the metal dipole load with the ratio of transverse dimensions to the longitudinal ones equal to 0.3 at three values of Z_L. Curve 1 corresponds to the waveform of the incident pulsed electromagnetic field, curves 2, 3, 4 correspond to the signal at the dipole load at $Z_L = 150\ \Omega$, $600\ \Omega$ and $2000\ \Omega$.

As it was supposed in calculations, resistance of the dipole material influences $\alpha(\omega)$ similar to $R'_\Sigma(\omega)$:

$$\alpha(\omega) = (R'_\Sigma(\omega) + R')/\rho.$$

Figure 3 presents the calculated signals at the dipole load with resistive material arms at the load impedance $Z_L = 600\ \Omega$. Curve 1 corresponds to the waveform of the incident electromagnetic field, curve 2 is the signal from the dipole with the arm material resistance $R' = 10^4\ \Omega/\text{m}$, curve 3 is the signal from the dipole with $R' = 2\cdot 10^4\ \Omega/\text{m}$. Calculations have shown that further increase of the arm material resistance approximates the waveform of the signal to the form of the incident electromagnetic field, however, the received signal amplitude diminishes. Thus, it is necessary to choose a compromise value of R' at which the output signal has small distortions and the level sufficient for recording.

The design of a dipole intended for application as the VRA element has been developed resulting from the investigations. The dipole arms are made as the plates of a

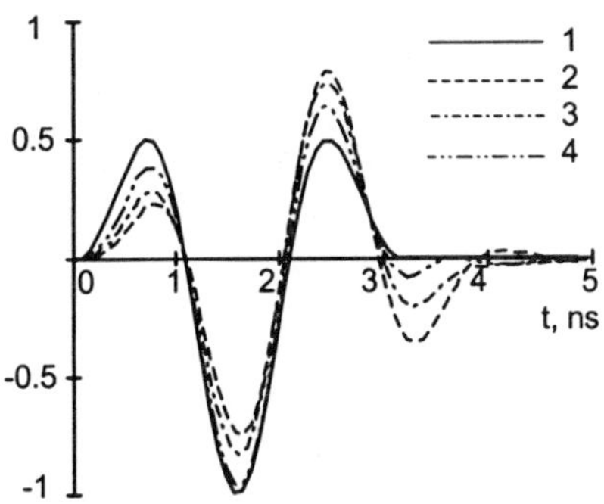

Figure 2. Calculated signal from metal dipole at different load impedances.

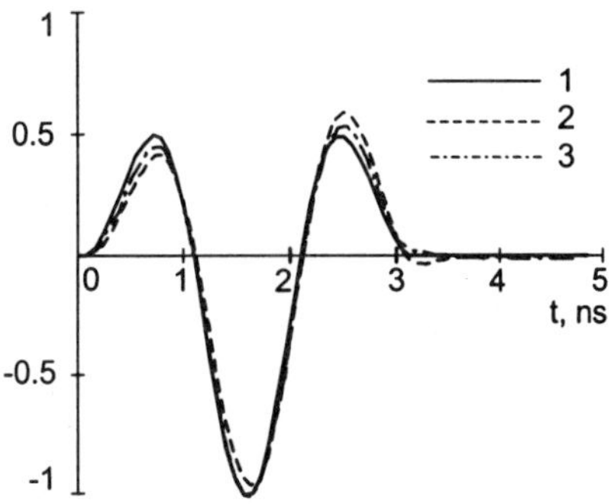

Figure 3. Calculated dipole signal at different resistance of the arm material.

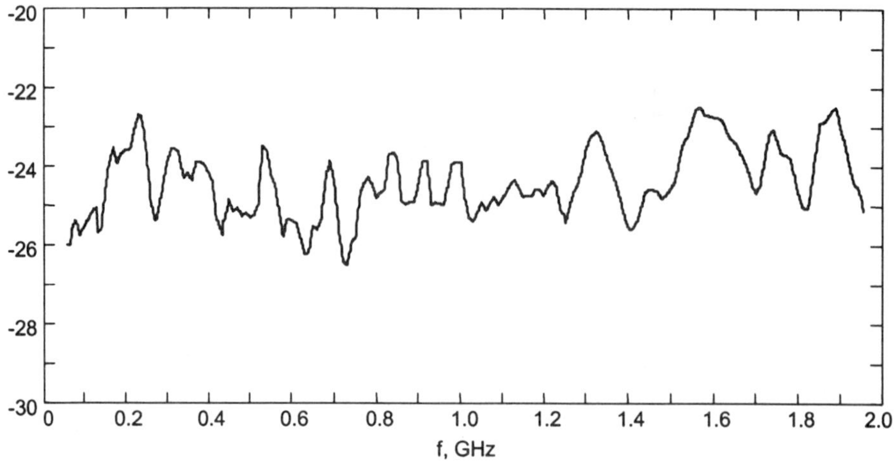

Figure 4. Dipole amplitude-frequency response.

low-conductivity material on the basis of carbon and metal contact pads. The signal from the dipole comes in to the balancing-matching unit with a 50-Ω output impedance through the transformer of impedances 4:1 and symmetric screened line of 150 Ω. Figure 4 presents the amplitude-frequency response of the dipole with a feeder. Having the length of 0.2 m and resistance of the arm material $R' = 2 \cdot 10^4$ Ω/m, the dipole had a frequency band of 0.07 ÷ 2.0 GHz at a –3 dB level. A dipole was used to record electromagnetic radiation pulses with the length from 1 ns to 3 ns.

3. VECTOR RECEIVING ANTENNA

VRA (figure 5) combines three independent channels two of which are formed by the identical dipole antennas – they are called horizontal X and Y. Their characteristics

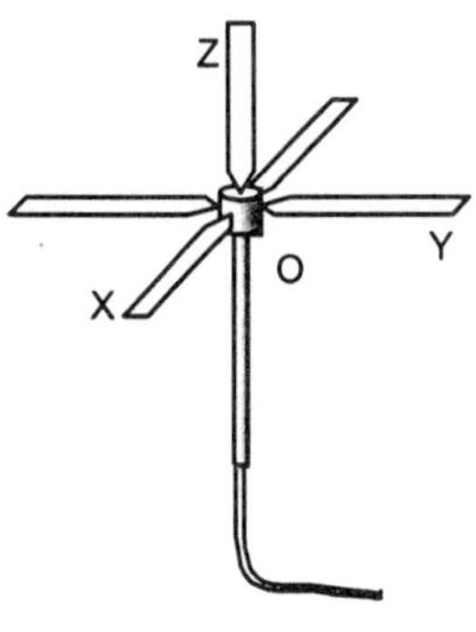

Figure 5. Vector receiving antenna.

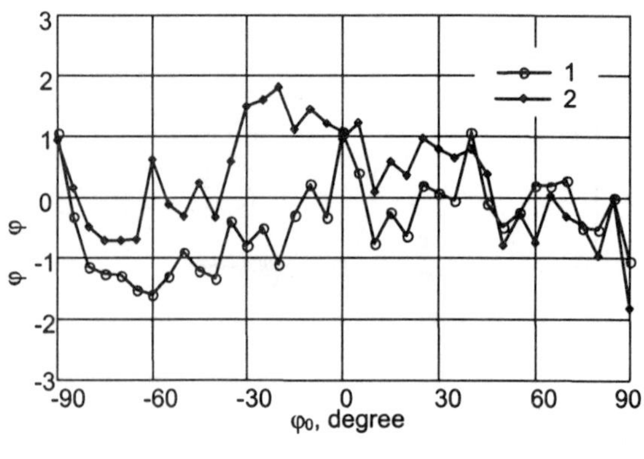

Figure 6. Measurement accuracy of vector **E** angle.

are presented above. The third channel - vertical Z represents a monopole made of a resistive material. To decrease the influence of the cables issuing from the area of connection of all VRA arms, horizontal dipole arms united by the in-phase current by means of balancing transformers are used as the counterbalance for the monopole.

Investigations on measurement of position of the electric field strength vector **E** at normal incidence of the linearly polarized pulsed electromagnetic field on the plane XOY were carried out. Figure 6 presents the measurement error of the vector **E** angular position depending on the angular position φ_0 of vector **E** in the incident field. In the angle calculation algorithm φ_m the signals of two channels are calibrated by time and amplitude. In figure 6, curve 1 corresponds to the angle measurement error by the first time lobe of the pulse, curve 2 – by the second time lobe. Angle measurement error doesn't exceed 2° and is comparable with the orientation accuracy of the transmitting and receiving antennas in the experimental setup.

4. INVESTIGATION OF RADIATED SIGNAL POLARIZATION STRUCTURE

PS measurement of a radiated pulsed electromagnetic field allows us to obtain additional information of the tuning of efficient transmitting antennas. Visualization of the vector **E** hodograph gives a notion concerning temporal and amplitude correlations of the main and cross-polarized components of radiated pulsed field. Voltage at the output of each VRA channel is proportional to the projections of the electric field strength **E**:

$$E_x(t)=\alpha_x U_x(t), \qquad E_y(t)=\alpha_y U_y(t), \qquad E_z(t)=\alpha_z U_z(t).$$

Let the plane XOY coincides with the incident wavefront plane. At each moment of time the coordinates of vector E end are determined in the following way:

$$|E_{XOY}| = \sqrt{\alpha_x^2 U_x^2 + \alpha_y^2 U_y^2} = \alpha_x \sqrt{U_x^2 + \frac{\alpha_y^2}{\alpha_x^2} U_y^2}\,, \qquad \varphi = arctg(\tfrac{U_y}{U_x}).$$

A set of these points during a pulse forms the godograph projection of vector **E** to the plane XOY. Analogous plotting is made for the planes YOZ and ZOX. There is no necessity to determine the effective length of the dipoles α_x, α_y, α_z. It is quite enough to know their ratios from calibration.

Radiation PS of a combined antenna[3] was investigated by means of VRA. A combined antenna optimized for excitation by a 2-ns long voltage bipolar pulse was used.

Radiation is linearly polarized in the direction of the pattern main maximum. Figure 7 presents the godograph projections of vector **E** to the plane XOY for the angles of 15°, 30°, 45° and 75° from the pattern maximum in horizontal plane. The signals from the dipoles X and Y are plotted in the abscissa and ordinate axes, respectively. Point «s» at the plots corresponds to the pulse start; other points are plotted in 0.5 ns. In the limits of the pattern width by half-power, radiation remains elliptically polarized with rotation of the vector **E** projection counter-clockwise for positive angles relative to the pattern maximum and clockwise for negative angles. Beyond the pattern width by half-power,

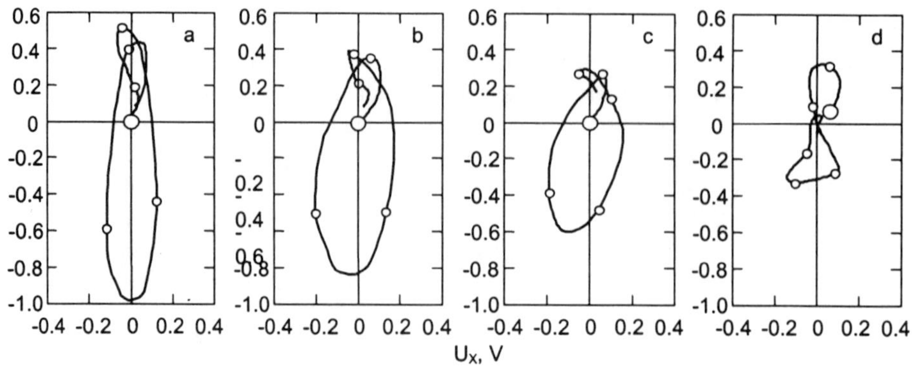

Figure 7. Godograph projection of vector **E** to the plane XOY at deviation from the combined antenna pattern maximum by 15° (a), 30° (b), 45° (c), 75° (d).

radiation PS becomes complicated, however, radiated power in these directions decreases essentially (figure 7 d).

5. INVESTIGATION OF REFLECTED SIGNAL POLARIZATION STRUCTURE

To investigate the pulsed electromagnetic field reflected from simple metal objects (plate, strip), a combined transmitting antenna excited by a 2-ns long bipolar voltage pulse was used. At the transmitting antenna orientation to the object by the pattern maximum the linear polarization was used, and at the object sounding at the angle of 30° from the pattern maximum the incident field was elliptically polarized with $\kappa=0.25$.

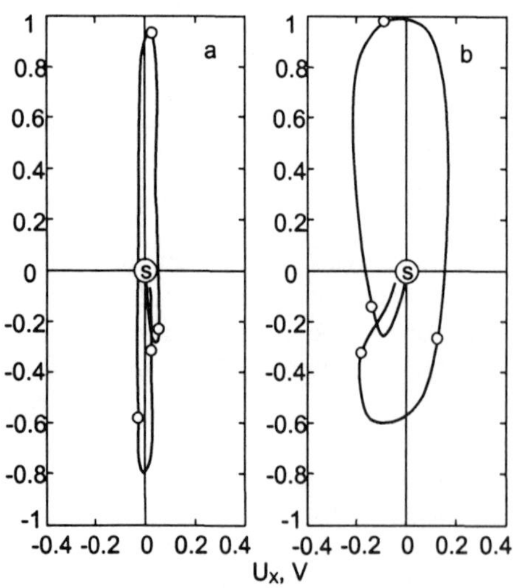

Figure 8. Reflection from a metal square plate.

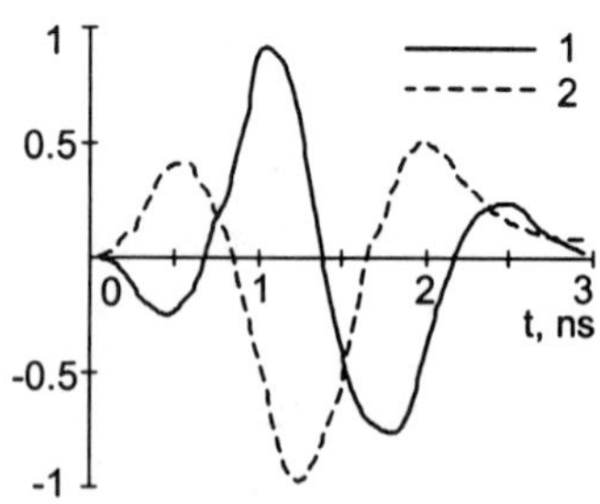

Figure 9. Reflected and radiated signals.

5.1. Reflection from metal square plate

In the experimental setup, a thin metal square plate (0.8×0.8 m) was placed so as to provide the wave mirror reflection from radiator to VRA. When the plate was sounded by the wave with linear polarization the reflected field was elliptic with $\kappa = 0.07$ (figure 8 a). Moreover, a change of the waveform of the reflected signal (curve 1 in figure 9) was observed in comparison with the incident field waveform (curve 2). At the sounded plate by the wave of elliptical polarization ($\kappa = 0.25$) the reflected signal preserved elliptical polarization, vector **E** rotation direction being changed (figure 8b).

5.2. Reflection from metal strip

A thin metal strip (0.6×0.15 m) was used in the experiment. It was placed so that directions to the radiator and VRA were symmetrical relative to the normal to the strip plane.

At sounding by the wave with linear polarization, reflection from the vertically placed strip coincides qualitatively with reflection from a large-dimensional plate (figure 10 b), but it is essentially less by the amplitude due to the scattering area decrease. At sounding of the strip inclined relative the vertical line by 45° into one or another side, appearance of a cross-polarized component with the change of polarization to the elliptical one is noted. Godographs of vector **E** reflected from the strip inclined by 45° and -45°are shown in figures 10 a and 10 c. Inclination of the godograph **E** to the corresponding side is seen as well as the change of the vector **E** rotation direction.

If a strip is sounded by the elliptically polarized wave ($\kappa = 0.25$) vertical and horizontal field components are distinguished at vertical and horizontal location of the strip, respectively, in the reflected signal structure.

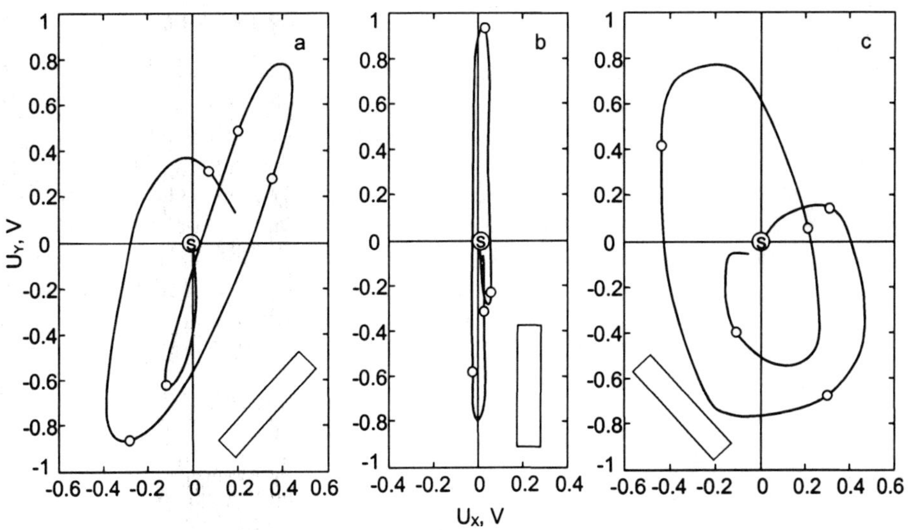

Figure 10. Reflection from metal strip.

5.3. Wave Arrival Direction Estimation

By means of VRA capable to record simultaneously and independently three projections of vector **E** of a pulsed electromagnetic field it is possible to estimate the arrival direction of the wave having polarization different from the linear one. By the signals recorded in the three VRA channels corresponding to vector **E** projections at the axis of a coordinate system related to VRA, godographs are plotted at the planes XOY, YOZ and ZOX. These godographs are the godograph projections of vector **E** plotted in the plane of the incident wavefront. Three godograph projections are approximated by ellipses, the coordinates of their large and small axes being estimated. Position of the large and small axes of the polarization ellipse approximating the godograph of vector **E** in the incident wavefront plane is estimated by the three determined projections of the large and small axes. Large and small axes of this ellipse determine the wavefront plane and the normal to this plane gives the wave arrival direction.

Table 1 presents the experimental results on estimation of the wave arrival of the pulsed electromagnetic radiation with elliptical polarization ($\kappa = 0.25$). Vertical θ_0 and azimuthal φ_0 angles of incidence were set while the angles θ_1 and φ_1 were estimated by the recorded signals at the VRA channels by means of the presented algorithm.

Table 1. Set and measured wave arrival angles

Vertical		Horizontal	
θ_0	θ_1	φ_0	φ_1
30°	34°	60°	58°
45°	43°	45°	44°
60°	57°	15°	12°

6. CONCLUSION

A method of calculation of a dipole antenna intended to receive ultrawideband electromagnetic pulses has been developed. Numerical simulation with the use of this method allowed estimating optimum dipole parameters and designing a dipole antenna providing small distortions of a recorded pulse. A vector receiving antenna intended to measure a polarization structure of electromagnetic pulses has been created on the basis of the developed dipole antenna. The VRA allowed recording the signals at the each dipole output simultaneously and independently that after processing allowed measuring instantaneous orientation of vector **E** with the accuracy of ±2° and determining the wave arrival direction if the wave had elliptical polarization. The VRA was used to measure the polarization structure of the radiated and reflected from objects ultrawideband electromagnetic pulses.

7. REFERENCES

1. A.I. Knyaz and V.A. Katorgin, Vector receiving antennas, *Zarubezhnaya radioelectronika* **8**, 36–42 (1984).
2. G.A. Lavrov and A.C. Knyazev, *Ground and Underground Antennas* (Sov. radio, Moscow, 1965).
3. V.I. Koshelev, Y.I. Buyanov, Y.A. Andreev, V.V. Plisko, and K.N. Sukhushin, in *Proc. IEEE Pulsed Power Plasma Science Conf.*, edited by R. Reinovsky and M. Newton (IEEE, 2001), pp. 1661–1664.

VERY HIGH VOLTAGE DIVIDER WITH LARGE BANDWIDTH

Joseph Wey and François Bieth*

1. INTRODUCTION

For the measurement of high electric voltages, resistors with the highest possible values are usually inserted between the two poles to be measured. Their resistances should fulfill as closely as possible the integral E·dl when using the electric current (j = σ·E) passing through the resistor. Two solutions exist: either measuring the current directly by means of a transformer, or measuring the potential on a small portion of this resistor, thus realizing a voltage divider. Very high frequencies (1 GHz) will cause geometrical constraints either to realize a pure resistance, or to compensate for parasitic impedances in the two arms of the divider. The other principles found in the literature, i.e. capacitive and even selfic dividers, are more the concern of antennae in high frequencies than of localized impedance. Generally speaking, in high frequencies, the elements of circuits (size > λ/10-λ/4) cannot be treated as simple local impedances, but it is necessary to take into account the propagation of waves.

A last possibility consists in measuring this voltage in a indirect way through the field E or H very locally in a transmission line with a perfectly defined impedance and coupled with the system to be measured. This principle requires that the relationship between the voltage and the fields are perfectly known, stable in frequency, in time and that the probe interferes only little with the transmission line.

2. OBJECTIVES

The study of pulsed generators to feed high-power microwave sources (vacuum tubes: Vircator, Milo, Klystron, Magnetron, etc.) or ultra-wide band systems (UWB antennae) requires the knowledge of certain physical quantities, such as the voltage time history for a

* J. Wey and F. Bieth, ISL, French German Research Institute, F-68301 Saint-Louis, France.

defined load. The main characteristics of this voltage are: rise time from 0.1 to 10 ns, width from 10 ns to several hundreds of ns and a peak amplitude from 100 kV to 1 MV. To avoid a mathematical treatment of the signal, the system will be chosen to transmit DC. Our bandwidth of interest covers DC up to 6 GHz. The generators studied are coaxial MARXs of the ISL/EMW type. Their output diameter (inner diameter of the external tube) is 137 mm.

3. CONTEXT

We consider a medium with no magnetic currents, no free charges, constant dielectric permittivity ε and the permeability of the vacuum (μ_0). We consider also quasi-stationary solutions (harmonic), which means:

$$\vec{E}(\vec{r}, t) = \vec{E}(\vec{r}) \cdot e^{j\omega t} \quad \text{and} \quad \vec{H}(\vec{r}, t) = \vec{H}(\vec{r}) \cdot e^{j\omega t}. \tag{1}$$

Maxwell's equations simplify to:

$$\vec{\nabla} \cdot \vec{H} = 0 \quad (2) \qquad \vec{\nabla} \wedge \vec{H} = (j\omega\varepsilon + \sigma) \cdot \vec{E} \quad (3)$$

$$\vec{\nabla} \cdot \vec{E} = 0 \quad (4) \qquad \vec{\nabla} \wedge \vec{E} = -j\omega\mu_0 \cdot \vec{H}. \quad (5)$$

$\vec{H}$ is the magnetic field, $\vec{E}$ the electrical one, σ the conductivity (0 in vacuum) and ω the pulsation. Because they can be derived from Eqs. (3) and (5), Eqs. (2) and (4) are superfluous in that case.

4. THEORETICAL PRINCIPLES

4.1. Basic principle

In order to realize an ideal measuring resistor, the system will be composed of a coaxial transmission line terminated with its characteristic impedance (Fig. 7), here 50 Ω, in order to be compatible with high-frequency measuring instruments. The line is coupled to the MARX generator and adjusts to its external output diameter. The maximum energy delivered by one (short) pulse of the MARX will be less than 1 kJ, so we have to choose the size of the 50 Ω resistor in order to withstand this energy adiabatically without noticeable changes. It is a tube with a wall thickness thinner than the skin depth for the highest frequency. The divider is realized by a 49.5 Ω and a 0.5 Ω part (Fig. 7). The central conductor of the N plug is connected to the interface between these 2 resistors. The coaxial profile surrounding the resistors should provide a progressive decrease of the measured voltage V(x) of the wave propagating along the resistor. The voltage between any point of the resistor and its base must be equal to that between this point and the corresponding point of the profile on the same wave front (Fig. 1 and Fig. 3).

The voltage V_{ab} between points a and b must not be confused with the difference of the potential Ψ between these 2 points:[1,2]

$$\Psi_b - \Psi_a = \int_a^b \left(\vec{E} + \frac{\partial}{\partial t}\vec{A}\right) \cdot d\vec{l} = V_{ab} + \int_a^b \frac{\partial}{\partial t}\vec{A} \cdot d\vec{l}. \tag{6}$$

The divider has to be verified and calibrated. Therefore we developed a "cone" adaptor equipped with a standard N plug to replace the MARX. In order to be able to verify the usable frequency bandwidth of a cone, a second cone is adapted to the output of the line.

4.2. First simple idea

By considering only transverse electromagnetic (TEM) plane waves, it can be deduced that everywhere on the surface of the resistor, the impedance of the line at the point x is terminated by the remaining resistance up to the end of the profile (Fig. 1).

The impedance of the profile at x is: $$Z(x) = \frac{60}{\sqrt{\varepsilon_r}} \cdot \ln\left[\frac{r(x)}{a}\right] . \tag{7}$$

The remaining resistance in x is: $$\mathcal{R}(x) = \frac{\mathcal{R}_0}{x_0} \cdot x . \tag{8}$$

This leads to the profile equation: $$r(x) = a \cdot \exp\left(\frac{\mathcal{R}_0 \cdot \sqrt{\varepsilon_r}}{60 \cdot x_0} \cdot x\right) . \tag{9}$$

A reference book like Meinke-Gundlach's[3] describes this principle without any details. This principle suffers from two errors on boundary conditions of the field E, although it can be considered to be valid in the generalized model of transmission lines:[4,5]

- considering the metal of the profile to be a perfect conductor, the field E on the profile must be perpendicular to it;
- the resistor not being a perfect conductor, the E field on its surface will not be perpendicular, but modified by a slight angle due to the field in the resistor.

Such a profile can only be connected to a cylindrical coaxial line with a break of angle of the outside profile, this line being connected to the measured MARX generator (Fig. 7).

4.3. Profile with spherical waves

4.3.1. General case and spherical approximation

In order to solve the Maxwell equations, we apply a coordinate transformation[6,7] to a curvilinear coordinate system (Fig. 2). By naming the new coordinates u, ϕ and s, by start-

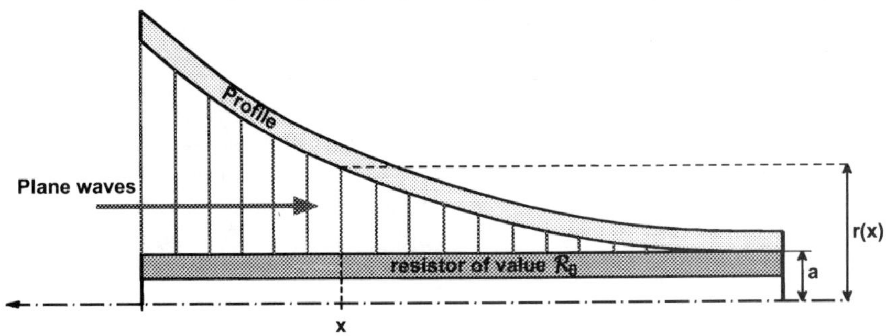

Figure 1. Profile with TEM plane waves.

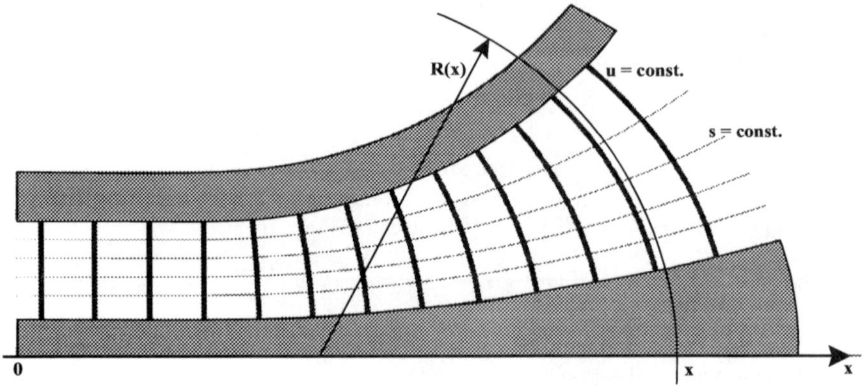

Figure 2. Grid of E-field lines and equipotentials in a curvilinear coordinate system.

ing from spherical coordinates θ, ϕ and r, the lines "u = const" correspond to lines "θ = const" and similarly s to r. In both cases, there is an axisymmetry on ϕ.

In the spherical system, the profile is a cone with a constant angle where we have the impedance of a conical transmission line:

$$Z = \frac{1}{2\pi} \cdot \sqrt{\frac{\mu_0}{\varepsilon}} \cdot \ln\left(\frac{\tan\frac{\theta_2}{2}}{\tan\frac{\theta_1}{2}}\right) \qquad \begin{array}{l}\theta_2 \text{ is the angle of the external profile,}\\ \theta_1 \text{ the angle of the internal one.}\end{array} \tag{10}$$

Here the electrical field has only one component (field lines are circles):

$$E_\theta = \frac{60}{\sqrt{\varepsilon_r}} \cdot \frac{I(r)}{2\pi \cdot r \cdot \sin\theta} \, . \tag{11}$$

In this cone, the wave fronts are spheres with the center of the cone as common center.

Our approximation is to say that we can find a "curved cone" (curvilinear coordinate system) where the wave fronts remain spherical with their center on the axis of symmetry (Ox), but are not necessarily concentric. Due to the fact that E-field lines are circles (in a plane of constant ϕ), the solutions will be locally identical to that of the conical line with constant angle at r = R(x), the radius of each bow crossing Ox in x.

Writing R(x) = x gives the conical line with constant angle.

Obviously, the bows are wave fronts and must obey to the propagation law: two lines "u", or bows, successively cut any lines "s" with the same length (the wave fronts are equidistant in a homogeneous medium). So, every time we find a model based on this spherical approximation, we will verify the results with the propagation law.

4.3.2. Resolution

The relations which allow us to determine the external profile are (Fig. 3) as follows:

1. the resistor is homogeneous ⇒ E_0 constant along it;

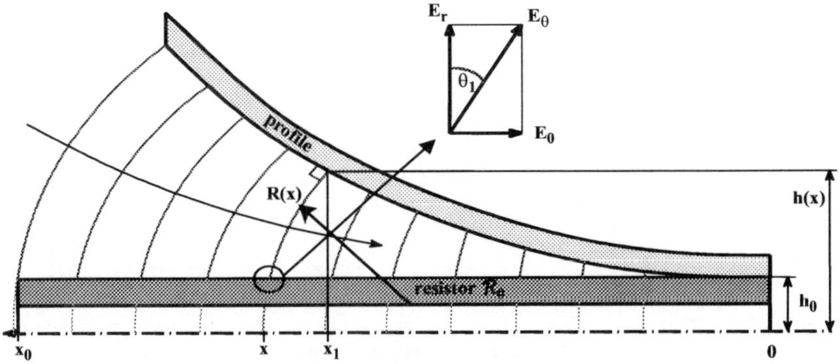

Figure 3. Profile with spherical waves.

2. the field conditions at walls (resistor and metal);
3. the voltage on the resistor (between x and 0) is equal to that between the resistor and external profile over the bow crossing Ox in x (0 is the end of the profile);
4. the impedance everywhere in the profile is equal to the rest of the resistance.

The outside profile will have as coordinates $[h(x_1), x_1]$. We have axisymmetry in ϕ around the axis Ox. On a bow R(x) crossing the outside of the resistor in x in the same plane ϕ, we can rewrite Eq. (11) with the variable x at the surface of the resistor:

$$\sin\theta \cdot E_\theta = \frac{60}{\sqrt{\varepsilon_r}} \cdot \frac{I(x)}{R(x)} = E_0(x) = \frac{\mathcal{R}_0 \cdot I(x)}{x_0} . \tag{12}$$

This leads to:

$$R(x) = \frac{60}{\mathcal{R}_0 \cdot \sqrt{\varepsilon_r}} \cdot x_0 = R_c = \text{constant.} \tag{13}$$

By applying Eq. (10), and writing: $Z(x) = \mathcal{R}_0 \cdot \dfrac{x}{x_0}$, (14)

the use of $\tan\dfrac{\theta}{2} = \dfrac{\sin\theta}{1 + \cos\theta}$ in Eq. (14) leads, after some manipulations, to:

$$\left(1 + \sqrt{1 - \frac{h^2(x)}{R^2}}\right) \cdot e^{\left(\frac{x}{R}\right)} - \left(1 + \sqrt{1 - \frac{h_0^2}{R^2}}\right) \cdot \frac{h(x)}{h_0} = 0 \tag{15}$$

with the coordinate on Ox (Fig. 3):

$$x_1(x) = x + \sqrt{R^2 - h^2(x)} - \sqrt{R^2 - h_0^2} . \tag{16}$$

We solve Eq. (15) with a numerical solver like MathCad and we compare the profile $h(x_1)$ with $h_e(x_1)$, an exponential one like Eq. (9):

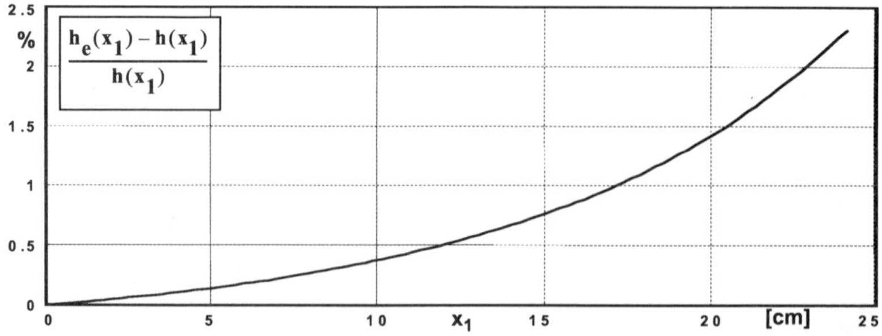

Figure 4. Relative difference between $h(x_1)$ and $h_e(x_1)$ in %.

$$h_e(x_1) = h_0 \cdot \exp\left(\frac{\mathcal{R}_0 \cdot \sqrt{\varepsilon_r}}{60 \cdot x_0} \cdot x_1\right) . \tag{17}$$

The relative difference between $h(x_1)$ and $h_e(x_1)$ is drawn in Figure 4. The profile $h(x_1)$ is an exact solution for wave fronts of constant sphericity.[8]

4.4. Transformation of plane waves into spherical waves

4.4.1. Principle of the TEM lens coupling to the divider

Figure 5 represents the complete principle of a TEM lens followed by the resistive coaxial divider. This lens allows to link the plane waves of the cylindrical coaxial line (outer diameter of 137 mm) with the divider (principle of Figure 7). The sphericity of the wave at the entrance of the divider is given by Eq. (13).

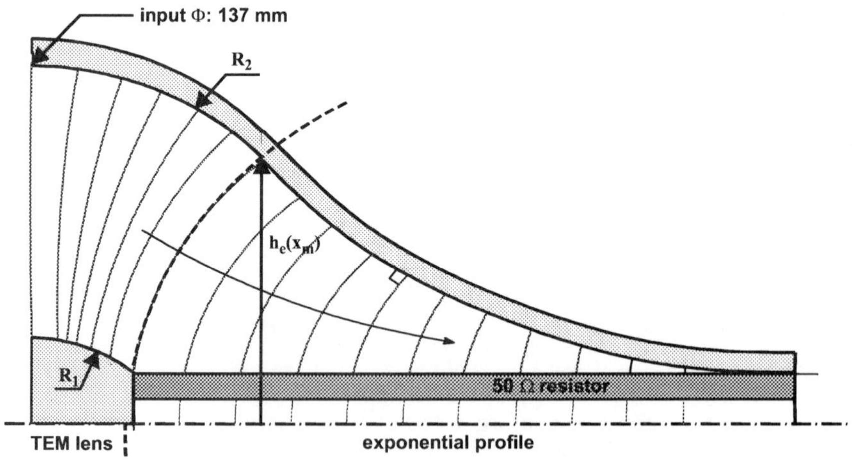

Figure 5. Principle of the coaxial divider preceded by the TEM lens.

4.4.2. Solving method

With the same approximation as in §4.3.1, we look for profiles of the conductors that allow spherical waves of progressive sphericity, starting with plane waves, up to the radius at the entrance of the divider. The impedance of the resulting line remains constant (Fig. 6).

The wave fronts have a sphericity R(t) and cross Ox (axis of symmetry) at t. We choose the internal profile $h_i(t)$, so that the angle θ_i of the slope varies linearly with t (factor K is a negative constant). The profile is arbitrarily defined as a circle, so we have:

$$\begin{cases} x_i(t) = \dfrac{\sin(K \cdot t)}{K} = t + R(t) \cdot [\cos(K \cdot t) - 1] & (18) \\ h_i(t) = h_{i0} + \dfrac{1 - \cos(K \cdot t)}{K} = R(t) \cdot \sin(K \cdot t) . & (19) \end{cases}$$

In Eqs. (18) and (19), the third member is derived from the geometrical construction of figure 6. By using Eq. (10), ($\theta_i = K \cdot t$), then we can express θ_e:

$$\theta_e(t) = 2 \cdot \operatorname{atan}\left[\exp\left(\frac{Z_0 \cdot \sqrt{\varepsilon_r}}{60}\right) \cdot \tan\left(\frac{K \cdot t}{2}\right)\right] . \quad (20)$$

Introducing Eq. (20) in the coordinates $h_e(t)$ and $x_e(t)$ of the external profile gives:

$$\begin{cases} h_e(t) = h_i(t) \cdot \dfrac{\sin(\theta_e(t))}{\sin(K \cdot t)} & (21) \\ x_e(t) = x_i(t) + \dfrac{h_i(t) \cdot \sin(K \cdot t)}{1 + \cos(K \cdot t)} - \dfrac{h_e(t) \cdot \sin(\theta_e(t))}{1 + \cos(\theta_e(t))} . & (22) \end{cases}$$

The parameter h_{e0} can be obtained from h_{i0} with the limit of Eq. (20) when t tends to zero:

$$Z(0) = \frac{60}{\sqrt{\varepsilon_r}} \cdot \ln\left(\frac{h_{i0}}{h_{e0}}\right) . \quad (23)$$

Eq. (23) is the expression for the impedance of the cylindrical coaxial line connected to the

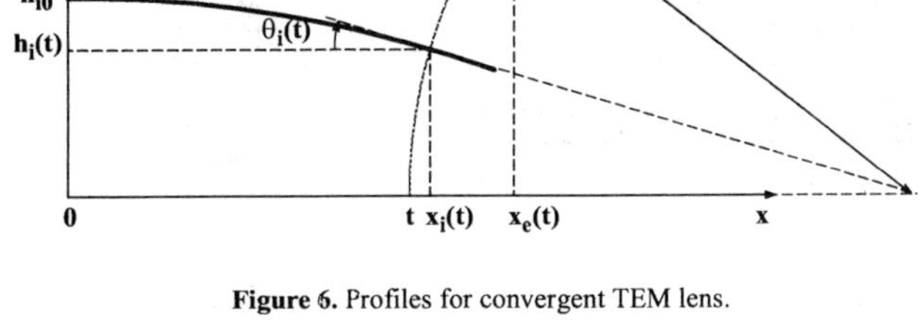

Figure 6. Profiles for convergent TEM lens.

divider (Fig. 7). Eqs. (21) and (22) describe a circle (only numerically proven) with the radius R_2. So we must verify ($K \cdot t_m$ being the input angle of the divider external profile):

$$\frac{(x_e(t))^2 + [h_e(t) - (h_{e0} + R_2)]^2}{R_2^2} - 1 = 0 \quad , \ \forall t \in [0, t_m] \, . \tag{24}$$

With $\varepsilon_r = 2.1$, $h_{i0} = 68.5$ mm, $Z = \mathcal{R}_0$ (50 Ω everywhere) and $x_0 = 250$ mm, we find: $1/K = R_1 = 349.55$ mm and $R_2 = 135.67$ mm. Eq. (24) is verified with an error of $\pm 10^{-15}$. But the paths are different between the internal (l_i) and external (l_e) profiles: $l_i = 32.126$ mm and $l_e = 41.421$ mm. Here the approximation with spherical waves is not as good. Due to the fact that the continuity of the TEM mode is imperfect, the appearance of high frequency modes has to be expected.

4.5. Cylindrical line and coupling telescope

4.5.1. Connecting the MARX generator

The divider with its TEM lens connects to a cylindrical line of 50 Ω plugged into the MARX generator having an outside profile diameter of 137 mm. Figure 7 presents the principle of the connection to the MARX. The insulating media are transformer oil ($\varepsilon_r = 2.1$ to 2.3) and PTFE ($\varepsilon_r = 2.1$).

4.5.2. Coupling telescope: principle

We developed adapting "cones" which are in fact telescopes allowing to adapt the N connector of 10 mm in diameter to the line of 137 mm in diameter. At first, they provide the magnification of the plane waves of the 10 mm line towards the 137 mm line. They will serve to characterize the line and the cones themselves and then to determine the transfer function of the coaxial divider (Fig. 8).

If the transfer function of the left-hand device in Figure 8 is close to unity, the transfer

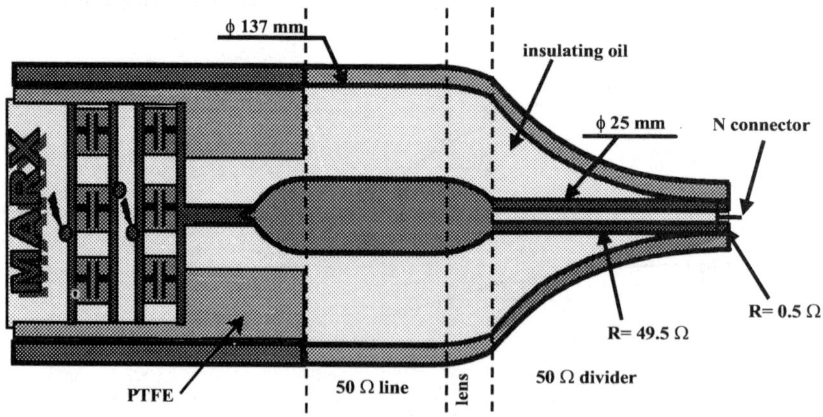

Figure 7. Principle of connection to the MARX generator.

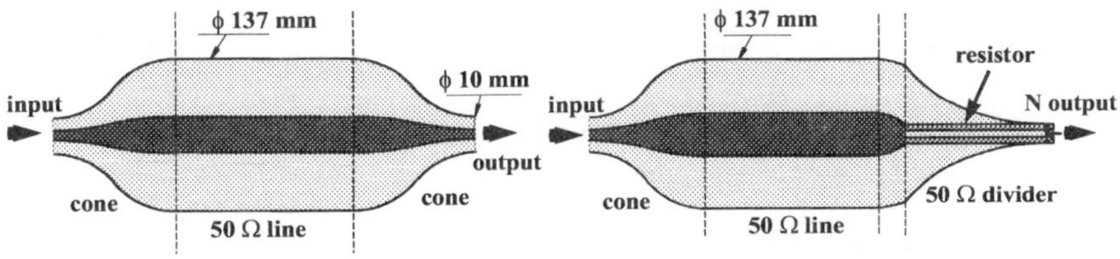

Figure 8. Principle of assembly. Left: 2 cones on the line, right: one cone and the divider.

function of the right-hand device will be equal to that of the resistive divider alone.

4.5.3. Coupling telescope: calculation method

The purpose being the magnification of plane waves (Fig. 9), one will apply the same principle as in §4.4.2. The telescope consists of 2 lenses:

- the left-hand one (Fig. 9) is divergent, the waves fronts are convex in the direction of propagation. The factor K will be positive (§4.4.2).
- the right-hand one is convergent (in the opposite direction). The factor K here is negative. The angle and height of the profile must be the same for both lenses in the center of the telescope. In fact, it will be sufficient to give the radius of the 2 lenses and the starting heights only for the internal profile.

In this case, for a total length of 121.68 mm of the telescope, we also find different internal (l_{int}) and external (l_{ext}) paths on the profiles: l_{int} = 123.63 mm and l_{ext} = 142.63 mm.

The difference, on the order of 13%, will provoke the appearance of modes at high frequencies. The telescope can be considered "too short" (121.68 mm) to fulfill the approximation. As for our first use the bandwidth is limited to 2 GHz, the results obtained are satisfying.

In Figure 9, it can be clearly seen that making a telescope with a monotonous rising profile is impossible, even if we choose a non-spherical transformation. The plane waves at the

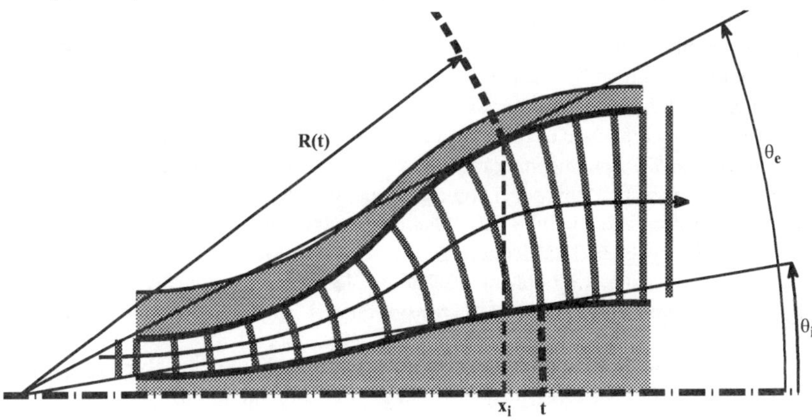

Figure 9. Principle of the coupling telescope.

input and at the output being parallel, the internal and external paths can never be equal, even by stretching the whole length of the telescope.

Neither the theory of non-uniform waveguides,[5] nor the M. Koch's solutions[4] will give us the right profiles. They only give us the interaction of the higher order modes which will be generated in the structure.

4.1. Skin effect in the resistor

We suppose that the current density is parallel to the axis of the resistor $\mathcal{R}_0$ (here a hollow cylinder) and constant on a perpendicular section to this axis. We obtain a good estimate of this hypothesis when the thickness of the resistor is thinner than the skin depth δ_c at frequency f. For a hollow tube of inner radius r_i and outer radius r_e, δ_c is always slightly superior to δ in the case of a semi-infinite plane:[9]

$$\delta(f) = \sqrt{\frac{1}{\pi f \sigma \mu_0}} \quad \text{and if } \delta < (r_e - r_i) \text{, we have: } \mathcal{R}(f) \approx \mathcal{R}_0 \cdot \frac{r_e^2 - r_i^2}{r_e^2 - [r_e - \delta(f)]^2} \tag{25}$$

If f_c, the cut-off frequency, is defined by: $\mathcal{R}(f_c) = \sqrt{2} \cdot \mathcal{R}_0$, (26)

$\mathcal{R}_0 = 50\ \Omega$, $r_i = 7.5$ mm, $r_e = 12.5$ mm and $x_0 = 250$ mm give: $\mathbf{f_c = 1.506}$ **GHz.**

4.2. Acceptable maximum energy

This type of carbon resistor can accept 170 kJ/l. If the calorific capacity is supposed to be in order of 4.18 kJ/°K·l and if the heating process is adiabatic, the maximum acceptable energy E_m for a resistor of dimensions given in §4.1 becomes: $\mathbf{E_m \approx 13}$ **kJ.**

5. EXPERIMENTAL RESULTS

5.1. Experimental set-up

The profiles in brass are manufactured with a computer-driven lathe. The divider and the lens coupled to the cylindrical line of 137 mm are made vacuum tight so that they can be filled with oil without bubbles (Fig. 10).

The dielectrics are transformer oil ($\varepsilon_r = 2.1$-2.3) and PTFE ($\varepsilon_r = 2.1$).

Both resistors are made of the same ceramics charged with carbon (adaptable σ). The outer radius r_e is 12.5 mm, the inner radius r_i is 7.5 mm. The length is calculated to be 147.5 mm and 2.5 mm (in fact, this precision cannot be obtained with a ceramic material). The theoretical division factor will be 100. It cannot be increased due to the mechanical weakness of the small resistor. The electrical contact to the metal parts is obtained by means of washers made of conducting elastomer. The whole is held together by a long insulating screw passing through the middle of the resistor tubes. Compared to carbon paper (papyex) and silver, these washers achieve the best electrical contact.

The resistors used do not have the theoretical values of 49.5 Ω and 0.5 Ω, because of manufacturing tolerances (10%). Furthermore, plunging them into insulating oil increases their resistive value (about 10%).

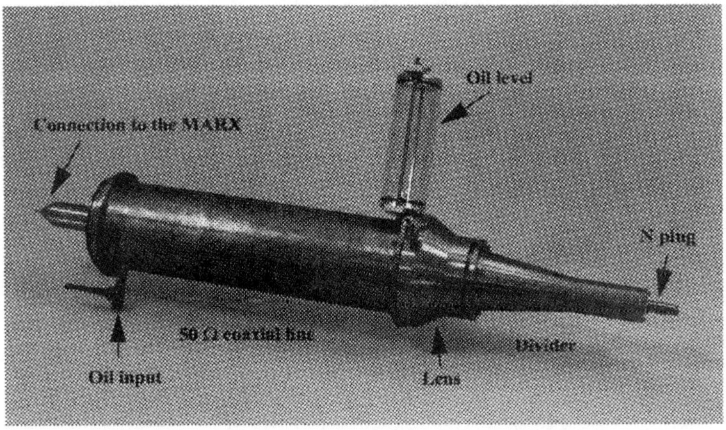

Figure 10. Picture of de high-voltage high-frequency divider.

5.2. Bandwidth of the coaxial line and 2 cones (telescopes)

We measured the transfer function with a vector analyzer (Anritsu MS4623A, 10 MHz - 6 GHz) and with the assembly of Figure 8. Figure 11 represents the module of its transfer function in transmission (S_{12}). Above 2 GHz, sharp absorbing peaks appear. They are due to higher order modes generated from the cones and, as the modes propagate more slowly than TEM, they interact in the vicinity of the resistor (mode interferences). In fact, the first mode (H_{11}) being able to appear in a coaxial line of 137 mm in diameter is, according to Ref.10, 761 MHz ($\varepsilon_r = 2.1$).

Due to the spherical profile of cones, the result (appearance of modes) is much better than in[4] where cones have constant angles.

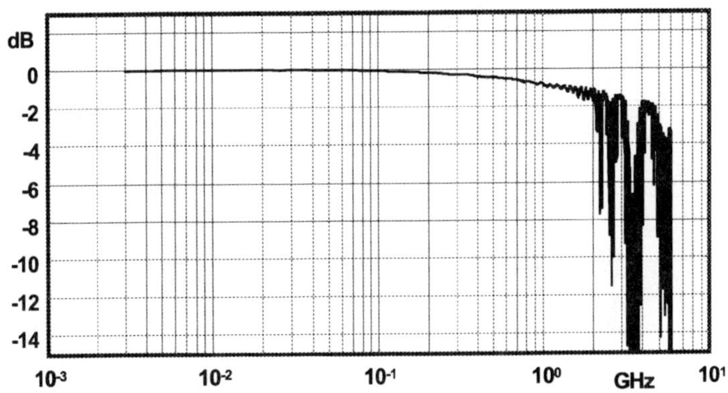

Figure 11. Bandwidth of the coaxial line equipped with the 2 cones (S_{12}).

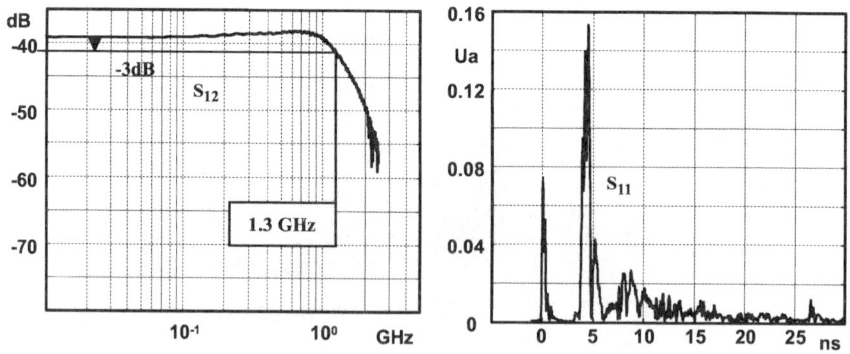

Figure 12. Bandwith (S_{12}) and reflexion (S_{11}) of the resistive divider.

5.3. Bandwidth of the resistive divider

As the division factor remains homogeneous, the frequency behavior (Fig. 12) cannot simply be explained by the result of §4.1. The reflection factor (S_{11}) will increase, on the other hand, with the total resistance.

In figure 12, the reflection is presented in time domain. The equivalent pulse of a flat band from 10 MHz to 3 GHz is about 0.2 ns half-width. Reflections remain lower than 15% or VSWR < 1.35. It is difficult to determine the origin of the reflection peaks, because, for cones, as well as for the resistor, they do not emanate from a punctual zone. The first peak (0 ns) is the result of a mismatch between oil and PTFE, the oil having an ε_r close to 2.3. The 15% peak, representing a total path (going back and forth) of about 1 m, is due to the entrance of the lens.

5.4. Field probe in the coaxial line

The coaxial line of 137 mm can be equipped with a field probe (Fig. 13). This probe, directly calibrated in the same geometry as the divider, will allow us a comparative measurement. The calibration results are in another report.[11] With such a probe, we are, for example, able to discriminate between transmitted and reflected waves by arranging 2 or 3 identical probes along the coaxial line.

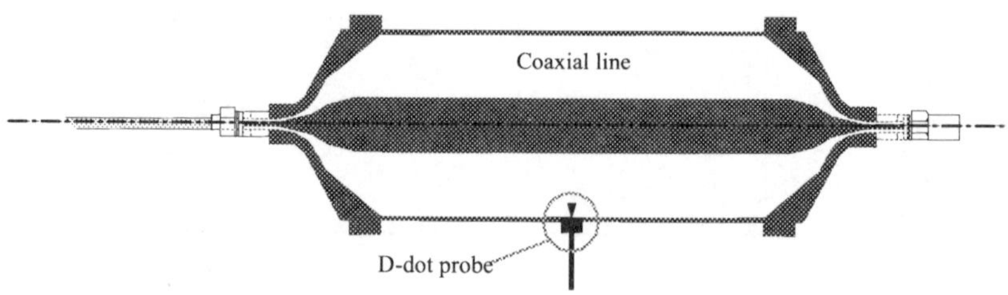

Figure 13. Assembly of a D-dot probe in the coaxial line.

This line, used as a system for antenna calibration, has a supplementary advantage over systems like or similar to GTEM cells.[11] The latter are lines terminated by an absorbent (pyramidal cones) and do not allow the measurement of the signal passing through the cell. This measurement is particularly important when modes appear, it informs us about the signal really transmitted to the antenna.

5.5. Measurements on MARX generator

The measurement of the signal at the output of the divider is carried out with a digital oscilloscope (Tektronix TDS694C, 10^{10} samples/s and 3 GHz bandwidth).

The figure 14 shows a rise time of 6 ns for the MARX generator. Oscillations on the signal are due to the mismatch between the MARX generator and the 50 Ω coaxial line.

The D-dot probe being derivative, its signal will be integrated numerically. Figure 14 shows that the divider answers perfectly linearly from 1 V to 150 kV, the signals of the probe and that of the divider having been only corrected by the calibration factors obtained with the vector analyzer at ~ 1 V. The superposing is near to perfect.

6. CONCLUSION

We demonstrated the capacities of our device to measure very high voltages (>120 kV) and very high frequencies up to 1.3 GHz (rise time: 0.27 ns). It is an original system compared to other systems found in the literature. The main restrictive factors are:

- carbon resistor: skin depth, contact quality, precision in ohmic value and in physical dimensions (length, diameter);
- the contact on the low resistor. We consider the circuit "small" compared to the wavelength. The wave propagation has to be taken into account.
- the profile of cones (telescopes) which limits us for the calibration and for the connection of the cylindrical line to the divisor (TEM lens). Modes appear beyond 2 GHz;
- the quality of dielectrics such as the insulating oil, the relative permittivity of

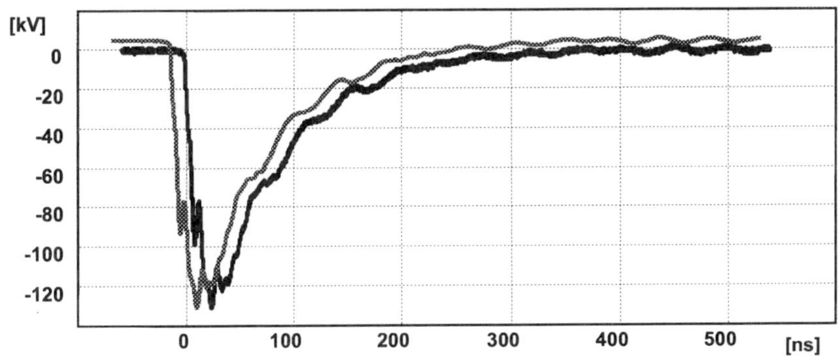

Figure 14. Comparison of D-dot (time-integrated) signal in grey with resistive divider signal in black. The D-dot signal is slightly shifted to distinguish it from the divider signal.

which can vary between 2.1 and 2.3. It provokes a mismatch of the line (reflections).

We also demonstrated the capacities of the system for the calibration of useful field probes for very high voltages. This use is particularly interesting, as it allows us to instrument the output of the generator while it feeds a real microwave source. We can now plan the measurement of transmission/reflection in order to assess the connection between the MARX generator and microwave tubes or antennae.

The use of more rigorous mathematical transformations [6,7] than that with spherical approximation could allow us to find ideal lenses or telescopes, but there is no way of determining them directly. It may be easier to use a computer simulation by "tinkering" with profiles in an intuitive way.

7. REFERENCES

1. J. Ph. Pérez, R. Carles, and R. Fleckinger, *Électromagnétisme: Vide et milieux matériels* (Masson, Paris, 1991).
2. R. Feynman, *Électromagnétisme 1* (InterEditions, Paris, 1995).
3. H. Meinke and F.W. Gundlach, *Taschenbuch der Hochfrequenztechnik - Band 2* (Springer-Verlag, Berlin, 1986).
4. M. Koch, *Analytische Feldberechnung in TEM-Zellen*, Dissertation an der Universität Hannover Deutschland (1998).
5. B.Z. Katsenelenbaum, L. Mercader Del Rìo, M. Pereyaslavets, M. Sorolla Aysa, and M. Thumm, *Theory of Non-uniform Waveguides: The cross-section method* (IEE, London, 1998).
6. M. Spiegel, *Vectoranalysis - Theorie und Anwendung* (McGraw-Hill, Hamburg, 1990).
7. A. Angot, *Compléments de Mathématiques à l'usage des Ingénieurs de l'Électrotechnique et des Télécommunications - 6e édition* (Masson, Paris, 1982).
8. J. Wey, F. Bieth, and P. Raymond, Diviseur ultra-large bande très haute tension, French German Research Institute F-68301 Saint-Louis France, ISL report R119 (2002).
9. H.E. Knoepfel, *Magnetic Fields: A Comprehensive Theoretical Treatise for Practical Use* (Wiley, New York, 2000).
10. N. Marcuvitz, *Waveguide Handbook* (McGraw-Hill, New York, 1951).
11. F. Bieth, J. Wey, and P. Raymond, Field Probes for HPM Measurements, Proceedings of the European Pulsed Power Symposium, Saint-Louis France, 22-24 October 2002.

AN UWB RADAR BASED STEALTHY 'LIE DETECTOR'

Enrico M. Staderini*

1. INTRODUCTION

The capability of remotely detecting breath rate and heart rate using radar techniques in the microwave region of the RF spectrum has been repeatedly demonstrated. Preliminary investigations[1-3], published as early as the 1970s, described conventional short range radar methods capable of monitoring thorax movements and breathing activity as well. In the last few years, with the introduction of UWB radars, remote, non-contact vital parameters monitoring revived. Interesting medical applications of UWB radars were first described in the cardiologic-pulmonary fields[4-6] and in ENT (ear-nose-throat) clinics.[7] A more in-depth and wider presentation of the possible medical uses of UWB radars has recently been presented.[8] In principle, an UWB radar pulse is able to pass through the human thorax (no matter of clothing) so as to be echoed back by the cardiac structures, namely the heart wall. The signal out of an UWB radar aimed at the subject's thorax is composed of two slowly varying periodic signals: the first conveying heart movements and the other carrying information on thorax displacement and respiration rate. As such data carry pathological and physiological meanings to the physician, a new, UWB-based, vital-parameters monitor has already been devised for a variety of applications in clinical, emergency and military medicine.[8]

Heart and breath rates, along with blood pressure and electro-dermal activity, are also of importance in the so called "polygraph testing" more commonly known as the lie detector test.[9] As a matter of fact, "lie detectors" don't actually detect lies and there is also no widespread agreement whether they can even detect deceptive behavior of an examinee. Scientific evidence and admissibility in court of polygraph testing remain controversial. Many authoritative papers have appeared in the literature, some of them lending credence to the limited scientific evidence[10] on polygraph testing, while others debunk the procedure,[11] even comparing it to astrology and tea-leaf reading.[12] Apart from

* Enrico M. Staderini, "Tor Vergata" University of Rome, Dept. of Biopathology and Imaging, Medical Physics Section, Via Montpellier, 1 00133 ROME Italy, staderini@med.uniroma2.it.

Ultra-Wideband Short-Pulse Electromagnetics 6
Edited by Mokole *et al.*, Kluwer Academic/Plenum Publishers, 2003

the scientific validity of polygraph testing *per se*, the added problem of countermeasures is actually a central one when the game becomes serious. Breathing and cardiac countermeasures are part of the training of every "professional" agent who might undergo polygraph testing by friend or foe. When properly mastered, countermeasures are much more "scientifically" effective than the polygraph itself.[13] To this end, an improved "lie detector" should be invisible so that no countermeasures can be taken.

An UWB radar-based "lie detector machine" would add the incredible feature of "stealthiness" to the examination. As a matter of fact, the subject under test might even be kept unaware of being polygraphed.

Present study has been mainly directed to obtaining a research instrument for fully legal psycho-physiological studies in a clinical setting. The current research is aimed at developing devices to assess, in real time and with no active participation of the subject, the mental stress during every day activities like using a personal computer or driving. Final results should enable the design of new and ergonomically improved man-machine interfaces or improved safety devices in car driving (falling asleep alerts).

By the way, the use of a conventional "lie detector" is at present believed not legally allowed in the U.S., neither at federal nor at state level jurisdictions. As a matter of fact, both the U.S. Army Criminal Investigation Command and the Naval Investigative Service, for example, need a *Polygraph Examination Statement of Consent*[14] or a *Polygraph Examination Waiver*,[15] signed by the prospective examinee, to administer a lie-detector test. In these cases, a stealthy polygraph would be actually unusable, not even for security screening purposes. Nevertheless, a stealthy "lie detector" might be used in covert operations or, as in our case, for psycho-neurophysiological studies where the system is used on purpose without the subject's awareness. In this situation a subject's activity to counteract, to fool or even to be conditioned by the machine should be avoided.

2. MATERIALS AND METHODS

The experimental setup is composed of an UWB radar device and an electrocardiograph (EKG) amplifier, for control purposes, connected to an analog-to-digital converter interfaced to a personal computer as depicted in Figure 1.

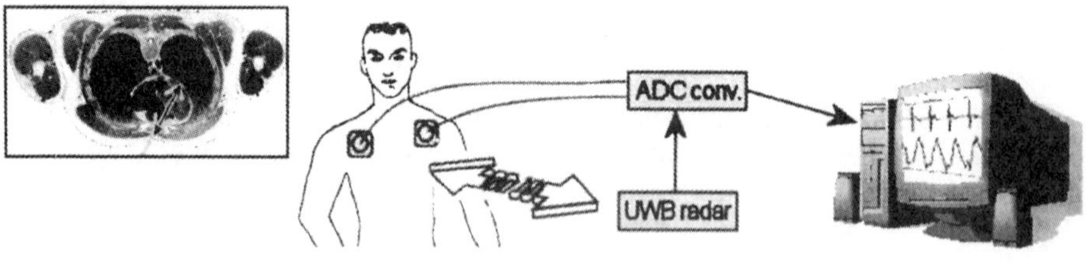

Figure 1. Schematic of the experimental setup for UWB vs. EKG comparison. Analog-to-digital converter (ADC conv.) includes a two-channel EKG for simultaneously detecting the electrical activity of the heart and the UWB field disturbance signal. In the box on the left, a slice from the Visible Human Project data[16] is shown with the possible radar beam path marked inside the thorax.

The analog-to-digital (ADC conv.) converter has been custom developed to integrate the UWB radar power supply, a dual-channel electrocardiograph (EKG) and a multi-channel, 10-bit resolution, 200-samples/sec/channel, analog-to-digital converter interfaced to the parallel port of a personal computer. The software was written in the Modula-2 programming language under the MS-DOS operating system. It dates back to the 1990s and has been used in our laboratory since then as a standard for research purposes on dedicated computers.

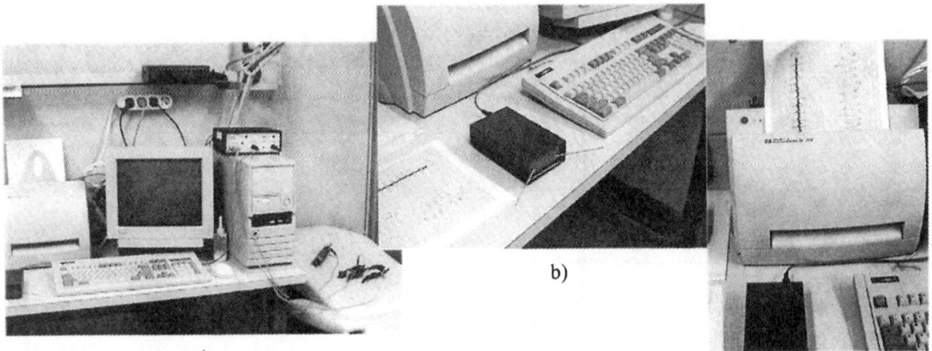

Figure 2. Experimental setup from left to right: a) hardware configuration on the bench; analog to digital converter and EKG amplifier on top of the personal computer; b) UWB radar prototype with dipole ribs protruding out of the black box; c) printer.

The UWB radar device used in this work is a replica with modifications of the micropower-impulse radar (MIR) originally developed at Lawrence Livermore National Laboratory (LLNL) in 1993[5, 6, 17]. Actual modifications and a detailed practical implementation of a prototype of the device has been presented elsewhere[18]. The UWB radar has been prototyped on a multi-hole board and emits 2 ns time-width pulses through a simple dipole antenna with considerable ringout. Although the assembly was all but adequate for a microwave system, nevertheless it was very successful in the preliminary investigations for a possible medical use. Pulses are emitted with a randomly variable pulse repetition delay (random dithering) having a mean repetition frequency of 2 MHz. Actual duty cycle is about 0.4% (not considering the ringing).

3. PRINCIPLE OF OPERATION

Figures 3 and 4 describe UWB radar signal generation from a moving reflector like the heart wall. In practice an analog signal is obtained thanks to the rise-time of the electromagnetic pulse. The rising shape of the reflected pulse is sampled at different time instants according to the time delay in the pulse round trip. Although it is more commonly known as a "radar" from the above rough presentation of its working principle, the kind of system presented here should be more appropriately named a "field disturbance sensor" or, generically, an "electromagnetic sensor."

In this regard, the UWB radar is definitely not a brand new biomedical microwave sensor with respect to the microwave CW devices described in the past, which used interferometry,[19] the Doppler effect,[20-22] backscattering,[23,24] or plain reflection.[25] As a matter of fact, the definition of "field disturbance sensor" has been used in the latest Federal Communications Commission (FCC) Report and Order on the revision of Part 15 of the Commission's Rules regarding Ultra-Wideband Transmission Systems.[26]

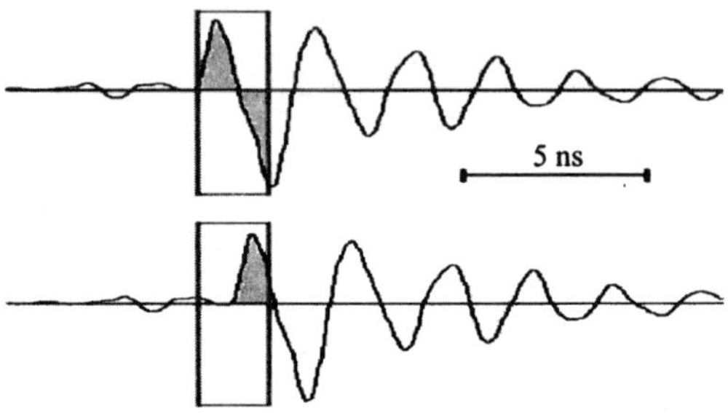

Figure 3. Example of UWB radar signal sampling for a UWB radar in monostatic mode $5 \cdot 10^{-2}$ m (2 in) away from the anterior thorax wall. Upper trace: echo from the heart wall in diastole; lower trace: echo from the heart wall in systole. Echo is delayed in systole as the heart wall is globally more distant (inside the thorax) from the antenna. Aperture time of the receiving sampler, in the boxes, is shown at the same time delay after transmitted pulse emission. Echo signal in the aperture time (averaged over many pulses) constitutes the signal output of the UWB radar.

In light of the cited FCC document the present device is all but legal as it doesn't comply with the rules. In effect, although no precise determination of frequency band of emission and no clear determination of spurious emissions has been carried out yet, nevertheless the device should be working in the low frequency band (below 960 MHz, as per section 15.509 of FCC rules) were medical imaging devices are not allowed.

According to the FCC, UWB based medical devices are only allowed in the high frequency band from 3,100 MHz to 10,600 MHz (as per section 15.513 of newly adopted FCC rules). However, the problem of compliance, with FCC or ERO rules and with any prior intellectual property issue, will be addressed at the time, if any, where the present research prototype will be engineered for industrial or commercial purposes.

A few more considerations would be useful for a better understanding of the operation and design trade-offs.

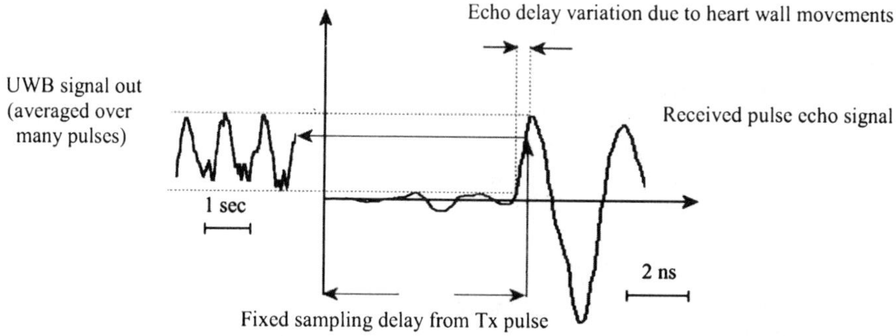

Figure 4. Genesis of the UWB heart signal exemplified considering an instantaneous sampling aperture. Although pulses are emitted with random time delay repetition, sampling is always synchronous with the transmitted pulse. UWB signal comes out of noise after thousands of averaged samplings (note the different time scales). Random dithering and subsequent averaging cancel heterodyning with other continuous wave (CW) signals present on the antenna.

From the signal generation scheme presented in Figure 4, it should be noted that a conversion is made of heart wall displacement to output voltage thanks to the rise time of the echo pulse. In fact, for a fixed sampling time delay from the transmitted pulse emission, the time-varying target surface displacement is responsible for a phase shift in the echo so that sampling occurs at different levels in the pulse rising period. This effect implies that there exists a minimum rise time in the pulse for completely acquiring the full dynamic range of the target moving surface (the heart wall). For example, estimating an electromagnetic field propagation speed in the tissues at about $180 \cdot 10^6$ m/s (some $400 \cdot 10^6$ miles per hour) and a maximum displacement of the heart wall during heart contractions in the range of $15 \cdot 10^{-3}$ m (half an inch), a minimum rise time, for the pulse, results at about $90 \cdot 10^{-12}$ sec or 90 ps. This extreme case is exemplified in Figure 4 for didactic purposes. A shorter rise time would have the effect of sensing a given surface more than once and the "onion effect" will be shown due to the presence of many pulses (ringing) in space at the same time. This might give an erroneous output signal presenting multiple waves per cardiac cycle as already hypothesized in a previous work.[5] It should effectively lead to the determination of a heart beat frequency erroneously multiple of the real one.[5]

It is often declared, when talking about UWB radio technology, that this technique is highly immune to multi-path fading or to its radar counterpart, the clutter. This is absolutely not the case with ringing pulses. As the emitted pulse is larger than irradiated structures, not only does diffraction play a major role, but the summation of echoes at the receiving antenna is of the highest importance. If the echoes which sum up at the input are coming back from still targets and only one is coming from the moving heart wall, then a simple and inoffensive, although quite high, d.c. level should be expected over the useful signal. In practice, echoes arriving from moving structures like lungs, blood vessels and thorax wall produce the effect of summing their signals (with various and time varying amplitudes) to the expected echo signal bouncing back from the target heart wall surface – no matter if the reflecting surfaces are before or behind the target. Furthermore, the echoes arrive at the antenna (in a true monostatic mode with a single

Tx-Rx radiator) when the emitted ringing pulse is not yet completely gone away. So the sampler is compelled to acquire part of the Tx pulse itself and this adds another exceptionally high d.c. level to the output signal. To cope with this effect, a high-pass filter has to be added in the amplifier chain as early as possible after the fast echo sampler. This feature only enables the reaching of the high gain needed to increase the signal to acceptable levels, without saturating the amplifier. From the preceding discussion, the rationale for naming this kind of UWB radar a "field disturbance sensor," more appropriately makes sense.

The sampling of its own emitted pulse, yet on the antenna when useful echo arrives, also makes it impossible to perform a fast scan of targets in space by varying the sampling delay. In fact, as a long time constant for the high-pass filter is chosen, so as to have a high pass at a low frequency with a good tracking of low speed of displacement targets, the scanning in space (that is the continuous variation of sampling delay from Tx pulse emission) should have to be set even slower. In this case only, indeed, the resulting spurious d.c. level variation (caused by the sampling of the emitted ringing pulse at different places) will be effectively canceled out by the high-pass filter. In the experiments performed thus far, not a single adequate trade-off was found in the latter regard. Consequently, scanning in space is considered impossible with the present device. This effect might limit seriously any possibility to consider an UWB radar device for a real-world medical (ultrasound-like) imaging scanner.

On the other end, a low-pass filter may also have to be used for cleaning the UWB radar signal from considerable noise. This limits the tracking of high speed of displacement targets as the heart valves are. By the way, technology improvements and trade-offs are possible in this respect.

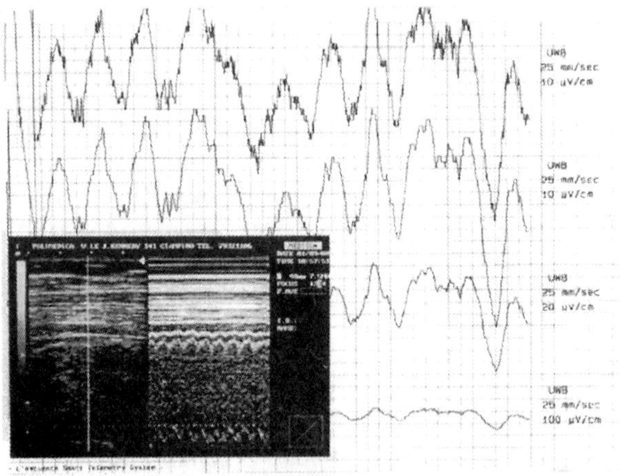

Figure 5. In the tracings the same UWB heart echo signal, with various degrees of low pass filtering and amplification, is plotted. Breathing modulation is visible due to movements of the chest wall. In the insert (bottom-left) an actual ultrasound echo image of the chest in the same subject is shown (SONOACE 1500 scanner from MEDISON Co.). On the right the M-mode tracing shows the heart wall displacement, approximately in the center of the image. Thorax movements due to breathing are not visible as the ultrasonic transducer, obviously, moves along with the chest. A good match is demonstrated between ultrasound M-mode echo and UWB M-mode echo.

As regards signal loss in the pulse round trip, it should be expected, on a pure time domain reflectometry theory,[27] that the return loss be due both to impedance mismatch and to attenuation into lossy media. Measurements and models, based on actual electromagnetic data of human tissues,[28] gave return loss figures, for a two way monostatic radar, in the order of 60 dB. Comparable results were derived from previous studies on UWB propagation through lossy media.[29] It appears that, for very short range radar operation, the R^{-2} dependence of power density on target distance R can be omitted and conventional two way radar equation, commonly found on reference handbooks,[30] seems at present not adequate to model radar propagation in this very special situation.

4. PHYSIOLOGICAL CONSIDERATIONS

From the previous description it appears that an UWB radar based monitor is able to detect heart and breath rate only. In conventional "lie detectors" attention is also given to monitor other signals capable of giving an insight into autonomic nervous system modifications which might be used to infer a deceptive behavior of the subject. Parasympathetic and sympathetic sections of the autonomic nervous system in humans play an important role in the control of heart rate, and from proper elaboration of the heart rate variability (HRV) signal, interesting data can be obtained.

HRV is obtained by measuring the time interval between successive heart beats. Using the EKG to detect heart beats, HRV is also known as "R-R signal" as time intervals are actually measured between successive R-waves in the EKG. Using UWB radar to detect heart beats, a suitable hardware or software detector measures time intervals between successive peaks or valleys in the radar signal (head to head or foot to foot in the UWB radar wave like that depicted in Figure 5).

HRV is fundamentally different from a signal derived by sampling a continuous process. HRV is by default a sampled signal whose sampling period is the signal itself. Although there is no sound physiological reason for considering HRV coming from a continuous process (that is, like the output of a delta-modulator), nevertheless it is common practice to derive a uniformly sampled HRV from crude HRV by interpolation (as if HRV might be defined for all points in time). In this way standard analyzing techniques can be used. HRV has been studied in the last twenty years in heart research laboratories worldwide and is now used, more and more frequently, on a clinical basis.

Continuous alteration in sympathetic-parasympathetic balance of the autonomic neural regulation makes the heart rhythm fluctuate around the mean heart rate. These fluctuations are periodic or quasi-periodic and they seem to originate from respiration regulation, blood pressure and thermoregulation. Parasympathetic (vagal) regulation of the heart is inhibited during inspiration[31] and it can be excluded using antivagal drugs like atropine.[32] Oscillations in blood pressure, due to baroreflex control loop, result in sympathetic activation.[33] Changes in peripheral blood vessel resistance due to thermoregulatory adjustments of peripheral blood flow are under the control of sympathetic activation too.[33] Parasympathetic stimuli make the heart slow down, while sympathetic activation makes the heart rate increase.

Under the hypothesis that mental stress could interact with sympatho-vagal system, commonly used "lie detectors" monitor the electrodermal activity. In our case HRV is monitored instead. Mental stress (for example in mental arithmetic) has been found to increase sympathetic activity and to decrease parasympathetic activity as detected using

HRV analysis.[34] Studies were also conducted to demonstrate the usefulness of HRV analysis to detect negative functional states (asthenia, depression, neurotic symptoms) in the course of psychological relaxation sessions.[35]

Keeping in mind the fact that no true and scientific "lie detector" exists or can be developed within the present state of the art, we can nevertheless consider HRV, detected with an UWB based radar, as a viable way of obtaining a signal related to mental stress. This might prove useful in situations where the autonomic nervous system of the subject is involved.

5. RESULTS

In Figure 6 a typical plotting of simultaneously recorded EKG and heart UWB radar signals is presented. The UWB radar signal, related to mechanical movement of the heart, follows the electrical event by a small amount of time as expected.

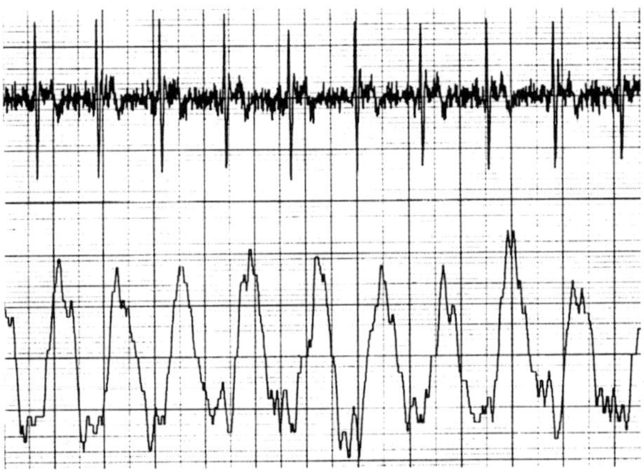

Figure 6. UWB radar tracing (bottom trace 20 μV/cm) versus high-pass filtered ECG (upper trace 400 μV/cm), time base 15 mm/sec.

An UWB radar signal of heart displacement is quite noisy, although very clearly and easily detected from a distance of $5\cdot10^{-2}$ m (2") from the thorax. In the initial experiments a thick plastic sheet was fixed to the table and the volunteer subject asked to press his chest onto the plastic sheet. In this way skin movements were strongly avoided and sufficient confidence was obtained that the UWB echoes are coming from inside the body. Subsequent signals were obtained with the antenna placed on the rear part of the thorax and the subject was asked to sit comfortably on a chair. The UWB radar worked at a distance of $15\cdot10^{-2}$ to $20\cdot10^{-2}$ m from the heart, with the back of the chair in between, as depicted in Figure 7.

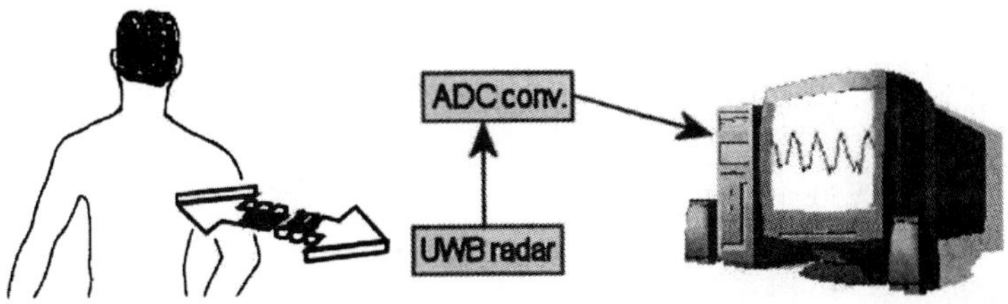

Figure 7. Schematic of the stealthy vital parameters monitor to be mounted into the back of a chair.

On the tracings in Figure 6, two different heart-cycle detectors implemented in software were used to extract the heart beat period from the EKG signal and the UWB echo signal. The EKG signal was high-pass filtered and the R wave was then detected using dynamic thresholding. The threshold is updated to the EKG signal whenever the latter is higher than the threshold. During time intervals in which the EKG signal is lower than the threshold, its level is decreased exponentially to cope with the varying R-wave amplitudes in successive beats. This method proved quite robust although somewhat *naïve*. Heart-beat period detection on UWB echo tracings used an algorithm previously used for arterial pressure analysis. Peaks and valleys of the signal are detected and the time lag between two consecutive valleys is considered the heart beat period. This kind of processing is quite sensitive to noise and it will be certainly abandoned in further studies. It was used at this moment because it was available, having been developed for other purposes.

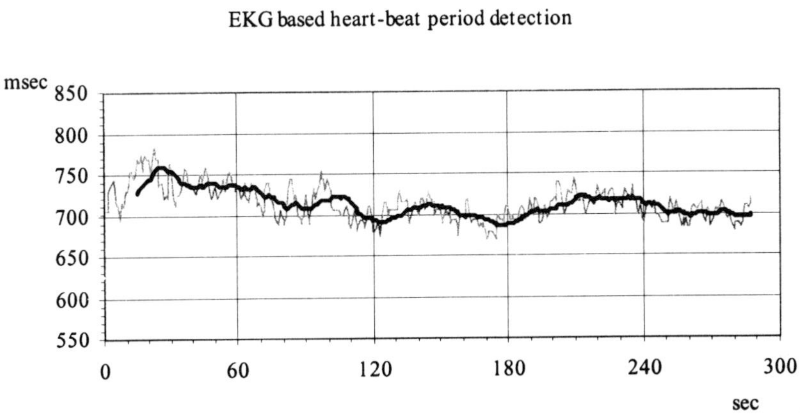

Figure 8. Estimated period of heart cycle from EKG tracing during five minutes (400 heart beats), which is overlayed with a 20-point, box-car, moving-average-regression curve (thicker curve).

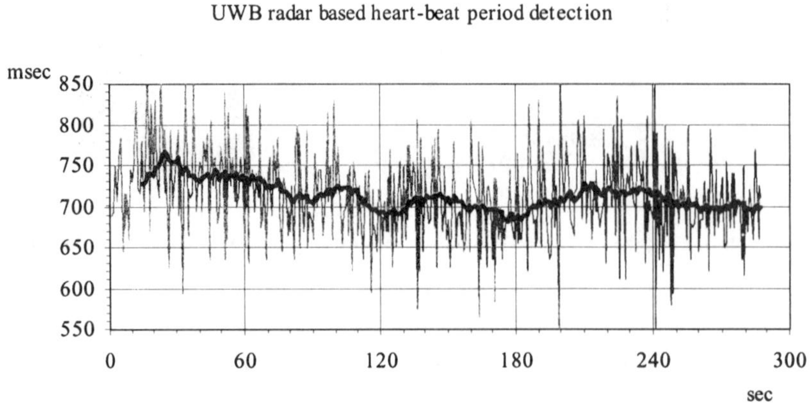

Figure 9. Estimated period of heart cycle from UWB radar echo tracing during five minutes (400 heart beats), which is overlayed with a 20-point, box-car, moving-average-regression curve (thicker curve). Data in this figure and in the preceding figure were recorded simultaneously on a subject at rest.

A simple statistical analysis of the data gave a mean heart period for EKG-detected beats of 713.1 ms and a mean heart period for UWB-detected beats of 712.9 ms. As both detection algorithms performed without errors for each heart beat (no heart beat was missed), the same mean value was expected. The standard deviation differed significantly in the two sets of data. In the EKG-derived data, the standard deviation was 21.06; while in the UWB-derived data, it was 56.95. The difference in standard deviation is attributable to the higher noise that contaminated UWB-derived HRV signal. A histogram comparison in Figure 10 demonstrates this phenomenon.

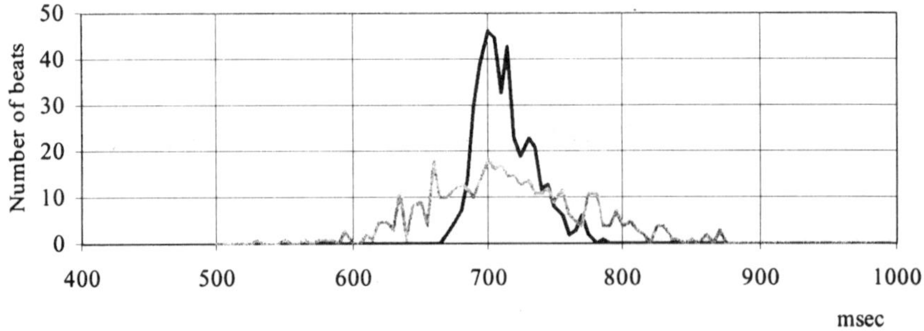

Figure 10. Histograms of heart periods detected using EKG (darker line) and UWB (lighter line). The histograms plot the number of heart beats having the same period (whether they were consecutive or not). Both distributions peak at the same mean value; however, noise in the UWB-echo tracing causes errors in period estimation and spreading of the relative histogram.

As explained in the previous paragraph, the activation of the autonomous nervous system should determine cyclic modifications on the heart period. Spontaneous respiration rate should account for variations intervening at breath rate, sympathetic blood pressure variations, and thermoregulation should also account for cyclic behavior at lower frequencies.

To verify a previous finding reported in the literature, EKG-derived HRV signal was Blackman-Harris windowed,[36] and the spectrum was estimated using the Fast Fourier Transform (FFT).

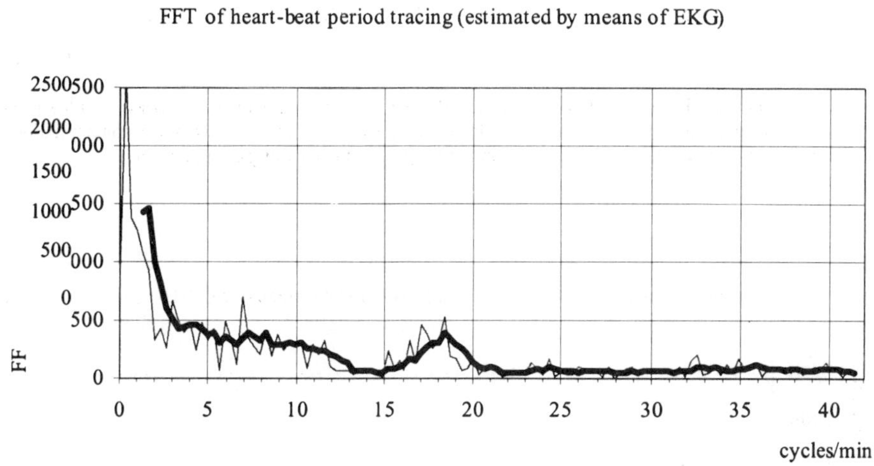

Figure 11. Spectrum of HRV signal derived from EKG (R-R tracing), which is overlayed with a box-car, 5-point, moving-average-regression curve for a clearer understanding.

A mean "sampling rate" was considered that was equal to the inverse of mean heart-beat period (= 1/0.713 Hz = 1.4 Hz). A 256-point FFT yielded a spectrum with an approximate resolution of 5.4 mHz or 0.33 cycles/min. The same FFT was performed on the UWB echo derived HRV signal. In Figure 11 the spectrum of the HRV signal is plotted with a superimposed, box-car, moving-average-regression curve. A variation related to breath rate is clearly visible in the 15-20 cycles/min band. Sympathetic activation is also visible in the 5-12 cycles/min frequency band. It is important to note that such periodic behavior of the HRV signal was also discernible to naked eye in the tracings reported in Figures 8 and 9 from which the spectra were estimated. The highly noisy, UWB-radar based, HRV signal is absolutely not suitable for FFT processing, as the Fourier Transform looses meaning with the uncorrelated noise build-up in the original signal (visible in Figure 12).

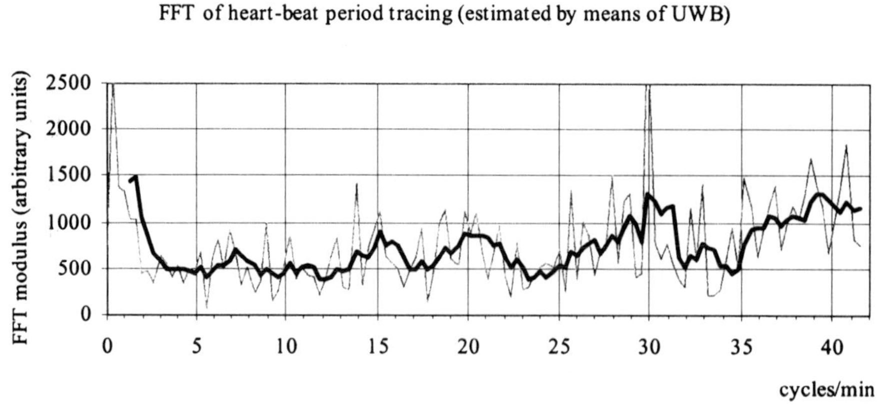

Figure 12. Spectrum of HRV signal derived from UWB radar echo tracing, which is overlayed with a box-car, 5-point, moving-average-regression curve for a clearer understanding. With heavy noise contamination, the Fourier Transform becomes statistically insignificant. The spectrum is meaningless in this case.

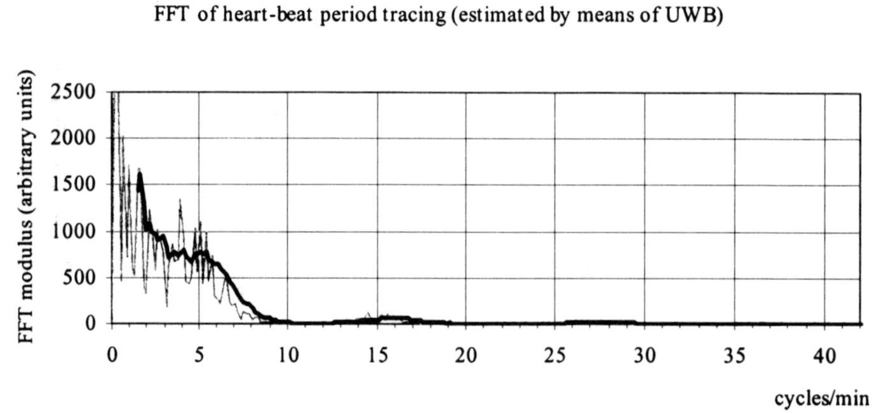

Figure 13. Spectrum of HRV signal derived from UWB radar echo tracing after Wavelet de-noising. After de-noising breath rate related peak is no more visible. Sympathetic activation is very low in the 3-10 cycles/min frequency band.

To cope with this problem, HRV frequency estimation is more commonly performed using autoregressive modeling (AR) or autoregressive moving average (ARMA) methods. Nonlinear methods are being used too. By the way, it is curious how distant bands in the spectrum are used in this work: from GHz in the microwave region of UWB radio to mHz in the heart-rate variation region.

In this initial study, the decision was made to filter out noise in the UWB-based HRV signal using Wavelet-based filtering[37] and then to use the FFT on the recovered signal. This procedure is somehow arbitrary, nevertheless it was quite effective in cleaning out noise while preserving the low-frequency spectral information. A discrete 256-point Wavelet transform was calculated, followed by a decimation of the higher resolution scales in the Wavelet domain. After taking the inverse Wavelet transform, a conventional 256-point FFT was used to estimate the spectrum. A more strict procedure is planned for future research.

To test the system, a math game was prepared and administered to a voluntary subject. Randomly generated arithmetic expressions were presented on the screen of a computer, and the subject was asked to do the calculation mentally and to enter the correct result as soon as possible. The game was made more challenging by programming more difficult calculations with the passage of time. After an initial period in which the subject feels he is adequate to the task and he is relaxing, an increase in difficulty of proposed calculations introduces anxiety and fear in the subject. Sympathetic activation arises and it is clearly visible in Figure 14 when compared to Figure 13.

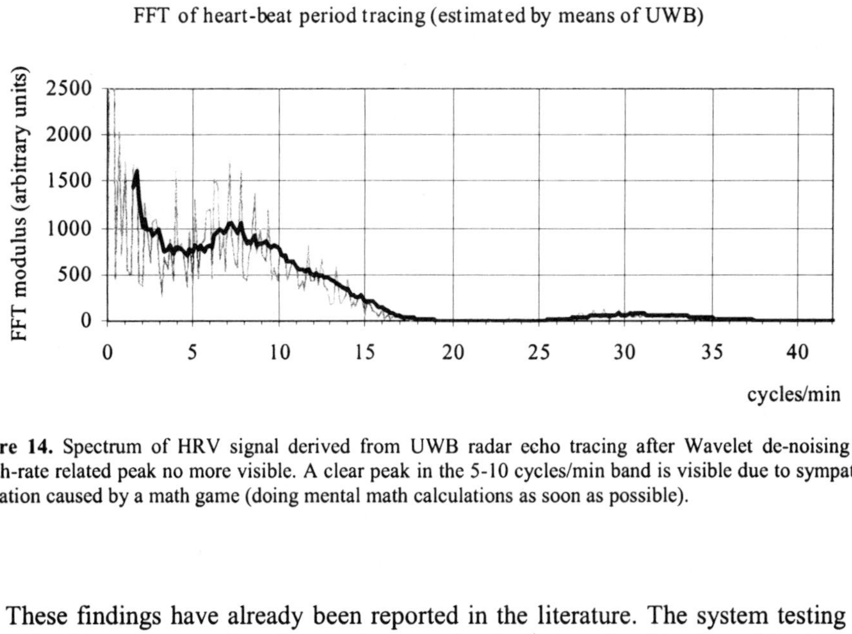

Figure 14. Spectrum of HRV signal derived from UWB radar echo tracing after Wavelet de-noising with breath-rate related peak no more visible. A clear peak in the 5-10 cycles/min band is visible due to sympathetic activation caused by a math game (doing mental math calculations as soon as possible).

These findings have already been reported in the literature. The system testing was started in this way to confirm the previous results. In fact, main objective of the task was to prove the equivalence of UWB-radar-derived HRV signal with the EKG-derived one. A straightforward application of this technique, thanks to its intrinsic remoteness and exceptional user-friendliness, is the monitoring of car drivers to prevent them from falling asleep during driving. A small UWB radar, which was attached on or built into the safety belt of the driver's seat, could sense the driver's heart rate and the HRV signal and could deliver a suitable alarm when necessary.

Figure 15. An UWB-radar-based, heart-rate monitor mounted into the driver's safety belt to prevent the driver from falling asleep.

The issue of safety, when using electromagnetic energy, is often a primary concern. Studies on the effects of UWB electromagnetic energy on humans, animal models and living cell cultures were performed in the past.[38-43] Unfortunately, their scope was primarily directed towards the assessment of the possible side effects on humans by UWB electromagnetic weapons. To this end, power in excess of a MW was used along with UWB electric fields in the range of 10s to 100s of kV/m. Although no final conclusion can be obtained from the limited published reports, nevertheless in our case, the power level and the electric field intensity are seven to nine orders of magnitude smaller, and no data at all is available from the literature in this regard.

Nowadays, the only argument that can be given in favor of the safety of UWB-radar-based, heart-rate monitors is that however dangerous cellular phones might be, medical UWB radars emit two to three orders of magnitude less power than common cellular phones. This means that a five-minute phone call on a mobile might equal ten hours of UWB-radar heart-rate monitoring.

6. CONCLUSIONS

The viability of monitoring sympathetic/parasympathetic activation due to mental stress in humans using the heart-rate-variability (HRV) signal detected with an UWB-radar-based, vital-signs monitor has been assessed. The system's strength is based on the possibility to maintain the subject under test unaware of being monitored so as to avoid any psychological discomfort. This leads to a more objective measure which, in the case of a proposed new "lie detector machine," adds the extra bonus of avoiding countermeasures when operated in a stealthy mode. The main weakness of the system lies in the presence of noise, which strongly degrades beat-to-beat time-period detection using an automatic software heart-cycle detector. Although advanced filtering techniques effectively help in overcoming noise disturbance, a better software detector might solve the problem in a more robust way.

Apart from "lie detectors," the proposed system and the signal processing software may be more appropriately used in neuro-psychological studies aimed at developing better man-machine interfaces or in detecting the degradation of human performances in dangerous or risky tasks like car driving, aircraft piloting or critical monitoring tasks.

7. ACKNOWLEDGEMENTS

The work of graduate students, Mr. Giorgio Cesarini (hardware and software setup) and Mrs. Elisa D'Agati (signal acquisition and analysis), are gratefully acknowledged. The physiotherapist, Mr. Stefano Mugnaini, provided useful hints in friendly discussions with the author about the methods and techniques used in the research.

This project was financially supported, in part, by the Italian National Research Council (C.N.R.) and STMicroelectronics S.r.l. (Italian company belonging to STMicroelectronics B.V. from Nederland).

8. REFERENCES

1. C.G. Caro and J.A. Bloice, Contactless apnoea detector based on radar, *Lancet* **2**(7731), 959–961 (1971).
2. C.I. Franks, B.H. Brown, and D.M. Johnston, Contactless respiration monitoring of infants, *Med. Biol. Eng* **14**(3), 306–318 (1976).
3. C.I. Franks, J.B. Watson, B.H. Brown, and E.F. Foster, Respiratory patterns and risk of sudden unexpected death in infancy, *Arch. Dis. Child* **55**(8), 595–604 (1980).
4. S.G. Azevedo and T.E. McEwan, Micropower Impulse Radar, *Energy and Technology Review*, UCRL-52000-96-1, 1-2, 16–13 (1996).
5. T.E. McEwan, Body monitoring and imaging apparatus and method, United States Patent 5,573,012 (1996).
6. T.E. McEwan, Body monitoring and imaging apparatus and method, United States Patent 5,766,208 (1998).
7. J.F. Holzrichter, G.C. Burnett, L.C. Ng, and W.A. Lea, Speech articulator measurements using low power EM-wave sensors [letter], *J. Acoust. Soc. Am.* **103**(1), 622–627 (1998).
8. E.M. Staderini, UWB radars in medicine, *IEEE AESS Systems Mag.*, 13–18 (January 2002).
9. K. Bonsor, How lie detectors work, http://www.howstuffworks.com/lie-detector.htm/printable (as visited on Jan 2002).
10. *Scientific Validity of Polygraph Testing: A Research Review and Evaluation*, Technical memorandum, Washington DC: U.S. Congress Office of Technology Assessment OTA-TM-H-15 (1983).
11. G.W. Maschke and G.J. Scalabrini, *The Lie Behind the Lie Detector,* http://antipolygraph.org (as downloaded on Jan 2002).
12. D.C. Richardson, Opening Statement before the U.S. Senate Committee on the Judiciary's Subcommittee on Administrative Oversight and the Courts, http://www.nopolygraph.com/drewtest.htm (as visited on Jan 2002).
13. D.G. Williams, *How to Sting the Polygraph* (Sting Publications, Chickasha OK, 1996).
14. U.S. Army criminal investigation command, Polygraph examination statement of consent, http://www.fas.org/sgp/othergov/polygraph/ota/consent2.jpg (as downloaded on Jan 2002).
15. Naval Investigative Service, Polygraph examination waiver, http://www.fas.org/sgp/othergov/polygraph/ota/consent1.jpg (as downloaded on Jan 2002).
16. Y.J. Chang, *The NPAC Visible Human Viewer*, Syracuse University, NY. http://www.npac.syr.edu/projects/vishuman/VisibleHuman.html (as downloaded on Jan 2002).
17. T.E. McEwan, Ultra-wideband radar motion sensor, United States Patent 5,361,070 (1994).
18. E.M. Staderini, Everything you always wanted to know about UWB radars..., Online Symposium for Electronics Engineers. Web broadcasted on Jan 16th, 2001. http://www.osee.net/whitepapers/paper4.pdf (as visited on Jan 2001).
19. D.W. Griffin, MW interferometers for biological studies, *Microwave J.*, 69–72 (May 1978).
20. K.H. Chan and J.C. Lin, An algorithm for extracting cardiopulmonary rates from chest movements, *IEEE/Seventh Annual Conference of the Engineering in Medicine and Biology Society*, 466–469 (1985).
21. J.Y. Lee and J.C. Lin, A microprocessor-based noninvasive arterial pulse-wave analyzer, *IEEE Trans. Biomedical Eng.* **BME-32**(6), 451–455 (1985).

22. W. Byrne, P. Flynn, R. Zapp, and M. Siegel, Adaptive filter processing in microwave remote heart monitors, *IEEE Trans. Biomedical Eng.* **BME-33**(7), 717–722 (1986).
23. K. Chen, D. Misra, H. Wang, H. Chuang, and E. Postow, An X-band microwave life-detection system, *IEEE Trans. Biomedical Eng.* **BME-33**(7), 697–701 (1986).
24. J.C. Lin, Microwave sensing of physiological movement and volume change: a review, *Bioelectromag.* **13**, 557–565 (1992).
25. K. Chen, Y. Huang, J. Zhang, and A. Normal, Microwave life-detection systems for searching human subjects under earthquake rubble or behind barrier, *IEEE Trans. Biomedical Eng.* **27**, 105–114 (2000).
26. *First Report and Order in the matter of Revision of Part 15 of the Commission's Rules Regarding Ultra-Wideband Transmission Systems*, ET Docket 98-153, Federal Communication Commission Washington, D.C. 20554; released April 22, 2002.
27. Hewlett Packard, *Time Domain Reflectometry Theory*, Application Note 1304-2.
28. C. Gabriel, Compilation of the Dielectric Properties of Body Tissues at RF and Microwave Frequencies, Report AL/OE-TR-1996-0037, Armstrong Laboratory (AFMC), Occupational and Environmental Health Directorate, Radiofrequency Radiation Division, Brooks Air Force Base TX 78235-5102 (1996).
29. M.A. Barnes, L.W. Fullerton, and P. Withington, Pulson Lossy Media Communication Link Project, Report to the FCC, Pulson Communication Inc., Atlanta GA 30339.
30. *Electronic Warfare and Radar Systems Engineering Handbook*, NAWCWPNS TP 8347, Naval Air Warfare Center, Weapons Division, Avionics Department, Electronic Warfare Division, Point Mugu CA 93042 (1999).
31. D. Eckberg, Human sinus arrhythmia as an index of vagal cardial outflow, *Respirat. Environ. Exercise Physiol.* **54**(4), 961–966 (1983).
32. S. Akselrod, D. Gordon, J. Madved, N. Snidman, D. Shannon, and R. Cohen, Hemodynamic regulation: investigation by spectral analysis, *Am. J. Physiol.* **249**, 867–875 (1985)
33. C. VanRavenswaaij-Arts, L. Kollée, J. Hopman, G. Stoelinga, and H. VanGeijn, Heart rate variability, review, *Ann. Intern. Med.* **118**(6), 436–447 (1993).
34. M. Pagani, R. Furlan, P. Pizzinelli, W. Crivellaro, S. Cerutti, and A. Malliani, Spectral analysis of R-R and arterial pressure variabilities to assess sympathovagal interaction during mental stress in humans, *J. Hypertens. Suppl.* **7**(6), S14-5 (1989).
35. V.A. Mashin and M.N. Mashina, Analysis of the heart rate variability in negative functional states in the course of psychological relaxation sessions, *Human Physiology* **26**(4), 420–425 (2000). Translated from Russian: originally published in *Fiziologiya Cheloveka* **26**(4), 48–54 (2000).
36. F.J. Harris, On the use of windows for harmonic analysis with the discrete fourier transform, *Proc. IEEE* (January 1978).
37. M. Akay (ed.), *Time Frequency and Wavelets in Biomedical Signal Processing* (Wiley, New York,1998), IEEE Press Series on Biomedical Engineering.
38. S.F. Cleary, F. Nickless, L.M. Liu, and R. Hoffman, Studies of exposure of rabbits to electromagnetic pulsed fields. *Bioelectromag.* **1**(3), 345–352 (1980).
39. J.A. D'Andrea, B.L. Cobb, and J.O. Lorge, Lack of behavioral effects in the rhesus monkey: high peak microwave pulses at 1.3 GHz, *Bioelectromag.* **10**(1), 65–76 (1989).
40. T.J. Walters, P.A. Mason, C.J. Sherry, C. Steffen, and J.H. Merritt, No detectable bioeffects following acute exposure to high peak power ultra-wide band electromagnetic radiation in rats, *Aviat. Space Environ. Med.* **66**(6), 562–569 (1995).
41. C.J. Sherry, D.W. Blick, T.J. Walters, G.C. Brown, and M.R. Murphy, Lack of behavioral effects in non-human primates after exposure to ultrawideband electromagnetic radiation in the microwave frequency range, *Radiat. Res.* **143**(1), 93–100 (1995).
42. J.H. Merritt, J.L. Kiel, and W.D. Hurt, Considerations for human exposure standards for fast-rise-time high-peak-power electromagnetic pulses, *Aviat. Space Environ. Med.* **66**(6), 586–595 (1995).
43. J.R. Jauchem, R.L. Seaman, H.M. Lehnert, S.P. Mathur, K.L. Ryan, M.R. Frei, and W.D. Hurt, Ultra-wideband electromagnetic pulses: lack of effects on heart rate and blood pressure during two-minute exposures of rats, *Bioelectromag.* **19**(5), 330–333 (1998).

ELECTROMAGNETIC COMPATIBILITY (EMC) OF ULTRAWIDEBAND (UWB) SHORT-PULSE SYSTEMS IN MILITARY ENVIRONMENTS

Lawrence S. Cohen and Arthur H. Light*

1. INTRODUCTION

Considerable literature has been presented describing proposed, experimental, and operational ultrawideband (UWB) short-pulse systems, both communications and radar. Nearly all of this literature has discussed the operation and electromagnetic compatibility (EMC) of these devices, either in commercial and residential environments or in relatively benign military environments similar to those frequently experienced by forward ground combat troops. In these environments, UWB systems have proved themselves to be highly capable and electromagnetically compatible. Much of the research has suggested that the excellent compatibility of such systems in test environments implies similar compatibility in high-intensity military and commercial environments, both as a potential interference source and as a potential interference victim.

This article considers several common high-intensity military and civilian environments, which have strong potential to interfere with UWB receivers and in which UWB transmitters may pose a significant source of interference to very sensitive receivers in close proximity to one another. For specific situations, we discuss the characteristics of the local environment which may be incompatible with UWB operation and the potential interference mechanisms which may have a detrimental effect on either the UWB system or other systems in the environment. In the case of an UWB receiver, we analyze the extreme interference-to-signal ratio (I/S) produced by some typical shipboard and aircraft transmitters in the same frequency bands in which current UWB systems are projected to operate. Furthermore, we investigate the potential of UWB

* Lawrence S. Cohen, Radar Division, Naval Research Laboratory, Washington, DC 20375. Arthur H. Light, Electronic Warfare Dept., Information Dominance, ITT Industries Advanced Engineering & Sciences, Alexandria, VA 22303.

transmitters interfering with narrowband receivers. We argue that UWB is not the panacea for the problems of spectrum overuse, even though it may have unique capabilities for particular situations.

2. BACKGROUND

On Valentine's Day 2002, the Federal Communications Commission (FCC) adopted its First Report and Order (R&O)[1] on the use and operation of ultrawideband (UWB) devices. The 25 April 2002 revision of Part 15 of the FCC[a] rules defines *UWB bandwidth* as:

> the frequency band bounded by the points that are 10 dB below the highest radiated emission, as based on the complete transmission system including the antenna. The upper boundary is designated f_H and the lower boundary is designated f_L. The frequency at which the highest radiated emission occurs is designated f_M.[a]

An *UWB transmitter* is defined as:

> an intentional radiator that, at any point in time, has a fractional bandwidth equal to or greater than 0.20 or has a UWB bandwidth equal to or greater than 500 MHz, regardless of the fractional bandwidth.[b]

Because an UWB transmission is normally carrierless, the FCC defined the *center frequency* f_C as $(f_H + f_L)/2$[c] and the *fractional bandwidth* as $2(f_H - f_L)/(f_H + f_L)$.[d]

The *effective isotropic radiated power (EIRP)* is defined as "the highest signal strength measured in any direction and at any frequency from the UWB device"[e] and is limited to the average levels shown in Table 1, when measured at a distance of 3 m from the UWB device. In addition, the UWB transmit device is limited to a peak EIRP "of the emissions contained within a 50 MHz bandwidth centered on the frequency at which the highest radiated emission occurs, f_M. That limit is 0 dBm EIRP."[f] Because UWB transmit devices operating at frequencies below 960 MHz have no FCC EIRP limit, these devices are limited to a radiated power of 200 μV/m[g] (46.02 dBμV/m or –49.18 dBm/m) when measured at a distance of 3 m by a receiver using a quasi-peak detector, as specified by the International Special Committee on Radio Interference (CISPR) of the International Electrotechnical Commission (IEC). The EIRP limits are shown in Table 1 and Figure 1. The FCC also specified three types of UWB classifications: (I) imaging systems; (II) vehicular radar systems; and (III) communications and measurement systems. Imaging systems are further divided into five classes: (1) ground penetrating radar (GPR) systems; (2) wall-imaging systems; (3) through-wall imaging systems; (4) medical systems; and (5) surveillance systems. These authors anticipate that all of these types and classes of UWB devices have already been or will soon be introduced into military environments.

---

[a] FCC Part 15.503(a).
[b] FCC Part 15.503(d).
[c] FCC Part 15.503(b).
[d] FCC Part 15.503(c).
[e] FCC Part 15.503(k).
[f] FCC Parts 15.209(a), 15.509(d), 15.511(d), 15.513(d), 15.515(d), 15.517(c), and 15.519(c).
[g] FCC Part 15.509(f).
[h] FCC Parts 15.209(a), 15.509(d), 15.511(d), 15.513(d), 15.515(d), 15.517(c), and 15.519(c).

Although the FCC does not have authority over the spectrum occupancy of equipment used by the military or other parts of the federal government, the National Telecommunications and Information Administration (NTIA) of the US Department of Commerce, which does have this authority,[3] usually allows the operation of commercial off-the-shelf (COTS) non-development activity (NDA) devices approved by the FCC for civilian use on a not-to-interfere (NTI) basis within the military. In the case of UWB devices, the NTIA's Institute for Telecommunications Sciences (ITS) has conducted extensive research into the measurement of the spectra of UWB devices and their potential to interfere with various government and military-operated aids for navigation and aircraft safety.[a,b]

This paper discusses the potential effects of UWB devices in the military environment, both to the UWB devices and from the UWB devices to equipment previously integrated into the military environment.

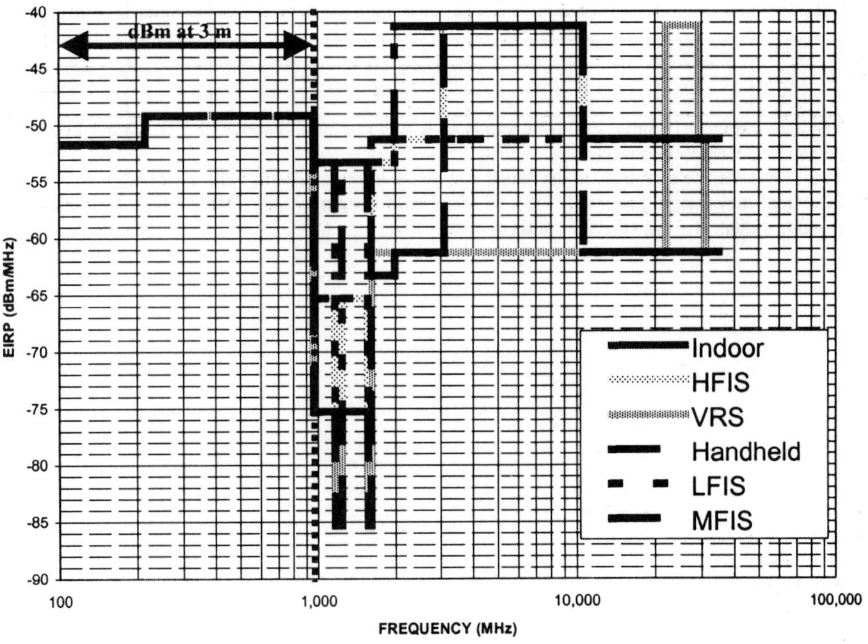

Figure 1. FCC peak EIRP limits for UWB systems.

Table 1. FCC limits on peak EIRP for UWB Devices

Frequency [MHz]	EIRP dBm/MHz*						dBm*,**		
	LFIS***	MFIS***	HFIS***	VRS***	Indoor	Hand-held	LFIS and HFIS	MFIS	VRS, Indoor, and Hand-held
216–960							-49.2[g]	-49.2[g]	-49.2[g]
960–1610	-65.3[h]	-53.3[i]	-65.3[j]	-75.3[k]	-75.3[l]	-75.3[m]			
1164–1240							-75.3[n]	-63.3[o]	-85.3[p]
1559–1610							-75.3[n]	-63.3[o]	-85.3[p]
1610–1990	-53.3[h]	-51.3[i]	-53.3[j]		-53.3[l]	-63.3[m]			
1610–22,000				-61.3[k]					
> 1990	-51.3[h]								
1990–3100			-51.3[j]		-51.3[l]	-61.3[m]			
1990–10,600		-41.3[i]							
3100–10,600			-41.3[j]		-41.3[l]	-41.3[m]			
> 10,600		-51.3[i]	-51.3[j]		-51.3[l]	-61.3[m]			
22,000–29,000				-41.3[k]					
29,000–31,000				-51.3[k]					
> 31,000				-61.3[k]					

* The levels in these columns are to be measured with a resolution bandwidth (RBW) > 1 kHz, but no bandwidth correction is allowed.

** These levels can be converted to dBV/m at 3 m by the formula dBV/m = dBm – 24.8.[q]

*** Low-frequency imaging system (LFIS), mid-frequency imaging system (MFIS), high-frequency imaging system (HFIS), vehicular radar system (VRS).

[i] Part 15.509(d).
[j] Part 15.511(d).
[k] Part 15.513(d).
[l] Part 15.515(d).
[m] Part 15.517(c).
[n] Part 15.519(c).
[o] Parts 15.509(e) and 15.513(e).
[p] Part 15.511(e).
[q] Parts 15.515(e), 15.517(d), and 15.519(d).
[r] Part 15.503(k), dBV = dBμV – 120.

3. DESCRIPTION OF UWB PULSE AND FREQUENCY OCCUPANCY

As noted by numerous authors, UWB traces its roots back to Marconi and his first spark gap transmitters. Many of the devices currently being demonstrated and described in the literature are ultra-high frequency (UHF: 300 MHz – 3 GHz) and super-high frequency (SHF: 3 - 30 GHz) "spark gap" devices.[6,7] Typically, the actual operating frequency and occupied bandwidth (BW) is a function both of the device's antenna frequency response and of the pulsed waveform's transient response. This transient response is a characteristic of the rapid discharge circuit generating the pulse. For these devices, the transmitted waveform is essentially only a repeating impulse. Some in the UWB community have, however, evolved their devices to where they can transmit extremely well shaped pulses consisting of only one or a very few, well-formed, full sine waves of a desired frequency.[8,9] For these devices, the occupied bandwidth is essentially a function of the transmitted waveform – the bandwidth is essentially twice the pulse width τ, divided by the transmitted frequency f_C ($BW \approx 2*\tau/f_C$).

Proponents of UWB devices claim that because of their extremely short pulses some of the advantages of UWB in the current RF environment are: low probability of interference to other users in the same RF space; high immunity to interference from other users of the same RF space; immunity to multipath fading; and the capability for high-frequency reuse.[6,8,10]

The UWB device does not cause significant interference because the energy of its transmitted pulse is spread over a very wide frequency band. With pulse widths on the order of nanoseconds (ns) or even only hundreds of picoseconds (ps), the main frequency lobe is spread over hundreds of megahertz (MHz) or gigahertz (GHz). Further, if an UWB device has a high data rate, in the tens or hundreds of MHz or higher, its line spectrum will have large gaps between its spectral lines. Even with a very noise-like dithered pulse, there will probably still be large gaps between these lines. Since the normal receiver looks at only a few tens of MHz, the normal receiver will only see a very small fraction of the UWB signal. Even better, there is a high probability that the normal receiver's channel will be located between the spectral lines and, consequently, will experience no interference at all.

Through the use of various techniques, such as pseudo-random noise-like (pseudo-noise [PN]) coding, PN time modulation, and correlating receivers, an UWB system can develop a very high processing gain (PG) and interference resistance (IR). It should be noted here that these figures of merit are only realized for coherent systems, such as those described by Petroff and Whitington[8] and XtremeSpectrum.[9] A measure of a system's PG is the ratio of its information bandwidth to its channel bandwidth. UWB systems have nearly the same information bandwidth as many conventional systems but much larger channel bandwidths; therefore, UWB systems have much higher PG/IR. Another method of determining PG is the product of duty factor (DF) and pulse integration. Petroff and Withington[8] have shown that some of their proposed devices have IR on the order of or greater than 50 dB; however,they also discussed similar systems with significantly less IR based on their data rates. Again, UWB systems score well here, but this measure is more difficult for the outsider to determine.

Multipath immunity is achieved through the combination of short pulses, relatively low data rates, small or well-defined dither, and comb filtering. This combination allows the receiver to look at only a very short window around an anticipated bit arrival. This permits the rejection of all the multipath returns, because they do not arrive within the acceptable look window. Not all proposed UWB modulation techniques have this advantage.

Several modulation techniques have been proposed for UWB systems: (1) pulse amplitude modulation (PAM); pulse position modulation (PPM), including dithered-pulse radar-type applications; (2) On-Off Keying (OOK); and (3) bi-polar phase shift keying (BPSK). At present, PPM appears to be the most widely used modulation technique in UWB systems.

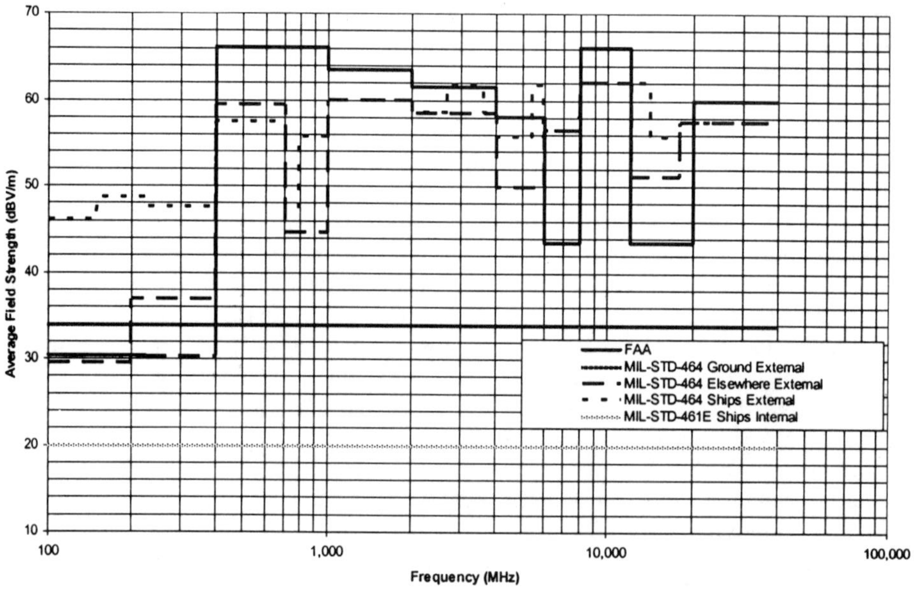

Figure 2. Typical average RF environments for military UWB systems.

4. COTS UWB SYSTEMS IN SELECTED ELECTROMAGNETIC ENVIRONMENTS

After having looked at the FCC rules for commercial UWB systems and some of the characteristics of COTS UWB systems, we now consider the use of COTS UWB devices in military environments, which have more stringent electromagnetic requirements. Not only must a military system survive an interaction with an UWB system, it must function without degradation. The five environmental descriptions of Fig. 2 that cover many of the environments in which a UWB would be expected to operate are discussed. While many of the environments for military UWB systems will resemble the usual US rural, suburban, and urban environments for which COTS UWB systems were developed, the environments described here will probably be much more common. The average field strengths in Fig. 2 are the maximum average levels that might be encountered in each of the five situations. In this context, average is defined as the product of the peak field strength times the duty factor of the RF-source equipment. The source equipment shall be at the closest distance to the victim equipment under normal circumstances, without considering the effect of the RF source's rotation rate. In some cases, the conditions may involve the victim equipment being in the main beam of a non-rotating CW transmitter.

The first, and overall most severe electromagnetic environment in terms of electric field intensity levels, is the one described in the US Federal Aviation Administration (FAA) Draft World RF Environment.[11] This environment (see Fig.2) allows worst-case RF environments in excess of 60 dBV/m through which a commercial aircraft would be expected to fly anywhere around the world. Although compiled by the FAA in 1989, it is still considered to be representative of the environments close to commercial airports around the world. Some of these levels are in good agreement with levels measured by NTIA ITS personnel studying the effects of UWB systems on FAA equipment.[4]

The next three situations are military environments, which must be considered during the procurement of equipment for use by the military, even if the equipment does not have to meet normal MIL-STDs because it is COTS.[c] All three are listed as external, or exposed to the weather, but the ground and "elsewhere" environments can largely be considered anywhere not inside a metal enclosure. Exceptions would be inside most military ships, aircraft, tanks, armored personnel carriers, and armored/shielded communications or command and control (C^2) vehicles or shelters.

The final environment is inside the metal hull of a naval ship as described by MIL-STD-461E.[13] This procurement standard is not applicable to COTS, but describes an environment that an UWB communication system procured for use on a ship would likely experience. Although it is the mildest environment of the set, the average field strength of 20 dBV/m greatly exceeds the threshold immunity levels expected by the European Union (EU) for equipment sold for use there. Typically, many devices have trouble meeting the EU requirements the first time and require additional hardening and re-testing.

Figure 3 compares the average field strengths of standard military systems (Fig. 2) to FCC EIRP limits (Fig. 1). The purpose of this comparison is to show the disparity between the allowed UWB power at a distance of 3 m from the antenna, about the maximum separation between a transmitter and a receiver in many interior compartments on a ship, and the expected military environments. In the interior of a ship, the anticipated peak average environment is 85 dB greater than the maximum received signal at the UWB receiver at a distance of 3 m. In many normal circumstances, the transmitter and the receiver may be as much as ten times as far apart.

These differences make it a near certainty that UWB systems in military environments will suffer interference from typical equipment currently operating in those environments. In many cases, even inside the hull of ships, this interference will be sufficient to render the UWB useless. This is not to say that UWB systems cannot be designed to operate successfully in these environments, but they will probably have to be designed for the environment as opposed to using COTS.

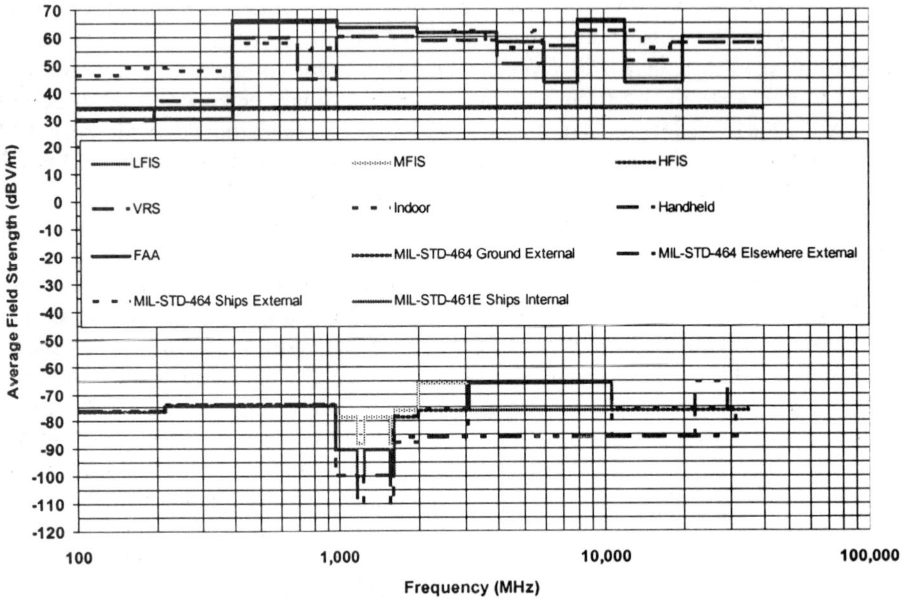

Figure 3. Comparison of UWB-allowed emission levels to military RF environments.

5. POTENTIAL FOR ELECTROMAGNETIC INTERFERENCE (EMI) TO SENSITIVE MILITARY EQUIPMENT

Studies by NTIA ITS[4] using UWB systems, available in 2000 and operating at the FCC Part 15.209 limit for radiation at frequencies above 1 GHz (–66 dBV/m measured at 3 m), showed that existing FAA radars were significantly affected by the operation of certain UWB systems within 3 km of the radar antenna. NTIA ITS studies of potential interference to Global Positioning System (GPS) receivers from simulated UWB emissions[5] demonstrated that UWB devices have a significant potential to interfere with GPS reception.

In response to the potential interference to navigation, the FCC imposed severe limitations on the operation of UWB systems in the radar L-band, 960 – 1610 MHz, which contains the Air Route Surveillance Radar (ARSR-4), the Airport Surveillance Radar (ASR-8), and the Air Traffic Control Beacon Interrogator (ATCBI-5). In addition, to protect GPS receivers, the FCC imposed more severe restrictions on the satellite-to-ground bands of 1164–1240 MHz and 1559–1610 MHz. These restrictions are shown in Table 1 and Figures 1 and 3. They are also beneficial to the military, which also operates a number of radars and navigational aids, such as Identification Friend or Foe (IFF) and Tactical Air Navigation (TACAN) in these bands.

In the rest of the 960–36,000 MHz band, the military has a large number of radars and communications systems, both point-to-point and satellite, that can potentially be affected by UWB systems. Most of these systems operate with sensitivities close to the background noise floor, −108 dBm/MHz. A comparison with the allowed UWB emission levels in the 2–10.6 GHz band shows that UWB systems are permitted to radiate up to −41 dBm/MHz, a difference of 67 dB. In order for a military receiver not to see the UWB, they will need to be separated by somewhat more than 2 km. Some military systems have enough processing gain and interference suppression techniques so that the output will not be affected regardless of the separation. Other receivers will be unable to process this interference out and will display the results as either false targets in the case of a radar or as increased bit error rate (BER) in the case of digital communications systems.

6. CONCLUSION

This paper has shown that the operation of COTS UWB systems in military environments has a strong potential to cause electromagnetic interference (EMI) both to an UWB system itself and to other sensitive equipment currently operating in those environments. In the case of interference to the UWB systems, there is the potential that the interference will be sufficient to render the operation of the UWB useless. In the case of the military receivers, many may suffer some degree of degradation, but most should be able to operate through the interference.

REFERENCES

1. Federal Communications Commission, *Revision of Part 15 of the Commission's Rules Regarding Ultra-Wideband Transmission Systems*, ET Docket 98-153, First Report and Order, adopted 14 February 2002, released 22 April 2002.
2. 47CFR15, *Radio Frequency Devices*, released 25 April 2002; http://www.fcc.gov/oet/info/rules/part15/part15_4_25_02.pdf.
3. NTIA2001, *Manual of Regulations and Procedures for Federal Radio Frequency Management*; http://www.ntia.doc.gov/osmhome/redbook/redbook.html.
4. W.A. Kissick (editor), *The Temporal and Spectral Characteristics of Ultrawideband Signals*, NTIA Report 01-383, January 2001; http://www.its.bldrdoc.gov/pub/ntia-rpt/01-383/01-383.pdf.
5. J.R. Hoffman, M.G. Cotton, R.G. Achatz, and R.N Statz, *Measurements to Determine Potential Interference to GPS Receivers from Ultrawideband Transmission Systems*, NTIA Report 01-384, February 2001; http://www.its.bldrdoc.gov/pub/ntia-rpt/01-384/report.pdf.
6. D.G. Leeper, Wireless data blaster, *Scientific American* **286**(5), 65–69 (May 2002).
7. R.J. Fontana, Recent applications of ultra wideband radar and communications systems, in: *Ultra-Wideband, Short-Pulse Electromagnetics 5*, edited by P.D. Smith and S.R. Cloude (Kluwer Academic/Plenum, New York, 2002), pp. 225–234.
8. P. Petroff and P. Withington, Time Modulated Ultra-Wideband (TM-UWB) Overview, Wireless Symposium/Portable by Design, white paper on web site, 25 February 2000; http://www.timedomain.com/Files/downloads/techpapers/PulsONOverview7_01.pdf.
9. XtremeSpectrum, Mono-phase and bi-phase ultra-wideband, white paper on web site; http://www.xtremespectrum.com/products/whitepapers.html.
10. M.Z. Win and R.A. Scholtz, Impulse radio: how it works, *IEEE Communications Letters* **2**(2), 36–38 (1998).
11. Electronics Industry Association (EIA), Minutes of Meeting #71, G-46, Electromagnetic Compatibility Committee, 26-27 October 1989 (unpublished).
12. MIL-STD-464, *Department of Defense Interface Standard Electromagnetic Environmental Effects Requirements for Systems*, 18 March 1997.
13. MIL-STD-461E, *Department of Defense Interface Standard Requirements for the Control of Electromagnetic Interference Characteristics of Subsystems and Equipment*, 20 August 1999.

DESIGN AND TESTING OF A 1- 3 GHz SFCW GROUND PENETRATING RADAR

Scott Peters, Ian Jeffrey, Colin Gilmore, Derek Collins, Joe LoVetri*

1. INTRODUCTION

At the present time no single landmine detection method is able to detect landmines with a confidence required to satisfy current regulations that deem a section of land suitable for human activity. In order to adequately guarantee human safety clearance requirements imposed by international bodies are stringent.[1] The scale of the problem requires a solution which is not only accurate and efficient but, given that most landmines are located in third world countries, inexpensive as well. In addition, the solution technique should require little technical expertise for its implementation. These requirements tend to diminish the prospects for ground penetrating radar (GPR) as a practicable humanitarian demining method. Given the gravity of the world-wide landmine problem, and the fact that a single modality solution is unlikely, it is worthwhile to explore the extent to which some of the problems with the GPR solution can be overcome. Making the GPR solution practicable requires that many of its associated problems be solved including the signal processing aspect of the method. In this paper, the design, construction, and testing of an inexpensive stepped-frequency continuous wave (SFCW) GPR is reported. Some of the signal processing aspects of the problem are dealt with in an accompanying paper in these proceedings.[2]

Typically, GPR is used in conjunction with imaging techniques which try to represent the material inhomogeneities beneath the ground and thereby detect landmines. Such imaging techniques require the use of a wide bandwidth of frequencies in order to obtain depth (or range) resolution. The range of frequencies is chosen so as to balance the conflicting requirements between the penetration depth and image resolution. That is, it is well known that for a lossy dielectric the attenuation constant is given by[3],

$$\alpha = 8.686\omega\sqrt{\frac{\mu\varepsilon_0\varepsilon'}{2}\left[\sqrt{1+\left(\frac{\varepsilon''}{\varepsilon'}\right)^2}-1\right]}\ \text{dB/m} \qquad (1)$$

* Department of Electrical and Computer Engineering, University of Manitoba, Winnipeg, Canada.

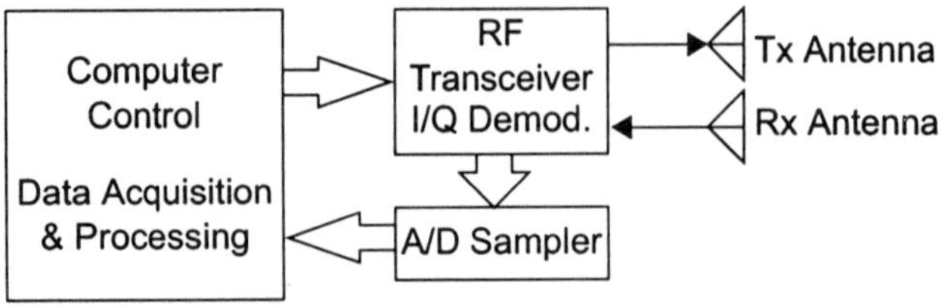

Figure 1. GPR system block diagram.

where ε_0 , $\hat{\varepsilon} = \varepsilon' + j\varepsilon''$, μ , and ω are the permittivity, complex dielectric constant, permeability, and angular frequency, respectively. Thus, at 2 GHz, for dry sandy soil having a permittivity $\varepsilon' = 2.55$ and loss tangent $\varepsilon''/\varepsilon' = 0.004$ the attenuation turns out to be 1.17 dB/m whereas for sandy soil with a 15% moisture content we might have $\varepsilon' = 11.58$ and $\varepsilon''/\varepsilon' = 0.087$ giving an attenuation of 54 dB/m. These high attenuations necessitate that the GPR system be designed with a large dynamic range (i.e., transmitted power minus sensitivity) in order to achieve detection at a sufficient depth. By carefully selecting an appropriate frequency range it is possible to detect landmines to the required depth while obtaining optimal resolution[4]. Frequencies anywhere from a few hundred megahertz to a few gigahertz have been used and tested in the past, at least in laboratory test situations.

Two prototype SFCW systems have been designed. The first prototype which is described is a 1.1-2.1 GHz GPR. A second prototype increases the bandwidth to 1.5-3 GHz. Both prototypes use off-the-shelf RF/microwave components which are low cost and readily available. Similar systems have been developed by Langman *et al.*[5] The decision to use a SFCW design as opposed to a pulse radar was made based on simplicity of design and cost. A comparison of the resolution which can be expected between pulse and CW radars has been presented by Langman and Inggs.[6] Given the availability of microwave components for use in wireless communication devices, the final cost of a production radar based on the prototype system described herein could easily be as low as a few hundred dollars. Initial tests have been performed in a laboratory based environment using a sand box as the ground medium. These tests reveal that the design has potential to accurately detect landmines in a controlled environment and give us confidence that a portable field unit can be successfully implemented.

2. 3. SYSTEM DESIGN

The ultimate goal of the GPR hardware is to measure and record both the magnitude and phase of a received scattered CW signal with respect to the transmitted signal. The transceiver part of the system handles the analogue transmission and receiving of the signals and obtains the required data using an I/Q demodulator. In order to make the task of data acquisition and calibration possible, the system was automated by a microcontroller that samples and stores received information for later offline processing. A simplified block diagram of the system is shown in Figure 1.

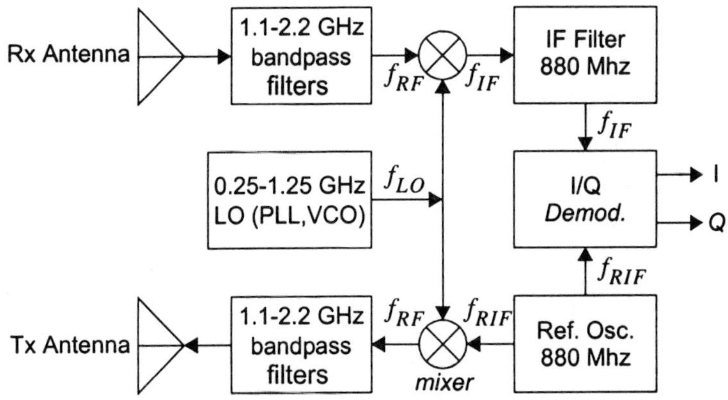

Figure 2. RF transceiver system block diagram.

2. 1. Microwave Transceiver - First Prototype

In order to keep the total system cost to a minimum and because RF components can be the most expensive parts of the GPR it was decided to use off-the-shelf components available for wireless communication devices where possible. Thus, for example, the IF frequency was chosen to be 880 MHz because quality low cost filters, mixers, and I/Q detectors are available at this frequency. A block diagram of the RF transceiver is given in Figure 2. As can be seen, the basic structure of the RF system is a super-heterodyne receiver, with I/Q demodulation. When the local oscillator scans from 0.25-1.25 GHz the output frequency of the GPR varies from 1.1-2.2 GHz. The I/Q demodulation is performed at a single frequency: 880 MHz. The I and Q signals are DC output signals and are sampled by a low cost A/D unit controlled by a microcontroller.

In the first prototype, to keep component costs to minimum, the LO was made up of four Voltage Controlled Oscillators (VCOs) with an RF switch for selection. The switch and tuning voltage for the VCOs are provided by the control system. The appropriate settings for the control of the LO are acquired via calibration with a spectrum analyser. Although this is a very inexpensive solution, it was found that fluctuations in the frequency and voltage output of the LO were difficult to control and as a result the spectral output of the system was not pure. In the second prototype, a phase-locked loop (PLL) was implemented to guarantee a precise frequency. Details of the PLL based LO are given below. In addition, bandpass filters were constructed by cascading inexpensive off-the-shelf lowpass and highpass filters. While the frequency response of such filters is far from ideal, fine tuning and calibration of the transceiver made the overall performance of the system acceptable. The first prototype design is not suitable for mass production because of the extensive calibration which is required. This deficiency was overcome in the second prototype design.

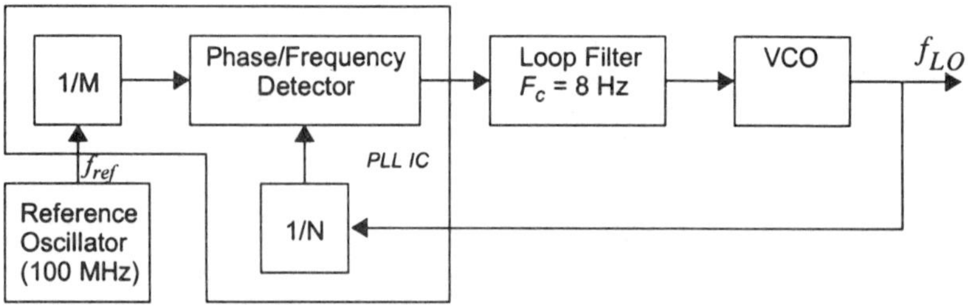

Figure 3. Block diagram of frequency synthesizer.

2. 2. Microwave Transceiver - Second Prototype

2. 2. 1. Frequency Synthesizer

As described above, a free running VCO as an implementation of an LO is not very practical. In addition to the necessary calibration required for each unit which is built, such a design is also very sensitive to noise, temperature, and even vibration. There are also many harmonics within the pass band of the system with amplitudes as high as 10dB below the main peak. Through the use of a phase locked loop, these factors can be reduced to acceptable levels. This is accomplished by incorporating feedback from the output signal and comparing this to a stable, low noise reference crystal oscillator.

The PLL IC used in this system has three major components: an *N*-divider to divide the feedback frequency, an *M*-divider to divide the reference frequency, and a phase detector. This is used in conjunction with a VCO and a loop filter to create the LO. The reference frequency is generally a very stable low noise oscillator such as a crystal (a 100 MHz crystal was used in this system). A block diagram of the PLL synthesizer is shown in Figure 3. The role of the dividers is to reduce the feedback frequency and the reference frequency to the same value in order to compare them with the phase detector. The phase detector outputs a voltage proportional to the difference in frequency and phase which is filtered to provide the VCO tuning voltage. The output frequency is set by changing the (digital) values of the frequency dividers so as to obtain the desired output. The output LO frequency is related to the reference frequency by

$$F_{LO} = \frac{N}{M} F_{ref} \tag{2}$$

where F is the frequency of signal f.

The role of the loop filter is to smooth the output waveform of the charge pump (in the phase detector) so as to reduce noise in the phase detector output. This noise results in a spectral broadening or increase phase noise in the output signal. For this reason the loop filter is a low-pass filter with a very low cut-off frequency. There is a trade off between cutoff frequency and response time of the system. A higher cutoff frequency will reduce the response time, but there will be an increase in levels of spurious noise and vice versa. The implementation of the loop filter in this system is a simple RC low-pass filter.

The initial design of the loop filter was for a cutoff frequency of 550 kHz (based on a maximum phase detector output of 55 MHz), however this did not remove enough spurious

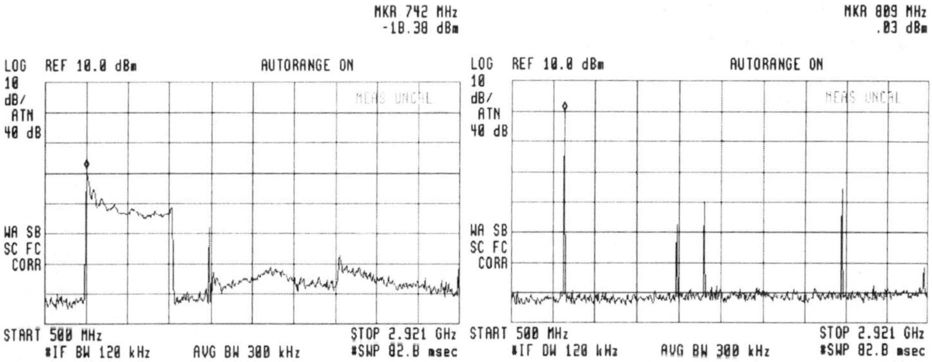

Figure 4. (a) shows the output spectrum with Fc = 500kHz, (b) shows the output spectrum with Fc = 8Hz.

oscillations of the charge pump output for a clean output signal. After some experimentation a cutoff frequency was found at 8 Hz which provided an acceptable relationship between spurious noise and response time. This is further illustrated in Figure 4 which shows output spectrums, both intended to be 800 MHz. The difference is only the cutoff frequency of the loop filter, in Figure 4-(a) $F_c = 550$ kHz, in (b) $F_c = 8$ Hz. It is clear that because of higher frequency oscillations in the tuning voltage, there is a spectral broadening in the output. It should also be noted that in the clean output spectrum, the peak is indeed at 800.025 MHz and not 809 MHz as indicated in the screen capture. This arises from shortfalls in the resolution of the spectrum analyser when attempting to view such a large bandwidth.

There are other advantages to this implementation as well. Firstly, the information required to tune the VCOs can be generated in real time because of the linear relationship between input data and output frequency. In the previous system, a table of tuning voltages was stored in the controller system requiring a large amount of memory. These tables also required a large amount of time to generate because the Frequency-Tuning Voltage characteristic of the VCOs is non-linear. Secondly, there is no need for analog interface circuitry to tune the VCOs, everything is controlled digitally which reduces system complexity and cost. This implementation adds very little to the overall cost of the system. The total increase in cost over the previous system is approximately \$30 (this includes only the extra components for the synthesizer circuitry, and assumes only small number of systems constructed, this amount would be reduced at higher production volumes).

2. 2. 2. RF Bandpass Filters

In the first prototype a pair of wideband filters from 1-2 GHz were used at both the output (transmit) and input (receive) ports of the system in order to eliminate noise and remove some of the VCO harmonics that occurred out of the desired system operating range. Unfortunately the filters were constructed by cascading a LP and HP filter which had a poor roll-off. In the second prototype, a set of five switched stripline filters were designed to cover the entire bandwidth of the system from 1.5 to 3 GHz. Each filter has a bandwidth of 500-600 MHz as shown in Table 1. Switching the filters allows for more of the mixing and VCO harmonics to be removed.

Table 1: Coupled stripline parameters.

Centre freq . (MHz)	Bandwidth (MHz)	W_{min}	S_{min}	Board Length (cm)
1800	500	.76	.209	15.4
2150	600	.7589	.209	13.1805
2450	600	.8074	.22	11.8115
2850	600	.8606	.25	10.4344
3250	600	.9041	.27	9.3963

The cross-sectional and longitudinal geometry of the coupled striplines is given in Figure 5. The even and odd mode characteristic impedances are given by[7]:

$$Z_c^e = \frac{Z_0}{4\sqrt{\varepsilon_r}}\frac{K(k_e')}{K(k_e)} \qquad Z_c^o = \frac{Z_0}{4\sqrt{\varepsilon_r}}\frac{K(k_0')}{K(k_0)} \tag{3}$$

where k_e and k_0 are functions of S, W and H given as

$$k_e = \tanh\left(\frac{\pi W}{4H}\right)\tanh\left(\pi\frac{W+S}{4H}\right) \qquad k_0 = \tanh\left(\frac{\pi W}{4H}\right)\coth\left(\pi\frac{W+S}{4H}\right) \tag{4}$$

and $k' = \sqrt{1-k^2}$. The ratio of quantities K and K' (elliptic integrals) can be approximated to one part in 10^5 by:

$$\frac{K(k)}{K(k')} = \frac{1}{\pi}\ln\left(2\frac{1+\sqrt{k}}{1-\sqrt{k}}\right) \quad \text{for} \quad 0.7 \le k \le 1 \tag{5}$$

and

$$\frac{K(k)}{K(k')} = \frac{1}{\pi}\ln\left[2\frac{1+\sqrt{k'}}{1-\sqrt{k'}}\right]^{-1} \quad \text{for} \quad k < 0.7 \tag{6}$$

Thus, by solving for $K(k')/K(k)$ in (3) and then substituting into (5) and (6) we can solve for k_e and k_o. The product $k_e k_o$ using (4) can be used to solve for the width W and the ratio k_e/k_o can then be used to solve for the required stripline spacing. The required even and odd mode characteristic impedances can be determined from the transformation of a low-pass filter prototype as described in various texts (see for example Collin[7] or

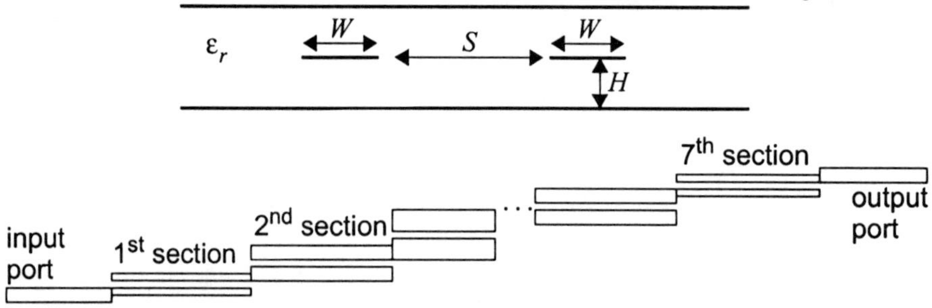

Figure 5. Cross-sectional geometry of stripline (top); longitudinal geometry (bottom).

Pozar). Once the filters were designed, they were verified using the *Ensemble* microwave circuit analysis program. results of the simulation are shown in Figure 6.

The design uses 6th order stripline filters using FR4-epoxy parameters $\varepsilon_r = 4.77$ and a board height of $H = 1.55$ mm. The dimensional characteristics obtained for each of the 5 filters designed are shown in Table 1. Given current standard PCB manufacturing tolerances of 0.2 mm for minimum spacing and line widths the numbers shown in the table mean that these filters can be easily manufactured using standard PCB manufacturing processes. The filters used in the prototype were manufactured using a numerically controlled milling machine to mill the surface of a PCB. The second substrate layer and top ground plane was then bonded to the PCB.

3. CONTROLLER SYSTEM

To properly sequence the RF side of the first prototype design, a system is needed to sequence the RF switch, tune the VCOs, and sample data from the IQ demodulator. There also needs to be a way for users to interact with the RADAR system. The main functions of the controller system are as follows: to write values to a Digital-to-Analog Converter (DAC) module in order to tune the VCOs, sample I and Q values from the RF section, and assert the proper control signals for the RF switch and user outputs. The main functions of the user interface allow users to initiate a scan, reset the system, select the amount of time per frequency step and select the number of frequencies that are stepped through. In addition, software was developed to retrieve data from the system and process it.

The PIC16C774 microcontroller chosen for this application has abundant input/output capabilities as well as an onboard 12-bit Analog-to-Digital Converter (ADC). There is 8 KB of memory in the system, which is enough space for four sets of scanned data. This data is then read with the PC software for offline processing.

In the second prototype, user control is moved entirely to the PC. The synthesizer is controlled with software developed in Visual Basic. The parallel port on the PC is used to set the switch control lines and chip enable lines of the synthesizer IC. Three lines of the parallel port are also used for the serial communication interface of the synthesizer.

The graphical user interface provides the same major functions listed above, but is now more configurable with respect to number of frequencies stepped, frequency step size, and step time. This also gives us the ability to implement further signal processing techniques

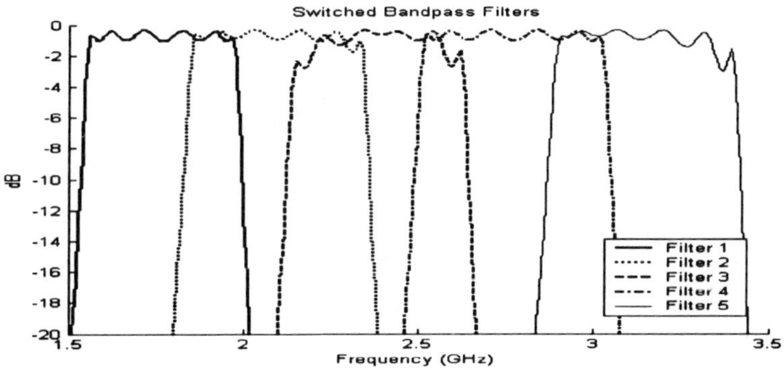

Figure 6. Bandpass filter transfer functions as simulated in Ensemble.

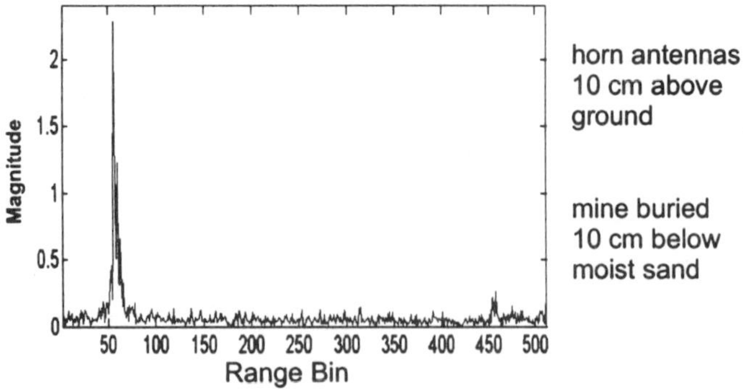

Figure 8. Typical range profile obtained with the system.

on the fly, such as averaging, and ground bounce removal. Sampling the data is now handled by a 12-bit ADC which can be readily controlled by the PC. The memory capacity of the system is greatly increased because of the PC based system.

4. LABORATORY TESTING

Two mock anti-personnel (AP) landmines were obtained from the Department of National Defence. These are shown in Figure 7 along side a 15 cm ruler. Normally such mines are buried 5-10 cm beneath the earth's surface. Thus the maximum penetration depth of a suitable GPR system must be at least 10 cm. These mines are constructed almost entirely of plastics and will therefore have a much smaller RADAR cross-section (i.e., response to RADAR signals) than a metallic object.The required resolving power of a GPR system for detecting such mines should be about 5 cm. Methods of increasing the resolution of our system are described in the accompanying paper in these proceedings.[2]

To limit the amount of uncontrollable variables present when testing the GPR system, testing was performed within an anechoic chamber. The test structure used to support and accurately position the transmit and receive antennas, as depicted in Figure 7, is a rigid platform constructed from wood and hard plastics to reduce reflections. Two ground media were used: Styrofoam and moist sand. A typical range profile which was obtained using the system is shown in Figure 8 Imaging of the ground was performed using Synthetic Aperture Radar (SAR) techniques. Information on the actual techniques which were obtained can be found in the accompanying paper[2].

Figure 7. Mock AP landmines (left), Test environment (right).

5. CONCLUSIONS AND FUTURE WORK

A low cost SFCW GPR system has been successfully designed, constructed and tested. The total cost for the system is under $2000, including antennas. The system performed better than expected, and detection of non-buried, buried and multiple AP landmines was achieved in a laboratory environment.

To increase the GPR resolution to the required 5 cm, the frequency range of the system will be increased. A bandwidth of 1-5 GHz is being considered; a frequency range that will give an increased resolution but maintain minimal penetration depth requirements. Lastly, the system will be constructed entirely out of surface mount components on a single PCB. This will aid greatly in reducing system inefficiencies and expense.

6. ACKNOWLEDGEMENTS

The authors wish to thank the University of Manitoba ECE technicians Al Symmons, Al McKay and Gord Toole, whose efforts in construction of the anechoic chamber, test structure and countless other tasks made the successful completion of this project possible. Much thanks to Eduardo Corral whose expertise and hard work in microcontroller design greatly eased the control system design and to Dr. Yoga Das of the Canadian Department of National Defence for his support and encouragement.

7. REFERENCES

1. "Clearance Requirements," *International Mine Action Standard, IMAS09.10* (UN Mine Action Service, 2001).
2. G. Thomas, E. Corral, J. LoVetri, "High Resolution Imaging Using a SFCW GPR," included in these proceedings.
3. S. Ramo, J.R. Whinnery, T. Van Duzer, *Fields and Waves in Communication Electronics*, 3rd Edition (John Wiley & Sons, Inc., New York, 1994).
4. David A. Noon, "Stepped-Frequency RADAR Design and Signal Processing Enhances Ground Penetrating RADAR Performance," Department of Electrical & Computer Engineering, (The University of Queensland and Cooperative Research Center for Sensor Signal and Information Processing, 1996).
5. A. Langman, S.P. Dimaio, B.E. Burns, and M.R. Inggs, "Development of a low cost SFCW ground penetrating radar," *International Geoscience and Remote Sensing Symposium, IGARSS '96*, "Remote Sensing for a Sustainable Future," vol. 4, pp. 2020-2022 (1996).
6. A. Langman, M.R. Inggs, "Pulse versus stepped frequency continuous wave modulation for ground penetrating radar," *2001 IEEE Int. Symp. on Geoscience and Remote Sensing, IGARSS '01*, vol. 3, pp. 1533 -1535, (2001).
7. R.E. Collin, *Foundations for Microwave Engineering*, 2nd Edition, (IEEE Press, New York, 2001).
8. D.M. Pozar, *Microwave Engineering*, 2nd Edition, (John Wiley & Sons, Inc., New York, 1998).

HIGH RESOLUTION IMAGING USING A SFCW GPR

Gabriel Thomas, Eduardo Corral, and Joe LoVetri*

1. INTRODUCTION

The radar group at the University of Manitoba has recently developed a stepped frequency continuous wave (SFCW) Ground Penetrating Radar (GPR) system that works in the range of 1-2 GHz[1] (see the companion paper by Peters et al.). The wideband signal used in this system offers a range resolution that is limited by the total number of frequencies sent and the frequency step size of the system. High resolution can be obtained by hardware means if the system is allowed to use more frequencies and the operational bandwidth is expanded in the range domain. In order to keep the cost of the system low and obtain more flexibility, it is shown in this paper that higher range resolution can be achieved by using signal processing algorithms thereby avoiding any modifications at the hardware level.

The use of superresolution in GPR systems has been suggested recently.[2] The majority of modern spectral estimation techniques used in radar signal processing which obtain higher resolution are based on the extrapolation of the signal.[3-5] The three main problems faced by these techniques are: i) splitting peaks in the final spectrum estimation encountered when non sinusoidal signals such as frequency modulated signals are analyzed; ii) the attenuation of prominent peaks; and iii) the appearance of spurious peaks. The attenuation and appearance of spurious peaks are mainly caused by a wrong selection of the parameters of the super resolution algorithm. Although there are techniques that can estimate the parameter values, there is still a margin of error and the final spectrum can yield misleading information. For this reason, both parametric and non-parametric high-resolution spectral estimation methods are explored in this paper. It is shown that a returned signal can be extrapolated up to three times its original length corresponding to a resolution improvement factor of a similar order. A qualitative measurement is taken by forming GPR range profiles of targets located at different range positions. It is shown that super resolution techniques improve the localization of a target.

*University of Manitoba, Winnipeg, Manitoba, Canada, R3T 5V6 [(thomas, ecorral, lovetri)@ee.umanitoba.ca].

2. GPR SYSTEM DESCRIPTION AND DATA COLLECTION

A complete hardware description of the GPR system can be found in Ref. 1. Here, the specifications of the system that must be considered in order to determine image resolution are described. The GPR system (the first prototype) used is able to generate a stepped frequency signal with a bandwidth of 1.022 GHz. The initial frequency f_0 is 1.13 GHz, the maximum frequency f_{N-1} is 2.152 GHz, the number of frequency steps used is $N = 512$; therefore, each frequency step is incremented by $\Delta f = 1.99$ MHz. During each step i, the GPR transmits a sinewave with frequency $f_i = f_0 + i\Delta f$, which can be expressed as

$$x_i(t) = A_i \cos(2\pi f_i t) \quad \text{for} \quad Ti \le t \le T(i+1) \quad \text{and} \quad i = 0,1,\ldots,N-1\,, \tag{1}$$

where T is the pulse duration, and A_i is the amplitude of the ith pulse. The signal received has the form

$$s_i(t) = A'_i \cos(2\pi f_i[t-\tau]) \quad \text{for} \quad Ti + \tau \le t \le T(i+1) + \tau\,, \tag{2}$$

where A'_i is the amplitude of the received pulse, τ is a delay associated with the range R of the target ($\tau = 2R/c$), and c is the propagation speed.

The received signal is mixed with a reference signal of the form of (1) after low pass filtering the resulting signal is[7]

$$sm_i = \frac{A'_i}{2}\cos[\psi_i] \quad \text{for} \quad Ti + \tau + \tau_s \le t \le T(i+1) + \tau + \tau_s\,, \tag{3}$$

in which the phase is given by

$$\psi_i = -2\pi\, f_i(\tau + \tau_s) \tag{4}$$

and $_s$ is the receiver transfer delay. It has been assumed that the reference signal has unit amplitude.

In-phase I and quadrature Q components of the signal are formed by mixing the received signal with a 90° shifted version of the transmitted signal yielding

$$sq_i = \frac{A'_i}{2}[\cos(\psi_i) + j\sin(\psi_i)] = \frac{A'_i}{2}\exp(j\psi_i)\,. \tag{5}$$

Each of these I and Q levels carries the frequency response of the object and is sampled via the available A/D converters in the GPR microcontroller. The 512 values of I and 512 values of Q are then transferred from the GPR system to a PC as data files. Once the I-Q data files are available at the PC, for each I and Q vector (512 values), the mean is computed and subtracted from each individual value eliminating with this any bias level caused during the A/D conversion.

Before running a scan with the target present, a calibration scan is made without any object. Once the calibration data is available, the object is placed in a specified location, and a scan is performed again. Once both records are available, the calibration data is subtracted from the measurements for clutter rejection purposes. The data generated at this point corresponds to the phase and magnitude differences between the transmitted and received signals at the GPR system and hence they carry the desired object information, such as presence and distance.

The receiving and transmitting horn type antennas were moved from left to right as indicated in Figure 1, covering a total cross range distance of 114 cm. The antennas were moved in steps of 3 cm and at every position a burst consisting of 512 pulses was generated. Figure 2 shows the intensity of a series of range profiles for a fixed range bin corresponding to the location of the target, which in this case is a metal plate of 15 cm diameter. The arc shapes of the range profiles in this perspective show the radiation extent in the cross-range direction. Thus, in cross range, resolution is limited by the radiation pattern of the mainlobe of the antenna. Therefore, the focus in this paper is to improve resolution of the range profiles and the resolution limits are discussed in the next section.

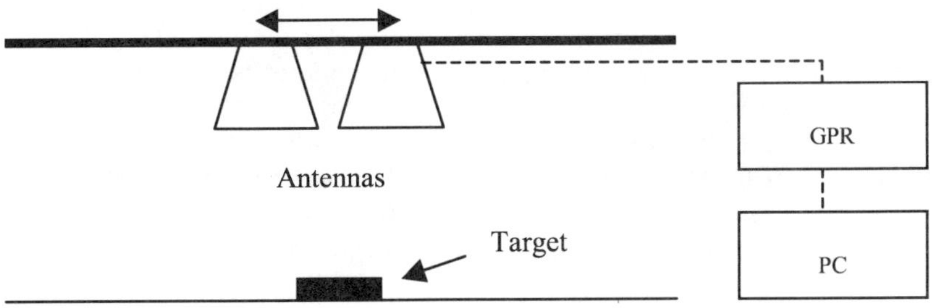

Figure 1. Physical set up of the GPR system.

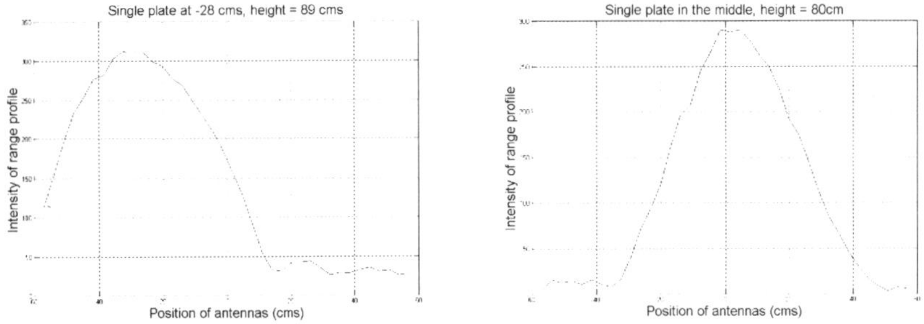

Figure 2. Intensity of range profiles for different positions of the GPR system for a target located at –28 cm with a target-to-antenna distance in the line of sight of 89 cm (left), and a target in the center with a distance of 80 cm (right).

3. RANGE RESOLUTION

In order to obtain range information, the Inverse Fourier Transform (IFT) of (5) is computed. This operation yields a range profile and can be calculated as

$$SQ(k) = \mathrm{IFT}\{sq_i\} = \sum_{i=0}^{N-1} sq_i \exp(j2\pi ki/N) \quad \text{for} \quad 0 \le k \le N-1. \tag{6}$$

The magnitude of (6) is given by[6]

$$|SQ(k)| = \left|\frac{A_i'}{2N}\right| \frac{\sin\left(\pi(k - N\Delta f(\tau + \tau_S))\right)}{\sin\left(\frac{\pi}{N}(k - N\Delta f(\tau + \tau_S))\right)}. \tag{7}$$

In order to calculate the range bin position that corresponds to a range equal to zero, the transmitter and receiver antennas were aligned together face to face and a range profile was obtained. The position of the obtained peak within the profile corresponds to the system delay only. That is = 0 in (7) and subsequent bins in the Fourier transform correspond to the delay of the target. Figure 3 (left) shows the range profile obtained. As it can be seen in this figure, range bin $k = 45$ corresponds to $R = 0$.

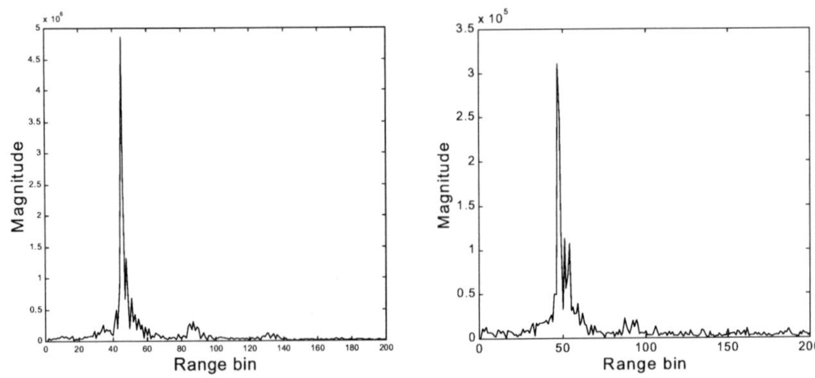

Figure 3. Left: position of the peak in (7) that corresponds to a target at $R = 0$. Right: range profile corresponding to a target at $R = 30$ cm.

For the system being considered, range resolution for a speed of propagation of $c = 3\times10^8$ m/s is given by[7]

$$\Delta R_s = \frac{c}{2\beta} = \frac{3\text{x}10^8}{2\left(1.022\text{x}10^9\right)} = 14.68 \text{ cm}, \tag{8}$$

where β is the total bandwidth of the signal. Thus, a peak in the range profile will jump 1 bin if the object is moved by approximately 15 cm away or towards the antennas. The range of a peak is then defined by

$$R = (k - 45)\Delta R_S \ . \tag{9}$$

A plot of a single range profile is shown in figure 3 (right). This profile was obtained with a disc made of aluminum, at a distance of R = 30 cm from the GPR antennas. The strong peak indicates the presence of a scattering object. Its range bin is k = 47 which is separated 2 bins from the bin that indicates a distance of zero, therefore confirming the real distance.

The distances of the object with respect to the GPR antennas were varied in order to verify the range resolution computed in (8). The disc was placed at 15, 30, 60, 90, 120 and 126 cm from the antennas. Figure 4 shows the 6 range profiles obtained.

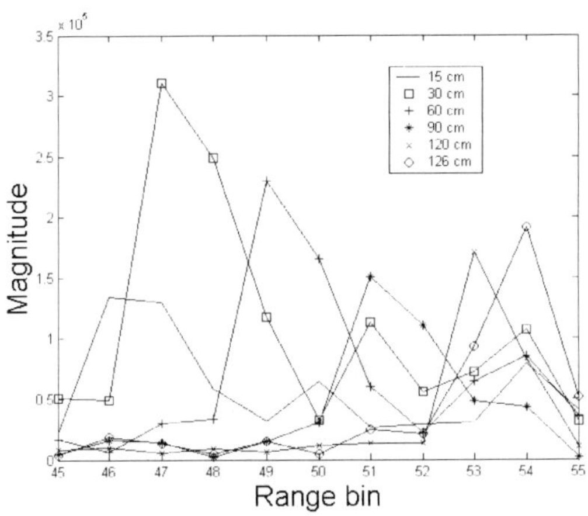

Figure 4. Range profiles of the same target located at 6 different positions shown in the legend.

Table 1 shows the range bin where each peak is located. As can be seen, range positions are not accurate.

Table 1. Range bins where peaks are located.

Distance (cm)	Range bin
15	46
30	47
60	49
90	51
120	52
126	54

4. HIGH-RESOLUTION IMAGING

As explained in the previous section, the current GPR system has a range resolution of 14.68 cm. In order to improve resolution, superresolution techniques based on data extrapolations using parametric and non-parametric power spectrum estimation were evaluated. Because the relevant information within the range profiles is located only in a few bins, low-pass filtering was applied prior to further processing. The filtering is introduced in order to help the extrapolation and power spectrum estimation steps with parametric and non-parametric methods. The filter used is a Hamming window finite impulse response low-pass filter with cut off normalized frequency $w_0 = 0.2$ rad and order 40. Once the signal is filtered, each burst is extrapolated by using superresolution techniques.

5. DATA EXTRAPOLATION

Let the data sequence $x(n)$ be represented as a linear regression of itself with η representing the error

$$x(n) = -\sum_{k=1}^{p} a_k x(n-k) + \eta \ . \tag{10}$$

This modeling of the data is called an autoregressive model of order p. There are different techniques for solving the autoregresive a_k coefficients. The technique used in this paper is based on Burg's minimization of the forward and backward errors in linear predictors. A complete presentation of the solution can be found in Marple.[8]

Once the a_k coefficients are found, the available samples are extended using the pth order forward linear predictor

$$\hat{x}(n) = -a_1 x(n-1) - a_2 x(n-2) - \ldots - a_p x(n-p) \ . \tag{11}$$

Previous samples are estimated by backward prediction using the conjugated coefficients. The model order used was fixed to 30 based on several tests. The data was extended up to

3 times. Therefore, the total bandwidth of the system increased by the same amount yielding a new resolution of 4.89 cm. The bin position for $R = 0$ is now 135. Figure 5 shows the high-resolution range profiles for the same data used in Figure 4. Table 2 shows the locations of the peaks within the range profiles using extrapolations. Once noticing the number of bins that separate the peaks, it can be seen that resolution has been increased by 3. However, the peaks are not clearly distinguishable.

Table 2. Range bins where peaks are located in Figure 5.

Distance (cm)	Range Bin
15	138
30	141
60	146
90	153
120	158
126	160

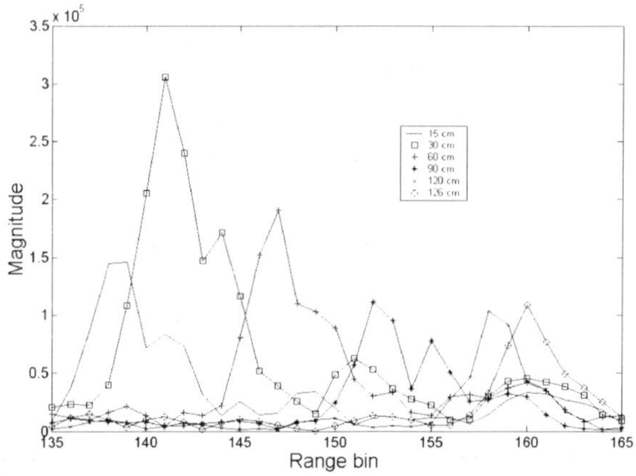

Figure 5. High-resolution range profiles using data extrapolation.

To further improve the spectral estimation,[9,10] a parametric method (in this case Burg again) is used to model the data into an AR filter which is used to generate an estimate of the power spectrum. After several preliminary tests, the chosen order was kept at 40. Figure 6 shows the results of the improved spectral estimations and Table 3 shows the peak positions within the range profiles. As it can be seen, the peaks become more distinguishable within the range profiles.

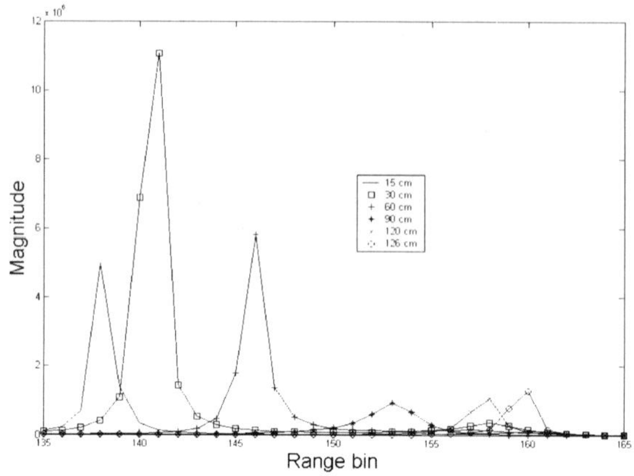

Figure 6. High resolution range profiles using data extrapolation followed by AR spectral estimation.

Table 3. Range bins where peaks are located in Figure 6.

Distance (cm)	Range Bin
15	138
30	140
60	146
90	152
120	158
126	159

Other power spectral estimation techniques such as Spatially Variant Apodization[11] and Adaptive Weighted Norm Extrapolations[5] approaches were also evaluated. These techniques are non-parametric and because of the limited space in this paper, they were not included. The results obtained were not as satisfactory as the ones shown here.

6. DETECTION OF BURIED OBJECTS

The GPR was used to detect buried objects in sand. It was found that the reference data subtraction yields false alarms because of variations in the sand made when digging a hole to bury the object. Hence, the imaging algorithm without the subtraction was applied. Figure 7 shows range images of two aluminum discs buried in sand at a depth of 10 cm, the distance between the antennas and the sand is 1cm, the discs are at ±24 cm from the center.

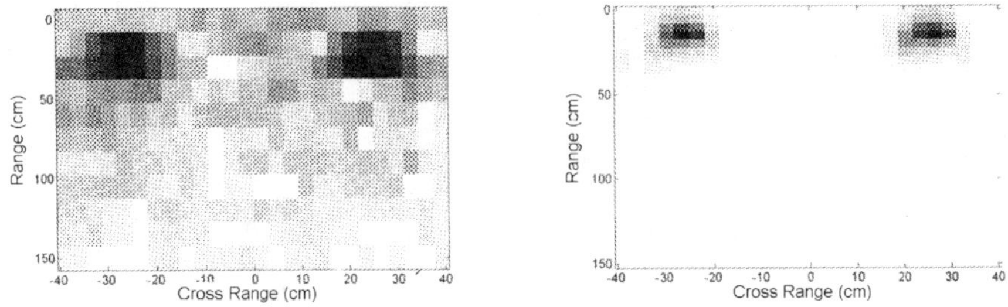

Figure 7. Left: Image of two discs buried in sand without superresolution. Right: Image of two discs buried in sand obtained by extrapolation and parametric PSD estimation.

Plastic mine detection was attempted, however the reflection from the sand was much stronger than the plastic and acceptable images were not obtained. Further research will be done to discriminate between sand and plastic mine reflections.

7. CONCLUSIONS

The existing GPR at the University of Manitoba successfully detects objects buried in sand. It will be used in the future for detection of buried land mines. In order to not only detect the land mines, but also be more accurate in their location, resolution of the current system was improved by signal processing means. The resolution offered by the hardware is approximately 15 cm and the superresolution techniques applied in this work improved the resolution by a factor of 3. It was found that filtering was helpful to obtain better range profiles when using alternative spectral estimation methods.

8. REFERENCES

1. S. Peters, I. Jeffrey, C. Gilmore, D. Collins, and J. LoVetri, Design and testing of a 1-3 GHz SFCW ground penetrating radar, in: *Ultra-Wideband, Short-Pulse Electromagnetics 6*, edited by E. Mokole, M. Kragalott, and K. Gerlach (Kluwer Academic / Plenum Publishers, New York, 2003), pp. 567–576.
2. S.M. Shrestha, I. Arai, T. Miwa, and Y. Tomizawa, Signal processing of ground penetrating radar using super resolution technique, *Proceedings of the 2001 IEEE Radar Conference*, 300–305 (2001).
3. S.R. DeGraaf, SAR imaging via modern 2D spectral estimation methods, *IEEE Trans. on Image Processing* 7(5), 729–761 (May 1998).
4. H.C. Stankwitz and M.R. Kozek, Superresolution for SAR/ISAR RCS measurements using spatially variant apodization, *Proc. AMTA Symp.*, Williamsburg, VA (Nov. 1995).

5. S.D. Cabrera, B.C. Flores, E. Rodriguez, and G. Thomas, Two-dimensional extrapolation and spectral estimation from arbitrary sampling configurations for SAR/ISAR imaging, *IEEE Proc. ASILOMAR*, Pacific Grove, CA (November 1994).
6. D.R. Wehner, *High-Resolution Radar* (Artech House, Norwood, 1995).
7. M. Chang, H. Lee, and I. Akiyama, Image reconstruction and enhancement for subsurface radar imaging using wavefield statistics, *Proceedings of the Twenty-Eighth Asilomar Conference on Signals, Systems and Computers* **2**, 1200–1204 (1994).
8. S.L. Marple, *Digital Spectral Analysis with Applications* (Prentice-Hall, Englewood Cliffs, 1987).
9. P. Millot and A. Berges, Ground based SAR imaging tool for design of buried mine detectors, *EUREL International Conference on the Detection of Abandoned Land Mines*, France (1996).
10. Y. Jiang, *ISAR Imaging using Autoregressive Estimator*, M.S. Thesis, The University of Texas at El Paso (1992).
11. H.C. Stankwitz, R.J. Dallaire, and J.R. Fienup, Non-linear apodization for sidelobe control in SAR imagery, *IEEE Trans. on Aerospace and Elect. Syst.* **31**(1), 267–279 (1995).

RADAR DETECTION AND IMAGING OF BURIED OBJECTS

K. Scheff and P. Hansen*

1. INTRODUCTION

An experimental ultrawideband (UWB) radar, designed and implemented by the Radar Division of the Naval Research Laboratory (NRL), is being used to develop techniques for the detection, imaging, and ultimately the identification of buried objects. This dual-frequency, dual-polarization, UWB radar has collected ultra-high resolution measurements (2-cm range resolution) of land mines and other objects in a laboratory setting. Initial free-space measurements are currently being used to develop and test imaging algorithms, as well as to explore the feasibility and limitations of various synthetic-aperture-radar (SAR) configurations. In this paper we describe the buried-object test bed and demonstrate the ability of the UWB radar to image several different types of objects. At the conclusion of the free-space measurements and analysis, the objects of interest will be buried and measured in various media.

Radar detection and identification of buried objects is currently a topic of great interest. `While the detection of buried objects has been more visible, the focus of our effort is imaging buried objects with enough resolution for reliable identification. To this end, the frequencies of interest tend toward the high end of the spectrum (2 to 12 GHz), which is traditionally used for buried object detection. Also ultra-wide bandwidths, defined as $\Delta f/f_{center} > 0.25$ (f_{center} is the center frequency and Δf is the instantaneous bandwidth), are used to obtain the desired resolution for object identification. To meet this experimental challenge, a specialized, UWB, ultra-high resolution, dual-polarized measurement radar system has been used to collect data suitable for imaging of objects of interest. This paper describes the experiment and provides sample images.

* Radar Division (Code 5340), Naval Research Laboratory, 4555 Overlook Avenue S.W., Washington DC 20375.

2. HARDWARE AND MEASUREMENT GEOMETRY

The UWB radar measurement system (Fig. 1) utilizes video impulse-excited X-band and S-band TWTs that can produce transmit pulses with 2-kW peak power and durations of 0.15 ns and 0.25 ns at the 3 dB power level. In this experiment, 30 dB of attenuation was inserted so that the peak power is 2W. Separate, ultrawideband (2-18 GHz), dual-polarized transmit and receive horn antennas are employed to maintain receiver isolation without the use of a duplexer. The receive system detects signals in coherent baseband quadrature form with a unique direct-sampling detector that uses a multiple sampling head, eight-bit, digital, sampling oscilloscope. The heads simultaneously sample both the rf in-phase and 90° phase-shifted receive signals.

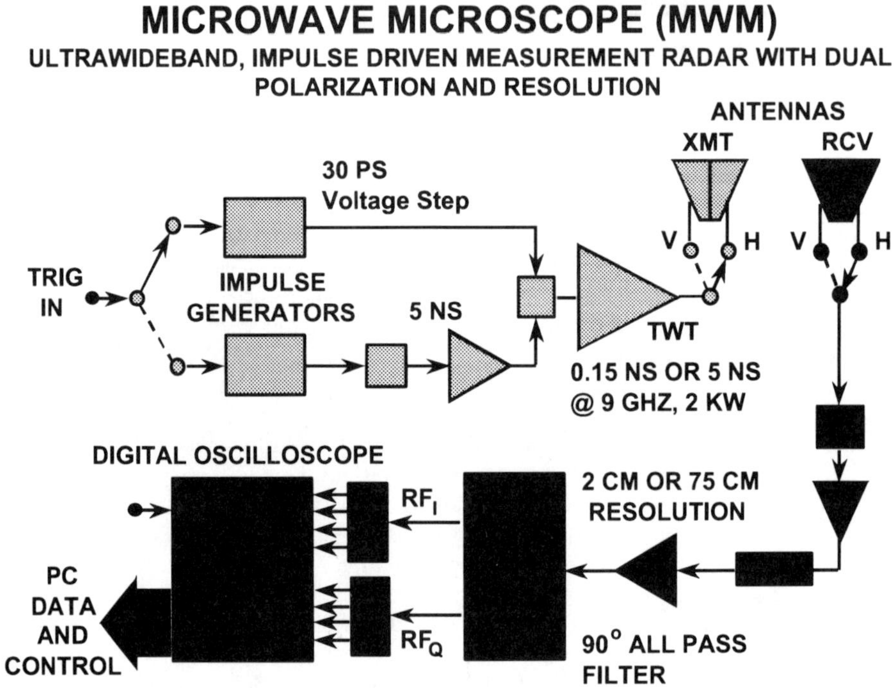

Figure 1. Diagram of UWB radar measurement system.

This detection technique supports an overall instantaneous receive bandwidth in excess 5 GHz. Since only one range point is sampled per pair of sampling heads per pulse, rapid sampling of single or multiple range points is achieved by using repetition rates as high as 100 kHz. Either the transmit frequency or the polarization state can be switched on a pulse-to-pulse basis. Range resolution of the received data is determined by the transmit pulse width and, when using a range scanning mode, by the range sample interval.

Synthetic aperture is achieved by positioning and directing the transmit and receive antennas over a 36-inch by 40-inch rectangle, with the object of interest centered on the y-axis and offset on the x-axis of the rectangle (Fig. 2). The receive antenna is positioned on a 1-inch xy grid spacing over the rectangle. The transmit antenna is moved along the x-axis only, but is fixed to the moving platform such that displacement in the x direction between the transmit and receive antennas remains constant. Both transmit and receive antennas are pointed at the object of interest, and the antennas are positioned such that polarization direction is preserved (that is, vertical polarization is in the x direction and horizontal polarization is in the y direction). The antennas are not in motion during data collection.

Data was collected at X-band (0.15-ns pulse duration) for a metal land mine, a plastic land mine, and a calibration sphere. The objects were placed on a shadowed stand 16 inches above the concrete and tile laboratory floor.

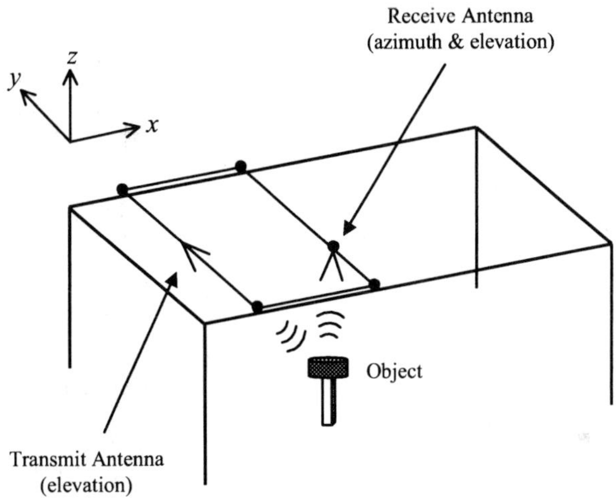

Figure 2. Collection geometry: two degrees of motion in the x and y directions; grid spacing is $\Delta x = \Delta y = 1$ inch. The transmit and receive antennas are placed on a rail system.

3. ANALYSIS

A simple algorithm can be used to form images in the time domain. Let (x,y,z), (x_t,y_t,z_t), and (x_r,y_r,z_r) be the coordinates of an arbitrary image point, the transmit position, and the receive position, respectively (Fig. 2). Images are formed in the time domain using the amplitude

$$A_{x,y,z} = \sqrt{I_{x,y,z}^2 + Q_{x,y,z}^2}$$

at (x,y,z), where

$$I_{x,y,z} = \sum_{t,r} V_I\left(\sqrt{(x-x_t)^2+(y-y_t)^2+(z-z_t)^2}+\sqrt{(x-x_r)^2+(y-y_r)^2+(z-z_r)^2}\right)$$

and

$$Q_{x,y,z} = \sum_{t,r} V_Q\left(\sqrt{(x-x_t)^2+(y-y_t)^2+(z-z_t)^2}+\sqrt{(x-x_r)^2+(y-y_r)^2+(z-z_r)^2}\right).$$

For each image point (x,y,z), $A_{x,y,z}$ is calculated by algebraically summing the in-phase (V_I) and quadrature (V_Q) components of the radar returns for each pair of transmit-receive antenna positions $\{(x_t,y_t,z_t),(x_r,y_r,z_r)\}$. This simple algorithm is possible because of the wide bandwidths (high resolution) of the transmitted signal and receive channel. Signals from objects out of the volume of interest can be range gated out to avoid degradation of the image. A typical $\{V_I,V_Q\}$ pair of signals is shown in Fig. 3. The first prominent reflection is from the object of interest, and the second reflection is from the floor.

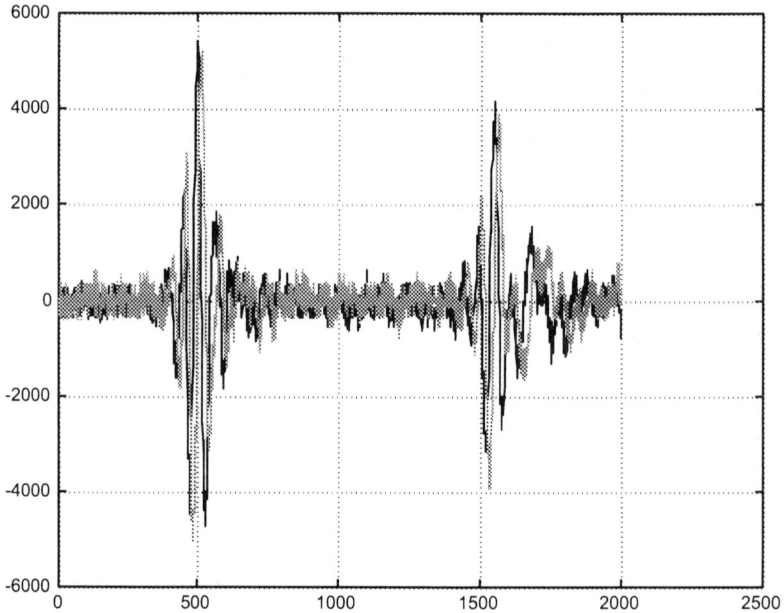

Figure 3. In-phase (black curve) and quadrature (gray curve) components of the received signal scattered from a 6-inch sphere located 16 inches above the laboratory floor. The first peak corresponds to the scatter from the sphere, and the second return is the reflection from the floor.

4. EXPERIMENTAL OBSERVATIONS

Various images of the three initial objects (vertical polarization only) show promise. Figure 4 shows a 6-inch calibration sphere. The side view displays the extent of the measurement geometry; that is, the extent of the 'skull cap' shape indicates that the measurement covers a small fraction of the sphere.

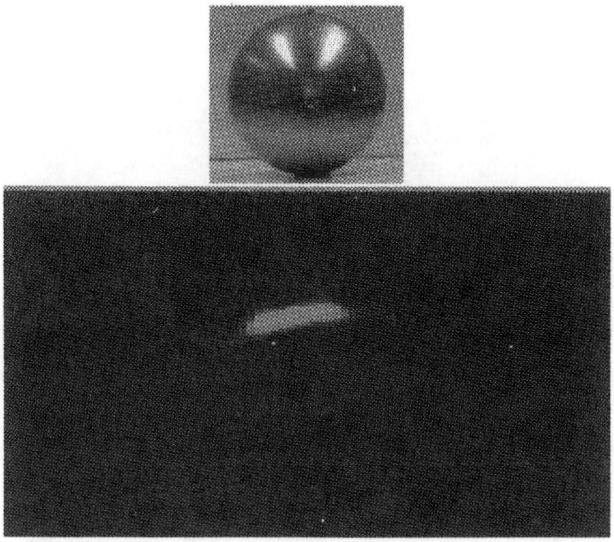

Figure 4. Side view of a sphere.

Figure 5 shows a cross-sectional image through the center of an inert, cylindrical, metallic, anti-tank mine with a photograph of the mine at the same physical scale. Note that the mine has corners where the top portion of the mine joins the main body and a rim around the base of the mine. The data collection was asymmetrical about the horizontal axis in this view, centered about one-third of the way across the mine when traveling left to right. Hence the rim about the base of the mine was not illuminated on the right side of the image, and the top of the mine is somewhat asymmetric in this image.

Figure 6 shows the same mine but from the top as viewed by our apparatus. Again the data collection was asymmetric in the horizontal view, but symmetric in the vertical aspect. The image is a top view of a horizontal cut through the mine at about the height where the top portion joins the main body. What is immediately apparent is the overall round shape of the mine, as well as some finer structure near the center of the mine. Also, the physical size is correct.

Figure 7 shows a side view of an inert plastic square mine and the corresponding image of it. The measurement geometry was as similar to the metallic mine as possible. Note the size and shape are correct in the image. Also, note significant returns from scatterers that appear to be inside to the mine. The vertical extent of the image is greater than the physical height of the mine, which could be due to multiple reflections within the mine or to the contents of the mine. Although the contents of the mine are unknown at present, detection and imaging the contents of non-metallic containers appear feasible.

Figure 8 is a top view of the plastic mine. The square outer boundary is discernible, as is some of the structure within the boundary. The asymmetry in the image is again a consequence our asymmetric measurement geometry with respect to the geometric center of the mine.

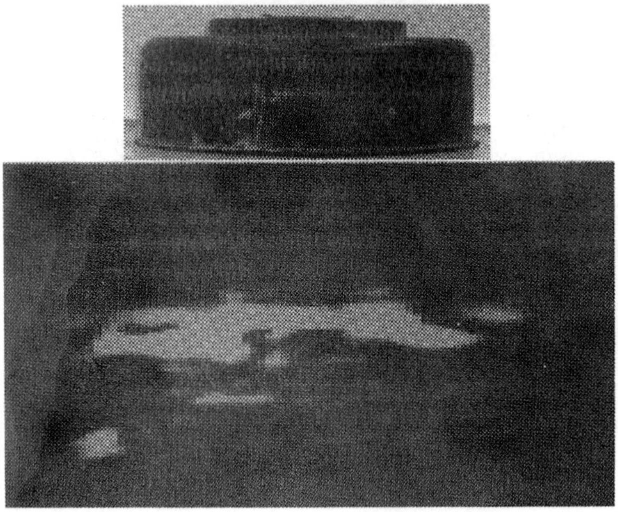

Figure 5. Side view of a metal anti-tank mine.

5. CONCLUSIONS

The quality of this free-space imagery suggests that object identification may be possible. While this imagery was performed using data from our measurement system at X-band with 2-cm resolution, our measurement system has also demonstrated 3-cm resolution at S-band. Moreover, S-band may be more conducive to the detection and imaging of buried objects, because S-band signals experience less attenuation while propagating through the medium than X-band signals. It should be noted that these results are probably representative of the highest microwave resolution that can technically be achieved, based on utilization of very finely controlled scanning and ultra-wide bandwidth. From this point, it might be pertinent to determine what degree of distortion (via bandwidth limitations and other factors) could be tolerated and still obtain a usable image.

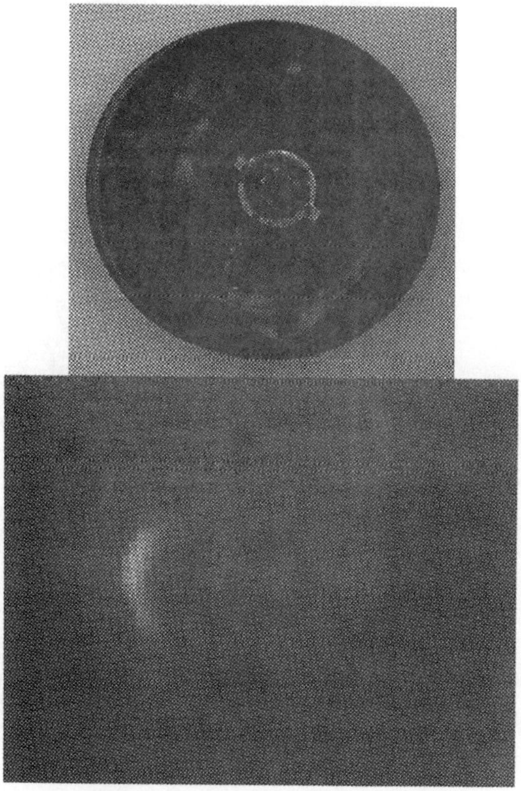

Figure 6. Top view of a metal anti-tank mine.

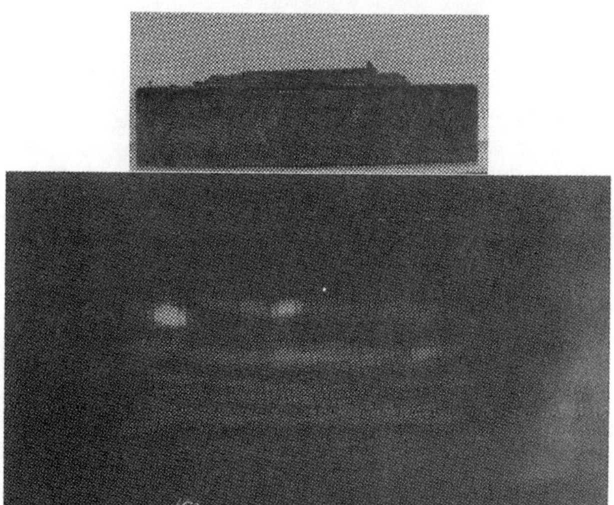

Figure 7. Side view of a plastic anti-tank mine.

Figure 8. Top view of a plastic anti-tank mine.

REFERENCES

1. P. Hansen, K. Scheff, E. Mokole, and E. Tomas, Dual frequency measurements of ocean forward scatter with an ultrawideband radar, *Proceedings of the IEEE 2001 National Radar Conference*, Atlanta GA, pp. 366–371 (May 2001).

AUTHOR INDEX

Abdalla, M., 381
Agee, F., 391
Andreev, Y., 357

Babin, A., 159
Bahar, E., 85
Balzovsky, E., 517
Baum, C., 23, 41, 281, 291, 299, 369, 381, 435, 445
Besieris, I., 193
Bieth, F., 525
Bleszynski, E., 143
Bleszynski, M., 143
Boag, A., 101, 113
Boryssenko, A., 239, 337
Bowen, L., 281
Broadbent, C., 131
Burger, J., 369, 381
Buyanov, Y., 357, 517

Cartwright, N., 123
Chamma, W., 495
Chumerin, P., 427
Chung, Y., 231
Clayton, M., 131
Cockreham, B., 381
Cohen, G., 251
Cohen, L., 555
Collins, C., 391
Collins, D., 565
Corral, E., 575
Crabbe, S., 509
Crisp, G., 473

Davanloo, F., 391
Didenko, A., 427

Farr, E., 281
Felsen, L., 11
Ferrieres, X., 251
Figotin, A., 159

Galdi, V., 11
Gaudet, J., 369, 401
Gaunaurd, G., 483
Gerlach, K., 75
Gilmore, C., 565
Glasgow, S., 1, 131
Gold, B., 345
Gooding-Williams, G., 473
Gwynne, J., 329

Hansen, P., 585
Hatfield, L., 409
Haustein, M., 409
Heyman, E., 101, 113, 203
Hovhannisyan, G., 131

Islam, H., 401

Jang, S., 445
Jaroszewicz, T., 143
Jeffrey, I., 565
Ji, Z., 435
Jung, B., 231, 435

Kayshap, S., 495
Kim, K., 445
Kluskens, M., 317
Kmec, M., 509
Koshelev, V., 63, 357, 517
Kragalott, M., 317, 345
Kristiansen, M., 409
Krompholz, H., 409
Kumru, C., 215

Lawry, D., 369
Lehr, J., 369
Lepelaars, E., 455
Letrou, C., 101, 113
Light, A., 555
Lomakin, V., 203
LoVetri, J., 565, 575

Maybeck, P., 465
Mclemore, D., 381
Mokole, E., 345
Monk, P., 251

Narayanan, R., 215
Novikov, S., 427

Oughstun, K., 123

Parent, M., 345
Pascoe, K., 465
Peatross, J., 1, 131
Pernet, S., 251
Peters, S., 565
Peyerl, P., 509
Pickles, W., 317
Plisko, V., 357
Pokryvailo, A., 415
Posner, F., 75
Prather, W., 369, 381

Rauschenbach, P., 509

Sachs, J., 509
Salo, G., 329
Samaddar, S., 345
Sarkar, T., 231, 435, 445
Sarychev, V., 63
Schamiloglu, E., 401
Schaubert, D., 239, 337
Scheff, K., 585
Seppanen, T., 51
Seydou, F., 51
Shaarawi, A., 193
Shapira, M., 415
Shipilov, S., 63
Shlivinski, A., 101, 113
Skipper, M., 381
Sorenson, G., 345
Spears, J., 409
Staderini, E., 539
Steinberg, B., 203
Stone, A., 291
Strifors, H., 483
Sukhushin, K., 357
Sullivan, A., 483

Thomas, G., 575
Thornhill, C., 473
Tijhuis, A., 223, 261, 455
Tkac, F., 509
Torres, R., 369, 381
Tran, T., 281, 369, 381
Tyo, J., 269
Tzeremes, G., 401

Van Beurden, M., 223
Van Riswick, J., 261
Vossen, S., 455

Ware, M., 1
Wey, J., 525
Wood, A., 465
Wood, W., 465

Xu, X., 215

Yakura, S., 175
Yankelevich, Y., 415
Yushkov, Y., 427

Zwamborn, P., 223, 455

SUBJECT INDEX

Absorption, 1
Analytic, 131
function, 113, 131
signal, 113
Antennas, 75, 239, 299, 329, 337, 357, 455, 555
aperture
distributions, 11
extended, 101, 113
illumination, 329
arrays, 239, 357, 369, 473, 495
dipole, 495
linear, 317
phased, 239
strip-line notch, 317
balun, 281
bandwidth, 317, 555
beamwidth, 75, 329, 369
bow tie, 337
center of rotation, 329
circular aperture, 193
coupling, 299, 317
cone (bi-, mono-), 337, 345
dipole, 23, 261, 299, 337, 357, 495, 517
effective length, 517
efficiency, 269, 329
electrically small, 299, 337, 345
flared notch, 317
flat dipole, 337
frequency response, 357
gain, 269, 281, 329
godograph projection, 517
half IRA, 369
high-voltage resistor, 281
horn, 575, 585
impedance matching, 269, 291, 337
impulse radiating (IRA), 269, 281, 291, 369
input impedance, 269, 281, 345, 357, 517
isolation, 299
orthogonal dipoles, 517
oscillator reflector, 269
passband, 357
pattern (energy, spectral, transient), 329, 337, 345, 357, 369, 517
polarization, 299, 329, 357, 517, 585
power spectral density, 337
pulse-compensated thin wire, 455
radiated field, 85, 101, 113, 345
co-polarized, 517
cross-polarized, 517
radome, 281
received field, 345
splitter, 281
stripline, 565
tapered slot, 239
TEM feed line, 269
TEM horn, 357
TEM mode, 291, 345, 517
TM mode, 345
vector potential, 261, 435
vector receiving antenna, 517
VSWR (voltage standing-wave ratio), 239, 317, 345, 357
Wu-King resistive loading, 455
Approximations,
asymptotic, 159
Born, 51
Hankel, 465
high-frequency, 85, 345
Kirchhoff, 85, 223
Asymptotics, 23, 123, 131, 159

Barycentric coordinates, 143
Basis,
half-basis functions, 143
orthonormal, 445
over-complete, 101, 113

Beam summation, 101, 113
Biorthogonal, 85, 101, 113
Bloch waves, 159
Branch cut, 123
B-scan, 455

Cauchy's theorem, 131
Causality, 123, 131, 143, 231, 445
Centrovelocity, 123
CMOS integral circuitry, 337
Compact range, 345
Complex source-point method, 11
Complex spectra, 63, 445
Complexity, 143
Conformal mapping, 23
Constitutive parameters, 455
  conductivity, 455
  permeability, 85, 175, 291, 455, 565
  permittivity, 85, 175, 455, 525, 565
Convolution, 113
  recursive, 175
Coupling, 381
Covariance, 465

D-dot probe, 525
Decimation, 113
Deconvolution, 495
Dielectric,
  layers, 11
Dirac delta, 85, 131
Dispersive media, 123, 131, 143
  Brillouin zone, 159
Dispersion, 337
  nonlinear, 175
  relation, 159
Dissipation, 131
Distribution, 131
  current, 203
  principal value, 131
  time-frequency, 483
  Weibull, 75
Divider (resistive, voltage), 525
Dyadic, 23, 41, 483

Eigenfrequency, 159
Eigenfunctions, 445
Electromagnetic compatibility (EMC), 555
  communications, 555
  effective isotropic radiated power, 555
  FCC imaging systems, 555
    GPR, 555
    wall imaging, 555
    through-wall imaging, 555
    medical, 555
    surveillance, 555
  FCC UWB classifications, 555
    communication & measurements, 555
    frequency occupancy, 555
    imaging systems, 555
    interference resistance, 555
    vehicular radar, 555
  fractional bandwidth, 555
  quasi-peak detector, 555
  radar, 555
Energy, 435
  free, 131
  integrated sidelobe, 329
  internal, 131
  total, 329
  total interaction, 131
Entire function, 41, 445
Entropy, 63

Fault integration, 203
Field,
  aperture, 269
  Gaussian, 1, 11
  integral equations, 435
    combined, electric, magnetic, 435
  macro-scale, 203
  micro-scale, 203
  tangential, 381
Filter,
  Butterworth, 483
  FIR, 509, 575
  low-pass filter, 575
  Volterra, 509
Four-wave interactions, 159
Fourier series, 231
Frames, 101, 113
  Weyl-Heisenberg, 101
  Gabor, 101, 113
  Radon, 113
  Wavelet, 101, 113
Frechet derivative, 51
Frequency-domain analysis, 11, 203, 337
Fresnel zones, 85

Gabor lattice, 11
Gauge, 261
Gaussian beam, 101, 113
Geometric optics, 269
Grazing angle, 75
Green's function, 143, 223, 239, 465, 483
Group delay, 1
Group velocity, 159

Heat, 131
Helmholtz equation, 51, 131, 143

Image processing, 113, 473, 483, 495, 509, 555, 575, 585
  high-resolution, 193, 575, 585
  back projection, 495
  target recognition, 465
  undersampling, 509
Impedance, 203, 525
Impulse response, 23, 63, 281
Inductance, 381
Integral equation, 41, 51, 143

EFIE (electric-field integral equation), 223, 239, 261, 435
Hallen-type, 261, 455
MFIE (magnetic-field integral equation), 435
MPIE (mixed-potential integral equation), 483
TDIE (time-domain integral equation), 143, 239, 251, 337
Interaction indices, 159
Inverse problem, 51
Iso-diffraction, 101, 113

Kronecker delta, 85

Laplace equation, 291
Lens, 369
artificial dielctric, 369
continually varying dielectric, 369
electromagnetic, 291
hyperboloidal, 369

Matrices,
admittance, 317
block diagonal, 251
block Hessenberg scheme, 143
block Toeplitz, 143
block triangular, 143
covariance, 465
Hessian, 159
Jacobian, 251
LU decomposition, 231
Mueller, 85
over-determined system, 159
sparse, 231
SVD (singular-value decomposition), 223
well-conditioned, 223
Maxwell's equations, 85, 143, 159, 175, 223, 251, 291, 345, 495, 525
Media,
anisotropic, 175, 203
Debye model, 175, 223
dispersive, 123, 193, 329
effective, 203
inhomogeneous, 51
Lorentz model, 1, 123, 131, 175
non-dispersive, 329
Multi-resolution, 203
decomposition, 203
homogenization, 203

Nano-structures, 85
Neumann series, 101, 113
Noise, 445, 465
additive, 465
Gaussian, 215, 465
white, 465
Nonlinear, 159
Numerical methods
absorbing boundary condition, 175, 231
alternating-direction implicit method, 231
boundary element, 381
conjugate gradient, 223
Courant-Friedrich-Levy stability condition, 231, 261
error (dispersion), 223, 231
extrapolation, 223
FEM (finite element), 51, 381
FDTD (finite-difference time-domain), 175, 231, 251, 261, 317, 345, 473, 495
Galerkin's method, 143, 231, 239, 435
Gauss-Lobatto quadrature, 251
Gaussian elimination, 223
Lagrange interpolating polynomials, 251
marching on in angle, 223
marching on in time, 143, 239, 261
Matrix Pencil, 435, 445, 465
MoM (method of moments), 143, 239, 337
perfectly matched layer, 175, 251
sensitivity analysis, 223
stability, 261
staggered Yee scheme, 175
WIPL-D, 445
Yee algorithm, 231, 251

Optics, 131, 159, 175
Optical phonon, 175
Optimization techniques, 281, 329, 369, 435, 445, 575
Burg's minimization, 575
dual-space, 51
energy, 435
ESPRIT, 465
least squares, 337
Newton, 51
quasi-Newton, 445
Oscillatory integrals, 159

Parabolic equation, 251
Perturbation, 85
Phase space, 101, 113
Photonic crystal, 159
Physical optics, 23, 85, 269
Polarimetry, 517
Polarization, 23, 41, 75, 85, 203, 317, 473, 517, 585
covariance matrix, 465
cross-pol coupling, 281
polarizability, 23
radar, 299, 473
Pole models (see SEM), 63
discrete, 63
continuous, 63
Potential problem, 291
Poynting
theorem, 131
vector, 1, 85, 123
Precursor (Brillouin, Sommerfeld), 1, 123
Processes (irreversible, reversible), 131
Propagation, 1, 11, 175, 555
Bloch modes, 159
Composite X waves, 193

diffraction, 193
dispersive, 1, 203
focusing, 193
Fresnel length, 23
full-wave, 85
main signal, 123
multi-layered structures, 193
multi-scale propagators, 101, 113
non-dispersive, 203
parallel-plate waveguide, 231
paraxial, 11
propagators, 11
ray tracing (quasi-real complex), 11
refraction, 193
superluminal, 193
TE mode, 51
TM mode, 51, 345
uniaxial, 175
Pulser, 391, 415, 427

Quasi-momentum, 159
Quasi-stationary, 525

Radar, 357, 473, 483, 495, 509, 517, 539, 555, 565, 575, 585
bandwidth, 215, 509, 555
bistatic, 495
biological, 539
analog-to-digital converter, 539
blood pressure, 539
breath rate, 539
electrocardiograph amplifier, 539
electro-dermal activity, 539
heart rate, 539
polygraph, 539
calibration, 509
clutter, 473, 509
detection, 455, 473, 509, 555, 585
digital, 509
impulse compression, 509
direction of arrival, 517
dynamic range, 565
field-programmable gate arrays (FPGA), 509
foliage penetration, 215
GPR (ground penetrating radar), 455, 473, 483, 509, 555, 565, 575, 585
loss tangent, 565
Microwave Microscope, 585
moving targets, 495
phase-locked loop, 565
point-spread function, 215
polarimetric, 585
power spectral density, 215
quasi-monostatic, 299
range, 215, 509
ambiguity, 509
resolution, 215, 509, 565
RCS (radar cross section), 23, 75
RF components, 509
low-temperature co-fired ceramic, 509
SAR (synthetic aperture radar), 215, 329, 369, 495
signal-to-interference ratio, 473
signal-to-noise ratio, 509
Silicon-Germanium technology, 509
spurious level, 509
time-bandwidth product, 445
Rayleigh-Sommerfeld formula, 193
RCS (radar cross section), 473
Reciprocity, 23, 85, 299
Reflection coefficient, 11, 85
Refractive index, 51, 123
Rellich lemma, 1, 51
Resolution, 329, 509, 565
cross-range, 329
down-range, 329, 565
high-range, 565
ultra-high range, 585
Resonance, 1, 203, 337, 381, 435, 445
Response
frequency domain, 357, 483
impulse, 509
time domain, 483
time-frequncy domain, 483
Riemann-Hilbert problem, 131

Scalar potential, 435
Scattering (target), 193, 251, 261, 445, 465, 473, 483, 495, 555
axial backscatter, 23
body of revolution, 483, 585
buried, 473, 483, 565, 575, 585
concrete floors, 495
conducting cube, 495
cone, 23
cylinders, 473, 483, 585
dielectric, 565
dielectric cube, 223
disk (aluminum, metal), 251, 575
early-time, 41, 445
flat plate, 223, 261, 517, 575
foliage, 215, 251
Kerr, 175
late-time, 41, 465
metal strip, 517
mine (plastic, metallic), 473, 509, 585
perfect conductor, 23, 223, 239
periscope, 75
reflectivity, 215
Raman, 175
random medium, 85
rough surface, 85
specular reflection, 473, 483
sea,
clutter, 75
spike, 75
state, 75
soil, 473, 483, 565
moisture content, 483
sphere, 585

walls, 495, 555
wires, 473
Seismogram, 455
SEM (singularity expansion method), 41, 63, 203, 445, 465
Signal processing, 75, 455, 483, 509, 555
annihilation filters, 465
E-pulse, K-pulse, S-pulse, 465
auto-correlation, 435, 509
auto-regressive model, 575
Butterworth filter, 483
clutter rejection, 575
correlation, 215
deconvolution, 495
Hamming window, 575
impulse response, 473, 509
Kalman filter, 465
linear (predictors, state space), 465, 575
maximum a posteriori (MAP) estimation, 465
Nyquist (frequency, sampling), 261, 509
optimal estimator, 465
profiles (range, cross-range), 575
signal components (I & Q), 575, 585
spread function, 483
spectral estimation, 575
transfer function, 465, 495
Simulation, 215, 337, 473
complex laminate, 203
Slow time, 159
Snell's law, 11
Sommerfeld integrals, 483
Sources, 369, 381
adiabatic heating, 525
amorphic diamond, 391
avalanche mode, 391
Blumlein (stacked), 391
coupling telescope, 525
delay time, 409
dividers (capacitive, voltage), 409, 415
dual-resonant transformer & coax, 381
explosive field emission, 409
gaseous carrier multiplication, 409
gain, 427
high-power, 357, 391, 401, 409
lifetime, 391, 401, 415
MARX generator, 525
Microwave, 525
multi-mode cavity, 427
peak power, 391
point-plane geometries, 409
pulse compressor, 415, 427
pulse width, 415, 427
quality factor, 427
relativistic magnetron, 427
repetition rate, 415
rise time, 391, 409, 415, 525
RMS jitter, 391
Rogowski coil, 415
shock-line pulsers, 369
Silicon avalanche shaper, 369
Silicon-nitride passivation layer, 401
solid-state, 369, 415
spark gap, 415
stability, 415
transfer function, 525
transients, 427
transition-matrix equations, 427
transition time, 415
traveling-wave tube (TWT), 427, 585
voltage attenuator, 415
voltage breakdown, 409, 415
wide band-gap material, 369
Special functions,
Associated Hermite, 445
Bessel, 345
complex error, 445
elliptic integrals, 565
Gamma, 23
Hankel, 51, 345
Hermite polynomials, 445
Kummer Hypergeometric, 23
Laguerre polynomials, 223
Legendre polynomials, 345
orthogonality, 203
prolate spheroidal, 435
spherical Hankel, 345
Spectral transmission-line equations, 203
Spectrum,
angular, 123
evanescent, 113
local, 101, 113
local time-dependent, 113
Susceptibility, 175
Switch, 369, 381, 391, 401
deposition (contact, encapsulating layer), 401
doping, 401
electron injection, 401
ferratron, 369
gas, 369, 409, 427
high power, 391
high-voltage coaxial Hydrogen, 381
junction recovery diode, 369
laser triggering, 369
liquid encapsulated Czochralski (LEC), 391, 401
photoconductive (GaAs, Si), 369, 391, 401
solid-state, 369
trigatron, 369
vertical gradient freeze, 401
wet etching, 401
Symmetry, 281, 299
dilation, 23
inversion, 159
spatial, 23
Toeplitz, 239
Wigner time reversal, 159

Taylor series, 175
Telegraphist's equation, 85

Time-domain analysis, 11, 23, 41, 113, 123, 203, 231, 239, 251, 317, 337, 445, 585
response, 281
Time-frequency analysis, 483
distributions, 483
Baraniuk-Jones optimal kernel, 483
Cohen class, 483
pseudo-Wigner, 483
plane, 445
Time invariance, 175
Transforms
discrete Fourier, 435
discrete local Radon, 113
fast Hadamard, 509
FFT (fast Fourier transform), 123, 143, 223, 483
fractional Fourier, 445
Fourier, 11, 175, 317, 337
Fourier-Laplace, 23, 41, 51
half Fourier, 445
Hilbert, 455, 495
inverse Fourier, 11, 575
Laplace, 299, 465
local windowed slant-stack, 113
slant-stack, 113, 203
Radon, 203
Transient, 203, 261, 337, 435
Translational invariance, 143
Transmission line, 381, 409, 525

Van Vleck formula, 473

Waveform,
bi-polar, 63, 357
bi-polar phase-shift keying, 555
Gaussian, 231, 261, 337, 435, 445, 455, 473, 483
LFM (linear frequency modulation), 215, 445
mono-polar, 357
M-sequence, 509
Nyquist, 435
on-off keying, 555
orthogonal, 435
pulse amplitude modulation, 555
pulse position modulation, 555
pseudo-noise coding, 555
pseudo-noise time modulation, 555
pseudo-random binary, 509
random noise, 215
rectangular envelope, 123
sine-modulated Gaussian, 223, 495
sinusoidal, 445
stepped frequency, 215
stepped-frequency CW, 565, 575
T-pulse, 435
triangular, 337
turn-on time, 445
wide-sense stationary, 215
Waves,
Bloch, 159
composite X, 193
lateral, 85
guided, 85
plane, 525
surface, 85
evanescent, 85
propagating, 85
TEM, 291, 345, 525
Wavelet (spatial, physical), 203
Wiener-Hopf factorization, 131
Wireless systems, 337